SCHAUM'S SOLVED PROBLEMS SERIES

2500 SOLVED PROBLEMS IN

DIFFERENTIAL EQUATIONS

by

Richard Bronson, Ph.D.
Fairleigh Dickinson University

McGRAW-HILL BOOK COMPANY
New York St. Louis San Francisco Auckland Bogotá Caracas Colorado Springs Hamburg Lisbon London Madrid Mexico Milan Montreal New Delhi Oklahoma City Panama Paris San Juan São Paulo Singapore Sydney Tokyo Toronto

Richard Bronson, Ph.D., *Professor of Mathematics and Computer Science at Fairleigh Dickinson University.*
Dr. Bronson, besides teaching, edits two mathematical journals and has written numerous technical papers. Among the books he has published are Schaum's Outlines in the areas of differential equations, operations research, and matrix methods.

Other Contributors to This Volume

Frank Ayres, Jr., Ph.D., *Dickinson College*

James Crawford, B.S., *Fairleigh Dickinson College*

Thomas M. Creese, Ph.D., *University of Kansas*

Robert M. Harlick, Ph.D., *University of Kansas*

Robert H. Martin, Jr., Ph.D., *North Carolina State University*

George F. Simmons, Ph.D., *Colorado College*

Murray R. Spiegel, Ph.D., *Rensselaer Polytechnic Institute*

C. Ray Wylie, Ph.D., *Furman University*

2500 Solved Problems in
DIFFERENTIAL EQUATIONS
INTERNATIONAL EDITION

1 2 3 4 5 6 7 8 9 0 BJEFCP 8 9 4 3 2 1 0 9

Library of Congress Cataloging-in-Publication Data

Bronson, Richard.
2500 solved problems in differential equations / by Richard Bronson.
p. cm. (Schaum's solved problems series)
ISBN 0-07-007979-X
1. Differential equations–Problems, exercises, etc. I. Title.
II. Series.
QA371.B83 1988
515.3'5'076 -dc19 88-17705
CIP

When ordering this title use ISBN 0-07-099158-8

Printed in Singapore

CONTENTS

To the Student

This collection of solved problems covers analytical techniques for solving differential equations. It is meant to be used as both a supplement for traditional courses in differential equations and a reference book for engineers and scientists interested in particular applications. The only prerequisite for understanding the material in this book is calculus.

The material within each chapter and the ordering of chapters are standard. The book begins with methods for solving first-order differential equations and continues through linear differential equations. In this latter category we include the methods of variation of parameters and undetermined coefficients, Laplace transforms, matrix methods, and boundary-value problems. Much of the emphasis is on second-order equations, but extensions to higher-order equations are also demonstrated.

Two chapters are devoted exclusively to applications, so readers interested in a particular type can go directly to the appropriate section. Problems in these chapters are cross-referenced to solution procedures in previous chapters. By utilizing this referencing system, readers can limit themselves to just those techniques that have value within a particular application.

CHAPTER 1
Basic Concepts

CLASSIFICATIONS

1.1 Determine which of the following are ordinary differential equations and which are partial differential equations:

(*a*) $\dfrac{d^2y}{dx^2} + 3\dfrac{dy}{dx} + 2y = 0$

(*b*) $\dfrac{\partial z}{\partial x} = z + x\dfrac{\partial z}{\partial y}$

▌ Equation (*a*) is an ordinary differential equation because it contains only ordinary (nonpartial) derivatives; (*b*) is a partial differential equation because it contains partial derivatives.

1.2 Determine which of the following are ordinary differential equations and which are partial differential equations:

(*a*) $xy' + y = 3$

(*b*) $y''' + 2(y'')^2 + y' = \cos x$

(*c*) $\dfrac{\partial^2 z}{\partial x^2} + \dfrac{\partial^2 z}{\partial y^2} = x^2 + y$

▌ Equations (*a*) and (*b*) are ordinary differential equations because they contain only ordinary derivatives; (*c*) is a partial differential equation because it contains at least one partial derivative.

1.3 Determine which of the following are ordinary differential equations and which are partial differential equations:

(*a*) $\dfrac{dy}{dx} = 5x + 3$

(*b*) $e^y \dfrac{d^2y}{dx^2} + 2\left(\dfrac{dy}{dx}\right)^2 = 1$

(*c*) $4\dfrac{d^3y}{dx^3} + (\sin x)\dfrac{d^2y}{dx^2} + 5xy = 0$

▌ All three equations are ordinary differential equations because each contains only ordinary derivatives.

1.4 Determine which of the following are ordinary differential equations and which are partial differential equations:

(*a*) $\left(\dfrac{d^2y}{dx^2}\right)^3 + 3y\left(\dfrac{dy}{dx}\right)^7 + y^3\left(\dfrac{dy}{dx}\right)^2 = 5x$

(*b*) $\dfrac{\partial^2 y}{\partial t^2} - 4\dfrac{\partial^2 y}{\partial x^2} = 0$

(*c*) $xy^2 + 3xy - 2x^3y = 7$

▌ Equation (*a*) is an ordinary differential equation, while (*b*) is a partial differential equation. Equation (*c*) is *neither*; since it contains no derivatives, it is not a differential equation of any type. It is an algebraic equation in x and y.

1.5 Determine which of the following are ordinary differential equations and which are partial differential equations:

(*a*) $(\sin x)y^2 + 2y = 3x^3 - 5$

(*b*) $e^{xy} - 2x + 3y^2 = 0$

(*c*) $(2x - 5y)^2 = 6$

▌ None of these equations is a differential equation, either ordinary or partial, because none of them involves derivatives.

1.6 Determine which of the following are ordinary differential equations and which are partial differential equations:

(*a*) $\dfrac{dy}{dx} = x + 5$

(*b*) $(y'')^2 + (y')^3 + 3y = x^2$

▌ Both are ordinary differential equations because each contains only ordinary derivatives.

1.7 Define *order* for an ordinary differential equation.

▌ The order of a differential equation is the order of the highest derivative appearing in the equation.

1.8 Define *degree* for an ordinary differential equation.

▌ If an ordinary differential equation can be written as a polynomial in the unknown function and its derivatives, then its degree is the power to which the highest-order derivative is raised.

1.9 Define *linearity* for an ordinary differential equation.

▌ An *n*th-order ordinary differential equation in the unknown function y and the independent variable x is *linear* if it has the form

$$b_n(x)\frac{d^ny}{dx^n} + b_{n-1}(x)\frac{d^{n-1}y}{dx^{n-1}} + \cdots + b_1(x)\frac{dy}{dx} + b_0(x)y = g(x)$$

The functions $b_j(x)$ $(j = 0, 1, 2, \ldots, n)$ and $g(x)$ are presumed known and depend only on the variable x. Differential equations that cannot be put into this form are *nonlinear*.

1.10 Determine the order, degree, linearity, unknown function, and independent variable of the ordinary differential equation $y'' - 5xy' = e^x + 1$.

▌ *Second order*: the highest derivative is the second. The unknown function is y, and the independent variable is x. *First degree*: the equation is written as a polynomial in the unknown function y and its derivatives, with the highest derivative (here the second) raised to the first power. *Linear*: in the notation of Problem 1.9, $b_2(x) = 1$, $b_1(x) = -5x$, $b_0(x) = 0$, and $g(x) = e^x + 1$.

1.11 Determine the order, degree, linearity, unknown function, and independent variable of the ordinary differential equation $y''' - 5xy' = e^x + 1$.

▌ *Third order*: the highest derivative is the third. The unknown function is y, and the independent variable is x. *First degree*: the equation is a polynomial in the unknown function y and its derivatives, with its highest derivative (here the third) raised to the first power. *Linear*: in the notation of Problem 1.9, $b_3(x) = 1$, $b_1(x) = -5x$, $b_2(x) = b_0(x) = 0$, and $g(x) = e^x + 1$.

1.12 Determine the order, degree, linearity, unknown function, and independent variable of the differential equation $y - 5xy' = e^x + 1$.

▌ *First order*: the highest derivative is the first. The unknown function is y, and the independent variable is x. *First degree*: the equation is a polynomial in the unknown function y and its derivative, with its highest derivative (here the first) raised to the first power. *Linear*: in the notation of Problem 1.9, $b_1(x) = -5x$, $b_0(x) = 1$, and $g(x) = e^x + 1$.

1.13 Determine the order, degree, linearity, unknown function, and independent variable of the differential equation $y - 5x(y')^2 = e^x + 1$.

▌ *First order*: the highest derivative is the first. The unknown function is y, and the independent variable is x. *Second degree*: the equation is a polynomial in the unknown function y and its derivative, with its highest derivative (here the first) raised to the second power. *Nonlinear*: the derivative of the unknown function is raised to a power other than the first.

1.14 Determine the order, degree, linearity, unknown function, and independent variable of the differential equation $y - 5x(y')^4 = e^x + 1$.

▌ *First order*: the highest derivative is the first. The unknown function is y, and the independent variable is x. *Fourth degree*: the equation is a polynomial in the unknown function y and its derivative, with its highest derivative (here the first) raised to the fourth power. *Nonlinear*: the derivative of the unknown function is raised to a power other than the first.

1.15 Determine the order, degree, linearity, unknown function, and independent variable of the differential equation $y'' - 5x(y')^4 = e^x + 1$.

▮ *Second order*: the highest derivative is the second. The unknown function is y, and the independent variable is x. *First degree*: the equation is a polynomial in the unknown function y and its derivatives, with its highest derivative (here the second) raised to the first power. *Nonlinear*: one of the derivatives of the unknown function is raised to a power other than the first.

1.16 Determine the order, degree, linearity, unknown function, and independent variable of the differential equation $(y'')^3 - 5x(y')^4 = e^x + 1$.

▮ *Second order*: the highest derivative is the second. The unknown function is y, and the independent variable is x. *Third degree*: the equation is a polynomial in the unknown function y and its derivatives, with the highest derivative (here the second) raised to the third power. *Nonlinear*: one of the derivatives of the unknown function is raised to a power other than the first.

1.17 Determine the order, degree, linearity, unknown function, and independent variable of the differential equation $(y''')^3 - 5x(y')^4 = e^x + 1$.

▮ *Third order*: the highest derivative is the third. The unknown function is y, and the independent variable is x. *Third degree*: the equation is a polynomial in the unknown function y and its derivatives, with the highest derivative (here the third) raised to the third power. *Nonlinear*: one of the derivatives of the unknown function is raised to a power other than the first.

1.18 Determine the order, degree, linearity, unknown function, and independent variable of the differential equation $y'' - 5x^2y' = e^x + 1$.

▮ *Second order*: the highest derivative is the second. The unknown function is y, and the independent variable is x. *First degree*: the equation is written as a polynomial in the unknown function y and its derivatives, with its highest derivative (here the second) raised to the first power. *Linear*: in the notation of Problem 1.9, $b_2(x) = 1$, $b_1(x) = 5x^2$, $b_0(x) = 0$, and $g(x) = e^x + 1$.

1.19 Determine the order, degree, linearity, unknown function, and independent variable of the differential equation $y'' - 5t^2y' = e^t + 1$.

▮ This problem is identical to Problem 1.18, with the single exception that now the independent variable is t.

1.20 Determine the order, degree, linearity, unknown function, and independent variable of the differential equation $z'' - 5x^2z' = e^x + 1$.

▮ This problem is identical to Problem 1.18, with the single exception that now the unknown function is z.

1.21 Determine the order, degree, linearity, unknown function, and independent variable of the ordinary differential equation

$$5x\frac{d^2y}{dx^2} + 3x^2\frac{dy}{dx} - (\sin x)y = 0$$

▮ *Second order*: the highest derivative is the second. The unknown function is y, and the independent variable is x. *First degree*: the equation is written as a polynomial in the unknown function y and its derivatives, with the highest derivative (here the second) raised to the first power. *Linear*: in the notation of Problem 1.9, $b_2(x) = 5x$, $b_1(x) = 3x^2$, $b_0(x) = -\sin x$, and $g(x) = 0$.

1.22 Determine the order, degree, linearity, unknown function, and independent variable of the differential equation

$$5x\frac{d^4y}{dx^4} + 3x^2\frac{dy}{dx} - (\sin x)y = 0$$

▮ *Fourth order*: the highest derivative is the fourth. The unknown function is y, and the independent variable is x. *First degree*: the equation is a polynomial in the unknown function y and its derivatives, with the highest derivative (here the fourth) raised to the first power. *Linear*: in the notation of Problem 1.9, $b_4(x) = 5x$, $b_3(x) = b_2(x) = 0$, $b_1(x) = 3x^2$, $b_0(x) = -\sin x$, and $g(x) = 0$.

1.23 Determine the order, degree, linearity, unknown function, and independent variable of the differential equation

$$5t\frac{d^4y}{dt^4} + 3t^2\frac{dy}{dt} - (\sin t)y = 0$$

▌ This problem is identical to Problem 1.22, with the single exception that now the independent variable is t rather than x.

1.24 Determine the order, degree, linearity, unknown function, and independent variable of the differential equation

$$5t\frac{d^4y}{dt^4} + 3t^2\left(\frac{dy}{dt}\right)^3 - (\sin t)y = 0$$

▌ *Fourth order*: the highest derivative is the fourth. The unknown function is y, and the independent variable is t. *First degree*: the equation is a polynomial in the unknown function y and its derivatives, with its highest derivative (here the fourth) raised to the first power. *Nonlinear*: one of the derivatives of the unknown function is raised to a power other than the first.

1.25 Determine the order, degree, linearity, unknown function, and independent variable of the differential equation

$$5t\frac{d^4y}{dt^4} + 3t^2\left(\frac{dy}{dt}\right)^3 - (\sin t)y^6 = 0$$

▌ *Fourth order*: the highest derivative is the fourth. The unknown function is y, and the independent variable is t. *First degree*: the equation is a polynomial in the unknown function y and its derivatives, with its highest derivative (here the fourth) raised to the first power. *Nonlinear*: one of the derivatives of the unknown function (as well as the unknown function itself) is raised to a power other than the first.

1.26 Determine the order, degree, linearity, unknown function, and independent variable of the differential equation

$$3t^2\left(\frac{dy}{dt}\right)^3 - (\sin t)y^6 = 0$$

▌ *First order*: the highest derivative is the first. The unknown function is y, and the independent variable is t. *Third degree*: the equation is a polynomial in the unknown function y and its derivative, with its derivative raised to the third power. *Nonlinear*: one of the derivatives of the unknown function y (as well as y itself) is raised to a power other than the first.

1.27 Determine the order, degree, linearity, unknown function, and independent variable of the differential equation

$$3t^2\left(\frac{dy}{dt}\right)^3 - (\sin t)\left(\frac{d^2y}{dt^2}\right)^6 = 0$$

▌ *Second order*: the highest derivative is the second. The unknown function is y, and the independent variable is t. *Sixth degree*: the equation is a polynomial in the unknown function y and its derivatives, with the highest derivative (here the second) raised to the sixth power. *Nonlinear*: at least one of the derivatives of the unknown function is raised to a power higher than the first.

1.28 Determine the order, degree, linearity, unknown function, and independent variable of the differential equation

$$3t^2\frac{d^3y}{dt^3} - (\sin t)\left(\frac{d^2y}{dt^2}\right)^6 = 0$$

▌ *Third order*: the highest derivative is the third. The unknown function is y, and the independent variable is t. *First degree*: the equation is a polynomial in the unknown function y and its derivatives, with its highest derivative (here the third) raised to the first power. *Nonlinear*: one of the derivatives of the unknown function y is raised to a power higher than the first.

1.29 Determine the order, degree, linearity, unknown function, and independent variable of the differential equation

$$3t^2\frac{d^3y}{dt^3} - (\sin t)\frac{d^2y}{dt^2} - (\cos t)y = 0$$

▌ *Third order*: the highest derivative is the third. The unknown function is y, and the independent variable is t. *First degree*: the equation is a polynomial in the unknown function y and its derivatives, with the highest derivative (here the third) raised to the first power. *Linear*: in the notation of Problem 1.9, $b_3(t) = 3t^2$, $b_2(t) = -\sin t$, $b_1(t) = 0$, $b_0(t) = -\cos t$, and $g(t) = 0$.

1.30 Determine the order, degree, linearity, unknown function, and independent variable of the differential equation

$$3t^2 \frac{d^3y}{dt^3} - (\sin t)\frac{d^2y}{dt^2} - \cos ty = 0$$

▌ *Third order*: the highest derivative is the third. The unknown function is y, and the independent variable is t. *No degree*: the equation cannot be written as a polynomial in the unknown function and its derivatives, because the unknown function y is an argument of the transcendental cosine function; degree is therefore undefined. *Nonlinear*: the unknown function is an argument of a transcendental function.

1.31 Determine the order, degree, linearity, unknown function, and independent variable of the differential equation $5\ddot{y} + 2e^t\dot{y} - 3y = t$.

▌ *Second order*: the highest derivative is the second. The unknown function is y, and the independent variable is t. *First degree*: the equation is a polynomial in the unknown function y and its derivatives, with the highest derivative (here the second) raised to the first power. *Linear*: in the notation of Problem 1.9, $b_2(t) = 5$, $b_1(t) = 2e^t$, $b_0(t) = -3$, and $g(t) = t$.

1.32 Determine the order, degree, linearity, unknown function, and independent variable of the differential equation $5\ddot{y} + 2e^{t\dot{y}} - 3y = t$.

▌ *Second order*: the highest derivative is the second. The unknown function is y, and the independent variable is t. *No degree*: the equation cannot be written as a polynomial in the unknown function y and its derivatives, because one of its derivatives (namely, $\dot{y}$) is an argument of the transcendental exponential function; degree is therefore undefined. *Nonlinear*: at least one derivative of the unknown function is an argument of a transcendental function.

1.33 Determine the order, degree, linearity, unknown function, and independent variable of the differential equation $5\dot{y} - 3\dot{y}y = t$.

▌ *First order*: the highest derivative is the first. The unknown function is y, and the independent variable is t. *First degree*: the equation is a polynomial in the unknown function y and its derivative, with its derivative raised to the first power. *Nonlinear*: the unknown function y is multiplied by its own derivative.

1.34 Determine the order, degree, linearity, unknown function, and independent variable of the differential equation $5\dot{y} - 3(\dot{y})^7y = t$.

▌ *First order*: the highest derivative is the first. The unknown function is y, and the independent variable is t. *Seventh degree*: the equation is a polynomial in the unknown function y and its derivative, with the highest power of its derivative being the seventh. *Nonlinear*: the unknown function y is multiplied by its own derivative; in addition, the derivative of y is raised to a power other than the first.

1.35 Determine the order, degree, linearity, unknown function, and independent variable of the differential equation $5\dot{y} - 3\dot{y}y^7 = t$.

▌ *First order*: the highest derivative is the first. The unknown function is y, and the independent variable is t. *First degree*: the equation is a polynomial in the unknown function y and its derivative, with the derivative raised to the first power. *Nonlinear*: the unknown function y is raised to a power other than the first (as well as being multiplied by its own derivative).

1.36 Determine the order, degree, linearity, unknown function, and independent variable of the differential equation $5\dot{z} - 3\dot{z}z^7 = t$.

▌ This problem is identical to Problem 1.35, with the single exception that now the unknown function is z.

1.37 Determine the order, degree, linearity, unknown function, and independent variable of the differential equation $t\ddot{y} + t^2\dot{y} - (\sin t)\sqrt{y} = t^2 - t + 1$.

▌ *Second order*: the highest derivative is the second. The unknown function is y, and the independent variable is t. *No degree*: because of the term $\sqrt{y}$, the equation cannot be written as a polynomial in y and its derivatives. *Nonlinear*: the unknown function y is raised to a power other than the first—in this case the one-half power.

1.38 Determine the order, degree, linearity, unknown function, and independent variable of the differential equation

$$5\left(\frac{d^4b}{dp^4}\right)^5 + 7\left(\frac{db}{dp}\right)^{10} + b^7 - b^5 = p$$

▌ *Fourth order.* The unknown function is b; the independent variable is p. *Fifth degree*: the highest (fourth) derivative is raised to the fifth power. *Nonlinear.*

1.39 Determine the order, degree, linearity, unknown function, and independent variable of the differential equation

$$s^2\frac{d^2t}{ds^2} + st\frac{dt}{ds} = s$$

▌ *Second order.* The unknown function is t; the independent variable is s. *First degree*: the equation is a polynomial in the unknown function t and its derivatives (with coefficients in s), and the second derivative is raised to the first power. *Nonlinear*: in the notation of Problem 1.9, $b_1 = st$, which depends on both s and t.

1.40 Determine the order, degree, linearity, unknown function, and independent variable for the differential equation

$$y\frac{d^2x}{dy^2} = y^2 + 1$$

▌ *Second order.* The unknown function is x; the independent variable is y. *First degree. Linear*: in the notation of Problem 1.9, $b_2(y) = y$, $b_1(y) = 0$, $b_0(y) = 0$, and $g(y) = y^2 + 1$.

1.41 Determine the order, degree, linearity, unknown function, and independent variable for the differential equation $(y'')^2 - 3yy' + xy = 0$.

▌ *Second order* because the highest derivative is the second, and *second degree* because this derivative is raised to the second power. The unknown function is y, and the independent variable is x. *Nonlinear* because one of the derivatives of y is raised to a power other than the first; in addition, the unknown function is multiplied by one of its own derivatives.

1.42 Determine the order, degree, linearity, unknown function, and independent variable for the differential equation $x^4y^{(4)} + xy^{(3)} = e^x$.

▌ *Fourth order* because the highest derivative is the fourth, and *first degree* because that derivative is raised to the first power. The unknown function is y, and the independent variable is x. *Linear*: in the notation of Problem 1.9, $b_4(x) = x^4$, $b_3(x) = x$, $b_2(x) = b_1(x) = b_0(x) = 0$, and $g(x) = e^x$.

1.43 Determine the order, degree, linearity, unknown function, and independent variable for the differential equation $y^{(4)} + xy^{(3)} + x^2y'' - xy' + \sin y = 0$.

▌ *Fourth order*: the highest derivative is the fourth. The unknown function is y, and the independent variable is x. *No degree* and *nonlinear* because the unknown function is the argument of a transcendental function, here the sine function.

1.44 Determine the order, degree, linearity, unknown function, and independent variable for the differential equation $t^2\ddot{s} - t\dot{s} = 1 - \sin t$.

▌ *Second order*: the highest derivative is the second. The unknown function is s, and the independent variable is t. *First degree*: the equation is a polynomial in the unknown function s and its derivatives, with its highest derivative (here the second) raised to the first power. *Linear*: in the notation of Problem 1.9, $b_2(t) = t^2$, $b_1(t) = -t$, $b_0(t) = 0$, and $g(t) = 1 - \sin t$.

1.45 Determine the order, degree, linearity, unknown function, and independent variable for the differential equation

$$\left(\frac{d^2r}{dy^2}\right)^2 + \frac{d^2r}{dy^2} + y\frac{dr}{dy} = 0$$

▌ *Second order*: the highest derivative is the second. The unknown function is r, and the independent variable is y. *Second degree*: the equation is a polynomial in the unknown function r and its derivatives, and the highest power of the highest derivative is the second. *Nonlinear*: one of the derivatives of the unknown function is raised to a power other than the first.

1.46 Determine the order, degree, linearity, unknown function, and independent variable for the differential equation $d^nx/dy^n = y^2 + 1$.

▌ For the derivative to make sense, n must be a nonnegative integer. If n is positive, then the equation is of *nth order* and *first degree* because this derivative is raised to the first power. The unknown function is x, and the independent variable is y. *Linear*: in the notation of Problem 1.9, $b_n(y) = 1$, $b_{n-1}(y) = b_{n-2}(y) = \cdots = b_1(y) = b_0(y) = 0$, and $g(y) = y^2 + 1$. If $n = 0$, the equation is algebraic.

1.47 Determine the order, degree, linearity, unknown function, and independent variable for the differential equation $(d^2y/dx^2)^{3/2} + y = x$.

▌ *Second order*: the highest derivative is the second. The unknown function is y, and the independent variable is x. *No degree* because the equation cannot be written as a polynomial in the unknown function and its derivatives; the 3/2 power precludes such a possibility. *Nonlinear*: a derivative of the unknown function is raised to a power other than the first.

1.48 Determine the order, degree, linearity, unknown function, and independent variable for the differential equation $d^7b/dp^7 = 3p$.

▌ *Seventh order* since the highest derivative is the seventh, and *first degree* since that derivative is raised to the first power. The unknown function is b, and the independent variable is p. *Linear*: in the notation of Problem 1.9, $b_7(p) = 1$, $b_6(p) = b_5(p) = \cdots = b_0(p) = 0$, and $g(p) = 3p$.

1.49 Determine the order, degree, linearity, unknown function, and independent variable for the differential equation $(dp/db)^7 = 3b$.

▌ *First order* since the highest derivative is the first, and *seventh degree* since that derivative is raised to the seventh power. The unknown function is p, and the independent variable is b. *Nonlinear* because one of the derivatives of the unknown function is raised to a power other than the first.

1.50 Must a linear ordinary differential equation always have a degree?

▌ Yes, and the degree is always 1 because the highest-order derivative is always raised to the first power.

1.51 If an ordinary differential equation has a degree, must it be linear?

▌ No. Counterexamples are provided by Problems 1.45 and 1.49.

FORMULATING PROPORTIONALITY PROBLEMS

1.52 Radium decomposes at a rate proportional to the amount present. Derive a differential equation for the amount of radium present at any time t.

▌ Let $R(t)$ denote the amount of radium present at time t. The decomposition rate is dR/dt, which is proportional to R. Thus, $dR/dt = kR$, where k is a constant of proportionality.

1.53 Bacteria are placed in a nutrient solution at time $t = 0$ and allowed to multiply. Under conditions of plentiful food and space, the bacteria population grows at a rate proportional to the population. Derive a differential equation for the approximate number of bacteria present at any time t.

▌ Let $N(t)$ denote the number of bacteria present in the nutrient solution at time t. The growth rate is dN/dt, which is proportional to N. Thus, $dN/dt = kN$, where k is a constant of proportionality.

1.54 One hundred grams of cane sugar in water is being converted into dextrose at a rate which is proportional to the unconverted amount. Find a differential equation expressing the rate of conversion after t minutes.

▌ Let q denote the number of grams converted in t minutes. Then $100 - q$ is the number of grams still unconverted, and the rate of conversion is given by $dq/dt = k(100 - q)$, k being the constant of proportionality.

1.55 Bacteria are placed in a nutrient solution at time $t = 0$ and allowed to multiply. Food is plentiful but space is limited, so ultimately the bacteria population will stabilize at a constant level M. Derive a differential equation for the approximate number of bacteria present at any time t if it is known that the growth rate of the bacteria is jointly proportional to both the number of bacteria present and the difference between M and the current population.

▮ Denote the number of bacteria present in the nutrient solution at time t by $N(t)$. The growth rate is dN/dt. Since this rate is jointly proportional to N and $(M - N)$, we have $dN/dt = kN(M - N)$, where k is a constant of proportionality.

1.56 Express the following proposition as a differential equation: the population P of a city increases at a rate which is jointly proportional to the current population and the difference between 200,000 and the current population.

▮ Let $P(t)$ denote the current population; then the rate of increase is dP/dt. Since this rate is jointly proportional to both P and $(200{,}000 - P)$, we have $dP/dt = kP(200{,}000 - P)$, where k is a constant of proportionality.

1.57 A bank pays interest to depositors at the rate of r percent per annum, compounded continuously. Derive a differential equation for the amount of money in an existing account at any time t, assuming no future withdrawals or additional deposits.

▮ Let $P(t)$ denote the amount in the account at time t. Then dP/dt, the change in P, is the interest received, which is the interest rate (converted to a decimal) times the current amount. Thus, $dP/dt = (r/100)P$.

1.58 When ethyl acetate in dilute aqueous solution is heated in the presence of a small amount of acid, it decomposes according to the equation

$$\underset{\text{(Ethyl acetate)}}{CH_3COOC_2H_5} + \underset{\text{(water)}}{H_2O} \longrightarrow \underset{\text{(acetic acid)}}{CH_3COOH} + \underset{\text{(ethyl alcohol)}}{C_2H_5OH}$$

Since this reaction takes place in dilute solution, the quantity of water present is so great that the loss of the small amount which combines with the ethyl acetate produces no appreciable change in the total amount. Hence, of the reacting substances only the ethyl acetate suffers a measurable change in concentration. A chemical reaction of this sort, in which the concentration of only one reacting substance changes, is called a *first-order reaction.* It is a law of physical chemistry that the rate at which a substance is used up in a first-order reaction is proportional to the amount of that substance instantaneously present. Find an expression for the concentration of ethyl acetate at any time t.

▮ Let Q be the amount of ethyl acetate present in the solution at time t, let V be the (constant) amount of water in which it is dissolved, and let C be the instantaneous concentration of the ethyl acetate. Then $Q = CV$, and, from the law governing first-order reactions,

$$\frac{dQ}{dt} = -kQ \quad \text{or} \quad \frac{d(CV)}{dt} = -k(CV)$$

or finally $dC/dt = -kC$.

PROBLEMS INVOLVING NEWTON'S LAW OF COOLING

1.59 Newton's law of cooling states that the rate at which a hot body cools is proportional to the difference in temperature between the body and the (cooler) surrounding medium. Derive a differential equation for the temperature of a hot body as a function of time if it is placed in a bath which is held at a constant temperature of 32 °F.

▮ Denote the temperature of the hot body at time t by $T(t)$, and assume it is placed in the bath at $t = 0$. The rate at which the body cools is dT/dt. Since this is proportional to $(T - 32)$, we have $dT/dt = k(T - 32)$, where k is a constant of proportionality.

1.60 A red-hot steel rod is suspended in air which remains at a constant temperature of 24 °C. Find a differential equation for the temperature of the rod as a function of time.

▮ Denote the temperature of the steel rod at time t by $T(t)$, and assume it is placed in the cooler medium at $t = 0$. The rate at which the rod cools is dT/dt. By Newton's law of cooling (see Problem 1.59), this rate is proportional to $(T - 24)$. Therefore, $dT/dt = k(T - 24)$, where k is a constant of proportionality.

PROBLEMS INVOLVING NEWTON'S SECOND LAW OF MOTION

1.61 Newton's second law of motion states that the time rate of change of the momentum of a body is equal to the net force acting on that body. Derive the differential equation governing the motion of a body when the only force acting on it is the force of gravity.

▌ Denote the mass of the body by m, and let $y(t)$ be the vertical distance to the body from some fixed reference height at any time t. Then the velocity of the body is dy/dt, the time rate of change of position. Its momentum is its mass times its velocity, or $m\frac{dy}{dt}$. The time rate of change of its momentum is $\frac{d}{dt}\left(m\frac{dy}{dt}\right) = m\frac{d^2y}{dt^2}$, if we assume its mass remains constant. The force of gravity is the only force acting on the body; it is given by mg, where g denotes the acceleration due to gravity (a constant 32 ft/s^2 or 9.8 m/s^2 close to the surface of the earth). Thus, the required equation is

$$m\frac{d^2y}{dt^2} = mg \qquad \text{or} \qquad \frac{d^2y}{dt^2} = g$$

1.62 Derive the differential equation governing the motion of a body that is subject to both the force of gravity and air resistance (which exerts a force that opposes and is proportional to the velocity of the body).

▌ This problem is similar to Problem 1.61, except now two forces act on the body in opposite directions. The force of gravity is mg, while the force due to air resistance is $-k\frac{dy}{dt}$, where k is a constant of proportionality. Thus the net force on the body is $mg - k\frac{dy}{dt}$, and it follows from Newton's second law of motion that $m\frac{d^2y}{dt^2} = mg - k\frac{dy}{dt}$.

1.63 Redo Problem 1.62 if the air resistance is replaced by a force that is proportional to the square of the velocity of the body.

▌ The new force is $-k(dy/dt)^2$, so the net force on the body is $mg - k(dy/dt)^2$. Newton's second law of motion now yields $m\frac{d^2y}{dt^2} = mg - k\left(\frac{dy}{dt}\right)^2$.

1.64 A particle of mass m moves along a straight line (the x axis) while subject to (1) a force proportional to its displacement x from a fixed point O in its path and directed toward O and (2) a resisting force proportional to its velocity. Write a differential equation for the motion of the particle.

▌ The first force may be represented by k_1x, and the second by $-k_2\frac{dx}{dt}$, where k_1 and k_2 are factors of proportionality. Newton's second law then yields $m\frac{d^2x}{dt^2} = k_1x - k_2\frac{dx}{dt}$.

1.65 A torpedo is fired from a ship and travels in a straight path just below the water's surface. Derive the differential equation governing the motion of the torpedo if the water retards the torpedo with a force proportional to its speed.

▌ Let $x(t)$ denote the distance of the torpedo from the ship at any time t. The velocity of the torpedo is dx/dt. The only force acting on the torpedo is the resisting force of the water, $k\frac{dx}{dt}$, where k is a constant of proportionality. If we assume the mass of the torpedo remains constant, then its time rate of change of momentum is $m\frac{d^2x}{dt^2}$. It follows from Newton's second law of motion (see Problem 1.61) that $m\frac{d^2x}{dt^2} = k\frac{dx}{dt}$.

1.66 Inside the earth, the force of gravity is proportional to the distance from the center. Assume that a hole is drilled through the earth from pole to pole, and a rock is dropped into the hole. Derive the differential equation for the motion of this rock.

▌ Let $s(t)$ denote the distance from the rock at any time t to the center of the earth. The force of gravity is ks, where k is a constant of proportionality, so Newton's second law of motion (see Problem 1.61) yields $m\frac{d^2s}{dt^2} = ks$.

1.67 A boat is being towed at the rate of 12 mi/h. At $t = 0$ the towing line is cast off and a man in the boat begins to row in the direction of motion, exerting a force of 20 lb. The combined weight of the man and the boat is

480 lb. The water resists the motion with a force equal to 1.75v lb, where v is the velocity of the boat in feet per second. Derive a differential equation governing the velocity of the boat.

▮ The boat moves along a straight line, which we take to be the x axis, with the positive direction being the direction of motion. Then $v = dx/dt$. For constant mass, Newton's second law (Problem 1.61) gives us

$$m\frac{dv}{dt} = \text{net force} = \text{forward force} - \text{resistance}$$

so that

$$\frac{480}{32}\frac{dv}{dt} = 20 - 1.75v \qquad \text{or} \qquad \frac{dv}{dt} + \frac{7}{60}v = \frac{4}{3}$$

We are also given the initial velocity of the boat, $v(0) = 12\text{ mi/h} = 12(5280)/(60)^2 = 17.6\text{ ft/s}$, which we would need to find the velocity at times after $t = 0$.

1.68 A mass is being pulled across the ice on a sled with a constant force. The resistance offered by the ice to the runners is negligible, but the resistance (in pounds) offered by the air is five times the velocity of the sled. Derive a differential equation for the velocity of the sled if the combined weight of the sled and the mass is 80 lb.

▮ We assume that the motion of the sled is along a straight line; we designate that line as the x axis, with the positive direction being the direction of motion. The velocity of the sled is then $v = dx/dt$. From Newton's second law of motion (see Problem 1.61), we have $m\frac{dv}{dt} = \text{forward force} - \text{resistance}$.

We denote the constant forward force by F, and $m = 80/32 = 2.5$ slugs. The differential equation is then

$$2.5\frac{dv}{dt} = F - 5v \qquad \text{or} \qquad \frac{dv}{dt} + 2v = \frac{2}{5}F$$

SPRING PROBLEMS

1.69 Hooke's law states that the restoring force of a spring is proportional to the displacement of the spring from its normal length. Use Hooke's law along with Newton's second law of motion to derive the differential equation governing the motion of the following system: A spring with a mass m attached to its lower end is suspended vertically from a mounting and allowed to come to rest in an equilibrium position. The system is then set in motion by releasing the mass with an initial velocity v_0 at a distance x_0 below its equilibrium position and simultaneously applying to the mass an external force $F(t)$ in the downward direction.

▮ For convenience, we choose the downward direction as the positive direction and take the origin to be the center of gravity of the mass in the equilibrium position (see Fig. 1.1). Furthermore, we assume that air resistance is present and is proportional to the velocity of the mass. Thus, at any time t, there are three forces acting on the system: (1) $F(t)$, measured in the positive direction; (2) a restoring force given by Hooke's law as $F_s = -kx$, where $k > 0$ is a constant of proportionality known as the *spring constant*; and (3) a force due to air resistance given by $F_a = -a\dot{x}$, where $a > 0$ is a constant of proportionality. Note that the restoring force F_s always acts in a direction that will tend to return the system to the equilibrium position: if the mass is below the equilibrium position, then x is positive and $-kx$ is negative; whereas if the mass is above the equilibrium position, then x is negative and $-kx$ is positive. Also note that because $a > 0$ the force F_a due to air resistance acts in the direction opposite the velocity and thus tends to retard, or *damp*, the motion of the mass.

It now follows from Newton's second law that $m\ddot{x} = -kx - a\dot{x} + F(t)$, or

$$\ddot{x} + \frac{a}{m}\dot{x} + \frac{k}{m}x = \frac{F(t)}{m} \qquad (1$$

Since the system starts at $t = 0$ with an initial velocity v_0 and from an initial position x_0, we have along with (*1*) the initial conditions $x(0) = x_0$ and $\dot{x}(0) = v_0$.

The force of gravity does not explicitly appear in (*1*), but it is present nonetheless. We automatically compensated for this force by measuring distance from the equilibrium position of the spring. If one wishes to exhibit gravity explicitly, then distance must be measured from the bottom end of the *natural length* of the spring. That is, the motion of a vibrating spring can be given by

$$\ddot{x} + \frac{a}{m}\dot{x} + \frac{k}{m}x = g + \frac{F(t)}{m}$$

if the origin, $x = 0$, is the terminal point of the unstretched spring before the mass m is attached.

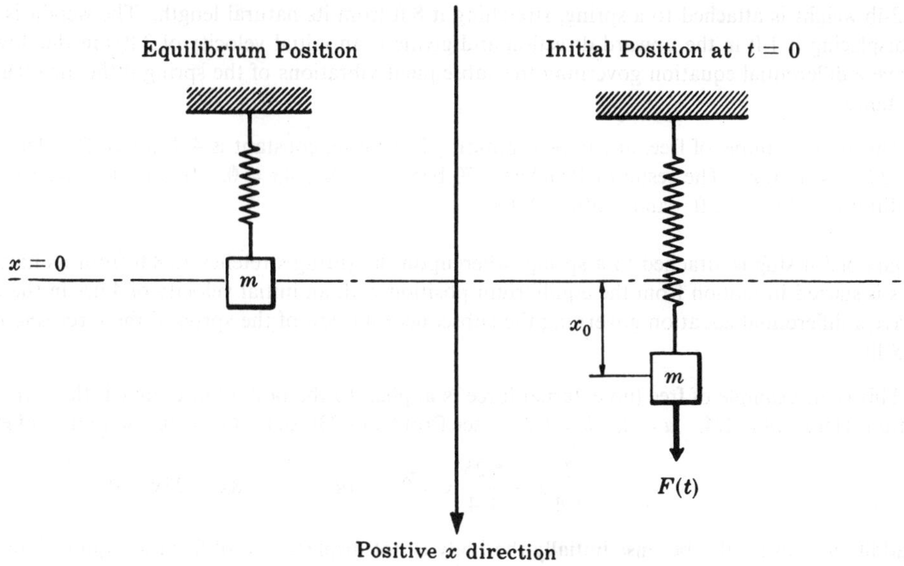

Fig. 1.1

1.70 Derive the differential equation governing the motion of the spring system shown in Fig. 1.1 if the vibrations are free and undamped.

▮ The vibrations are free if no external force is applied to the spring, and they are undamped if air resistance is zero. With $F(t) = 0$ and $a = 0$, *(1)* of Problem 1.69 reduces to $\ddot{x} + (k/m)x = 0$.

1.71 A steel ball weighing 128 lb is suspended from a spring, whereupon the spring stretches 2 ft from its natural length. What is the value of the spring constant?

▮ The applied force responsible for the 2-ft displacement is the weight of the ball, 128 lb. Thus, $F = -128$ lb. Hooke's law then gives $-128 = -k(2)$, or $k = 64$ lb/ft.

1.72 A 32-lb weight is attached to a spring, stretching it 8 ft from its natural length. What is the value of the spring constant?

▮ The applied force responsible for the 8-ft displacement is the 32-lb weight. At equilibrium, this force is balanced by the restoring force of the spring, so by Hooke's law $-32 = -k(8)$, or $k = 4$ lb/ft.

1.73 A mass of 1/4 slug is attached to a spring, whereupon the spring stretches 1.28 ft from its natural length. What is the value of spring constant?

▮ The applied force responsible for the 1.28-ft displacement is the weight of the attached body, which is $(1/4)(32) = 8$ lb. At equilibrium, this force is balanced by the restoring force of the spring, so by Hooke's law $-8 = -k(1.28)$, or $k = 6.25$ lb/ft.

1.74 A 10-kg mass is attached to a spring, stretching it 0.7 m from its natural length. What is the value of the spring constant?

▮ The applied force responsible for the 0.7-m displacement is the weight of the attached body, which is $10(9.8) = 9.8$ N. At equilibrium, this force is balanced by the restoring force of the spring, so by Hooke's law $-98 = k(0.7)$, from which $k = 140$ N/m.

1.75 A steel ball weighing 128 lb is suspended from a spring, whereupon the spring stretches 2 ft from its natural length. The ball is started in motion with no initial velocity by displacing it 6 in above the equilibrium position. Derive a differential equation governing the subsequent vibrations of the spring if there is no air resistance.

▮ This is an example of free, undamped motion. The spring constant was determined in Problem 1.71 to be $k = 64$ lb/ft; the weight of the ball is $mg = 128$ lb, so $m = 128/32 = 4$ slugs. With these values, the result of Problem 1.70 becomes $\ddot{x} + (64/4)x = 0$, or $\ddot{x} + 16x = 0$. In addition, we have the initial conditions $x(0) = -1/2$ ft (the minus sign is required because the ball is initially displaced *above* the equilibrium position, which is in the *negative* direction) and $\dot{x}(0) = 0$.

1.76 A 32-lb weight is attached to a spring, stretching it 8 ft from its natural length. The weight is started in motion by displacing it 1 ft in the upward direction and giving it an initial velocity of 2 ft/s in the downward direction. Derive a differential equation governing the subsequent vibrations of the spring if the air offers negligible resistance.

▮ This is an example of free, undamped motion. The spring constant is 4 lb/ft (see Problem 1.72), and $m = 32/32 = 1$ slug. The result of Problem 1.70 becomes $\ddot{x} + 4x = 0$. In addition, we have the initial conditions $x(0) = -1$ ft and $\dot{x}(0) = 2$ ft/s.

1.77 A mass of 1/4 slug is attached to a spring, whereupon the spring stretches 1.28 ft from its natural length. The mass is started in motion from the equilibrium position with an initial velocity of 4 ft/s in the downward direction. Derive a differential equation governing the subsequent motion of the spring if the force due to air resistance is $-2\dot{x}$ lb.

▮ This is an example of free (no external force is applied to the body) but damped (there is air resistance) motion. Here $m = 1/4$, $a = 2$, $k = 6.25$ (see Problem 1.73), and $F(t) = 0$, so that (*1*) of Problem 1.69 becomes

$$\ddot{x} + \frac{2}{1/4}\dot{x} + \frac{6.25}{1/4}x = 0 \qquad \text{or} \qquad \ddot{x} + 8\dot{x} + 25x = 0$$

In addition, $x(0) = 0$, because initially the body is not displaced at all from its equilibrium position, and $\dot{x}(0) = 4$ ft/s.

1.78 A 10-kg mass is attached to a spring, stretching it 0.7 m from its natural position. The mass is started in motion from the equilibrium position with an initial velocity of 1 m/s in the upward direction. Derive a differential equation governing the subsequent motion of the spring if the force due to air resistance is $-90\dot{x}$ N.

▮ Here $m = 10$, $a = 90$, $k = 140$ (see Problem 1.74), and $F(t) = 0$, so that (*1*) of Problem 1.69 becomes

$$\ddot{x} + \frac{90}{10}\dot{x} + \frac{140}{10}x = 0 \qquad \text{or} \qquad \ddot{x} + 9\dot{x} + 14x = 0$$

In addition, $x(0) = 0$ (the mass starts at the equilibrium position) and $\dot{x}(0) = -1$ (the initial velocity is in the upward, or negative, direction).

1.79 Redo Problem 1.78 if, in addition, an external force $5 \sin t$ (in newtons) is applied to the system.

▮ The constants m, a, and k remain as before, but now $F(t) = 5 \sin t$ and (*1*) of Problem 1.69 becomes

$$\ddot{x} + \frac{90}{10}\dot{x} + \frac{140}{10}x = \frac{5 \sin t}{10} \qquad \text{or} \qquad \ddot{x} + 9\dot{x} + 14x = \tfrac{1}{2}\sin t$$

Vibrations subject to external forces are called *forced* vibrations.

1.80 A 128-lb weight is attached to a spring having a spring constant of 64 lb/ft. The weight is started in motion with no initial velocity by displacing it 6 in above the equilibrium position and by simultaneously applying to the weight an external force $F(t) = 8 \sin 4t$. Derive a differential equation governing the subsequent vibrations of the spring if there is no air resistance.

▮ This is an example of forced (there is an applied external force) but undamped (there is no air resistance) motion. Here $m = 128/32 = 4$ slugs, $k = 64$ lb/ft, $a = 0$, and $F(t) = 8 \sin 4t$ lb, so (*1*) of Problem 1.69 becomes

$$\ddot{x} + \frac{64}{4}x = \frac{8 \sin 4t}{4} \qquad \text{or} \qquad \ddot{x} + 16x = 2 \sin 4t$$

The initial conditions are $x(0) = -\frac{1}{2}$ ft and $\dot{x}(0) = 0$.

ELECTRIC CIRCUIT PROBLEMS

1.81 Kirchoff's loop law states that the algebraic sum of the voltage drops in a simple closed electric circuit is zero. Use this law to derive a differential equation for the current I in a simple circuit consisting of a resistor, a capacitor, an inductor, and an electromotive force (usually a battery or a generator) connected in series.

▮ The circuit is shown in Fig. 1.2, where R is the resistance in ohms, C is the capacitance in farads, L is the inductance in henries, $E(t)$ is the electromotive force (emf) in volts, and I is the current in amperes. It is known

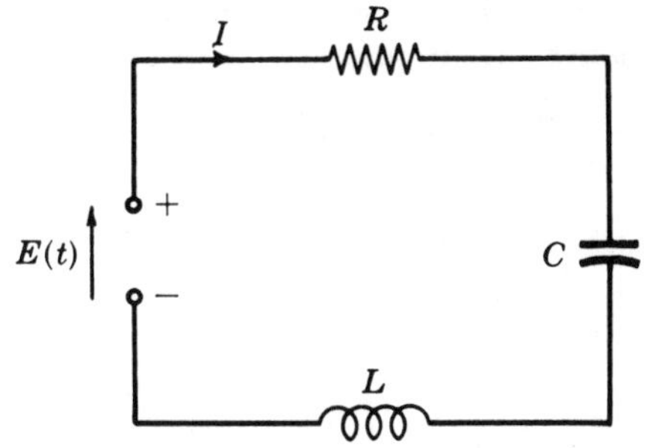

Fig. 1.2

that the voltage drops across a resistor, a capacitor, and an inductor are respectively RI, $\frac{1}{C}q$, and $L\frac{dI}{dt}$, where q is the charge on the capacitor. The voltage drop across an emf is $-E(t)$. Thus, from Kirchhoff's loop law, we have

$$RI + L\frac{dI}{dt} + \frac{1}{C}q - E(t) = 0 \qquad (1)$$

The relationship between q and I is $I = dq/dt$. Differentiating (1) with respect to t and using this relation, we obtain $R\frac{dI}{dt} + L\frac{d^2I}{dt^2} + \frac{1}{C}I - \frac{dE(t)}{dt} = 0$, which may be rewritten as

$$\frac{d^2I}{dt^2} + \frac{R}{L}\frac{dI}{dt} + \frac{1}{LC}I = \frac{1}{L}\frac{dE(t)}{dt} \qquad (2)$$

1.82 Derive a differential equation for the charge on the capacitor in the series RCL circuit of Fig. 1.2.

▮ From the last problem we have $I = dq/dt$ and so $dI/dt = d^2q/dt^2$. Substituting these equalities into (1) of Problem 1.81 and rearranging, we obtain

$$\frac{d^2q}{dt^2} + \frac{R}{L}\frac{dq}{dt} + \frac{1}{LC}q = \frac{1}{L}E(t) \qquad (1)$$

1.83 A simple series RCL circuit has $R = 180\ \Omega$, $C = 1/280$ F, $L = 20$ H, and applied voltage $E(t) = 10\sin t$. Derive a differential equation for the charge on the capacitor at any time t.

▮ Substituting the given quantities into (1) of Problem 1.82, we get

$$\frac{d^2q}{dt^2} + \frac{180}{20}\frac{dq}{dt} + \frac{1}{20(1/280)}q = \frac{1}{20}(10\sin t) \qquad \text{or} \qquad \ddot{q} + 9\dot{q} + 14q = \tfrac{1}{2}\sin t$$

1.84 A simple series RCL circuit has $R = 10\ \Omega$, $C = 10^{-2}$ F, $L = 1/2$ H, and applied voltage $E = 12$ V. Derive a differential equation for the amount of charge on the capacitor at any time t.

▮ Substituting the given quantities into (1) of Problem 1.82, we obtain

$$\frac{d^2q}{dt^2} + \frac{10}{1/2}\frac{dq}{dt} + \frac{1}{(1/2)(10^{-2})}q = \frac{1}{1/2}(12) \qquad \text{or} \qquad \ddot{q} + 20\dot{q} + 200q = 24$$

1.85 Find a differential equation for the current in the circuit of Problem 1.84.

▮ Substituting the values given in Problem 1.84 into (2) of Problem 1.81, we obtain

$$\frac{d^2I}{dt^2} + \frac{10}{1/2}\frac{dI}{dt} + \frac{1}{(1/2)(10^{-2})}I = \frac{1}{1/2}\frac{d(12)}{dt} \qquad \text{or} \qquad \frac{d^2I}{dt^2} + 20\frac{dI}{dt} + 200I = 0$$

1.86 A simple series RCL circuit has $R = 6\ \Omega$, $C = 0.02$ F, $L = 0.1$ H, and no applied voltage. Derive a differential equation for the current in the circuit at any time t.

▮ Substituting the given quantities into (2) of Problem 1.81, we obtain

$$\frac{d^2I}{dt^2} + \frac{6}{0.1}\frac{dI}{dt} + \frac{1}{(0.1)(0.02)}I = \frac{1}{0.1}\frac{d(0)}{dt} \qquad \text{or} \qquad \frac{d^2I}{dt^2} + 60\frac{dI}{dt} + 500I = 0$$

1.87 Use Kirchoff's loop law to derive a differential equation for the current in a simple circuit consisting of a resistor, an inductor, and an electromotive force connected in series (a series *RL* circuit).

▮ The circuit is similar to the one in Fig. 1.2, but without the capacitor. The voltage drops are given in Problem 1.81, so it follows from Kirchoff's law that

$$RI + L\frac{dI}{dt} - E(t) = 0 \quad \text{or} \quad \frac{dI}{dt} + \frac{R}{L}I = \frac{1}{L}E(t) \tag{1}$$

1.88 A simple series *RL* circuit has an emf given by 3 sin 2*t* (in volts), a resistance of 10 Ω, and an inductance of 0.5 H. Derive a differential equation for the current in the system.

▮ Substituting the given quantities into (*1*) of Problem 1.87, we obtain

$$\frac{dI}{dt} + \frac{10}{0.5}I = \frac{1}{0.5}(3\sin 2t) \quad \text{or} \quad \frac{dI}{dt} + 20I = 6\sin 2t$$

1.89 Derive a differential equation for the current *I* in a series *RL* circuit having a resistance of 10 Ω, an inductance of 4 H, and no applied electromotive force.

▮ Here $E(t) = 0$, $R = 10$, and $L = 4$, so (*1*) of Problem 1.87 becomes $\dfrac{dI}{dt} + \dfrac{10}{4}I = 0.$

1.90 Use Kirchoff's loop law to derive a differential equation for the charge on the capacitor of a circuit consisting of a resistor, a capacitor, and an electromotive force (emf) connected in series (a series *RC* circuit).

▮ The circuit is similar to the one in Fig. 1.2, but without the inductor. The voltage drops are as given in Problem 1.81, so it follows from Kirchoff's law that $RI + q/C - E(t) = 0$. Since $I = dq/dt$, this may be rewritten as

$$\frac{dq}{dt} + \frac{1}{RC}q = \frac{1}{R}E(t) \tag{1}$$

1.91 A series *RC* circuit has an emf given by 400 cos 2*t* (in volts), a resistance of 100 Ω, and a capacitance of 0.01 F. Find a differential equation for the charge on the capacitor.

▮ Substituting the given quantities into (*1*) of Problem 1.90, we obtain

$$\frac{dq}{dt} + \frac{1}{100(0.01)}q = \frac{1}{100}(400\cos 2t) \quad \text{or} \quad \dot{q} + q = 4\cos 2t \tag{1}$$

1.92 Derive a differential equation for the current in the circuit of the previous problem.

▮ Differentiating (*1*) of that problem with respect to time, we obtain $\dfrac{d}{dt}\left(\dfrac{dq}{dt}\right) + \dfrac{dq}{dt} = -8\sin 2t.$ Using the relationship $I = dq/dt$, we find that $dI/dt + I = -8\sin 2t.$

1.93 Derive a differential equation for the charge on the capacitor of a series *RC* circuit having a resistance of 10 Ω, a capacitance of 10^{-3} F, and an emf of 100 sin 120π*t*.

▮ Substituting these quantities into (*1*) of Problem 1.90, we obtain $\dot{q} + 100q = 10\sin 120\pi t.$

GEOMETRICAL PROBLEMS

1.94 Derive a differential equation for the orthogonal trajectories of the one-parameter family of curves in the *xy* plane defined by $F(x, y, c) = 0$, where *c* denotes the parameter.

▮ The orthogonal trajectories consist of a second family of curves having the property that each curve in this new family intersects at right angles every curve in the original family. Thus, at every point of intersection, the slope of the tangent of each curve in the new family must be the negative reciprocal of the slope of the tangent of each curve in the original family. To get the slope of the tangent, we differentiate $F(x, y, c) = 0$ implicitly with respect to *x*, and then eliminate *c* by solving for it in the equation $F(x, y, c) = 0$ and substituting for it in the derived equation. This gives an equation connecting *x*, *y*, and *y*′, which we solve for *y*′ to obtain a differential equation of the form $dy/dx = f(x, y)$. The orthogonal trajectories are then the solutions of

$$\frac{dy}{dx} = -\frac{1}{f(x, y)} \tag{1}$$

For many families of curves, one cannot explicitly solve for dy/dx and obtain a differential equation of the form $dy/dx = f(x, y)$. We do not consider such curves in this book.

1.95 Derive a differential equation for the orthogonal trajectories of the family of curves $y = cx^2$.

▮ The family of curves is a set of parabolas symmetric about the y axis with vertices at the origin. In the notation of Problem 1.94, we have $F(x, y, c) = y - cx^2 = 0$. Implicitly differentiating the given equation with respect to x, we obtain $dy/dx = 2cx$. To eliminate c, we observe, from the given equation, that $c = y/x^2$; hence, $dy/dx = 2y/x$. We have found $f(x, y) = 2y/x$, so (1) of Problem 1.94 becomes $\dfrac{dy}{dx} = \dfrac{-x}{2y}$.

1.96 Derive a differential equation for the orthogonal trajectories of the family of curves $x^2 + y^2 = cx$.

▮ The family of curves is a set of circles centered at $(c/2, 0)$. In the notation of Problem 1.94, we have $F(x, y, c) = x^2 + y^2 - cx$. Implicitly differentiating the given equation with respect to x, we obtain $2x + 2y\dfrac{dy}{dx} = c$. Eliminating c between this equation and $x^2 + y^2 - cx = 0$ gives $\dfrac{dy}{dx} = \dfrac{y^2 - x^2}{2xy}$. We have found $f(x, y) = (y^2 - x^2)/2xy$, so (1) of Problem 1.94 becomes $\dfrac{dy}{dx} = \dfrac{2xy}{x^2 - y^2}$.

1.97 Derive a differential equation for the orthogonal trajectories of the family of curves $x^2 + y^2 = c^2$.

▮ This family of curves is a set of circles with centers at the origin and radii c. In the notation of Problem 1.94, we have $F(x, y, c) = x^2 + y^2 - c^2$. Implicitly differentiating the given equation with respect to x, we get $2x + 2yy' = 0$ or $dy/dx = -x/y$. Since $f(x, y) = -x/y$, (1) of Problem 1.94 becomes $dy/dx = y/x$.

1.98 Derive a differential equation for the orthogonal trajectories of the family of curves $y = ce^x$.

▮ In the notation of Problem 1.94, we have $F(x, y, c) = y - ce^x$. Implicitly differentiating this equation with respect to x, we obtain $y' - ce^x = 0$. Since $y = ce^x$, it follows that $y' - y = 0$ or $y' = y$. Here $f(x, y) = y$, so (1) of Problem 1.94 becomes $dy/dx = -1/y$.

1.99 Derive a differential equation for the orthogonal trajectories of the family of curves $xy = C$.

▮ In the notation of Problem 1.94, we have $F(x, y, C) = xy - C$. Implicitly differentiating this equation with respect to x, we get $y + xy' = 0$ or $y' = -y/x$. Here $f(x, y) = -y/x$, so (1) of Problem 1.94 becomes $dy/dx = x/y$.

1.100 Derive a differential equation for the orthogonal trajectories of the cardioid $\rho = C(1 + \sin\theta)$, expressed in polar coordinates.

▮ Differentiating with respect to θ to obtain $\dfrac{d\rho}{d\theta} = C\cos\theta$, solving for $C = \dfrac{1}{\cos\theta}\dfrac{d\rho}{d\theta}$, and substituting for C in the given equation lead to the differential equation of the given family: $\dfrac{d\rho}{d\theta} = \dfrac{\rho\cos\theta}{1 + \sin\theta}$. In polar coordinates, the differential equation of the orthogonal trajectories is obtained by replacing $d\rho/d\theta$ by $-\rho^2\, d\theta/d\rho$, which gives us

$$-\frac{d\theta}{d\rho} = \frac{\cos\theta}{\rho(1 + \sin\theta)} \quad \text{or} \quad \frac{d\rho}{\rho} + (\sec\theta + \tan\theta)\,d\theta = 0$$

1.101 A curve is defined by the condition that at each of its points (x, y), its slope dy/dx is equal to twice the sum of the coordinates of the point. Express the condition by means of a differential equation.

▮ The differential equation representing the condition is $dy/dx = 2(x + y)$.

1.102 A curve is defined by the condition that the sum of the x and y intercepts of its tangents is always equal to 2. Express the condition by means of a differential equation.

The equation of the tangent at (x, y) on the curve is $Y - y = \frac{dy}{dx}(X - x)$, and the x and y intercepts are, respectively, $X = x - y\frac{dx}{dy}$ and $Y = y - x\frac{dy}{dx}$. The differential equation representing the condition is

$$X + Y = x - y\frac{dx}{dy} + y - x\frac{dy}{dx} = 2 \quad \text{or} \quad x\left(\frac{dy}{dx}\right)^2 - (x + y - 2)\frac{dy}{dx} + y = 0$$

PRIMITIVES

1.103 Define *essential constants* in the context of a relationship between two variables.

If a relationship between two variables involves n arbitrary constants, then those constants are essential if they cannot be replaced by a smaller number of constants.

1.104 Show that only one arbitrary constant is essential in the relationship $y = x^2 + A + B$ involving the variables x and y.

Since $A + B$ is no more than a single arbitrary constant, only one essential constant is involved.

1.105 Show that only one arbitrary constant is essential in the relationship $y = Ae^{x+B}$ involving the variables x and y.

Since $y = Ae^{x+B} = Ae^x e^B$, and Ae^B is no more than a single arbitrary constant, only one essential constant is required.

1.106 Show that only one arbitrary constant is essential in the relationship $y = A + \ln Bx$ involving the variables x and y.

Since $y = A + \ln Bx = A + \ln B + \ln x$, and $(A + \ln B)$ is no more than a single constant, only one essential constant is involved.

1.107 Define *primitive* in the context of a relation between two variables.

A primitive is a relationship between two variables which contains only essential arbitrary constants. Examples are $y = x^4 + C$ and $y = Ax^2 + Bx$, involving the variables x and y.

1.108 Describe a procedure for obtaining a differential equation from a primitive.

In general, a primitive involving n essential arbitrary constants will give rise to a differential equation of order n, free of arbitrary constants. This equation is obtained by eliminating the n constants between the $n + 1$ equations consisting of the primitive and the n equations obtained by differentiating the primitive n times with respect to the independent variable.

1.109 Obtain the differential equation associated with the primitive $y = Ax^2 + Bx + C$.

Since there are three arbitrary constants, we consider the four equations

$$y = Ax^2 + Bx + C \qquad \frac{dy}{dx} = 2Ax + B \qquad \frac{d^2y}{dx^2} = 2A \qquad \frac{d^3y}{dx^3} = 0$$

The last of these, being free of arbitrary constants and of the proper order, is the required equation.

Note that the constants could not have been eliminated between the first three of the above equations. Note also that the primitive can be obtained readily from the differential equation by integration.

1.110 Obtain the differential equation associated with the primitive $x^2y^3 + x^3y^5 = C$.

Differentiating once with respect to x, we obtain

$$\left(2xy^3 + 3x^2y^2\frac{dy}{dx}\right) + \left(3x^2y^5 + 5x^3y^4\frac{dy}{dx}\right) = 0$$

or

$$\left(2y + 3x\frac{dy}{dx}\right) + xy^2\left(3y + 5x\frac{dy}{dx}\right) = 0 \qquad \text{for } xy \neq 0$$

as the required equation. Written in differential notation, these equations are

$$(2xy^3\,dx + 3x^2y^2\,dy) + (3x^2y^5\,dx + 5x^3y^4\,dy) = 0 \tag{1}$$

and

$$(2y\,dx + 3x\,dy) + xy^2(3y\,dx) + 5x\,dy) = 0 \tag{2}$$

Note that the primitive can be obtained readily from (*1*) by integration but not so readily from (*2*). To obtain the primitive when (*2*) is given, it is necessary to determine the factor xy^2 which was removed from (1).

1.111 Obtain the differential equation associated with the primitive $y = A\cos ax + B\sin ax$, A and B being arbitrary constants and a being a fixed constant.

▌ Here $$\frac{dy}{dx} = -Aa\sin ax + Ba\cos ax$$

and $$\frac{d^2y}{dx^2} = -Aa^2\cos ax - Ba^2\sin ax = -a^2(A\cos ax + B\sin ax) = -a^2y$$

The required differential equation is $d^2y/dx^2 + a^2y = 0$.

1.112 Obtain the differential equation associated with the primitive $y = Ae^{2x} + Be^x + C$.

▌ Here $$\frac{dy}{dx} = 2Ae^{2x} + Be^x \qquad \frac{d^2y}{dx^2} = 4Ae^{2x} + Be^x \qquad \frac{d^3y}{dx^3} = 8Ae^{2x} + Be^x$$

Then $$\frac{d^3y}{dx^3} - \frac{d^2y}{dx^2} = 4Ae^{2x} \quad \text{and} \quad \frac{d^2y}{dx^2} - \frac{dy}{dx} = 2Ae^{2x} \quad \text{so} \quad \frac{d^3y}{dx^3} - \frac{d^2y}{dx^2} = 2\left(\frac{d^2y}{dx^2} - \frac{dy}{dx}\right)$$

The required equation is $\dfrac{d^3y}{dx^3} - 3\dfrac{d^2y}{dx^2} + 2\dfrac{dy}{dx} = 0$.

1.113 Obtain the differential equation associated with the primitive $y = C_1e^{3x} + C_2e^{2x} + C_3e^x$.

▌ Here

$$\frac{dy}{dx} = 3C_1e^{3x} + 2C_2e^{2x} + C_3e^x \qquad \frac{d^2y}{dx^2} = 9C_1e^{3x} + 4C_2e^{2x} + C_3e^x \qquad \frac{d^3y}{dx^3} = 27C_1e^{3x} + 8C_2e^{2x} + C_3e^x$$

The elimination of the constants by elementary methods is somewhat tedious. If three of the equations are solved for C_1, C_2, and C_3 using determinants, and these are substituted in the fourth equation, the result may be put in the form (called the *eliminant*):

$$\begin{vmatrix} e^{3x} & e^{2x} & e^x & y \\ 3e^{3x} & 2e^{2x} & e^x & y' \\ 9e^{3x} & 4e^{2x} & e^x & y'' \\ 27e^{3x} & 8e^{2x} & e^x & y''' \end{vmatrix} = e^{6x}\begin{vmatrix} 1 & 1 & 1 & y \\ 3 & 2 & 1 & y' \\ 9 & 4 & 1 & y'' \\ 27 & 8 & 1 & y''' \end{vmatrix} = e^{6x}(-2y''' + 12y'' - 22y' + 12y) = 0$$

The required differential equation is $\dfrac{d^3y}{dx^3} - 6\dfrac{d^2y}{dx^2} + 11\dfrac{dy}{dx} - 6y = 0$.

1.114 Obtain the differential equation associated with the primitive $y = Cx^2 + C^2$.

▌ Since $dy/dx = 2Cx$, we have

$$C = \frac{1}{2x}\frac{dy}{dx} \quad \text{and} \quad y = Cx^2 + C^2 = \frac{1}{2x}\frac{dy}{dx}x^2 + \frac{1}{4x^2}\left(\frac{dy}{dx}\right)^2$$

The required differential equation is $\left(\dfrac{dy}{dx}\right)^2 + 2x^3\dfrac{dy}{dx} - 4x^2y = 0$.

(*Note*: The primitive involves one arbitrary constant of degree two, and the resulting differential equation is of order 1 and degree 2.)

1.115 Find the differential equation of the family of circles of fixed radius r with centers on the x axis.

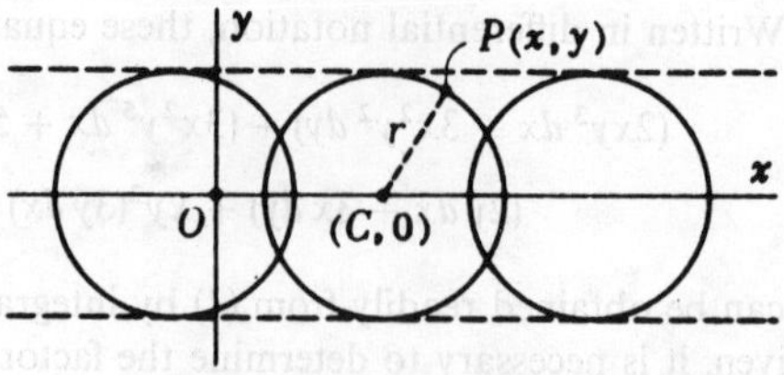

Fig. 1.3

▮ The equation of the family (see Fig. 1.3) is $(x-C)^2+y^2=r^2$, C being an arbitrary constant. Then

$$x-C+y\frac{dy}{dx}=0 \quad \text{so} \quad x-C=-y\frac{dy}{dx}$$

and the differential equation is $y^2\left(\frac{dy}{dx}\right)^2+y^2=r^2$.

1.116 Find the differential equation of the family of parabolas with foci at the origin and axes along the x axis.

▮ The equation of the family of parabolas is $y^2=4A(A+x)$. (See Figs 1.4 and 1.5.) Then $yy'=2A$, $A=\frac{1}{2}yy'$, and $y^2=2yy'(\frac{1}{2}yy'+x)$. The required equation is $y\left(\frac{dy}{dx}\right)^2+2x\frac{dy}{dx}-y=0$.

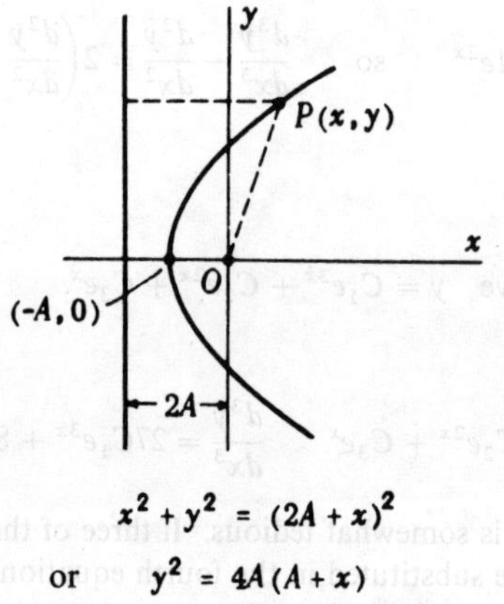

Fig. 1.4

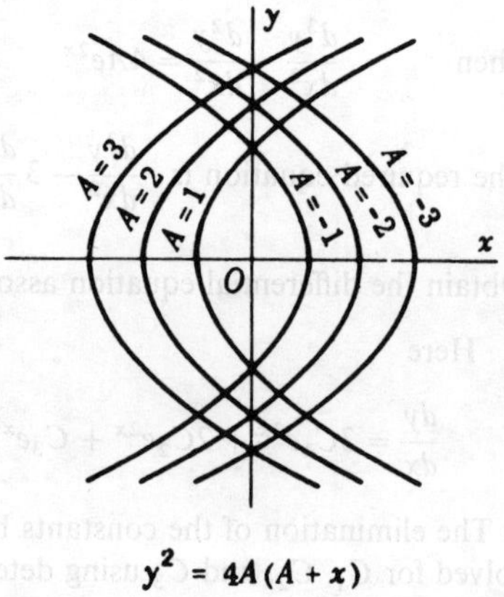

Fig. 1.5

1.117 Form the differential equation representing all tangents to the parabola $y^2=2x$.

▮ At any point (A, B) on the parabola, the equation of the tangent is $y-B=(x-A)/B$ or, since $A=\frac{1}{2}B^2$, $By=x+\frac{1}{2}B^2$. Eliminating B between this and $By'=1$, which is obtained by differentiation with respect to x, we get as the required differential equation $2x(y')^2-2yy'+1=0$.

CHAPTER 2
Solutions

VALIDATING SOLUTIONS

2.1 Determine whether $y(x) = 3e^x$ is a solution of $y' + y = 0$.

▌ Differentiating $y(x)$, we get $y'(x) = 3e^x$. Then $y' + y = 3e^x + 3e^x = 6e^x \neq 0$. Since $y(x)$ does not satisfy the differential equation anywhere, it is not a solution.

2.2 Determine whether $y(x) = 5$ is a solution of $y' + y = 0$.

▌ Differentiating $y(x)$, we get $y'(x) = 0$. Then $y' + y = 0 + 5 = 5 \neq 0$. Since $y(x)$ does not satisfy the differential equation anywhere, it is not a solution.

2.3 Determine whether $y(x) = \cos x$ is a solution to $y' + y = 0$.

▌ Differentiating $y(x)$, we get $y'(x) = -\sin x$. Then $y' + y = -\sin x + \cos x$, which is not identically zero on any interval. Because $y(x)$ does not satisfy the differential equation on any interval, it is not a solution.

Note that $y' + y$ is zero wherever $\sin x = \cos x$, which occurs at infinitely many discrete points. There is, however, no *interval* on which $\sin x = \cos x$, so there is no interval on which the differential equation is satisfied.

2.4 Determine whether $y = 3e^{-x}$ is a solution of $y' + y = 0$.

▌ Differentiating $y(x)$, we get $y'(x) = -3e^{-x}$. Then $y' + y = -3e^{-x} + 3e^{-x} = 0$. Thus $y(x) = 3e^{-x}$ satisfies the differential equation for all values of x and is a solution on the interval $(-\infty, \infty)$.

2.5 Determine whether $y = 5e^{-x}$ is a solution of $y' + y = 0$.

▌ Differentiating $y(x)$, we get $y'(x) = -5e^{-x}$. Then $y' + y = -5e^{-x} + 5e^{-x} = 0$. Thus $y(x) = 5e^{-x}$ satisfies the differential equation for all values of x and is a solution on the interval $(-\infty, \infty)$.

2.6 Show that $y(x) = Ce^{-x}$ is a solution of $y' + y = 0$ on the interval $(-\infty, \infty)$ for any arbitrary constant C.

▌ Differentiating $y(x)$, we get $y'(x) = -Ce^{-x}$. Then $y' + y = -Ce^{-x} + Ce^{-x} = 0$ for all real values of x.

2.7 Determine whether $y(x) = 2$ is a solution of $y' + y^2 = 0$.

▌ Differentiating $y(x)$, we get $y'(x) = 0$. Then $y' + y^2 = 0 + (2)^2 = 4 \neq 0$. Thus $y(x)$ does not satisfy the differential equation anywhere and is not a solution.

2.8 Determine whether $y = e^x$ is a solution of $y' + y^2 = 0$.

▌ Differentiating $y(x)$, we get $y'(x) = e^x$. Then $y' + y^2 = e^x + (e^x)^2 = e^x + e^{2x} \neq 0$. Thus $y(x)$ does not satisfy the differential equation anywhere and is not a solution.

2.9 Determine whether $y = -x$ is a solution of $y' + y^2 = 0$.

▌ Differentiating $y(x)$, we get $y'(x) = -1$. Then $y' + y^2 = -1 + (-x)^2 = x^2 - 1$, which is not identically zero on any interval. Since $y(x)$ does not satisfy the differential equation on any interval, it is not a solution.

Note that $x^2 - 1$ is zero at ± 1; but for $y(x)$ to be a solution, $x^2 - 1$ would have to be zero on some interval—and that is not the case.

2.10 Determine whether $y = 1/x$ is a solution of $y' + y^2 = 0$.

▌ Differentiating $y(x)$, we get $y'(x) = -1/x^2$. Then $y' + y^2 = -\dfrac{1}{x^2} + \left(\dfrac{1}{x}\right)^2 = 0$ for all nonzero x. Since the differential equation is satisfied whenever $x \neq 0$, $y(x)$ is a solution on any interval that does not include the origin.

2.11 Determine whether $y = 2/x$ is a solution of $y' + y^2 = 0$.

▮ Differentiating $y(x)$, we get $y' = -2/x^2$. Then $y' + y^2 = -\frac{2}{x^2} + \left(\frac{2}{x}\right)^2 = \frac{2}{x^2} \neq 0$. Since $y(x)$ does not satisfy the differential equation anywhere, it is not a solution.

2.12 Determine whether $y = 1/(x-2)$ is a solution of $y' + y^2 = 0$.

▮ Here $y'(x) = -1/(x-2)^2$, so $y' + y^2 = \frac{-1}{(x-2)^2} + \left(\frac{1}{x-2}\right)^2 = 0$ for all $x \neq 2$. Since the differential equation is satisfied whenever $x \neq 2$, $y(x)$ is a solution on any interval that does not include $x = 2$.

2.13 Show that $y = 1/(x+k)$ is a solution to $y' + y^2 = 0$ on any interval that does not include the point $x = k$, where k denotes an arbitrary constant.

▮ Differentiating $y(x)$, we get $y'(x) = -1/(x+k)^2$. Then

$$y' + y^2 = \frac{-1}{(x+k)^2} + \left(\frac{1}{x+k}\right)^2 = 0 \qquad \text{for all } x \neq k$$

2.14 Determine whether $y = e^{2x}$ is a solution of $y'' - 4y = 0$.

▮ Differentiating y twice, we find $y' = 2e^{2x}$ and $y'' = 4e^{2x}$. Then $y'' - 4y = 4e^{2x} - 4(e^{2x}) = 0$, so y is a solution to the differential equation everywhere.

2.15 Determine whether $y = e^{2x}$ is a solution of $y'' + 4y = 0$.

▮ As in the previous problem, $y'' = 4e^{2x}$; then $y'' + 4y = 4e^{2x} + 4(e^{2x}) = 8e^{2x} \neq 0$, so y is not a solution.

2.16 Determine whether $y = \sin 2x$ is a solution of $y'' + y = 0$.

▮ Differentiating y twice, we find $y' = 2\cos 2x$ and $y'' = -4\sin 2x$. Then $y'' + 4y = -4\sin 2x + 4(\sin 2x) = 0$, so $y = \sin 2x$ is a solution to the differential equation everywhere.

2.17 Determine whether $y = 2\sin x$ is a solution of $y'' + y = 0$.

▮ Differentiating y, we obtain $y' = 2\cos x$ and $y'' = -2\sin x$. Then $y'' + 4y = -2\sin x + 4(2\sin x) = 6\sin x$, which is zero only for integral multiples of π. Since $6\sin x$ is not identically zero on any interval, $y = 2\sin x$ is not a solution to the differential equation.

2.18 Determine whether $y = 2\cos 2x$ is a solution of $y'' + y = 0$.

▮ Differentiating y, we find $y' = -4\sin 2x$ and $y'' = -8\cos 2x$. Then $y'' + y = -8\cos 2x + 4(2\cos 2x) = 0$, so $y = 2\cos 2x$ is a solution to the differential equation everywhere.

2.19 Determine whether $y(x) = 0$ is a solution of $y'' + 4y = 0$.

▮ For the identically zero function, $y' = y'' = 0$; hence $y'' + 4y = 0 + 4(0) = 0$. It follows that $y(x)$ is a solution to this differential equation everywhere.

2.20 Show that $y(x) = c_1 \sin 2x + c_2 \cos 2x$ is a solution of $y'' + 4y = 0$ for all values of the arbitrary constants c_1 and c_2.

▮ Differentiating y, we find

$$y' = 2c_1 \cos 2x - 2c_2 \sin 2x \qquad \text{and} \qquad y'' = -4c_1 \sin 2x - 4c_2 \cos 2x$$

Hence,
$$\begin{aligned} y'' + 4y &= (-4c_1 \sin 2x - 4c_2 \cos 2x) + 4(c_1 \sin 2x + c_2 \cos 2x) \\ &= (-4c_1 + 4c_1)\sin 2x + (-4c_2 + 4c_2)\cos 2x = 0 \end{aligned}$$

Thus, $y = c_1 \sin 2x + c_2 \cos 2x$ satisfies the differential equation for all values of x and is a solution on the interval $(-\infty, \infty)$.

2.21 Determine whether $y = e^{-2t}$ is a solution of $\dddot{y} - 4\ddot{y} - 4\dot{y} + 16y = 0$.

▮ Differentiating y, we obtain $\dot{y} = -2e^{-2t}$, $\ddot{y} = 4e^{-2t}$, and $\dddot{y} = -8e^{-2t}$. Then

$$\dddot{y} - 4\ddot{y} - 4\dot{y} + 16y = -8e^{-2t} - 4(4e^{-2t}) - 4(-2e^{-2t}) + 16(e^{-2t}) = 0$$

Thus, y is a solution of the given differential equation for all values of t on the interval $(-\infty, \infty)$.

2.22 Determine whether $y = e^{2t}$ is a solution of $\dddot{y} - 4\ddot{y} - 4\dot{y} + 16y = 0$.

▮ Differentiating y, we obtain $\dot{y} = 2e^{2t}$, $\ddot{y} = 4e^{2t}$, and $\dddot{y} = 8e^{2t}$. Then

$$\dddot{y} - 4\ddot{y} - 4\dot{y} + 16y = 8e^{2t} - 4(4e^{2t}) - 4(2e^{2t}) + 16(e^{2t}) = 0$$

Since y satisfies the differential equation everywhere, it is a solution everywhere.

2.23 Determine whether $y = e^{3t}$ is a solution of $\dddot{y} - 4\ddot{y} - 4\dot{y} + 16y = 0$.

▮ Differentiating $y(t) = e^{3t}$, we obtain $\dot{y} = 3e^{3t}$, $\ddot{y} = 9e^{3t}$, and $\dddot{y} = 27e^{3t}$. Then

$$\dddot{y} - 4\ddot{y} - 4\dot{y} + 16y = 27e^{3t} - 4(9e^{3t}) - 4(3e^{3t}) + 16(e^{3t}) = -5e^{3t} \neq 0$$

Therefore, y is not a solution.

2.24 Determine whether $y = e^{4t}$ is a solution of $\dddot{y} - 4\ddot{y} - 4\dot{y} + 16y = 0$.

▮ Differentiating $y(t) = e^{4t}$, we obtain $\dot{y} = 4e^{4t}$, $\ddot{y} = 16e^{4t}$, and $\dddot{y} = 64e^{4t}$. Then

$$\dddot{y} - 4\ddot{y} - 4\dot{y} + 16y = 64e^{4t} - 4(16e^{4t}) - 4(4e^{4t}) + 16(e^{4t}) = 0$$

Thus $y(t)$ is a solution everywhere.

2.25 Determine whether $y = -0.5e^{4t}$ is a solution of $\dddot{y} - 4\ddot{y} - 4\dot{y} + 16y = 0$.

▮ Differentiating $y(t) = -0.5e^{4t}$, we obtain $\dot{y} = -2e^{4t}$, $\ddot{y} = -8e^{4t}$, and $\dddot{y} = -32e^{4t}$. Then

$$\dddot{y} - 4\ddot{y} - 4\dot{y} + 16y = -32e^{4t} - 4(-8e^{4t}) - 4(-2e^{4t}) + 16(-0.5e^{4t}) = 0$$

Thus $y(t)$ is a solution everywhere.

2.26 Show that $y(t) = c_1e^{2t} + c_2e^{-2t} + c_3e^{4t}$ is a solution of $\dddot{y} - 4\ddot{y} - 4\dot{y} + 16y = 0$ for all values of the arbitrary constants c_1, c_2, and c_3.

▮ Differentiating $y(t)$, we get

$$\dot{y} = 2c_1e^{2t} - 2c_2e^{-2t} + 4c_3e^{4t}$$
$$\ddot{y} = 4c_1e^{2t} + 4c_2e^{-2t} + 16c_3e^{4t}$$

and

$$\dddot{y} = 8c_1e^{2t} - 8c_2e^{-2t} + 64c_3e^{4t}$$

Then

$$\begin{aligned}\dddot{y} - 4\ddot{y} - 4\dot{y} + 16y &= 8c_1e^{2t} - 8c_2e^{-2t} + 64c_3e^{4t} - 4(4c_1e^{2t} + 4c_2e^{-2t} + 16c_3e^{4t}) \\ &\quad - 4(2c_1e^{2t} - 2c_2e^{-2t} + 4c_3e^{4t}) + 16(c_1e^{2t} + c_2e^{-2t} + c_3e^{4t}) \\ &= 0\end{aligned}$$

Thus $y(t)$ is a solution for all values of t.

2.27 Determine whether $x(t) = -\frac{1}{2}t$ is a solution of $\dot{x} - 2x = t$.

▮ Differentiating $x(t)$, we get $\dot{x} = -\frac{1}{2}$. Then $\dot{x} - 2x = -\frac{1}{2} - 2(-\frac{1}{2}t) = t - \frac{1}{2}$, which is never equal to t, the right-hand side of the differential equation. Therefore, $x(t)$ is not a solution.

2.28 Determine whether $x(t) = -\frac{1}{4}$ is a solution of $\dot{x} - 2x = t$.

▮ Differentiating $x(t)$, we get $\dot{x} = 0$, so $\dot{x} - 2x = 0 - 2(-\frac{1}{4}) = \frac{1}{2}$. This is equal to t, the right-hand side of the differential equation, only when $t = \frac{1}{2}$. Since $x(t)$ does not satisfy the differential equation on any interval, it is not a solution.

2.29 Determine whether $x(t) = -\frac{1}{2}t - \frac{1}{4}$ is a solution of $\dot{x} - 2x = t$.

▮ Differentiating $x(t)$, we obtain $\dot{x} = -\frac{1}{2}$. Then $\dot{x} - 2x = -\frac{1}{2} - 2(-\frac{1}{2}t - \frac{1}{4}) = t$. Therefore, $x(t)$ is a solution for all values of t in the interval $(-\infty, \infty)$.

2.30 Determine whether $x(t) = Ae^{2t}$ is a solution of $\dot{x} - 2x = t$ for any value of the arbitrary constant A.

▌ Differentiating $x(t)$, we get $\dot{x} = 2Ae^{2t}$, so $\dot{x} - 2x = 2Ae^{2t} - 2Ae^{2t} = 0$. This is equal to t, the right-hand side of the differential equation, only at $t = 0$. Since $x(t)$ satisfies the differential equation only at a single point, it is not a solution anywhere.

2.31 Determine whether $x(t) = -\frac{1}{2}t - \frac{1}{4} + Ae^{2t}$ is a solution of $\dot{x} - 2x = t$ for any value of the arbitrary constant A.

▌ Differentiating $x(t)$, we obtain $\dot{x} = -\frac{1}{2} + 2Ae^{2t}$. Then

$$\dot{x} - 2x = -\tfrac{1}{2} + 2Ae^{2t} - 2(-\tfrac{1}{2}t - \tfrac{1}{4} + Ae^{2t}) = t$$

so $x(t)$ is a solution everywhere.

2.32 Determine whether $y(x) = 2e^{-x} + xe^{-x}$ is a solution of $y'' + 2y' + y = 0$.

▌ Differentiating $y(x)$, we obtain

$$y'(x) = -2e^{-x} + e^{-x} - xe^{-x} = -e^{-x} - xe^{-x} \qquad \text{and} \qquad y''(x) = e^{-x} - e^{-x} + xe^{-x} = xe^{-x}$$

Substituting these values into the differential equation gives

$$y'' + 2y' + y = xe^{-x} + 2(-e^{-x} - xe^{-x}) + (2e^{-x} + xe^{-x}) = 0$$

Thus, $y(x)$ is a solution everywhere.

2.33 Determine whether $y(x) \equiv 1$ is a solution of $y'' + 2y' + y = x$.

▌ From $y(x) \equiv 1$ it follows that $y'(x) \equiv 0$ and $y''(x) \equiv 0$. Substituting these values into the differential equation, we obtain $y'' + 2y' + y = 0 + 2(0) + 1 = 1 \neq x$. Thus, $y(x) \equiv 1$ is not a solution.

2.34 Show that $y(x) \equiv 0$ is the only solution of $(y')^2 + y^2 = 0$ on the entire interval $(-\infty, \infty)$.

▌ By direct substitution, we find that $y(x) \equiv 0$ satisfies the differential equation identically for all values of x in $(-\infty, \infty)$ and is, therefore, a solution. Any other function must be nonzero at some point (otherwise it would not be different from the given function), and at such a point its square must be positive. Therefore, for such a function, the left side of the differential equation must be positive at that point, because it is a sum of squares; it then cannot equal zero, the right side of the differential equation. It follows that any nonzero function cannot satisfy the differential equation at some point on $(-\infty, \infty)$ and thus cannot be a solution over the entire interval.

2.35 Determine whether $y = x^2 - 1$ is a solution of $(y')^4 + y^2 = -1$.

▌ Note that the left side of the differential equation must be nonnegative for every real function $y(x)$ and any x, since it is the sum of terms raised to the second and fourth powers, while the right side of the equation is negative. Since no function $y(x)$ will satisfy this equation, the given differential equation has no solution.

2.36 Show that $y = \ln x$ is a solution of $xy'' + y' = 0$ on $\mathscr{I} = (0, \infty)$ but is not a solution on $\mathscr{I} = (-\infty, \infty)$.

▌ On $(0, \infty)$ we have $y' = 1/x$ and $y'' = -1/x^2$. Substituting these values into the differential equation, we obtain

$$xy'' + y' = x\left(-\frac{1}{x^2}\right) + \frac{1}{x} = 0$$

Thus, $y = \ln x$ is a solution on $(0, \infty)$. However, $y = \ln x$ cannot be a solution on $(-\infty, \infty)$ because the logarithm is undefined for negative numbers and zero.

2.37 Show that $y = 1/(x^2 - 1)$ is a solution of $y' + 2xy^2 = 0$ on $\mathscr{I} = (-1, 1)$, but not on any larger interval containing $\mathscr{I}$.

▌ On $(-1, 1)$, $y = 1/(x^2 - 1)$ and its derivative $y' = -2x/(x^2 - 1)^2$ are well-defined functions. Substituting these values into the differential equation, we have

$$y' + 2xy^2 = -\frac{2x}{(x^2 - 1)^2} + 2x\left(\frac{1}{x^2 - 1}\right)^2 = 0$$

Thus, $y = 1/(x^2 - 1)$ is a solution on $\mathscr{I} = (-1, 1)$. However, $1/(x^2 - 1)$ is not defined at $x = \pm 1$ and therefore cannot be a solution on any interval containing either of these two points.

PRIMITIVES

2.38 Explain what is meant by a primitive associated with a differential equation.

▌ A primitive associated with a differential equation of order n is a primitive (see Problem 1.107) that contains n arbitrary constants and is a solution of the differential equation.

2.39 Show that $y = C_1 \sin x + C_2 x$ is a primitive associated with the differential equation $(1 - x\cot x)\dfrac{d^2y}{dx^2} - x\dfrac{dy}{dx} + y = 0.$

▌ We substitute $y = C_1 \sin x + C_2 x$, $y' = C_1 \cos x + C_2$, and $y'' = -C_1 \sin x$ in the differential equation to obtain

$$(1 - x\cot x)(-C_1 \sin x) - x(C_1 \cos x + C_2) + (C_1 \sin x + C_2 x) = -C_1 \sin x + C_1 x \cos x - C_1 x \cos x - C_2 x + C_1 \sin x + C_2 x = 0$$

so y is a solution. In addition, the order of the differential equation (2) equals the number of arbitrary constants.

2.40 Show that $y = C_1 e^x + C_2 xe^x + C_3 e^{-x} + 2x^2e^x$ is a primitive associated with the differential equation $\dfrac{d^3y}{dx^3} - \dfrac{d^2y}{dx^2} - \dfrac{dy}{dx} + y = 8e^x.$

▌ We have

$$\begin{aligned} y &= C_1e^x + C_2xe^x + C_3e^{-x} + 2x^2e^x \\ y' &= (C_1 + C_2)e^x + C_2xe^x - C_3e^{-x} + 2x^2e^x + 4xe^x \\ y'' &= (C_1 + 2C_2)e^x + C_2xe^x + C_3e^{-x} + 2x^2e^x + 8xe^x + 4e^x \end{aligned}$$

and

$$y''' = (C_1 + 3C_2)e^x + C_2xe^x - C_3e^{-x} + 2x^2e^x + 12xe^x + 12e^x$$

and $y''' - y'' - y' + y = 8e^x$. Also, the order of the differential equation and the number of arbitrary constants are both 3.

2.41 Show that $y = 2x + Ce^x$ is a primitive of the differential equation $\dfrac{dy}{dx} - y = 2(1 - x)$.

▌ We substitute $y = 2x + Ce^x$ and $y' = 2 + Ce^x$ in the differential equation to obtain $2 + Ce^x - (2x + Ce^x) = 2 - 2x$. Furthermore, the order of the differential equation equals the number of arbitrary constants (2).

2.42 Show that $y = C_1e^x + C_2e^{2x} + x$ is a primitive of the differential equation $\dfrac{d^2y}{dx^2} - 3\dfrac{dy}{dx} + 2y = 2x - 3.$

▌ We substitute $y = C_1e^x + C_2e^{2x} + x$, $y' = C_1e^x + 2C_2e^{2x} + 1$, and $y'' = C_1e^x + 4C_2e^{2x}$ in the differential equation to obtain

$$C_1e^x + 4C_2e^{2x} - 3(C_1e^x + 2C_2e^{2x} + 1) + 2(C_1e^x + C_2e^{2x} + x) = 2x - 3$$

Moreover, the order of the differential equation and the number of arbitrary constants in y are both 2.

2.43 Show that $(y - C)^2 = Cx$ is a primitive of the differential equation $4x\left(\dfrac{dy}{dx}\right)^2 + 2x\dfrac{dy}{dx} - y = 0.$

▌ Here $2(y - C)\dfrac{dy}{dx} = C$, so that $\dfrac{dy}{dx} = \dfrac{C}{2(y - C)}$. Then $4x(y')^2 + 2xy' - y$ becomes

$$4x\frac{C^2}{4(y - C)^2} + 2x\frac{C}{2(y - C)} - y = \frac{C^2x + Cx(y - C) - y(y - C)^2}{(y - C)^2} = \frac{y[Cx - (y - C)^2]}{(y - C)^2} = 0$$

Furthermore, the order of the differential equation (1) is the same as the number of arbitrary constants in the proposed primitive.

2.44 Determine whether $y = c_1e^{-x} + \frac{1}{4}e^{3x}$ is a primitive of $y'' - y' - 2y = e^{3x}$.

▌ One can show by direct substitution that y is a solution of the differential equation. However, since y contains only one arbitrary constant whereas the order of the differential equation is 2, y is not a primitive of the differential equation.

2.45 Determine whether $y = c_1xe^x + c_2x^2e^x + \frac{1}{6}x^2e^x - 1$ is a primitive of $y''' - 3y'' + 3y' - y = e^x + 1$.

▌ By direct substitution we can show that y is a solution of the differential equation. However, since y contains only 2 arbitrary constants whereas the order of the differential equation is 3, y is not a primitive.

2.46 Determine whether $y = 3e^{2x}$ is a primitive of $y'' - 2y' + y = 3e^{2x}$.

▌ Since y contains no arbitrary constants while the order of the differential equation is 2, y cannot be a primitive of the differential equation.

2.47 Determine whether $y = A$ is a primitive of $y' + y = 0$.

▌ Substituting $y = A$ and its derivative $y' = 0$ into the left side of the differential equation, we obtain $y' + y = 0 + A = A$, which equals 0, the right side of the differential equation, only if $A = 0$. If $A \neq 0$, then $y = A$ is not a solution. Since $y = A$ is not a solution for arbitrary A, y is not a primitive.

2.48 Determine whether $y = Ax$ is a primitive for $y' - 3y = 0$.

▌ Substituting $y = Ax$ and $y' = A$ into the left side of the differential equation, we obtain $y' - 3y = A - 3Ax = A(1 - 3x)$. If $y = Ax$ is a primitive, it must satisfy the differential equation for all values of A. Thus, $A(1 - 3x)$ must be zero for all A, but it is zero only when $x = \frac{1}{3}$. That means $y = Ax$ does not satisfy the differential equation on any *interval*; for that reason it is not a solution and, therefore, not a primitive.

2.49 Determine whether $y = Cx + 2C^2$ is a primitive for $2\left(\frac{dy}{dx}\right)^2 + x\frac{dy}{dx} - y = 0$.

▌ The derivative of y is $y' = C$. Then

$$2\left(\frac{dy}{dx}\right)^2 + x\frac{dy}{dx} - y = 2C^2 + xC - (Cx + 2C^2) = 0$$

so y is a solution for all values of the arbitrary constant C. Since y contains only the one arbitrary constant C and the differential equation is of order 1, y is a primitive for the differential equation.

2.50 Show that $y = -\frac{1}{8}x^2$ is a particular solution of $2\left(\frac{dy}{dx}\right)^2 + x\frac{dy}{dx} - y = 0$.

▌ Here $y' = -\frac{1}{4}x$, so

$$2\left(\frac{dy}{dx}\right)^2 + x\frac{dy}{dx} - y = 2\left(-\frac{1}{4}x\right)^2 + x\left(-\frac{1}{4}x\right) - \left(-\frac{1}{8}x^2\right) = 0$$

2.51 Use the results of Problems 2.49 and 2.50 to show that not every particular solution of a differential equation can be generated from a primitive of that differential equation by selecting specific values for the arbitrary constants.

▌ We have shown that a primitive for $2\left(\frac{dy}{dx}\right)^2 + x\frac{dy}{dx} - y = 0$ is $y = Cx + 2C^2$ with arbitrary constant C, while a particular solution is the parabola $y = -\frac{1}{8}x^2$. The primitive represents a family of straight lines, and clearly the equation of a parabola cannot be obtained by manipulating the arbitrary constant C. (A solution that cannot be generated from a primitive is called a *singular* solution of the differential equation.)

2.52 Determine graphically a relationship between the primitive $y = Cx - C^2$ and the singular solution $y = x^2/4$ of the differential equation $y = xy' - (y')^2$.

▌ Referring to Fig. 2.1, we see that $y = Cx - C^2$ represents a family of straight lines tangent to the parabola $y = x^2/4$. The parabola is the *envelope* of the family of straight lines.

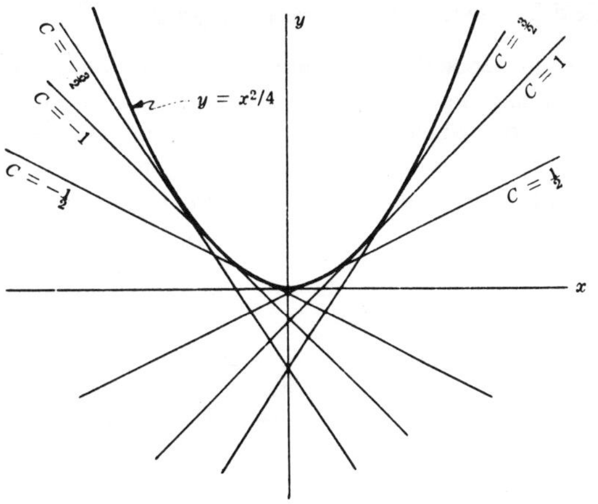

Fig. 2.1

DIRECTION FIELDS

2.53 Construct a direction field for the first-order differential equation $y' = y - t$.

▌ At each point in the (t, y) plane, we compute dy/dt by substituting y and t into the right-hand side of the differential equation; then through the point, we graph a short line segment having the derivative as its slope. In particular, at $(0, 0)$ we have $y' = 0 - 0 = 0$; at $(0, 1)$, $y' = 1 - 0 = 1$; at $(1, 0)$, $y' = 0 - 1 = -1$; at $(1, 1)$, $y' = 1 - 1 = 0$; and at $(-1, -1)$, $y' = -1 - (-1) = 0$. The direction field for these and other points is shown in Fig. 2.2.

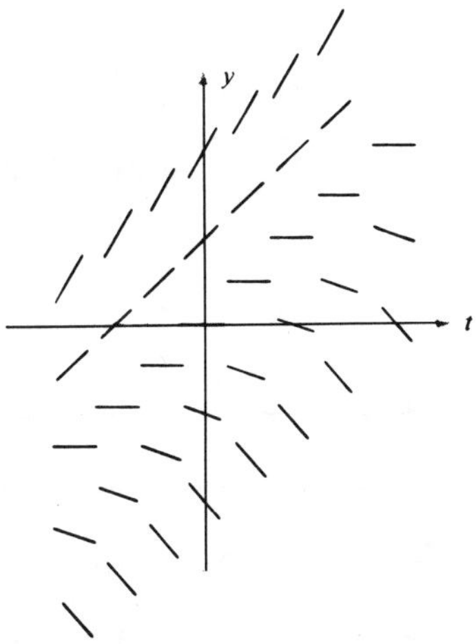

Fig. 2.2

2.54 Graph the solution curves that pass through the direction field obtained in the previous problem.

▌ The curves are shown in Fig. 2.3.

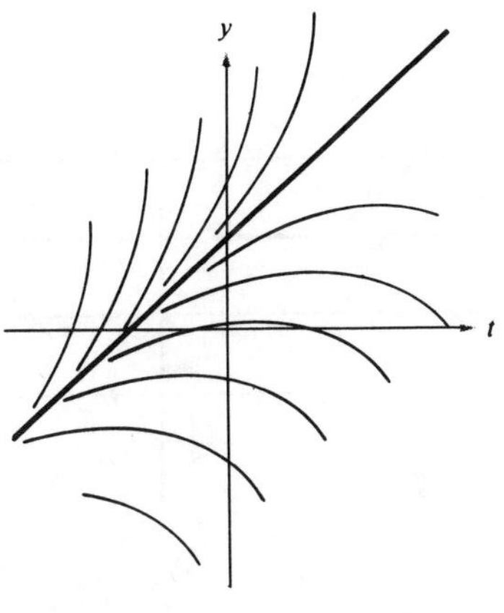

Fig. 2.3

2.55 Construct a direction field for the first-order differential equation $y' = 5y(y - 1)$.

▌ Since y' in this case is independent of t, for any given y_0 the slopes of the solutions at (t, y_0) are the same for all t. Noting that the right-hand side of this equation is zero when y is 0 or 1, positive when y is in $(-\infty, 0) \cup (1, \infty)$, and negative when y is on $(0, 1)$, we can readily verify that Fig. 2.4 gives a reasonable indication of the direction field. For example, if $y = \frac{1}{2}$ then $y' = 5(\frac{1}{2})(-\frac{1}{2}) = -\frac{5}{4}$, so the solutions have slope $-\frac{5}{4}$ when they cross the line $y = \frac{1}{2}$. Also, if $y = \frac{1}{4}$ then $y' = -\frac{15}{16}$, and if $y = -\frac{1}{4}$ then $y' = \frac{25}{16}$ (these values are indicated in Fig. 2.4).

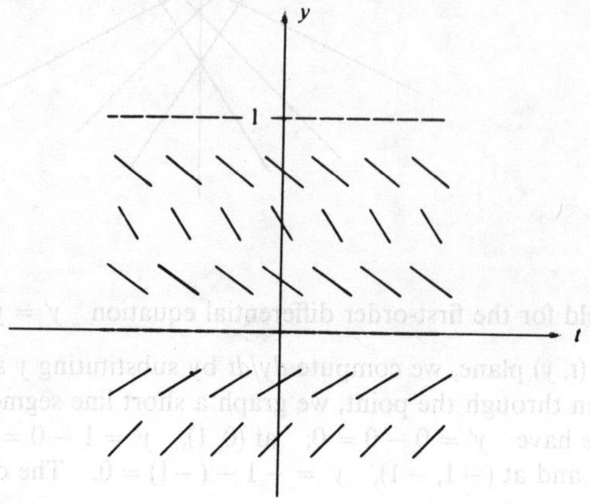

Fig. 2.4

2.56 Sketch a direction field for $y' = y - 1$.

▌ The derivative is independent of t and depends only on y. At $(t, y) = (0, 0)$, $y' = 0 - 1 = -1$; at $(1, 0)$, $y' = 0 - 1 = -1$; at $(2, 0)$, $y' = 0 - 1 = -1$; at $(1, 1)$, $y' = 1 - 1 = 0$; at $(2, 1)$, $y' = 1 - 1 = 0$; and at $(2, 2)$, $y' = 2 - 1 = 1$. The direction field at these points and others is shown in Fig. 2.5.

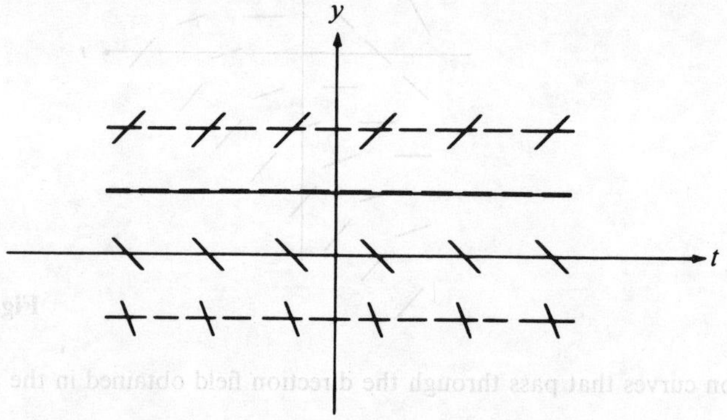

Fig. 2.5

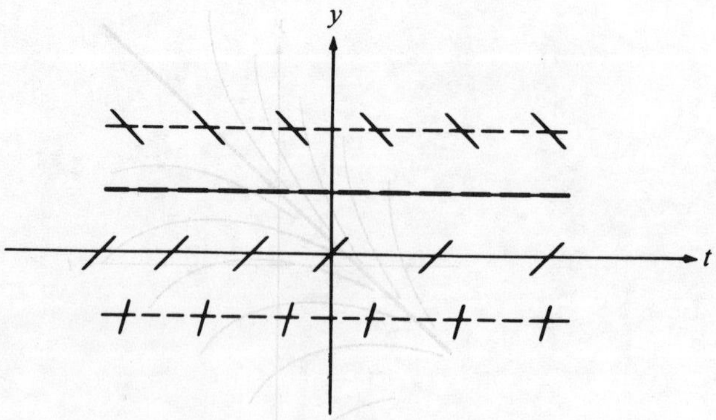

Fig. 2.6

2.57 Sketch a direction field for $y' = 1 - y$.

▌ At $(t, y) = (0, 0)$, $y' = 1 - 0 = 1$; at $(1, 0)$, $y' = 1 - 0 = 1$; at $(1, 1)$, $y' = 1 - 1 = 0$; at $(2, 1)$, $y' = 1 - 1 = 0$; at $(2, 2)$, $y' = 1 - 2 = -1$; at $(-2, -1)$, $y' = 1 - (-1) = 2$. The direction field at these points and others is shown in Fig. 2.6.

2.58 Sketch a direction field for $y' = y^3 - y^2$.

▌ At $(t, y) = (0, 0)$, $y' = 0^3 - 0^2 = 0$; at $(0, 1)$, $y' = 1^3 - 1^2 = 0$; at $(0, 2)$, $y' = 2^3 - 2^2 = 4$; at $(0, -2)$, $y' = (-2)^3 - (-2)^2 = -12$, at $(1, -1)$, $y' = (-1)^3 - (-1)^2 = -2$. The direction field at these points and others is shown in Fig. 2.7.

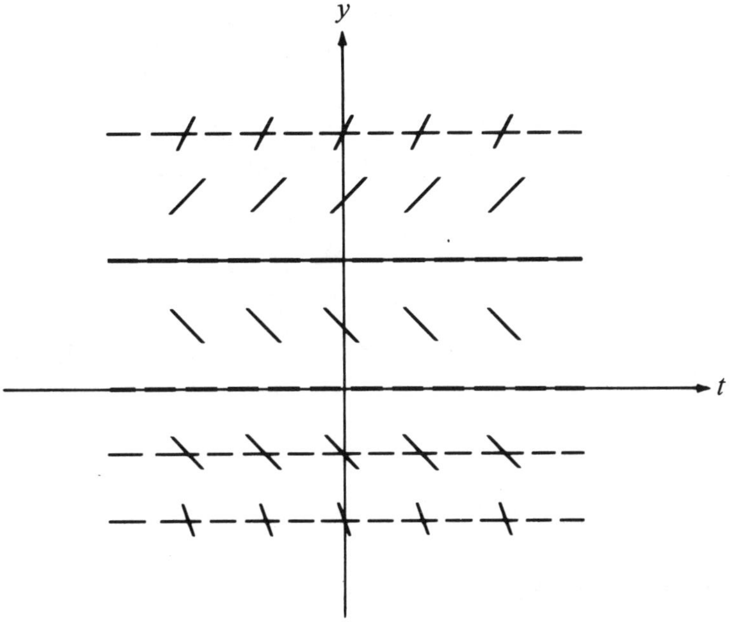

Fig. 2.7

2.59 Sketch a direction field for $y' = 1 - y^2$.

▌ At $(t, y) = (0, 0)$, $y' = 1 - 0^2 = 1$; at $(1, 1)$, $y' = 1 - 1^2 = 0$, at $(1, 2)$, $y' = 1 - 2^2 = -3$; at $(0, -1)$, $y' = 1 - (-1)^2 = 0$; and at $(-1, -2)$, $y' = 1 - (-2)^2 = -3$. The direction field at these points and others is shown in Fig. 2.8.

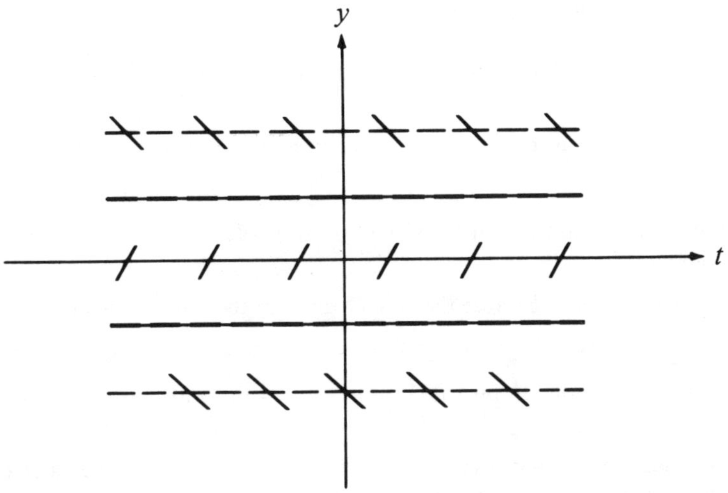

Fig. 2.8

2.60 Sketch a direction field for $y' = 2x$, along with some of the solution curves that pass through it.

The direction field and three curves are shown in Fig. 2.9.

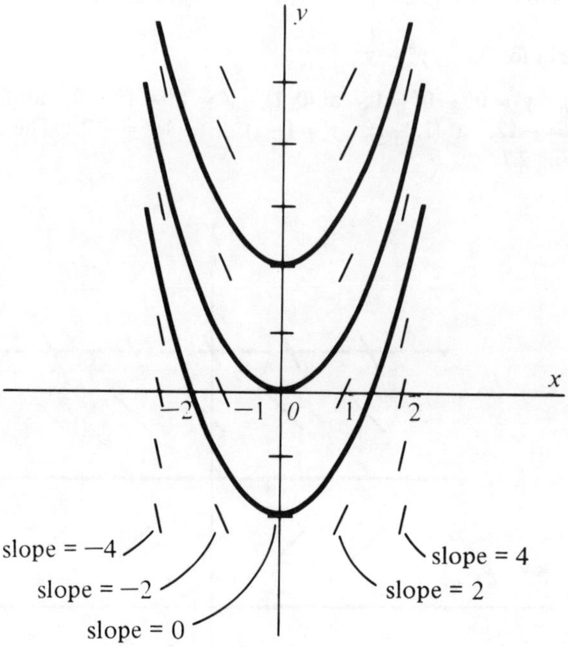

Fig. 2.9

INITIAL AND BOUNDARY CONDITIONS

2.61 Determine whether the conditions on $y(x)$ given by $y(0) = 1$, $y'(0) = 2$ are boundary conditions or initial conditions.

They are initial conditions because they are given at the same value of the independent variable, here at $x = 0$.

2.62 Determine whether the conditions on $y(x)$ given by $y(1) = 0$, $y'(2) = 0$ are boundary or initial conditions.

They are boundary conditions because they are not both given at the same value of the independent variable. One is given at $x = 1$, and the other at $x = 2$.

2.63 Determine whether the conditions on $y(t)$ given by $y(3) = 0$, $y'(3) = 0$, $y''(3) = 1$ are boundary or initial conditions.

They are initial conditions because they are all given at the same value of the independent variable, here at $t = 3$.

2.64 Determine whether the conditions on $x(t)$ given by $x(\pi) = 1$, $x'(\pi) = 2$, $x''(\pi) = 3$, $x'''(\pi) = 4$ are boundary or initial conditions.

They are initial conditions because they are all given at the same value of the independent variable, here at $t = \pi$.

2.65 Determine whether the conditions on $x(t)$ given by $x(0) = 0$, $x'(0) = 0$, $x''(\pi) = 0$ are boundary or initial conditions.

They are boundary conditions because they are not all given at the same value of the independent variable. Two are given at $t = 0$ while the third is given at $t = \pi$.

2.66 Determine whether the conditions on $s(t)$ given by $s(5) = s(10) = 15$ are boundary or initial conditions.

▌ They are boundary conditions because they are not both given at the same value of the independent variable. One is given at $t = 5$, and the other at $t = 10$.

2.67 Determine whether a single condition is a boundary or initial condition.

▌ A single subsidiary condition is an initial condition because it satisfies the criterion that all conditions are prescribed at the same value of the independent variable.

2.68 Determine whether the conditions on $y(x)$ given by $y(-7.5) = 0$, $y'(-7.5) = 1$, $y''(-7.5) = 0$, $y^{(3)}(-7.5) = 1$, $y^{(4)}(-7.5) = 0$, and $y^{(5)}(-7.5) = 1$ are boundary or initial conditions.

▌ They are initial conditions because they are all specified at the same value of the independent variable, here $x = -7.5$.

2.69 Determine C so that $y(x) = 2x + Ce^x$ will satisfy the condition $y(0) = 3$.

▌ When $x = 0$ and $y = 3$, we have $3 = 2(0) + Ce^0$ and $C = 3$. Then $y = 2x + 3e^x$.

2.70 Determine C so that $(y - C)^2 = Cx$ will satisfy the condition $y(1) = 2$.

▌ When $x = 1$ and $y = 2$, we have $(2 - C)^2 = C$ and $C = 1, 4$. Thus, $(y - 1)^2 = x$ and $(y - 4)^2 = 4x$.

2.71 Determine C_1 and C_2 so that $y = x + C_1e^x + C_2e^{2x}$ will satisfy the boundary conditions $y(0) = 0$ and $y(1) = 0$.

▌ When $x = 0$ and $y = 0$, we have $C_1 + C_2 = 0$. When $x = 1$ and $y = 0$, we have $C_1e + C_2e^2 = -1$. Then $C_1 = -C_2 = \dfrac{1}{e^2 - e}$ and the required equation is $y = x + \dfrac{e^x - e^{2x}}{e^2 - e}$.

2.72 Determine c_1 and c_2 so that $y(x) = c_1 \sin 2x + c_2 \cos 2x + 1$ will satisfy the conditions $y(\pi/8) = 0$ and $y'(\pi/8) = \sqrt{2}$.

▌ Note that

$$y(\pi/8) = c_1 \sin\frac{\pi}{4} + c_2 \cos\frac{\pi}{4} + 1 = c_1\left(\frac{\sqrt{2}}{2}\right) + c_2\left(\frac{\sqrt{2}}{2}\right) + 1$$

To satisfy the condition $y(\pi/8) = 0$, we require $c_1(\frac{1}{2}\sqrt{2}) + c_2(\frac{1}{2}\sqrt{2}) + 1 = 0$, or equivalently,

$$c_1 + c_2 = -\sqrt{2} \tag{1}$$

Since $y'(x) = 2c_1 \cos 2x - 2c_2 \sin 2x$,

$$y'(\pi/8) = 2c_1 \cos\frac{\pi}{4} - 2c_2 \sin\frac{\pi}{4} = 2c_1\left(\frac{\sqrt{2}}{2}\right) - 2c_2\left(\frac{\sqrt{2}}{2}\right) = \sqrt{2}c_1 - \sqrt{2}c_2$$

To satisfy the condition $y'(\pi/8) = \sqrt{2}$, we require $\sqrt{2}c_1 - \sqrt{2}c_2 = \sqrt{2}$, or equivalently,

$$c_1 - c_2 = 1 \tag{2}$$

Solving (1) and (2) simultaneously, we obtain $c_1 = -\frac{1}{2}(\sqrt{2} - 1)$ and $c_2 = -\frac{1}{2}(\sqrt{2} + 1)$.

2.73 Determine c_1 and c_2 so that $y(x) = c_1e^{2x} + c_2e^x + 2\sin x$ will satisfy the conditions $y(0) = 0$ and $y'(0) = 1$.

▌ Because $\sin 0 = 0$, $y(0) = c_1 + c_2$. To satisfy the condition $y(0) = 0$, we require

$$c_1 + c_2 = 0 \tag{1}$$

From $y'(x) = 2c_1e^{2x} + c_2e^x + 2\cos x$, we have $y'(0) = 2c_1 + c_2 + 2$. To satisfy the condition $y'(0) = 1$, we require $2c_1 + c_2 + 2 = 1$, or

$$2c_1 + c_2 = -1 \tag{2}$$

Solving (1) and (2) simultaneously, we obtain $c_1 = -1$ and $c_2 = 1$.

PARTICULAR SOLUTIONS

2.74 Find the solution to the initial-value problem $y' + y = 0$; $y(3) = 2$, if the general solution to the differential equation is known to be $y(x) = c_1 e^{-x}$, where c_1 is an arbitrary constant.

▌ Since $y(x)$ is a solution of the differential equation for every value of c_1, we seek that value of c_1 which will also satisfy the initial condition. Note that $y(3) = c_1 e^{-3}$. To satisfy the initial condition $y(3) = 2$, it is sufficient to choose c_1 so that $c_1 e^{-3} = 2$, that is, to choose $c_1 = 2e^3$. Substituting this value for c_1 into $y(x)$, we obtain $y(x) = 2e^3 e^{-x} = 2e^{3-x}$ as the solution of the initial-value problem.

2.75 Find a solution to the initial-value problem $y'' + 4y = 0$; $y(0) = 0$, $y'(0) = 1$, if the general solution to the differential equation is known to be $y(x) = c_1 \sin 2x + c_2 \cos 2x$.

▌ Since $y(x)$ is a solution of the differential equation for all values of c_1 and c_2, we seek those values of c_1 and c_2 that will also satisfy the initial conditions. Note that $y(0) = c_1 \sin 0 + c_2 \cos 0 = c_2$. To satisfy the first initial condition, $y(0) = 0$, we choose $c_2 = 0$. Furthermore, $y'(x) = 2c_1 \cos 2x - 2c_2 \sin 2x$; thus, $y'(0) = 2c_1 \cos 0 - 2c_2 \sin 0 = 2c_1$. To satisfy the second initial condition, $y'(0) = 1$, we choose $2c_1 = 1$, or $c_1 = \frac{1}{2}$. Substituting these values of c_1 and c_2 into $y(x)$, we obtain $y(x) = \frac{1}{2}\sin 2x$ as the solution of the initial-value problem.

2.76 Find a solution to the boundary-value problem $y'' + 4y = 0$; $y(\pi/8) = 0$, $y(\pi/6) = 1$, if the general solution to the differential equation is $y(x) = c_1 \sin 2x + c_2 \cos 2x$.

▌ Note that
$$y(\pi/8) = c_1 \sin\frac{\pi}{4} + c_2 \cos\frac{\pi}{4} = c_1\left(\frac{\sqrt{2}}{2}\right) + c_2\left(\frac{\sqrt{2}}{2}\right)$$

To satisfy the condition $y(\pi/8) = 0$, we require
$$c_1\left(\frac{\sqrt{2}}{2}\right) + c_2\left(\frac{\sqrt{2}}{2}\right) = 0 \qquad (1)$$

Furthermore,
$$y(\pi/6) = c_1 \sin\frac{\pi}{3} + c_2 \cos\frac{\pi}{3} = c_1\frac{\sqrt{3}}{2} + \frac{c_2}{2}$$

To satisfy the second condition, $y(\pi/6) = 1$, we require
$$\tfrac{1}{2}\sqrt{3}c_1 + \tfrac{1}{2}c_2 = 1 \qquad (2)$$

Solving (1) and (2) simultaneously, we find $c_1 = -c_2 = 2/(\sqrt{3} - 1)$. Substituting these values into $y(x)$, we obtain $y(x) = \dfrac{2}{\sqrt{3}-1}(\sin 2x - \cos 2x)$ as the solution of the boundary-value problem.

2.77 Find a solution to the boundary-value problem $y'' + 4y = 0$; $y(0) = 1$, $y(\pi/2) = 2$, if the general solution to the differential equation is known to be $y(x) = c_1 \sin 2x + c_2 \cos 2x$.

▌ Since $y(0) = c_1 \sin 0 + c_2 \cos 0 = c_2$, we must choose $c_2 = 1$ to satisfy the condition $y(0) = 1$. Since $y(\pi/2) = c_1 \sin \pi + c_2 \cos \pi = -c_2$, we must choose $c_2 = -2$ to satisfy the condition $y(\pi/2) = 2$. Thus, to satisfy both boundary conditions simultaneously, we must require c_2 to equal both 1 and -2, which is impossible. Therefore, this problem does not have a solution.

2.78 Find a solution to the initial-value problem $y'' + y = 0$; $y(0) = 1$, $y'(0) = 2$, if the general solution to the differential equation is $y(x) = A \sin x + B \cos x$, where A and B are arbitrary constants.

▌ At $x = 0$, $y(0) = A \sin 0 + B \cos 0 = B$, so we must choose $B = 1$ to satisfy the condition $y(0) = 1$. Furthermore, $y'(x) = A \cos x - B \sin x$, so $y'(0) = A \cos 0 - B \sin 0 = A$. To satisfy the condition $y'(0) = 2$, we must choose $A = 2$. Then $y(x) = 2 \sin x + \cos x$ is the solution to the initial-value problem.

2.79 Rework Problem 2.78 if the subsidiary conditions are $y(\pi/2) = 1$, $y'(\pi/2) = 2$.

▌ At $x = \pi/2$, $y(\pi/2) = A \sin(\pi/2) + B \cos(\pi/2) = A$, so we must choose $A = 1$ to satisfy the condition $y(\pi/2) = 1$. Also $y'(\pi/2) = A \cos(\pi/2) - B \sin(\pi/2) = -B$, so we must choose $B = -2$ to satisfy the condition $y'(\pi/2) = 2$. Then $y(x) = \sin x - 2 \cos x$.

2.80 Rework Problem 2.78 if the subsidiary conditions are $y(0) = 1$, $y(\pi/2) = 1$.

▌ The problem is now a boundary-value problem because the subsidiary conditions are specified at different values of the independent variable x. At $x = 0$, $y(0) = A\sin 0 + B\cos 0 = B$, so we must choose $B = 1$ to satisfy the first subsidiary condition. At $x = \pi/2$, $y(\pi/2) = A\sin(\pi/2) + B\cos(\pi/2) = A$, so we must choose $A = 1$ to satisfy the second subsidiary condition. Then $y(x) = \sin x + \cos x$.

2.81 Rework Problem 2.78 if the subsidiary conditions are $y'(0) = 1$, $y'(\pi/2) = 1$.

▌ The problem is now a boundary-value problem. For the given $y(x)$, we have $y'(x) = A\cos x - B\sin x$. At $x = 0$, $y'(0) = A\cos 0 - B\sin 0 = A$, so we must choose $A = 1$ if we are to satisfy the first boundary condition. At $x = \pi/2$, we have $y'(\pi/2) = A\cos(\pi/2) - B\sin(\pi/2) = -B$, so we must choose $B = -1$ to satisfy the second boundary condition. Then $y(x) = \sin x - \cos x$.

2.82 Rework Problem 2.78 if the subsidiary conditions are $y(0) = 1$, $y'(\pi) = 1$.

▌ The problem is now a boundary-value problem. With $y(x) = A\sin x + B\cos x$ and $y'(x) = A\cos x - B\sin x$, we have $y(0) = A\sin 0 + B\cos 0 = B$ and $y'(\pi) = A\cos\pi - B\sin\pi = -A$. To satisfy the first condition we must choose $B = 1$; to satisfy the second condition we must choose $A = -1$. Then $y(x) = -\sin x + \cos x$.

2.83 Rework Problem 2.78 if the subsidiary conditions are $y(0) = 0$, $y(\pi) = 2$.

▌ Since $y(0) = A\sin 0 + B\cos 0 = B$, we must choose $B = 0$ to satisfy the subsidiary condition $y(0) = 0$. Since $y(\pi) = A\sin\pi + B\cos\pi = -B$, we must choose $B = -2$ to satisfy the condition $y(\pi) = 2$. Thus, to satisfy both conditions simultaneously, we must require B to equal both 0 and -2, which is impossible. Therefore, this boundary-value problem does not possess a solution.

2.84 Rework Problem 2.78 if the subsidiary conditions are $y(0) = y'(0) = 0$.

▌ At $x = 0$, $y(0) = A\sin 0 + B\cos 0 = B$, so we must choose $B = 0$ to satisfy the first initial condition. Furthermore, $y'(0) = A\cos 0 - B\sin 0 = A$, so we must choose $A = 0$ to satisfy the second initial condition. Substituting these values into the general solution, we get $y(x) = 0\sin x + 0\cos x = 0$ as the solution to the initial-value problem.

2.85 Rework Problem 2.78 if the subsidiary conditions are $y(\pi/4) = 0$, $y(\pi/6) = 1$.

▌ At $x = \pi/4$, we have $y(\pi/4) = A\sin(\pi/4) + B\cos(\pi/4) = A(\sqrt{2}/2) + B(\sqrt{2}/2)$. Thus, to satisfy the condition $y(\pi/4) = 0$, we require

$$A\frac{\sqrt{2}}{2} + B\frac{\sqrt{2}}{2} = 0 \tag{1}$$

Furthermore, $y(\pi/6) = A\sin(\pi/6) + B\cos(\pi/6) = A(\frac{1}{2}) + B(\sqrt{3}/2)$. To satisfy the condition $y(\pi/6) = 1$, then, we require

$$\frac{A}{2} + B\frac{\sqrt{3}}{2} = 1 \tag{2}$$

Solving (1) and (2) simultaneously, we determine $A = \dfrac{-2}{\sqrt{3}-1}$ and $B = \dfrac{2}{\sqrt{3}-1}$. Substituting these values into the general solution, we obtain $y(x) = \dfrac{2}{\sqrt{3}-1}(-\sin x + \cos x)$.

2.86 Rework Problem 2.78 if the subsidiary conditions are $y(0) = 0$, $y'(\pi/2) = 1$.

▌ At $x = 0$, $y(0) = A\sin 0 + B\cos 0 = B$, so we must choose $B = 0$ to satisfy the first boundary condition. At $x = \pi/2$, $y'(\pi/2) = A\cos(\pi/2) - B\sin(\pi/2) = -B$, so we must choose $B = -1$ to satisfy the second boundary condition. Thus, we must have B equal to both 0 and -1 simultaneously, which is impossible. Therefore, the boundary-value problem does not have a solution.

2.87 Rework Problem 2.78 if the subsidiary conditions are $y(0) = 1$, $y(\pi) = -1$.

▌ At $x = 0$, $y(0) = A\sin 0 + B\cos 0 = B$, so we must choose $B = 1$ to satisfy the first boundary condition. At $x = \pi$, $y(\pi) = A\sin\pi + B\cos\pi = -B$, so we must choose $B = 1$ to satisfy the second boundary condition. Thus, $B = 1$ is sufficient to satisfy both boundary conditions, with no restrictions placed on A. The solution is $y = A \sin x + \cos x$, where A is an orbitrary constant.

2.88 A general solution to a certain differential equation is $y(x) = c_1 e^{-x} + c_2 e^{2x} - 2x^2 + 2x - 3$, where c_1 and c_2 are arbitrary constants. Find a particular solution which also satisfies the initial conditions $y(0) = 1$, $y'(0) = 4$.

▮ Since
$$y = c_1 e^{-x} + c_2 e^{2x} - 2x^2 + 2x - 3 \tag{1}$$
we have
$$y' = -c_1 e^{-x} + 2c_2 e^{2x} - 4x + 2 \tag{2}$$
Applying the first initial condition to (1), we obtain
$$c_1 e^{-(0)} + c_2 e^{2(0)} - 2(0)^2 + 2(0) - 3 = 1 \qquad \text{or} \qquad c_1 + c_2 = 4 \tag{3}$$
Applying the second initial condition to (2), we obtain
$$-c_1 e^{-(0)} + 2c_2 e^{2(0)} - 4(0) + 2 = 4 \qquad \text{or} \qquad -c_1 + 2c_2 = 2 \tag{4}$$
Solving (3) and (4) simultaneously, we find that $c_1 = 2$ and $c_2 = 2$. Substituting these values into (1), we obtain the solution of the initial-value problem as $y = 2e^{-x} + 2e^{2x} - 2x^2 + 2x - 3$.

2.89 A general solution to a certain differential equation is $y(x) = c_1 e^x + c_3 x e^x + x e^x \ln |x|$, where c_1 and c_3 are arbitrary constants. Find a particular solution which also satisfies the initial conditions $y(1) = 0$, $y'(1) = 1$.

▮ Since
$$y = c_1 e^x + c_3 x e^x + x e^x \ln |x| \tag{1}$$
we have
$$y' = c_1 e^x + c_3 e^x + c_3 x e^x + e^x \ln |x| + x e^x \ln |x| + e^x \tag{2}$$
Applying the first initial condition to (1), we obtain $c_1 e^1 + c_3(1)e^1 + (1)e^1 \ln 1 = 0$ or (since $\ln 1 = 0$),
$$c_1 e + c_3 e = 0 \tag{3}$$
Applying the second initial condition to (2), we obtain $c_1 e^1 + c_3 e^1 + c_3(1)e^1 + e^1 \ln 1 + (1)e^1 \ln 1 + e^1 = 1$, or
$$c_1 e + 2c_3 e = 1 - e \tag{4}$$
Solving (3) and (4) simultaneously, we find that $c_1 = -c_3 = (e - 1)/e$. Substituting these values into (1), we obtain the solution of the initial-value problem as $y = e^{x-1}(e - 1)(1 - x) + x e^x \ln |x|$.

2.90 A general solution to a certain differential equation is $y = e^{-2x}(c_1 \cos 2x + c_2 \sin 2x) + \frac{7}{65} \sin x - \frac{4}{65} \cos x$, where c_1 and c_2 are arbitrary constants. Find a particular solution which also satisfies the initial conditions $y(0) = 1$, $y'(0) = 0$.

▮ For y as given, we have
$$y' = -2e^{-2x}(c_1 \cos 2x + c_2 \sin 2x) + e^{-2x}(-2c_1 \sin 2x + 2c_2 \cos 2x) + \frac{7}{65} \cos x + \frac{4}{65} \sin x$$
Applying the first initial condition to y, we obtain $c_1 = 69/65$. Applying the second initial condition to y' gives $-2c_1 + 2c_2 = -7/65$, so that $c_2 = 131/130$. Substituting these values for c_1 and c_2, we obtain the solution of the initial-value problem as
$$y = e^{-2x}\left(\frac{69}{65} \cos 2x + \frac{131}{130} \sin 2x\right) + \frac{7}{65} \sin x - \frac{4}{65} \cos x$$

2.91 The general solution to a certain third-order differential equation is $y = c_1 e^x + c_2 e^{2x} + c_3 e^{3x}$, where c_1, c_2, and c_3 are arbitrary constants. Find a particular solution which also satisfies the initial conditions $y(\pi) = 0$, $y'(\pi) = 0$, $y''(\pi) = 1$.

▮ We have
$$y = c_1 e^x + c_2 e^{2x} + c_3 e^{3x}$$
$$y' = c_1 e^x + 2c_2 e^{2x} + 3c_3 e^{3x} \tag{1}$$
and
$$y'' = c_1 e^x + 4c_2 e^{2x} + 9c_3 e^{3x}$$
Applying each initial condition separately, we obtain
$$c_1 e^{\pi} + c_2 e^{2\pi} + c_3 e^{3\pi} = 0$$
$$c_1 e^{\pi} + 2c_2 e^{2\pi} + 3c_3 e^{3\pi} = 0$$
$$c_1 e^{\pi} + 4c_2 e^{2\pi} + 9c_3 e^{3\pi} = 1$$

Solving these equations simultaneously, we find $c_1 = \frac{1}{2}e^{-\pi}$, $c_2 = -e^{-2\pi}$, and $c_3 = \frac{1}{2}e^{-3\pi}$. Substituting these values into the first equation of (*1*), we obtain $y = \frac{1}{2}e^{(x-\pi)} - e^{2(x-\pi)} + \frac{1}{2}e^{3(x-\pi)}$.

2.92 Solve the initial-value problem $x'' - x' - 2x = e^{3t}$; $x(0) = 1$, $x'(0) = 2$, if the general solution to the differential equation is $x(t) = c_1e^{-t} + c_2e^{2t} + \frac{1}{4}e^{3t}$, where c_1 and c_2 are arbitrary constants.

▌ The first initial condition yields $x(0) = c_1e^{-0} + c_2e^{2(0)} + \frac{1}{4}e^{3(0)} = 1$, which may be rewritten as

$$c_1 + c_2 = \tfrac{3}{4} \tag{1}$$

Furthermore, $x'(t) = -c_1e^{-t} + 2c_2e^{2t} + \frac{3}{4}e^{3t}$, so the second initial condition yields $x'(0) = -c_1e^{-0} + 2c_2e^{2(0)} + \frac{3}{4}e^{3(0)} = 2$, which may be rewritten as

$$-c_1 + 2c_2 = \tfrac{5}{4} \tag{2}$$

Solving (*1*) and (*2*) simultaneously, we find $c_1 = \frac{1}{12}$ and $c_2 = \frac{2}{3}$. Thus, $x(t) = \frac{1}{12}e^{-t} + \frac{2}{3}e^{2t} + \frac{1}{4}e^{3t}$ is a solution to the initial-value problem.

2.93 Rework Problem 2.92 if the initial conditions are $x(0) = 2$, $x'(0) = 1$.

▌ With $x(t) = c_1e^{-t} + c_2e^{2t} + \frac{1}{4}e^{3t}$ we have $x'(t) = -c_1e^{-t} + 2c_2e^{2t} + \frac{3}{4}e^{3t}$. The initial conditions then yield $x(0) = c_1e^{-0} + c_2e^{2(0)} + \frac{1}{4}e^{3(0)} = 2$ and $x'(0) = -c_1e^{-0} + 2c_2e^{2(0)} + \frac{3}{4}e^{3(0)} = 1$, which may be rewritten as

$$\begin{aligned} c_1 + c_2 &= \tfrac{7}{4} \\ -c_1 + 2c_2 &= \tfrac{1}{4} \end{aligned} \tag{1}$$

Solving system (*1*), we obtain $c_1 = \frac{13}{12}$ and $c_2 = \frac{2}{3}$, and the solution to the initial-value problem is $x(t) = \frac{13}{12}e^{-t} + \frac{2}{3}e^{2t} + \frac{1}{4}e^{3t}$.

2.94 Rework Problem 2.92 if the initial conditions are $x(1) = 2$, $x'(1) = 1$.

▌ Applying the initial conditions to the expressions for $x(t)$ and $x'(t)$ as determined in the previous problem, we obtain $x(1) = c_1e^{-1} + c_2e^{2(1)} + \frac{1}{4}e^{3(1)} = 2$ and $x'(1) = -c_1e^{-1} + 2c_2e^{2(1)} + \frac{3}{4}e^{3(1)} = 1$, which may be rewritten as

$$\begin{aligned} c_1e^{-1} + c_2e^2 &= 2 - \tfrac{1}{4}e^3 \\ -c_1e^{-1} + 2c_2e^2 &= 1 - \tfrac{3}{4}e^3 \end{aligned} \tag{1}$$

Solving system (*1*), we obtain $c_1 = e + \frac{1}{12}e^4$ and $c_2 = e^{-2} - \frac{1}{3}e$. The solution to the initial-value problem is then $x(t) = (e + \frac{1}{12}e^4)e^{-t} + (e^{-2} - \frac{1}{3}e)e^{2t} + \frac{1}{4}e^{3t}$.

2.95 A general solution to a certain second-order differential equation is $x(t) = c_1e^t + c_2e^{-t} + 4\sin t$, where c_1 and c_2 are arbitrary constants. Find a particular solution which also satisfies the initial conditions $x(0) = 1$, $\dot{x}(0) = -1$.

▌ For $x(t)$ as given, we have $\dot{x}(t) = c_1e^t - c_2e^{-t} + 4\cos t$. The initial conditions then yield

$$\begin{aligned} x(0) &= c_1e^0 + c_2e^{-(0)} + 4\sin 0 = 1 \qquad \text{or} \qquad c_1 + c_2 = 1 \\ \dot{x}(0) &= c_1e^0 - c_2e^{-(0)} + 4\cos 0 = -1 \qquad \text{or} \qquad c_1 - c_2 = -5 \end{aligned} \tag{1}$$

Solving system (*1*), we obtain $c_1 = -2$ and $c_2 = 3$. Substituting these values into the general solution, we get the particular solution $x(t) = -2e^t + 3e^{-t} + 4\sin t$.

2.96 A general solution to a certain second-order differential equation is $x(t) = c_1t + c_2 + t^2 - 1$, where c_1 and c_2 are arbitrary constants. Find a particular solution which also satisfies the initial conditions $x(1) = 1$, $\dot{x}(1) = 2$.

▌ For $x(t)$ as given, we have $\dot{x}(t) = c_1 + 2t$. The initial conditions yield

$$\begin{aligned} x(1) &= c_1(1) + c_2 + (1)^2 - 1 = 1 \qquad \text{or} \qquad c_1 + c_2 = 1 \\ \dot{x}(1) &= c_1 + 2(1) = 2 \qquad \text{or} \qquad c_1 = 0 \end{aligned} \tag{1}$$

Solving system (*1*), we obtain $c_1 = 0$ and $c_2 = 1$. Substituting these values into the general solution, we get the particular solution $x(t) = (0)t + 1 + t^2 - 1 = t^2$.

2.97 A general solution to a certain second-order differential equation is $z(t) = Ae^t + Bte^t + t^2e^t$, where A and B are arbitrary constants. Find a particular solution which also satisfies the initial conditions $z(1) = 1$, $\dot{z}(1) = -1$.

▮ For the given $z(t)$, we have $\dot{z}(t) = Ae^t + B(e^t + te^t) + (2te^t + t^2e^t)$. The initial conditions then yield $z(1) = Ae + Be + e = 1$ and $\dot{z}(1) = Ae + B(e + e) + (2e + e) = -1$, which may be rewritten as

$$\begin{aligned} A + B &= e^{-1} - 1 \\ A + 2B &= -e^{-1} - 3 \end{aligned} \qquad (1)$$

Solving system (1), we obtain $A = 1 + 3e^{-1}$ and $B = -2 - 2e^{-1}$. Substituting these values into the general solution, we get the particular solution $z(t) = (1 + 3e^{-1})e^t + (-2 - 2e^{-1})te^t + t^2e^t$.

2.98 A general solution to a particular third-order differential equation is $z(t) = A + Bt + Ct^2 + 2t^3$, where A, B, and C are arbitrary constants. Find a particular solution which also satisfies the initial conditions $z(1) = \dot{z}(1) = \ddot{z}(1) = 0$.

▮ For $z(t)$ as given, we have $\dot{z}(t) = B + 2Ct + 6t^2$ and $\ddot{z}(t) = 2C + 12t$. The initial conditions yield

$$\begin{aligned} z(1) &= A + B(1) + C(1)^2 + 2(1)^3 = 0 &\quad\text{or}\quad A + B + C &= -2 \\ \dot{z}(1) &= B + 2C(1) + 6(1)^2 = 0 &\quad\text{or}\quad B + 2C &= -6 \\ \ddot{z}(1) &= 2C + 12(1) = 0 &\quad\text{or}\quad 2C &= -12 \end{aligned} \qquad (1)$$

Solving system (1), we obtain $A = -2$, $B = 6$, and $C = -6$. Substituting these values into the general solution, we get the particular solution $z(t) = -2 + 6t - 6t^2 + 2t^3$.

2.99 A general solution to a third-order differential equation is $z(t) = Ae^{2t} + Be^{-2t} + Ce^{-3t}$, where A, B, and C are arbitrary constants. Find a particular solution which also satisfies the initial conditions $z(0) = 0$, $z'(0) = 9$, $z''(0) = -5$.

▮ For $z(t)$ as given, we have $z'(t) = 2Ae^{2t} - 2Be^{-2t} - 3Ce^{-3t}$ and $z''(t) = 4Ae^{2t} + 4Be^{-2t} + 9Ce^{-3t}$. The initial conditions yield

$$\begin{aligned} z(0) &= Ae^{2(0)} + Be^{-2(0)} + Ce^{-3(0)} = 0 &\quad\text{or}\quad A + B + C &= 0 \\ z'(0) &= 2Ae^{2(0)} - 2Be^{-2(0)} - 3Ce^{-3(0)} = 9 &\quad\text{or}\quad 2A - 2B - 3C &= 9 \\ z''(0) &= 4Ae^{2(0)} + 4Be^{-2(0)} + 9Ce^{-3(0)} = -5 &\quad\text{or}\quad 4A + 4B + 9C &= -5 \end{aligned} \qquad (1)$$

Solving system (1), we obtain $A = 2$, $B = -1$, and $C = -1$, so the particular solution is $z(t) = 2e^{2t} - e^{-2t} - e^{-3t}$.

2.100 A general solution to a fourth-order differential equation is $y(s) = Ae^s + Be^{-s} + Ce^{2s} + De^{3s} + s^2 + 2$, where A, B, C, and D are arbitrary constants. Find a particular solution which also satisfies the initial conditions $y(0) = 1$, $y'(0) = y''(0) = 4$, $y'''(0) = 10$.

▮ For $y(s)$ as given, we have

$$\begin{aligned} y'(s) &= Ae^s - Be^{-s} + 2Ce^{2s} + 3De^{3s} + 2s \\ y''(s) &= Ae^s + Be^{-s} + 4Ce^{2s} + 9De^{3s} + 2 \\ y'''(s) &= Ae^s - Be^{-s} + 8Ce^{2s} + 27De^{3s} \end{aligned}$$

Consequently, the initial conditions yield

$$\begin{aligned} y(0) &= Ae^{(0)} + Be^{-(0)} + Ce^{2(0)} + De^{3(0)} + (0)^2 + 2 = 1 \\ y'(0) &= Ae^{(0)} - Be^{-(0)} + 2Ce^{2(0)} + 3De^{3(0)} + 2(0) = 4 \\ y''(0) &= Ae^{(0)} + Be^{-(0)} + 4Ce^{2(0)} + 9De^{3(0)} + 2 = 4 \\ y'''(0) &= Ae^{(0)} - Be^{-(0)} + 8Ce^{2(0)} + 27De^{3(0)} = 10 \end{aligned}$$

which may be rewritten

$$\begin{aligned} A + B + C + D &= -1 \\ A - B + 2C + 3D &= 4 \\ A + B + 4C + 9D &= 2 \\ A - B + 8C + 27D &= 10 \end{aligned} \qquad (1)$$

System (1) has as its solution $A = D = 0$, $B = -2$, and $C = 1$, so the particular solution to the initial-value problem is $y(s) = -2e^{-s} + e^{2s} + s^2 + 2$.

2.101 A general solution to a fourth-order differential equation is $y(\theta) = A + B\theta + C\theta^2 + D\theta^3 + \theta^4$, where A, B, C, and D are arbitrary constants. Find a particular solution which also satisfies the initial conditions $y(-1) = 0$, $y'(-1) = 1$, $y''(-1) = 2$, $y'''(-1) = 0$.

▌ For the given function $y(\theta)$, we have $y'(\theta) = B + 2C\theta + 3D\theta^2 + 4\theta^3$, $y''(\theta) = 2C + 6D\theta + 12\theta^2$, and $y'''(\theta) = 6D + 12\theta$. Applying the initial conditions, we obtain

$$\begin{aligned} y(-1) &= A + B(-1) + C(-1)^2 + D(-1)^3 + (-1)^4 = 0 \\ y'(-1) &= B + 2C(-1) + 3D(-1)^2 + 4(-1)^3 = 1 \\ y''(-1) &= 2C + 6D(-1) + 12(-1)^2 = 2 \\ y'''(-1) &= 6D + 12(-1) = 0 \end{aligned}$$

which may be rewritten as

$$\begin{aligned} A - B + C - D &= -1 \\ B - 2C + 3D &= 5 \\ 2C - 6D &= -10 \\ 6D &= 12 \end{aligned} \tag{1}$$

System (*1*) has as its solution $A = B = C = 1$, and $D = 2$, so the particular solution to the initial-value problem is $y(\theta) = 1 + \theta + \theta^2 + 2\theta^3 + \theta^4$.

2.102 Find a particular solution to the initial-value problem $y' = \dfrac{x^2 + y^2}{xy}$; $y(1) = -2$, if it is known that a general solution to the differential equation is given implicitly by $y^2 = x^2 \ln x^2 + kx^2$, where k is an arbitrary constant.

▌ Applying the initial condition to the general solution, we obtain $(-2)^2 = 1^2 \ln(1^2) + k(1^2)$, or $k = 4$. (Recall that $\ln 1 = 0$). Thus, the solution to the initial-value problem is

$$y^2 = x^2 \ln x^2 + 4x^2 \qquad \text{or} \qquad y = -\sqrt{x^2 \ln x^2 + 4x^2}$$

The negative square root is taken so as to be consistent with the initial condition. That is, we cannot choose the positive square root, since then $y(1) = \sqrt{1^2 \ln(1^2) + 4(1^2)} = 2$, which violates the initial condition.

2.103 Find a particular solution to the initial-value problem $y' = e^x/y$; $y(0) = 1$, if it is known that a general solution to the differential equation is given implicitly by $y^2 = 2e^x + k$, where k is an arbitrary constant.

▌ Applying the initial condition, we obtain $1^2 = 2e^0 + k$, or $k = -1$. Thus, the solution to the initial-value problem is

$$y^2 = 2e^x - 1 \qquad \text{or} \qquad y = \sqrt{2e^x - 1}$$

[Note that we cannot choose the negative square root, since then $y(0) = -1$, which violates the initial condition.]

To ensure that y remains real, we must restrict x so that $2e^x - 1 \ge 0$. To guarantee that y' exists [note that $y'(x) = dy/dx = e^x/y$], we must restrict x so that $2e^x - 1 \ne 0$. Together these conditions imply that $2e^x - 1 > 0$, or $x > \ln \frac{1}{2}$.

2.104 Find a particular solution to the initial-value problem $\dot{y} = y^5 \sin t$; $y(0) = 1$, if it is known that the general solution to the differential equation is given implicitly by $1/y^4 = 4\cos t + c_1$, where c_1 is an arbitrary constant.

▌ Applying the initial condition, we obtain $1/1^4 = 4\cos 0 + c_1$, or $c_1 = -3$. Substituting this value into the general solution and solving explicitly for y, we obtain $y(t) = \left(\dfrac{1}{4\cos t - 3}\right)^{1/4}$. This equation makes sense only if $4\cos t - 3 > 0$. Also, since y must be defined on an *interval containing 0* (a solution to an initial-value problem is defined on an *interval* that contains the initial point), we see that if θ is the number in $(0, \pi/2)$ such that $\cos\theta = \frac{3}{4}$ (that is, $\theta = \text{Arccos}\,\frac{3}{4}$), then the solution y is indeed defined on $(-\theta, \theta)$. Moreover, $y(t) \to +\infty$ as $t \to \theta^-$ and as $t \to -\theta^+$.

2.105 Find a particular solution to the initial-value problem $x\cos x + (1 - 6y^5)y' = 0$ $y(\pi) = 0$; if the general solution to the differential equation is given implicitly by $x\sin x + \cos x + y - y^6 = c$, where c is an arbitrary constant.

▌ Applying the initial condition to the general solution, we find $\pi\sin\pi + \cos\pi + 0 + 0^2 = c$, or $c = -1$. A particular solution is then $x\sin x + \cos x + y - y^6 = -1$. We can rearrange this equation to

$x \sin x + \cos x + 1 = y^6 - y$, but since we cannot solve either equation for y explicitly, we must be content with the solution in implicit form.

SIMPLIFYING SOLUTIONS

2.106 Verify and reconcile the fact that $y = c_1 \cos x + c_2 \sin x$ and $y = A\cos(x + B)$ are primitives of $\dfrac{d^2y}{dx^2} + y = 0$.

▌ From $y = c_1 \cos x + c_2 \sin x$, we obtain first $y' = -c_1 \sin x + c_2 \cos x$ and then

$$y'' = -c_1 \cos x - c_2 \sin x = -y \qquad \text{or} \qquad \frac{d^2y}{dx^2} + y = 0$$

From $y = A\cos(x + B)$, we obtain first $y' = -A\sin(x + B)$ and then, again, $y'' = -A\cos(x + B) = -y$. To reconcile the two primitives we write

$$y = A\cos(x + B) = A(\cos x \cos B - \sin x \sin B) = (A\cos B)\cos x + (-A\sin B)\sin x = c_1 \cos x + c_2 \sin x$$

2.107 Show that $\ln x^2 + \ln(y^2/x^2) = A + x$ may be written as $y^2 = Be^x$.

▌ Since we have $\ln x^2 + \ln \dfrac{y^2}{x^2} = \ln\left(x^2 \dfrac{y^2}{x^2}\right) = \ln y^2 = A + x$, we may write $y^2 = e^{A+x} = e^A e^x = Be^x$.

2.108 Show that $\operatorname{Arcsin} x - \operatorname{Arcsin} y = A$ may be written as $x\sqrt{1 - y^2} - y\sqrt{1 - x^2} = B$.

▌ We first let $\sin(\operatorname{Arcsin} x - \operatorname{Arcsin} y) = \sin A = B$. Then, for a difference of angles, we have

$$\sin(\operatorname{Arcsin} x)\cos(\operatorname{Arcsin} y) - \cos(\operatorname{Arcsin} x)\sin(\operatorname{Arcsin} y) = x\sqrt{1 - y^2} - y\sqrt{1 - x^2} = B$$

2.109 Show that $\ln(1 + y) + \ln(1 + x) = A$ may be written as $xy + x + y = c$.

▌ We first note that $\ln(1 + y) + \ln(1 + x) = \ln[(1 + y)(1 + x)] = A$. Then

$$(1 + y)(1 + x) = xy + x + y + 1 = e^A = B \qquad \text{and} \qquad xy + x + y = B - 1 = c.$$

2.110 Show that $\sinh y + \cosh y = cx$ may be written as $y = \ln x + A$.

▌ By definition, $\sinh y + \cosh y = \frac{1}{2}(e^y - e^{-y}) + \frac{1}{2}(e^y + e^{-y}) = e^y = cx$. Then $y = \ln c + \ln x = A + \ln x$.

CHAPTER 3

Separable First-Order Differential Equations

SOLUTIONS WITH RATIONAL FUNCTIONS

3.1 Define *separable* as applied to differential equations.

A first-order differential equation is separable if it can be expressed in the form $\frac{dy}{dx} = -\frac{A(x)}{B(y)}$, where $A(x)$ is a function only of x, and $B(y)$ is a function only of y. Such equations have the differential form $A(x)\,dx + B(y)\,dy = 0$. The variables x and y may be replaced by any two other variables without affecting separability.

3.2 Prove that every solution of the separable differential equation $A(x)\,dx + B(y)\,dy = 0$ is given by $\int A(x)\,dx + \int B(y)\,dy = c$, where c represents an arbitrary constant.

Rewrite the differential equation as $A(x) + B(y)y' = 0$. If $y(x)$ is a solution, it must satisfy this equation identically in x; hence, $A(x) + B[y(x)]y'(x) = 0$. Integrating both sides of this last equation with respect to x, we obtain

$$\int A(x)\,dx + \int B[y(x)]y'(x)\,dx = c$$

In the second integral, make the change of variables $y = y(x)$, so that $dy = y'(x)\,dx$. The result of this substitution is $\int A(x)\,dx + \int B(y)\,dy = c$.

The two integrals may be, for all practical purposes, impossible to evaluate. In such a case, numerical techniques may have to be used to obtain an approximate solution. Even if the indicated integrations can be performed, it may not be algebraically possible to solve for y explicitly in terms of x. In that case, the solution is left in implicit form.

3.3 Solve $x\,dx - y^2\,dy = 0$.

For this differential equation, $A(x) = x$ and $B(y) = -y^2$. The solution is $\int x\,dx + \int(-y^2)\,dy = c$, which, after the indicated integrations are performed, becomes $x^2/2 - y^3/3 = c$. Solving for y explicitly, we obtain the solution as $y = (\frac{3}{2}x^2 + k)^{1/3}$, where $k = -3c$.

3.4 Solve $y' = y^2x^3$.

We first rewrite this equation in the differential form $x^3\,dx - (1/y^2)\,dy = 0$. Then $A(x) = x^3$ and $B(y) = -1/y^2$. The solution is $\int x^3\,dx + \int(-1/y^2)\,dy = c$ or, after the indicated integrations, $x^4/4 + 1/y = c$. Solving explicitly for y, we obtain the solution as $y = \frac{-4}{x^4 + k}$, where $k = -4c$.

3.5 Solve $x\,dx + y\,dy = 0$.

The variables are separated, so integrating each term gives $\int x\,dx + \int y\,dy = c$ or $\frac{1}{2}x^2 + \frac{1}{2}y^2 = c$. Solving explicitly for y, we obtain the two expressions $y = \sqrt{k - x^2}$ and $y = -\sqrt{k - x^2}$, where $k = 2c$.

3.6 Solve $x\,dx - y^3\,dy = 0$.

The variables are separated, so integrating each term gives $\int x\,dx - \int y^3\,dy = c$ or $\frac{1}{2}x^2 - \frac{1}{4}y^4 = c$. Solving explicitly for y, we obtain the two expressions $y = (k + 2x^2)^{1/4}$ and $y = -(k + 2x^2)^{1/4}$, where $k = -4c$.

3.7 Solve $y^4y' = x + 1$.

This equation may be rewritten in the differential form $y^4\,dy - (x + 1)\,dx = 0$, which is separable. The solution is $\int y^4\,dy - \int(x + 1)\,dx = c$ or, after the indicated integrations are performed, $\frac{1}{5}y^5 - \frac{1}{2}x^2 - x = c$. This may be solved explicitly for y to yield $y = (\frac{5}{2}x^2 + 5x + k)^{1/5}$, where $k = 5c$.

3.8 Solve $y' = (x + 1)/y$.

▌ This equation may be rewritten as $yy' = x + 1$ or, in differential form, as $y\,dy - (x + 1)\,dx = 0$. The differential equation is separable and has the solution $\int y\,dy - \int (x + 1)\,dx = c$. Performing the indicated integrations, we get $\frac{1}{2}y^2 - \frac{1}{2}x^2 - 2x = c$. Solving for y explicitly, we obtain the two expressions $y = \sqrt{x^2 + 4x + k}$ and $y = -\sqrt{x^2 + 4x + k}$, where $k = 2c$.

3.9 Solve $\dot{y} = y^2$.

▌ Separating variables gives, in differential form, $y^{-2}\,dy = dt$. This has the solution $\int y^{-2}\,dt = \int dt + c$ or $-1/y = t + c$. Thus, $y = -1/(t + c)$.

3.10 Solve $dy/dt = y^2t^2$.

▌ This equation may be rewritten in the differential form $\dfrac{1}{y^2}\,dy - t^2\,dt = 0$, which is separable. The solution is $\displaystyle\int \frac{1}{y^2}\,dy - \int t^2\,dt = c$ or, after the indicated integrations are performed, $-\dfrac{1}{y} - \dfrac{1}{3}t^3 = c$. Solving explicitly for y, we get $y = -\dfrac{3}{t^3 + k}$, where $k = 3c$.

3.11 Solve $dz/dt = z^3t^2$.

▌ This equation may be rewritten in the differential form $\dfrac{1}{z^3}\,dz - t^2\,dt = 0$, which is separable. The solution is $\displaystyle\int \frac{1}{z^3}\,dz - \int t^2\,dt = c$ or, after the indicated integrations are performed, $-\dfrac{1}{2z^2} - \dfrac{1}{3}t^3 = c$. Solving explicitly for z, we obtain $z = \pm\left(\dfrac{1}{k - \frac{2}{3}t^3}\right)^{1/2}$, where $k = -2c$.

3.12 Solve $\dfrac{1}{x^4}\,dx + dt = 0$.

▌ This equation is separable. Integrating term by term, we obtain $-\frac{1}{3}x^{-3} + t = c$. Solving explicitly for x (assuming it is the unknown function), we obtain $x(t) = (3t + k)^{-1/3}$, where $k = -3c$,

3.13 Solve $s\,ds + s^3(\theta^2 - 3)\,d\theta = 0$ for $s(\theta)$.

▌ We can rewrite this equation as $s^{-2}\,ds + (\theta^2 - 3)\,d\theta = 0$, which has as its solution $\int s^{-2}\,ds + \int (\theta^2 - 3)\,d\theta = c$ Performing the indicated integrations, we obtain $-s^{-1} + \frac{1}{3}\theta^3 - 3\theta = c$. Solving explicitly for s, we find $s = (\frac{1}{3}\theta^3 - 3\theta - c)^{-1}$.

3.14 Solve $\dot{x} = x^2t + x^2$.

▌ This equation may be rewritten as $dx/dt = x^2(t + 1)$ or, in differential form, as $\dfrac{1}{x^2}\,dx - (t + 1)\,dt = 0$. Integrating term by term, we obtain $-\dfrac{1}{x} - \frac{1}{2}t^2 - t = c$, which may be explicitly solved for $x(t)$, giving $x = -(\frac{1}{2}t^2 + t + c)^{-1}$.

3.15 Solve $x\dot{x} = (t - 1)^2$.

▌ This equation may be rewritten as $x\dfrac{dx}{dt} = (t - 1)^2$, which has the differential form $x\,dx - (t - 1)^2\,dt = 0$. Integrating term by term, we obtain $\frac{1}{2}x^2 - \frac{1}{3}(t - 1)^3 = c$ or, explicitly, $x = \pm[k + \frac{2}{3}(t - 1)^3]^{1/2}$, where $k = 2c$

3.16 Solve $x^{-2}\dot{x} = (t + 3)^3$.

▌ This equation may be rewritten as $x^{-2}\dfrac{dx}{dt} = (t + 3)^3$, which has the differential form $x^{-2}\,dx - (t + 3)^3\,dt = 0$. Integrating term by term, we obtain $-x^{-1} - \frac{1}{4}(t + 3)^4 = c$. Thus $x = -1/[c + \frac{1}{4}(t + 3)^4]$.

3.17 Solve $x^3\,dx + (y+1)^2\,dy = 0$.

▌ The variables are separated. Integrating term by term, we get $\dfrac{x^4}{4} + \dfrac{(y+1)^3}{3} = c$. Solving explicitly for y, we obtain $y = -1 + (k - \frac{3}{4}x^4)^{1/3}$, where $k = 3c$.

3.18 Solve $x^2\,dx + (y-3)^4\,dy = 0$.

▌ The variables are separated. Integrating term by term, we get $\dfrac{x^3}{3} + \dfrac{(y-3)^5}{5} = c$. Solving explicitly for y, we obtain $y = 3 + (k - \frac{5}{3}x^3)^{1/5}$, where $k = 5c$.

3.19 Solve $\dfrac{dx}{dt} = \dfrac{t-1}{x^2 - 4x + 4}$.

▌ This equation may be rewritten in the differential form $(x-2)^2\,dx - (t-1)\,dt = 0$, which is separable. Integrating term by term, we get $\dfrac{(x-2)^3}{3} - \dfrac{(t-1)^2}{2} = c$. Solving explicitly for x, we obtain $x = 2 + [\frac{3}{2}(t-1)^2 + k]^{1/3}$, where $k = 3c$.

3.20 Solve $\dfrac{ds}{dt} = \dfrac{s^2 + 6s + 9}{t^2}$.

▌ This equation can be written in the differential form $\dfrac{ds}{(s+3)^2} - \dfrac{dt}{t^2} = 0$, which is separable. Integrating term by term, we get $-(s+3)^{-1} + t^{-1} = c$. Solving explicitly for s, we obtain $s = -3 + t/(1 - ct)$.

3.21 Solve $\dfrac{ds}{dt} = \dfrac{t^2}{s^2 + 6s + 9}$.

▌ This equation can be written in the differential form $(s+3)^2\,ds - t^2\,dt = 0$, which is separable. Integrating term by term, we get $\frac{1}{3}(s+3)^3 - \frac{1}{3}t^3 = c$. Solving explicitly for s, we obtain $s = -3 + (t^3 + k)^{1/3}$, where $k = 3c$.

3.22 Solve $\dfrac{dr}{dt} = \dfrac{r^3 + 3r^2 + 3r + 1}{x^2 - 2x + 1}$.

▌ This equation can be written in the differential form $\dfrac{dr}{(r+1)^3} - \dfrac{dx}{(x-1)^2} = 0$, which is separable. Integrating term by term, we get $-\frac{1}{2}(r+1)^{-2} + (x-1)^{-1} = c$. Solving explicitly for $r(x)$, we obtain $r = -1 + \left[\dfrac{2 - 2c(x-1)}{x-1}\right]^{-1/2}$.

3.23 Solve $y' = x/(y+2)$.

▌ We may rewrite this equation as $(y+2)y' = x$, and in differential form as $(y+2)\,dy - x\,dx = 0$. The variables are separated, so term-by-term integration produces the solution $\frac{1}{2}y^2 + 2y - \frac{1}{2}x^2 = c$ or $y^2 + 4y + (k - x^2) = 0$, where $k = -2c$. To solve explicitly for y we use the quadratic formula, getting

$$y = \frac{-4 \pm \sqrt{16 - 4(k - x^2)}}{2} = -2 \pm \sqrt{4 - (k - x^2)} = -2 \pm \sqrt{d + x^2} \qquad \text{where} \qquad d = 4 - k$$

3.24 Solve $y' = x^2/(y-1)$.

▌ We rewrite this equation first as $(y-1)y' - x^2 = 0$ and then in the differential form $(y-1)\,dy - x^2\,dx = 0$. Integrating term by term, we obtain as the solution of this separable equation $\frac{1}{2}y^2 - y - \frac{1}{3}x^3 = c$ or $y^2 - 2y + (k - \frac{2}{3}x^3) = 0$, where $k = -2c$. To solve explicitly for y we use the quadratic formula, getting

$$y = \frac{2 \pm \sqrt{4 - 4(k - \frac{2}{3}x^3)}}{2} = 1 \pm \sqrt{1 - (k - \tfrac{2}{3}x^3)} = 1 \pm \sqrt{d + \tfrac{2}{3}x^3} \qquad \text{where} \qquad d = 1 - k$$

3.25 Solve $\dfrac{dy}{dt} = \dfrac{t+1}{y-1}$.

This equation may be rewritten in the differential form $(y-1)\,dy - (t+1)\,dt = 0$, which is separable. Integrating term by term, we obtain the solution $\frac{1}{2}y^2 - y - \frac{1}{2}t^2 - t = c$ or $y^2 - 2y + (k - t^2 - 2t) = 0$, where $k = -2c$. To solve explicitly for y we use the quadratic formula, getting

$$y = \frac{2 \pm \sqrt{4 - 4(k - t^2 - 2t)}}{2} = 1 \pm \sqrt{1 - (k - t^2 - 2t)} = 1 \pm \sqrt{t^2 - 2t + d} \qquad \text{where} \qquad d = 1 - k$$

3.26 Solve $\dfrac{dz}{dt} = \dfrac{t^2+2}{z+3}$.

This equation may be rewritten in the differential form $(z+3)\,dz - (t^2+2)\,dt = 0$, which is separable. Integrating term by term, we obtain the solution $\frac{1}{2}z^2 + 3z - \frac{1}{3}t^3 - 2t = c$ or $z^2 + 6z + (k - \frac{2}{3}t^3 - 4t) = 0$, where $k = -2c$. To solve explicitly for z we use the quadratic formula, getting

$$z = \frac{-6 \pm \sqrt{36 - 4(k - \frac{2}{3}t^3 - 4t)}}{2} = -3 \pm \sqrt{9 - (k - \tfrac{2}{3}t^3 - 4t)} = -3 \pm \sqrt{\tfrac{2}{3}t^3 + 4t + d} \qquad \text{where} \qquad d = 9 - k$$

3.27 Solve $y' = \dfrac{x+1}{y^4+1}$.

This equation may be written in differential form as $(x+1)\,dx + (-y^4 - 1)\,dy = 0$, which is separable. The solution is $\int (x+1)\,dx + \int (-y^4 - 1)\,dy = c$ or, after integration, $\dfrac{x^2}{2} + x - \dfrac{y^5}{5} - y = c$. Since it is algebraically impossible to solve this equation explicitly for y, the solution must be left in implicit form.

3.28 Solve $y' = \dfrac{x^2+7}{y^9 - 3y^4}$.

In differential form this equation is $(y^9 - 3y^4)\,dy - (x^2+7)\,dx = 0$, which is separable. Its solution is $\int (y^9 - 3y^4)\,dy - \int (x^2+7)\,dx = c$ or, after integration, $\frac{1}{10}y^{10} - \frac{3}{5}y^5 - \frac{1}{3}x^3 + 7x = c$. Since it is algebraically impossible to solve this equation explicitly for y, the solution must be left in implicit form.

SOLUTIONS WITH LOGARITHMS

3.29 Solve $y' = 5y$.

This equation may be written in the differential form $5\,dx - (1/y)\,dy = 0$. Then the solution is $\int 5\,dx + \int -\dfrac{1}{y}\,dy = c$ or, after integration, $5x - \ln|y| = c$.

To solve for y explicitly, we first rewrite the solution as $\ln|y| = 5x - c$ and then take the exponentials of both sides. Thus, $e^{\ln|y|} = e^{5x-c}$. Noting that $e^{\ln|y|} = |y|$, we obtain $|y| = e^{5x}e^{-c}$, or $y = \pm e^{-c}e^{5x}$. The solution is given explicitly by $y = ke^{5x}$, $k = \pm e^{-c}$.

Note that the presence of the term $-1/y$ in the differential form of the differential equation requires the restriction $y \neq 0$ in our derivation of the solution. This restriction is equivalent to the restriction $k \neq 0$, since $y = ke^{5x}$. However, by inspection, $y \equiv 0$ is a solution of the differential equation as originally given. Thus, $y = ke^{5x}$ is the solution for all k.

3.30 Solve $y' = Ay$, where A denotes a constant.

In differential form this equation is $(1/y)\,dy - A\,dx = 0$. Its solution is $\int (1/y)\,dy - \int A\,dx = c$ or, after integration, $\ln|y| - Ax = c$, which may be rewritten as $\ln|y| = Ax + c$. Taking the exponentials of both sides of this last equation and noting that $e^{\ln|y|} = |y|$, we obtain $|y| = e^{Ax+c} = e^c e^{Ax}$. Thus, $y = \pm e^c e^{Ax} = ke^{Ax}$, where $k = \pm e^c$.

3.31 Solve $y' = xy$.

The differential form of this equation is $(1/y)\,dy - x\,dx = 0$. Integrating term by term, we obtain the solution $\ln|y| - \frac{1}{2}x^2 = c$, which we rewrite as $\ln|y| = c + \frac{1}{2}x^2$. Taking the exponentials of both sides of this equation, we get $|y| = e^{c + x^2/2} = e^c e^{x^2/2}$, so $y = ke^{x^2/2}$, where $k = \pm e^c$.

3.32 Solve $dy/dt = y(t-2)$.

▌ In differential form this equation is $(1/y)\,dy - (t-2)\,dt = 0$. The solution is $\int (1/y)\,dy - \int (t-2)\,dt = c$. The indicated integrations result in $\ln|y| - \frac{1}{2}(t-2)^2 = c$, which may be rewritten as $\ln|y| = c + \frac{1}{2}(t-2)^2$. Taking the exponentials of both sides of this equation, we get $|y| = e^{c+(t-2)^2/2} = e^c e^{(t-2)^2/2}$, so that $y = ke^{(t-2)^2/2}$, where $k = \pm e^c$.

3.33 Solve $dy/dt = -2yt^2$.

▌ In differential form this equation is $(1/y)\,dy + 2t^2\,dt = 0$. Integrating term by term, we obtain the solution $\ln|y| + \frac{2}{3}t^3 = c$ or, rewritten, $\ln|y| = c - \frac{2}{3}t^3$. Taking the exponentials of both sides of this last equation, we get $|y| = e^{c-(2/3)t^3} = e^c e^{-2t^3/3}$, so $y = ke^{-2t^3/3}$, where $k = \pm e^c$.

3.34 Solve $dy/dt = y/t$.

▌ In differential form this equation is $(1/y)\,dy - (1/t)\,dt = 0$. Integrating term by term, we obtain the solution $\ln|y| - \ln|t| = c$, or, rewritten, $\ln|y/t| = c$. Taking the exponentials of both sides of this last equation, we get $|y/t| = e^c$, so $y/t = \pm e^c$ and $y = kt$, where $k = \pm e^c$.

3.35 Solve $\dfrac{dz}{dt} = \dfrac{z+1}{t}$.

▌ In differential form this equation is $\dfrac{1}{z+1}\,dz - \dfrac{1}{t}\,dt = 0$. Integrating term by term, we obtain the solution $\ln|z+1| - \ln|t| = c$, which we rewrite as $\ln\left|\dfrac{z+1}{t}\right| = c$. Taking the exponentials of both sides of this last equation, we get $\left|\dfrac{z+1}{t}\right| = e^c$ or $\dfrac{z+1}{t} = \pm e^c$. Then $z = kt - 1$, where $k = \pm e^c$.

3.36 Solve $dy/dx + 3y = 8$.

▌ We have $dy/dx = 8 - 3y$. Separating the variables gives $\dfrac{dy}{8-3y} = dx$, with solution $\displaystyle\int \frac{dy}{8-3y} = \int dx + c$. The indicated integrations produce $-\frac{1}{3}\ln|8-3y| = x + c$, which were written as $\ln|8-3y| = -3x - 3c$. Taking the exponentials of both sides gives us $|8-3y| = e^{-3x-3c} = e^{-3c}e^{-3x}$ or $8 - 3y = \pm e^{-3c}e^{-3x}$, so that $y = \frac{8}{3} + ke^{-3x}$, where $k = \pm\frac{1}{3}e^{-3c}$.

3.37 Solve $dy/dx - 5y = 3$.

▌ Separating the variables, we obtain the differential form $\dfrac{1}{3+5y}\,dy - dx = 0$. Integrating term by term gives $\frac{1}{5}\ln|3+5y| - x = c$, so that $\ln|3+5y| = 5c + 5x$. Taking the exponentials of both sides gives $|3+5y| = e^{5c+5x} = e^{5c}e^{5x}$ or $3 + 5y = \pm e^{5c}e^{5x}$. Then $y = -\frac{3}{5} + ke^{5x}$, where $k = \pm\frac{1}{5}e^{5c}$.

3.38 Solve $(5-t)\,dx + (x+3)\,dt = 0$ for $x(t)$.

▌ We first rewrite the differential equation as $\dfrac{1}{x+3}\,dx - \dfrac{1}{t-5}\,dt = 0$, which is separable. The solution, obtained by integrating term by term, is $\ln|x+3| - \ln|5-t| = c$ or $\ln\left|\dfrac{x+3}{t-5}\right| = c$. Taking the exponentials of both sides of this last equation, we get $\left|\dfrac{x+3}{t-5}\right| = e^c$. Hence, $x = -3 + k(t-5)$, where $k = \pm e^c$.

3.39 Solve $(5-t)\,dx - (x+3)\,dt = 0$ for $x(t)$.

▌ We rewrite the differential equation as $\dfrac{1}{x+3}\,dx + \dfrac{1}{t-5}\,dt = 0$, which is separable. Integrating term by term, we obtain the solution $\ln|x+3| + \ln|t-5| = c$ or $\ln|(x+3)(t-5)| = c$. Taking the exponentials of both sides of this last equation, we get $|(x+3)(t-5)| = e^c$. Hence, $(x+3)(t-5) = \pm e^c$ and $x = -3 + \dfrac{k}{t-5}$, where $k = \pm e^c$.

3.40 Solve $dy/dt = -y/t$.

▮ In differential form this equation is $(1/y)\,dy + (1/t)\,dt = 0$. Integrating term by term, we get $\ln|y| + \ln|t| = c$ or $\ln|yt| = c$. Taking the exponentials of both sides gives us $|yt| = e^c$ or $yt = \pm e^c$. Thus $y = k/t$, where $k = \pm e^c$.

3.41 Solve $y\,dy + (y^2 + 1)\,dx = 0$.

▮ We rewrite this equation as $\dfrac{y}{y^2+1}\,dy + dx = 0$, which is separable. Integrating term by term, we get $\frac{1}{2}\ln(1 + y^2) + x = c$ or, rewritten, $\ln(1 + y^2) = 2c - 2x$. Taking the exponentials of both sides, we obtain $1 + y^2 = e^{2c-2x} = e^{2c}e^{-2x}$. Then $y^2 = -1 + ke^{-2x}$, where $k = e^{2c}$. Solving explicitly for y, we obtain as the solution $y = \pm(-1 + ke^{-2x})^{1/2}$.

3.42 Solve $y^2x\,dy + (y^3 - 1)\,dx = 0$.

▮ We rewrite this equation as $\dfrac{y^2}{y^3-1}\,dy + \dfrac{1}{x}\,dx = 0$, which is separable. Integrating term by term, we get $\frac{1}{3}\ln|y^3 - 1| + \ln|x| = c$, which we may rewrite as $\ln|y^3 - 1| + 3\ln|x| = 3c$ and then as $\ln|(y^3 - 1)x^3| = 3c$. Taking exponentials gives $|(y^3 - 1)x^3| = e^{3c}$, so that $(y^3 - 1)x^3 = \pm e^{3c}$. The solution in implicit form is then $y^3 - 1 = kx^{-3}$, where $k = \pm e^{3c}$. Solving explicitly for y, we obtain $y = (1 + kx^{-3})^{1/3}$.

3.43 Solve $(y^2 + 1)(x + 1)\,dy = (y^3 + 3y)\,dx$.

▮ We rewrite this equation as $\dfrac{y^2+1}{y^3+3y}\,dy - \dfrac{1}{x+1}\,dx = 0$, which is separable. Integrating term by term, we get $\frac{1}{3}\ln|y^3 + 3y| - \ln|x + 1| = c$, from which we find $\ln\left|\dfrac{y^3+3y}{(x+1)^3}\right| = 3c$. Taking the exponentials of both sides gives $\left|\dfrac{y^3+3y}{(x+1)^3}\right| = e^{3c}$, so that $\dfrac{y^3+3y}{(x+1)^3} = \pm e^{3c}$. This yields $y^3 + 3y = k(x + 1)^3$, where $k = \pm e^{3c}$, as the solution in implicit form.

3.44 Solve $(4x + xy^2)\,dx + (y + x^2y)\,dy = 0$ for $y(x)$.

▮ This equation can be written as $x(4 + y^2)\,dx + y(1 + x^2)\,dy = 0$ and then separated into $\dfrac{x\,dx}{1+x^2} + \dfrac{y\,dy}{4+y^2} = 0$. Integrating term by term gives $\frac{1}{2}\ln(1 + x^2) + \frac{1}{2}\ln(4 + y^2) = c$, which we rewrite as $\ln[(1 + x^2)(4 + y^2)] = 2c$ or $(1 + x^2)(4 + y^2) = e^{2c}$. Thus the general solution is $(1 + x^2)(4 + y^2) = k$, where $k = e^{2c}$.

3.45 Solve $(y^3 + y)(t^2 + 1)\,dy = (ty^4 + 2y^2t)\,dt$ for $y(t)$.

▮ This equation can be written as $\dfrac{y^3+y}{y^4+2y^2}\,dy - \dfrac{t}{t^2+1}\,dt = 0$, which is separable. Integration yields $\frac{1}{4}\ln(y^4 + 2y^2) - \frac{1}{2}\ln(t^2 + 1) = c$, which we rewrite as $\ln(y^4 + 2y^2) - \ln(t^2 + 1)^2 = 4c$ or $\ln\dfrac{y^4+2y^2}{(t^2+1)^2} = 4c$. After taking the exponentials of both sides, we have $\dfrac{y^4+2y^2}{(t^2+1)^2} = e^{4c}$, which yields the implicit solution $y^4 + 2y^2 = k(t^2 + 1)^2$, where $k = e^{4c}$.

3.46 Solve $xv\,dv - (1 + v^2)\,dx = 0$.

▮ We first separate the variables and rewrite the differential equation as $\dfrac{v}{1+v^2}\,dv - \dfrac{1}{x}\,dx = 0$. Then, integrating term by term yields $\frac{1}{2}\ln(1 + v^2) - \ln|x| = c$, from which $\ln\dfrac{1+v^2}{x^2} = 2c$, so that $\dfrac{1+v^2}{x^2} = e^{2c}$. Thus, in implicit form, $1 + v^2 = kx^2$, where $k = e^{2c}$.

3.47 Solve $2xv\,dv + (v^2 - 1)\,dx = 0$.

▮ We first separate the variables and rewrite the differential equation as $\frac{2v}{v^2 - 1}\,dv + \frac{1}{x}\,dx = 0$. Then, integrating term by term yields $\ln|v^2 - 1| + \ln|x| = c$, so that $\ln|x(v^2 - 1)| = c$. Taking exponentials gives $|x(v^2 - 1)| = e^c$, from which $x(v^2 - 1) = \pm e^c$. Thus, in implicit form, $v^2 - 1 = k/x$, where $k = \pm e^c$.

3.48 Solve $3xv\,dv + (2v^2 - 1)\,dx = 0$.

▮ This differential equation may be rewritten as $\frac{3v}{2v^2 - 1}\,dv + \frac{1}{x}\,dx = 0$, which is separable. Integrating term by term, we obtain $\frac{3}{4}\ln|2v^2 - 1| + \ln|x| = c$, so that $\ln|2v^2 - 1|^3 + \ln|x|^4 = 4c$, and $\ln|(2v^2 - 1)^3x^4| = 4c$. Exponentiation gives $(2v^2 - 1)^3x^4 = \pm e^{4c}$, and the solution, in implicit form, is $2v^2 - 1 = k/x^{4/3}$, where $k = \pm e^{4c/3}$.

3.49 Solve $4xv\,dv + (3v^2 - 1)\,dx = 0$.

▮ This differential equation may be rewritten as $\frac{4v}{3v^2 - 1}\,dv + \frac{1}{x}\,dx = 0$, which is separable. Integrating term by term yields $\frac{2}{3}\ln|3v^2 - 1| + \ln|x| = c$, which we rewrite as $\ln|3v^2 - 1|^2 + \ln|x|^3 = 3c$. Then $\ln|(3v^2 - 1)^2x^3| = 3c$, so that $(3v^2 - 1)^2x^3 = \pm e^{3c}$ and the solution, in implicit form, is $3v^2 - 1 = kx^{-3/2}$, where $k = \pm e^{3c/2}$.

3.50 Solve $x(v^2 - 1)\,dv + (v^3 - 3v)\,dx = 0$.

▮ This differential equation may be rewritten as $\frac{v^2 - 1}{v^3 - 3v}\,dv + \frac{1}{x}\,dx = 0$, which is separable. Integrating term by term yields $\frac{1}{3}\ln|v^3 - 3v| + \ln|x| = c$, which we rewrite as $\ln|v^3 - 3v| + \ln|x|^3 = 3c$. Then $\ln|x^3(v^3 - 3v)| = 3c$, so that $|x^3(v^3 - 3v)| = e^{3c}$. In implicit form, the solution is then $x^3(v^3 - 3v) = \pm e^{3c}$.

3.51 Solve $(1 + x^3)\,dy = x^2y\,dx$ for $y(x)$.

▮ We rewrite this equation as $\frac{1}{y}\,dy - \frac{x^2}{1 + x^3}\,dx = 0$. Integration yields $\ln|y| - \frac{1}{3}\ln|1 + x^3| = c$ or $3\ln|y| - \ln|1 + x^3| = 3c$, so that $\ln\left|\frac{y^3}{1 + x^3}\right| = 3c$. Exponentiation gives us $\frac{y^3}{1 + x^3} = \pm e^{3c}$, so that, in implicit form, $y^3 = k(1 + x^3)$, where $k = \pm e^{3c}$.

3.52 Solve $(t^2 + 4)\,dx + 3t(1 - 2x)\,dt = 0$ for $x(t)$.

▮ We rewrite this equation as $\frac{1}{1 - 2x}\,dx + \frac{3t}{t^2 + 4} = 0$. Integrating term by term, we obtain $-\frac{1}{2}\ln|1 - 2x| + \frac{3}{2}\ln(t^2 + 4) = c$, which is equivalent to $\ln\left|\frac{(t^2 + 4)^3}{1 - 2x}\right| = 2c$. Taking exponentials, we get $\frac{(t^2 + 4)^3}{1 - 2x} = \pm e^{2c} = k$ so that $(t^2 + 4)^3 = k(1 - 2x)$, and in explicit form

$$x = \frac{1}{2}\left[1 - \frac{(t^2 + 4)^3}{k}\right] = \frac{1}{2} - A(t^2 + 4)^3 \qquad \text{where} \qquad A = \frac{1}{2k}$$

3.53 Solve $z\dot{z} = (z^2 + 1)/(t + 1)^2$.

▮ We rewrite this equation as $\frac{z}{z^2 + 1}\,dz - \frac{1}{(t + 1)^2}\,dt = 0$. Integrating term by term yields

$$\frac{1}{2}\ln(z^2 + 1) + \frac{1}{t + 1} = c \qquad \text{or} \qquad \ln(z^2 + 1) = 2c - \frac{2}{t + 1}$$

Then $$z^2 + 1 = e^{2c - 2/(t+1)} = e^{2c}e^{-2/(t+1)} = ke^{-2/(t+1)}$$

and $$z = \pm(-1 + ke^{-2/(t+1)})^{1/2}.$$

3.54 Solve $(2z+1)\dot{z} = 4t(z^2+z)$.

▌ This equation has the differential form $\dfrac{2z+1}{z^2+z}\,dz - 4t\,dt = 0$. Integrating term by term, we get $\ln|z^2+z| - 2t^2 = c$, which we may write as $\ln|z^2+z| = c + 2t^2$. Then exponentiation gives $|z^2+z| = e^{c+2t^2} = e^c e^{2t^2}$, from which $z^2 + z + ke^{2t^2} = 0$, where $k = \pm e^c$. We may solve for z explicitly, using the quadratic formula to obtain $z = \dfrac{-1 \pm \sqrt{1-4ke^{2t^2}}}{2}$.

3.55 Solve $\dfrac{1}{x}\,dx - \dfrac{v^3}{v^4+1}\,dv = 0$.

▌ This equation is separable. Integrating term by term, we get $\ln|x| - \frac{1}{4}\ln(v^4+1) = c$, so that $\ln x^4 - \ln(v^4+1) = 4c$ and $\ln\dfrac{x^4}{v^4+1} = 4c$. By taking exponentials we get $\dfrac{x^4}{v^4+1} = e^{4c}$, which gives us $x^4 = k(v^4+1)$, where $k = e^{4c}$.

3.56 Solve $\dfrac{1}{x}\,dx - \dfrac{3v^2}{1-2v^3}\,dv = 0$.

▌ This equation is separable. Integrating term by term, we get $\ln|x| + \frac{1}{2}\ln|1-2v^3| = c$, from which we obtain $\ln x^2 + \ln|1-2v^3| = 2c$. Writing this as $\ln|x^2(1-2v^3)| = 2c$, we then find that $x^2(1-2v^3) = \pm e^{2c} = k$.

3.57 Solve $y\dfrac{du}{dy} = \dfrac{u^2}{1-u}$.

▌ In differential form, this equation is $\dfrac{1-u}{u^2}\,du - \dfrac{1}{y}\,dy = 0$, which is separable. Integrating term by term, and noting that

$$\int \frac{1-u}{u^2}\,du = \int\left(\frac{1}{u^2} - \frac{1}{u}\right)du = -\frac{1}{u} - \ln|u|$$

we have as the solution $-\dfrac{1}{u} - \ln|u| - \ln|y| = c$, which may be simplified to $\dfrac{1}{u} = -c - \ln|uy|$.

3.58 Solve $x(1+\sqrt{v})\,dv + v\sqrt{v}\,dx = 0$.

▌ Separating the variables, we obtain $\dfrac{1+\sqrt{v}}{v\sqrt{v}}\,dv + \dfrac{1}{x}\,dx = 0$. Integrating term by term, after noting that

$$\int \frac{1+\sqrt{v}}{v\sqrt{v}}\,dv = \int\left(v^{-3/2} + \frac{1}{v}\right)dv = -2v^{-1/2} + \ln|v|$$

we get $-\dfrac{2}{\sqrt{v}} + \ln|v| + \ln|x| = c$, so that, in implicit form, $-\dfrac{2}{\sqrt{v}} + \ln|vx| = c$.

3.59 Solve $xv\,dv - (1 + 2v^2 + v^4)\,dx = 0$.

▌ Separating the variables, we obtain $\dfrac{v}{(1+v^2)^2}\,dv - \dfrac{1}{x}\,dx = 0$. Term-by-term integration yields $-\frac{1}{2}(1+v^2)^{-1} - \ln|x| = c$, which we rewrite as $-(1+v^2)^{-1} = \ln x^2 + \ln|k|$, where $k = e^{2c}$. Then $-(1+v^2)^{-1} = \ln|kx^2|$, so that, in implicit form, $(1+v^2)\ln|kx^2| = -1$.

3.60 Solve $dy/dx = y - y^2$.

▌ We rewrite this equation as $\dfrac{1}{y-y^2}\,dy - dx = 0$, which is separable. The solution is $\displaystyle\int \frac{1}{y-y^2}\,dy - \int dx = c$. Now, using partial-fractions techniques, we have for the leftmost term,

$$\int \frac{dy}{y-y^2} = \int \frac{dy}{y(1-y)} = \int\left(\frac{1}{y} + \frac{1}{1-y}\right)dy = \ln|y| - \ln|1-y| = \ln\left|\frac{y}{1-y}\right|$$

The solution thus becomes $\ln\left|\dfrac{y}{1-y}\right| - x = c$ or, after rearrangement and exponentiation, $\left|\dfrac{y}{1-y}\right| = e^{c+x} = e^c e^x$, so that $\dfrac{y}{1-y} = \pm e^c e^x = ke^x$. To solve explicitly for y, we write $y = ke^x(1-y)$, from which $y = \dfrac{ke^x}{1+ke^x}$.

3.61 Solve $4x\,dy - y\,dx = x^2\,dy$.

▌ This equation may be rewritten as $y\,dx + (x^2 - 4x)\,dy = 0$ or $\dfrac{1}{x(x-4)}\,dx + \dfrac{1}{y}\,dy = 0$, which is separable. The solution is $\int \dfrac{1}{x(x-4)}\,dx + \int \dfrac{1}{y}\,dy = c$. For the leftmost term, the method of partial fractions gives

$$\int \frac{1}{x(x-4)}\,dx = \int \frac{1/4}{x-4}\,dx - \int \frac{1/4}{x}\,dx = \frac{1}{4}\ln|x-4| - \frac{1}{4}\ln|x|$$

The solution then becomes $\frac{1}{4}\ln|x-4| - \frac{1}{4}\ln|x| + \ln|y| = c$ or $\ln|x-4| - \ln|x| + \ln y^4 = 4c$, so that $\ln\left|\dfrac{(x-4)y^4}{x}\right| = 4x$. Exponentiation then yields $\dfrac{(x-4)y^4}{x} = \pm e^{4c}$, from which $y = \pm\left(\dfrac{kx}{x-4}\right)^{1/4}$, where $k = \pm e^{4c}$.

3.62 Solve $\dfrac{dy}{dx} = \dfrac{4y}{x(y-3)}$.

▌ We rewrite this equation in the differential form $\dfrac{y-3}{y}\,dy - \dfrac{4}{x}\,dx = 0$, which has as its solution $\int \dfrac{y-3}{y}\,dy - \int \dfrac{4}{x}\,dx = c$. By the method of partial fractions,

$$\int \frac{y-3}{y}\,dy = \int \left(1 - \frac{3}{y}\right) dy = y - 3\ln|y|$$

so the solution becomes $y - 3\ln|y| - 4\ln|x| = c$. Thus $\ln|y^3x^4| = y - c$ and, after exponentiation, $|y^3x^4| = e^{y-c} = e^{-c}e^y$, so that $y^3x^4 = ke^y$, where $k = \pm e^{-c}$.

3.63 Solve $x^2(y+1)\,dx + y^2(x-1)\,dy = 0$.

▌ This equation may be rewritten as $\dfrac{x^2}{x-1}\,dx + \dfrac{y^2}{y+1}\,dy = 0$, with solution $\int \dfrac{x^2}{x-1}\,dx + \int \dfrac{y^2}{y+1}\,dy = c$. Since

$$\int \frac{x^2}{x-1}\,dx = \int \left(x + 1 + \frac{1}{x-1}\right) dx = \frac{1}{2}x^2 + x + \ln|x-1|$$

and

$$\int \frac{y^2}{y+1}\,dy = \int \left(y - 1 + \frac{1}{y+1}\right) dy = \frac{1}{2}y^2 - y + \ln|y+1|$$

the solution becomes

$$\tfrac{1}{2}x^2 + x + \ln|x-1| + \tfrac{1}{2}y^2 - y + \ln|y+1| = c$$

or

$$x^2 + 2x + y^2 - 2y + 2\ln|x-1| + 2\ln|y+1| = 2c$$

or

$$(x^2 + 2x + 1) + (y^2 - 2y + 1) + 2\ln|(x-1)(y+1)| = 2c + 2$$

or

$$(x+1)^2 + (y-1)^2 + 2\ln|(x-1)(y+1)| = k$$

where $k = 2c + 2$.

3.64 Solve $\dfrac{dy}{dt} = y^2 - y^3$.

▌ This equation may be rewritten in the differential form $\frac{1}{y^2 - y^3}\,dy - dt = 0$, with solution $\int \frac{1}{y^2 - y^3}\,dy - \int dt = c$. By the method of partial fractions,

$$\frac{1}{y^2 - y^3} = \frac{1}{y^2(1 - y)} = \frac{1}{y^2} + \frac{1}{y} + \frac{1}{1 - y}$$

so after the indicated integrations the solution becomes $-\frac{1}{y} + \ln|y| - \ln|1 - y| - t = c$ or, rearranged, $\ln\left|\frac{y}{1 - y}\right| - \frac{1}{y} = t + c$.

3.65 Solve $(t + 1)y\dot{y} = 1 - y^2$.

▌ This equation may be rewritten in the differential form $\frac{y}{y^2 - 1}\,dy + \frac{1}{t + 1}\,dt = 0$, with solution $\int \frac{y}{y^2 - 1}\,dy + \int \frac{1}{t + 1}\,dt = c$. Integrating directly, we have $\frac{1}{2}\ln|y^2 - 1| + \ln|t + 1| = c$, so that $\ln|y^2 - 1| + \ln(t + 1)^2 = 2c$ or $\ln|(y^2 - 1)(t + 1)^2| = 2c$. Exponentiation then yields $|(y^2 - 1)(t + 1)^2| = e^{2c}$ or $(y^2 - 1)(t + 1)^2 = \pm e^{2c}$, from which $y^2 = 1 + k/(t + 1)^2$, where $k = \pm e^{2c}$.

3.66 Solve $y\frac{du}{dy} = -\frac{u + u^5}{2 + u^4}$.

▌ This equation may be rewritten in the differential form $\frac{1}{y}\,dy + \frac{2 + u^4}{u + u^5}\,du = 0$, which has as its solution $\int \frac{1}{y}\,dy + \int \frac{2 + u^4}{u + u^5}\,du = c$. By the method of partial fractions, $\frac{2 + u^4}{u + u^5} = \frac{2 + u^4}{u(1 + u^4)} = \frac{2}{u} - \frac{u^3}{1 + u^4}$. Thus, after the indicated integrations, the solution becomes $\ln|y| + 2\ln|u| - \frac{1}{4}\ln(1 + u^4) = c$. Multiplication by 4 gives $4\ln|y| + 8\ln|u| - \ln(1 + u^4) = 4c$, which may be written as $\ln\frac{y^4u^8}{1 + u^4} = 4c$. Exponentiation yields $\frac{y^4u^8}{1 + u^4} = e^{4c}$, which we write as $y^4u^8 = k(1 + u^4)$, where $k = e^{4c}$.

3.67 Solve $v + x\frac{dv}{dx} = \frac{2x^2v}{x^2 - x^2v^2}$.

▌ This equation may be rewritten as $x\frac{dv}{dx} = -\frac{v(v^2 + 1)}{v^2 - 1}$ or $\frac{1}{x}\,dx + \frac{v^2 - 1}{v(v^2 + 1)}\,dv = 0$. Using the method of partial fractions, we can expand this to

$$\frac{1}{x}\,dx + \left(-\frac{1}{v} + \frac{2v}{v^2 + 1}\right)dv = 0$$

The solution to this separable equation is $\ln|x| - \ln|v| + \ln(v^2 + 1) = c$, so that $\ln\left|\frac{x(v^2 + 1)}{v}\right| = c$ and $\frac{x(v^2 + 1)}{v} = \pm e^c$. Then $x(v^2 + 1) = kv$, where $k = \pm e^c$.

3.68 Solve $y\frac{du}{dy} = -\frac{u^2 + u}{u + 2}$ for $u(y)$.

▌ In differential form, this equation is $\frac{u + 2}{u^2 + u}\,du + \frac{1}{y}\,dy = 0$, which is separable. Using partial-fraction techniques, find that

$$\int \frac{u + 2}{u^2 + u}\,du = \int \frac{u + 2}{u(u + 1)}\,du = \int\left(\frac{2}{u} - \frac{1}{u + 1}\right)du = 2\ln|u| - \ln|u + 1| = \ln\left|\frac{u^2}{u + 1}\right|$$

so the solution to the differential equation becomes $\ln\left|\dfrac{u^2}{u+1}\right| + \ln|y| = c$, from which we obtain $\dfrac{yu^2}{u+1} = k$, where $k = \pm e^c$, or $yu^2 = k(u+1)$. To solve for u explicitly, we rewrite this last equation as $u^2 - (k/y)u - (k/y) = 0$ and then use the quadratic formula to obtain

$$u = \frac{k/y \pm \sqrt{(k/y)^2 + 4(k/y)}}{2} = \frac{A^2 \pm A\sqrt{1+4y}}{2y} \qquad \text{where} \qquad A = \pm\sqrt{k}$$

3.69 Solve $y(u^2+2)\,du - (u^3-u)\,dy = 0$.

Separating the variables gives us $\dfrac{u^2+2}{u^3-u}\,du - \dfrac{1}{y}\,dy = 0$. By the method of partial fractions, we have, for the leftmost term,

$$\int \frac{u^2+2}{u^3-u}\,du = \int \frac{u^2+2}{u(u-1)(u+1)}\,du = \int\left(-\frac{2}{u} + \frac{3/2}{u-1} + \frac{3/2}{u+1}\right)du = -2\ln|u| + \frac{3}{2}\ln|u-1| + \frac{3}{2}\ln|u+1|$$

Then the solution to the differential equation is

$$-2\ln|u| + \frac{3}{2}\ln|u-1| + \frac{3}{2}\ln|u+1| - \ln|y| = c$$

or

$$-4\ln|u| + 3\ln|u-1| + 3\ln|u+1| - 2\ln|y| = 2c$$

so that

$$\ln\left|\frac{(u-1)^3(u+1)^3}{u^4y^2}\right| = 2c \qquad \text{and} \qquad \frac{(u-1)^3(u+1)^3}{u^4y^2} = \pm e^{2c}$$

This yields $(u^2-1)^3 = ku^4y^2$, where $k = \pm e^{2c}$.

3.70 Solve $y(u^2+2)\,du + (u^3+u)\,dy = 0$.

Separating the variables gives us $\dfrac{u^2+2}{u^3+u}\,du + \dfrac{1}{y}\,dy = 0$. By the method of partial fractions, we have, for the leftmost term,

$$\int \frac{u^2+2}{u^3+u}\,du = \int \frac{u^2+2}{u(u^2+1)}\,du = \int\left(\frac{-u}{u^2+1} + \frac{2}{u}\right)du = -\frac{1}{2}\ln(u^2+1) + 2\ln|u|$$

Then the solution to the differential equation is $-\frac{1}{2}\ln(u^2+1) + 2\ln|u| + \ln|y| = c$, which may be simplified to $\ln\dfrac{u^4y^2}{u^2+1} = 2c$. Exponentiation gives $\dfrac{u^4y^2}{u^2+1} = e^{2c}$, so that $u^4y^2 = k(u^2+1)$, where $k = e^{2c}$.

3.71 Solve $y(u^2+1)\,du + (u^3-3u)\,dy = 0$.

Separating the variables gives $\dfrac{u^2+1}{u^3-3u}\,du + \dfrac{1}{y}\,dy = 0$. Using the method of partial fractions, we find that

$$\begin{aligned}\int \frac{u^2+1}{u^3-3u}\,du &= \int \frac{u^2+1}{u(u-\sqrt{3})(u+\sqrt{3})}\,du \\ &= \int\left(\frac{-1/3}{u} + \frac{2/3}{u-\sqrt{3}} + \frac{2/3}{u+\sqrt{3}}\right)du \\ &= -\frac{1}{3}\ln|u| + \frac{2}{3}\ln|u-\sqrt{3}| + \frac{2}{3}\ln|u+\sqrt{3}|\end{aligned}$$

Then the solution to the differential equation is $-\frac{1}{3}\ln|u| + \frac{2}{3}\ln|u-\sqrt{3}| + \frac{2}{3}\ln|u+\sqrt{3}| + \ln|y| = c$. After multiplication by 3 and rearrangement, this solution becomes $\ln\left|\dfrac{(u-\sqrt{3})^2(u+\sqrt{3})^2y^3}{u}\right| = 3c$, and exponentiation gives $\dfrac{(u-\sqrt{3})^2(u+\sqrt{3})^2y^3}{u} = \pm e^{3c}$, which we may write as $(u^2-3)^2y^3 = ku$, where $k = \pm e^{3c}$.

3.72 Solve $x(v^2-1)\,dv + (v^3-4v)\,dx = 0$.

▌ Separating the variables gives us $\frac{v^2-1}{v^3-4v}\,dv + \frac{1}{x}\,dx = 0$. Using the method of partial fractions, we find that

$$\int \frac{v^2-1}{v^3-4v}\,dv = \int \frac{v^2-1}{v(v-2)(v+2)}\,dv = \int \left(\frac{1/4}{v} + \frac{3/8}{v-2} + \frac{3/8}{v+2}\right) dv = \frac{1}{4}\ln|v| + \frac{3}{8}\ln|v-2| + \frac{3}{8}\ln|v+2|$$

Then the solution to the differential equation is $\frac{1}{4}\ln|v| + \frac{3}{8}\ln|v-2| + \frac{3}{8}\ln|v+2| + \ln|x| = c$. Multiplication by 8 and rearrangement yield $\ln|v^2(v-2)^3(v+2)^3x^8| = 8c$, and exponentiation gives $v^2(v-2)^3(v+2)^3x^8 = \pm e^{8c}$, which may be written as $v^2(v^2-4)^3x^8 = k$, where $k = \pm e^{8c}$.

3.73 Solve $x(v^2+1)\,dv + (v^3-2v)\,dx = 0$.

▌ Separating the variables gives us $\frac{v^2+1}{v^3-2v}\,dv + \frac{1}{x}\,dx = 0$. Using the method of partial fractions, we find that

$$\int \frac{v^2+1}{v^3-2v}\,dv = \int \frac{v^2+1}{v(v-\sqrt{2})(v+\sqrt{2})}\,dv = \int \left(-\frac{1/2}{v} + \frac{3/4}{v-\sqrt{2}} + \frac{3/4}{v+\sqrt{2}}\right) dv$$

$$= -\frac{1}{2}\ln|v| + \frac{3}{4}\ln|v-\sqrt{2}| + \frac{3}{4}\ln|v+\sqrt{2}|$$

The solution to the differential equation is, therefore, $-\frac{1}{2}\ln|v| + \frac{3}{4}\ln|v-\sqrt{2}| + \frac{3}{4}\ln|v+\sqrt{2}| + \ln|x| = c$. Multiplication by 4 and rearrangement yield $\ln\left|\frac{(v-\sqrt{2})^3(v+\sqrt{2})^3x^4}{v^2}\right| = 4c$, and exponentiation gives $(v-\sqrt{2})^3(v+\sqrt{2})^3x^4 = kv^2$, where $k = \pm e^{4c}$, which we may write as $(v^2-2)^3x^4 = kv^2$.

3.74 Solve $x(v^2-1)\,dv + (v^3+2v)\,dx = 0$.

▌ Separating the variables gives us $\frac{v^2-1}{v^3+2v}\,dv + \frac{1}{x}\,dx = 0$. Then, by the method of partial fractions,

$$\int \frac{v^2-1}{v^3+2v}\,dv = \int \frac{v^2-1}{v(v^2+2)}\,dv = \int \left(\frac{-1/2}{v} + \frac{(3/2)v}{v^2+2}\right) dv = -\frac{1}{2}\ln|v| + \frac{3}{4}\ln(v^2+2)$$

Thus, the solution to the differential equation is $-\frac{1}{2}\ln|v| + \frac{3}{4}\ln(v^2+2) + \ln|x| = c$. After multiplication by 4 and rearrangement, this becomes $\ln\frac{(v^2+2)^3x^4}{v^2} = 4c$, so that exponentiation gives $(v^2+2)^3x^4 = kv^2$, where $k = e^{4c}$.

SOLUTIONS WITH TRANSCENDENTAL FUNCTIONS

3.75 Solve $dy/dx = y^2 + 1$.

▌ By separating the variables we obtain $\frac{1}{1+y^2}\,dy - dx = 0$, which has the solution $\int \frac{1}{1+y^2}\,dy - \int dx = c$. The integrations yield $\arctan y - x = c$, from which $y = \tan(x+c)$.

3.76 Solve $dy/dx = 2y^2 + 3$.

▌ By separating the variables we obtain $\frac{1}{2y^2+3}\,dy - dx = 0$, which has the solution $\int \frac{1}{2y^2+3}\,dy - \int dx = c$. The integrations yield $\frac{1}{\sqrt{6}}\arctan\sqrt{\tfrac{2}{3}}\,y - x = c$, or $\arctan\sqrt{\tfrac{2}{3}}\,y = \sqrt{6}(x+c)$. Then $\sqrt{\tfrac{2}{3}}\,y = \tan\sqrt{6}(x+c)$, so that $y = \sqrt{\tfrac{3}{2}}\tan\sqrt{6}(x+c)$.

3.77 Solve $dy/dx = y^2 + 2y + 2$.

▌ We separate the variables to obtain $\frac{1}{y^2+2y+2}\,dy - dx = 0$. Now, since

$$\int \frac{dy}{y^2+2y+2} = \int \frac{dy}{1+(y+1)^2} = \arctan(y+1)$$

the solution to the differential equation is $\arctan(y+1) - x = c$; hence, $y = -1 + \tan(x+c)$.

3.78 Solve $dy/dx - y^2 + 6y = 13$.

▌ We first write the equation as $dy = (y^2 - 6y + 13)\,dx$ and then as $\dfrac{1}{y^2 - 6y + 13}\,dy - dx = 0$. Since

$$\int \frac{1}{y^2 - 6y + 13}\,dy = \int \frac{1}{(y-3)^2 + 4}\,dy = \frac{1}{4}\int \frac{1}{1 + \frac{1}{4}(y-3)^2}\,dy = \frac{1}{2}\arctan\frac{y-3}{2}$$

the solution to the differential equation is $\dfrac{1}{2}\arctan\dfrac{y-3}{2} - x = c$ or $\arctan\dfrac{y-3}{2} = 2(x+c)$. Thus, $\frac{1}{2}(y-3) = \tan(2x+k)$, where $k = 2c$, or $y = 3 + 2\tan(2x+k)$.

3.79 Solve $\dfrac{dy}{dx} = \dfrac{y^3 + 4y}{x(y^2 + 2y + 4)}$.

▌ Separating the variables gives us $\dfrac{y^2 + 2y + 4}{y^3 + 4y}\,dy - \dfrac{1}{x}\,dx = 0$. Then, by the method of partial fractions, we obtain

$$\int \frac{y^2 + 2y + 4}{y^3 + 4y}\,dy = \int \frac{y^2 + 2y + 4}{y(y^2 + 4)}\,dy = \int\left(\frac{1}{y} + \frac{2}{y^2 + 4}\right)dy = \ln|y| + \arctan\frac{y}{2}$$

Then the solution to the differential equation is $\ln|y| + \arctan(y/2) - \ln|x| = c$ or, rearranged, $\arctan(y/2) = \ln|kx/y|$, where $k = \pm e^c$.

3.80 Solve $\dfrac{dy}{dx} = \dfrac{y^3 + 9y}{x^2(2y^2 + 10y + 9)}$.

▌ Separating the variables gives us $\dfrac{2y^2 + 10y + 9}{y^3 + 9y}\,dy - \dfrac{1}{x^2}\,dx = 0$. And, by the method of partial fractions,

$$\int \frac{2y^2 + 10y + 9}{y^3 + 9y}\,dy = \int \frac{2y^2 + 10y + 9}{y(y^2 + 9)}\,dy = \int\left(\frac{1}{y} + \frac{y + 10}{y^2 + 9}\right)dy$$

$$= \int\left(\frac{1}{y} + \frac{y}{y^2 + 9} + \frac{10}{y^2 + 9}\right)dy = \ln|y| + \frac{1}{2}\ln(y^2 + 9) + \frac{10}{3}\arctan\frac{y}{3}$$

The solution to the differential equation is, then, $\ln|y| + \dfrac{1}{2}\ln(y^2 + 9) + \dfrac{10}{3}\arctan\dfrac{y}{3} - \dfrac{1}{x} = c$, which we write first as

$$2\ln|y| + \ln(y^2 + 9) + \ln|k| + \frac{20}{3}\arctan\frac{y}{3} = \frac{2}{x} \qquad \text{where} \qquad k = \pm e^{-2c}$$

and then as $\ln|ky^2(y^2 + 9)| + \frac{20}{3}\arctan(y/3) = 2/x$.

3.81 Solve $(v^2 + 2v + 2)\,dx + x(v - 1)\,dv = 0$.

▌ Separating the variables gives us $\dfrac{v - 1}{v^2 + 2v + 2}\,dv + \dfrac{1}{x}\,dx = 0$. Then, by the method of partial fractions, we have

$$\int \frac{v - 1}{v^2 + 2v + 2}\,dv = \int \frac{(v+1) - 2}{(v+1)^2 + 1}\,dv = \int \frac{v + 1}{(v+1)^2 + 1}\,dv - 2\int \frac{1}{1 + (v+1)^2}\,dv$$

$$= \frac{1}{2}\ln[(v+1)^2 + 1] - 2\arctan(v+1)$$

The solution to the differential equation is thus $\frac{1}{2}\ln(v^2 + 2v + 2) - 2\arctan(v+1) + \ln|x| = c$, which we multiply by 2 and rearrange to obtain $\ln[x^2(v^2 + 2v + 2)] - 4\arctan(v+1) = k$, where $k = 2c$.

3.82 Solve $(u + 2e^u)\,dy + y(1 + 2e^u)\,du = 0$.

▌ Separating the variables gives us $\dfrac{1}{y}\,dy + \dfrac{1 + 2e^u}{u + 2e^u}\,du = 0$. Integrating term by term then yields $\ln|y| + \ln|u + 2e^u| = c$, which we rearrange to $\ln|y(u + 2e^u)| = c$. Exponentiation then gives the solution as $y(u + 2e^u) = k$, where $k = \pm e^c$.

3.83 Solve $dy/dx = \sec y \tan x$.

▌ Separating the variables gives us $dy/\sec y = \tan x\, dx$, which we rewrite as $\cos y\, dy = (\sin x/\cos x)\, dx$. Then integration yields

$$\int \cos y\, dy = -\int \frac{d(\cos x)}{\cos x} \quad \text{or} \quad \sin y = -\ln|\cos x| + c.$$

3.84 Solve $dy/dx = \tan y$.

▌ Since $\tan y = \sin y/\cos y$, the differential equation may be rewritten as $\dfrac{\cos y}{\sin y}\, dy - dx = 0$. Integrating term by term, we obtain the solution $\ln|\sin y| - x = c$, which we rewrite as $\ln|\sin y| = c + x$. Then $\sin y = ke^x$, where $k = \pm e^c$, so that $y = \arcsin ke^x$.

3.85 Solve $\dot{y} = e^y$.

▌ Separating the variables gives us $e^{-y}\, dy - dt = 0$, and integrating term by term yields $-e^{-y} - t = c$. Hence $e^{-y} = k - t$, where $k = -c$, and $y = -\ln(k - t)$.

3.86 Solve $\dot{y} = te^y$.

▌ Separating the variables gives us $e^{-y}\, dy - t\, dt = 0$, and integrating term by term yields $-e^{-y} - \frac{1}{2}t^2 = c$. Hence $y = -\ln(k - \frac{1}{2}t^2)$, where $k = -c$.

3.87 Solve $\dot{y} = ye^t$.

▌ Separating the variables gives us $\dfrac{1}{y}\, dy - e^t\, dt = 0$, and integrating term by term yields $\ln|y| - e^t = c$. Then $\ln|y| = c + e^t$, so that exponentiation yields $|y| = e^{c+e^t} = e^c e^{e^t}$. Thus, $y = ke^{e^t}$, where $k = \pm e^c$.

3.88 Solve $\dot{y} = y^5 \sin t$.

▌ Separating variables yields $\dfrac{dy}{y^5} = \sin t\, dt$ and integration gives us $\displaystyle\int \frac{dy}{y^5} = \int \sin t\, dt + c$. Therefore, $-\frac{1}{4}y^{-4} = -\cos t + c$, or $\dfrac{1}{y^4} = 4\cos t + k$, where $k = -4c$.

3.89 Solve $x\, dv - \sqrt{1 - v^2}\, dx = 0$.

▌ Separating the variables gives us $\dfrac{1}{\sqrt{1 - v^2}}\, dv - \dfrac{1}{x}\, dx = 0$, and then integrating term by term yields $\arcsin v - \ln|x| = c$ as the solution in implicit form.

HOMOGENEOUS EQUATIONS

3.90 Define *homogeneous* with regard to first-order differential equations.

▌ A first-order differential equation in standard form $\dfrac{dy}{dx} = f(x, y)$ is *homogeneous* if $f(tx, ty) = f(x, y)$ for every real number t in some nonempty interval.

Note: The word *homogeneous* has an entirely different meaning in the general context of linear differential equations. (See Chapter 8.)

3.91 Determine whether the equation $y' = (y + x)/x$ is homogeneous.

▌ The equation is homogeneous because

$$f(tx, ty) = \frac{ty + tx}{tx} = \frac{t(y + x)}{tx} = \frac{y + x}{x} = f(x, y)$$

3.92 Determine whether the equation $y' = y^2/x$ is homogeneous.

▌ The equation is not homogeneous because

$$f(tx, ty) = \frac{(ty)^2}{tx} = \frac{t^2y^2}{tx} = t\frac{y^2}{x} \neq f(x, y)$$

3.93 Determine whether the equation $y' = \dfrac{2xye^{x/y}}{x^2 + y^2 \sin(x/y)}$ is homogeneous.

▌ The equation is homogeneous because

$$f(tx, ty) = \frac{2(tx)(ty)e^{tx/ty}}{(tx)^2 + (ty)^2 \sin(tx/ty)} = \frac{t^2 2xy\, e^{x/y}}{t^2x^2 + t^2y^2 \sin(x/y)} = \frac{2xy\, e^{x/y}}{x^2 + y^2 \sin(x/y)} = f(x, y)$$

3.94 Determine whether $y' = (x^2 + y)/x^3$ is homogeneous.

▌ The equation is not homogeneous because

$$f(tx, ty) = \frac{(tx)^2 + ty}{(tx)^3} = \frac{t^2x^2 + ty}{t^3x^3} = \frac{tx^2 + y}{t^2x^3} \neq f(x, y)$$

3.95 Determine whether $y' = \dfrac{2y^4 + x^4}{xy^3}$ is homogeneous.

▌ The equation is homogeneous because

$$f(tx, ty) = \frac{2(ty)^4 + (tx)^4}{(tx)(ty)^3} = \frac{2t^4y^4 + t^4x^4}{t^4xy^3} = \frac{2y^4 + x^4}{xy^3} = f(x, y)$$

3.96 Determine whether $y' = \dfrac{2xy}{x^2 - y^2}$ is homogeneous.

▌ The equation is homogeneous because

$$f(tx, ty) = \frac{2(tx)(ty)}{(tx)^2 - (ty)^2} = \frac{2t^2xy}{t^2(x^2 - y^2)} = \frac{2xy}{x^2 - y^2} = f(x, y)$$

3.97 Determine whether $y' = \dfrac{x^2 + y^2}{xy}$ is homogeneous.

▌ The equation is homogeneous because

$$f(tx, ty) = \frac{(tx)^2 + (ty)^2}{(tx)(ty)} = \frac{t^2(x^2 + y^2)}{t^2xy} = \frac{x^2 + y^2}{xy} = f(x, y)$$

3.98 Determine whether $y' = (y - x)/x$ is homogeneous.

▌ The equation is homogeneous because

$$f(tx, ty) = \frac{ty - tx}{tx} = \frac{t(y - x)}{tx} = \frac{y - x}{x} = f(x, y)$$

3.99 Determine whether $y' = (2y + x)/x$ is homogeneous.

▌ The equation is homogeneous because

$$f(tx, ty) = \frac{2(ty) + tx}{tx} = \frac{t(2y + x)}{tx} = \frac{2y + x}{x} = f(x, y)$$

3.100 Determine whether $y' = \dfrac{x^2 + 2y^2}{xy}$ is homogeneous.

▌ The equation is homogeneous because

$$f(tx, ty) = \frac{(tx)^2 + 2(ty)^2}{(tx)(ty)} = \frac{t^2(x^2 + 2y^2)}{t^2xy} = \frac{x^2 + 2y^2}{xy} = f(x, y)$$

3.101 Determine whether $y' = \dfrac{2x + y^2}{xy}$ is homogeneous.

▌ The equation is not homogeneous because

$$f(tx, ty) = \frac{2(tx) + (ty)^2}{(tx)(ty)} = \frac{2tx + t^2y^2}{t^2xy} = \frac{2x + ty^2}{txy} \neq f(x, y)$$

3.102 Determine whether $y' = \dfrac{2xy}{y^2 - x^2}$ is homogeneous.

▌ The equation is homogeneous because

$$f(tx, ty) = \frac{2(tx)(ty)}{(ty)^2 - (tx)^2} = \frac{2t^2xy}{t^2(y^2 - x^2)} = \frac{2xy}{y^2 - x^2} = f(x, y)$$

3.103 Determine whether $y' = \dfrac{x^2 + y^2}{2xy}$ is homogeneous.

▌ The equation is homogeneous because

$$f(tx, ty) = \frac{(tx)^2 + (ty)^2}{2(tx)(ty)} = \frac{t^2(x^2 + y^2)}{2t^2xy} = \frac{x^2 + y^2}{2xy} = f(x, y)$$

3.104 Determine whether $y' = \dfrac{y}{x + \sqrt{xy}}$ is homogeneous.

▌ The equation is homogeneous for $t > 0$, because then

$$f(tx, ty) = \frac{ty}{tx + \sqrt{(tx)(ty)}} = \frac{ty}{tx + |t|\sqrt{xy}} = \frac{y}{x + \sqrt{xy}} = f(x, y)$$

3.105 Determine whether $y' = \dfrac{y^2}{xy + (xy^2)^{1/3}}$ is homogeneous.

▌ The equation is not homogeneous because

$$f(tx, ty) = \frac{(ty)^2}{(tx)(ty) + [(tx)(ty)^2]^{1/3}} = \frac{t^2y^2}{t^2xy + (t^3xy^2)^{1/3}} = \frac{t^2y^2}{t^2xy + t(xy^2)^{1/3}} = \frac{ty^2}{txy + (xy^2)^{1/3}} \neq f(x, y)$$

3.106 Determine whether $y' = \dfrac{x^4 + 3x^2y^2 + y^4}{x^3y}$ is homogeneous.

▌ The equation is homogeneous because

$$f(tx, ty) = \frac{(tx)^4 + 3(tx)^2(ty)^2 + (ty)^4}{(tx)^3(ty)} = \frac{t^4x^4 + 3t^4x^2y^2 + t^4y^4}{t^4x^3y} = \frac{x^4 + 3x^2y^2 + y^4}{x^3y} = f(x, y)$$

3.107 What is a homogeneous function of degree n?

▌ A function $g(x, y)$ of two variables is a homogeneous function of degree n if $g(tx, ty) = t^n g(x, y)$ for all real numbers t in some nonempty interval.

3.108 Determine whether $g(x, y) = xy + y^2$ is homogeneous and, if so, find its degree.

▌ The function is homogeneous of degree 2 because

$$g(tx, ty) = (tx)(ty) + (ty)^2 = t^2(xy + y^2)$$

3.109 Determine whether $g(x, y) = x + y\sin(y/x)^2$ is homogeneous and, if so, find its degree.

▌ The function is homogeneous of degree 1 because

$$g(tx, ty) = tx + ty\sin\left(\frac{ty}{tx}\right)^2 = t\left[x + y\sin\left(\frac{y}{x}\right)^2\right]$$

.110 Determine whether $g(x, y) = x^3 + xy^2e^{x/y}$ is homogeneous and, if so, find its degree.

▌ The function is homogeneous of degree 3 because

$$g(tx, ty) = (tx)^3 + (tx)(ty)^2e^{tx/ty} = t^3(x^3 + xy^2e^{x/y})$$

.111 Determine whether $g(x, y) = x + xy$ is homogeneous and, if so, find its degree.

▌ The function is not homogeneous because

$$g(tx, ty) = tx + (tx)(ty) = tx + t^2xy$$

.112 Determine whether $g(x, y) = \sqrt{x^2 - y^2}$ is homogeneous and, if so, find its degree.

▌ The function is homogeneous of degree 1 because, for $t > 0$,

$$g(tx, ty) = \sqrt{(tx)^2 - (ty)^2} = \sqrt{t^2(x^2 - y^2)} = |t|\sqrt{x^2 - y^2} = |t|g(x, y) = tg(x, y)$$

.113 Determine whether $g(x, y) = 2x \sinh(y/x) + 3y \cosh(y/x)$ is homogeneous and, if so, find its degree.

▌ The function is homogeneous of degree 1 because

$$g(tx, ty) = 2(tx) \sinh \frac{ty}{tx} + 3(ty) \cosh \frac{ty}{tx} = 2tx \sinh \frac{y}{x} + 3ty \cosh \frac{y}{x} = tg(x, y)$$

.114 Determine whether $g(x, y) = \sqrt{x + y}$ is homogeneous and, if so, find its degree.

▌ The function is homogeneous of degree 1/2 because

$$g(tx, ty) = \sqrt{tx + ty} = \sqrt{t(x + y)} = \sqrt{t}\sqrt{x + y} = t^{1/2}g(x, y)$$

.115 Determine whether $g(x, y) = x\sqrt{x + y}$ is homogeneous and, if so, find its degree.

▌ The function is homogeneous of degree 3/2 because

$$g(tx, ty) = tx\sqrt{tx + ty} = tx\sqrt{t(x + y)} = t\sqrt{t}x\sqrt{x + y} = t^{3/2}g(x, y)$$

.116 Determine whether $g(x, y) = x \sin(y/x^2)$ is homogeneous and, if so, find its degree.

▌ The function is not homogeneous because

$$g(tx, ty) = tx \sin \frac{ty}{(tx)^2} = tx \sin \frac{y}{tx^2} \neq t^n g(x, y)$$

for any real value of n.

.117 Determine whether $g(x, y) = x^3 \sin(x^2/y^2)$ is homogeneous and, if so, find its degree.

▌ The function is homogeneous of degree 3 because

$$g(tx, ty) = (tx)^3 \sin \frac{(tx)^2}{(ty)^2} = t^3x^3 \sin \frac{t^2x^2}{t^2y^2} = t^3x^3 \sin \frac{x^2}{y^2} = t^3g(x, y)$$

OLUTIONS OF HOMOGENEOUS EQUATIONS

.118 Show that the differential equation $M(x, y)\,dx + N(x, y)\,dy = 0$ is homogeneous if $M(x, y)$ and $N(x, y)$ are homogeneous functions of the same degree.

▌ The differential equation may be rewritten as $\dfrac{dy}{dx} = -\dfrac{M(x, y)}{N(x, y)}$. If $M(x, y)$ and $N(x, y)$ are homogeneous of degree n, then

$$f(tx, ty) = -\frac{M(tx, ty)}{N(tx, ty)} = -\frac{t^nM(x, y)}{t^nN(x, y)} = -\frac{M(x, y)}{N(x, y)} = f(x, y)$$

.119 Prove that if $y' = f(x, y)$ is homogeneous, then the differential equation can be rewritten as $y' = g(y/x)$, where $g(y/x)$ depends only on the quotient y/x.

▌ We know that $f(x, y) = f(tx, ty)$. Since this equation is valid for all t in some interval, it must be true, in particular, for $t = 1/x$. Thus, $f(x, y) = f(1, y/x)$. If we now define $g(y/x) = f(1, y/x)$, we then have $y' = f(x, y) = f(1, y/x) = g(y/x)$ as required.

3.120 Show that the transformation $y = vx$; $dy/dx = v + x\,dv/dx$ converts a homogeneous differential equation into a separable one.

▌ From the previous problem, we know that the homogeneous differential equation $y' = f(x, y)$ can be written as $y' = g(y/x)$. Substituting for y' and y/x in this equation, we get $v + x\dfrac{dv}{dx} = g(v)$, which may be rewritten as

$$[v - g(v)]\,dx + x\,dv = 0 \qquad \text{or} \qquad \frac{1}{x}\,dx + \frac{1}{v - g(v)}\,dv = 0$$

This last equation is separable.

3.121 Prove that if $y' = f(x, y)$ is homogeneous, then the differential equation can be rewritten as $y' = h(x/y)$, where $h(x/y)$ depends only on the quotient x/y.

▌ We have $f(x, y) = f(tx, ty)$. Since this equation is valid for all t in some interval, it must be true in particular for $t = 1/y$. Thus, $f(x, y) = f(x/y, 1)$. If we now define $h(x/y) = f(x/y, 1)$, we have $y' = f(x, y) = f(x/y, 1) = h(x/y)$ as required.

3.122 Show that the transformation $x = yu$; $\dfrac{dx}{dy} = u + y\dfrac{du}{dy}$ converts a homogeneous differential equation into a separable one.

▌ From the previous problem, we know that the homogeneous equation $y' = f(x, y)$ can be written as $y' = h(x/y)$, which is equivalent to the differential equation $\dfrac{dx}{dy} = \dfrac{1}{h(x/y)}$. Substituting for dx/dy and x/y in this last equation, we get $u + y\dfrac{du}{dy} = \dfrac{1}{h(u)}$, which may be rewritten as

$$\left[u - \frac{1}{h(u)}\right]dy + y\,du = 0 \qquad \text{or} \qquad \frac{1}{u - 1/h(u)}\,du + \frac{1}{y}\,dy = 0$$

The last equation is separable.

3.123 Solve $y' = (y + x)/x$.

▌ This differential equation is homogeneous (see Problem 3.91). Using the substitution $y = vx$; $\dfrac{dy}{dx} = v + x\dfrac{dv}{dx}$, we obtain $v + x\dfrac{dv}{dx} = \dfrac{xv + x}{x}$, which can be simplified to $x\dfrac{dv}{dx} = 1$ or $\dfrac{1}{x}\,dx - dv = 0$. This last equation is separable; its solution is $v = \ln|x| - c$, or $v = \ln|kx|$, where $k = \pm e^{-c}$. Finally, substituting $v = y/x$, we obtain the solution to the given differential equation as $y = x\ln|kx|$.

3.124 Solve $y' = \dfrac{2y^4 + x^4}{xy^3}$.

▌ This differential equation is homogeneous (see Problem 3.95). Using the substitution $y = vx$; $\dfrac{dy}{dx} = v + x\dfrac{dv}{dx}$, we obtain $v + x\dfrac{dv}{dx} = \dfrac{2(xv)^4 + x^4}{x(xv)^3}$, which can be simplified to

$$x\frac{dv}{dx} = \frac{v^4 + 1}{v^3} \qquad \text{or} \qquad \frac{1}{x}\,dx - \frac{v^3}{v^4 + 1}\,dv = 0$$

This last equation is separable; its solution was found in Problem 3.55 to be $x^4 = k(v^4 + 1)$. Since $v = y/x$, the solution becomes $x^4 = k[(y/x)^4 + 1]$, or $x^8 = k(y^4 + x^4)$ in implicit form.

3.125 Rework Problem 3.124 using the transformation suggested in Problem 3.122.

▮ We first rewrite the differential equation as $\frac{dx}{dy} = \frac{xy^3}{2y^4 + x^4}$. Then, using the substitution $x = yu$; $\frac{dx}{dy} = u + y\frac{du}{dy}$, we have $u + y\frac{du}{dy} = \frac{(yu)y^3}{2y^4 + (yu)^4}$, which can be simplified to

$$y\frac{du}{dy} = -\frac{u + u^5}{2 + u^4} \qquad \text{or} \qquad \frac{1}{y}dy + \frac{2 + u^4}{u + u^5}du = 0$$

This last equation is separable; its solution was found in Problem 3.66 to be $y^4u^8 = k(1 + u^4)$. Since $u = x/y$, the solution becomes $y^4(x/y)^8 = k[1 + (x/y)^4]$ or, on simplification, $x^8 = k(y^4 + x^4)$ as before.

3.126 Solve $y' = (x^2 + y^2)/xy$.

▮ This differential equation is homogeneous (see Problem 3.97). Using the substitution $y = vx$; $\frac{dy}{dx} = v + x\frac{dv}{dx}$, we obtain $v + x\frac{dv}{dx} = \frac{x^2 + (xv)^2}{x(xv)}$, which can be simplified to

$$x\frac{dv}{dx} = \frac{1}{v} \qquad \text{or} \qquad \frac{1}{x}dx - v\,dv = 0$$

The solution to this separable equation is $\ln|x| - v^2/2 = c$ or, equivalently, $v^2 = \ln x^2 + k$, where $k = -2c$. Substituting $v = y/x$, we find that the solution to the given differential equation is $y^2 = x^2 \ln x^2 + kx^2$.

3.127 Solve $y' = 2xy/(x^2 - y^2)$.

▮ This differential equation is homogeneous (see Problem 3.96). Using the substitution $y = vx$; $\frac{dy}{dx} = v + x\frac{dv}{dx}$, we obtain $v + x\frac{dv}{dx} = \frac{2x(xv)}{x^2 - (xv)^2}$. This last equation is separable and has as its solution $x(v^2 + 1) = kv$ (see Problem 3.67). Since $v = y/x$, this solution may be rewritten as $x[(y/x)^2 + 1] = k(y/x)$ or, after simplification, $y^2 + x^2 = ky$.

3.128 Solve $y' = (y - x)/x$.

▮ This differential equation is homogeneous (see Problem 3.98). Using the substitution $y = vx$; $\frac{dy}{dx} = v + x\frac{dv}{dx}$, we obtain $v + x\frac{dv}{dx} = \frac{vx - x}{x}$, which may be simplified to

$$x\frac{dv}{dx} = -1 \qquad \text{or} \qquad dv + \frac{1}{x}dx = 0$$

This last equation is separable and has as its solution $v + \ln|x| = c$. If we set $c = \ln|k|$ (that is, $k = \pm e^c$) and substitute $v = y/x$, the solution becomes $y/x = \ln|k| - \ln|x| = \ln|k/x|$, or $y = x\ln|k/x|$.

3.129 Rework Problem 3.128 using the substitution suggested in Problem 3.122.

▮ We first write the differential equation as $\frac{dx}{dy} = \frac{x}{y - x}$. Then, using the substitution $x = yu$; $\frac{dx}{dy} = u + y\frac{du}{dy}$, we obtain $u + y\frac{du}{dy} = \frac{yu}{y - yu}$. This equation may be simplified to $y\frac{du}{dy} = \frac{u^2}{1 - u}$, which is separable and has as its solution $\frac{1}{u} = -c - \ln|uy|$ (see Problem 3.57). If we set $c = -\ln|k|$ and substitute $u = x/y$, the solution becomes $\frac{1}{x/y} = \ln|k| - \ln|(x/y)y| = \ln|k/x|$, or $y = x\ln|k/x|$ as before.

3.130 Solve $y' = \frac{2xye^{(x/y)^2}}{y^2 + y^2e^{(x/y)^2} + 2x^2e^{(x/y)^2}}$.

▮ Noting the (x/y) term in the exponential, we shall try the substitution $u = x/y$, which is equivalent to the substitution $x = uy$. To do so, we rewrite the differential equation as $\frac{dx}{dy} = \frac{y^2 + y^2e^{(x/y)^2} + 2x^2e^{(x/y)^2}}{2xye^{(x/y)^2}}$, and then use the substitution $x = uy$; $\dfrac{dx}{dy} = u + y\dfrac{du}{dy}$ to obtain

$$y\frac{du}{dy} = \frac{1 + e^{u^2}}{2ue^{u^2}} \qquad \text{or} \qquad \frac{1}{y}\,dy - \frac{2ue^{u^2}}{1 + e^{u^2}}\,du = 0$$

This equation is separable; its solution is $\ln|y| - \ln(1 + e^{u^2}) = c$, which can be rewritten as $y = k(1 + e^{u^2})$, where $c = \ln|k|$. Substituting $u = x/y$ into this result, we obtain the solution of the given differential equation as $y = k[1 + e^{(x/y)^2}]$.

3.131 Solve $y' = (2y + x)/x$.

▌ This differential equation is homogeneous (see Problem 3.99). Using the substitution $y = vx$; $\dfrac{dy}{dx} = v + x\dfrac{dv}{dx}$, we obtain $v + x\dfrac{dv}{dx} = \dfrac{2vx + x}{x}$, which may be simplified to $x\dfrac{dv}{dx} = v + 1$. This last equation is separable and has as its solution $v = kx - 1$ (see Problem 3.35, with z and t replaced by v and x, respectively). Since $v = y/x$, the solution becomes $y/x = kx - 1$ or $y = kx^2 - x$.

3.132 Rework Problem 3.131 using the substitution suggested in Problem 3.122.

▌ We first write the differential equation as $\dfrac{dx}{dy} = \dfrac{x}{2y + x}$. Then, using the substitution $x = yu$; $\dfrac{dx}{dy} = u + y\dfrac{du}{dy}$, we obtain $u + y\dfrac{du}{dy} = \dfrac{uy}{2y + uy}$, which may be simplified to $y\dfrac{du}{dy} = \dfrac{u}{2 + u} - u = -\dfrac{u^2 + u}{u + 2}$. This last equation is separable and has as its solution in implicit form $yu^2 = k(u + 1)$ (see Problem 3.68). Since $u = x/y$, that solution becomes $y(x/y)^2 = k(x/y + 1)$ or $x^2 = k(x + y)$. Setting $A = 1/k$, we may rewrite this as $y = Ax^2 - x$, which is identical to the solution obtained in the previous problem except for the letter designating the arbitrary constant.

3.133 Solve $y' = (x^2 + 2y^2)/xy$.

▌ This differential equation is homogeneous (see Problem 3.100). Using the substitution $y = vx$; $\dfrac{dy}{dx} = v + x\dfrac{dv}{dx}$, we obtain $v + x\dfrac{dv}{dx} = \dfrac{x^2 + 2(vx)^2}{x(vx)}$, which may be simplified to $x\dfrac{dv}{dx} = \dfrac{1 + v^2}{v}$ or, in differential form, $xv\,dv - (1 + v^2)\,dx = 0$. This last equation is separable and has as its solution $1 + v^2 = kx^2$ (see Problem 3.46). Since $v = y/x$, the solution becomes $1 + (y/x)^2 = kx^2$ or $y^2 = kx^4 - x^2$.

3.134 Rework Problem 3.133 using the substitution suggested in Problem 3.122.

▌ We first write the differential equation as $\dfrac{dx}{dy} = \dfrac{xy}{x^2 + 2y^2}$. Then, using the substitution $x = yu$; $\dfrac{dx}{dy} = u + y\dfrac{du}{dy}$, we obtain $u + y\dfrac{du}{dy} = \dfrac{(yu)y}{(yu)^2 + 2y^2}$, which may be simplified to $y\dfrac{du}{dy} = \dfrac{u}{u^2 + 2} - u = -\dfrac{u^3 + u}{u^2 + 2}$. This last equation is separable and has as its solution in implicit form $u^4y^2 = k(u^2 + 1)$ (see Problem 3.70). Since $u = x/y$, the solution becomes $(x/y)^4y^2 = k[(x/y)^2 + 1]$ or $x^4 = k(x^2 + y^2)$. Setting $A = 1/k$, we may rewrite this solution as $Ax^4 = x^2 + y^2$, which is algebraically identical to the solution obtained in the previous problem.

3.135 Solve $y' = (x^2 + y^2)/2xy$.

▌ This differential equation is homogeneous (see Problem 3.103). Using the substitution $y = vx$; $\dfrac{dy}{dx} = v + x\dfrac{dv}{dx}$, we obtain $v + x\dfrac{dv}{dx} = \dfrac{x^2 + (vx)^2}{2x(vx)}$, which may be simplified to $x\dfrac{dv}{dx} = \dfrac{1 - v^2}{2v}$ or, in differential form, $2xv\,dv + (v^2 - 1)\,dx = 0$. This last equation is separable and has as its solution $v^2 - 1 = k/x$ (see Problem 3.47). Since $v = y/x$, the solution becomes $(y/x)^2 - 1 = k/x$ or $y^2 = x^2 + kx$.

3.136 Solve $y' = (x^2 + y^2)/3xy$.

▌ This differential equation is almost identical to that of the previous problem. The same substitution reduces it to $v + x\dfrac{dv}{dx} = \dfrac{x^2 + (vx)^2}{3x(vx)}$, which may be simplified to $x\dfrac{dv}{dx} = \dfrac{1 - 2v^2}{3v}$ or, in differential form,

$3xv\,dv + (2v^2 - 1)\,dx = 0$. This last equation is separable and has as its solution $2v^2 - 1 = k/x^{4/3}$ (see Problem 3.48). Since $v = y/x$, the solution becomes $2y^2 - x^2 = kx^{2/3}$.

3.137 Solve $y' = (x^2 + y^2)/4xy$.

This differential equation is almost identical to those of the two previous problems. The same substitution reduces it to $v + x\dfrac{dv}{dx} = \dfrac{x^2 + (vx)^2}{4x(vx)}$, which may be simplified to $x\dfrac{dv}{dx} = \dfrac{1 - 3v^2}{4v}$ or, in differential form, $4xv\,dv + (3v^2 - 1)\,dx = 0$. This last equation is separable and has as its solution $3v^2 - 1 = kx^{-3/2}$ (see Problem 3.49). Since $v = y/x$, the solution becomes $3y^2 - x^2 = kx^{1/2}$.

3.138 Rework the previous problem using the substitution suggested in Problem 3.122.

We first write the differential equation as $\dfrac{dx}{dy} = \dfrac{4xy}{x^2 + y^2}$. Then, using the substitution $x = yu$; $\dfrac{dx}{dy} = u + y\dfrac{du}{dy}$, we obtain $u + y\dfrac{du}{dy} = \dfrac{4(yu)y}{(yu)^2 + y^2}$, which may be simplified to $y\dfrac{du}{dy} = \dfrac{4u}{u^2 + 1} - u = \dfrac{-u^3 + 3u}{u^2 + 1}$. This last equation is separable and has as its solution in implicit form $(u^2 - 3)^2 y^3 = ku$ (see Problem 3.71). Setting $u = x/y$, we obtain $[(x/y)^2 - 3]^2 y^3 = k(x/y)$, which may be simplified to $(x^2 - 3y^2)^2 = kx$ and then to $3y^2 - x^2 = Ax^{1/2}$, where $A = \pm\sqrt{k}$.

3.139 Solve $y' = 2xy/(y^2 - x^2)$.

This differential equation is homogeneous (see Problem 3.102). Using the substitution $y = vx$; $\dfrac{dy}{dx} = v + x\dfrac{dv}{dx}$, we obtain $v + x\dfrac{dv}{dx} = \dfrac{2x(vx)}{(vx)^2 - x^2}$, which may be simplified to $x\dfrac{dv}{dx} = \dfrac{2v}{v^2 - 1} - v = \dfrac{-v^3 + 3v}{v^2 - 1}$ or, in differential form, $x(v^2 - 1)\,dv + (v^3 - 3v)\,dx = 0$. This last equation is separable and has as its solution $x^3(v^3 - 3v) = k$ (see Problem 3.50). Since $v = y/x$, the solution becomes $y^3 - 3yx^2 = k$.

3.140 Solve $y' = 3xy/(y^2 - x^2)$.

This differential equation is similar to that of the previous problem. The same substitution reduces it to $v + x\dfrac{dv}{dx} = \dfrac{3x(vx)}{(vx)^2 - x^2}$, which may be simplified to $x\dfrac{dv}{dx} = \dfrac{-v^3 + 4v}{v^2 - 1}$ or, in differential form, $x(v^2 - 1)\,dv + (v^3 - 4v)\,dx = 0$. This last equation is separable and has as its solution $v^2(v^2 - 4)^3 x^8 = k$ (see Problem 3.72). Since $v = y/x$, the solution becomes $y^2(y^2 - 4x^2)^3 = k$.

3.141 Solve $y' = 3xy/(y^2 + x^2)$.

This equation is similar to that of the previous problem. Using the substitution $y = vx$; $\dfrac{dy}{dx} = v + x\dfrac{dv}{dx}$, we obtain $v + x\dfrac{dv}{dx} = \dfrac{3x(vx)}{(vx)^2 + x^2}$, which may be simplified to $x\dfrac{dv}{dx} = \dfrac{-v^3 + 2v}{v^2 + 1}$ or, in differential form, $x(v^2 + 1)\,dv + (v^3 - 2v)\,dx = 0$. This last equation is separable and has as its solution $(v^2 - 2)^3 x^4 = kv^2$ (see Problem 3.73). Since $v = y/x$, the solution becomes $(y^2 - 2x^2)^3 = ky^2$.

3.142 Solve $y' = 3xy/(x^2 - y^2)$.

This problem is similar to Problem 3.139. The same substitution reduces the equation to $v + x\dfrac{dv}{dx} = \dfrac{3x(vx)}{x^2 - (vx)^2}$, which may be simplified to $x\dfrac{dv}{dx} = \dfrac{v^3 + 2v}{1 - v^2}$ or, in differential form, $x(v^2 - 1)\,dv + (v^3 + 2v)\,dx = 0$. This last equation is separable and has as its solution $(v^2 + 2)^3 x^4 = kv^2$ (see Problem 3.74). Since $v = y/x$, this solution becomes $(y^2 + 2x^2)^3 = ky^2$.

3.143 Solve $y' = \dfrac{y}{x + \sqrt{xy}}$.

This differential equation is homogeneous (see Problem 3.104). Using the substitution $y = vx$; $\dfrac{dy}{dx} = v + x\dfrac{dv}{dx}$, we obtain $v + x\dfrac{dv}{dx} = \dfrac{vx}{x + \sqrt{x(vx)}}$, which may be simplified to $x\dfrac{dv}{dx} = \dfrac{-v\sqrt{v}}{1 + \sqrt{v}}$ or, in differential form, $x(1+\sqrt{v})\,dv + v\sqrt{v}\,dx = 0$. The solution to this last equation is $-2/\sqrt{v} + \ln|vx| = c$ (see Problem 3.58). Since $v = y/x$, the solution becomes $-2\sqrt{x/y} + \ln|y| = c$.

3.144 Solve $y' = \dfrac{x^4 + 3x^2y^2 + y^4}{x^3y}$.

▌ This differential equation is homogeneous (see Problem 3.106). Using the substitution $y = vx$; $\dfrac{dy}{dx} = v + x\dfrac{dv}{dx}$, we obtain

$$v + x\frac{dv}{dx} = \frac{x^4 + 3x^2(vx)^2 + (vx)^4}{x^3(vx)} \qquad \text{or} \qquad x\frac{dv}{dx} = \frac{1 + 2v^2 + v^4}{v}$$

or, in differential form, $xv\,dv - (1 + 2v^2 + v^4)\,dx = 0$. The solution to this last equation is $(1 + v^2)\ln|kx|^2 = -1$ (see Problem 3.59). Since $v = y/x$, this solution becomes $(x^2 + y^2)\ln|kx^2| = -x^2$, or $y^2 = -x^2\left(1 + \dfrac{1}{\ln|kx^2|}\right)$.

3.145 Solve $(x^3 + y^3)\,dx - 3xy^2\,dy = 0$.

▌ This equation is homogeneous of degree 3. We use the transformation $y = vx$; $dy = v\,dx + x\,dv$, to obtain $x^3[(1 + v^3)\,dx - 3v^2(v\,dx + x\,dv)] = 0$, which we simplify to $(1 - 2v^3)\,dx - 3v^2x\,dv = 0$ and then write as $\dfrac{dx}{x} - \dfrac{3v^2\,dv}{1 - 2v^3} = 0$. The solution to this equation is $x^2(1 - 2v^3) = k$ (see Problem 3.56). Since $v = y/x$, we have $x^2[1 - 2(y/x)^3] = k$ or $x^3 - 2y^3 = kx$.

3.146 Solve $x\,dy - y\,dx - \sqrt{x^2 - y^2}\,dx = 0$.

▌ The equation is homogeneous of degree 1. Using the transformation $y = vx$; $dy = v\,dx + x\,dv$ and dividing by x, we get $v\,dx + x\,dv - v\,dx - \sqrt{1 - v^2}\,dx = 0$ or $x\,dv - \sqrt{1 - v^2}\,dx = 0$, which we write as $\dfrac{dv}{\sqrt{1 - v^2}} - \dfrac{dx}{x} = 0$. The solution to this equation is $\arcsin v - \ln|x| = c$ (see Problem 3.89). Since $v = y/x$, we have $\arcsin(y/x) = c + \ln|x| = \ln|kx|$, where $c = \ln k$.

3.147 Solve $[2x\sinh(y/x) + 3y\cosh(y/x)]\,dx - 3x\cosh(y/x)\,dy = 0$.

▌ This equation is homogeneous of degree 1. Using the transformation $y = vx$; $dy = v\,dx + x\,dv$ and dividing by x, we obtain $2\sinh v\,dx - 3x\cosh v\,dv = 0$. Separating the variables yields $2\dfrac{dx}{x} - 3\dfrac{\cosh v}{\sinh v}\,dv = 0$. Integrating, we get $2\ln x - 3\ln\sinh v = \ln c$, so that $x^2 = c\sinh^3 v$. Since $v = y/x$, this becomes $x^2 = c\sinh^3(y/x)$.

3.148 Solve $(2x + 3y)\,dx + (y - x)\,dy = 0$.

▌ This equation is homogeneous of degree 1. The transformation $y = vx$; $dy = v\,dx + x\,dv$ reduces it to

$$(2 + 3v)\,dx + (v - 1)(v\,dx + x\,dv) = 0 \qquad \text{or} \qquad (v^2 + 2v + 2)\,dx + x(v - 1)\,dv = 0$$

The solution to this last equation is $\ln[x^2(v^2 + 2v + 2)] - 4\arctan(v + 1) = k$ (see Problem 3.81). Since $v = y/x$, the solution becomes $\ln(y^2 + 2xy + 2x^2) - 4\arctan\dfrac{x + y}{x} = k$.

3.149 Solve $(1 + 2e^{x/y})\,dx + 2e^{x/y}(1 - x/y)\,dy = 0$.

▌ This equation is homogeneous of degree zero. The appearance of the quantity x/y throughout the equation suggests the substitution $x = uy$; $\dfrac{dx}{dy} = u + y\dfrac{du}{dy}$ or, equivalently, $dx = u\,dy + y\,du$. This transforms the differential equation into $(1 + 2e^u)(u\,dy + y\,du) + 2e^u(1 - u)\,dy = 0$, which we simplify to $(u + 2e^u)\,dy + y(1 + 2e^u)\,du = 0$. The solution to this last equation is $y(u + 2e^u) = k$ (see Problem 3.82). Since $u = x/y$, the solution becomes $x + 2ye^{x/y} = k$.

MISCELLANEOUS TRANSFORMATIONS

3.150 Solve $dy/dx = (y - 4x)^2$.

▮ The transformation $y - 4x = v;\ dy = 4\,dx + dv$ reduces this equation to $4\,dx + dv = v^2\,dx$ or $dx - \dfrac{dv}{v^2 - 4} = 0$. Then integration gives $x + \dfrac{1}{4}\ln\dfrac{v+2}{v-2} = c_1$, so that $\ln\dfrac{v+2}{v-2} = \ln c - 4x$. Exponentiation then yields $\dfrac{v+2}{v-2} = ce^{-4x}$, and substitution for v yields $\dfrac{y - 4x + 2}{y - 4x - 2} = ce^{-4x}$.

3.151 Solve $\tan^2(x + y)\,dx - dy = 0$.

▮ The transformation $x + y = v;\ dy = dv - dx$ reduces this equation to

$$\tan^2 v\,dx - (dv - dx) = 0 \qquad \text{or} \qquad dx - \frac{dv}{1 + \tan^2 v} = 0 \qquad \text{or} \qquad dx - \cos^2 v\,dv = 0$$

Integrating gives $x - \frac{1}{2}v - \frac{1}{4}\sin 2v = c_1$ which, after substitution for v and simplification, becomes $2(x - y) = c + \sin 2(x + y)$.

3.152 Solve $(2 + 2x^2y^{1/2})y\,dx + (x^2y^{1/2} + 2)x\,dy = 0$.

▮ The transformation $x^2y^{1/2} = v;\ y = \dfrac{v^2}{x^4};\ dy = \dfrac{2v}{x^4}\,dv - \dfrac{4v^2}{x^5}\,dx$ reduces the given equation to

$$(2 + 2v)\frac{v^2}{x^4}\,dx + x(v + 2)\left(\frac{2v}{x^4}\,dv - \frac{4v^2}{x^5}\,dx\right) = 0 \qquad \text{or} \qquad v(3 + v)\,dx - x(v + 2)\,dv = 0$$

Then the method of partial fractions gives $\dfrac{dx}{x} - \dfrac{2}{3}\dfrac{dv}{v} - \dfrac{1}{3}\dfrac{dv}{v + 3} = 0$, and integration yields $3\ln x - 2\ln v - \ln(v + 3) = \ln c_1$, from which $x^3 = c_1v^2(v + 3)$. Finally, substitution gives $1 = c_1xy(x^2y^{1/2} + 3)$ or $xy(x^2y^{1/2} + 3) = k$.

3.153 Solve $(2x^2 + 3y^2 - 7)x\,dx - (3x^2 + 2y^2 - 8)y\,dy = 0$.

▮ The transformation $x^2 = u;\ y^2 = v$ reduces this equation to $(2u + 3v - 7)\,du - (3u + 2v - 8)\,dv = 0$. Then the transformation $u = s + 2;\ v = t + 1$ yields the homogeneous equation $(2s + 3t)\,ds - (3s + 2t)\,dt = 0$, and the transformation $s = rt;\ ds = r\,dt + t\,dr$ yields $2(r^2 - 1)\,dt + (2r + 3)t\,dr = 0$. Separating the variables, we get $2\dfrac{dt}{t} + \dfrac{2r + 3}{r^2 - 1}\,dr = 2\dfrac{dt}{t} - \dfrac{1}{2}\dfrac{dr}{r + 1} + \dfrac{5}{2}\dfrac{dr}{r - 1} = 0$. Then integration yields $4\ln t - \ln(r + 1) + 5\ln(r - 1) = \ln C$. Exponentiation and successive substitutions then yield

$$\frac{t^4(r - 1)^5}{r + 1} = \frac{(s - t)^5}{s + t} = \frac{(u - v - 1)^5}{u + v - 3} = \frac{(x^2 - y^2 - 1)^5}{x^2 + y^2 - 3} = c$$

so that $(x^2 - y^2 - 1)^5 = c(x^2 + y^2 - 3)$.

3.154 Solve $x^2(x\,dx + y\,dy) + y(x\,dy - y\,dx) = 0$.

▮ Here $x\,dx + y\,dy = \frac{1}{2}d(x^2 + y^2)$ and $x\,dy - y\,dx = x^2d(y/x)$ suggest the transformation $x^2 + y^2 = \rho^2$; $y/x = \tan\theta$, or $x = \rho\cos\theta;\ y = \rho\sin\theta;\ dx = -\rho\sin\theta\,d\theta + \cos\theta\,d\rho;\ dy = \rho\cos\theta\,d\theta + \sin\theta\,d\rho$. The given equation then takes the form $\rho^2\cos^2\theta(\rho\,d\rho) + \rho\sin\theta(\rho^2\,d\theta) = 0$ or $d\rho + \tan\theta\sec\theta\,d\theta = 0$. Integration gives $\rho + \sec\theta = c_1$, so that $\sqrt{x^2 + y^2}\,\dfrac{x + 1}{x} = c_1$, which may be written $(x^2 + y^2)(x + 1)^2 = cx^2$.

3.155 Solve $y(xy + 1)\,dx + x(1 + xy + x^2y^2)\,dy = 0$.

▮ The transformation $xy = v;\ dy = \dfrac{x\,dv - v\,dx}{x^2}$ reduces this equation to

$\dfrac{v}{x}(v + 1)\,dx + x(1 + v + v^2)\dfrac{x\,dv - v\,dx}{x^2} = 0$, which can be simplified to $v^3\,dx - x(1 + v + v^2)\,dv = 0$.

Separating the variables yields $\frac{dx}{x} - \frac{dv}{v^3} - \frac{dv}{v^2} - \frac{dv}{v} = 0$. Integration then gives $\ln x + \frac{1}{2v^2} + \frac{1}{v} - \ln v = c_1$, so that $2v^2 \ln (v/x) - 2v - 1 = cv^2$. Finally, substitution of $v = xy$ yields $2x^2y^2 \ln y - 2xy - 1 = cx^2y^2$.

3.156 Solve $(y - xy^2)\,dx - (x + x^2y)\,dy = 0$ or, rewritten, $y(1 - xy)\,dx - x(1 + xy)\,dy = 0$.

▌ The transformation $xy = v$; $dy = \frac{x\,dv - v\,dx}{x^2}$ reduces this equation to

$$\frac{v}{x}(1 - v)\,dx - x(1 + v)\frac{x\,dv - v\,dx}{x^2} = 0 \qquad \text{or} \qquad 2v\,dx - x(1 + v)\,dv = 0$$

Then $2\frac{dx}{x} - \frac{1 + v}{v}\,dv = 0$, and integration gives $2 \ln x - \ln v - v = \ln c$, from which $x^2/v = ce^v$, and $x = cye^{xy}$.

3.157 Solve $(1 - xy + x^2y^2)\,dx + (x^3y - x^2)\,dy = 0$ or, rewritten, $y(1 - xy + x^2y^2)\,dx + x(x^2y^2 - xy)\,dy = 0$.

▌ The transformation $xy = v$; $dy = \frac{x\,dv - v\,dx}{x^2}$ reduces this equation to

$$\frac{v}{x}(1 - v + v^2)\,dx + x(v^2 - v)\frac{x\,dv - v\,dx}{x^2} = 0 \qquad \text{or} \qquad v\,dx + x(v^2 - v)\,dv = 0$$

Then $dx/x + (v - 1)\,dv = 0$, and integration gives $\ln x + \frac{1}{2}v^2 - v = c$, from which $\ln x = xy - \frac{1}{2}x^2y^2 + c$.

3.158 Solve $(x + y)\,dx + (3x + 3y - 4)\,dy = 0$.

▌ The expressions $(x + y)$ and $(3x + 3y)$ suggest the transformation $x + y = t$. We use $y = t - x$; $dy = dt - dx$ to obtain $t\,dx + (3t - 4)(dt - dx) = 0$ or, rewritten, $(4 - 2t)\,dx + (3t - 4)\,dt = 0$, in which the variables are separable. We then have $2\,dx + \frac{3t - 4}{2 - t}\,dt = 2\,dx - 3\,dt + \frac{2}{2 - t}\,dt = 0$. Integration yields $2x - 3t - 2\ln(2 - t) = c_1$, and after substitution for t we have $2x - 3(x + y) - 2\ln(2 - x - y) = c_1$, from which $x + 3y + 2\ln(2 - x - y) = c$.

3.159 Solve $(2x - 5y + 3)\,dx - (2x + 4y - 6)\,dy = 0$.

▌ We first solve $2x - 5y + 3 = 0$ and $2x + 4y - 6 = 0$ simultaneously to obtain $x = h = 1$, $y = k = 1$. Then the transformation

$$x = x' + h = x' + 1; \qquad dx = dx'$$
$$y = y' + k = y' + 1; \qquad dy = dy'$$

reduces the given equation to $(2x' - 5y')\,dx' - (2x' + 4y')\,dy' = 0$, which is homogeneous of degree 1. (Note that this latter equation can be written without computing the transformation.)

Using the transformation $y' = vx'$; $dy' = v\,dx' + x'\,dv$, we obtain $(2 - 5v)\,dx' - (2 + 4v)(v\,dx' + x'\,dv) = 0$ or $(2 - 7v - 4v^2)\,dx' - x'(2 + 4v)\,dv = 0$, which we separate into $\frac{dx'}{x'} + \frac{4}{3}\frac{dv}{4v - 1} + \frac{2}{3}\frac{dv}{v + 2} = 0$. Integrating, we get $\ln x' + \frac{1}{3}\ln(4v - 1) + \frac{2}{3}\ln(v + 2) = \ln c_1$, or $x'^3(4v - 1)(v + 2)^2 = c$.

Replacing v by y'/x' gives us $(4y' - x')(y' + 2x')^2 = c$, and replacing x' by $x - 1$ and y' by $y - 1$ yields the primitive $(4y - x - 3)(y + 2x - 3)^2 = c$.

3.160 Solve $(x - y - 1)\,dx + (4y + x - 1)\,dy = 0$.

▌ Solving $x - y - 1 = 0$ and $4y + x - 1 = 0$ simultaneously, we obtain $x = h = 1$, $y = k = 0$. The transformation

$$x = x' + h = x' + 1; \qquad dx = dx'$$
$$y = y' + k = y'; \qquad dy = dy'$$

reduces the given equation to $(x' - y')\,dx' + (4y' + x')\,dy' = 0$, which is homogeneous of degree 1. [Note that this transformation $x - 1 = x'$, $y = y'$ could have been obtained by inspection, that is, by examining the terms $(x - y - 1)$ and $(4y + x - 1)$.]

Using the transformation $y' = vx'$; $dy' = v\,dx' + x'\,dv$, we obtain $(1 - v)\,dx' + (4v + 1)(v\,dx' + x'\,dv) = 0$. Then

$$\frac{dx'}{x'} + \frac{4v+1}{4v^2+1}\,dv = \frac{dx'}{x'} + \frac{1}{2}\frac{8v}{4v^2+1}\,dv + \frac{dv}{4v^2+1} = 0$$

Integration gives $\ln x' + \frac{1}{2}\ln(4v^2 + 1) + \frac{1}{2}\arctan 2v = c$, which we rewrite as $\ln(x')^2(4v^2 + 1) + \arctan 2v = c$. Substitution for v then gives $\ln(4y'^2 + x'^2) + \arctan(2y'/x') = c$, and substitution for x' and y' yields $\ln[4y^2 + (x-1)^2] + \arctan\dfrac{2y}{x-1} = c$.

INITIAL-VALUE PROBLEMS

3.161 Discuss how to solve the initital-value problem $A(x)\,dx + B(y)\,dy = 0$; $y(x_0) = y_0$.

▌ The solution may be obtained by first solving the separable differential equation for its general solution and then applying the initial condition to evaluate the arbitrary constant. Alternatively, the solution may be obtained directly from

$$\int_{x_0}^{x} A(x)\,dx + \int_{y_0}^{y} B(y)\,dy = 0 \tag{1}$$

This last approach may not determine the solution *uniquely*; that is, the integrations may produce many solutions (in implicit form), of which only one will satisfy the initial-value problem.

3.162 Solve $e^x\,dx - y\,dy = 0$; $y(0) = 1$.

▌ The solution to the differential equation is $\int e^x\,dx + \int(-y)\,dy = c$ or, after evaluation, $y^2 = 2e^x + k$, $k = -2c$. Applying the initial condition (see Problem 2.103), we find that $k = -1$ and that the solution to the initial-value problem is $y = \sqrt{2e^x - 1}$, $x > \ln\frac{1}{2}$.

3.163 Use (*1*) of Problem 3.161 to solve the previous problem.

▌ Here $x_0 = 0$ and $y_0 = 1$, so (*1*) of Problem 3.161 becomes $\int_0^x e^x\,dx + \int_1^y(-y)\,dy = 0$. Evaluating these integrals, we get

$$e^x\Big|_0^x + \frac{-y^2}{2}\Big|_1^y = 0 \qquad \text{or} \qquad e^x - e^0 + \frac{-y^2}{2} - (-\tfrac{1}{2}) = 0$$

Thus, $y^2 = 2e^x - 1$ and, as before, $y = \sqrt{2e^x - 1}$, $x > \ln\frac{1}{2}$.

3.164 Solve $x\cos x\,dx + (1 - 6y^5)\,dy = 0$; $y(\pi) = 0$.

▌ Here $x_0 = \pi$, $y_0 = 0$, $A(x) = x\cos x$, and $B(y) = 1 - 6y^5$. Substituting these values into (*1*) of Problem 3.161, we obtain $\int_\pi^x x\cos x\,dx + \int_0^y(1 - 6y^5)\,dy = 0$. Evaluating these integrals (the first one by integration by parts), we find

$$x\sin x\Big|_\pi^x + \cos x\Big|_\pi^x + (y - y^6)\Big|_0^y = 0 \qquad \text{or} \qquad x\sin x + \cos x + 1 = y^6 - y$$

Since we cannot solve this last equation for y explicitly, we leave the solution in implicit form. (See also Problem 2.105).

3.165 Solve $\sin x\,dx + y\,dy = 0$; $y(0) = -2$.

▌ The differential equation is separable, so we have $\int_0^x \sin x\,dx + \int_{-2}^y y\,dy = 0$. Evaluating these integrals, we get

$$-\cos x\Big|_0^x + \tfrac{1}{2}y^2\Big|_{-2}^y = 0 \qquad \text{or} \qquad -\cos x + 1 + \tfrac{1}{2}y^2 - 2 = 0$$

from which $y^2 = 2 + 2\cos x$, or $y = -\sqrt{2 + 2\cos x}$. The negative square root is chosen to be consistent with the initial condition.

3.166 Solve $(x^2 + 1)\,dx + (1/y)\,dy = 0$; $y(-1) = 1$.

▌ The differential equation is separable, so we have $\int_{-1}^x (x^2 + 1)\,dx + \int_1^y (1/y)\,dy = 0$, from which

$$(\tfrac{1}{3}x^3 + x)\Big|_{-1}^x + \ln|y|\Big|_1^y = 0 \qquad \text{or} \qquad \tfrac{1}{3}x^3 + x - (-\tfrac{1}{3} - 1) + \ln|y| - \ln 1 = 0$$

Then $\ln|y| = -(x^3 + 3x + 4)/3$, and $y = e^{-(x^3+3x+4)/3}$. The plus sign in front of the exponential is consistent with the initial condition.

3.167 Solve $xe^{x^2}\,dx + (y^5 - 1)\,dy = 0$; $y(0) = 0$.

▮ The differential equation is separable, so we have $\int_0^x xe^{x^2}\,dx + \int_0^y (y^5 - 1)\,dy = 0$. The indicated integrations give

$$\tfrac{1}{2}e^{x^2}\Big|_0^x + (\tfrac{1}{6}y^6 - y)\Big|_0^y = 0 \qquad \text{or} \qquad \tfrac{1}{2}e^{x^2} - \tfrac{1}{2} + \tfrac{1}{6}y^6 - y - 0 = 0$$

from which we obtain $y^6 - 6y + 3(e^{x^2} - 1) = 0$, which is the solution in implicit form.

3.168 Solve $y' = (x^2y - y)/(y + 1)$; $y(3) = -1$.

▮ Separating the variables, we find that the differential equation has the form $\dfrac{y+1}{y}\,dy - (x^2 - 1)\,dx = 0$. The solution to the initial-value problem then is $\displaystyle\int_{-1}^{y}\left(1 + \frac{1}{y}\right)dy - \int_3^x (x^2 - 1)\,dx = 0$. The indicated integrations give

$$(y + \ln|y|)\Big|_{-1}^{y} - (\tfrac{1}{3}x^3 - x)\Big|_3^x = 0 \qquad \text{or} \qquad y + \ln|y| - (-1 + \ln 1) - (\tfrac{1}{3}x^3 - x) + 9 - 3 = 0$$

from which we obtain $y + \ln|y| = \frac{1}{3}x^3 - x - 7$.

3.169 Solve $y' + 3y = 8$; $y(0) = 2$.

▮ The solution to the differential equation was shown in Problem 3.36 to be $y(x) = \frac{8}{3} + ke^{-3x}$. Applying the initial condition, we get $2 = y(0) = \frac{8}{3} + ke^{-3(0)}$, so that $k = -\frac{2}{3}$. Thus, the solution to the initial-value problem is $y(x) = \frac{8}{3} - \frac{2}{3}e^{-3x}$.

3.170 Solve the preceding problem if the initial condition is $y(0) = 4$.

▮ The solution to the differential equation remains the same. Applying the new initial condition, we get $4 = y(0) = \frac{8}{3} + ke^{-3(0)}$, so that $k = \frac{4}{3}$. Thus, the solution to this initial-value problem is $y(x) = \frac{8}{3} + \frac{4}{3}e^{-3x}$.

3.171 Solve Problem 3.169 if the initial condition is $y(1) = 0$.

▮ The solution to the differential equation remains the same. Applying the new initial condition, we get $0 = y(1) = \frac{8}{3} + ke^{-3(1)}$, so that $k = -\frac{8}{3}e^3$. The solution to this initial-value problem is then $y(x) = \frac{8}{3} - \frac{8}{3}e^3e^{-3x} = \frac{8}{3}(1 - e^{-3(x-1)})$.

3.172 Solve Problem 3.169 if the initial condition is $y(-2) = 1$.

▮ The solution to the differential equation remains as before. Applying the new initial condition, we get $1 = y(-2) = \frac{8}{3} + ke^{-3(-2)}$, so that $k = -\frac{5}{3}e^{-6}$. The solution to this initial-value problem is then $y(x) = \frac{8}{3} - \frac{5}{3}e^{-6}e^{-3x} = \frac{1}{3}(8 - 5e^{-3(x+2)})$.

3.173 Solve Problem 3.169 if the initial condition is $y(-2) = -1$.

▮ The solution to the differential equation remains as before. Applying the new initial condition, we get $-1 = y(-2) = \frac{8}{3} + ke^{-3(-2)}$, so that $k = -\frac{11}{3}e^{-6}$. The solution to this initial-value problem is then $y(x) = \frac{8}{3} - \frac{11}{3}e^{-6}e^{-3x} = \frac{1}{3}(8 - 11e^{-3(x+2)})$.

3.174 Solve Problem 3.169 if the initial condition is $y(4) = -3$.

▮ The solution to the differential equation remains as before. Applying the new initial condition, we get $-3 = y(4) = \frac{8}{3} + ke^{-3(4)}$, so $k = -\frac{17}{3}e^{12}$. The solution to the initial-value problem is then $y(x) = \frac{8}{3} - \frac{17}{3}e^{12}e^{-3x} = (8 - 17e^{-3(x-4)})/3$.

3.175 Solve $dy/dt = y^5 \sin t$; $y(0) = 1$.

▮ The general solution to the differential equation was shown in Problem 3.88 to be $1/y^4 = 4\cos t + c_1$. As a result of Problem 2.104 we have $y(t) = \left(\dfrac{1}{4\cos t - 3}\right)^{1/4}$, where $-\arccos\dfrac{3}{4} < t < \arccos\dfrac{3}{4}$, as the solution to the initial-value problem.

3.176 Solve the previous problem if the initial condition is $y(0) = \frac{1}{2}$.

▮ The general solution to the differential equation remains as before. Applying the initial condition, we get $\dfrac{1}{(1/2)^4} = 4\cos 0 + c_1$, so $c_1 = 2^4 - 4 = 12$. The solution to the initial-value problem becomes $1/y^4 = 4\cos t + 12$ or, explicitly, $y(t) = \left(\dfrac{1}{4\cos t + 12}\right)^{1/4}$. The solution y is defined *for all t*, since $4\cos t + 12$ is always positive.

3.177 Solve $y' + y = 0$; $y(3) = 2$.

▮ The general solution to the differential equation was shown in Problem 3.30 to be $y = ke^{-x}$. (Here, A of Problem 3.30 is equal to -1.) As a result of Problem 2.74, we have $y(x) = 2e^{3-x}$ as the solution to the initial-value problem.

3.178 Solve $\dot{y} = y^2$; $y(0) = 4$.

▮ The general solution to the differential equation was shown in Problem 3.9 to be $y(t) = -1/(t + c)$. Applying the initial condition, we get $4 = y(0) = -1/(0 + c)$, so that $c = -\frac{1}{4}$. The solution to the initial-value problem then is $y(t) = -1/(t - \frac{1}{4}) = -4/(4t - 1)$.

3.179 Solve the previous problem if the initial condition is $y(-1) = -2$.

▮ The solution to the differential equation remains the same. Applying the new initial condition, we get $-2 = y(-1) = -1/(-1 + c)$, so that $c = \frac{3}{2}$. The solution to the initial-value problem is then $y(t) = -1/(t + \frac{3}{2}) = -2/(2t + 3)$.

Observe that this solution is not valid in any interval containing $t = -\frac{3}{2}$. Since a solution to an initial-value problem must be valid in some interval containing the initial time, in this case $t = -1$, it follows that the above solution is valid only on the interval $(-\frac{3}{2}, \infty)$. By similar reasoning, the solution to the previous problem is valid only on $(-\infty, \frac{1}{4})$.

3.180 Solve Problem 3.178 if the initial condition is $y(0) = -2$.

▮ The general solution to the differential equation remains as before. Since $y(0) = -2$, we have $c = \frac{1}{2}$. Then $y(t) = -1/(t + \frac{1}{2}) = -2/(2t + 1)$ is the solution to the initial-value problem. Since this solution is defined only for $t \neq -\frac{1}{2}$, and since 0 must be in the interval for which y is defined, it follows that $(-\frac{1}{2}, \infty)$ is the interval of definition for y.

3.181 Solve $dz/dt = z^3t^2$; $z(2) = 3$.

▮ The solution to the differential equation was found in Problem 3.11 to have the form $-\dfrac{1}{2z^2} - \dfrac{1}{3}t^3 = c$.

Applying the initial condition, we get $-\dfrac{1}{2(3)^2} - \dfrac{1}{3}(2)^3 = c$, so that $c = -\frac{49}{18}$ and the solution becomes, in explicit form, $z = \left(\dfrac{1}{\frac{49}{9} - \frac{2}{3}t^3}\right)^{1/2}$, where the positive square root is chosen consistent with the initial condition.

3.182 Rework the previous problem if the initial condition is $z(2) = -3$.

▮ The solution to the differential equation remains the same. Applying the initial condition, we get $-\dfrac{1}{2(-3)^2} - \dfrac{1}{3}(2)^3 = c$, so again $c = -\frac{49}{18}$. Now, however, the solution to the initial-value problem in explicit form becomes $z = -\left(\dfrac{1}{\frac{49}{9} - \frac{2}{3}t^3}\right)^{1/2}$, where the negative square root is chosen to be consistent with the initial condition.

3.183 Solve $dy/dt = y(t - 2)$; $y(2) = 5$.

▮ The solution to the differential equation was found in Problem 3.32 to be $y(t) = ke^{(t-2)^2/2}$. Applying the initial condition, we get $5 = ke^{(2-2)^2/2} = ke^0 = k$, so the solution to the initial-value problem is $y(t) = 5e^{(t-2)^2/2}$.

3.184 Solve $dy/dt = -2yt^2$; $y(2) = 3$.

▌ The solution to the differential equation was found in Problem 3.33 to be $y(t) = ke^{2t^3}$. Applying the initial condition, we get $3 = y(2) = ke^{2(2)^3} = ke^{16}$, so $k = 3e^{-16}$. Then the solution to the initial-value problem is $y(t) = 3e^{-16}e^{2t^3} = 3e^{2(t-2)^3}$.

3.185 Solve $4x\,dy - y\,dx = x^2\,dy$; $y(5) = -1$.

▌ The solution to the differential equation was found in Problem 3.61 to have the form $(x-4)y^4/x = k$. Applying the initial condition, we get $(5-4)(-1)^4/5 = k$, so $k = \frac{1}{5}$ and the solution to the initial-value problem is $y(x) = -\left[\dfrac{x}{5(x-4)}\right]^{1/4}$, where the negative fourth root is taken consistent with the initial condition.

3.186 Solve $x^2(y+1)\,dx + y^2(x-1)\,dy = 0$; $y(-1) = 2$.

▌ The solution to the differential equation was found in Problem 3.63 to have the form $(x+1)^2 + (y-1)^2 + 2\ln|(x-1)(y+1)| = k$. Applying the initial condition, we get $k = (-1+1)^2 + (2-1)^2 + 2\ln|(-1-1)(2+1)| = 1 + 2\ln 6$. The solution to the initial-value problem is then

$$(x+1)^2 + (y-1)^2 + 2\ln|(x-1)(y+1)| = 1 + 2\ln 6 \quad \text{or} \quad (x+1)^2 + (y-1)^2 + \ln\frac{(x-1)^2(y+1)^2}{36} = 1$$

3.187 Solve $dy/dt = y^2 - y^3$; $y(1) = 2$.

▌ The solution to the differential equation was found in Problem 3.64 to be $-1/y + \ln|y| - \ln|1-y| - t = c$. Applying the initial condition, we get $-\frac{1}{2} + \ln 2 - \ln 1 - 1 = c$, so $c \approx -0.80685$. The solution to the initial-value problem is thus $-1/y + \ln|y| - \ln|1-y| = -0.80685$.

3.188 Solve the preceding problem if the initial condition is $y(2) = 0$.

▌ The solution obtained in Problem 3.64 is the solution to the differential equation only when $y \neq 0$ and $y \neq 1$, because the partial-fraction decomposition used to generate the solution is undefined at these two values of y. Here $y_0 = 0$, so we are in one of these special cases. By inspection, we note that two constant solutions to the differential equation are $y \equiv 0$ and $y \equiv 1$. Since the first of these also satisfies the initial condition, it is the solution to the initial-value problem.

3.189 Solve $dy/dx = y - y^2$; $y(0) = 2$.

▌ The solution to the differential equation was found in Problem 3.60 to be $y(x) = \dfrac{ke^x}{1+ke^x}$. Applying the initial condition, we get $2 = y(0) = \dfrac{ke^0}{1+ke^0} = \dfrac{k}{1+k}$, so $k = -2$. The solution to the initial-value problem is then $y(x) = \dfrac{-2e^x}{1-2e^x} = \dfrac{2}{2-e^{-x}}$.

3.190 Solve the previous problem if the initial condition is $y(0) = 1$.

▌ The solution obtained in Problem 3.60 is the solution to the differential equation only when $y \neq 0$ and $y \neq 1$, because the partial-fraction decomposition used to obtain the solution is undefined at these two values of y. Here $y_0 = 1$, so we are in one of these special cases. By inspection, we note that two constant solutions to the differential equation are $y \equiv 0$ and $y \equiv 1$. Since the latter solution also satisfies the initial condition, it is the solution to the initial-value problem.

3.191 Solve $dy/dt = 2ty^2$; $y(0) = y_0$.

▌ If $y_0 \neq 0$, then by separation of variables we have $\dfrac{y'}{y^2} = 2t$ and hence $\displaystyle\int_{y_0}^{y} \frac{dy}{y^2} = \int_0^t 2t\,dt$. The integrations result in $-\dfrac{1}{y} + \dfrac{1}{y_0} = t^2$ or $\dfrac{1}{y} = \dfrac{1}{y_0} - t^2$, so that $y(t) = \dfrac{y_0}{1 - y_0 t^2}$ as long as $y_0 t^2 \neq 1$. If $y_0 > 0$ then $y(t) \longrightarrow +\infty$ as $t \longrightarrow 1/\sqrt{y_0}$, and so solutions to this equation "blow up" in finite time whenever the initial condition is positive. Note, however, that if $y_0 < 0$, then y exists for all $t \geq 0$ and $y(t) \longrightarrow 0$ as $t \longrightarrow +\infty$.

Note also that if $y_0 = 0$, then the solution becomes $y(t) = 0$, which is the solution to $dy/dt = 2ty^2$; $y(0) = 0$. Thus, $y(t)$ as found above solves the initial-value problem for all values of y_0.

3.192 Solve $y' = (y + x)/x$; $y(-1) = -2$.

▌ The solution to the differential equation was found in Problem 3.123 to be $y = x \ln |kx|$. Applying the initial condition, we get $-2 = y(-1) = -1 \ln |k(-1)|$, so $\ln |k| = 2$ and the solution to the initial-value problem is $y(x) = x \ln |kx| = x(\ln |k| + \ln |x|) = x(2 + \ln |x|)$.

3.193 Solve $y' = (x^2 + y^2)/xy$; $y(1) = -2$.

▌ The solution to the differential equation was found in Problem 3.126 to be $y^2 = x^2 \ln x^2 + kx^2$. Applying the initial condition, we obtain $(-2)^2 = (1)^2 \ln (1)^2 + k(1)^2$, or $k = 4$. Thus, the solution to the initial-value problem is $y^2 = x^2 \ln x^2 + 4x^2$ or $y = -\sqrt{x^2 \ln x^2 + 4x^2}$. The negative square root is taken consistent with the initial condition.

3.194 Solve $y' = (x^2 + y^2)/2xy$; $y(1) = -2$.

▌ The solution to the differential equation was found in Problem 3.135 to be $y^2 = x^2 + kx$. Applying the initial condition, we get $(-2)^2 = (1)^2 + k(1)$, or $k = 3$. The solution to the initial-value problem is then $y^2 = x^2 + 3x$ or, explicitly, $y = -\sqrt{x^2 + 3x}$, where the negative square root is chosen consistent with the initial condition.

3.195 Solve $y' = 2xy/(y^2 - x^2)$; $y(4) = 0$.

▌ The solution to the differential equation was found in Problem 3.139 to be $y^3 - 3yx^2 = k$. Applying the initial condition, we get $k = (0)^3 - 3(0)(4)^2 = 0$, so the solution to the initial-value problem is $y^3 - 3yx^2 = 0$.

3.196 Solve $y' = \dfrac{x^4 + 3x^2y^2 + y^4}{x^3y}$; $y(2) = 1$.

▌ The general solution to the differential equation was found in Problem 3.144 to be $y^2 = -x^2\left(1 + \dfrac{1}{\ln |kx^2|}\right)$.

Applying the initial condition, we get $(1)^2 = -(2)^2\left(1 + \dfrac{1}{\ln |k(2)^2|}\right)$, so $\ln |4k| = -\dfrac{4}{5}$. Then the general solution becomes

$$y^2 = -x^2\left[1 + \frac{1}{\ln |(4k)(x^2/4)|}\right] = -x^2\left[1 + \frac{1}{\ln |4k| + \ln (x^2/4)}\right] = -x^2\left[1 + \frac{1}{-4/5 + \ln (x^2/4)}\right]$$

or, explicitly, $y = \left[-x^2\left(1 + \dfrac{5}{-4 + 5\ln (x^2/4)}\right)\right]^{1/2}$

CHAPTER 4
Exact First-Order Differential Equations

TESTING FOR EXACTNESS

4.1 Define *exact* as regards a differential equation.

▌ A differential equation $M(x, y)\,dx + N(x, y)\,dy = 0$ is exact if there exists a function $g(x, y)$ such that $dg(x, y) = M(x, y)\,dx + N(x, y)\,dy$.

4.2 Develop a test for determining whether a first-order differential equation is exact.

▌ If $M(x, y)$ and $N(x, y)$ are continuous functions and have continuous first partial derivatives on some rectangle of the (x, y) plane, then the differential equation $M(x, y)\,dx + N(x, y)\,dy = 0$ is exact if and only if $\dfrac{\partial M(x, y)}{\partial y} = \dfrac{\partial N(x, y)}{\partial x}$.

4.3 Determine whether the differential equation $2xy\,dx + (1 + x^2)\,dy = 0$ is exact.

▌ Here $M(x, y) = 2xy$ and $N(x, y) = 1 + x^2$. Since $\dfrac{\partial M}{\partial y} = \dfrac{\partial N}{\partial x} = 2x$, the differential equation is exact.

4.4 Determine whether the differential equation $(x + \sin y)\,dx + (x \cos y - 2y)\,dy = 0$ is exact.

▌ Here $M(x, y) = x + \sin y$ and $N(x, y) = x \cos y - 2y$. Since $\dfrac{\partial M}{\partial y} = \cos y = \dfrac{\partial N}{\partial x}$, the differential equation is exact.

4.5 Determine whether the differential equation $xe^{xy}\,dx + ye^{xy}\,dy = 0$ is exact.

▌ Here $\dfrac{\partial M}{\partial y} = \dfrac{\partial(xe^{xy})}{\partial y} = x^2e^{xy}$ and $\dfrac{\partial N}{\partial x} = \dfrac{\partial(ye^{xy})}{\partial x} = y^2e^{xy}$. Since these two partial derivatives are not equal, the differential equation is *not* exact.

4.6 Determine whether the differential equation $(xy + x^2)\,dx + (-1)\,dy = 0$ is exact.

▌ Here, $M(x, y) = xy + x^2$ and $N(x, y) = -1$; hence $\dfrac{\partial M}{\partial y} = x$ and $\dfrac{\partial N}{\partial x} = 0$. Since $\dfrac{\partial M}{\partial y} \neq \dfrac{\partial N}{\partial x}$, the equation is *not* exact.

4.7 Determine whether the differential equation $(2xy + x)\,dx + (x^2 + y)\,dy = 0$ is exact.

▌ Here $M(x, y) = 2xy + x$ and $N(x, y) = x^2 + y$. Since $\dfrac{\partial M}{\partial y} = 2x = \dfrac{\partial N}{\partial x}$, the equation is exact.

4.8 Determine whether the differential equation $(y + 2xy^3)\,dx + (1 + 3x^2y^2 + x)\,dy = 0$ is exact.

▌ Here $M(x, y) = y + 2xy^3$ and $N(x, y) = 1 + 3x^2y^2 + x$. Since $\dfrac{\partial M}{\partial y} = 1 + 6xy^2 = \dfrac{\partial N}{\partial x}$, the equation is exact.

4.9 Determine whether the differential equation $ye^{xy}\,dx + xe^{xy}\,dy = 0$ is exact.

▌ Here $M(x, y) = ye^{xy}$ and $N(x, y) = xe^{xy}$. Since $\dfrac{\partial M}{\partial y} = e^{xy} + xye^{xy} = \dfrac{\partial N}{\partial x}$, the equation is exact.

4.10 Determine whether the differential equation $\sin x \cos y\,dx - \sin y \cos x\,dy = 0$.

▌ Here $M(x, y) = \sin x \cos y$ and $N(x, y) = -\sin y \cos x$. Since the partial derivative $\dfrac{\partial M}{\partial y} = -\sin x \sin y$ is not equal to the partial derivative $\dfrac{\partial N}{\partial x} = \sin y \sin x$, the equation is *not* exact.

4.11 Determine whether the differential equation $y\,dx + x\,dy = 0$ is exact.

Here $M(x, y) = y$ and $N(x, y) = x$. Since $\dfrac{\partial M}{\partial y} = 1 = \dfrac{\partial N}{\partial x}$, the equation is exact.

4.12 Determine whether the differential equation $(x - y)\,dx + (x + y)\,dy = 0$.

Here $M(x, y) = x - y$ and $N(x, y) = x + y$. Since $\dfrac{\partial M}{\partial y} = -1$ and $\dfrac{\partial N}{\partial x} = 1$ are not equal, the equation is *not* exact.

4.13 Determine whether the differential equation $(y \sin x + xy \cos x)\,dx + (x \sin x + 1)\,dy = 0$ is exact.

Here $M(x, y) = y \sin x + xy \cos x$ and $N(x, y) = x \sin x + 1$. Since $\dfrac{\partial M}{\partial y} = \sin x + x \cos x = \dfrac{\partial N}{\partial x}$, the equation is exact.

4.14 Determine whether the differential equation $x^2\,dy + y^2\,dx = 0$ is exact.

Here $M(x, y) = y^2$ [recall that $M(x, y)$ is the coefficient of dx] and $N(x, y) = x^2$. Since the partial derivatives $\dfrac{\partial M}{\partial y} = 2y$ and $\dfrac{\partial N}{\partial x} = 2x$ are not equal, the equation is *not* exact.

4.15 Determine whether the differential equation $x \sin y\,dy + \frac{1}{2}x^2 \cos y\,dx = 0$ is exact.

Here $M(x, y) = \frac{1}{2}x^2 \cos y$ and $N(x, y) = x \sin y$. Since the partial derivatives $\dfrac{\partial M}{\partial y} = -\frac{1}{2}x^2 \sin y$ and $\dfrac{\partial N}{\partial x} = \sin y$ are not equal, the equation is *not* exact.

4.16 Determine whether the differential equation $(3x^4y^2 - x^2)\,dy + (4x^3y^3 - 2xy)\,dx = 0$ is exact.

Here $M(x, y) = 4x^3y^3 - 2xy$ and $N(x, y) = 3x^4y^2 - x^2$. Since $\dfrac{\partial M}{\partial y} = 12x^3y^2 - 2x = \dfrac{\partial N}{\partial x}$, the equation is exact.

4.17 Determine whether the differential equation $e^{x^3}(3x^2y - x^2)\,dx + e^{x^3}\,dy = 0$ is exact.

Here $M(x, y) = 3x^2ye^{x^3} - x^2e^{x^3}$ and $N(x, y) = e^{x^3}$. Since $\dfrac{\partial M}{\partial y} = 3x^2e^{x^3} = \dfrac{\partial N}{\partial x}$, the equation is exact.

4.18 Determine whether the differential equation given in the previous problem is exact after it is divided by the nonzero quantity e^{x^3}.

The new equation is $(3x^2y - x^2)\,dx + dy = 0$, in which $M(x, y) = 3x^2y - x^2$ and $N(x, y) = 1$. Since $\dfrac{\partial M}{\partial y} = 3x^2$ and $\dfrac{\partial N}{\partial x} = 0$ are not equal, the new differential equation is *not* exact.

4.19 Determine whether the differential equation $\dfrac{xy - 1}{x^2y}\,dx - \dfrac{1}{xy^2}\,dy = 0$ is exact.

Here $M(x, y) = x^{-1} - x^{-2}y^{-1}$ and $N(x, y) = -x^{-1}y^{-2}$. Since $\dfrac{\partial M}{\partial y} = x^{-2}y^{-2} = \dfrac{\partial N}{\partial x}$, the equation is exact.

4.20 Determine whether the differential equation $(2x^3y + 4y^3)\,dy + 3x^2y^2\,dx = 0$ is exact.

Here $M(x, y) = 3x^2y^2$ [recall that $M(x, y)$ is the coefficient of dx] and $N(x, y) = 2x^3y + 4y^3$. Since $\dfrac{\partial M}{\partial y} = 6x^2y = \dfrac{\partial N}{\partial x}$, the equation is exact.

4.21 Determine whether the differential equation $(2t^3 + 3y)\,dt + (3t + y - 1)\,dy = 0$ is exact.

▌ Here $M(t, y) = 2t^3 + 3y$ and $N(t, y) = 3t + y - 1$. Since $\dfrac{\partial M}{\partial y} = 3 = \dfrac{\partial N}{\partial t}$, the equation is exact.

4.22 Determine whether the differential equation $(t^2 - y)\,dt - t\,dy = 0$ is exact.

▌ Here $M(t, y) = t^2 - y$ and $N(t, y) = -t$. Since $\dfrac{\partial M}{\partial y} = -1 = \dfrac{\partial N}{\partial t}$, the equation is exact.

4.23 Determine whether the differential equation $y(t - 2)\,dt - t^2\,dy = 0$ is exact.

▌ Here $M(t, y) = y(t - 2)$ and $N(t, y) = -t^2$. Since $\dfrac{\partial M}{\partial y} = t - 2$ and $\dfrac{\partial N}{\partial t} = -2t$ are not equal, the equation is *not* exact.

4.24 Determine whether the differential equation $dt - \sqrt{a^2 - t^2}\,dy = 0$ is exact when a denotes a constant.

▌ Here $M(t, y) = 1$ and $N(t, y) = -\sqrt{a^2 - t^2}$. Since $\dfrac{\partial M}{\partial y} = 0 \neq \dfrac{\partial N}{\partial t} = \dfrac{2t}{\sqrt{a^2 - t^2}}$, the equation is *not* exact.

4.25 Determine whether the differential equation $(3e^{3t}y - 2t)\,dt + e^{3t}\,dy = 0$ is exact.

▌ Here $M(t, y) = 3e^{3t}y - 2t$ and $N(t, y) = e^{3t}$. Since $\dfrac{\partial M}{\partial y} = 3e^{3t} = \dfrac{\partial N}{\partial t}$, the equation is exact.

4.26 Determine whether the differential equation $(\cos y + y\cos t)\,dt + (\sin t - t\sin y)\,dy = 0$ is exact.

▌ Here $M(t, y) = \cos y + y\cos t$ and $N(t, y) = \sin t - t\sin y$. Since $\dfrac{\partial M}{\partial y} = -\sin y + \cos t = \dfrac{\partial N}{\partial t}$, the equation is exact.

4.27 Determine whether the differential equation for $x(t)$ defined by $(2t + 3x + 4)\,dt + (3t + 4x + 5)\,dx = 0$ is exact.

▌ With t as the independent variable and x as the dependent variable, we have $M(t, x) = 2t + 3x + 4$ and $N(t, x) = 3t + 4x + 5$. Then $\dfrac{\partial M}{\partial x} = 3 = \dfrac{\partial N}{\partial t}$, so the equation is exact.

4.28 Determine whether the differential equation for $x(t)$ defined by $(6t^5x^3 + 4t^3x^5)\,dt + (3t^6x^2 + 5t^4x^4)\,dx = 0$ is exact.

▌ With t as the independent variable and x as the dependent variable, we have $M(t, x) = 6t^5x^3 + 4t^3x^5$ and $N(t, x) = 3t^6x^2 + 5t^4x^4$. Then $\dfrac{\partial M}{\partial x} = 18t^5x^2 + 20t^3x^4 = \dfrac{\partial N}{\partial t}$, so the equation is exact.

4.29 Determine whether the differential equation for $x(t)$ defined by $(2t + 3x + 4)\,dx + (3t + 4x + 5)\,dt = 0$ is exact.

▌ With t as the independent variable and x as the dependent variable, we have $M(t, x) = 3t + 4x + 5$, because M is always the coefficient of the differential of the independent variable; then also $N(t, x) = 2t + 3x + 4$. Since $\dfrac{\partial M}{\partial x} = 4$ and $\dfrac{\partial N}{\partial t} = 2$ are not equal, the equation is *not* exact.

4.30 Determine whether the differential equation for $x(t)$ defined by $2t(xe^{t^2} - 1)\,dt + e^{t^2}\,dx = 0$ is exact.

▌ With t as the independent variable and x as the dependent variable, we have $M(t, x) = 2t(xe^{t^2} - 1)$ and $N(t, x) = e^{t^2}$. Then $\dfrac{\partial M}{\partial x} = 2te^{t^2} = \dfrac{\partial N}{\partial t}$, so the equation is exact.

4.31 Determine whether the differential equation for $x(t)$ defined by $2t(xe^{t^2} - 1)\,dx + e^{t^2}\,dt = 0$ is exact.

▌ With t as the independent variable and x as the dependent variable, we have $M(t, x) = e^{t^2}$, because $M(t, x)$

is the coefficient of the differential of the independent variable; then also $N(t, x) = 2t(xe^{t^2} - 1)$. Now

$$\frac{\partial M}{\partial x} = 0 \quad \text{and} \quad \frac{\partial N}{\partial t} = 2(xe^{t^2} - 1) + (2t)(2txe^{t^2})$$

Since these partial derivatives are not equal, the equation is not exact.

4.32 Determine whether the differential equation for $x(t)$ defined by $\left(3 + \frac{x^2}{t^2}\right) dt - 2\frac{x}{t}\, dx = 0$ is exact.

▮ Here $M(t, x) = 3 + \frac{x^2}{t^2}$ and $N(t, x) = -2\frac{x}{t}$. Since $\frac{\partial M}{\partial x} = 2xt^{-2} = \frac{\partial N}{\partial t}$, the equation is exact.

4.33 Determine whether the differential equation for $z(t)$ defined by $(t^2 + z^2)\, dt + (2tz - z)\, dz = 0$ is exact.

▮ Here $M(t, z) = t^2 + z^2$ and $N(t, z) = 2tz - z$. Since $\frac{\partial M}{\partial z} = 2z = \frac{\partial N}{\partial t}$, the equation is exact.

4.34 Determine whether the differential equation for $z(t)$ defined by $(t + z\cos t)\, dt + (\sin t - 3z^2 + 5)\, dz = 0$ is exact.

▮ Here $M(t, z) = t + z\cos t$ and $N(t, z) = \sin t - 3z^2 + 5$. Since $\frac{\partial M}{\partial z} = \cos t = \frac{\partial N}{\partial t}$, the equation is exact.

4.35 Determine whether the differential equation for $u(v)$ defined by $2(u^2 + uv - 3)\, du + (u^2 + 3v^2 - v)\, dv = 0$ is exact.

▮ Here $M(v, u) = u^2 + 3v^2 - v$, because we associate M with the coefficient of the differential of the independent variable; then $N(v, u) = 2(u^2 + uv - 3)$. Since $\frac{\partial M}{\partial u} = 2u = \frac{\partial N}{\partial v}$, the equation is exact.

4.36 Determine whether the differential equation for $u(v)$ defined by $(4v^3u^3 + 1/v)\, dv + (3v^4u^2 - 1/u)\, du = 0$ is exact.

▮ Here $M(v, u) = 4v^3u^3 + \frac{1}{v}$ and $N(v, u) = 3v^4u^2 - \frac{1}{u}$. Since $\frac{\partial M}{\partial u} = 12v^3u^2 = \frac{\partial N}{\partial v}$, the equation is exact.

4.37 Determine whether the differential equation for $v(u)$ defined by $(v^2e^{uv^2} + 4u^3)\, du + (2uve^{uv^2} - 3v^2)\, dv = 0$ is exact.

▮ Here u is the independent variable and v is the dependent variable, so $M(u, v) = v^2e^{uv^2} + 4u^3$ and $N(u, v) = 2uve^{uv^2} - 3v^2$. Since $\frac{\partial M}{\partial v} = 2ve^{uv^2} + 2v^3ue^{uv^2} = \frac{\partial N}{\partial u}$, the equation is exact.

4.38 Determine whether the differential equation for $\rho(\theta)$ defined by $(1 + e^{2\theta})\, d\rho + 2\rho e^{2\theta}\, d\theta = 0$ is exact.

▮ Here $M(\theta, \rho) = 2\rho e^{2\theta}$ and $N(\theta, \rho) = 1 + e^{2\theta}$. Since $\frac{\partial M}{\partial \rho} = 2e^{2\theta} = \frac{\partial N}{\partial \theta}$, the equation is exact.

4.39 Determine whether the differential equation $(t\sqrt{t^2 + y^2} - y)\, dt + (y\sqrt{t^2 + y^2} - t)\, dy = 0$ is exact.

▮ Here $M(t, y) = t\sqrt{t^2 + y^2} - y$ and $N(t, y) = y\sqrt{t^2 + y^2} - t$. Since $\frac{\partial M}{\partial y} = ty(t^2 + y^2)^{-1/2} - 1 = \frac{\partial N}{\partial t}$, the equation is exact.

4.40 Determine whether the differential equation $y' = \frac{2 + ye^{xy}}{2y - xe^{xy}}$ is exact.

▮ Rewriting this equation in differential form, we obtain $(2 + ye^{xy})\,dx + (xe^{xy} - 2y)\,dy = 0$. Here, $M(x, y) = 2 + ye^{xy}$ and $N(x, y) = xe^{xy} - 2y$. Since $\dfrac{\partial M}{\partial y} = e^{xy} + xye^{xy} = \dfrac{\partial N}{\partial x}$, the differential equation is exact.

4.41 Determine whether the differential equation $dy/dx = y/x$ is exact.

▮ In differential form, this equation may be written as $dy - \dfrac{y}{x}\,dx = 0$ or $\dfrac{y}{x}\,dx - dy = 0$, which is not exact. The original differential equation also has the differential form $y\,dx - x\,dy = 0$, but this equation also is not exact. If, however, we write the original differential equation as $-(1/x)\,dx + (1/y)\,dy = 0$, then $M(x, y) = -1/x$ and $N(x, y) = 1/y$, so that $\dfrac{\partial M}{\partial y} = 0 = \dfrac{\partial N}{\partial x}$ and the equation is exact. Thus, a differential equation has many differential forms, some of which may be exact. Exactness is a property of differential equations *in differential form* (see Problem 4.1).

4.42 Determine whether the differential equation $dy/dx = -y/x$ is exact.

▮ This equation has the differential form $\dfrac{y}{x}\,dx + dy = 0$, which is not exact. If, however, we write the original equation as $y\,dx + x\,dy = 0$, then we have $M(x, y) = y$ and $N(x, y) = x$, so that $\dfrac{\partial M}{\partial y} = 1 = \dfrac{\partial N}{\partial x}$ and the equation is exact.

SOLUTIONS OF EXACT EQUATIONS

4.43 Develop a method for solving an exact differential equation.

▮ If the differential equation $M(x, y)\,dx + N(x, y)\,dy = 0$ is exact, then it follows from Problem 4.1 that there exists a function $g(x, y)$ such that

$$dg(x, y) = M(x, y)\,dx + N(x, y)\,dy \tag{1}$$

But also

$$dg(x, y) = \frac{\partial g(x, y)}{\partial x}\,dx + \frac{\partial g(x, y)}{\partial y}\,dy \tag{2}$$

so $g(x, y)$ must satisfy the equations

$$\frac{\partial g(x, y)}{\partial x} = M(x, y) \qquad \text{and} \qquad \frac{\partial g(x, y)}{\partial y} = N(x, y) \tag{3}$$

It follows from (*1*) that the exact differential equation may be written as $dg(x, y) = 0 = 0\,dx$. Integrating this with respect to x and noting that y is itself a function of x, we obtain the solution to the exact differential equation in implicit form as

$$g(x, y) = c \tag{4}$$

where c denotes an arbitrary constant. The function $g(x, y)$ is obtained by solving (*3*).

4.44 Solve $2xy\,dx + (1 + x^2)\,dy = 0$.

▮ This differential equation is exact (see Problem 4.3). We must determine a function $g(x, y)$ that satisfies (*3*) of Problem 4.43. Substituting $M(x, y) = 2xy$ into (*3*) of Problem 4.43, we obtain $\partial g/\partial x = 2xy$. Integrating both sides of this equation with respect to x, we find

$$\int \frac{\partial g}{\partial x}\,dx = \int 2xy\,dx \qquad \text{or} \qquad g(x, y) = x^2y + h(y) \tag{1}$$

Note that when we integrate with respect to x, the constant (*with respect to* x) of integration can depend on y.

We now determine $h(y)$. Differentiating $g(x, y)$ of (*1*) with respect to y, we obtain $\partial g/\partial y = x^2 + h'(y)$. Then, substituting this equation along with $N(x, y) = 1 + x^2$ into (*3*) of Problem 4.43, we obtain

$$x^2 + h'(y) = 1 + x^2 \qquad \text{or} \qquad h'(y) = 1$$

Integrating this last equation with respect to y, we obtain $h(y) = y + c_1$ (c_1 constant). Substituting this expression into (*1*) yields $g(x, y) = x^2y + y + c_1$. Thus, the solution to the differential equation, which is given implicitly by $g(x, y) = c$, is $x^2y + y = c_2$, where $c_2 = c - c_1$. Solving for y explicitly, we obtain the solution as $y = \dfrac{c_2}{x^2 + 1}$.

4.45 Solve $(x + \sin y)\,dx + (x \cos y - 2y)\,dy = 0$.

▌ It follows from Problem 4.4 that this equation is exact. With $M(x, y) = x + \sin y$ and $N(x, y) = x \cos y - 2y$, we seek a function $g(x, y)$ that satisfies (3) of Problem 4.43. Substituting $M(x, y)$ in (3) of Problem 4.43, we obtain $\partial g/\partial x = x + \sin y$. Integrating both sides of this equation with respect to x, we find

$$\int \frac{\partial g}{\partial x}\,dx = \int (x + \sin y)\,dx$$

or

$$g(x, y) = \tfrac{1}{2}x^2 + x \sin y + h(y) \tag{1}$$

To find $h(y)$, we differentiate (1) with respect to y, obtaining $\partial g/\partial y = x \cos y + h'(y)$, and then substitute this result along with $N(x, y) = x \cos y - 2y$ into (3) of Problem 4.43. Thus we find

$$x \cos y + h'(y) = x \cos y - 2y \qquad \text{or} \qquad h'(y) = -2y$$

from which it follows that $h(y) = -y^2 + c_1$. Substituting this $h(y)$ into (1), we obtain $g(x, y) = \frac{1}{2}x^2 + x \sin y - y^2 + c_1$. The solution of the differential equation is then given implicitly by $g(x, y) = c$, or by $\frac{1}{2}x^2 + x \sin y - y^2 = c_2$, where $c_2 = c - c_1$.

4.46 Solve $(2xy + x)\,dx + (x^2 + y)\,dy = 0$.

▌ This equation is exact, with $M(x, y) = 2xy + x$ and $N(x, y) = x^2 + y$ (see Problem 4.7). We require $\partial g/\partial x = M(x, y)$, so $\partial g/\partial x = 2xy + x$. Integrating this with respect to x, we obtain

$$g(x, y) = \int \frac{\partial g}{\partial x}\,dx = \int (2xy + x)\,dx = x^2y + \tfrac{1}{2}x^2 + h(y) \tag{1}$$

To find $h(y)$, we first differentiate (1) with respect to y, obtaining $\partial g/\partial y = x^2 + h'(y)$. Since we require $\partial g/\partial y = N(x, y)$, it follows that $x^2 + h'(y) = x^2 + y$ or $h'(y) = y$. Upon integration, we find that $h(y) = \frac{1}{2}y^2 + c_1$, so (1) becomes $g(x, y) = x^2y + \frac{1}{2}x^2 + \frac{1}{2}y^2 + c_1$. The solution to the differential equation is then $g(x, y) = c$, or $x^2y + \frac{1}{2}x^2 + \frac{1}{2}y^2 = c_2$ where $c_2 = c - c_1$.

4.47 Solve $(y + 2xy^3)\,dx + (1 + 3x^2y^2 + x)\,dy = 0$.

▌ This equation is exact, with $M(x, y) = y + 2xy^3$ and $N(x, y) = 1 + 3x^2y^2 + x$ (see Problem 4.8). We require $\partial g/\partial x = M(x, y)$, so $\partial g/\partial x = y + 2xy^3$. Integrating this with respect to x, we obtain

$$g(x, y) = \int \frac{\partial g}{\partial x}\,dx = \int (y + 2xy^3)\,dx = xy + x^2y^3 + h(y) \tag{1}$$

To find $h(y)$, we first differentiate (1) with respect to y, obtaining $\partial g/\partial y = x + 3x^2y^2 + h'(y)$. Since we require $\partial g/\partial y = N(x, y)$, it follows that

$$x + 3x^2y^2 + h'(y) = 1 + 3x^2y^2 + x \qquad \text{or} \qquad h'(y) = 1$$

Upon integrating this last equation with respect to y, we find $h(y) = y + c_1$, so (1) becomes $g(x, y) = xy + x^2y^3 + y + c_1$. The solution to the differential equation is then $g(x, y) = c$, or $xy + x^2y^3 + y = c_2$ where $c_2 = c - c_1$.

4.48 Solve $ye^{xy}\,dx + xe^{xy}\,dy = 0$.

▌ This equation is exact, with $M(x, y) = ye^{xy}$ and $N(x, y) = xe^{xy}$ (see Problem 4.9). We require $\partial g/\partial x = M(x, y)$, so $\partial g/\partial x = ye^{xy}$. Integrating this with respect to x, we obtain

$$g(x, y) = \int \frac{\partial g}{\partial x}\,dx = \int ye^{xy}\,dx = e^{xy} + h(y) \tag{1}$$

To find $h(y)$, we first differentiate (1) with respect to y, obtaining $\partial g/\partial y = xe^{xy} + h'(y)$. Since we require $\partial g/\partial y = N(x, y)$, it follows that $xe^{xy} + h'(y) = xe^{xy}$, or $h'(y) = 0$. Upon integrating this last equation with respect to y, we find $h(y) = c_1$, a constant, so (1) becomes $g(x, y) = e^{xy} + c_1$. The solution to the differential equation is $g(x, y) = c$, or $e^{xy} = c_2$ where $c_2 = c - c_1$.

4.49 Solve $3x^2y^2\,dx + (2x^3y + 4y^3)\,dy = 0$.

▌ This equation is exact, with $M(x, y) = 3x^2y^2$ and $N(x, y) = 2x^3y + 4y^3$ (see Problem 4.20). We require $\partial g/\partial x = M(x, y)$, so $\partial g/\partial x = 3x^2y^2$. Integrating this with respect to x, we obtain

$$g(x, y) = \int \frac{\partial g}{\partial x}\,dx = \int 3x^2y^2\,dx = x^3y^2 + h(y) \tag{1}$$

To find $h(y)$, we first differentiate (1) with respect to y, obtaining $\partial g/\partial y = 2x^3y + h'(y)$. Since we require $\partial g/\partial y = N(x, y)$, it follows that

$$2x^3y + h'(y) = 2x^3y + 4y^3 \qquad \text{or} \qquad h'(y) = 4y^3$$

Upon integrating this last equation with respect to y, we find $h(y) = y^4 + c_1$, so (1) becomes $g(x, y) = x^3y^2 + y^4 + c_1$. The solution to the differential equation is $g(x, y) = c$, or $x^3y^2 + y^4 = c_2$ where $c_2 = c - c_1$.

4.50 Solve $y\,dx + x\,dy = 0$.

▌ This equation is exact, with $M(x, y) = y$ and $N(x, y) = x$ (see Problem 4.11). We require $\partial g/\partial x = M(x, y)$, so $\partial g/\partial x = y$. Integrating with respect to x, we obtain

$$g(x, y) = \int \frac{\partial g}{\partial x}\,dx = \int y\,dx = xy + h(y) \tag{1}$$

To find $h(y)$, we first differentiate (1) with respect to y, obtaining $\partial g/\partial y = x + h'(y)$. Since we require $\partial g/\partial y = N(x, y)$, it follows that $x + h'(y) = x$ or $h'(y) = 0$. Upon integrating this last equation with respect to y, we find $h(y) = c_1$, so (1) becomes $g(x, y) = xy + c_1$. The solution to the differential equation is $g(x, y) = c$, or $xy = c_2$ where $c_2 = c - c_1$.

4.51 Solve $(y \sin x + xy \cos x)\,dx + (x \sin x + 1)\,dy = 0$.

▌ This equation is exact, with $M(x, y) = y \sin x + xy \cos x$ and $N(x, y) = x \sin x + 1$ (see Problem 4.13). We require $\partial g/\partial x = M(x, y)$, so $\partial g/\partial x = y \sin x + xy \cos x$. Integrating this with respect to x, we obtain

$$g(x, y) = \int \frac{\partial g}{\partial x}\,dx = \int (y \sin x + xy \cos x)\,dx = -y \cos x + (xy \sin x + y \cos x) = xy \sin x + h(y) \tag{1}$$

It follows that $\partial g/\partial y = x \sin x + h'(y)$. We require $\partial g/\partial y = N(x, y)$, so we have

$$x \sin x + h'(y) = x \sin x + 1 \qquad \text{or} \qquad h'(y) = 1$$

Upon integrating this last equation with respect to y, we find $h(y) = y + c_1$, so (1) becomes $g(x, y) = xy \sin x + y + c_1$. The solution to the differential equation, $g(x, y) = c$, may then be written as $xy \sin x + y + c_1 = c$, or $y = c_2 - xy \sin x$ where $c_2 = c - c_1$.

4.52 Solve $(3x^4y^2 - x^2)\,dy + (4x^3y^3 - 2xy)\,dx = 0$.

▌ This equation is exact, with $M(x, y) = 4x^3y^3 - 2xy$ and $N(x, y) = 3x^4y^2 - x^2$ (see Problem 4.16). Then we have

$$g(x, y) = \int M(x, y)\,dx = \int (4x^3y^3 - 2xy)\,dx = x^4y^3 - x^2y + h(y) \tag{1}$$

from which $\partial g/\partial y = 3x^4y^2 - x^2 + h'(y)$. Now since $\partial g/\partial y = N(x, y) = 3x^4y^2 - x^2$, we have $h'(y) = 0$. Then $h(y) = c_1$, and (1) becomes $g(x, y) = x^4y^3 - x^2y + c_1$. The solution to the differential equation is then $x^4y^3 - x^2y = k$, where k is an arbitrary constant.

4.53 Solve $e^{x^3}(3x^2y - x^2)\,dx + e^{x^3}\,dy = 0$.

▌ This equation is exact, with $M(x, y) = 3x^2ye^{x^3} - x^2e^{x^3}$ and $N(x, y) = e^{x^3}$ (see Problem 4.17). Then

$$g(x, y) = \int \frac{\partial g}{\partial x}\,dx = \int M(x, y)\,dx = \int (3x^2ye^{x^3} - x^2e^{x^3})\,dx = ye^{x^3} - \tfrac{1}{3}e^{x^3} + h(y) \tag{1}$$

from which $\partial g/\partial y = e^{x^3} + h'(y)$. Since this last must equal $N(x, y) = e^{x^3}$, we have $h'(y) = 0$, so that $h(y) = c_1$. Then (1) becomes $g(x, y) = ye^{x^3} - \frac{1}{3}e^{x^3} + c_1$, and the solution to the differential equation is $ye^{x^3} - \frac{1}{3}e^{x^3} = k$ or $y = ke^{-x^3} + \frac{1}{3}$, where k is an arbitrary constant.

4.54 Solve $\dfrac{xy-1}{x^2y}\,dx - \dfrac{1}{xy^2}\,dy = 0.$

▮ This equation is exact, with $M(x, y) = x^{-1} - x^{-2}y^{-1}$ and $N(x, y) = -x^{-1}y^{-2}$ (see Problem 4.19). Then

$$g(x, y) = \int \frac{\partial g}{\partial x}\,dx = \int M(x, y)\,dx = \int (x^{-1} - x^{-2}y^{-1})\,dx = \ln|x| + x^{-1}y^{-1} + h(y) \tag{1}$$

from which we may write $\partial g/\partial y = -x^{-1}y^{-2} + h'(y) = N(x, y) = -x^{-1}y^{-2}$. This gives us $h'(y) = 0$, from which $h(y) = c_1$. Then (1) becomes $g(x, y) = \ln|x| + x^{-1}y^{-1} + c_1$, and the solution to the differential equation is $\ln|x| + x^{-1}y^{-1} = k$, where k is arbitrary. This solution may also be written as $\ln|x| + \ln|C| = -x^{-1}y^{-1}$ where $k = -\ln|C|$, from which $\ln|Cx| = -1/xy$ or $y = -1/(x \ln|Cx|)$.

4.55 Solve $(2t^3 + 3y)\,dt + (3t + y - 1)\,dy = 0.$

▮ This equation is exact, with $M(t, y) = 2t^3 + 3y$ and $N(t, y) = 3t + y - 1$ (see Problem 4.21). Then

$$g(t, y) = \int \frac{\partial g}{\partial t}\,dt = \int M(t, y)\,dt = \int (2t^3 + 3y)\,dt = \frac{1}{2}t^4 + 3ty + h(y) \tag{1}$$

from which we may write $\partial g/\partial y = 3t + h'(y) = N(t, y) = 3t + y - 1$. This yields $h'(y) = y + 1$, from which $h(y) = \frac{1}{2}y^2 + y + c_1$. Then (1) becomes $g(t, y) = \frac{1}{2}t^4 + 3ty + \frac{1}{2}y^2 + y + c_1$, and the solution to the differential equation is $\frac{1}{2}t^4 + 3ty + \frac{1}{2}y^2 + y = k$ where k is an arbitrary constant. If we rewrite the solution as $y^2 + (6t + 2)y + (t^4 + C) = 0$ where $C = -2k$, then we can use the quadratic formula to solve for y explicitly, obtaining

$$y = \frac{-(6t + 2) \pm \sqrt{(6t + 2)^2 - 4(t^4 + C)}}{2} = -(3t + 1) \pm \sqrt{9t^2 + 6t - t^4 + K} \qquad K = 1 - C$$

4.56 Solve $(t^2 - y)\,dt - t\,dy = 0.$

▮ This equation is exact, with $M(t, y) = t^2 - y$ and $N(t, y) = -t$ (see Problem 4.22). Then

$$g(t, y) = \int \frac{\partial g}{\partial t}\,dt = \int M(t, y)\,dt = \int (t^2 - y)\,dt = \frac{1}{3}t^3 - ty + h(y) \tag{1}$$

from which we may write $\partial g/\partial y = -t + h'(y) = N(t, y) = -t$. This yields $h'(y) = 0$, from which $h(y) = c_1$. Then (1) becomes $g(t, y) = \frac{1}{3}t^3 - ty + c_1$, and the solution is $\frac{1}{3}t^3 - ty = k$ or, explicitly, $y = \frac{1}{3}t^2 - k/t$, with k arbitrary.

4.57 Solve $(3e^{3t}y - 2t)\,dt + e^{3t}\,dy = 0.$

▮ This equation is exact, with $M(t, y) = 3e^{3t}y - 2t$ and $N(t, y) = e^{3t}$ (see Problem 4.25). Then

$$g(t, y) = \int \frac{\partial g}{\partial t}\,dt = \int M(t, y)\,dt = \int (3e^{3t}y - 2t)\,dt = e^{3t}y - t^2 + h(y) \tag{1}$$

from which we may write $\partial g/\partial y = e^{3t} + h'(y) = N(t, y) = e^{3t}$. This yields $h'(y) = 0$, so that $h(y) = c_1$. Then (1) becomes $g(t, y) = e^{3t}y - t^2 + c_1$, and the solution is $e^{3t}y - t^2 = k$ or $y = (t^2 + k)e^{-3t}$, with k arbitrary.

4.58 Solve $(\cos y + y\cos t)\,dt + (\sin t - t\sin y)\,dy = 0.$

▮ This equation is exact, with $M(t, y) = \cos y + y\cos t$ and $N(t, y) = \sin t - t\sin y$ (see Problem 4.26). Then

$$g(t, y) = \int \frac{\partial g}{\partial t}\,dt = \int M(t, y)\,dt = \int (\cos y + y\cos t)\,dt = t\cos y + y\sin t + h(y) \tag{1}$$

from which we may write $\partial g/\partial y = -t\sin y + \sin t + h'(y)$. Since this must equal $N(t, y) = \sin t - t\sin y$, we have $h'(y) = 0$, so that $h(y) = c_1$. Then (1) becomes $g(t, y) = t\cos y + y\sin t + c_1$, and the solution is, implicitly, $t\cos y + y\sin t = k$.

4.59 Solve $(t\sqrt{t^2 + y^2} - y)\,dt + (y\sqrt{t^2 + y^2} - t)\,dy = 0.$

▮ This equation is exact, with $M(t, y) = t\sqrt{t^2 + y^2} - y$ and $N(t, y) = y\sqrt{t^2 + y^2} - t$ (see Problem 4.39). Then

$$g(t, y) = \int \frac{\partial g}{\partial t}\,dt = \int M(t, y)\,dt = \int [t(t^2 + y^2)^{1/2} - y]\,dt = \frac{1}{3}(t^2 + y^2)^{3/2} - ty + h(y) \tag{1}$$

from which we may write $\partial g/\partial y = y(t^2 + y^2)^{1/2} - t + h'(y) = N(t, y) = y(t^2 + y^2)^{1/2} - t$. This yields $h'(y) = 0$, from which $h(y) = c_1$. Then (1) becomes $g(t, y) = \frac{1}{3}(t^2 + y^2)^{3/2} - ty + c_1$, and the solution is, implicitly, $(t^2 + y^2)^{3/2} - 3ty = k$.

4.60 Solve $(2t + 3x + 4)\,dt + (3t + 4x + 5)\,dx = 0$.

▌ We presume that t is the independent variable and x is the dependent variable, so we are seeking $x(t)$. Then this equation is exact, with $M(t, x) = 2t + 3x + 4$ and $N(t, x) = 3t + 4x + 5$ (see Problem 4.27). Now

$$g(t, x) = \int \frac{\partial g}{\partial t}\,dt = \int M(t, x)\,dt = \int (2t + 3x + 4)\,dt = t^2 + 3tx + 4t + h(x) \tag{1}$$

from which we write $\partial g/\partial x = 3t + h'(x) = N(t, x) = 3t + 4x + 5$. This yields $h'(x) = 4x + 5$, from which $h(x) = 2x^2 + 5x + c_1$. Then (1) becomes $t^2 + 3tx + 4t + 2x^2 + 5x + c_1$, and the solution is, implicitly, $t^2 + 3tx + 4t + 2x^2 + 5x = k$. This solution may be rewritten as $2x^2 + (3t + 5)x + (t^2 + 4t - k) = 0$ and then solved explicitly for x with the quadratic formula, to yield $x = \dfrac{-(3t + 5) \pm \sqrt{(3t + 5)^2 - 8(t^2 + 4t - k)}}{4}$.

4.61 Solve $(6t^5x^3 + 4t^3x^5)\,dt + (3t^6x^2 + 5t^4x^4)\,dx = 0$.

▌ We presume that t is the independent variable, x is the dependent variable, and we want $x(t)$. Then this equation is exact, with $M(t, x) = 6t^5x^3 + 4t^3x^5$ and $N(t, x) = 3t^6x^2 + 5t^4x^4$ (see Problem 4.28). Now

$$g(t, x) = \int \frac{\partial g}{\partial t}\,dt = \int M(t, x)\,dt = \int (6t^5x^3 + 4t^3x^5)\,dt = t^6x^3 + t^4x^5 + h(x) \tag{1}$$

from which we may write $\partial g/\partial x = 3t^6x^2 + 5t^4x^4 + h'(x) = N(t, x) = 3t^6x^2 + 5t^4x^5$. This yields $h'(x) = 0$, from which $h(x) = c_1$. Then (1) becomes $g(t, x) = t^6x^3 + t^4x^5 + c_1$, and the solution is, implicitly, $t^6x^3 + t^4x^5 = k$.

4.62 Solve $2t(xe^{t^2} - 1)\,dt + e^{t^2}\,dx = 0$.

▌ We presume that t is the independent variable, x is the dependent variable, and we want $x(t)$. Then this equation is exact, with $M(t, x) = 2t(xe^{t^2} - 1)$ and $N(t, x) = e^{t^2}$ (see Problem 4.30). Now

$$g(t, x) = \int \frac{\partial g}{\partial t}\,dt = \int M(t, x)\,dt = \int 2t(xe^{t^2} - 1)\,dt = xe^{t^2} - t^2 + h(x) \tag{1}$$

from which we write $\partial g/\partial x = e^{t^2} + h'(x) = N(t, x) = e^{t^2}$. This yields $h'(x) = 0$, from which $h(x) = c_1$. Then (1) becomes $g(t, x) = xe^{t^2} - t^2 + c_1$; the solution is $xe^{t^2} - t^2 = k$ or, explicitly, $x(t) = (t^2 + k)e^{-t^2}$.

4.63 Solve $(t^2 + z^2)\,dt + (2tz - z)\,dz = 0$ for $z(t)$.

▌ This equation is exact, with $M(t, z) = t^2 + z^2$ and $N(t, z) = 2tz - z$ (see Problem 4.33). Then

$$g(t, z) = \int \frac{\partial g}{\partial t}\,dt = \int M(t, z)\,dt = \int (t^2 + z^2)\,dt = \frac{1}{3}t^3 + tz^2 + h(z) \tag{1}$$

from which we have $\partial g/\partial z = 2tz + h'(z) = N(t, z) = 2tz - z$. This yields $h'(z) = -z$, from which $h(z) = -\frac{1}{2}z^2 + c_1$. Then (1) becomes $g(t, z) = \frac{1}{3}t^3 + tz^2 - \frac{1}{2}z^2 + c_1$; the solution is $2t^3 + 6tz^2 - 3z^2 = k$ or, explicitly, $z(t) = \pm\left(\dfrac{k - 2t^3}{6t - 3}\right)^{1/2}$.

4.64 Solve $\left(3 + \dfrac{x^2}{t^2}\right)dt - 2\dfrac{x}{t}\,dx = 0$.

▌ This equation is exact, with $M(t, x) = 3 + \dfrac{x^2}{t^2}$ and $N(t, x) = -2\dfrac{x}{t}$ (see Problem 4.32). Then

$$g(t, x) = \int \frac{\partial g}{\partial t}\,dt = \int M(t, x)\,dt = \int \left(3 + \frac{x^2}{t^2}\right)dt = 3t - \frac{x^2}{t} + h(x) \tag{1}$$

from which we may write $\dfrac{\partial g}{\partial x} = -\dfrac{2x}{t} + h'(x) = N(t, x) = -2\dfrac{x}{t}$. This yields $h'(x) = 0$, so that

$h(x) = c_1$. Then (1) becomes $g(t, x) = 3t - \dfrac{x^2}{t} + c_1$; the solution to the differential equation is $3t - \dfrac{x^2}{t} = k$ or, explicitly, $x(t) = \pm\sqrt{3t^2 - kt}$.

4.65 Solve $(t + z\cos t)\,dt + (\sin t - 3z^2 + 5)\,dz = 0$ for $z(t)$.

▌ This equation is exact, with $M(t, z) = t + z\cos t$ and $N(z, t) = \sin t - 3z^2 + 5$ (see Problem 4.34). Then

$$g(t, z) = \int \frac{\partial g}{\partial t}\,dt = \int M(t, z)\,dt = \int (t + z\cos t)\,dt = \frac{1}{2}t^2 + z\sin t + h(z) \qquad (1)$$

from which we write $\partial g/\partial z = \sin t + h'(z) = N(t, z) = \sin t - 3z^2 + 5$. This yields $h'(z) = -3z^2 + 5$, so that $h(z) = -z^3 + 5z + c_1$. Then (1) becomes $g(t, z) = \frac{1}{2}t^2 + z\sin t - z^3 + 5z + c_1$, and the solution is, implicitly, $\frac{1}{2}t^2 + z\sin t - z^3 + 5z = k$.

4.66 Solve $2(u^2 + uv - 3)\,du + (u^2 + 3v^2 - v)\,dv$ for $u(v)$.

▌ This equation is exact, with $M(v, u) = u^2 + 3v^2 - v$ and $N(v, u) = 2u^2 + 2uv - 6$ (see Problem 4.35). Then

$$g(v, u) = \int \frac{\partial g}{\partial v}\,dv = \int M(v, u)\,dv = \int (u^2 + 3v^2 - v)\,dv = vu^2 + v^3 - \frac{1}{2}v^2 + h(u) \qquad (1)$$

from which we may write $\partial g/\partial u = 2uv + h'(u) = N(v, u) = 2u^2 + 2uv - 6$. This yields $h'(u) = 2u^2 - 6$, from which $h(u) = \frac{2}{3}u^3 - 6u + c_1$. Then (1) becomes $g(v, u) = vu^2 + v^3 - \frac{1}{2}v^2 + \frac{2}{3}u^3 - 6u + c_1$; the solution is, after fractions are cleared, $4u^3 + 6v^3 + 6vu^2 - 3v^2 - 36u = k$.

4.67 Solve $(4v^3u^3 + 1/v)\,dv + (3v^4u^2 - 1/u)\,du = 0$ for $u(v)$.

▌ This equation is exact, with $M(v, u) = 4v^3u^3 + \dfrac{1}{v}$ and $N(v, u) = 3v^4u^2 - \dfrac{1}{u}$ (see Problem 4.36). Then

$$g(v, u) = \int \frac{\partial g}{\partial v}\,dv = \int M(v, u)\,dv = \int \left(4v^3u^3 + \frac{1}{v}\right)dv = v^4u^3 + \ln|v| + h(u) \qquad (1)$$

from which we may write $\partial g/\partial u = 3v^4u^2 + h'(u) = N(v, u) = 3v^4u^2 - 1/u$. This yields $h'(u) = -1/u$, from which $h(u) = -\ln|u| + c_1$. Then (1) becomes $g(v, u) = v^4u^3 + \ln|v| - \ln|u| + c_1$; the solution is, implicitly, $v^4u^3 + \ln|v/u| = k$.

4.68 Solve $(v^2e^{uv^2} + 4u^3)\,du + (2uve^{uv^2} - 3v^2)\,dv = 0$ for $v(u)$.

▌ This equation is exact, with $M(u, v) = v^2e^{uv^2} + 4u^3$ and $N(u, v) = 2uve^{uv^2} - 3v^2$ (see Problem 4.37). Then

$$g(u, v) = \int \frac{\partial g}{\partial u}\,du = \int M(u, v)\,du = \int (v^2e^{uv^2} + 4u^3)\,du = e^{uv^2} + u^4 + h(v) \qquad (1)$$

from which we write $\partial g/\partial v = 2uve^{uv^2} + h'(v) = N(u, v) = 2uve^{uv^2} - 3v^2$. This yields $h'(v) = -3v^2$, from which $h(v) = -v^3 + c_1$. Then (1) becomes $g(u, v) = e^{uv^2} + u^4 - v^3 + c_1$; the solution is, implicitly, $e^{uv^2} + u^4 - v^3 = k$.

4.69 Solve $(1 + e^{2\theta})\,d\rho + 2\rho e^{2\theta}\,d\theta = 0$ for $\rho(\theta)$.

▌ This equation is exact, with $M(\theta, \rho) = 2\rho e^{2\theta}$ and $N(\theta, \rho) = 1 + e^{2\theta}$ (see Problem 4.38). Then

$$g(\theta, \rho) = \int \frac{\partial g}{\partial \theta}\,d\theta = \int M(\theta, \rho)\,d\theta = \int 2\rho e^{2\theta}\,d\theta = \rho e^{2\theta} + h(\rho) \qquad (1)$$

from which we write $\partial g/\partial \rho = e^{2\theta} + h'(\rho) = N(\theta, \rho) = 1 + e^{2\theta}$. This yields $h'(\rho) = 1$, so that $h(\rho) = \rho + c_1$. Then (1) becomes $\rho e^{2\theta} + \rho + c_1$. The solution is $\rho e^{2\theta} + \rho = k$ or, explicitly, $\rho = k/(1 + e^{2\theta})$.

4.70 Solve $y' = \dfrac{2 + ye^{xy}}{2y - xe^{xy}}$.

▎It was shown in Problem 4.40 that this equation is exact in the differential form $(2 + ye^{xy})\,dx + (xe^{xy} - 2y)\,dy = 0$. Here $M(x, y) = 2 + ye^{xy}$ and $N(x, y) = xe^{xy} - 2y$. Since $\partial g/\partial x = M(x, y)$, we have

$$g(x, y) = \int \frac{\partial g}{\partial x}\,dx = \int (2 + ye^{xy})\,dx = 2x + e^{xy} + h(y) \qquad (1)$$

from which we may write $\partial g/\partial y = xe^{xy} + h'(y)$; then equating this to $N(x, y)$ yields $xe^{xy} + h'(y) = xe^{xy} - 2y$, from which $h'(y) = -2y$. It follows that $h(y) = -y^2 + c_1$. Then (1) becomes $g(x, y) = 2x + e^{xy} - y^2 + c_1$, and the solution to the differential equation is given implicitly by $2x + e^{xy} - y^2 = c_2$, where $c_2 = c - c_1$.

4.71 Solve $dy/dx = y/x$.

▎It was shown in Problem 4.41 that in the differential form $-(1/x)\,dx + (1/y)\,dy = 0$, this equation is exact with $M(x, y) = -1/x$ and $N(x, y) = 1/y$. Then

$$g(x, y) = \int \frac{\partial g}{\partial x}\,dx = \int M(x, y)\,dx = \int -\frac{1}{x}\,dx = -\ln|x| + h(y) \qquad (1)$$

from which we may write $\partial g/\partial y = h'(y) = N(x, y) = 1/y$. This yields $h'(y) = 1/y$, so that $h(y) = \ln|y| + c_1$. Then (1) becomes $g(x, y) = -\ln|x| + \ln|y| + c_1$, or $g(x, y) = \ln|y/x| + c_1$. The solution to the differential equation is $\ln|y/x| = k$, or $y = Cx$ where $C = \pm e^k$.

4.72 Solve $dy/dx = -y/x$.

▎It was shown in Problem 4.42 that in the differential form $y\,dx + x\,dy = 0$, this equation is exact with $M(x, y) = y$ and $N(x, y) = x$. Then

$$g(x, y) = \int \frac{\partial g}{\partial x}\,dx = \int M(x, y)\,dx = \int y\,dx = xy + h(y) \qquad (1)$$

from which we may write $\partial g/\partial y = x + h'(y) = N(x, y) = x$. This yields $h'(y) = 0$, so that $h(y) = c_1$. Then (1) becomes $g(x, y) = xy + c_1$, so the solution to the differential equation is $xy = k$ or, explicitly, $y = k/x$.

4.73 Solve $-\dfrac{y}{x^2}\,dx + \dfrac{1}{x}\,dy = 0$.

▎Here $M(x, y) = -\dfrac{y}{x^2}$ and $N(x, y) = \dfrac{1}{x}$, and since $\dfrac{\partial M}{\partial y} = -\dfrac{1}{x^2} = \dfrac{\partial N}{\partial x}$ the differential equation is exact. Then

$$g(x, y) = \int \frac{\partial g}{\partial x}\,dx = \int M(x, y)\,dx = \int \left(\frac{y}{x^2}\right)dx = \frac{y}{x} + h(y) \qquad (1)$$

from which we write $\dfrac{\partial g}{\partial y} = \dfrac{1}{x} + h'(y) = N(x, y) = \dfrac{1}{x}$. This yields $h'(y) = 0$, from which $h(y) = c_1$. Then (1) becomes $g(x, y) = y/x + c_1$, and the solution to the differential equation is $y/x = k$ or $y = kx$.

4.74 Solve $\dfrac{1}{y}\,dx - \dfrac{x}{y^2}\,dy = 0$.

▎Here $M(x, y) = \dfrac{1}{y}$ and $N(x, y) = -\dfrac{x}{y^2}$, so $\dfrac{\partial M}{\partial y} = -\dfrac{1}{y^2} = \dfrac{\partial N}{\partial x}$ and the equation is exact. Then

$$g(x, y) = \int \frac{\partial g}{\partial x}\,dx = \int M(x, y)\,dx = \int \frac{1}{y}\,dx = \frac{x}{y} + h(y) \qquad (1)$$

from which we can write $\dfrac{\partial g}{\partial y} = -\dfrac{x}{y^2} + h'(y) = N(x, y) = -\dfrac{x}{y^2}$. This yields $h'(y) = 0$, so that $h(y) = c_1$. Then (1) becomes $g(x, y) = x/y + c_1$, so the solution to the differential equation is $x/y = k$, or $y = Cx$ where $C = 1/k$.

4.75 Solve $\dfrac{-y}{x^2+y^2}\,dx + \dfrac{x}{x^2+y^2}\,dy = 0.$

▌ Here $M(x, y) = \dfrac{-y}{x^2+y^2}$ and $N(x, y) = \dfrac{x}{x^2+y^2}$, so $\dfrac{\partial M}{\partial y} = \dfrac{y^2 - x^2}{(x^2+y^2)^2} = \dfrac{\partial N}{\partial x}$ and the equation is exact. Then

$$g(x, y) = \int \frac{\partial g}{\partial x}\,dx = \int M(x, y)\,dx = \int \frac{-y}{x^2+y^2}\,dx = \arctan\frac{y}{x} + h'(y) \tag{1}$$

from which we may write $\dfrac{\partial g}{\partial y} = \dfrac{x}{x^2+y^2} + h'(y) = N(x, y) = \dfrac{x}{x^2+y^2}$. This yields $h'(y) = 0$, from which $h(y) = c_1$. Then (*1*) becomes $g(x, y) = \arctan(y/x) + c_1$, and the solution to the differential equation is $\arctan(y/x) = k$. This may be rewritten as $y/x = \tan k$, or as $y = Cx$ where $C = \tan k$.

4.76 Solve $(1/x)\,dx + (1/y)\,dy = 0.$

▌ Here $M(x, y) = 1/x$ and $N(x, y) = 1/y$, so $\dfrac{\partial M}{\partial y} = 0 = \dfrac{\partial N}{\partial x}$ and the equation is exact. Then

$$g(x, y) = \int \frac{\partial g}{\partial x}\,dx = \int M(x, y)\,dx = \int \frac{1}{x}\,dx = \ln|x| + h(y) \tag{1}$$

from which we write $\partial g/\partial y = h'(y) = N(x, y) = 1/y$. This yields $h'(y) = 1/y$, from which $h(y) = \ln|y| + c_1$. Then (*1*) becomes $g(x, y) = \ln|x| + \ln|y| + c_1 = \ln|xy| + c_1$, and the solution to the differential equation is $\ln|xy| = k$, or $y = c/x$ where $c = \pm e^k$.

4.77 Solve $xy^2\,dx + x^2y\,dy = 0.$

▌ Here $M(x, y) = xy^2$ and $N(x, y) = x^2y$, so $\dfrac{\partial M}{\partial y} = 2xy = \dfrac{\partial N}{\partial x}$ and the equation is exact. Then

$$g(x, y) = \int \frac{\partial g}{\partial x}\,dx = \int M(x, y)\,dx = \int xy^2\,dx = \frac{1}{2}x^2y^2 + h(y) \tag{1}$$

from which we may write $\partial g/\partial y = x^2y + h'(y) = N(x, y) = x^2y$. This yields $h'(y) = 0$, so that $h(y) = c_1$. Then (*1*) becomes $g(x, y) = \frac{1}{2}x^2y^2 + c_1$, so the solution to the differential equation is $\frac{1}{2}x^2y^2 = k$ or, explicitly, $y = c/x$ where $c = \pm\sqrt{2k}$.

4.78 Solve $x^{-n}y^{-n+1}\,dx + x^{-n+1}y^{-n}\,dy = 0$, for real numbers $n \neq 1$.

▌ Here $M(x, y) = x^{-n}y^{-n+1}$ and $N(x, y) = x^{-n+1}y^{-n}$, so $\dfrac{\partial M}{\partial y} = (-n+1)x^{-n}y^{-n} = \dfrac{\partial N}{\partial x}$ and the equation is exact. Then

$$g(x, y) = \int \frac{\partial g}{\partial x}\,dx = \int M(x, y)\,dx = \int x^{-n}y^{-n+1}\,dx = \frac{x^{-n+1}y^{-n+1}}{-n+1} + h(y) \tag{1}$$

from which we may write $\partial g/\partial y = x^{-n+1}y^{-n} + h'(y) = N(x, y) = x^{-n+1}y^{-n}$. This yields $h'(y) = 0$, so that $h(y) = c_1$. Then (*1*) becomes $g(x, y) = \dfrac{-(xy)^{-n+1}}{n-1} + c_1$, and the solution to the differential equation is $\dfrac{-(xy)^{-n+1}}{n-1} = k$. This may be rewritten as $(xy)^{n-1} = \dfrac{-1}{k(n-1)}$, or as $y = \dfrac{C}{x}$ where $C = \left[\dfrac{-1}{k(n-1)}\right]^{1/(n-1)}$

4.79 Solve $\dfrac{x}{x^2+y^2}\,dx + \dfrac{y}{x^2+y^2}\,dy = 0.$

▌ Here $M(x, y) = \dfrac{x}{x^2+y^2}$ and $N(x, y) = \dfrac{y}{x^2+y^2}$, so $\dfrac{\partial M}{\partial y} = \dfrac{-2xy}{(x^2+y^2)^2} = \dfrac{\partial N}{\partial x}$ and the equation is exact. Then

$$g(x, y) = \int \frac{\partial g}{\partial x}\,dx = \int M(x, y)\,dx = \int \frac{x}{x^2 + y^2}\,dx = \frac{1}{2}\ln(x^2 + y^2) + h(y) \tag{1}$$

from which we may write $\dfrac{\partial g}{\partial y} = \dfrac{y}{x^2 + y^2} + h'(y) = N(x, y) = \dfrac{y}{x^2 + y^2}$. This yields $h'(y) = 0$, from which $h(y) = c_1$. Then (1) becomes $g(x, y) = \frac{1}{2}\ln(x^2 + y^2) + c_1$; the solution to the differential equation is $\frac{1}{2}\ln(x^2 + y^2) = k$, which we may rewrite as $x^2 + y^2 = e^{2k}$, or explicitly as $y = \pm\sqrt{C - x^2}$ where $C = e^{2k}$.

4.80 Solve $\dfrac{x}{(x^2 + y^2)^n}\,dx + \dfrac{y}{(x^2 + y^2)^n}\,dy = 0$ for real numbers $n \neq 1$.

▌ Here $M(x, y) = \dfrac{x}{(x^2 + y^2)^n}$ and $N(x, y) = \dfrac{y}{(x^2 + y^2)^n}$, so $\dfrac{\partial M}{\partial y} = \dfrac{-2nxy}{(x^2 + y^2)^{n+1}} = \dfrac{\partial N}{\partial x}$ and the equation is exact. Then

$$g(x, y) = \int \frac{\partial g}{\partial x}\,dx = \int M(x, y)\,dx = \int \frac{x}{(x^2 + y^2)^n}\,dx = \frac{-1}{2(n-1)(x^2 + y^2)^{n-1}} + h(y) \tag{1}$$

from which we may write $\dfrac{\partial g}{\partial y} = \dfrac{y}{(x^2 + y^2)^n} + h'(y) = N(x, y) = \dfrac{y}{(x^2 + y^2)^n}$. This yields $h'(y) = 0$, so that $h(y) = c_1$. Then (1) becomes $g(x, y) = \dfrac{-1}{2(n-1)(x^2 + y^2)^{n-1}} + c_1$, and the solution to the differential equation is $\dfrac{-1}{2(n-1)(x^2 + y^2)^{n-1}} = k$. This may be rewritten as $(x^2 + y^2)^{n-1} = \dfrac{-1}{2k(n-1)}$, and then explicitly as $y = \pm\sqrt{c - x^2}$ where $c = \left[\dfrac{-1}{2k(n-1)}\right]^{1/(n-1)}$.

4.81 Solve $ax^{a-1}y^b\,dx + bx^a y^{b-1}\,dy = 0$ for nonzero values of the real constants a and b.

▌ Here $M(x, y) = ax^{a-1}y^b$ and $N(x, y) = bx^a y^{b-1}$, so $\dfrac{\partial M}{\partial y} = abx^{a-1}y^{b-1} = \dfrac{\partial N}{\partial x}$ and the equation is exact. Then

$$g(x, y) = \int \frac{\partial g}{\partial x}\,dx = \int M(x, y)\,dx = \int ax^{a-1}y^b\,dx = x^a y^b + h(y) \tag{1}$$

from which we may write $\partial g/\partial y = bx^a y^{b-1} + h'(y) = N(x, y) = bx^a y^{b-1}$. This yields $h'(y) = 0$, from which $h(y) = c_1$. Then (1) becomes $g(x, y) = x^a y^b + c_1$; the solution to the differential equation is $x^a y^b = k$, or $y = Cx^{-a/b}$ where $C = k^{1/b}$.

INTEGRATING FACTORS

4.82 Define *integrating factor* for a differential equation of the form $M(x, y)\,dx + N(x, y)\,dy = 0$.

▌ A function $I(x, y)$ is an integrating factor for such a differential equation if $I(x, y)[M(x, y)\,dx + N(x, y)\,dy] = 0$ is exact.

4.83 Determine whether $-1/x^2$ is an integrating factor for $y\,dx - x\,dy = 0$.

▌ Multiplying the given differential equation by $-1/x^2$ yields

$$\frac{-1}{x^2}(y\,dx - x\,dy) = 0 \qquad \text{or} \qquad \frac{-y}{x^2}\,dx + \frac{1}{x}\,dy = 0$$

This last equation is exact (see Problem 4.73); hence $-1/x^2$ is an integrating factor for the equation.

4.84 Determine whether $-1/xy$ is an integrating factor for $y\,dx - x\,dy = 0$.

▌ Multiplying the given differential equation by $-1/xy$ yields

$$\frac{-1}{xy}(y\,dx - x\,dy) = 0 \qquad \text{or} \qquad -\frac{1}{x}\,dx + \frac{1}{y}\,dy = 0$$

This last equation is exact; hence $-1/xy$ is an integrating factor for the equation.

4.85 Determine whether $1/y^2$ is an integrating factor for $y\,dx - x\,dy = 0$.

▌ Multiplying the given differential equation by $1/y^2$ yields

$$\frac{1}{y^2}(y\,dx - x\,dy) = 0 \qquad \text{or} \qquad \frac{1}{y}\,dx - \frac{x}{y^2}\,dy = 0$$

This last equation is exact (see Problem 4.74); hence $1/y^2$ is an integrating factor for the equation.

4.86 Determine whether $-1/(x^2 + y^2)$ is an integrating factor for $y\,dx - x\,dy = 0$.

▌ Multiplying the given differential equation by $-1/(x^2 + y^2)$ yields

$$\frac{-1}{x^2 + y^2}(y\,dx - x\,dy) = 0 \qquad \text{or} \qquad \frac{-y}{x^2 + y^2}\,dx + \frac{x}{x^2 + y^2}\,dy = 0$$

This last equation is exact (see Problem 4.75); hence $-1/(x^2 + y^2)$ is an integrating factor for the equation.

4.87 Show that $1/xy$ is an integrating factor for $y\,dx + x\,dy = 0$.

▌ Multiplying the given differential equation by $1/xy$ yields

$$\frac{1}{xy}(y\,dx + x\,dy) = 0 \qquad \text{or} \qquad \frac{1}{x}\,dx + \frac{1}{y}\,dy = 0$$

Since this last equation is exact (see Problem 4.76), $1/xy$ is an integrating factor.

4.88 Show that xy is an integrating factor for $y\,dx + x\,dy = 0$.

▌ Multiplying the given differential equation by xy yields

$$xy(y\,dx + x\,dy) = 0 \qquad \text{or} \qquad xy^2\,dx + x^2y\,dy = 0$$

Since this last equation is exact (see Problem 4.77), xy is an integrating factor.

4.89 Show that $1/(xy)^n$ is an integrating factor for $y\,dx + x\,dy = 0$, for any real number n.

▌ Multiplying the given differential equation by $1/(xy)^n$ yields

$$\frac{1}{(xy)^n}(y\,dx + x\,dy) = 0 \qquad \text{or} \qquad x^{-n}y^{-n+1}\,dx + x^{-n+1}y^{-n}\,dy = 0$$

Since this last equation is exact for all real values of n (see Problems 4.76 and 4.78), $1/(xy)^n$ is an integrating factor.

4.90 Show that $(x^2 + y^2)^{-n}$ is an integrating factor for $x\,dx + y\,dy = 0$, for any real number n.

▌ Multiplying the given differential equation by $(x^2 + y^2)^{-n}$ yields $\dfrac{x}{(x^2 + y^2)^n}\,dx + \dfrac{y}{(x^2 + y^2)^n}\,dy = 0$. Since this last equation is exact (see Problems 4.79 and 4.80), $(x^2 + y^2)^{-n}$ is an integrating factor. Observe that if $n = 0$, then $(x^2 + y^2)^{-0} = 1$ is an integrating factor, which implies that the differential equation is exact in its original form.

4.91 Show that $x^{a-1}y^{b-1}$ is an integrating factor for $ay\,dx + bx\,dy = 0$ for any real-valued constants a and b.

▌ Multiplying the given differential equation by $x^{a-1}y^{b-1}$ yields

$$ax^{a-1}y^b\,dx + bx^ay^{b-1}\,dy = 0 \tag{1}$$

Here $M(x, y) = ax^{a-1}y^b$ and $N(x, y) = bx^ay^{b-1}$. Since $\dfrac{\partial M}{\partial y} = abx^{a-1}y^{b-1} = \dfrac{\partial N}{\partial x}$, (1) is exact; hence $x^{a-1}y^{b-1}$ is an integrating factor for the original differential equation.

4.92 Show that $\dfrac{1}{Mx - Ny}$, for $Mx - Ny$ not identically zero, is an integrating factor for the equation $M\,dx + N\,dy = yf_1(xy)\,dx + xf_2(xy)\,dy = 0$. Investigate the case $Mx - Ny = 0$.

▌ Multiplying the given equation by $\dfrac{1}{Mx - Ny}$ yields $\dfrac{yf_1(xy)}{xy[f_1(xy) - f_2(xy)]}\,dx + \dfrac{xf_2(xy)}{xy[f_1(xy) - f_2(xy)]}\,dy = 0.$ This equation is exact because

$$\frac{\partial}{\partial y}\left[\frac{f_1}{x(f_1 - f_2)}\right] = \frac{x(f_1 - f_2)\dfrac{\partial f_1}{\partial y} - f_1 x\left(\dfrac{\partial f_1}{\partial y} - \dfrac{\partial f_2}{\partial y}\right)}{x^2(f_1 - f_2)^2} = \frac{-f_2\dfrac{\partial f_1}{\partial y} + f_1\dfrac{\partial f_2}{\partial y}}{x(f_1 - f_2)^2}$$

$$\frac{\partial}{\partial x}\left[\frac{f_2}{y(f_1 - f_2)}\right] = \frac{y(f_1 - f_2)\dfrac{\partial f_2}{\partial x} - f_2 y\left(\dfrac{\partial f_1}{\partial x} - \dfrac{\partial f_2}{\partial x}\right)}{y^2(f_1 - f_2)^2} = \frac{f_1\dfrac{\partial f_2}{\partial x} - f_2\dfrac{\partial f_1}{\partial x}}{y(f_1 - f_2)^2}$$

and
$$\frac{\partial}{\partial y}\left[\frac{f_1}{x(f_1 - f_2)}\right] - \frac{\partial}{\partial x}\left[\frac{f_2}{y(f_1 - f_2)}\right] = \frac{f_2\left(-y\dfrac{\partial f_1}{\partial y} + x\dfrac{\partial f_1}{\partial x}\right) + f_1\left(y\dfrac{\partial f_2}{\partial y} - x\dfrac{\partial f_2}{\partial x}\right)}{xy(f_1 - f_2)^2}$$

This last is identically zero because $y\,\partial f(xy)/\partial y = x\,\partial f(xy)/\partial x$.

If $Mx - Ny = 0$, then $M/N = y/x$ and the differential equation reduces to $x\,dy + y\,dx = 0$, with solution $xy = C$.

4.93 If $M\,dx + N\,dy = 0$ has an integrating factor μ which depends only on x, show that $\mu = e^{\int f(x)\,dx}$, where $f(x) = (M_y - N_x)/N$, $M_y = \partial M/\partial y$, and $N_x = \partial N/\partial x$. Write the condition that μ depends only on y.

▌ By hypothesis, $\mu M\,dx + \mu N\,dy = 0$ is exact. Then $\dfrac{\partial(\mu M)}{\partial y} = \dfrac{\partial(\mu N)}{\partial x}$. Since μ depends only on x, this last equation can be written

$$\mu\frac{\partial M}{\partial y} = \mu\frac{\partial N}{\partial x} + N\frac{d\mu}{dx} \qquad \text{or} \qquad N\frac{d\mu}{dx} = \mu\left(\frac{\partial M}{\partial y} - \frac{\partial N}{\partial x}\right)$$

Thus
$$\frac{d\mu}{\mu} = \frac{M_y - N_x}{N}\,dx = f(x)\,dx$$

and integration yields $\ln\mu = \int f(x)\,dx$, so that $\mu = e^{\int f(x)\,dx}$. If we interchange M and N, x and y, then we see that there will be an integrating factor μ depending only on y if $(N_x - M_y)/M = g(y)$ and that in this case $\mu = e^{\int g(y)\,dy}$.

4.94 Develop a table of integrating factors.

▌ From the results of Problems 4.83 through 4.91, we obtain the first two columns of Table 4.1. The last column follows from Problems 4.71 through 4.81, where in each case we have suppressed the c_1 term in $g(x, y)$ for simplicity.

SOLUTION WITH INTEGRATING FACTORS

4.95 Solve $(y^2 - y)\,dx + x\,dy = 0$.

▌ No integrating factor is immediately apparent. Note, however, that if terms are strategically regrouped, the differential equation can be rewritten as

$$-(y\,dx - x\,dy) + y^2\,dx = 0 \qquad (1)$$

The group of terms in parentheses has many integrating factors (see Table 4.1). Trying each integrating factor separately, we find that the only one that makes the entire equation exact is $I(x, y) = 1/y^2$. Using this integrating factor, we can rewrite (*1*) as

$$-\frac{y\,dx - x\,dy}{y^2} + 1\,dx = 0 \qquad (2)$$

Since (*2*) is exact, it can be solved by the method of Problem 4.43. Alternatively, we note from Table 4.1 that (*2*) can be rewritten as $-d(x/y) + 1\,dx = 0$, or as $d(x/y) = 1\,dx$. Integrating, we obtain the solution $\dfrac{x}{y} = x + c$ or $y = \dfrac{x}{x + c}$.

TABLE 4.1

Group of Terms	Integrating Factor $I(x, y)$	Exact Differential
$y\,dx - x\,dy$	$-\dfrac{1}{x^2}$	$\dfrac{x\,dy - y\,dx}{x^2} = d\left(\dfrac{y}{x}\right)$
$y\,dx - x\,dy$	$\dfrac{1}{y^2}$	$\dfrac{y\,dx - x\,dy}{y^2} = d\left(\dfrac{x}{y}\right)$
$y\,dx - x\,dy$	$-\dfrac{1}{xy}$	$\dfrac{x\,dy - y\,dx}{xy} = d\left(\ln\left\|\dfrac{y}{x}\right\|\right)$
$y\,dx - x\,dy$	$-\dfrac{1}{x^2+y^2}$	$\dfrac{x\,dy - y\,dx}{x^2+y^2} = d\left(\arctan\dfrac{y}{x}\right)$
$y\,dx + x\,dy$	$\dfrac{1}{xy}$	$\dfrac{y\,dx + x\,dy}{xy} = d(\ln\|xy\|)$
$y\,dx + x\,dy$	$\dfrac{1}{(xy)^n},\quad n > 1$	$\dfrac{y\,dx + x\,dy}{(xy)^n} = d\left[\dfrac{-1}{(n-1)(xy)^{n-1}}\right]$
$y\,dy + x\,dx$	$\dfrac{1}{x^2+y^2}$	$\dfrac{y\,dy + x\,dx}{x^2+y^2} = d[\frac{1}{2}\ln(x^2+y^2)]$
$y\,dy + x\,dx$	$\dfrac{1}{(x^2+y^2)^n},\quad n > 1$	$\dfrac{y\,dy + x\,dx}{(x^2+y^2)^n} = d\left[\dfrac{-1}{2(n-1)(x^2+y^2)^{n-1}}\right]$
$ay\,dx + bx\,dy$ (a, b constants)	$x^{a-1}y^{b-1}$	$x^{a-1}y^{b-1}(ay\,dx + bx\,dy) = d(x^a y^b)$

4.96 Solve $(y - xy^2)\,dx + (x + x^2y^2)\,dy = 0$.

▌ No integrating factor is immediately apparent. Note, however, that the differential equation can be rewritten as

$$(y\,dx + x\,dy) + (-xy^2\,dx + x^2y^2\,dy) = 0 \qquad (1)$$

The first group of terms has many integrating factors (see Table 4.1). One of these factors, namely $I(x, y) = 1/(xy)^2$, is an integrating factor for the entire equation. Multiplying (*1*) by $1/(xy)^2$ yields $\dfrac{y\,dx + x\,dy}{(xy)^2} + \dfrac{-xy^2\,dx + x^2y^2\,dy}{(xy)^2} = 0$ or, equivalently,

$$\frac{y\,dx + x\,dy}{(xy)^2} = \frac{1}{x}\,dx - 1\,dy \qquad (2)$$

From Table 4.1, $\dfrac{y\,dx + x\,dy}{(xy)^2} = d\left(\dfrac{-1}{xy}\right)$ so that (*2*) can be rewritten as $d\left(\dfrac{-1}{xy}\right) = \dfrac{1}{x}\,dx - 1\,dy$. Integrating both sides of this last equation, we find $\dfrac{-1}{xy} = \ln|x| - y + c$, which is the solution in implicit form.

4.97 Solve $y' = (y + 1)/x$.

▌ Rewriting the equation in differential form, we obtain $(y + 1)\,dx - x\,dy = 0$, or

$$(y\,dx - x\,dy) + 1\,dx = 0 \qquad (1)$$

The first group of terms in (*1*) has many integrating factors (see Table 4.1); one of them, $I(x, y) = -1/x^2$, is an integrating factor for the entire equation. Multiplying (*1*) by $I(x, y)$ yields $\dfrac{y\,dx - x\,dy}{-x^2} + \left(-\dfrac{1}{x^2}\right)dx = 0$, which we write as $d\left(\dfrac{y}{x}\right) + d\left(\dfrac{1}{x}\right) = 0$. Integrating this last equation, we get $\dfrac{y}{x} + \dfrac{1}{x} = c$ or $y = cx - 1$ as the solution to the differential equation.

4.98 Solve $y' = y/(x-1)$.

▌ Rewriting the equation in differential form, we obtain $y\,dx - (x-1)\,dy = 0$, or

$$(y\,dx - x\,dy) + 1\,dy = 0 \tag{1}$$

The first group of terms in (1) has many integrating factors (see Table 4.1); one of them, $I(x, y) = 1/y^2$, is an integrating factor for the entire equation. Multiplying (1) by $I(x, y)$ yields $\dfrac{y\,dx - x\,dy}{y^2} + \dfrac{1}{y^2}\,dy = 0$, which we write as $d\left(\dfrac{x}{y}\right) + d\left(-\dfrac{1}{y}\right) = 0$. Integrating this last equation, we obtain, as the solution, $\dfrac{x}{y} - \dfrac{1}{y} = c$, or $y = k(x-1)$ where $k = 1/c$.

4.99 Solve $y' = (x^2 + y + y^2)/x$.

▌ We rewrite the equation as $(x^2 + y + y^2)\,dx - x\,dy = 0$, or

$$(y\,dx - x\,dy) + (x^2 + y^2)\,dx = 0 \tag{1}$$

The first group of terms in (1) has many integrating factors (see Table 4.1), one of which, $I(x, y) = -1/(x^2 + y^2)$, is also an integrating factor for the entire equation. Multiplying (1) by $I(x, y)$ yields $\dfrac{-y\,dx + x\,dy}{x^2 + y^2} - 1\,dx = 0$, which we write as $d\left(\arctan \dfrac{y}{x}\right) - d(x) = 0$. Integrating this last equation, we obtain as the solution $\arctan(y/x) - x = c$ or, explicitly, $y = x\tan(x + c)$.

4.100 Solve $y' = -y(1 + x^3y^3)/x$.

▌ We rewrite the equation as $y(1 + x^3y^3)\,dx + x\,dy = 0$, or

$$(y\,dx + x\,dy) + x^3y^3\,dx = 0 \tag{1}$$

The first group of terms in (1) has many integrating factors (see Table 4.1), one of which, $I(x, y) = 1/(xy)^3$, is also an integrating factor for the entire equation. Multiplying (1) by $I(x, y)$ yields $\dfrac{y\,dx + x\,dy}{(xy)^3} + 1\,dx = 0$, which we write as $d\left[\dfrac{-1}{2(xy)^2}\right] + d(x) = 0$. Integrating this last equation, we obtain as the solution $\dfrac{-1}{2(xy)^2} + x = c$ or, explicitly, $y = \pm 1/x\sqrt{2(x-c)}$.

4.101 Solve $(y + x^4)\,dx - x\,dy = 0$.

▌ We rewrite the equation as $y\,dx - x\,dy + x^4\,dx = 0$. The combination $y\,dx - x\,dy$ suggests several integrating factors (see Table 4.1), but only $1/x^2$ leads to favorable results, i.e., to

$$\frac{y\,dx - x\,dy}{x^2} + x^2\,dx = 0 \qquad \text{or} \qquad -d\left(\frac{y}{x}\right) + d\left(\frac{x^3}{3}\right) = 0$$

Then integration gives $-\dfrac{y}{x} + \dfrac{x^3}{3} = c$ or, explicitly, $y = \frac{1}{3}x^4 - cx$.

4.102 Solve $(x^3 + xy^2 - y)\,dx + x\,dy = 0$.

▌ We rewrite the equation as $x(x^2 + y^2)\,dx + x\,dy - y\,dx = 0$ and multiply by the integrating factor $I(x, y) = 1/(x^2 + y^2)$ to obtain

$$x\,dx + \frac{x\,dy - y\,dx}{x^2 + y^2} = 0 \qquad \text{or} \qquad d\left(\frac{x^2}{2}\right) + d\left(\tan^{-1}\frac{y}{x}\right) = 0$$

Then integration gives $\dfrac{x^2}{2} + \tan^{-1}\dfrac{y}{x} = c$ or, explicitly, $y = x\tan(c - \frac{1}{2}x^2)$.

4.103 Solve $x\,dy + y\,dx - 3x^3y^2\,dy = 0$.

▌ The terms $x\,dy + y\,dx$ suggest $I(x, y) = 1/(xy)^k$, and the last term requires $k = 3$. Upon multiplication by the integrating factor $1/(xy)^3$, the equation becomes $\dfrac{x\,dy + y\,dx}{x^3y^3} - \dfrac{3}{y}\,dy = 0$, whose primitive is $\dfrac{-1}{2x^2y^2} - 3\ln y = C_1$. Then $6\ln y = \ln C - \dfrac{1}{x^2y^2}$ or $y^6 = Ce^{-1/(x^2y^2)}$.

4.104 Solve $x\,dx + y\,dy + 4y^3(x^2 + y^2)\,dy = 0$.

▌ The last term suggests $I(x, y) = 1/(x^2 + y^2)$ as an integrating factor, and multiplication by $I(x, y)$ yields $\dfrac{x\,dx + y\,dy}{x^2 + y^2} + 4y^3\,dy = 0$, which is exact. Its primitive is $\frac{1}{2}\ln(x^2 + y^2) + y^4 = \ln C_1$ or $(x^2 + y^2)e^{2y^4} = C$.

4.105 Solve $x\,dy - y\,dx - (1 - x^2)\,dx = 0$.

▌ Here $1/x^2$ is the integrating factor, since all other possibilities suggested by $x\,dy - y\,dx$ render the last term inexact. Multiplication by $1/x^2$ yields

$$\frac{x\,dy - y\,dx}{x^2} - \left(\frac{1}{x^2} - 1\right)dx = 0 \qquad \text{or} \qquad d\left(\frac{y}{x}\right) + d\left(\frac{1}{x}\right) + d(x) = 0$$

Integration yields the solution $\dfrac{y}{x} + \dfrac{1}{x} + x = C$ or $y + x^2 + 1 = Cx$.

4.106 Solve $(x + x^4 + 2x^2y^2 + y^4)\,dx + y\,dy = 0$ or $x\,dx + y\,dy + (x^2 + y^2)^2\,dx = 0$.

▌ An integrating factor suggested by the form of the equation is $I(x, y) = \dfrac{1}{(x^2 + y^2)^2}$. Using it, we get $\dfrac{x\,dx + y\,dy}{(x^2 + y^2)^2} + dx = 0$, whose primitive is $-\dfrac{1}{2(x^2 + y^2)} + x = C_1$ or $(C + 2x)(x^2 + y^2) = 1$.

4.107 Solve $y' = -y(1 + x^4y)/x$.

▌ This equation has the differential form $(y\,dx + x\,dy) + x^4y^2\,dx = 0$. An integrating factor for the first group of terms that also renders the entire equation exact is $I(x, y) = 1/(xy)^2$. Using it, we get

$$\frac{y\,dx + x\,dy}{(xy)^2} + x^2\,dx = 0 \qquad \text{or} \qquad d\left(\frac{-1}{xy}\right) + d\left(\frac{1}{3}x^3\right) = 0$$

Integrating yields $\dfrac{-1}{xy} + \dfrac{1}{3}x^3 = c$ or, explicitly, $y = (\frac{1}{3}x^4 - cx)^{-1}$.

4.108 Solve $y' = -y/(y^3 + x^2y - x)$.

▌ This equation has the differential form $(y\,dx - x\,dy) + y(x^2 + y^2)\,dy = 0$. An integrating factor for the first group of terms that also renders the entire equation exact is $I(x, y) = -1/(x^2 + y^2)$. Using it, we get

$$\frac{x\,dy - y\,dx}{x^2 + y^2} - y\,dy = 0 \qquad \text{or} \qquad d\left(\arctan\frac{y}{x}\right) - d\left(\frac{1}{2}y^2\right) = 0$$

Integrating yields the solution $\arctan(y/x) - \frac{1}{2}y^2 = c$.

4.109 Solve $xy^2\,dx + (x^2y^2 + x^2y)\,dy = 0$.

▌ Rearranging gives us $(xy)(y\,dx + x\,dy) + (xy)^2\,dy = 0$, from which we find $(y\,dx + x\,dy) + xy\,dy = 0$. An integrating factor for the first group of terms that also renders the entire equation exact is $I(x, y) = 1/xy$. Using it, we get

$$\frac{x\,dy + y\,dx}{xy} + 1\,dy = 0 \qquad \text{or} \qquad d(\ln|xy|) + d(y) = 0$$

Integrating yields, as the solution in implicit form, $\ln|xy| + y = c$.

4.110 Solve $(x^3y^2 - y)\,dx + (x^2y^4 - x)\,dy = 0$.

▌ Rearranging yields $(x\,dy + y\,dx) - x^3y^2\,dx - x^2y^4\,dy = 0$. An integrating factor for the first group of terms that also renders the entire equation exact is $I(x, y) = 1/(xy)^2$. Using it, we get

$$\frac{x\,dy + y\,dx}{(xy)^2} - x\,dx - y^2\,dy = 0 \qquad \text{or} \qquad d\left(\frac{-1}{xy}\right) - d\left(\frac{1}{2}x^2\right) - d\left(\frac{1}{3}y^3\right) = 0$$

Integrating, we obtain as the solution in implicit form $\dfrac{-1}{xy} - \dfrac{1}{2}x^2 - \dfrac{1}{3}y^3 = c$, or $3x^3y + 2xy^4 + kxy = -6$ where $k = 6c$.

4.111 Solve $y' = \dfrac{3yx^2}{x^3 + 2y^4}$.

▌ Rewriting the equation in differential form, we have $(3yx^2)\,dx + (-x^3 - 2y^4)\,dy = 0$. No integrating factor is immediately apparent, but we can rearrange this equation as $x^2(3y\,dx - x\,dy) - 2y^4\,dy = 0$. The group in parentheses is of the form $ay\,dx + bx\,dy$, where $a = 3$ and $b = -1$, which has an integrating factor x^2y^{-2}. Since the expression in parentheses is already multiplied by x^2, we try $I(x, y) = y^{-2}$. Multiplying by y^{-2} yields

$$x^2y^{-2}(3y\,dx - x\,dy) - 2y^2\,dy = 0$$

which can be simplified (see Table 4.1) to $d(x^3y^{-1}) = 2y^2\,dy$. Integration then yields $x^3y^{-1} = \frac{2}{3}y^3 + c$ as the solution in implicit form.

4.112 Solve $y' = -\frac{1}{3}(y/x)$.

▌ Rewriting the equation in differential form gives us $y\,dx + 3x\,dy = 0$, which is of the form $ay\,dx + bx\,dy = 0$ with $a = 1$ and $b = 3$. An integrating factor is $I(x, y) = x^{1-1}y^{3-1} = y^2$. Multiplying by $I(x, y)$, we get $y^3\,dx + 3xy^2 = 0$ or $d(xy^3) = 0$. Integrating then yields $xy^3 = c$ or, explicitly, $y = (c/x)^{1/3}$.

4.113 Solve $y' = -(2y^4 + 1)/4xy^3$.

▌ Rewriting the equation in differential form gives $4xy^3\,dy + (2y^4 + 1)\,dx = 0$, which we rearrange to $y^3(2y\,dx + 4x\,dy) + 1\,dx = 0$. The terms in parentheses here are of the form $ay\,dx + bx\,dy$ with $a = 2$ and $b = 4$, which suggests the integrating factor $I(x, y) = x^{2-1}y^{4-1} = xy^3$. Since the expression in parentheses is already multiplied by y^3, we try $I(x, y) = x$. Multiplying by x yields

$$(2xy^4\,dx + 4x^2y^3\,dy) + x\,dx = 0 \qquad \text{or} \qquad d(x^2y^4) + d(\tfrac{1}{2}x^2) = 0$$

Integrating, we obtain the solution in implicit form as $x^2y^4 + \frac{1}{2}x^2 = c$ or, explicitly, as $y = \pm(cx^{-2} - \frac{1}{2})^{1/4}$.

4.114 Solve $y' = 2xy - x$.

▌ Rewriting this equation in differential form, we have $(-2xy + x)\,dx + dy = 0$. No integrating factor is immediately apparent. Note, however, that for this equation $M(x, y) = -2xy + x$ and $N(x, y) = 1$, so that

$$\frac{1}{N}\left(\frac{\partial M}{\partial y} - \frac{\partial N}{\partial x}\right) = \frac{(-2x) - (0)}{1} = -2x$$

is a function of x alone. Then from Problem 4.93, we have $I(x) = e^{\int(-2x)\,dx} = e^{-x^2}$ as an integrating factor. Multiplying by e^{-x^2} yields

$$(-2xye^{-x^2} + xe^{-x^2})\,dx + e^{-x^2}\,dy = 0$$

which is exact. To solve this equation, we compute

$$g(x, y) = \int(-2xye^{-x^2} + xe^{-x^2})\,dx = ye^{-x^2} - \tfrac{1}{2}e^{-x^2} + h(y)$$

from which we write $\partial g/\partial y = e^{-x^2} + h'(y) = N(x, y) = e^{-x^2}$. This yields $h'(y) = 0$, from which $h(y) = c_1$. Then $g(x, y) = ye^{-x^2} - \frac{1}{2}e^{-x^2} + c_1$. The solution to the original equation is $ye^{-x^2} - \frac{1}{2}e^{-x^2} = k$ or $y = ke^{x^2} + \frac{1}{2}$.

4.115 Solve $y^2\,dx + xy\,dy = 0$.

▌ Here $M(x, y) = y^2$ and $N(x, y) = xy$; hence,

$$\frac{1}{M}\left(\frac{\partial N}{\partial x} - \frac{\partial M}{\partial y}\right) = \frac{y - 2y}{y^2} = -\frac{1}{y}$$

is a function of y alone. From Problem 4.93, then, $I(y) = e^{-\int (1/y)\,dy} = e^{-\ln y} = 1/y$. Multiplying the given differential equation by $I(y)$ yields the exact equation $y\,dx + x\,dy = 0$, which has the solution $y = c/x$.

An alternative method would be first to divide the given differential equation by xy^2 and then to note that the resulting equation is separable.

4.116 Solve $y' = 1/x^2(1 - 3y)$.

We rewrite this equation in the differential form $(3x^2y - x^2)\,dx + 1\,dy = 0$, with $M(x, y) = 3x^2y - x^2$ and $N(x, y) = 1$. Then $\dfrac{\partial M/\partial y - \partial N/\partial x}{N} = \dfrac{3x^2 - 0}{1} = 3x^2$ is a function only of x, so it follows from Problem 4.93 that an integrating factor is $I(x) = e^{\int 3x^2\,dx} = e^{x^3}$. Multiplying by $I(x)$, we obtain $e^{x^3}(3x^2y - x^2)\,dx + e^{x^3}\,dy = 0$, which is exact. Its solution (see Problem 4.53) is $y = ke^{-x^3} + \frac{1}{3}$.

4.117 Solve $y' = 1/(2xy)$.

In differential form this equation is $1\,dx - 2xy\,dy = 0$, with $M(x, y) = 1$ and $N(x, y) = -2xy$. Then $\dfrac{\partial N/\partial x - \partial M/\partial y}{M} = \dfrac{-2y - 0}{1} = -2y$ is a function only of y, so it follows from Problem 4.93 that an integrating factor is $I(y) = e^{\int -2y\,dy} = e^{-y^2}$. Multiplying by it, we obtain $e^{-y^2}\,dx - 2xye^{-y^2}\,dy = 0$ or $d(xe^{-y^2}) = 0$. Integrating this last equation yields $xe^{-y^2} = c$, or $x = ce^{y^2}$. This may be rewritten as $y^2 = \ln|kx|$, where $k = 1/c$, or as $y = \pm\sqrt{\ln|kx|}$.

4.118 Solve $y' = \dfrac{-3y}{x(2 + y^3)}$.

In differential form, this equation is $3y\,dx + x(2 + y^3)\,dy = 0$, with $M(x, y) = 3y$ and $N(x, y) = x(2 + y^3)$. Then $\dfrac{\partial N/\partial x - \partial M/\partial y}{M} = \dfrac{(2 + y^3) - 3}{3y} = \dfrac{1}{3}y^2 - \dfrac{1}{3y}$ is a function only of y, so it follows from Problem 4.93 that an integrating factor is

$$I(y) = e^{\int (y^2/3 - 1/3y)\,dy} = e^{y^3/9 - (1/3)\ln|y|} = e^{(-1/3)\ln|y|}e^{y^3/9} = e^{\ln|y^{-1/3}|}e^{y^3/9} = y^{-1/3}e^{y^3/9}$$

Multiplying by it, we obtain $3y^{2/3}e^{y^3/9}\,dx + (2xy^{-1/3}e^{y^3/9} + xy^{8/3}e^{y^3/9})\,dy = 0$, or $d(3xy^{2/3}e^{y^3/9}) = 0$. Integrating yields the solution in implicit form as $3xy^{2/3}e^{y^3/9} = c$, or after both sides are raised to the ninth power, $x^9y^6e^{y^3} = k$ where $k = (c/3)^9$.

4.119 Solve $(x^2 + y^2 + x)\,dx + xy\,dy = 0$.

Because $\dfrac{\partial M}{\partial y} = 2y$ and $\dfrac{\partial N}{\partial x} = y$, the equation is not exact. However, $\dfrac{1}{N}\left(\dfrac{\partial M}{\partial y} - \dfrac{\partial N}{\partial x}\right) = \dfrac{2y - y}{xy} = \dfrac{1}{x} = f(x)$, and $e^{\int f(x)\,dx} = e^{\int dx/x} = e^{\ln x} = x$ is an integrating factor. Multiplying by the integrating factor, we obtain

$$(x^3 + xy^2 + x^2)\,dx + x^2y\,dy = 0 \qquad \text{or} \qquad x^3\,dx + x^2\,dx + (xy^2\,dx + x^2y\,dy) = 0$$

Then, noting that $xy^2\,dx + x^2y\,dy = d(\frac{1}{2}x^2y^2)$, we integrate to obtain $\dfrac{x^4}{4} + \dfrac{x^3}{3} + \dfrac{1}{2}x^2y^2 = C_1$ or $3x^4 + 4x^3 + 6x^2y^2 = c$.

4.120 Solve $(2xy^4e^y + 2xy^3 + y)\,dx + (x^2y^4e^y - x^2y^2 - 3x)\,dy = 0$.

Here $\dfrac{\partial M}{\partial y} = 8xy^3e^y + 2xy^4e^y + 6xy^2 + 1$ and $\dfrac{\partial N}{\partial x} = 2xy^4e^y - 2xy^2 - 3$, so the equation is not exact. However, $\dfrac{\partial M}{\partial y} - \dfrac{\partial N}{\partial x} = 8xy^3e^y + 8xy^2 + 4$ and $-\dfrac{1}{M}\left(\dfrac{\partial M}{\partial y} - \dfrac{\partial N}{\partial x}\right) = -\dfrac{4}{y}$. Then $I(y) = e^{-4\int dy/y} = e^{-4\ln y} = 1/y^4$ is an integrating factor; upon multiplication by $I(y)$ the equation takes the form $\left(2xe^y + 2\dfrac{x}{y} + \dfrac{1}{y^3}\right)dx + \left(x^2e^y - \dfrac{x^2}{y^2} - 3\dfrac{x}{y^4}\right)dy = 0$, which is exact.

Now $g(x, y) = \displaystyle\int \left(2xe^y + 2\frac{x}{y} + \frac{1}{y^3}\right)dx = x^2e^y + \frac{x^2}{y} + \frac{x}{y^3} + h(y)$, from which we may write

$\frac{\partial g}{\partial y} = x^2e^y - \frac{x^2}{y^2} - 3\frac{x}{y^4} + h'(y) = x^2e^y - \frac{x^2}{y^2} - 3\frac{x}{y^4}$. This yields $h'(y) = 0$, so that $h(y) = c$ and the primitive is $x^2e^y + \frac{x^2}{y} + \frac{x}{y^3} = C$.

4.121 Solve $(2x^3y^2 + 4x^2y + 2xy^2 + xy^4 + 2y)\,dx + 2(y^3 + x^2y + x)\,dy = 0$.

Here $\frac{\partial M}{\partial y} = 4x^3y + 4x^2 + 4xy + 4xy^3 + 2$ and $\frac{\partial N}{\partial x} = 2(2xy + 1)$; so the equation is not exact. However, $\frac{1}{N}\left(\frac{\partial M}{\partial y} - \frac{\partial N}{\partial x}\right) = 2x$, and an integrating factor is $I(x) = e^{\int 2x\,dx} = e^{x^2}$. When $I(x)$ is introduced, the given equation becomes $(2x^3y^2 + 4x^2y + 2xy^2 + xy^4 + 2y)e^{x^2}\,dx + 2(y^3 + x^2y + x)e^{x^2}\,dy = 0$, which is exact.

Now

$$g(x, y) = \int (2x^3y^2 + 4x^2y + 2xy^2 + xy^4 + 2y)e^{x^2}\,dx = \int (2xy^2 + 2x^3y^2)e^{x^2}\,dx + \int (2y + 4x^2y)e^{x^2}\,dx + \int xy^4e^{x^2}\,dx$$
$$= x^2y^2e^{x^2} + 2xye^{x^2} + \tfrac{1}{2}y^4e^{x^2} + h(y)$$

from which we may write $\partial g/\partial y = 2x^2ye^{x^2} + 2xe^{x^2} + 2y^3e^{x^2} + h'(y) = 2(y^3 + x^2y + x)e^{x^2}$. Thus $h'(y) = 0$ and $h(y) = c$, and the primitive is $(2x^2y^2 + 4xy + y^4)e^{x^2} = C$.

4.122 Solve $y' = \frac{xy^2 - y}{x}$.

Rewriting this equation in differential form yields $y(1 - xy)\,dx + x(1)\,dy = 0$. Based on Problem 4.92, we choose $I(x, y) = \frac{1}{x[y(1 - xy)] - yx} = \frac{-1}{(xy)^2}$ as an integrating factor. Multiplying by $I(x, y)$, we obtain $\frac{xy - 1}{x^2y}\,dx - \frac{1}{xy^2}\,dy = 0$, which is exact. Its solution is $y = -1/(x \ln|Cx|)$ (see Problem 4.54).

4.123 Solve $y(x^2y^2 + 2)\,dx + x(2 - 2x^2y^2)\,dy = 0$.

The given equation is of the form $yf_1(xy)\,dx + xf_2(xy)\,dy = 0$, so $\frac{1}{Mx - Ny} = \frac{1}{3x^3y^3}$ is an integrating factor (see Problem 4.92). When it is introduced, the equation becomes $\frac{x^2y^2 + 2}{3x^3y^2}\,dx + \frac{2 - 2x^2y^2}{3x^2y^3}\,dy = 0$, which is exact.

Now

$$g(x, y) = \int \frac{x^2y^2 + 2}{3x^3y^2}\,dx = \int \left(\frac{1}{3x} + \frac{2}{3x^3y^2}\right) dx = \frac{1}{3}\ln x - \frac{1}{3x^2y^2} + h(y)$$

from which we may write $\frac{\partial g}{\partial y} = \frac{2}{3x^2y^3} + h'(y) = \frac{2 - 2x^2y^2}{3x^2y^3}$. This yields $h'(y) = -2/3y$, and so $h(y) = -\frac{2}{3}\ln y$. The primitive is then $\frac{1}{3}\ln x - \frac{1}{3x^2y^2} - \frac{2}{3}\ln y = \ln C_1$, and $x = Cy^2e^{1/x^2y^2}$.

4.124 Solve $y(2xy + 1)\,dx + x(1 + 2xy - x^3y^3)\,dy = 0$.

The given equation is of the form $yf_1(xy)\,dx + xf_2(xy)\,dy = 0$, so $\frac{1}{Mx - Ny} = \frac{1}{x^4y^4}$ is an integrating factor (see Problem 4.92). When it is introduced, the equation becomes $\left(\frac{2}{x^3y^2} + \frac{1}{x^4y^3}\right) dx + \left(\frac{1}{x^3y^4} + \frac{2}{x^2y^3} - \frac{1}{y}\right) dy = 0$, which is exact.

Now $g(x, y) = \int \left(\frac{2}{x^3y^2} + \frac{1}{x^4y^3}\right) dx = -\frac{1}{x^2y^2} - \frac{1}{3x^3y^3} + h(y)$, from which we may write $\frac{\partial g}{\partial y} = \frac{2}{x^2y^3} + \frac{1}{x^3y^4} + h'(y) = \frac{1}{x^3y^4} + \frac{2}{x^2y^3} - \frac{1}{y}$. This yields $h'(y) = -1/y$, so that $h(y) = -\ln y$. The primitive is then $-\ln y - \frac{1}{x^2y^2} - \frac{1}{3x^3y^3} = C_1$, and $y = Ce^{-(3xy+1)/(3x^3y^3)}$.

4.125 Obtain an integrating factor by inspection for $(2xy^4e^y + 2xy^3 + y)\,dx + (x^2y^4e^y - x^2y^2 - 3x)\,dy = 0$.

▌ When the given equation is written in the form

$$y^4(2xe^y\,dx + x^2e^y\,dy) + 2xy^3\,dx - x^2y^2\,dy + y\,dx - 3x\,dy = 0$$

the leftmost term is the product of y^4 and an exact differential. This suggests $1/y^4$ as a possible integrating factor. To show that it is an integrating factor, we may verify that it produces an exact equation.

4.126 Obtain an integrating factor by inspection for $(x^2y^3 + 2y)\,dx + (2x - 2x^3y^2)\,dy = 0$.

▌ When the given equation is written in the form $2(y\,dx + x\,dy) + x^2y^3\,dx - 2x^3y^2\,dy = 0$, the term $(y\,dx + x\,dy)$ suggests $1/(xy)^k$ as a possible integrating factor. An examination of the remaining terms shows that each will be an exact differential if $k = 3$. Thus, $1/(xy)^3$ is an integrating factor.

4.127 Obtain an integrating factor by inspection for $(2xy^2 + y)\,dx + (x + 2x^2y - x^4y^3)\,dy = 0$.

▌ When the given equation is written in the form $(x\,dy + y\,dx) + 2xy(x\,dy + y\,dx) - x^4y^3\,dy = 0$, the first two terms suggest $1/(xy)^k$ as an integrating factor. The third term will be an exact differential if $k = 4$; thus, $1/(xy)^4$ is an integrating factor.

4.128 Solve $x^2\dfrac{dy}{dx} + xy + \sqrt{1 - x^2y^2} = 0$.

▌ In differential form, this equation is $x(x\,dy + y\,dx) + \sqrt{1 - x^2y^2}\,dx = 0$. The integrating factor $\dfrac{1}{x\sqrt{1 - x^2y^2}}$ reduces it to $\dfrac{x\,dy + y\,dx}{\sqrt{1 - x^2y^2}} + \dfrac{dx}{x} = 0$, whose primitive is $\arcsin xy + \ln x = C$.

4.129 Solve $\dfrac{dy}{dx} = \dfrac{y - xy^2 - x^3}{x + x^2y + y^3}$.

▌ In differential form, this equation is $(x^3 + xy^2 - y)\,dx + (y^3 + x^2y + x)\,dy = 0$. When it is rewritten as $(x^2 + y^2)(x\,dx + y\,dy) + x\,dy - y\,dx = 0$, the terms $x\,dy - y\,dx$ suggest several possible integrating factors. By trial and error we determine that $1/(x^2 + y^2)$ reduces the given equation to $x\,dx + y\,dy + \dfrac{x\,dy - y\,dx}{x^2 + y^2} = x\,dx + y\,dy + \dfrac{(x\,dy - y\,dx)/x^2}{1 + (y/x)^2} = 0$. Its primitive is $\frac{1}{2}x^2 + \frac{1}{2}y^2 + \arctan(y/x) = C_1$ or $x^2 + y^2 + 2\arctan(y/x) = C$.

4.130 Solve $x(4y\,dx + 2x\,dy) + y^3(3y\,dx + 5x\,dy) = 0$.

▌ Suppose that the effect of multiplying the given equation by $x^\alpha y^\beta$ is to produce an equation

$$(4x^{\alpha+1}y^{\beta+1}\,dx + 2x^{\alpha+2}y^\beta\,dy) + (3x^\alpha y^{\beta+4}\,dx + 5x^{\alpha+1}y^{\beta+3}\,dy) = 0 \tag{1}$$

each of whose two parenthesized terms is an exact differential. Then the first term of (1) is proportional to

$$d(x^{\alpha+2}y^{\beta+1}) = (\alpha + 2)x^{\alpha+1}y^{\beta+1}\,dx + (\beta + 1)x^{\alpha+2}y^\beta\,dy \tag{2}$$

That is,
$$\frac{\alpha + 2}{4} = \frac{\beta + 1}{2} \quad \text{or} \quad \alpha - 2\beta = 0 \tag{3}$$

Also, the second term of (1) is proportional to

$$d(x^{\alpha+1}y^{\beta+4}) = (\alpha + 1)x^\alpha y^{\beta+4}\,dx + (\beta + 4)x^{\alpha+1}y^{\beta+3}\,dy \tag{4}$$

That is,
$$\frac{\alpha + 1}{3} = \frac{\beta + 4}{5} \quad \text{or} \quad 5\alpha - 3\beta = 7 \tag{5}$$

Solving (3) and (5) simultaneously, we find $\alpha = 2$ and $\beta = 1$. When these substitutions are made in (1), the equation becomes $(4x^3y^2\,dx + 2x^4y\,dy) + (3x^2y^5\,dx + 5x^3y^4\,dy) = 0$. Its primitive is $x^4y^2 + x^3y^5 = C$.

4.131 Solve $(8y\,dx + 8x\,dy) + x^2y^3(4y\,dx + 5x\,dy) = 0$.

▌ Suppose that the effect of multiplying the given equation by $x^\alpha y^\beta$ is to produce an equation

$$(8x^\alpha y^{\beta+1}\,dx + 8x^{\alpha+1}y^\beta\,dy) + (4x^{\alpha+2}y^{\beta+4}\,dx + 5x^{\alpha+3}y^{\beta+3}\,dy) = 0 \tag{1}$$

each of whose two parenthesized parts is an exact differential. The first part is proportional to

$$d(x^{\alpha+1}y^{\beta+1}) = (\alpha+1)x^{\alpha}y^{\beta+1}\,dx + (\beta+1)x^{\alpha+1}y^{\beta}\,dy \tag{2}$$

That is,

$$\frac{\alpha+1}{8} = \frac{\beta+1}{8} \qquad \text{or} \qquad \alpha-\beta=0 \tag{3}$$

The second part of (1) is proportional to

$$d(x^{\alpha+3}y^{\beta+4}) = (\alpha+3)x^{\alpha+2}y^{\beta+4}\,dx + (\beta+4)x^{\alpha+3}y^{\beta+3}\,dy \tag{4}$$

That is,

$$\frac{\alpha+3}{4} = \frac{\beta+4}{5} \qquad \text{or} \qquad 5\alpha-4\beta=1 \tag{5}$$

Solving (3) and (5) simultaneously, we find $\alpha = 1$ and $\beta = 1$. When these substitutions are made in (1), the equation becomes $(8xy^2\,dx + 8x^2y\,dy) + (4x^3y^5\,dx + 5x^4y^4\,dy) = 0$. Its primitive is $4x^2y^2 + x^4y^5 = C$. [*Note*: In this and the previous problem it was not necessary to write statements (2) and (4) since, after a little practice, (3) and (5) may be obtained directly from (1).]

4.132 Solve $x^3y^3(2y\,dx + x\,dy) - (5y\,dx + 7x\,dy) = 0$.

▌ Multiplying the given equation by $x^{\alpha}y^{\beta}$ yields

$$(2x^{\alpha+3}y^{\beta+4}\,dx + x^{\alpha+4}y^{\beta+3}\,dy) - (5x^{\alpha}y^{\beta+1}\,dx + 7x^{\alpha+1}y^{\beta}\,dy) = 0 \tag{1}$$

If the first parenthesized term of (1) is to be exact, then $\dfrac{\alpha+4}{2} = \dfrac{\beta+4}{1}$ and $\alpha - 2\beta = 4$. If the second part of (1) is to be exact, then $\dfrac{\alpha+1}{5} = \dfrac{\beta+1}{7}$ and $7\alpha - 5\beta = -2$. Solving these two equations simultaneously, we find $\alpha = -8/3$ and $\beta = -10/3$. Then (1) becomes

$$(2x^{1/3}y^{2/3}\,dx + x^{4/3}y^{-1/3}\,dy) - (5x^{-8/3}y^{-7/3}\,dx + 7x^{-5/3}y^{-10/3}\,dy) = 0$$

Each of its two terms is exact, and its primitive is $\frac{3}{2}x^{4/3}y^{2/3} + 3x^{-5/3}y^{-7/3} = C_1$. This may be rewritten as $x^{4/3}y^{2/3} + 2x^{-5/3}y^{-7/3} = C$ or $x^3y^3 + 2 = Cx^{5/3}y^{7/3}$.

4.133 Solve $\dfrac{dy}{dt} = \dfrac{y^2+t^2+t}{-y}$.

▌ This equation has the differential form $(y^2+t^2+t)\,dt + y\,dy = 0$, or $(t\,dt + y\,dy) + (t^2+y^2)\,dt = 0$. Multiplication by the integrating factor $1/(t^2+y^2)$ (see Table 4.1) yields

$$\frac{t\,dt + y\,dy}{t^2+y^2} + 1\,dt = 0 \qquad \text{or} \qquad d[\tfrac{1}{2}\ln(t^2+y^2)] + d(t) = 0$$

Integrating, we obtain the solution $\frac{1}{2}\ln(t^2+y^2) + t = c$, which we may rewrite as $t^2 + y^2 = e^{(2c-2t)} = e^{2c}e^{-2t} = ke^{-2t}$, or as $y = \pm(ke^{-2t} - t^2)^{1/2}$.

4.134 Solve $\dfrac{dy}{dt} = \dfrac{3t-2y}{t}$.

▌ This equation has the differential form $(2y-3t)\,dt + t\,dy = 0$, or $(2y\,dt + t\,dy) - 3t\,dt = 0$. In the latter equation, the terms in parentheses have the form $ay\,dt + bt\,dy$ with $a = 2$ and $b = 1$, which suggests the integrating factor $t^{2-1}y^{1-1} = t$. Multiplying by it, we get

$$(2yt\,dt + t^2\,dy) - 3t^2 = 0 \qquad \text{or} \qquad d(t^2y) - d(t^3) = 0$$

Integrating yields the solution $t^2y - t^3 = c$, or $y = (t^3+c)/t^2$.

4.135 Solve $\dfrac{dy}{dt} = \dfrac{y^2-t}{2yt}$.

▌ This equation has the differential form $2yt\,dy + (t-y^2)\,dt = 0$, or $y(-y\,dt + 2t\,dy) + t\,dt = 0$. In the latter equation, the terms in parentheses have the form $ay\,dt + bt\,dy$ with $a = -1$ and $b = 2$, which suggests the integrating factor $t^{-1-1}y^{2-1} = t^{-2}y$. Since the expression in parentheses is already multiplied by y, we try $I(t, y) = t^{-2}$. Multiplying by it, we get

$$(-y^2t^{-2}\,dt + 2yt^{-1}\,dy) + t^{-1}\,dt = 0 \qquad \text{or} \qquad d(t^{-1}y^2) + d(\ln|t|) = 0$$

Integrating yields $t^{-1}y^2 + \ln|t| = c$, which we may write as $t^{-1}y^2 = -\ln|t| - \ln|k|$ where $c = -\ln|k|$, and then as $y = \pm\sqrt{-t\ln|kt|}$.

4.136 Solve $\dfrac{dx}{dt} = \dfrac{3t^2(t^2+x^2)+t}{x}$.

▮ This equation has the differential form $x\,dt - [3t^2(t^2+x^2)+t]\,dt = 0$, or $(x\,dt - t\,dx) - 3t^2(t^2+x^2)\,dt = 0$. Multiplying the latter equation by $I(t, x) = -1/(t^2+x^2)$, we get

$$\frac{t\,dx - x\,dt}{t^2+x^2} - 3t^2\,dt = 0 \qquad \text{or} \qquad d\left(\arctan\frac{x}{t}\right) - d(t^3) = 0$$

Integrating yields $\arctan(x/t) - t^3 = c$, or $x = t\tan(t^3 + c)$.

4.137 Solve $\dfrac{dx}{dt} = \dfrac{x + \ln t}{t}$.

▮ This equation has the differential form $-t\,dx + (x + \ln t)\,dt = 0$, or $(x\,dt - t\,dx) + \ln t\,dt = 0$. Multiplying the latter equation by $I(t, x) = -1/t^2$ (see Table 4.1), we get

$$\frac{t\,dx - x\,dt}{t^2} - \frac{\ln t}{t^2}\,dt = 0 \qquad \text{or} \qquad d\left(\frac{x}{t}\right) - \frac{\ln t}{t^2}\,dt = 0$$

Integrating yields $\dfrac{x}{t} + \dfrac{1}{t}\ln t + \dfrac{1}{t} = c$, or $x = ct - 1 - \ln t$.

4.138 Solve $\dfrac{dx}{dt} = \dfrac{3t^2+x^2}{2xt}$.

▮ This equation has the differential form $(3t^2+x^2)\,dt - 2xt\,dx = 0$, or $x(x\,dt - 2t\,dx) + 3t^2\,dt = 0$. The terms in parentheses in the latter equation have the form $ax\,dt + bt\,dx$ with $a = 1$ and $b = -2$, which suggests the integrating factor $t^{1-1}x^{-2-1} = x^{-3}$. Since the expression in parentheses is already multiplied by x, we try $I(t, x) = x^{-4}$. Multiplying by it, we get $(x^{-2}\,dt - 2x^{-3}t\,dx) + 3t^2x^{-4}\,dt = 0$, which is *not* exact even though the first two terms can be expressed as $d(x^{-2}t)$.

In the first differential form above, however, we have $M(t, x) = 3t^2 + x^2$ and $N(t, x) = -2xt$, and $\dfrac{\partial M/\partial x - \partial N/\partial t}{N} = \dfrac{2x-(-2x)}{-2xt} = -\dfrac{2}{t}$, a function only of t. It follows from Problem 4.93 (with y replaced by x and x replaced by t) that an integrating factor is $I(t) = e^{\int -(2/t)\,dt} = e^{-2\ln|t|} = e^{\ln t^{-2}} = t^{-2}$. Multiplying the equation in differential form by $I(t)$ yields $\left(3 + \dfrac{x^2}{t^2}\right)dt - 2\dfrac{x}{t}\,dx = 0$, which is exact. Its solution is $x = \pm\sqrt{3t^2 - kt}$ (see Problem 4.64).

4.139 Solve $\dfrac{dx}{dt} = \dfrac{-x}{t(t^2x+1)}$.

▮ This equation has the differential form $x\,dt + t(t^2x+1)\,dx = 0$, or $(x\,dt + t\,dx) + t^3x\,dx = 0$. An integrating factor for the terms in parentheses in the latter equation is $1/(tx)^n$ for any n; taking $n = 3$ gives an integrating factor for the entire equation. Multiplying by it, we get

$$\frac{x\,dt - t\,dx}{(xt)^3} + \frac{1}{x^2}\,dx = 0 \qquad \text{or} \qquad d\left[\frac{-1}{2(xt)^2}\right] + d\left(-\frac{1}{x}\right) = 0$$

Integrating yields $\dfrac{-1}{2x^2t^2} - \dfrac{1}{x} = c$ or $ct^2x^2 + t^2x + \frac{1}{2} = 0$, which may be solved for x explicitly with the quadratic formula.

INITIAL-VALUE PROBLEMS

4.140 Solve $\dfrac{dy}{dx} + \dfrac{2xy}{1+x^2} = 0$; $y(2) = 3$.

▮ The solution to this differential equation in differential form was found in Problem 4.44 to be $y = c_2/(x^2+1)$. Applying the initial condition, we get $3 = c_2/[(2)^2+1]$, from which $c_2 = 15$. Thus, the solution to the initial-value problem is $y = 15/(x^2+1)$.

4.141 Solve the preceding problem if the initial condition is $y(0) = -1$.

▌ The solution to the differential equation remains $y = c_2/(x^2 + 1)$. Applying the new initial condition, we get $-1 = c_2/[(0)^2 + 1]$, from which $c_2 = -1$. The solution to the initial-value problem is now $y = -1/(x^2 + 1)$.

4.142 Solve $\dfrac{dy}{dx} = \dfrac{x + \sin y}{2y - x\cos y}$; $y(2) = \pi$.

▌ The solution to this differential equation in differential form was found in Problem 4.45 to be $\frac{1}{2}x^2 + x\sin y - y^2 = c_2$. Applying the initial condition, we get $c_2 = \frac{1}{2}(2)^2 + 2\sin\pi - \pi^2 = 2 - \pi^2$. The solution to the initial-value problem is $\frac{1}{2}x^2 + x\sin y - y^2 = 2 - \pi^2$.

4.143 Solve $(x\sin x + 1)\,dy/dx + (\sin x + x\cos x)y = 0$; $y(\pi/2) = 3$.

▌ The solution to this differential equation in differential form was found in Problem 4.51 to be $y = c_2 - xy\sin x$. Applying the initial condition, we get $3 = c_2 - (\pi/2)(3)\sin(\pi/2)$, from which $c_2 = 3 + \frac{3}{2}\pi$. The solution to the initial-value problem is $y = 3 + \frac{3}{2}\pi - xy\sin x$.

4.144 Solve $t\dfrac{dy}{dt} + y = t^2$; $y(-1) = 2$.

▌ The solution to this differential equation in differential form was found in Problem 4.56 to be $y = \frac{1}{3}t^2 - k/t$. Applying the initial condition, we get $2 = \frac{1}{3}(-1)^2 - k/(-1)$ or $k = \frac{5}{3}$. The solution to the initial-value problem is $y = \frac{1}{3}(t^2 - 5/t)$.

4.145 Solve $e^{3t}\dfrac{dy}{dt} + 3e^{3t}y = 2t$; $y(2) = 1$.

▌ The solution to this differential equation in differential form was found in Problem 4.57 to be $y = (t^2 + k)e^{-3t}$. Applying the initial condition, we get $1 = (2^2 + k)e^{-3(2)}$, from which $k = e^6 - 4$. The solution to the initial-value problem is $y = (t^2 + e^6 - 4)e^{-3t}$.

4.146 Solve $\dfrac{dy}{dt} + \dfrac{y - t\sqrt{t^2 + y^2}}{t - y\sqrt{t^2 + y^2}} = 0$; $y(4) = 3$.

▌ The solution to this differential equation in differential form was found in Problem 4.59 to be $(t^2 + y^2)^{3/2} - 3ty = k$. Applying the initial condition, we get $(4^2 + 3^2)^{3/2} - 3(4)(3) = k$, from which $k = 125 - 36 = 89$. The solution to the initial-value problem is given implicitly by $(t^2 + y^2)^{3/2} - 3ty = 89$.

4.147 Solve $(3t^6x^2 + 5t^4x^4)\dfrac{dx}{dt} + 6t^5x^3 + 4t^3x^5 = 0$; $x(0) = 0$.

▌ The solution to this differential equation in differential form was found in Problem 4.61 to be $t^6x^3 + t^4x^5 = k$. Applying the initial condition, we get $0^60^3 + 0^40^5 = k$, or $k = 0$. The solution to the initial-value problem is $t^6x^3 + t^4x^5 = 0$, which may be written as $t^4x^3(t^2 + x^2) = 0$. Since t is the independent variable which must take on all values in some interval that includes the initial time, it follows that either $x^3 \equiv 0$ or $t^2 + x^2 \equiv 0$. The last equation is impossible, so $x^3 \equiv 0$ or $x(t) \equiv 0$ is the only solution

4.148 Solve $x\dfrac{dy}{dx} = y - y^2$; $y(-1) = 2$.

▌ The solution to this differential equation in differential form was found in Problem 4.95 to be $y = x/(x + c)$. Applying the initial condition, we get $2 = (-1)/(-1 + c)$, from which $c = \frac{1}{2}$. The solution to the initial-value problem is $y = x/(x + \frac{1}{2})$.

4.149 Solve the preceding problem if the initial condition is $y(0) = 2$.

▌ The solution to the differential equation remains as before. Applying the new initial condition, we get $2 = 0/(0 + c)$, which cannot be solved for c. This initial-value problem has no solution.

4.150 Solve $y' = y/(x - 1)$; $y(0) = -5$.

▌ The solution to this differential equation was found in Problem 4.98 to be $y = k(x - 1)$. Applying the initial condition, we get $-5 = k(0 - 1)$, from which $k = 5$. The solution to the initial-value problem is $y = 5(x - 1)$.

4.151 Solve $y' = (x^2 + y + y^2)/x$; $y(2) = 2$.

▮ The solution to this differential equation was found in Problem 4.99 to be $\arctan(y/x) - x = c$. Applying the initial condition, we get $\arctan(2/2) - 2 = c$ or $c = \frac{1}{4}\pi - 2$. The solution to the initial-value problem is $y = x\tan(x + \frac{1}{4}\pi - 2)$.

4.152 Solve $y' = -y(1 + x^4y)/x$; $y(-1) = -2$.

▮ The solution to this differential equation was found in Problem 4.107 to be $-\dfrac{1}{xy} + \dfrac{1}{3}x^3 = c$. Applying the initial condition, we get $c = \dfrac{-1}{(-1)(-2)} + \dfrac{1}{3}(-1)^3 = -\dfrac{5}{6}$. The solution to the initial-value problem is $y = 1/(\frac{1}{3}x^4 + \frac{5}{6}x)$.

4.153 Solve $3xy' + y = 0$; $y(-1) = 2$.

▮ The solution to this differential equation was found in Problem 4.112 to be $xy^3 = c$. Applying the initial condition, we get $c = (-1)(2)^3 = -8$. The solution to the initial-value problem is $y = (-8/x)^{1/3}$.

4.154 Solve $\dfrac{dy}{dt} = \dfrac{y^2 + t^2 + t}{-y}$; $y(-2) = -1$.

▮ The solution to this differential equation was found in Problem 4.133 to be $t^2 + y^2 = ke^{-2t}$. Applying the initial condition, we get $(-2)^2 + (-1)^2 = ke^{-2(-2)}$, from which $k = 5e^{-4}$. The solution to the initial-value problem is $y = -(5e^{-4}e^{-2t} - t^2)^{1/2} = -(5e^{-2(t+2)} - t^2)^{1/2}$, where the negative square root is taken consistent with the initial condition.

4.155 Solve $\dfrac{dy}{dt} = \dfrac{3t - 2y}{t}$; $y(-5) = 3$.

▮ The solution to this differential equation was found in Problem 4.134 to be $t^2y - t^3 = c$. Applying the initial condition, we get $(-5)^2(3) - (-5)^3 = c$ or $c = 200$. The solution to the initial-value problem is $y = (t^3 + 200)/t^2$.

4.156 Solve $\dfrac{dx}{dt} = \dfrac{x + \ln t}{t}$; $x(1) = 100$.

▮ The solution to this differential equation was found in Problem 4.137 to be $x = ct - 1 - \ln t$. Applying the initial condition, we get $100 = c(1) - 1 - \ln 1$, from which $c = 101$. The solution to the initial-value problem is $x = 101t - 1 - \ln t$.

4.157 Solve $\dfrac{dx}{dt} = \dfrac{3t^2 + x^2}{2xt}$; $x(2) = -4$.

▮ The solution to this differential equation was found in Problem 4.138 to be $x = \pm\sqrt{3t^2 - kt}$. Applying the initial condition after squaring both sides of this last equation, we get $(-4)^2 = 3(2)^2 - k(2)$, from which $k = -2$. The solution to the initial-value problem is $x = -\sqrt{3t^2 + 2t}$, where we have chosen the negative square root consistent with the initial condition.

CHAPTER 5
Linear First-Order Differential Equations

HOMOGENEOUS EQUATIONS

5.1 Show that $I(x, y) = e^{\int p(x)\,dx}$ is an integrating factor for $y' + p(x)y = 0$, where $p(x)$ denotes an integrable function.

▌ Multiplying the differential equation by $I(x, y)$ gives

$$e^{\int p(x)\,dx}y' + p(x)e^{\int p(x)\,dx}y = 0 \tag{1}$$

which is exact. In fact, (1) is equivalent to $\dfrac{d}{dx}(ye^{\int p(x)\,dx}) = 0$.

5.2 Find the general solution to the first-order differential equation $y' + p(x)y = q(x)$ if both $p(x)$ and $q(x)$ are integrable functions of x.

▌ We multiply the differential equation by $I(x, y) = e^{\int p(x)\,dx}$. Then, using the results of the previous problem, we rewrite it as $\dfrac{d}{dx}(e^{\int p(x)\,dx}y) = e^{\int p(x)\,dx}q(x)$. Integrating both sides with respect to x gives us

$$\int \frac{d}{dx}(e^{\int p(x)\,dx}y)\,dx = \int e^{\int p(x)\,dx}q(x)\,dx \qquad \text{or} \qquad e^{\int p(x)\,dx}y + c_1 = \int e^{\int p(x)\,dx}q(x)\,dx \tag{1}$$

Finally, setting $c_1 = -c$ and solving (1) for y, we obtain

$$y = ce^{-\int p(x)\,dx} + e^{-\int p(x)\,dx}\int e^{\int p(x)\,dx}q(x)\,dx \tag{2}$$

5.3 Solve $y' - 5y = 0$.

▌ Here $p(x) = -5$ and $I(x, y) = e^{\int(-5)\,dx} = e^{-5x}$. Multiplying the differential equation by $I(x, y)$, we obtain

$$e^{-5x}y' - 5e^{-5x}y = 0 \qquad \text{or} \qquad \frac{d}{dx}(ye^{-5x}) = 0$$

Integration yields $ye^{-5x} = c$, or $y = ce^{5x}$.

5.4 Solve $y' + 100y = 0$.

▌ Here $p(x) = 100$ and $I(x, y) = e^{\int 100\,dx} = e^{100x}$. Multiplying the differential equation by $I(x, y)$, we obtain

$$e^{100x}y' + 100e^{100x}y = 0 \qquad \text{or} \qquad \frac{d}{dx}(ye^{100x}) = 0$$

Integration yields $ye^{100x} = c$, or $y = ce^{-100x}$.

5.5 Solve $dy/dt = y/2$ for $y(t)$.

▌ We rewrite the equation as $\dfrac{dy}{dt} - \frac{1}{2}y = 0$. Then $p(t) = -\frac{1}{2}$ and $I(t, y) = e^{\int -(1/2)\,dt} = e^{-t/2}$. Multiplying the differential equation by $I(t, y)$, we obtain

$$e^{-t/2}\frac{dy}{dt} - \frac{1}{2}e^{-t/2}y = 0 \qquad \text{or} \qquad \frac{d}{dt}(ye^{-t/2}) = 0$$

Integration yields $ye^{-t/2} = c$, or $y(t) = ce^{t/2}$.

5.6 Solve $\dfrac{dQ}{dt} + \dfrac{1}{20}Q = 0$.

▌ Here $p(t) = 1/20$ and $I(t, Q) = e^{\int (1/20)\,dt} = e^{t/20}$. Multiplying the differential equation by $I(t, Q)$, we obtain

$$e^{t/20}\frac{dQ}{dt} + \frac{1}{20}e^{t/20}Q = 0 \qquad \text{or} \qquad \frac{d}{dt}(Qe^{t/20}) = 0$$

Integrating yields $Qe^{t/20} = c$, or $Q = ce^{-t/20}$.

5.7 Solve $dQ/dt + 0.04Q = 0$.

▌ Here $p(t) = 0.04$ and $I(t, Q) = e^{\int 0.04\,dt} = e^{0.04t}$. Multiplying the differential equation by $I(t, Q)$, we obtain

$$e^{0.04t}\frac{dQ}{dt} + 0.04e^{0.04t}Q = 0 \qquad \text{or} \qquad \frac{d}{dt}(e^{0.04t}Q) = 0$$

Integrating yields $Qe^{0.04t} = c$, or $Q = ce^{-0.04t}$.

5.8 Solve $dN/dt = kN$ for $N(t)$ if k denotes a constant.

▌ We rewrite the equation as $dN/dt - kN = 0$. Then $p(t) = -k$ and $I(t, N) = e^{\int -k\,dt} = e^{-kt}$. Multiplying the differential equation by $I(t, N)$, we obtain

$$e^{-kt}\frac{dN}{dt} - ke^{-kt}N = 0 \qquad \text{or} \qquad \frac{d}{dt}(Ne^{-kt}) = 0$$

Integrating yields $Ne^{-kt} = c$, or $N(t) = ce^{kt}$.

5.9 Solve $y' + 2xy = 0$.

▌ Here $p(x) = 2x$ and $I(x, y) = e^{\int 2x\,dx} = e^{x^2}$. Multiplying the differential equation by $I(x, y)$, we get

$$e^{x^2}y' + 2xe^{x^2}y = 0 \qquad \text{or} \qquad \frac{d}{dx}(ye^{x^2}) = 0$$

Integrating yields $ye^{x^2} = c$, or $y = ce^{-x^2}$.

.10 Solve $y' - 3xy = 0$.

▌ Here $p(x) = -3x$ and $I(x, y) = e^{\int -3x\,dx} = e^{-3x^2/2}$. Multiplying the differential equation by $I(x, y)$, we get

$$e^{-3x^2/2}y' - 3xe^{-3x^2/2}y = 0 \qquad \text{or} \qquad \frac{d}{dx}(ye^{-3x^2/2}) = 0$$

Integrating yields $ye^{-3x^2/2} = c$, or $y = ce^{3x^2/2}$.

.11 Solve $y' - 3x^2y = 0$.

▌ Here $p(x) = -3x^2$ and $I(x, y) = e^{\int -3x^2\,dx} = e^{-x^3}$. Multiplying the differential equation by $I(x, y)$, we get

$$e^{-x^3}y' - 3x^2e^{-x^3} = 0 \qquad \text{or} \qquad \frac{d}{dx}(ye^{-x^3}) = 0$$

Integrating yields $ye^{-x^3} = c$, or $y = ce^{x^3}$.

.12 Solve $dy/dt + t^3y = 0$.

▌ Here $p(t) = t^3$ and $I(t, y) = e^{\int t^3\,dt} = e^{t^4/4}$. Multiplying the differential equation by $I(t, y)$, we find

$$e^{t^4/4}y' + t^3e^{t^4/4}y = 0 \qquad \text{or} \qquad \frac{d}{dt}(ye^{t^4/4}) = 0$$

Integrating yields $ye^{t^4/4} = c$, or $y = ce^{-t^4/4}$.

.13 Solve $dy/dt + (t - 1)y = 0$.

▌ Here $p(t) = t - 1$ and $I(t, y) = e^{\int (t-1)\,dt} = e^{t^2/2 - t}$. Multiplying the differential equation by $I(t, y)$, we find

$$e^{t^2/2-t}\frac{dy}{dt} + (t - 1)e^{t^2/2-t}y = c \qquad \text{or} \qquad \frac{d}{dt}(ye^{t^2/2-t}) = 0$$

Integrating yields $ye^{t^2/2-t} = c$, or $y = ce^{t-t^2/2}$.

5.14 Solve $dy/dt + e^t y = 0$.

▌ Here $p(t) = e^t$ and $I(t, y) = e^{\int e^t dt} = e^{e^t}$. Multiplying the differential equation by $I(t, y)$, we get

$$e^{e^t}\frac{dy}{dt} + e^t e^{e^t} y = 0 \qquad \text{or} \qquad \frac{d}{dt}(ye^{e^t}) = 0$$

Integration yields $ye^{e^t} = c$, or $y = ce^{-e^t}$.

5.15 Solve $dx/d\theta - x \sin\theta = 0$.

▌ Here $p(\theta) = -\sin\theta$ and $I(\theta, x) = e^{\int -\sin\theta\, d\theta} = e^{\cos\theta}$. Multiplying the differential equation by $I(\theta, x)$, we get

$$e^{\cos\theta}\frac{dx}{d\theta} - xe^{\cos\theta}\sin\theta = 0 \qquad \text{or} \qquad \frac{d}{d\theta}(xe^{\cos\theta}) = 0$$

Integration yields $xe^{\cos\theta} = c$, or $x = ce^{-\cos\theta}$.

5.16 Solve $\dfrac{dx}{dt} + \dfrac{1}{t}x = 0$.

▌ Here $p(t) = 1/t$ and $I(t, x) = e^{\int (1/t)dt} = e^{\ln|t|} = |t|$. Multiplying the differential equation by $I(t, x)$, we get $|t|\dfrac{dx}{dt} + \dfrac{|t|}{t}x = 0$. When $t > 0$, $|t| = t$ and this equation becomes $t\dfrac{dx}{dt} + x = 0$. When $t < 0$, $|t| = -t$ and the differential equation becomes $-t\dfrac{dx}{dt} - x = 0$, which reduces to the same equation. Thus $t\dfrac{dx}{dt} + x = 0$ is appropriate for all $t \neq 0$. It may be rewritten in differential form as $\dfrac{d}{dt}(tx) = 0$, and integration yields $tx = c$, or $x = c/t$.

5.17 Solve $\dfrac{dx}{dt} + \dfrac{2}{t}x = 0$.

▌ Here $p(t) = 2/t$ and $I(t, x) = e^{\int (2/t)dt} = e^{2\ln|t|} = e^{\ln t^2} = t^2$. Multiplying the differential equation by $I(t, x)$, we obtain $t^2\dfrac{dx}{dt} + 2tx = 0$ or $\dfrac{d}{dt}(t^2x) = 0$. Integration yields $t^2x = c$, or $x = c/t^2$.

5.18 Solve $\dfrac{dN}{dt} + \dfrac{5}{t}N = 0$.

▌ Here $p(t) = 5/t$ and $I(t, N) = e^{\int (5/t)dt} = e^{5\ln|t|} = e^{\ln|t^5|} = |t^5|$. Multiplying the differential equation by $I(t, N)$, we get $|t^5|\dfrac{dN}{dt} + \dfrac{5|t^5|}{t}N = 0$. Using the logic of Problem 5.16, we can show that this becomes $t^5\dfrac{dN}{dt} + 5t^4N = 0$ for all $t \neq 0$. This last equation may be rewritten in the differential form $\dfrac{d}{dt}(t^5N) = 0$, and integration yields $t^5N = c$, or $N = ct^{-5}$.

5.19 Solve $\dfrac{dN}{dt} - \dfrac{5}{t}N = 0$.

▌ Here $p(t) = -5/t$ and $I(t, N) = e^{\int (-5/t)dt} = e^{-5\ln|t|} = e^{\ln|t^{-5}|} = |t^{-5}|$. Multiplying the differential equation by $I(t, N)$ and simplifying field $t^{-5}\dfrac{dN}{dt} - 5t^{-4}N = 0$ or $\dfrac{d}{dt}(t^{-5}N) = 0$. Integration then gives $t^{-5}N = c$, or $N = ct^5$.

5.20 Solve $\dfrac{dN}{dt} - \dfrac{5}{t^2}N = 0$.

▌ Here $p(t) = -5/t^2$ and $I(t, N) = e^{\int (-5/t^2)dt} = e^{5t^{-1}}$. Multiplying the differential equation by $I(t, N)$ and simplifying, we get

$$e^{5t^{-1}}\frac{dN}{dt} - \frac{5}{t^2}e^{5t^{-1}}N = 0 \qquad \text{or} \qquad \frac{d}{dt}(Ne^{5t^{-1}}) = 0$$

Integration yields $Ne^{5/t} = c$, or $N = ce^{-5/t}$.

5.21 Solve $\dfrac{du}{dx} = \dfrac{2xu}{x^2+2}$.

▌ We rewrite the equation as $\dfrac{du}{dx} - \dfrac{2x}{x^2+2}u = 0$, from which $p(x) = -\dfrac{2x}{x^2+2}$ and

$$I(x, u) = e^{\int[-2x/(x^2+2)]\,dx} = e^{-\ln(x^2+2)} = e^{\ln(x^2+2)^{-1}} = \frac{1}{x^2+2}$$

Multiplying the rewritten equation by $I(x, u)$, we obtain

$$\frac{1}{x^2+2}\frac{du}{dx} - \frac{2x}{(x^2+2)^2}u = 0 \qquad \text{or} \qquad \frac{d}{dx}\left(\frac{u}{x^2+2}\right) = 0$$

Integration yields $\dfrac{u}{x^2+2} = c$, or $u = c(x^2+2)$.

5.22 Solve $\dfrac{du}{dx} = \dfrac{u}{x^2+1}$.

▌ We rewrite the equation as $\dfrac{du}{dx} - \dfrac{1}{x^2+1}u = 0$, from which $p(x) = -1/(x^2+1)$ and $I(x, u) = e^{\int[-1/(x^2+1)]\,du} = e^{-\arctan x}$. Multiplying the rewritten equation by $I(x, u)$, we obtain

$$e^{-\arctan x}\frac{du}{dx} - \frac{1}{x^2+1}e^{-\arctan x}u = 0 \qquad \text{or} \qquad \frac{d}{dx}(ue^{-\arctan x}) = 0$$

Integrating yields $ue^{-\arctan x} = c$, or $u = ce^{\arctan x}$.

5.23 Solve $e^x\,dT/dx + (e^x - 1)T = 0$.

▌ We rewrite this equation as $\dfrac{dT}{dx} + \dfrac{e^x - 1}{e^x}T = 0$, and then as $\dfrac{dT}{dx} + (1 - e^{-x})T = 0$. Now $p(x) = 1 - e^{-x}$ and $I(x, T) = e^{\int(1-e^{-x})\,dx} = e^{x+e^{-x}}$. Multiplying the rewritten equation by $I(x, T)$, we get

$$e^{x+e^{-x}}\frac{dT}{dx} + (1 - e^{-x})e^{x+e^{-x}}T = 0 \qquad \text{or} \qquad \frac{d}{dt}(Te^{x+e^{-x}}) = 0$$

Integrating yields $Te^{x+e^{-x}} = c$, or $T = ce^{-(x+e^{-x})}$.

5.24 Solve $(e^x - 1)\,dT/dx + e^xT = 0$.

▌ We rewrite this equation as $\dfrac{dT}{dx} + \dfrac{e^x}{e^x - 1}T = 0$. Now $p(x) = \dfrac{e^x}{e^x - 1}$ and

$$I(x, T) = e^{\int e^x/(e^x-1)\,dx} = e^{\ln|e^x-1|} = |e^x - 1|.$$

Multiplying the rewritten equation by $I(x, T)$ and simplifying, we get

$$(e^x - 1)\frac{dT}{dx} + e^xT = 0 \qquad \text{or} \qquad \frac{d}{dx}[T(e^x - 1)] = 0$$

(The equation on the left is the differential equation in its original form. Thus, some work could have been saved if the original equation had been recognized as being exact.) Integrating yields $T(e^x - 1) = c$, or $T = c/(e^x - 1)$.

5.25 Solve $(\sin\theta)\,dT/d\theta = T\cos\theta$.

▌ We rewrite this equation as $\dfrac{dT}{d\theta} - \dfrac{\cos\theta}{\sin\theta}T = 0$. Then $p(\theta) = -\cos\theta/\sin\theta$ and $I(\theta, T) = e^{\int -(\cos\theta/\sin\theta)\,d\theta} = e^{-\ln|\sin\theta|} = e^{\ln|\sin^{-1}\theta|} = \left|\dfrac{1}{\sin\theta}\right|$. Multiplying the rewritten equation by $I(\theta, T)$

and simplifying, we get

$$\frac{1}{\sin\theta}\frac{dT}{d\theta} - \frac{\cos\theta}{\sin^2\theta}T = 0 \quad \text{or} \quad \frac{d}{d\theta}\left(\frac{T}{\sin\theta}\right) = 0$$

Integrating yields $T/\sin\theta = c$ or $T = c\sin\theta$.

5.26 Solve $dT/d\theta + T\sec\theta = 0$.

▮ Here $p(\theta) = \sec\theta$ and $I(\theta, T) = e^{\int \sec\theta\, d\theta} = e^{\ln|\sec\theta + \tan\theta|} = |\sec\theta + \tan\theta|$. Multiplying the differential equation by $I(\theta, T)$ and simplifying, we get

$$(\sec\theta + \tan\theta)\frac{dT}{d\theta} + (\sec^2\theta + \sec\theta\tan\theta)T = 0 \quad \text{or} \quad \frac{d}{dt}[T(\sec\theta + \tan\theta)] = 0$$

Integrating yields $T(\sec\theta + \tan\theta) = c$ or $T = c/(\sec\theta + \tan\theta)$.

5.27 Solve $dz/dt + z\ln t = 0$.

▮ Here $p(t) = \ln t$ and (via integration by parts) $I(t, z) = e^{\int \ln t\, dt} = e^{t\ln t - t}$. Multiplying the differential equation by $I(t, z)$, we obtain

$$e^{t\ln t - t}\frac{dz}{dt} + ze^{t\ln t - t}\ln t = 0 \quad \text{or} \quad \frac{d}{dt}(ze^{t\ln t - t}) = 0$$

Integrating yields $ze^{t\ln t - t} = c$, or $z = ce^{t - t\ln t}$.

5.28 Solve $\dfrac{dz}{dx} + \dfrac{2z}{x^2 + x} = 0$.

▮ Here $p(x) = 2/(x^2 + x)$ and (via partial fractions)

$$I(x, z) = e^{\int 2/(x^2+x)\,dx} = e^{\int [2/x - 2/(x+1)]\,dx} = e^{2\ln|x| - 2\ln|x+1|} = e^{\ln x^2 - \ln(x+1)^2} = e^{\ln[x^2/(x+1)^2]} = \frac{x^2}{(x+1)^2}$$

Multiplying the differential equation by $I(x, z)$ and simplifying, we obtain

$$\frac{x^2}{(x+1)^2}\frac{dz}{dx} + \frac{2x}{(x+1)^3}z = 0 \quad \text{or} \quad \frac{d}{dx}\left[\frac{zx^2}{(x+1)^2}\right] = 0$$

Integrating yields $\dfrac{zx^2}{(x+1)^2} = c$, or $z = c(x+1)^2/x^2 = c(1 + 1/x)^2$.

NONHOMOGENEOUS EQUATIONS

5.29 Solve $y' - 3y = 6$.

▮ Here $p(x) = -3$. Then $\int p(x)\,dx = \int -3\,dx = -3x$, from which $I(x, y) = e^{-3x}$. Multiplying the differential equation by $I(x, y)$, we obtain

$$e^{-3x}y' - 3e^{-3x}y = 6e^{-3x} \quad \text{or} \quad \frac{d}{dx}(ye^{-3x}) = 6e^{-3x}$$

Integrating both sides of this last equation with respect to x yields $ye^{-3x} = \int 6e^{-3x}\,dx = -2e^{-3x} + c$, or $y = ce^{3x} - 2$.

5.30 Solve $y' + 6y = 3$.

▮ Here $p(x) = 6$ and $I(x, y) = e^{\int 6\,dx} = e^{6x}$. Multiplying the differential equation by $I(x, y)$, we get

$$e^{6x}y' + 6e^{6x}y = 3e^{6x} \quad \text{or} \quad \frac{d}{dx}(ye^{6x}) = 3e^{6x}$$

Integration yields $ye^{6x} = \frac{1}{2}e^{6x} + c$, or $y = \frac{1}{2} + ce^{-6x}$.

5.31 Solve $dI/dt + 50I = 5$.

▌ Here $p(t) = 50$ and $I(t, I) = e^{\int 50\,dt} = e^{50t}$. Multiplying the differential equation by this integrating factor, we get

$$e^{50t}\frac{dI}{dt} + 50e^{50t}I = 5e^{50t} \qquad \text{or} \qquad \frac{d}{dt}(Ie^{50t}) = 5e^{50t}$$

Integrating yields $Ie^{50t} = \frac{1}{10}e^{50t} + c$, or $I = \frac{1}{10} + ce^{-50t}$.

5.32 Solve $dq/dt + 10q = 20$.

▌ Here $p(t) = 10$ and $I(t, q) = e^{\int 10\,dt} = e^{10t}$. Multiplying the differential equation by $I(t, q)$, we get

$$e^{10t}\frac{dq}{dt} + 10e^{10t}q = 20e^{10t} \qquad \text{or} \qquad \frac{d}{dt}(qe^{10t}) = 20e^{10t}$$

Integrating yields $qe^{10t} = 2e^{10t} + c$, or $q = 2 + ce^{-10t}$.

5.33 Solve $dI/dt + \frac{20}{3}I = 6$.

▌ Here $p(t) = \frac{20}{3}$ and $I(t, I) = e^{\int (20/3)\,dt} = e^{(20/3)t}$. Multiplying the differential equation by this integrating factor, we get

$$e^{(20/3)t}\frac{dI}{dt} + e^{(20/3)t}\frac{20}{3}I = 6e^{(20/3)t} \qquad \text{or} \qquad \frac{d}{dt}(Ie^{(20/3)t}) = 6e^{(20/3)t}$$

Integrating yields $Ie^{(20/3)t} = \frac{9}{10}e^{(20/3)t} + c$, or $I = \frac{9}{10} + ce^{-(20/3)t}$.

5.34 Solve $\dot{q} + 10q = \frac{1}{2}$.

▌ Here $p(t) = 10$ and $I(t, I) = e^{\int 10\,dt} = e^{10t}$. Multiplying the differential equation by $I(t, I)$, we get

$$e^{10t}\dot{q} + 10e^{10t}q = \tfrac{1}{2}e^{10t} \qquad \text{or} \qquad \frac{d}{dt}(qe^{10t}) = \tfrac{1}{2}e^{10t}$$

Integrating yields $qe^{10t} = \frac{1}{20}e^{10t} + c$, or $q = \frac{1}{20} + ce^{-10t}$.

5.35 Solve $dv/dt + \frac{1}{4}v = 32$.

▌ Here $p(t) = \frac{1}{4}$ and $I(t, v) = e^{\int (1/4)\,dt} = e^{t/4}$. Multiplying the differential equation by $I(t, v)$, we get

$$e^{t/4}\frac{dv}{dt} + \frac{1}{4}e^{t/4}v = 32e^{t/4} \qquad \text{or} \qquad \frac{d}{dt}(ve^{t/4}) = 32e^{t/4}$$

Integrating yields $ve^{t/4} = 128e^{t/4} + c$, or $v = 128 + ce^{-t/4}$.

5.36 Solve $dv/dt + 25v = 9.8$.

▌ Here $p(t) = 25$ and $I(t, v) = e^{\int 25\,dt} = e^{25t}$. Multiplying the differential equation by $I(t, v)$, we get

$$e^{25t}\frac{dv}{dt} + 25e^{25t}v = 9.8e^{25t} \qquad \text{or} \qquad \frac{d}{dt}(ve^{25t}) = 9.8e^{25t}$$

Integrating yields $ve^{25t} = 0.392e^{25t} + c$, or $v = 0.392 + ce^{-25t}$.

5.37 Solve $\dfrac{dv}{dt} + \dfrac{k}{m}v = \pm g$ for k, m, and g constant.

▌ Here $p(t) = k/m$ and $I(t, v) = e^{\int (k/m)\,dt} = e^{kt/m}$. Multiplying the differential equation by $I(t, v)$, we get

$$e^{kt/m}\frac{dv}{dt} + \frac{k}{m}e^{kt/m}v = \pm ge^{kt/m} \qquad \text{or} \qquad \frac{d}{dt}(ve^{kt/m}) = \pm ge^{kt/m}$$

Integrating yields $ve^{kt/m} = \pm\dfrac{mg}{k}e^{kt/m} + c$, or $v = ce^{-kt/m} \pm \dfrac{mg}{k}$.

5.38 Solve $\dot{T} + kT = 100k$ for k constant.

▌ Here $p(t) = k$ and $I(t, T) = e^{\int k\,dt} = e^{kt}$. Multiplying the differential equation by $I(t, T)$, we get

$$Te^{kt} + ke^{kt}T = 100ke^{kt} \quad \text{or} \quad d(Te^{kt}) = 100ke^{kt}$$

Integrating yields $Te^{kt} = \int 100ke^{kt}\,dt = 100e^{kt} + c$, so $T = 100 + ce^{-kt}$.

5.39 Solve $\dot{T} + kT = ak$ for a and k constant.

▌ Here $p(t) = k$, so $I(t, T) = e^{\int k\,dt} = e^{kt}$. Multiplying the differential equation by $I(t, T)$, we get

$$Te^{kt} + ke^{kt}T = ake^{kt} \quad \text{or} \quad d(Te^{kt}) = ake^{kt}$$

Integrating yields $Te^{kt} = \int ake^{kt}\,dt = ae^{kt} + c$; therefore, $T = a + ce^{-kt}$.

5.40 Solve $dv/dt = -\frac{1}{2}$.

▌ Here $p(t) = 0$, so $I(t, v) = e^{\int 0\,dt} = e^0 = 1$, which indicates that the differential equation can be integrated directly. Doing so, we obtain $v = -\frac{1}{2}t + c$.

5.41 Solve $dv/dt = g$ for g constant.

▌ Here $p(t) = 0$ and $I(t, v) = 1$, as in the previous problem, so we may integrate the differential equation directly with respect to time. Doing so, we obtain $v = gt + c$.

5.42 Solve $y' - 2xy = x$.

▌ Here, $p(x) = -2x$ and $I(x, y) = e^{\int p(x)\,dx} = e^{-x^2}$. Multiplying the differential equation by $I(x, y)$, we obtain

$$e^{-x^2}y' - 2xe^{-x^2}y = xe^{-x^2} \quad \text{or} \quad \frac{d}{dx}(ye^{-x^2}) = xe^{-x^2}$$

Integrating yields $ye^{-x^2} = \int xe^{-x^2}\,dx = -\frac{1}{2}e^{-x^2} + c$, or $y = ce^{x^2} - \frac{1}{2}$.

5.43 Solve $dy/dx + 2xy = 4x$.

▌ Here $p(x) = 2x$ and $\int p(x)\,dx = \int 2x\,dx = x^2$, so $I(x, y) = e^{x^2}$ is an integrating factor. Multiplication and integration then yield $ye^{x^2} = \int 4xe^{x^2}\,dx = 2e^{x^2} + c$, or $y = 2 + ce^{-x^2}$.

5.44 Solve $y' + y = \sin x$.

▌ Here $p(x) = 1$; hence $I(x, y) = e^{\int 1\,dx} = e^x$. Multiplying the differential equation by $I(x, y)$, we obtain

$$e^xy' + e^xy = e^x \sin x \quad \text{or} \quad \frac{d}{dx}(ye^x) = e^x \sin x$$

To integrate the right side, we use integration by parts twice, and the result of integration is $ye^x = \frac{1}{2}e^x(\sin x - \cos x) + c$, or $y = ce^{-x} + \frac{1}{2}\sin x - \frac{1}{2}\cos x$.

5.45 Solve $y' + (4/x)y = x^4$.

▌ Here $p(x) = 4/x$; hence $I(x, y) = e^{\int p(x)\,dx} = e^{\ln x^4} = x^4$. Multiplying the differential equation by $I(x, y)$, we find

$$x^4y' + 4x^3y = x^8 \quad \text{or} \quad \frac{d}{dx}(yx^4) = x^8$$

Integrating with respect to x yields $yx^4 = \frac{1}{9}x^9 + c$, or $y = c/x^4 + \frac{1}{9}x^5$.

5.46 Solve $x\,dy/dx - 2y = x^3 \cos 4x$.

▌ We write the equation as $\dfrac{dy}{dx} - \dfrac{2}{x}y = x^2 \cos 4x$. Then $p(x) = -2/x$ and an integrating factor is $e^{\int(-2/x)\,dx} = e^{-2\ln x} = e^{\ln x^{-2}} = x^{-2}$. Multiplying by x^{-2}, we have

$$x^{-2}\frac{dy}{dx} - 2x^{-3}y = \cos 4x \quad \text{or} \quad \frac{d}{dx}(x^{-2}y) = \cos 4x$$

Then by integrating we find $x^{-2}y = \frac{1}{4}\sin 4x + c$, or $y = \frac{1}{4}x^2 \sin 4x + cx^2$.

5.47 Solve $x\dfrac{dy}{dx} = y + x^3 + 3x^2 - 2x$.

▮ We rewrite the equation as $\dfrac{dy}{dx} - \dfrac{1}{x}y = x^2 + 3x - 2$. Then $p(x) = -1/x$ and $\int p(x)\,dx = -\int \dfrac{dx}{x} = -\ln|x|$ so $e^{-\ln|x|} = \left|\dfrac{1}{x}\right|$ is an integrating factor. Then we have

$$y\frac{1}{x} = \int \frac{1}{x}(x^2 + 3x - 2)\,dx = \int \left(x + 3 - \frac{2}{x}\right)dx = \frac{1}{2}x^2 + 3x - 2\ln x + c_1$$

or $2y = x^3 + 6x^2 - 4x\ln x + cx$.

5.48 Solve $\dfrac{dQ}{dt} + \dfrac{3}{100 - t}Q = 2$.

▮ Here $p(t) = 3/(100 - t)$ and

$$I(t, Q) = e^{\int 3/(100-t)\,dt} = e^{-3\ln|100-t|} = e^{\ln|(100-t)^{-3}|} = |(100 - t)^{-3}|$$

Multiplying the differential equation by $I(t, Q)$, we get $|(100 - t)^{-3}|\dfrac{dQ}{dt} + \dfrac{3|(100 - t)^{-3}|}{100 - t}Q = |(100 - t)^{-3}|2$. By reasoning similar to that in Problem 5.16, we can show that this reduces to $(100 - t)^{-3}\dfrac{dQ}{dt} + 3(100 - t)^{-4}Q = 2(100 - t)^{-3}$ for all $t \neq 100$. This last equation may be written as $\dfrac{d}{dt}[(100 - t)^{-3}Q] = 2(100 - t)^{-3}$. Integrating then yields $(100 - t)^{-3}Q = (100 - t)^{-2} + c$, or $Q = 100 - t + c(100 - t)^3$.

5.49 Solve $\dfrac{dQ}{dt} + \dfrac{2}{10 + 2t}Q = 4$.

▮ Here $p(t) = \dfrac{2}{10 + 2t}$ and $I(t, Q) = e^{\int 2/(10+2t)\,dt} = e^{\ln|10+2t|} = |10 + 2t|$. Multiplying the differential equation by $I(t, Q)$ and simplifying, we get

$$(10 + 2t)\frac{dQ}{dt} + 2Q = 4(10 + 2t) \qquad \text{or} \qquad \frac{d}{dt}[(10 + 2t)Q] = 40 + 8t$$

Integrating yields $(10 + 2t)Q = 40t + 4t^2 + c$, or $Q = \dfrac{40t + 4t^2 + c}{10 + 2t}$.

5.50 Solve $\dfrac{dQ}{dt} + \dfrac{2}{20 - t}Q = 4$.

▮ Here $p(t) = \dfrac{2}{20 - t}$ and $I(t, Q) = e^{\int 2/(20-t)\,dt} = e^{-2\ln|20-t|} = e^{\ln(20-t)^{-2}} = (20 - t)^{-2}$. Multiplying the differential equation by $I(t, Q)$, we get

$$(20 - t)^{-2}\frac{dQ}{dt} + 2(20 - t)^{-3}Q = 4(20 - t)^{-2} \qquad \text{or} \qquad \frac{d}{dt}[(20 - t)^{-2}Q] = 4(20 - t)^{-2}$$

Integrating yields $(20 - t)^{-2}Q = 4(20 - t)^{-1} + c$, or $Q = 4(20 - t) + c(20 - t)^2$.

5.51 Solve $dI/dt + 20I = 6\sin 2t$.

▮ Here $p(t) = 20$ and $I(t, I) = e^{\int 20\,dt} = e^{20t}$. Multiplying the differential equation by this integrating factor, we get

$$e^{20t}\frac{dI}{dt} + 20e^{20t}I = 6e^{20t}\sin 2t \qquad \text{or} \qquad \frac{d}{dt}(Ie^{20t}) = 6e^{20t}\sin 2t$$

Integrating (and noting that the right side requires integration by parts twice), we obtain $Ie^{20t} = (\frac{30}{101}\sin 2t - \frac{3}{101}\cos 2t)e^{20t} + c$, or $I = \frac{30}{101}\sin 2t - \frac{3}{101}\cos 2t + ce^{-20t}$.

5.52 Solve $dq/dt + q = 4\cos 2t$.

▌ Here $p(t) = 1$ and $I(t, q) = e^{\int 1\,dt} = e^t$. Multiplying the differential equation by $I(t, q)$, we get

$$e^t \frac{dq}{dt} + e^t q = 4e^t \cos 2t \qquad \text{or} \qquad \frac{d}{dt}(qe^t) = 4e^t \cos 2t$$

Integrating both sides of this equation (with two integrations by parts required for the right side), we obtain $qe^t = \frac{8}{5}e^t \sin 2t + \frac{4}{5}e^t \cos 2t + c$, or $q = \frac{8}{5}\sin 2t + \frac{4}{5}\cos 2t + ce^{-t}$.

5.53 Solve $\dfrac{dI}{dt} + 5I = \dfrac{110}{3}\sin 120\pi t$.

▌ Here $p(t) = 5$ and $I(t, I) = e^{\int 5\,dt} = e^{5t}$. Multiplying the differential equation by this integrating factor, we get

$$e^{5t}\frac{dI}{dt} + 5e^{5t}I = \frac{110}{3}e^{5t}\sin 120\pi t \qquad \text{or} \qquad \frac{d}{dt}(e^{5t}I) = \frac{110}{3}e^{5t}\sin 120\pi t$$

Then
$$Ie^{5t} = \frac{110}{3}\int e^{5t}\sin 120\pi t\,dt = \frac{110}{3}e^{5t}\frac{5\sin 120\pi t - 120\pi\cos 120\pi t}{25 + 14{,}400\pi^2} + c$$

or
$$I = \frac{22}{3}\frac{\sin 120\pi t - 24\pi\cos 120\pi t}{1 + 576\pi^2} + ce^{-5t}$$

5.54 Solve $\dot{q} + 100q = 10\sin 120\pi t$.

▌ Here $p(t) = 100$ and $I(t, q) = e^{\int 100\,dt} = e^{100t}$. Multiplying the differential equation by $I(t, q)$, we get

$$qe^{100t} + 100e^{100t}q = 10e^{100t}\sin 120\pi t \qquad \text{or} \qquad \frac{d}{dt}(qe^{100t}) = 10e^{100t}\sin 120\pi t$$

Then
$$qe^{100t} = 10\int e^{100t}\sin 120\pi t\,dt = 10e^{100t}\frac{100\sin 120\pi t - 120\pi\cos 120\pi t}{10{,}000 + 14{,}400\pi^2} + A$$
$$= e^{100t}\frac{10\sin 120\pi t - 12\pi\cos 120\pi t}{100 + 144\pi^2} + A$$

or
$$q = \frac{10\sin 120\pi t - 12\pi\cos 120\pi t}{100 + 144\pi^2} + Ae^{-100t}$$

5.55 Solve $dQ/dt + 0.04Q = 3.2e^{-0.04t}$.

▌ Here $p(t) = 0.04$ and $I(t, Q) = e^{\int 0.04\,dt} = e^{0.04t}$. Multiplying the differential equation by $I(t, Q)$, we get $e^{0.04t}\dfrac{dQ}{dt} + 0.04e^{0.04t}Q = 3.2$, or $\dfrac{d}{dt}(Qe^{0.04t}) = 3.2$. Integrating yields $Qe^{0.04t} = 3.2t + c$, or $Q = 3.2te^{-0.04t} + ce^{-0.04t}$.

5.56 Solve $dv/dx - xv = -x$.

▌ Here $p(x) = -x$ and $I(x, v) = e^{\int -x\,dx} = e^{-x^2/2}$. Multiplying the differential equation by $I(x, v)$, we get

$$e^{-x^2/2}\frac{dv}{dx} - xe^{-x^2/2}v = -xe^{-x^2/2} \qquad \text{or} \qquad \frac{d}{dx}(ve^{-x^2/2}) = -xe^{-x^2/2}$$

Integrating yields $ve^{-x^2/2} = e^{-x^2/2} + c$, or $v = 1 + ce^{x^2/2}$.

5.57 Solve $\dfrac{dv}{dx} - \dfrac{2}{x}v = \dfrac{2}{3}x^4$.

▌ Here $p(x) = -2/x$ and $I(x, v) = e^{\int(-2/x)dx} = e^{-2\ln|x|} = e^{\ln(1/x^2)} = 1/x^2$. Multiplying the differential equation by $I(x, v)$, we get $\dfrac{1}{x^2}\dfrac{dv}{dx} - \dfrac{2}{x^3}v = \dfrac{2}{3}x^2$ or $\dfrac{d}{dx}\left(\dfrac{v}{x^2}\right) = \dfrac{2}{3}x^2$. Integrating then yields $\dfrac{v}{x^2} = \dfrac{2}{9}x^3 + c$, or $v = \dfrac{2}{9}x^5 + cx^2$.

5.58 Solve $v' + \frac{1}{2}xv = 3x$.

▌ The integrating factor here is $I(x, v) = e^{\int (x/2)\,dx} = e^{x^2/4}$. Multiplying the differential equation by it, we obtain

$$e^{x^2/4}v' + \tfrac{1}{2}xe^{x^2/4}v = 3xe^{x^2/4} \qquad \text{or} \qquad \frac{d}{dx}(ve^{x^2/4}) = 3xe^{x^2/4}$$

Integrating yields $ve^{x^2/4} = 6e^{x^2/4} + c$, or $v = 6 + ce^{-x^2/4}$.

5.59 Solve $v' - \dfrac{2}{x}v = -2$.

▌ The integrating factor here is $I(x, v) = 1/x^2$, the same as in Problem 5.57. Multiplying by it, we obtain $\dfrac{1}{x^2}v' - \dfrac{2}{x^3}v = \dfrac{-2}{x^2}$ or $\dfrac{d}{dx}\left(\dfrac{v}{x^2}\right) = \dfrac{-2}{x^2}$. Integrating then yields $\dfrac{v}{x^2} = \dfrac{2}{x} + c$, or $v = 2x + cx^2$.

5.60 Solve $v' - \dfrac{5}{x}v = -5x^2$.

▌ The integrating factor here is $I(x, v) = e^{\int (-5/x)\,dx} = e^{-5\ln|x|} = e^{\ln|x^{-5}|} = |x^{-5}|$. Multiplying the differential equation by it and simplifying, we obtain $x^{-5}v' - 5x^{-6}v = -5x^{-3}$ or $\dfrac{d}{dx}(vx^{-5}) = -5x^{-3}$. Then $vx^{-5} = \int -5x^{-3}\,dx = \frac{5}{2}x^{-2} + c$, and $v = \frac{5}{2}x^3 + cx^5$.

5.61 Solve $v' - v = -e^x$.

▌ The integrating factor here is $I(x, v) = e^{\int -1\,dx} = e^{-x}$. Multiplying the differential equation by it, we get $v'e^{-x} - ve^{-x} = -1$ or $\dfrac{d}{dx}(ve^{-x}) = -1$. Then $ve^{-x} = \int (-1)\,dx = -x + c$, and $v = (c - x)e^x$.

5.62 Solve $\dfrac{dv}{dt} + \dfrac{1}{t}v = \cos t$.

▌ The integrating factor here is $I(t, x) = e^{\int (1/t)\,dt} = e^{\ln|t|} = |t|$. Multiplying the differential equation by it and simplifying, we get $tv' + v = t\cos t$ or $\dfrac{d}{dt}(tv) = t\cos t$. Then $tv = \int t\cos t\,dt = t\sin t + \cos t + c$, and $v = \sin t + \dfrac{1}{t}\cos t + \dfrac{c}{t}$.

5.63 Solve $\dfrac{dv}{dt} + \dfrac{3}{2t}v = 6t$.

▌ The integrating factor here is $I(t, x) = e^{\int (3/2t)\,dt} = e^{(3/2)\ln|t|} = e^{\ln|t^{3/2}|} = |t^{3/2}|$. Multiplying the differential equation by it and simplifying, we get

$$t^{3/2}\frac{dv}{dt} + \frac{3}{2}t^{1/2}v = 6t^{5/2} \qquad \text{or} \qquad \frac{d}{dt}(vt^{3/2}) = 6t^{5/2}$$

Then $vt^{3/2} = \int 6t^{5/2}\,dt = \dfrac{12}{7}t^{7/2} + c$, and $v = \dfrac{12}{7}t^2 + ct^{-3/2}$.

5.64 Solve $\dfrac{dv}{dt} + \dfrac{2}{t}v = 4$.

▌ The integrating factor here is $I(t, v) = e^{\int (2/t)\,dt} = e^{2\ln|t|} = e^{\ln t^2} = t^2$. Multiplying the differential equation by it, we obtain $t^2\dfrac{dv}{dt} + 2tv = 4t^2$ or $\dfrac{d}{dt}(vt^2) = 4t^2$. Then $vt^2 = \int 4t^2\,dt = \frac{4}{3}t^3 + c$, and $v = \frac{4}{3}t + ct^{-2}$.

5.65 Solve $(x - 2)\,dy/dx = y + 2(x - 2)^3$.

▌ We rewrite the equation as $\dfrac{dy}{dx} - \dfrac{1}{x - 2}y = 2(x - 2)^2$. Then $\int p(x)\,dx = -\int \dfrac{dx}{x - 2} = -\ln|x - 2|$, and an integrating factor is $e^{-\ln|x-2|} = \left|\dfrac{1}{x-2}\right|$. Multiplication by it and integration yield

$$y\frac{1}{x-2} = 2\int (x-2)^2 \frac{1}{x-2}\,dx = 2\int (x-2)\,dx = (x-2)^2 + c \qquad \text{or} \qquad y = (x-2)^3 + c(x-2)$$

5.66 Solve $dy/dx + y\cot x = 5e^{\cos x}$.

▮ An integrating factor is $e^{\int \cot x\,dx} = e^{\ln|\sin x|} = |\sin x|$, and multiplication by it and integration yield $y \sin x = 5\int e^{\cos x} \sin x\,dx = -5e^{\cos x} + c$. Therefore, $y = \dfrac{-5e^{\cos x} + c}{\sin x}$.

5.67 Solve $x^3\,dy/dx + (2 - 3x^2)y = x^3$.

▮ We rewrite this equation as $\dfrac{dy}{dx} + \dfrac{2-3x^2}{x^3}y = 1$. Then we have $\displaystyle\int \frac{2-3x^2}{x^3}\,dx = -\frac{1}{x^2} - 3\ln x$, and an integrating factor is $\dfrac{1}{x^3 e^{1/x^2}}$. Multiplication by it and integration yield

$$\frac{y}{x^3 e^{1/x^2}} = \int \frac{dx}{x^3 e^{1/x^2}} = \frac{1}{2e^{1/x^2}} + c_1 \qquad \text{or} \qquad 2y = x^3 + cx^3 e^{1/x^2}$$

5.68 Solve $dy/dx - 2y\cot 2x = 1 - 2x\cot 2x - 2\csc 2x$.

▮ An integrating factor is $e^{-\int 2\cot 2x\,dx} = e^{-\ln|\sin 2x|} = |\csc 2x|$. Then

$$y\csc 2x = \int (\csc 2x - 2x\cot 2x \csc 2x - 2\csc^2 2x)\,dx = x\csc 2x + \cot 2x + c$$

or $y = x + \cos 2x + c\sin 2x$.

5.69 Solve $y\ln y\,dx + (x - \ln y)\,dy = 0$.

▮ With x taken as the dependent variable, this equation may be put in the form $\dfrac{dx}{dy} + \dfrac{1}{y\ln y}x = \dfrac{1}{y}$. Then $e^{\int dy/(y\ln y)} = e^{\ln(\ln y)} = \ln y$ is an integrating factor. Multiplication by it and integration yield $x\ln y = \displaystyle\int \ln y\frac{dy}{y} = \frac{1}{2}\ln^2 y + K$, and the solution is $2x\ln y = \ln^2 y + c$.

5.70 Solve $dv/dx + 2v\cos x = \sin^2 x\cos x$.

▮ Here $e^{2\int \cos x\,dx} = e^{2\sin x}$ is an integrating factor. Then multiplication by it and integration yield

$$ve^{2\sin x} = \int e^{2\sin x}\sin^2 x\cos x\,dx = \tfrac{1}{2}e^{2\sin x}\sin^2 x - \tfrac{1}{2}e^{2\sin x}\sin x + \tfrac{1}{4}e^{2\sin x} + c$$

or $v = \frac{1}{2}\sin^2 x - \frac{1}{2}\sin x + \frac{1}{4} + ce^{-2\sin x}$.

5.71 Solve $dv/dx + v = 4\sin x$.

▮ The integrating factor here is $I(x, v) = e^{\int 1\,dx} = e^x$. Then multiplication by it and integration give

$$ve^x = 4\int e^x \sin x\,dx = 2e^x(\sin x - \cos x) + c \qquad \text{or} \qquad v = 2(\sin x - \cos x) + ce^{-x}$$

5.72 Solve $dv/dx - v = -x$.

▮ Using the integrating factor e^{-x}, we obtain

$$ve^{-x} = \int -xe^{-x}\,dx = xe^{-x} + e^{-x} + c \qquad \text{or} \qquad v = x + 1 + ce^x$$

5.73 Solve $\dfrac{dv}{dx} - \dfrac{2}{x}v = \dfrac{-1}{x^2}$.

▮ Using the integrating factor $I(x, v) = e^{\int(-2/x)\,dx} = e^{-2\ln|x|} = e^{\ln x^{-2}} = 1/x^2$, we obtain

$$\frac{v}{x^2} = \int(-x^{-4})\,dx = \frac{x^{-3}}{3} + c \qquad \text{or} \qquad v = \frac{1}{3}x^{-1} + cx^2$$

BERNOULLI EQUATIONS

5.74 Develop a method for obtaining nontrivial solutions to the Bernoulli equation, $\frac{dy}{dx} + p(x)y = q(x)y^n$, for $n \neq 0, 1$.

Observe that the trivial solution $y \equiv 0$ is always a solution. To find others, set $v = y^{-n+1}$. Then $y = v^{1/(-n+1)}$ and $\frac{dy}{dx} = \frac{1}{-n+1} v^{n/(-n+1)} \frac{dv}{dx}$. Substituting these relationships into the Bernoulli equation yields

$$\frac{1}{-n+1} v^{n/(-n+1)} \frac{dv}{dx} + p(x)v^{1/(-n+1)} = q(x)v^{n/(-n+1)} \qquad \text{or} \qquad \frac{dv}{dx} - (n-1)p(x)v = -(n-1)q(x)$$

This last equation is linear and may be solved by the method of Problem 5.2.

5.75 Solve $y' + xy = xy^2$.

This is a Bernoulli equation with $p(x) = q(x) = x$ and $n = 2$. Setting $v = y^{-2+1} = y^{-1} = 1/y$, we have $y = 1/v$ and $\frac{dy}{dx} = \frac{-1}{v^2}\frac{dv}{dx}$. The original differential equation then becomes

$$\frac{-1}{v^2}\frac{dv}{dx} + x\frac{1}{v} = x\frac{1}{v^2} \qquad \text{or} \qquad \frac{dv}{dx} - xv = -x$$

This equation is linear, and its solution is $v = 1 + ce^{x^2/2}$ (see Problem 5.56). Since $y = 1/v$, we have $y = \frac{1}{1 + ce^{x^2/2}}$ as a set of nontrivial solutions to the original differential equation.

5.76 Solve $\frac{dy}{dx} - \frac{3}{x}y = x^4 y^{1/3}$.

This is a Bernoulli equation with $p(x) = -3/x$, $q(x) = x^4$, and $n = 1/3$. Setting $v = y^{-1/3+1} = y^{2/3}$, we have $y = v^{3/2}$ and $\frac{dy}{dx} = \frac{3}{2}v^{1/2}\frac{dv}{dx}$. The original differential equation thus becomes

$$\frac{3}{2}v^{1/2}\frac{dv}{dx} - \frac{3}{x}v^{3/2} = x^4 v^{1/2} \qquad \text{or} \qquad \frac{dv}{dx} - \frac{2}{x}v = \frac{2}{3}x^4$$

This last equation is linear, and its solution is $v = \frac{2}{9}x^5 + cx^2$ (see Problem 5.57). Thus, for the original equation, $y^{2/3} = \frac{2}{9}x^5 + cx^2$ or, explicitly, $y = \pm(\frac{2}{9}x^5 + cx^2)^{3/2}$.

5.77 Solve $\frac{dy}{dx} - y = xy^5$.

This is a Bernoulli equation with $p(x) = -1$, $q(x) = x$, $n = 5$. The transformation $y^{-4} = v$, $y^{-5}\frac{dy}{dx} = -\frac{1}{4}\frac{dv}{dx}$ reduces it to $\frac{dv}{dx} + 4v = -4x$, for which an integrating factor is $e^{4\int dx} = e^{4x}$. Then $ve^{4x} = -4\int xe^{4x}\,dx = -xe^{4x} + \frac{1}{4}e^{4x} + c$, so that, for the original equation,

$$y^{-4}e^{4x} = -xe^{4x} + \frac{1}{4}e^{4x} + c \qquad \text{or} \qquad \frac{1}{y^4} = -x + \frac{1}{4} + ce^{-4x}$$

5.78 Solve $xy' + y = xy^3$.

This is a Bernoulli equation with $n = 3$. Setting $v = y^{-3+1} = y^{-2}$, we have $y = v^{-1/2}$ and $y' = -\frac{1}{2}v^{-3/2}v'$. The original differential equation then becomes $-\frac{1}{2}v^{-3/2}v' + \frac{1}{x}v^{-1/2} = v^{-3/2}$, or $v' - \frac{2}{x}v = -2$. This last equation is linear, and its solution is $v = 2x + cx^2$ (see Problem 5.59). Then for the original equation, $y^{-2} = 2x + cx^2$ or $y = \pm\sqrt{\frac{1}{2x + cx^2}}$.

5.79 Solve $y' + xy = 6x\sqrt{y}$.

▮ This is a Bernoulli equation with $n = \frac{1}{2}$. Setting $v = y^{-1/2+1} = y^{1/2}$, we have $y = v^2$ and $y' = 2vv'$. The original differential equation then becomes $2vv' + xv^2 = 6xv$, or $v' + \frac{1}{2}xv = 3x$. The solution to this last equation is given in Problem 5.58 as $v = 6 + ce^{-x^2/4}$. Then, for the original equation, $y^{1/2} = 6 + ce^{-x^2/4}$, or $y = (6 + ce^{-x^2/4})^2$.

5.80 Solve $y' + y = y^2$.

▮ This is a Bernoulli equation with $n = 2$. Setting $v = y^{-2+1} = y^{-1}$, we have $y = v^{-1}$ and $y' = -v^{-2}v'$. The original differential equation then becomes $-v^{-2}v' + v^{-1} = v^{-2}$, or $v' - v = -1$. This last equation is linear with integrating factor $I(x, v) = e^{\int -1\,dx} = e^{-x}$. Multiplying by it, we obtain

$$e^{-x}v' - e^{-x}v = -e^{-x} \qquad \text{or} \qquad \frac{d}{dx}(ve^{-x}) = -e^{-x}$$

Then integration yields $ve^{-x} = \int(-e^{-x})\,dx = e^{-x} + c$, so that $v = 1 + ce^x$. Thus, for the original equation, $y^{-1} = 1 + ce^x$, and $y = (1 + ce^x)^{-1}$.

5.81 Solve $y' + y = y^{-2}$.

▮ This is a Bernoulli equation with $n = -2$. Setting $v = y^{2+1} = y^3$, we have $y = v^{1/3}$ and $y' = \frac{1}{3}v^{-2/3}v'$. The original differential equation thus becomes $\frac{1}{3}v^{-2/3}v' + v^{1/3} = v^{-2/3}$ or $v' + 3v = 3$. This last equation is linear with integrating factor $e^{\int 3\,dx} = e^{3x}$. Multiplying by it, we get

$$e^{3x}v' + 3e^{3x}v = 3e^{3x} \qquad \text{or} \qquad \frac{d}{dx}(ve^{3x}) = 3e^{3x}$$

Then integration yields $ve^{3x} = \int 3e^{3x}\,dx = e^{3x} + c$, so that $v = 1 + ce^{-3x}$. Thus, for the original equation, $y^3 = 1 + ce^{-3x}$, or $y = (1 + ce^{-3x})^{1/3}$.

5.82 Solve $x\,dy + y\,dx = x^3y^6\,dx$.

▮ We rewrite the equation first as $xy' + y = x^3y^6$ and then as $y' + \dfrac{1}{x}y = x^2y^6$, to obtain a Bernoulli equation with $n = 6$. Setting $v = y^{-6+1} = y^{-5}$, we have $y = v^{-1/5}$ and $y' = -\dfrac{1}{5}v^{-6/5}v'$. Our equation then becomes $-\dfrac{1}{5}v^{-6/5}v' + \dfrac{1}{x}v^{-1/5} = x^2v^{-6/5}$, or $v' - \dfrac{5}{x}v = -5x^2$. The solution to this last equation is $v = \frac{5}{2}x^3 + cx^5$ (see Problem 5.60); hence $y = v^{-1/5} = (\frac{5}{2}x^3 + cx^5)^{-1/5}$.

5.83 Solve $dy + y\,dx = y^2e^x\,dx$.

▮ This equation may be rewritten as $y' + y = y^2e^x$, which is a Bernoulli equation with $n = 2$. Setting $v = y^{-2+1} = y^{-1}$, we have $y = v^{-1}$ and $y' = -v^{-2}v'$. The rewritten equation then becomes $-v^{-2}v' + v^{-1} = v^{-2}e^x$, or $v' - v = -e^x$. The solution to this last equation is $v = (c - x)e^x$ (see Problem 5.61), so $y = v^{-1} = e^{-x}/(c - x)$.

5.84 Solve $\dfrac{dx}{dt} - \dfrac{1}{2t}x = \left(-\dfrac{1}{2}\cos t\right)x^3$.

▮ This is a Bernoulli equation in the dependent variable x and the independent variable t, with $n = 3$. Setting $v = x^{-3+1} = x^{-2}$, we have $x = v^{-1/2}$ and $\dfrac{dx}{dt} = -\dfrac{1}{2}v^{-3/2}\dfrac{dv}{dt}$. The differential equation thus becomes

$$-\frac{1}{2}v^{-3/2}\frac{dv}{dt} - \frac{1}{2t}v^{-1/2} = \left(-\frac{1}{2}\cos t\right)v^{-3/2} \qquad \text{or} \qquad \frac{dv}{dt} + \frac{1}{t}v = \cos t$$

The solution to this last equation is $v = \sin t + \dfrac{1}{t}\cos t + \dfrac{c}{t}$ (see Problem 5.62). Thus,

$$x = v^{-1/2} = \left(\sin t + \frac{1}{t}\cos t + \frac{c}{t}\right)^{-1/2}.$$

5.85 Solve $\dfrac{dx}{dt} - \dfrac{1}{2t}x = -2tx^4$.

▌ This is a Bernoulli equation for $x(t)$ with $n = 4$. Setting $v = x^{-4+1} = x^{-3}$, we have $x = v^{-1/3}$ and $\dfrac{dx}{dt} = -\dfrac{1}{3}v^{-4/3}\dfrac{dv}{dt}$. The differential equation becomes

$$-\frac{1}{3}v^{-4/3}\frac{dv}{dt} - \frac{1}{2t}v^{-1/3} = -2tv^{-4/3} \qquad \text{or} \qquad \frac{dv}{dt} + \frac{3}{2t}v = 6t$$

The solution to this last equation is $v = \frac{12}{7}t^2 + ct^{-3/2}$ (see Problem 5.63). Therefore, $x = v^{-1/3} = (\frac{12}{7}t^2 + ct^{-3/2})^{-1/3}$.

5.86 Solve $\dfrac{dz}{dt} - \dfrac{1}{2t}z = -z^5$.

▌ This is a Bernoulli equation for $z(t)$ with $n = 5$. Setting $v = z^{-5+1} = z^{-4}$, we have $z = v^{-1/4}$ and $\dfrac{dz}{dt} = -\dfrac{1}{4}v^{-5/4}\dfrac{dv}{dt}$. The differential equation becomes

$$-\frac{1}{4}v^{-5/4}\frac{dv}{dt} - \frac{1}{2t}v^{-1/4} = -v^{-5/4} \qquad \text{or} \qquad \frac{dv}{dt} + \frac{2}{t}v = 4$$

The solution to this last equation is $v = \frac{4}{3}t + ct^{-2}$ (see Problem 5.64), so $z = v^{-1/4} = (\frac{4}{3}t + ct^{-2})^{-1/4}$.

5.87 Solve $\dfrac{dy}{dx} + 2xy + xy^4 = 0$, or $y^{-4}\dfrac{dy}{dx} + 2xy^{-3} = -x$.

▌ The transformation $y^{-3} = v$; $-3y^{-4}\dfrac{dy}{dx} = \dfrac{dv}{dx}$ reduces either equation to $\dfrac{dv}{dx} - 6xv = 3x$. Using the integrating factor $e^{-\int 6x\,dx} = e^{-3x^2}$, we obtain

$$ve^{-3x^2} = \int 3xe^{-3x^2}\,dx = -\frac{1}{2}e^{-3x^2} + c \qquad \text{or} \qquad \frac{1}{y^3} = -\frac{1}{2} + ce^{3x^2}$$

5.88 Solve $\dfrac{dy}{dx} + \dfrac{1}{3}y = \dfrac{1}{3}(1-2x)y^4$, or $y^{-4}\dfrac{dy}{dx} + \dfrac{1}{3}y^{-3} = \dfrac{1}{3}(1-2x)$.

▌ The transformation $y^{-3} = v$; $-3y^{-4}\dfrac{dy}{dx} = \dfrac{dv}{dx}$ reduces either equation to $\dfrac{dv}{dx} - v = 2x - 1$, for which e^{-x} is an integrating factor. Then integrating by parts gives

$$ve^{-x} = \int (2x-1)e^{-x}\,dx = -2xe^{-x} - e^{-x} + c \qquad \text{or} \qquad \frac{1}{y^3} = -1 - 2x + ce^x$$

5.89 Solve $\dfrac{dy}{dx} + y = y^2(\cos x - \sin x)$, or $y^{-2}\dfrac{dy}{dx} + y^{-1} = \cos x - \sin x$.

▌ The transformation $y^{-1} = v$; $-y^{-2}\dfrac{dy}{dx} = \dfrac{dv}{dx}$ reduces either equation to $\dfrac{dv}{dx} - v = \sin x - \cos x$, for which e^{-x} is an integrating factor. Then multiplication and integration give

$$ve^{-x} = \int (\sin x - \cos x)e^{-x}\,dx = -e^{-x}\sin x + c \qquad \text{or} \qquad \frac{1}{y} = -\sin x + ce^x$$

5.90 Solve $x\,dy - [y + xy^3(1 + \ln x)]\,dx = 0$, or $y^{-3}\dfrac{dy}{dx} - \dfrac{1}{x}y^{-2} = 1 + \ln x$.

▌ The transformation $y^{-2} = v$; $-2y^{-3}\dfrac{dy}{dx} = \dfrac{dv}{dx}$ reduces either equation to $\dfrac{dv}{dx} + \dfrac{2}{x}v = -2(1 + \ln x)$, for which $e^{\int 2\,dx/x} = x^2$ is an integrating factor. Then multiplication and integration give

$$vx^2 = -2\int (x^2 + x^2\ln x)\,dx = -\frac{4}{9}x^3 - \frac{2}{3}x^3\ln x + c \qquad \text{or} \qquad \frac{x^2}{y^2} = -\frac{2}{3}x^3\left(\frac{2}{3} + \ln x\right) + c$$

MISCELLANEOUS TRANSFORMATIONS

5.91 Develop a method for solving the differential equation $f'(y)\dfrac{dy}{dx} + f(y)P(x) = Q(x)$ for $y(x)$.

Set $v = f(y)$ so that $\dfrac{dv}{dx} = f'(y)\dfrac{dy}{dx}$. Then the given differential equation may be written as $\dfrac{dv}{dx} + P(x)v = Q(x)$, which is linear and may be solved by the method developed in Problem 5.2.

5.92 Show that the Bernoulli equation is a special case of the differential equation described in the previous problem.

The Bernoulli equation, $\dfrac{dy}{dx} + p(x)y = y^n q(x)$, may be written as

$$(-n+1)y^{-n}\frac{dy}{dx} + (-n+1)p(x)y^{-n+1} = (-n+1)q(x)$$

Set $Q(x) = (-n+1)q(x)$ and $P(x) = (-n+1)p(x)$; then the Bernoulli equation has the form $(-n+1)y^{-n}\dfrac{dy}{dx} + y^{-n+1}P(x) = Q(x)$, which is identical to the differential equation described in the previous problem for the special case $f(y) = y^{-n+1}$.

5.93 Solve $\sin y \dfrac{dy}{dx} = (\cos x)(2\cos y - \sin^2 x)$.

We rewrite this equation as $-\sin y \dfrac{dy}{dx} + (\cos y)(2\cos x) = \sin^2 x \cos x$, which has the form required by Problem 5.91 with $f(y) = \cos y$, $P(x) = 2\cos x$, and $Q(x) = \sin^2 x \cos x$. The substitution $v = \cos y$ transforms the rewritten equation into $\dfrac{dv}{dx} + (2\cos x)v = \sin^2 x \cos x$, which is linear. Its solution is $v = \frac{1}{2}\sin^2 x - \frac{1}{2}\sin x + \frac{1}{4} + ce^{-2\sin x}$ (see Problem 5.70). The solution to the original equation is, implicitly, $\cos y = \frac{1}{2}\sin^2 x - \frac{1}{2}\sin x + \frac{1}{4} + ce^{-2\sin x}$.

5.94 Solve $\dfrac{dy}{dx} + 1 = 4e^{-y}\sin x$.

We rewrite this equation as $e^y\dfrac{dy}{dx} + e^y = 4\sin x$, which has the form required by Problem 5.91 with $f(y) = e^y$, $P(x) = 1$, and $Q(x) = 4\sin x$. The substitution $v = e^y$ transforms it to $\dfrac{dv}{dx} + v = 4\sin x$, whose solution is given in Problem 5.71 as $v = 2(\sin x - \cos x) + ce^{-x}$. Then the solution to the original equation is, implicitly, $e^y = 2(\sin x - \cos x) + ce^{-x}$ and, explicitly, $y = \ln[2(\sin x - \cos x) + ce^{-x}]$.

5.95 Solve $x^2 \cos y \dfrac{dy}{dx} = 2x\sin y - 1$.

We write this equation as $\cos y \dfrac{dy}{dx} + \sin y\left(-\dfrac{2}{x}\right) = \dfrac{-1}{x^2}$, which has the form required by Problem 5.91 with $f(y) = \sin y$, $P(x) = -\dfrac{2}{x}$, and $Q(x) = \dfrac{-1}{x^2}$. The substitution $v = \sin y$ transforms it to $\dfrac{dv}{dx} - \dfrac{2}{x}v = \dfrac{-1}{x^2}$, whose solution is given in Problem 5.73 as $v = \frac{1}{3}x^{-1} + cx^2$. Then the solution to the original equation is, implicitly, $\sin y = \frac{1}{3}x^{-1} + cx^2$ and, explicitly, $y = \arcsin\dfrac{1 + kx^3}{3x}$, where $k = 3c$.

5.96 Solve $\sin y \dfrac{dy}{dx} = (\cos y)(1 - x\cos y)$.

We rewrite this equation as $\dfrac{\sin y}{\cos^2 y}\dfrac{dy}{dx} - \dfrac{1}{\cos y} = -x$, which has the form required by Problem 5.91 with

$f(y) = 1/\cos y$, $P(x) = -1$, and $Q(x) = -x$. The substitution $v = 1/\cos y$ transforms it to $\dfrac{dv}{dx} - v = -x$, whose solution is given in Problem 5.72 as $v = x + 1 + ce^x$. Then the solution to the original equation is, implicitly, $1/\cos y = x + 1 + ce^x$ and, explicitly, $y = \operatorname{arcsec}(x + 1 + ce^x)$.

5.97 Solve $x\dfrac{dy}{dx} - y + 3x^3y - x^2 = 0$, or $x\,dy - y\,dx + 3x^3y\,dx - x^2\,dx = 0$.

▌ Here $(x\,dy - y\,dx)$ suggests the transformation $y/x = v$. Then $\dfrac{x\,dy - y\,dx}{x^2} + 3x^2\dfrac{y}{x}\,dx - dx = 0$ is reduced to $\dfrac{dv}{dx} + 3x^2v = 1$, for which e^{x^3} is an integrating factor. Multiplication and integration then yield $ve^{x^3} = \int e^{x^3}\,dx + c$ or $y = xe^{-x^3}\int e^{x^3}\,dx + cxe^{-x^3}$. The indefinite integral here cannot be evaluated in terms of elementary functions.

5.98 Solve $(4r^2s - 6)\,dr + r^3\,ds = 0$, or $(r\,ds + s\,dr) + 3s\,dr = (6/r^2)\,dr$.

▌ The first term of the second equation suggests the substitution $rs = t$, which reduces the equation to $dt + 3\dfrac{t}{r}\,dr = \dfrac{6}{r^2}\,dr$, or $\dfrac{dt}{dr} + \dfrac{3}{r}t = \dfrac{6}{r^2}$. Then r^3 is an integrating factor, and the solution is $tr^3 = r^4s = 3r^2 + c$, or $s = \dfrac{3}{r^2} + \dfrac{c}{r^4}$.

5.99 Solve $x\sin\theta\,d\theta + (x^3 - 2x^2\cos\theta + \cos\theta)\,dx = 0$, or $-\dfrac{x\sin\theta\,d\theta + \cos\theta\,dx}{x^2} + 2\cos\theta\,dx = x\,dx$.

▌ The substitution $xy = \cos\theta$; $dy = -\dfrac{x\sin\theta\,d\theta + \cos\theta\,dx}{x^2}$, reduces the second equation to $dy + 2xy\,dx = x\,dx$, or $\dfrac{dy}{dx} + 2xy = x$. An integrating factor is e^{x^2}, and the solution is

$$ye^{x^2} = \frac{\cos\theta}{x}e^{x^2} = \int e^{x^2}x\,dx = \frac{1}{2}e^{x^2} + K \qquad \text{or} \qquad 2\cos\theta = x + cxe^{-x^2}$$

INITIAL-VALUE PROBLEMS

5.100 Solve $y' - 5y = 0$; $y(0) = 3$.

▌ The solution to the differential equation is given in Problem 5.3 as $y = ce^{5x}$. Applying the initial condition directly, we have $3 = ce^{5(0)} = c$, so the solution to the initial-value problem is $y = 3e^{5x}$.

5.101 Solve $y' - 5y = 0$; $y(3) = 0$.

▌ The solution to the differential equation is the same as in the previous problem. Applying the initial condition, we get $0 = ce^{3(3)}$, or $c = 0$. The solution to the initial-value problem is $y \equiv 0$.

5.102 Solve $y' - 5y = 0$; $y(3) = 4$.

▌ The solution to the differential equation is the same as in Problem 5.100. Applying the initial condition, we find that $4 = ce^{3(3)} = ce^9$, or $c = 4e^{-9}$. The solution to the initial-value problem is $y = 4e^{-9}e^{3x} = 4e^{3(x-3)}$.

5.103 Solve $y' - 5y = 0$; $y(\pi) = 2$.

▌ The solution to the differential equation is the same as in Problem 5.100. Applying the initial condition, we obtain $2 = ce^{3(\pi)}$, or $c = 2e^{-3\pi}$. The solution to the initial-value problem is $y = 2e^{-3\pi}e^{3x} = 2e^{3(x-\pi)}$.

5.104 Solve $y' + 2xy = 0$; $y(3) = 4$.

▌ The solution to the differential equation is given in Problem 5.9 as $y = ce^{-x^2}$. Applying the initial condition directly, we have $4 = ce^{-(3)^2} = ce^{-9}$, or $c = 4e^9$. The solution to the initial-value problem is $y = 4e^9e^{-x^2} = 4e^{-(x^2-9)}$.

5.105 Solve $y' + 2xy = 0$; $y(-2) = 3$.

▮ The solution to the differential equation is the same as in the previous problem. Applying the initial condition, we have $3 = ce^{-(-2)^2} = ce^{-4}$, or $c = 3e^4$. The solution to the initial-value problem is $y = 3e^4e^{-x^2} = 3e^{-(x^2-4)}$.

5.106 Solve $dy/dt + (t-1)y = 0$; $y(1) = 5$.

▮ The solution to the differential equation is given in Problem 5.13 as $y = ce^{t-t^2/2}$. Applying the initial condition, we get $5 = ce^{1-1^2/2} = ce^{1/2}$, or $c = 5e^{-1/2}$. The solution to the initial-value problem is $y = 5e^{-1/2}e^{t-t^2/2} = 5e^{t-t^2/2-1/2}$.

5.107 Solve $dy/dt + (t-1)y = 0$; $y(-3) = 0$.

▮ The solution to the differential equation is the same as in the previous problem. Applying the new initial condition, we obtain $0 = ce^{(-3)-(-3)^2/2} = ce^{-15/2}$, or $c = 0$. The solution to the initial-value problem is $y \equiv 0$.

5.108 Solve $\dfrac{dN}{dt} - \dfrac{5}{t}N = 0$; $N(1) = 1000$.

▮ The solution to the differential equation is given in Problem 5.19 as $N = ct^5$. Applying the initial condition, we have $1000 = c(1)^5 = c$, so the solution to the initial-value problem is $N = 1000t^5$.

5.109 Solve $\dfrac{dN}{dt} - \dfrac{5}{t}N = 0$; $N(2) = 1000$.

▮ The solution to the differential equation is the same as in the previous problem. Applying the initial condition, we obtain $1000 = c(2)^5 = 32c$, or $c = 31.25$. The solution to the initial-value problem is $N = 31.25t^5$.

5.110 Solve $y' - 3y = 6$; $y(0) = 1$.

▮ The solution to the differential equation is given in Problem 5.29 as $y = ce^{3x} - 2$. Applying the initial condition directly, we have $1 = ce^{3(0)} - 2 = c - 2$, or $c = 3$. The solution to the initial-value problem is $y = 3e^{3x} - 2$.

5.111 Solve $y' - 3y = 6$; $y(1) = 0$.

▮ The solution to the differential equation is the same as in the previous problem. Applying the initial condition directly, we have $0 = ce^{3(1)} - 2$, so that $2 = ce^3$ or $c = 2e^{-3}$. The solution to the initial-value problem is $y = 2e^{-3}e^{3x} - 2 = 2e^{3(x-1)} - 2$.

5.112 Solve $y' - 3y = 6$; $y(-5) = 4$.

▮ The solution to the differential equation is the same as in Problem 5.110. Applying the initial condition, we find that $4 = ce^{3(-5)} - 2$, so that $6 = ce^{-15}$ or $c = 6e^{15}$. The solution to the initial-value problem is $y = 6e^{15}e^{3x} - 2 = 6e^{3(x+5)} - 2$.

5.113 Solve $dq/dt + 10q = 20$; $q(0) = 2$.

▮ The solution to the differential equation is given in Problem 5.32 as $q = 2 + ce^{-10t}$. Applying the initial condition, we get $2 = 2 + ce^{-10(0)}$, or $c = 0$. The solution to the initial-value problem is $q \equiv 2$.

5.114 Solve $dq/dt + 10q = 20$; $q(0) = 500$.

▮ The solution to the differential equation is the same as in the previous problem. Applying the initial condition to it, we find that $500 = 2 + ce^{-10(0)} = 2 + c$, or $c = 498$. The solution to the initial-value problem is $q = 2 + 498e^{-10t}$.

5.115 Solve $dq/dt + 10q = 20$; $q(4) = 500$.

▮ The solution to the differential equation is the same as in Problem 5.113. Applying the initial condition, we have $500 = 2 + ce^{-10(4)}$, so that $498 = ce^{-40}$ or $c = 498e^{40}$. The solution to the initial-value problem is $q = 2 + 498e^{40}e^{-10t} = 2 + 498e^{-10(t-4)}$.

5.116 Solve $dv/dt + 25v = 9.8$; $v(0) = 5$.

▌ The solution to the differential equation is given in Problem 5.36 as $v = 0.392 + ce^{-25t}$. Applying the initial condition, we get $5 = 0.392 + ce^{-25(0)}$, or $c = 4.608$. The solution to the initial-value problem is $v = 0.392 + 4.608e^{-25t}$.

5.117 Solve $dv/dt + 25v = 9.8$; $v(0.1) = 5$.

▌ The solution to the differential equation is the same as in the previous problem. Applying the new initial condition, we get $5 = 0.392 + ce^{-25(0.1)}$, so that $4.608 = ce^{-2.5}$ or $c = 4.608e^{2.5} = 56.137$. The solution to the initial-value problem is $v = 0.392 + 56.137e^{-25t}$.

5.118 Solve $y' + y = \sin x$; $y(\pi) = 1$.

▌ From Problem 5.44 the solution to the differential equation is $y = ce^{-x} + \frac{1}{2}\sin x - \frac{1}{2}\cos x$. Applying the initial condition directly, we obtain $1 = ce^{-\pi} + \frac{1}{2}$, or $c = \frac{1}{2}e^{\pi}$. Thus $y = \frac{1}{2}e^{\pi}e^{-x} + \frac{1}{2}\sin x - \frac{1}{2}\cos x = \frac{1}{2}(e^{\pi - x} + \sin x - \cos x)$.

5.119 Solve $x\,dy/dx - 2y = x^3 \cos 4x$; $y(\pi) = 1$.

▌ The solution to the differential equation is given in Problem 5.46 as $y = \frac{1}{4}x^2 \sin 4x + cx^2$. Applying the initial condition, we obtain $1 = \frac{1}{4}\pi^2 \sin 4\pi + c\pi^2 = c\pi^2$, or $c = 1/\pi^2$. The solution to the initial-value problem is $y = \frac{1}{4}x^2 \sin 4x + (x/\pi)^2$.

5.120 Solve $x\,dy/dx - 2y = x^3 \cos 4x$; $y(1) = \pi$.

▌ The solution to the differential equation is the same as in the previous problem. Applying the new initial condition, we find that $\pi = \frac{1}{4}(1^2)\sin 4 + c(1^2)$, or $c = \pi - \frac{1}{4}\sin 4 = 3.331$. The solution to the initial-value problem is $y = \frac{1}{4}x^2 \sin 4x + 3.331x^2$.

5.121 Solve $y' + xy = xy^2$; $y(0) = 1$.

▌ The solution to the differential equation is given in Problem 5.75 as $y = \dfrac{1}{1 + ce^{x^2/2}}$. Applying the initial condition, we find that $1 = \dfrac{1}{1 + c}$, or $c = 0$. The solution to the initial-value problem is $y \equiv 1$.

5.122 Solve $y' + xy = xy^2$; $y(1) = 0$.

▌ Applying this initial condition to the solution found in Problem 5.75, we have $0 = \dfrac{1}{1 + ce^{1/2}}$, which has no solution. Thus, there is no value of c that will satisfy the initial condition. However, a Bernoulli equation also admits the trivial solution $y \equiv 0$, and since this solution does satisfy the initial condition, it is the solution to this initial-value problem.

5.123 Solve $y' + xy = 6x\sqrt{y}$; $y(0) = 0$.

▌ In Problem 5.79 we found a nontrivial solution to the differential equation to be $y = (6 + ce^{-x^2/4})^2$. Applying the initial condition, we obtain $0 = (6 + ce^0)^2$, or $c = -6$. One solution to the initial-value problem is thus $y = 36(1 - e^{-x^2/4})^2$. The trivial solution to the Bernoulli equation, $y \equiv 0$, also satisfies the initial condition, so it is a second solution to the initial-value problem.

5.124 Solve $x\,dy + y\,dx = x^3y^6\,dx$; $y(1) = 5$.

▌ A nontrivial solution to this Bernoulli equation is given in Problem 5.82 as $y = (\frac{5}{2}x^3 + cx^5)^{-1/5}$. Applying the initial condition, we obtain $5 = (\frac{5}{2} + c)^{-1/5}$, so that $5^{-5} = \frac{5}{2} + c$ or $c = -2.49968$. The solution to the initial-value problem is $y = (2.5x^3 - 2.49968x^5)^{-1/5}$.

5.125 Solve $x\,dy + y\,dx = x^3y^6\,dx$; $y(1) = 0$.

▌ A nontrivial solution to the differential equation is given in the previous problem. Applying the initial condition, we have $0 = (\frac{5}{2} + c)^{-1/5}$ or $0 = \dfrac{1}{5/2 + c}$, which has no solution. However, the trivial solution, $y \equiv 0$, does satisfy the initial condition; hence it is the solution to the initial-value problem.

CHAPTER 6

Applications of First-Order Differential Equations

POPULATION GROWTH PROBLEMS

6.1 A certain population of bacteria is known to grow at a rate proportional to the amount present in a culture that provides plentiful food and space. Initially there are 250 bacteria, and after seven hours 800 bacteria are observed in the culture. Find an expression for the approximate number of bacteria present in the culture at any time t.

▌ The differential equation governing this system was determined in Problem 1.53 to be $dN/dt = kN$, where $N(t)$ denotes the number of bacteria present and k is a constant of proportionality. Its solution is $N = ce^{kt}$ (see Problem 5.8).

At $t = 0$, we are given $N = 250$. Applying this initial condition, we get $250 = ce^{k(0)} = c$, so the solution becomes $N = 250e^{kt}$.

At $t = 7$, we are given $N = 800$. Substituting this condition and solving for k, we get $800 = 250e^{k(7)}$, or $k = \frac{1}{7}\ln\frac{800}{250} = 0.166$. Now the solution becomes

$$N = 250e^{0.166t} \tag{1}$$

which is an expression for the approximate number of bacteria present at any time t measured in hours.

6.2 Determine the approximate number of bacteria that will be present in the culture described in the previous problem after 24 h.

▌ We require N at $t = 24$. Substituting $t = 24$ into (1) of the previous problem, we obtain $N = 250e^{0.166(24)} = 13{,}433$.

6.3 Determine the amount of time it will take for the bacteria described in Problem 6.1 to increase to 2500.

▌ We seek a value of t corresponding to $N = 2500$. Substituting $N = 2500$ into (1) of Problem 6.1 and solving for t, we find $2500 = 250e^{0.166t}$, so that $10 = e^{0.166t}$ and $t = (\ln 10)/0.166 = 13.9$ h.

6.4 A bacteria culture is known to grow at a rate proportional to the amount present. After one hour, 1000 bacteria are observed in the culture; and after four hours, 3000. Find an expression for the number of bacteria present in the culture at any time t.

▌ As in Problem 6.1, the differential equation governing this system is $dN/dt = kN$, where $N(t)$ denotes the number of bacteria present and k is a constant of proportionality, and its solution is $N = ce^{kt}$.

At $t = 1$, $N = 1000$; hence, $1000 = ce^{k}$.

At $t = 4$, $N = 3000$; hence, $3000 = ce^{4k}$.

Solving these two equations for k and c, we find $k = \frac{1}{3}\ln 3 = 0.366$ and $c = 1000e^{-0.366} = 694$. Substituting these values of k and c into the solution yields $N = 694e^{0.366t}$ as the number of bacteria present at any time t.

6.5 In the previous problem, determine the number of bacteria originally in the culture.

▌ We require N at $t = 0$. Substituting $t = 0$ into the result of the previous problem, we obtain $N = 694e^{(0.366)(0)} = 694$.

6.6 A bacteria culture is known to grow at a rate proportional to the amount present. Find an expression for the approximate number of bacteria in such a culture if the initial number is 300 and if it is observed that the population has increased by 20 percent after 2 h.

▌ As in Problem 6.1, the differential equation governing this system is $dN/dt = kN$, where k is a constant of proportionality, and its solution is $N = ce^{kt}$.

At $t = 0$, we are given $N = 300$. Applying this initial condition, we get $300 = ce^{k(0)} = c$, so the solution becomes $N = 300e^{kt}$.

At $t = 2$, the population has grown by 20 percent or 60 bacteria and stands at $300 + 60 = 360$. Substituting this condition and solving for k, we get $360 = 300e^{k(2)}$ or $k = \frac{1}{2}\ln\frac{360}{300} = 0.09116$. The number of bacteria present at any time t is thus

$$N = 300e^{0.09116t} \qquad t \text{ in hours} \tag{1}$$

6.7 Determine the number of bacteria that will be present in the culture of the previous problem after 24 h.

▌ We require N at $t = 24$. Substituting this value of t into (1) of the previous problem, we obtain $N = 300e^{0.09116(24)} = 2675$.

6.8 Determine the number of bacteria present in the culture of Problem 6.6 after 1 week.

▌ We require N at $t = 7(24) = 168$ h. Substituting this value of t into (1) of Problem 6.6, we obtain $N = 300e^{0.09116(168)} = 1.34 \times 10^9$.

6.9 Determine the amount of time it will take the culture described in Problem 6.6 to double its original population.

▌ We seek the value of t associated with $N = 2(300) = 600$. Substituting $N = 600$ into (1) of Problem 6.6 and then solving for t, we get $600 = 300e^{0.09116t}$, or $t = \dfrac{\ln(600/300)}{0.09116} = 7.6$ h.

6.10 A certain culture of bacteria grows at a rate that is proportional to the number present. If it is found that the number doubles in 4 h, how many may be expected at the end of 12 h?

▌ Let x denote the number of bacteria present at time t hours. Then $\dfrac{dx}{dt} = kx$, or $\dfrac{dx}{x} = k\,dt$.

First Solution: Integrating the second equation, we have $\ln x = kt + \ln c$, so that $x = ce^{kt}$. Assuming that $x = x_0$ at time $t = 0$, we have $c = x_0$ and $x = x_0e^{kt}$; at time $t = 4$, we have $x = 2x_0$; then $2x_0 = x_0e^{4k}$ and $e^{4k} = 2$. Now when $t = 12$, $x = x_0e^{12k} = x_0(e^{4k})^3 = x_0(2^3) = 8x_0$; that is, there are eight times the original number.

Second Solution: Again we integrate the second equation, this time between the limits $t = 0$, $x = x_0$ and $t = 4$, $x = 2x_0$. We write $\int_{x_0}^{2x_0} \dfrac{dx}{x} = k\int_0^4 dt$, from which $\ln 2x_0 - \ln x_0 = 4k$ so that $4k = \ln 2$. Now if we integrate between the limits $t = 0$, $x = x_0$ and $t = 12$, $x = x$, we get $\int_{x_0}^{x} \dfrac{dx}{x} = k\int_0^{12} dt$, from which $\ln\dfrac{x}{x_0} = 12k = 3(4k) = 3\ln 2 = \ln 8$. Then $x = 8x_0$, as before.

6.11 If, in the previous problem, there are 10^4 bacteria at the end of 3 h and 4×10^4 at the end of 5 h, how many were there to start?

▌ ***First Solution:*** When $t = 3$, $x = 10^4$; hence, the equation $x = ce^{kt}$ of the previous problem becomes $10^4 = ce^{3k}$, and so $c = \dfrac{10^4}{e^{3k}}$. Also, when $t = 5$, $x = 4 \times 10^4$; hence, $4 \times 10^4 = ce^{5k}$ and so $c = \dfrac{4 \times 10^4}{e^{5k}}$. Equating these values of c gives us $\dfrac{10^4}{e^{3k}} = \dfrac{4 \times 10^4}{e^{5k}}$ from which $e^{2k} = 4$ and $e^k = 2$. Thus, the original number is $c = \dfrac{10^4}{e^{3k}} = \dfrac{10^4}{8}$ bacteria.

Second Solution: Integrating the differential equation of the previous problem between the limits $t = 3$, $x = 10^4$ and $t = 5$, $x = 4 \times 10^4$ gives us $\int_{10^4}^{4\times 10^4} \dfrac{dx}{x} = k\int_3^5 dt$, from which $\ln 4 = 2k$, and $k = \ln 2$. Integrating between the limits $t = 0$, $x = x_0$ and $t = 3$, $x = 10^4$ gives us $\int_{x_0}^{10^4} \dfrac{dx}{x} = k\int_0^3 dt$, from which $\ln\dfrac{10^4}{x_0} = 3k = 3\ln 2 = \ln 8$. Then $x_0 = \dfrac{10^4}{8}$ as before.

6.12 In a culture of yeast the amount of active ferment grows at a rate proportional to the amount present. If the amount doubles in 1 h, how many times the original amount may be anticipated at the end of $2\frac{3}{4}$ h?

▌ Let $N(t)$ denote the amount of yeast present at time t. Then $dN/dt = kN$, where k is a constant of proportionality. The solution to this equation is given in Problem 5.8 as $N = ce^{kt}$. If we designate the initial amount of yeast as N_0, then $N = N_0$ at $t = 0$, and it follows that $N_0 = ce^{k(0)} = c$. We may then rewrite the solution as $N = N_0e^{kt}$.

After 1 h, the amount present is $N = 2N_0$; applying this condition and solving for k, we find $2N_0 = N_0e^{k(1)}$, so that $e^k = 2$ and $k = \ln 2 = 0.693$. Thus, the amount of yeast present at any time t is $N = N_0e^{0.693t}$. After 2.75 h the amount will be $N = N_0e^{0.693(2.75)} = 6.72N_0$. This represents a 6.72-fold increase over the original amount.

6.13 The rate at which yeast cells multiply is proportional to the number present. If the original number doubles in 2 h, in how many hours will it triple?

▌Let $N(t)$ denote the number of yeast cells present at time t. Then it follows from the previous problem that $N = N_0e^{kt}$, where N_0 designates the initial number present and k is a constant of proportionality. At $t = 2$, we know that $N = 2N_0$. Substituting this condition into the equation and solving for k, we get $2N_0 = N_0e^{k(2)}$, from which $e^{2k} = 2$, or $k = \frac{1}{2}\ln 2 = 0.3466$. Thus, the number of yeast cells in this culture at any time t is $N = N_0e^{0.3466t}$.

We seek t for which $N = 3N_0$; Substituting for N and solving for t, we obtain $3N_0 = N_0e^{0.3466t}$, from which $t = \dfrac{\ln(3N_0/N_0)}{0.3466} = 3.17$ h.

6.14 Bacteria are placed in a nutrient solution and allowed to multiply. Food is plentiful but space is limited, so competition for space will force the bacteria population to stabilize at some constant level M. Determine an expression for the population at time t if the growth rate of the bacteria is jointly proportional to the number of bacteria present and the difference between M and the current population.

▌Let $N(t)$ denote the number of bacteria present at time t. The differential equation governing this system was determined in Problem 1.55 to be $\dfrac{dN}{dt} = kN(M - N)$, where k is a constant of proportionality. If we rewrite this equation in the differential form $\dfrac{1}{N(M-N)}\,dN - k\,dt = 0$, we see it is separable. Integrating term by term and noting that by partial fractions $\dfrac{1}{N(M-N)} = \dfrac{1/M}{N} + \dfrac{1/M}{M-N}$, we get

$$\frac{1}{M}\ln N - \frac{1}{M}\ln(M-N) - kt = c, \qquad \text{or} \qquad \frac{1}{M}\ln\frac{N}{M-N} = c + kt,$$

from which

$$\frac{N}{M-N} = e^{cM+kMt} = e^{cM}e^{kMt} = Ce^{kMt} \qquad \text{where } C = e^{cM}$$

Solving for N, we obtain $N = \dfrac{CM}{C + e^{-kMt}}$. If we now denote the initial population by N_0, then at $t = 0$ this becomes $N_0 = \dfrac{CM}{C+1}$, and $C = \dfrac{N_0}{M - N_0}$. Thus, the solution can be written

$$N = \frac{MN_0}{N_0 + (M - N_0)e^{-kMt}} \qquad (1)$$

which is an expression for the bacteria population at any time t. Equation (1) is often referred to as the *logistics equation.*

6.15 The population of a certain country is known to increase at a rate proportional to the number of people presently living in the country. If after 2 years the population has doubled, and after 3 years the population is 20,000, find the number of people initially living in the country.

▌Let N denote the number of people living in the country at any time t, and let N_0 denote the number of people initially living in the country. Then, $\dfrac{dN}{dt} - kN = 0$, which has the solution $N = ce^{kt}$. At $t = 0$, $N = N_0$; hence, it follows from that $N_0 = ce^{k(0)}$, or that $c = N_0$. Thus, the solution becomes $N = N_0e^{kt}$.

At $t = 2$, $N = 2N_0$. Substituting these values, we get $2N_0 = N_0e^{2k}$, from which $k = \frac{1}{2}\ln 2 = 0.347$. Thus, the solution finally becomes $N = N_0e^{0.347t}$.

At $t = 3$, $N = 20{,}000$. Substituting these values, we obtain $20{,}000 = N_0e^{(0.347)(3)} = N_0(2.832)$, or $N_0 = 7062$.

6.16 If the population of a country doubles in 50 years, in how many years will it treble under the assumption that the rate of increase is proportional to the number of inhabitants?

▌ Let y denote the population at time t years, and y_0 the population at time $t = 0$. Then $\dfrac{dy}{dt} = ky$, or $\dfrac{dy}{y} = k\,dt$, where k is a proportionality factor.

First Solution: Integrating the second equation gives us $\ln y = kt + \ln c$, or $y = ce^{kt}$. Let $y = y_0$ at time $t = 0$; then $c = y_0$ and $y = y_0e^{kt}$.

At $t = 50$, we know $y = 2y_0$. Then we have $2y_0 = y_0e^{50k}$ or $e^{50k} = 2$. When $y = 3y_0$, $y = y_0e^{kt}$ gives $3 = e^{kt}$. Then $3^{50} = e^{50kt} = (e^{50k})^t = 2^t$, and so $t = 79$ years.

Second Solution: Integrating this time between the limits $t = 0$, $y = y_0$ and $t = 50$, $y = 2y_0$ gives us $\int_{y_0}^{2y_0} \dfrac{dy}{y} = k\int_0^{50} dt$, from which $\ln 2y_0 - \ln y_0 = 50k$, and so $50k = \ln 2$. Also, integrating between the limits $t = 0$, $y = y_0$ and $t = t$, $y = 3y_0$ gives us $\int_{y_0}^{3y_0} \dfrac{dy}{y} = k\int_0^t dt$, from which $\ln 3 = kt$. Then $50\,\ln 3 = 50kt = t\ln 2$, and $t = \dfrac{50\ln 3}{\ln 2} = 79$ years.

DECAY PROBLEMS

6.17 A certain radioactive material is known to decay at a rate proportional to the amount present. If initially there is 100 mg of the material present and if after 2 years it is observed that 5 percent of the original mass has decayed, find an expression for the mass at any time t.

▌ Let $N(t)$ denote the amount of material present at time t. The differential equation governing this system is $dN/dt = kN$, and its solution is $N = ce^{kt}$ (see Problem 5.8). At $t = 0$, we are given $N = 100$. Applying this initial condition, we get $100 = ce^{k(0)} = c$. Thus, the solution becomes $N = 100e^{kt}$.

At $t = 2$, 5 percent of the original mass of 100 mg, or 5 mg, has decayed. Hence, at $t = 2$, $N(2) = 100 - 5 = 95$. Substituting this condition in the equation $N = 100e^{kt}$ and solving for k, we get $95 = 100e^{k(2)}$, or $k = \dfrac{1}{2}\ln\dfrac{95}{100} = -0.0256$. The amount of radioactive material present at any time t is, therefore,

$$N = 100e^{-0.0256t} \qquad t \text{ in years} \qquad (1)$$

6.18 In the previous problem, determine the time necessary for 10 percent of the original mass to decay.

▌ We require t when N has decayed to 90 percent of its original mass. Since the original mass was 100 mg, we seek the value of t corresponding to $N = 90$. Substituting $N = 90$ into (*1*) of the previous problem gives us $90 = 100e^{-0.0256t}$, so that $-0.0256t = \ln 0.9$, and $t = -(\ln 0.9)/0.0256 = 4.12$ years.

6.19 A certain radioactive material is known to decay at a rate proportional to the amount present. If initially there is 50 mg of the material present and after 2 h it is observed that the material has lost 10 percent of its original mass, find an expression for the mass of the material remaining at any time t.

▌ Let N denote the amount of material present at time t. Then $dN/dt - kN = 0$ and, as in Problem 6.17, $N = ce^{kt}$. At $t = 0$, we are given $N = 50$. Therefore, $50 = ce^{k(0)}$, or $c = 50$. Thus, we now have $N = 50e^{kt}$.

At $t = 2$, 10 percent of the original mass of 50 mg, or 5 mg, has decayed. Hence, at $t = 2$, $N = 50 - 5 = 45$. Substituting these values into the last equation and solving for k, we get $45 = 50e^{2k}$, or $k = \dfrac{1}{2}\ln\dfrac{45}{50} = -0.053$. The amount of mass present at any time t is therefore

$$N = 50e^{-0.053t} \qquad t \text{ in hours} \qquad (1)$$

6.20 In the previous problem, determine the mass of the material after 4 h.

▌ We require N at $t = 4$. Substituting $t = 4$ into (*1*) of the previous problem and then solving for N, we find that $N = 50e^{(-0.053)(4)} = 50(0.809) = 40.5$ mg.

6.21 Determine the time at which the mass described in Problem 6.19 has decayed to one-half its initial mass.

▌ We require t when $N = 50/2 = 25$. Substituting $N = 25$ into (*1*) of Problem 6.19 and solving for t, we find $25 = 50e^{-0.053t}$, so that $-0.053t = \ln\frac{1}{2}$ and $t = 13$ h. (The time required to reduce a decaying material to one-half its original mass is called the *half-life* of the material. For this material the half-life is 13 h.)

6.22 A certain radioactive material is known to decay at a rate proportional to the amount present. If after 1 h it is observed that 10 percent of the material has decayed, find the half-life of the material.

▌ Let $N(t)$ denote the amount of the material present at time t. Then $dN/dt = kt$, where k is a constant of proportionality. The solution to this equation is given in Problem 5.8 as $N = ce^{kt}$. If we designate the initial mass as N_0, then $N = N_0$ at $t = 0$, and we have $N_0 = ce^{k(0)} = c$. Thus, the solution becomes $N = N_0e^{kt}$.

At $t = 1$, 10 percent of the original mass N_0 has decayed, so 90 percent remains. Hence $N = 0.9N_0$ at $t = 1$. Substituting this condition and solving for k, we get $0.9N_0 = N_0e^{k(1)}$, from which $0.9 = e^k$, and $k = \ln 0.9 = -0.105$. The amount of radioactive material present at any time t is thus

$$N = N_0e^{-0.105t} \qquad t \text{ in hours} \tag{1}$$

The half-life is the time associated with $N = \frac{1}{2}N_0$. Substituting this value into (*1*) and solving for t, we obtain $\frac{1}{2}N_0 = N_0e^{-0.105t}$, so that $-0.105t = \ln\frac{1}{2}$ and $t = 6.60$ h.

6.23 Find the half-life of a radioactive substance if three-quarters of it is present after 8 h.

▌ Let $N(t)$ denote the amount of material present at time t. Then it follows from the previous problem that $N = N_0e^{kt}$, where N_0 denotes the initial amount of material and k is a constant of proportionality. If three-quarters of the initial amount is present after 8 h, it follows that $\frac{3}{4}N_0 = N_0e^{k(8)}$, from which $e^{8k} = \frac{3}{4}$ and $k = \frac{1}{8}\ln\frac{3}{4} = -0.03596$. Thus the amount of material present at any time t is $N = N_0e^{-0.03596t}$.

We require t when $N = \frac{1}{2}N_0$. Substituting this value into the previous equation and solving for t, we get $\frac{1}{2}N_0 = N_0e^{-0.03596t}$, from which $e^{-0.03596t} = \frac{1}{2}$ and $t = (\ln\frac{1}{2})/(-0.03596) = 19.3$ h.

6.24 Radium decomposes at a rate proportional to the amount present. If half the original amount disappears in 1600 years, find the percentage lost in 100 years.

▌ Let $R(t)$ denote the amount of radium present at time t. It follows from Problem 1.52 that $dR/dt = kR$, where k is a constant of proportionality. Solving this equation, we get $R = ce^{kt}$. If we designate the initial amount as R_0 (at $t = 0$) and apply this condition, we find $R_0 = ce^{k(0)} = c$, so the solution becomes $R = R_0e^{kt}$.

Since the half-life of radium is 1600 years, we have the condition $R = \frac{1}{2}R_0$ when $t = 1600$. Applying this condition to the last equation and then solving for k give $\frac{1}{2}R_0 = R_0e^{k(1600)}$, from which $\frac{1}{2} = e^{1600k}$, and $k = (\ln\frac{1}{2})/1600 = -0.0004332$. The amount of radium present at any time t is thus

$$R = R_0e^{-0.0004332t} \qquad t \text{ in years} \tag{1}$$

The amount present after 100 years will be $R = R_0e^{-0.0004332(100)} = 0.958R_0$, so the percent decrease from the initial amount R_0 is $\dfrac{R_0 - 0.958R_0}{R_0}\,100 = 4.2$ percent.

6.25 A certain radioactive material is known to decay at a rate proportional to the amount present. If initially $\frac{1}{2}$ g of the material is present and 0.1 percent of the original mass has decayed after 1 week, find an expression for the mass at any time t.

▌ Let $N(t)$ denote the amount of material present at time t. Then $dN/dt = kt$, where k is a constant of proportionality, and the solution to this equation is $N = ce^{kt}$ (see Problem 5.8). Since $N = \frac{1}{2}$ at $t = 0$, we have $\frac{1}{2} = ce^{k(0)} = c$, so the solution becomes $N = \frac{1}{2}e^{kt}$.

If we take the time unit to be 1 week, then 0.1 percent of the initial mass has decayed at $t = 1$, and 99.9 percent remains. Thus, at $t = 1$, we have $N = 0.999(\frac{1}{2}) = 0.4995$ g. Applying this condition to the last equation, we get $0.4995 = \frac{1}{2}e^{k(1)}$, from which $e^k = 0.999$, so that $k = \ln 0.999 = -0.001$. The amount of radioactive material present at any time t is thus

$$N = \tfrac{1}{2}e^{-0.001t} \qquad t \text{ in weeks} \tag{1}$$

6.26 Determine the half-life of the material described in the previous problem.

▌ The half-life is the time t associated with the decay of one-half the original mass. Here the original mass is $\frac{1}{2}$ g, so we seek the time when $N = \frac{1}{4}$. Substituting this value into (1) of the previous problem and solving for t, we obtain $\frac{1}{4} = \frac{1}{2}e^{-0.001t}$, from which $-0.001t = \ln\frac{1}{2}$ and $t = 693$ weeks. The half-life is 693 weeks or 13.3 years.

6.27 Rework Problem 6.25 using a time unit of 1 day.

▌ Our work through the derivation of the equation $N = \frac{1}{2}e^{kt}$ in Problem 6.25 remains valid. Now after 1 week, or 7 days, the mass has been reduced to 0.4995 g, so $N = 0.4995$ at $t = 7$. Applying this condition, we get $0.4995 = \frac{1}{2}e^{k(7)}$, from which $e^{7k} = 0.999$ and $k = \frac{1}{7}\ln 0.999 = -0.0001429$. The amount of radioactive material present at any time t is thus

$$N = \tfrac{1}{2}e^{-0.0001429t} \qquad t \text{ in days} \tag{1}$$

6.28 Use the result of the previous problem to determine the half-life of the material.

▌ Since the original mass is $\frac{1}{2}$ g, we require t when $N = \frac{1}{4}$. Substituting this value into (1) of the previous problem, we obtain $\frac{1}{4} = \frac{1}{2}e^{-0.0001429t}$, from which $-0.0001429t = \ln\frac{1}{2}$ and $t = 4850$. The half-life is 4850 days or 13.3 years (as we found in Problem 6.26).

6.29 A certain radioactive material is known to decay at a rate proportional to the amount present. If initially 500 mg of the material is present and after 3 years 20 percent of the original mass has decayed, find an expression for the mass at any time t.

▌ Let $R(t)$ denote the amount of radioactive material present at time t. Then $dR/dt = kR$, where k is a constant of proportionality. The solution to this equation is given in Problem 5.8 (with R replacing N) as $R = ce^{kt}$. At $t = 0$, $N = 500$; applying this condition yields $500 = ce^{k(0)} = c$, so the solution becomes $R = 500e^{kt}$.

If we take the time unit to be 1 year, then 20 percent of the original mass has decayed at $t = 3$, and 80 percent remains. Thus, at $t = 3$, $R = 0.8(500) = 400$. Applying this condition to the last equation, we get $400 = 500e^{k(3)}$, so that $e^{3k} = 0.8$ and $k = \frac{1}{3}\ln 0.8 = -0.07438$. The amount of radioactive material present at any time t is then

$$R = 500e^{-0.07438t} \qquad t \text{ in years} \tag{1}$$

6.30 For the material described in the previous problem, determine the amount remaining after 25 years.

▌ We require R when $t = 25$. Substituting $t = 25$ into (1) of the previous problem, we obtain $R = 500e^{-0.07438(25)} = 77.9$ mg.

6.31 For the material described in Problem 6.29, determine the amount remaining after 200 weeks.

▌ Since the time unit in Problem 6.29 is 1 year, we require R when $t = 200/52 = 3.846$ years. Substituting $t = 3.846$ into (1) of Problem 6.29, we get $R = 500e^{-0.07438(3.846)} = 375.6$ mg.

6.32 Determine the amount of time required for the material of Problem 6.29 to decay to 30 percent of its original amount.

▌ We require t when $R = 0.3(500) = 150$. Substituting $R = 150$ into (1) of Problem 6.29, we obtain $150 = 500e^{-0.07438t}$, from which $-0.07438t = \ln\frac{150}{500}$ and $t = 16.2$ years.

6.33 Determine the amount of time required for the material of Problem 6.29 to decay to 250 mg.

▌ We require t when $R = 250$. Substituting $R = 250$ into (1) of Problem 6.29, we obtain $250 = 500e^{-0.07438t}$, from which $-0.07438t = \ln\frac{250}{500}$ and $t = 9.3$ years. Note that 9.3 years is the half-life of the material.

6.34 After 2 days, 10 g of a radioactive chemical is present. Three days later, 5 g is present. How much of the chemical was present initially, assuming the rate of disintegration is proportional to the amount present?

▌ Let $N(t)$ denote the amount of chemical present at time t. Then $dN/dt = kN$, where k is a constant of proportionality, and the solution to this equation is $N = ce^{kt}$. Measuring time in units of 1 day, we have

$N = 10$ at $t = 2$; hence, $10 = ce^{2k}$. Moreover, $N = 5$ at $t = 5$ (3 days later); so $5 = ce^{5t}$. Solving these last two equations simultaneously for k and c, we find $2 = e^{-3k}$, so that $k = -\dfrac{\ln 2}{3} = -0.231$, and $c = 10e^{-2(-0.231)} = 15.87$. Substituting these values of c and k into $N = ce^{kt}$, we obtain $N = 15.87e^{-0.231t}$ as an expression for the amount of radioactive chemical present at any time t. At $t = 0$, this amount is $N = 15.87e^{-0.231(0)} = 15.87$ g.

6.35 Under certain conditions it is observed that the rate at which a solid substance dissolves varies directly as the product of the amount of undissolved solid present in the solvent and the difference between the saturation concentration and the instantaneous concentration of the substance. If 40 kg of solute is dumped into a tank containing 120 kg of solvent and at the end of 12 min the concentration is observed to be 1 part in 30, find the amount of solute in solution at any time t. The saturation concentration is 1 part of solute in 3 parts of solvent.

▌ If Q is the amount of the material in solution at time t, then $40 - Q$ is the amount of undissolved material present at that time, and $Q/120$ is the corresponding concentration. Hence, according to the given information,

$$\frac{dQ}{dt} = k(40 - Q)\left(\frac{1}{3} - \frac{Q}{120}\right) = \frac{k}{120}(40 - Q)^2$$

This is a simple separable equation for which we have $\dfrac{dQ}{(40 - Q)^2} = \dfrac{k}{120}\,dt$, with solution $\dfrac{1}{40 - Q} = \dfrac{k}{120}t + c$. Since $Q = 0$ when $t = 0$, we find that $c = \frac{1}{40}$. Also, when $t = 12$, $Q = \frac{1}{30}(120) = 4$, so we have

$$\frac{1}{40 - 4} = \frac{k}{120}12 + \frac{1}{40} \qquad \text{from which} \qquad \frac{k}{120} = \frac{1}{4320}$$

Then the solution becomes $\dfrac{1}{40 - Q} = \dfrac{t}{4320} + \dfrac{1}{40}$, from which we find that $Q = 40 - \dfrac{4320}{t + 108}$.

6.36 A certain chemical dissolves in water at a rate proportional to the product of the amount undissolved and the difference between the concentration in a saturated solution and the concentration in the actual solution. In 100 g of a saturated solution it is known that 50 g of the substance is dissolved. If when 30 g of the chemical is agitated with 100 g of water, 10 g is dissolved in 2 h, how much will be dissolved in 5 h?

▌ Let x denote the number of grams of the chemical undissolved after t hours. At that time the concentration of the actual solution is $\dfrac{30 - x}{100}$, and that of a saturated solution is $\dfrac{50}{100}$. Then

$$\frac{dx}{dt} = kx\left(\frac{50}{100} - \frac{30 - x}{100}\right) = kx\frac{x + 20}{100} \qquad \text{or} \qquad \frac{dx}{x} - \frac{dx}{x + 20} = \frac{k}{5}dt.$$

Integrating the latter equation between $t = 0$, $x = 30$ and $t = 2$, $x = 30 - 10 = 20$, we get $\int_{30}^{20} \frac{dx}{x} - \int_{30}^{20} \frac{dx}{x + 20} = \frac{k}{5}\int_0^2 dt$, from which $k = \frac{5}{2}\ln\frac{5}{6} = -0.46$.

Integrating now between $t = 0$, $x = 30$ and $t = 5$, $x = x$, we get $\int_{30}^{x} \frac{dx}{x} - \int_{30}^{x} \frac{dx}{x + 20} = \frac{k}{5}\int_0^5 dt$, from which $\ln\dfrac{5x}{3(x + 20)} = k = -0.46$. Then $\dfrac{x}{x + 20} = \dfrac{3}{5}e^{-0.46} = 0.38$, and $x = 12$. Thus, the amount dissolved after 5 h is $30 - 12 = 18$ g.

6.37 Chemical A dissolves in solution at a rate proportional to both the instantaneous amount of undissolved chemical and the difference in concentration between the actual solution C_a and saturated solution C_s. A porous inert solid containing 10 lb of A is agitated with 100 gal of water, and after an hour 4 lb of A is dissolved. If a saturated solution contains 0.2 lb of A per gallon, find the amount of A which is undissolved after 2 h.

▌ Let x lb of A be undissolved after t hours. Then

$$\frac{dx}{dt} = kx(C_s - C_a) = kx\left(0.2 - \frac{10 - x}{100}\right) = \frac{kx(x + 10)}{100} = cx(x + 10)$$

where $c = k/100$. Separating the variables and integrating, we get

$$\int \frac{dx}{x(x+10)} = \frac{1}{10}\int\left(\frac{1}{x} - \frac{1}{x+10}\right)dx = \frac{1}{10}\ln\frac{x}{x+10} = ct + C_1$$

Using the conditions $t = 0$, $x = 10$ and $t = 1$, $x = 6$, we find $x = \dfrac{5(3/4)^t}{1-(1/2)(3/4)^t}$. When $t = 2$ h, $x = 3.91$ lb of A undissolved.

6.38 Find the time required to dissolve 80 percent of the chemical A described in the previous problem.

▮ Substituting $x = 0.8(10) = 2$ into the last equation of the previous problem and solving for t, we have $2 = \dfrac{5(3/4)^t}{1-(1/2)(3/4)^t}$, from which we find $6(\frac{3}{4})^t = 2$ or $(\frac{3}{4})^t = \frac{1}{3}$. Then $t = (\ln\frac{1}{3})/(\ln\frac{3}{4}) = 3.82$ h.

COMPOUND-INTEREST PROBLEMS

6.39 A depositor places \$10,000 in a certificate of deposit account which pays 7 percent interest per annum, compounded continuously. How much will be in the account after 2 years?

▮ Let $P(t)$ denote the amount of money in the account at time t. The differential equation governing the growth of the account was determined in Problem 1.57 to be $dP/dt = 0.07P$ for an annual interest rate of 7 percent. This equation is linear and separable; its solution is $P = ce^{0.07t}$. At $t = 0$, the initial principal is $P = \$10{,}000$. Applying this condition, we find $10{,}000 = ce^{0.07(0)} = c$, so the solution becomes $P = 10{,}000e^{0.07t}$.

We require P when $t = 2$ years. Substituting this value of t into the last equation, we find $P = 10{,}000e^{0.07(2)} = \$11{,}502.74$.

6.40 How much will the depositor of the previous problem have after 5 years if the interest rate remains constant over that time?

▮ Substituting $t = 5$ into the solution derived in the previous problem, we obtain $P = 10{,}000e^{0.07(0.5)} = \$14{,}190.68$.

6.41 A woman places \$2000 in an account for her child upon his birth. Assuming no additional deposits or withdrawals, how much will the child have at his eighteenth birthday if the bank pays 5 percent interest per annum, compounded continuously, for the entire time period?

▮ Let $P(t)$ denote the amount of money in the account at time t. The differential equation governing the growth of the money was determined in Problem 1.57 to be $dP/dt = 0.05P$ for an annual interest rate of 5 percent. The solution to this differential equation is $P = ce^{0.05t}$. At $t = 0$, we have $P = \$2000$, so $2000 = ce^{0.05(0)} = c$, and the solution becomes $P = 2000e^{0.05t}$.

We require the principal at $t = 18$. Substituting this value of t into the last equation, we obtain $P = 2000e^{0.05(18)} = \$4919.21$.

6.42 How long will it take for the initial deposit to double under the conditions described in the previous problem?

▮ We seek the value of t corresponding to $P = \$4000$. Substituting this quantity into the solution derived in the previous problem, we obtain $4000 = 2000e^{0.05t}$, from which $t = \dfrac{\ln(4000/2000)}{0.05} = 13.86$ years.

6.43 Solve Problem 6.41 if the interest rate is 6.5 percent.

▮ For this new interest rate, the solution derived in Problem 6.41 becomes $P = 2000e^{0.065t}$. Then, at $t = 18$, we have $P = 2000e^{0.065(18)} = \$6443.99$.

6.44 Solve Problem 6.41 if the interest rate is $9\frac{1}{4}$ percent.

▮ For this new interest rate, the solution derived in Problem 6.41 becomes $P = 2000e^{0.0925t}$. Then, at $t = 18$, we have $P = 2000e^{0.0925(18)} = \$10{,}571.35$.

6.45 A man places \$700 in an account that accrues interest continuously. Assuming no additional deposits and no withdrawals, how much will be in the account after 10 years if the interest rate is a constant $7\frac{1}{2}$ percent for the first 6 years and a constant $8\frac{1}{4}$ percent for the last 4 years?

▮ For the first 6 years, the differential equation governing the growth is given by Problem 1.57 as $dP/dt = 0.075P$, which has as its solution $P = ce^{0.075t}$ $(0 \le t \le 6)$. At $t = 0$, $P = 700$; hence $700 = ce^{0.075(0)} = c$, and the solution becomes $P = 700e^{0.075t}$. At the end of 6 years, the account will have grown to $P = 700e^{0.075(6)} = \$1097.82$. This amount also represents the beginning balance for the 4-year period.

Over the next 4 years, the growth of the account is governed by the differential equation $dP/dt = 0.0825P$, which has as its solution $P = Ce^{0.0825t}$ $(6 \le t \le 10)$. At $t = 6$, $P = 1097.82$; hence $1097.82 = Ce^{0.0825(6)}$, and $C = 1097.82e^{-0.495} = 669.20$. The solution thus becomes $P = 669.20e^{0.0825t}$, and at year 10 the account will have grown to $P = 669.20e^{0.0825(10)} = \$1527.03$.

6.46 How long will it take a bank deposit to double if interest is compounded continuously at a constant rate of 4 percent per annum?

▮ The differential equation governing the growth of the account is $dP/dt = 0.04P$ (see Problem 1.57); this equation has as its solution $P = ce^{0.04t}$. If we denote the initial deposit as P_0, we have $P_0 = ce^{0.04(0)} = c$, and the solution becomes $P = P_0e^{0.04t}$.

We seek t corresponding to $P = 2P_0$. Substituting this value into the last equation and solving for t, we obtain $2P_0 = P_0e^{0.04t}$, from which $t = \dfrac{\ln(2P_0/P_0)}{0.04} = 17.33$ years.

6.47 How long will it take a bank deposit to double if interest is compounded continuously at a constant rate of 8 percent per annum?

▮ With this new interest rate, the solution derived in the previous problem becomes $P = P_0e^{0.08t}$. We seek t corresponding to $2P_0$; hence we write $2P_0 = P_0e^{0.08t}$, so that $2 = e^{0.08t}$ and $t = (\ln 2)/0.08 = 8.66$ years.

6.48 A woman plans to place a single sum in a certificate of deposit account with a guaranteed interest rate of $6\frac{1}{4}$ percent for 5 years. How much should she deposit if she wants the account to be worth \$25,000 at the end of the 5-year period?

▮ The differential equation governing the growth of this account was determined in Problem 1.57 and is $dP/dt = 0.0625P$; its solution is $P = ce^{0.0625t}$. Since we want $P = 25{,}000$ at $t = 5$, we have $25{,}000 = ce^{0.0625(5)}$, from which $c = 25{,}000e^{-0.3125} = 18{,}290.39$. Thus the solution becomes $P = 18{,}290.39e^{0.0625t}$. At $t = 0$, the initial amount must be $P = 18{,}290.39e^{0.0625(0)} = \$18{,}290.39$.

6.49 A man currently has \$12,000 and plans to invest it in an account that accrues interest continuously. What interest rate must he receive, if his goal is to have \$15,000 in $2\frac{1}{2}$ years?

▮ Let $P(t)$ denote the amount in the account at any time t, and let r represent the interest rate (which is presumed fixed for the entire period). The differential equation governing the growth of the account is given in Problem 1.57 as $dP/dt = (r/100)P$, which has as its solution $P = ce^{(r/100)t}$. At $t = 0$, $P = 12{,}000$; hence $12{,}000 = ce^{(r/100)(0)} = c$, so the solution becomes $P = 12{,}000e^{(r/100)t}$.

We require r corresponding to $P = 15{,}000$ and $t = 2.5$. Substituting these values into the last equation and solving for r, we obtain $15{,}000 = 12{,}000e^{(r/100)(2.5)}$, which reduces to $1.25 = e^{r/40}$ and yields $r = 40 \ln 1.25 = 8.926$ percent.

6.50 What interest rate must the man in the previous problem receive if his goal is \$16,000 in 3 years?

▮ Substituting $P = 16{,}000$ and $t = 3$ into the solution derived in the previous problem, we obtain $16{,}000 = 12{,}000e^{(r/100)3}$, from which we find that $r = \dfrac{100}{3} \ln \dfrac{16{,}000}{12{,}000} = 9.589$ percent.

COOLING AND HEATING PROBLEMS

6.51 Newton's law of cooling states that *the time rate of change of the temperature of a body is proportional to the temperature difference between the body and its surrounding medium.* Using Newton's law of cooling, derive a differential equation for the cooling of a hot body surrounded by a cool medium.

▮ Let T denote the temperature of the body, and let T_m denote the temperature of the surrounding medium. Then the time rate of change of the temperature of the body is dT/dt, and Newton's law of cooling can be formulated as $dT/dt = -k(T - T_m)$, or as

$$\frac{dT}{dt} + kT = kT_m \qquad (1)$$

where k is a positive constant of proportionality. Since k is chosen positive, the minus sign in Newton's law is required to make dT/dt negative for a cooling process. Note that in such a process, T is greater than T_m; thus $T - T_m$ is positive.

6.52 A metal bar at a temperature of 100°F is placed in a room at a constant temperature of 0°F. If after 20 min the temperature of the bar is 50°F, find an expression for the temperature of the bar at any time.

▌ The surrounding medium is the room, which is being held at a constant temperature of 0°F, so $T_m = 0$ and (1) of Problem 6.51 becomes $\dfrac{dT}{dt} + kT = 0$. This equation is linear; it also has the differential form $\dfrac{1}{T}\,dT + k\,dt = 0$, which is separable. Solving either form, we get $T = ce^{-kt}$. Since $T = 100$ at $t = 0$ (the temperature of the bar is initially 100°F), it follows that $100 = ce^{-k(0)}$ or $100 = c$. Substituting this value into the solution, we obtain $T = 100e^{-kt}$.

At $t = 20$, we are given that $T = 50$; hence, the last equation becomes $50 = 100e^{-20k}$, from which $k = \dfrac{-1}{20}\ln\dfrac{50}{100} = \dfrac{-1}{20}(-0.693) = 0.035$. The temperature of the bar at any time t is then

$$T = 100e^{-0.035t} \tag{1}$$

6.53 Find the time it will take for the bar in the previous problem to reach a temperature of 25°F.

▌ We require t when $T = 25$. Substituting $T = 25$ into (1) of the previous problem, we obtain $25 = 100e^{-0.035t}$ or $-0.035t = \ln\frac{1}{4}$. Solving, we find that $t = 39.6$ min.

6.54 Determine the temperature of the bar described in Problem 6.52 after 10 min.

▌ We require T when $t = 10$. Substituting $t = 10$ into (1) of Problem 6.52, we find that $T = 100e^{(-0.035)(10)} = 100(0.705) = 70.5$°F.

It should be noted that since Newton's law is valid only for small temperature differences, the above calculations represent only a first approximation to the physical situation.

6.55 A body at a temperature of 50°F is placed outdoors where the temperature is 100°F. If after 5 min the temperature of the body is 60°F, find an expression for the temperature of the body at any time.

▌ With $T_m = 100$ (the surrounding medium is the outside air), (1) of Problem 6.51 becomes $dT/dt + kT = 100k$. This equation is linear and has as its solution $T = ce^{-kt} + 100$ (see Problem 5.38). Since $T = 50$ when $t = 0$, it follows that $50 = ce^{-k(0)} + 100$, or $c = -50$. Substituting this value into the solution, we obtain $T = -50e^{-kt} + 100$.

At $t = 5$, we are given that $T = 60$; hence, from the last equation, $60 = -50e^{-5k} + 100$. Solving for k, we obtain $-40 = -50e^{-5k}$, so that $k = -\frac{1}{5}\ln\frac{40}{50} = -\frac{1}{5}(-0.223) = 0.045$. Substituting this value, we obtain the temperature of the body at any time t as

$$T = -50e^{-0.045t} + 100 \tag{1}$$

6.56 Determine how long it will take the body in the previous problem to reach a temperature of 75°F.

▌ We require t when $T = 75$. Substituting $T = 75$ into (1) of the previous problem, we have $75 = -50e^{-0.045t} + 100$ or $e^{-0.045t} = \frac{1}{2}$. Solving for t, we find $-0.045t = \ln\frac{1}{2}$, or $t = 15.4$ min.

6.57 Determine the temperature of the body described in Problem 6.55 after 20 min.

▌ We require T when $t = 20$. Substituting $t = 20$ into (1) of Problem 6.55 and then solving for T, we find $T = -50e^{(-0.045)(20)} + 100 = -50(0.41) + 100 = 79.5$°F.

6.58 A body at an unknown temperature is placed in a room which is held at a constant temperature of 30°F. If after 10 min the temperature of the body is 0°F and after 20 min the temperature of the body is 15°F, find an expression for the temperature of the body at time t.

▌ Here the temperature of the surrounding medium, T_m, is held constant at 30°F, so (1) of Problem 6.51 becomes $dT/dt + kT = 30k$. The solution to this differential equation is given by Problem 5.39 (with $a = 30$) as $T = ce^{-kt} + 30$.

At $t = 10$, we are given that $T = 0$. Hence, $0 = ce^{-10k} + 30$ or $ce^{-10k} = -30$.
At $t = 20$, we are given that $T = 15$. Hence, $15 = ce^{-20k} + 30$ or $ce^{-20k} = -15$.

Solving these last two equations for k and c, we find $k = \frac{1}{10}\ln 2 = 0.069$ and $c = -30e^{10k} = -30(2) = -60$. Substituting these values into the solution, we obtain, for the temperature of the body at any time t,

$$T = -60e^{-0.069t} + 30 \tag{1}$$

6.59 Find the initial temperature of the body described in the previous problem, just as it is placed into the room.

▌ We require T at $t = 0$. Substituting $t = 0$ into (1) of the previous problem, we find that $T = -60e^{(-0.069)(0)} + 30 = -60 + 30 = -30$°F.

6.60 A body at a temperature of 0°F is placed in a room whose temperature is kept at 100°F. If after 10 min the temperature of the body is 25°F, find an expression for the temperature of the body at time t.

▌ Here the temperature of the surrounding medium is the temperature of the room, which is held constant at $T_m = 100$. Thus, (1) of Problem 6.51 becomes $dT/dt + kT = 100k$; its solution is given by Problem 5.38 as $T = 100 + ce^{kt}$. At $t = 0$, we have $T = 0$; hence $0 = 100 + ce^{k(0)} = 100 + c$. Thus $c = -100$ and the solution becomes $T = 100 - 100e^{kt}$.

At $t = 10$, we have $T = 25$; hence $25 = 100 - 100e^{k(10)}$, or $e^{10k} = 0.75$, so that $k = -0.02877$. Thus, the last equation becomes $T = 100 - 100e^{-0.02877t}$.

6.61 Find the time needed for the body described in the previous problem to reach a temperature of 50°F.

▌ We require t when $T = 50$. Substituting $T = 50$ into the result of the previous problem and solving for t, we find $50 = 100 - 100e^{-0.02877t}$ or $e^{-0.02877t} = 0.5$. Then $t = (\ln 0.5)/(-0.02877) = 24.1$ min.

6.62 Find the temperature of the body described in Problem 6.60 after 20 min.

▌ Substituting $t = 20$ into the result of Problem 6.60, we have $T = 100 - 100e^{-0.02877(20)} = 43.75$°F.

6.63 A body at a temperature of 50°F is placed in an oven whose temperature is kept at 150°F. If after 10 min the temperature of the body is 75°F, find an expression for the temperature of the body at time t.

▌ Here the temperature of the surrounding medium is the temperature of the oven, which is held constant at $T_m = 150$°F. Thus (1) of Problem 6.51 becomes $dT/dt + kt = 150k$; its solution is given by Problem 5.39 (with $a = 150$) as $T = 150 + ce^{-kt}$. At $t = 0$, we have $T = 50$. Hence $50 = 150 + ce^{-k(0)}$, so $c = -100$ and the solution becomes $T = 150 - 100e^{-kt}$.

At $t = 10$, we have $T = 75$. Hence $75 = 150 - 100e^{-k(10)}$ or $e^{-10k} = 0.75$, so that $k = 0.02877$ and the last equation becomes $T = 150 - 100e^{-0.02877t}$.

6.64 Find the time required for the body described in the previous problem to reach a temperature of 100°F.

▌ Substituting $T = 100$ into the result of the previous problem, we find that $100 = 150 - 100e^{-0.02877t}$ or $e^{-0.02877t} = 0.50$. Then $t = (\ln 0.50)/(-0.02877) = 24.1$ min.

6.65 Find the time required for the body described in Problem 6.63 to reach a temperature of 70°F.

▌ Substituting $T = 70$ into the result of Problem 6.63, we find that $70 = 150 - 100e^{-0.02877t}$. Then $t = (\ln 0.80)/(-0.02877) = 7.76$ min.

6.66 Find the time required for the body described in Problem 6.63 to reach a temperature of 200°F.

▌ Since a body can never reach a temperature higher than that of the surrounding medium, which here is $T_m = 150$°, the body of Problem 6.63 can never attain a temperature of 200°F.

6.67 A body whose temperature is initially 100°C is allowed to cool in air whose temperature remains at a constant 20°C. Find a formula which gives the temperature of the body as a function of time t if it is observed that after 10 min the body has cooled to 40°C.

▌ If we let T denote the instantaneous temperature of the body in degrees Celsius and t denote the time in minutes since the body began to cool, then the rate of cooling is $\dfrac{dT}{dt} = k(T - 20)$. This equation can be solved either as a separable equation or as a linear equation. Regarding it as a separable equation, we rearrange it to $\dfrac{dT}{T - 20} = k\,dt$. Integration yields $\ln(T - 20) = kt + \ln|c|$, from which we write $\dfrac{T - 20}{c} = e^{kt}$ and find that

$T = 20 + ce^{kt}$. Since $T = 100$ when $t = 0$, it follows that $c = 80$, so that

$$T = 20 + 80e^{kt} \tag{1}$$

To determine the value of k, we use the fact that $T = 40$ when $t = 10$. Under this condition, (1) becomes $40 = 20 + 80e^{10k}$, or $e^{10k} = \frac{1}{4}$. We can solve approximately for k by writing $10k = -\ln 4 \approx -1.386$, so that $k \approx -0.1386$. The instantaneous temperature of the body is then given by $T = 20 + 80e^{-0.1386t}$.

.68 Find an alternative expression for the temperature in the previous problem.

▌ Rather than solving approximately for k, we can solve explicitly for e^k, obtaining $e^k = (\frac{1}{4})^{1/10}$. Then (1) can be written as $T = 20 + 80(e^k)^t = 20 + 80(\frac{1}{4})^{t/10}$.

.69 Find an alternative expression for the solution to Problem 6.52.

▌ Upon applying the second boundary condition in that problem, we obtained $50 = 100e^{-20k}$, which can be written as $\frac{1}{2} = (e^{-k})^{20}$ to give us $e^{-k} = (\frac{1}{2})^{1/20}$. Then the result of Problem 6.52 becomes $T = 100e^{-kt} = 100(e^{-k})^t = 100(\frac{1}{2})^{t/20}$.

.70 Rework Problem 6.53 using the expression obtained in the previous problem.

▌ We require t when $T = 25$. Substituting $T = 25$ into the result of the previous problem, we obtain $25 = 100(\frac{1}{2})^{t/20}$, which we rewrite as $(t/20)\ln\frac{1}{2} = \ln 0.25$. Then $t = (20\ln 0.25)/(\ln\frac{1}{2}) = 40$ min. The difference between this answer and the one obtained in Problem 6.53 is due to round-off error in computing k in Problem 6.52.

.71 Find an alternative expression for the solution to Problem 6.55.

▌ Upon applying the second boundary condition in that problem, we obtained $-40 = -50e^{-5k}$, which can be written as $0.8 = (e^{-k})^5$ to give $e^{-k} = (0.8)^{1/5}$. Then the result of Problem 6.55 becomes $T = -50(e^{-k})^t + 100 = -50(0.8)^{t/5} + 100$.

.72 Rework Problem 6.57 using the expression obtained in the previous problem.

▌ We require T when $t = 20$. Substituting $t = 20$ into the result of the previous problem, we find $T = -50(0.8)^4 + 100 = 79.5°F$.

.73 According to Newton's law of cooling, the rate at which a substance cools in air is proportional to the difference between the temperature of the substance and that of the air. If the temperature of the air is 30° and the substance cools from 100° to 70° in 15 min, find when the temperature will be 40°.

▌ Let T be the temperature of the substance at time t minutes. Then $\dfrac{dT}{dt} = -k(T - 30)$ or $\dfrac{dT}{T - 30} = -k\,dt$. (*Note*: The use of $-k$ is optional. We shall find that k is positive here; but if we used $+k$, we would find k to be equally negative.)

Integrating between the limits $t = 0$, $T = 100$ and $t = 15$, $T = 70$, we obtain $\displaystyle\int_{100}^{70}\frac{dT}{T-30} = -k\int_0^{15} dt$, so that $\ln 40 - \ln 70 = -15k = \ln\frac{4}{7}$ and $15k = \ln\frac{7}{4} = 0.56$.

Integrating between the limits $t = 0$, $T = 100$ and $t = t$, $T = 40$, we obtain $\displaystyle\int_{100}^{40}\frac{dT}{T-30} = -k\int_0^t dt$, so that $\ln 10 - \ln 70 = -kt$. Multiplying by -15 and rearranging, we obtain $15kt = 15\ln 7$, from which $t = (15\ln 7)/0.56 = 52$ min.

LOW PROBLEMS

.74 A tank initially holds V_0 gal of brine that contains a lb of salt. Another brine solution, containing b lb of salt per gallon, is poured into the tank at the rate of e gal/min while, simultaneously, the well-stirred solution leaves the tank at the rate of f gal/min (see Fig. 6.1). Find a differential equation for the amount of salt in the tank at any time t.

▌ Let Q denote the amount (in pounds) of salt in the tank at any time. The time rate of change of Q, dQ/dt, equals the rate at which salt enters the tank minus the rate at which salt leaves the tank. Salt enters the tank at the rate of be lb/min. To determine the rate at which salt leaves the tank, we first calculate the volume of brine in

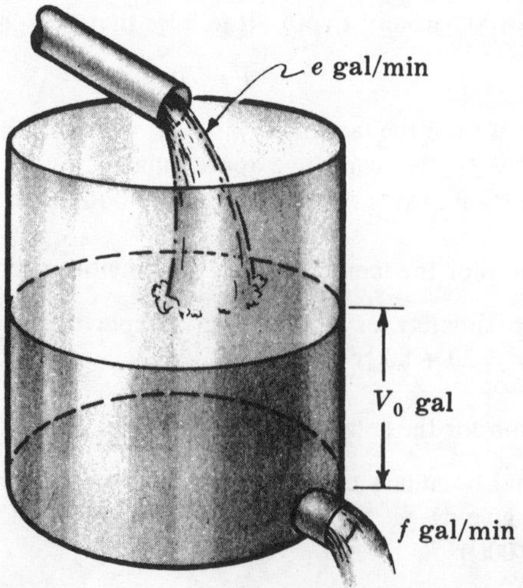

Fig. 6.1

the tank at any time t, which is the initial volume V_0 plus the volume of brine added et minus the volume of brine removed ft. Thus, the volume of brine in the tank at any time is $V_0 + et - ft$. The concentration of salt in the tank at any time is then $Q/(V_0 + et - ft)$, from which it follows that salt leaves the tank at the rate of $f[Q/(V_0 + et - ft)]$ lb/min. Thus, $dQ/dt = be - f[Q/(V_0 + et - ft)]$, so that

$$\frac{dQ}{dt} + \frac{f}{V_0 + (e-f)t} Q = be \qquad (1)$$

At $t = 0$, $Q = a$, so we also have the initial condition $Q(0) = a$.

6.75 A tank initially holds 100 gal of a brine solution containing 20 lb of salt. At $t = 0$, fresh water is poured into the tank at the rate of 5 gal/min, while the well-stirred mixture leaves the tank at the same rate. Find the amount of salt in the tank at any time t.

Here, $V_0 = 100$, $a = 20$, $b = 0$, $e = f = 5$, and (1) of Problem 6.74 becomes $\frac{dQ}{dt} + \frac{1}{20}Q = 0$. The solution to this differential equation is given in Problem 5.6 as $Q = ce^{-t/20}$. At $t = 0$, we are given that $Q = a = 20$. Substituting these values into the last equation, we find that $c = 20$, so that the solution can be rewritten as $Q = 20e^{-t/20}$.

Note that as $t \to \infty$, $Q \to 0$ as it should, since only fresh water is being added.

6.76 A tank initially holds 100 gal of a brine solution containing 1 lb of salt. At $t = 0$ another brine solution containing 1 lb of salt per gallon is poured into the tank at the rate of 3 gal/min, while the well-stirred mixture leaves the tank at the same rate. Find the amount of salt in the tank at any time t.

Here $V_0 = 100$, $a = 1$, $b = 1$, and $e = f = 3$; hence, (1) of Problem 6.74 becomes $\frac{dQ}{dt} + 0.03Q = 3$. The solution to this linear differential equation is $Q = ce^{-0.03t} + 100$.

At $t = 0$, $Q = a = 1$. Substituting these values into the last equation, we find $1 = ce^0 + 100$, or $c = -99$. Then the solution can be rewritten as $Q = -99e^{-0.03t} + 100$.

6.77 Find the time at which the mixture described in the previous problem contains 2 lb of salt.

We require t when $Q = 2$. Substituting $Q = 2$ into the result of the previous problem, we obtain $2 = -99e^{-0.03t} + 100$ or $e^{-0.03t} = \frac{98}{99}$, from which $t = -\frac{1}{0.03}\ln\frac{98}{99} = 0.338$ min.

6.78 A 50-gal tank initially contains 10 gal of fresh water. At $t = 0$, a brine solution containing 1 lb of salt per gallon is poured into the tank at the rate of 4 gal/min, while the well-stirred mixture leaves the tank at the rate of 2 gal/min. Find the amount of time required for overflow to occur.

Here $a = 0$, $b = 1$, $e = 4$, $f = 2$, and $V_0 = 10$. From Problem 6.74, the volume of brine in the tank at any time t is $V_0 + et - ft = 10 + 2t$. We require t when $10 + 2t = 50$; hence, $t = 20$ min.

6.79 Find the amount of salt in the tank described in the previous problem at the moment of overflow.

For this problem, (1) of Problem 6.74 becomes $\frac{dQ}{dt} + \frac{2}{10+2t}Q = 4$. This is a linear equation whose solution is given in Problem 5.49 as $Q = \frac{40t + 4t^2 + c}{10 + 2t}$.

At $t = 0$, $Q = a = 0$. Substituting these values into the last equation, we find that $c = 0$. We require Q at the moment of overflow, which is $t = 20$. Thus, $Q = \frac{40(20) + 4(20)^2}{10 + 2(20)} = 48$ lb.

6.80 A tank initially holds 10 gal of fresh water. At $t = 0$, a brine solution containing $\frac{1}{2}$ lb of salt per gallon is poured into the tank at a rate of 2 gal/min, while the well-stirred mixture leaves the tank at the same rate. Find the amount of salt in the tank at any time t.

Here $V_0 = 10$, $a = 0$, $b = \frac{1}{2}$, and $e = f = 2$. Hence (1) of Problem 6.74 becomes $\frac{dQ}{dt} + \frac{1}{5}Q = 1$; its solution is $Q = ce^{-t/5} + 5$ (see Problem 5.39 with $k = \frac{1}{5}$, $a = 5$, and T replaced by Q).

At $t = 0$, $Q = a = 0$; hence $0 = ce^{-0/5} + 5 = c + 5$, and $c = -5$. Thus the last equation becomes $Q = -5e^{-t/5} + 5$, which represents the amount of salt in the tank at any time t.

6.81 Determine the concentration of salt in the tank described in the previous problem at any time t.

The volume V of liquid in the tank remains constant at 10 gal. The concentration is $Q/V = -\frac{1}{2}e^{-t/5} + \frac{1}{2}$.

6.82 A tank initially holds 80 gal of a brine solution containing $\frac{1}{8}$ lb of salt per gallon. At $t = 0$, another brine solution containing 1 lb of salt per gallon is poured into the tank at the rate of 4 gal/min, while the well-stirred mixture leaves the tank at the rate of 8 gal/min. Find the amount of salt in the tank at any time t.

Here $V_0 = 80$, $a = \frac{1}{8}(80) = 10$, $b = 1$, $e = 4$, and $f = 8$. Then (1) of Problem 6.74 becomes

$$\frac{dQ}{dt} + \frac{8}{80 + (4 - 8)t}Q = 1(4) \qquad \text{or} \qquad \frac{dQ}{dt} + \frac{2}{20 - t}Q = 4$$

The solution of this equation is given in Problem 5.50 as $Q = 4(20 - t) + c(20 - t)^2$. Applying the initial condition $Q(0) = a = 10$, we get $10 = 4(20) + c(20)^2$, so that $c = -7/40$. Therefore, the amount of salt in the tank at time t is $Q = 4(20 - t) - \frac{7}{40}(20 - t)^2$.

6.83 Determine when the tank described in the previous problem will be empty.

We seek t corresponding to a volume $V = 0$. From Problem 6.74, we have $V = 0 = 80 + 4t - 8t$, so that $t = 20$ min.

6.84 Determine when the tank described in Problem 6.82 will hold 40 gal of solution.

We seek t corresponding to a volume $V = 40$. From Problem 6.74, we have $V = 40 = 80 + 4t - 8t$, so that $t = 10$ min.

6.85 Find the amount of salt in the tank described in Problem 6.82 when the tank contains exactly 40 gal of brine.

The amount of salt in the tank at any time t is given in Problem 6.82 as $Q = 4(20 - t) - \frac{7}{40}(20 - t)^2$. From Problem 6.84, the tank will contain 40 gal of solution when $t = 10$. At that time, $Q = 4(20 - 10) - \frac{7}{40}(20 - 10)^2 = 22.5$ lb.

6.86 Determine when the tank described in Problem 6.82 will contain the most salt.

The amount of salt in the tank at any time t is given in Problem 6.82 as

$$Q = 4(20 - t) - \tfrac{7}{40}(20 - t)^2 \tag{1}$$

Since d^2Q/dt^2 is always negative, the maximum value of Q occurs when $dQ/dt = 0$. Setting the derivative of (1) equal to zero, we get $-4 + \frac{14}{40}(20 - t) = 0$, from which $t = 8.57$ min. At that time, there will be 22.857 lb of salt in the tank.

6.87 A tank contains 100 gal of brine made by dissolving 80 lb of salt in water. Pure water runs into the tank at the rate of 4 gal/min, and the mixture, kept uniform by stirring, runs out at the same rate. Find the amount of salt in the tank at any time t.

▌ Here $V_0 = 100$, $a = 80$, $b = 0$, and $e = f = 4$. Then (*1*) of Problem 6.74 becomes $dQ/dt + 0.04Q = 0$, which has as its solution $Q = ce^{-0.04t}$ (see Problem 5.7). Applying the initial condition $Q(0) = a = 80$, we obtain $80 = c^{-0.04(0)} = c$, so the amount of salt in the tank at time t is $Q = 80e^{-0.04t}$.

6.88 Find the concentration of salt in the tank described in the previous problem at any time t.

▌ Since the outflow equals the inflow of liquid, the volume of liquid in the tank remains a constant $V = 100$. From the result of the previous problem, it follows that the concentration is $C = Q/V = 0.8e^{-0.04t}$.

6.89 Assume that the outflow of the tank described in Problem 6.87 runs into a second tank which contains 100 gal of pure water initially. The mixture in the second tank is kept uniform by constant stirring and is allowed to run out at the rate of 4 gal/min. Determine the amount of salt in the second tank at any time t.

▌ For the second tank, $V_0 = 100$, $a = 0$, $b = 0.8e^{-0.04t}$ (see the previous problem), and $e = f = 4$. Then (*1*) of Problem 6.74 becomes $dQ/dt + 0.04Q = 3.2e^{-0.04t}$, which has as its solution $Q = 3.2te^{-0.04t} + ce^{-0.04t}$ (see Problem 5.55). At $t = 0$, $Q = a = 0$; hence $0 = 3.2(0)e^{-0.04(0)} + ce^{-0.04(0)} = c$. The amount of salt in the second tank at any time t is thus $Q = 3.2te^{-0.04t}$.

6.90 Determine the amount of salt in each of the two tanks described in Problems 6.87 and 6.89 after 1 h.

▌ Using the results of the two problems with $t = 60$ min, we have $Q = 80e^{-0.04(60)} = 7.26$ lb of salt in the first tank, and $Q = 3.2(60)e^{-0.04(60)} = 17.42$ lb of salt in the second tank.

6.91 Determine when the amounts of salt in the tanks described in Problems 6.87 and 6.89 will be equal.

▌ We equate the results of the two problems to obtain $80e^{-0.04t} = 3.2te^{-0.04t}$, from which $t = 80/3.2 = 25$ min.

6.92 A tank contains 100 gal of brine made by dissolving 60 lb of salt in water. Salt water containing 1 lb of salt per gallon runs in at the rate of 2 gal/min, and the mixture, kept uniform by stirring, runs out at the rate of 3 gal/min. Find the amount of salt in the tank at the end of 1 h.

▌ Here $V_0 = 100$, $a = 60$, $b = 1$, $e = 2$, and $f = 3$. Then (*1*) of Problem 6.74 becomes $\dfrac{dQ}{dt} + \dfrac{3}{100 - t}Q = 2$, which has as its solution $Q = 100 - t + c(100 - t)^3$ (see Problem 5.48).

At $t = 0$, $Q = a = 60$; hence $60 = 100 + c(100)^3$, so that $c = -0.00004$ and the solution becomes $Q = (100 - t) - 0.00004(100 - t)^3$. At $t = 60$ min, this equation yields $Q = (100 - 60) - 0.00004(100 - 60)^3 = 37.44$ lb.

6.93 A cylindrical tank contains 40 gal of a salt solution containing 2 lb of salt per gallon. A salt solution of concentration 3 lb/gal flows into the tank at 4 gal/min. How much salt is in the tank at any time if the well-stirred mixture flows out at 4 gal/min?

▌ Let the tank contain A lb of salt after t minutes. Then

Rate of change of amount of salt = rate of entrance − rate of exit

$$\frac{dA}{dt}\,\frac{\text{lb}}{\text{min}} = 3\,\frac{\text{lb}}{\text{gal}} \times 4\,\frac{\text{gal}}{\text{min}} - \frac{A\text{ lb}}{40\text{ gal}} \times 4\,\frac{\text{gal}}{\text{min}}$$

Solving the equation $\dfrac{dA}{dt} = 12 - \dfrac{A}{10}$ subject to $A = 40(2) = 80$ at $t = 0$, we find $A = 120 - 40e^{-t/10}$.

6.94 A right circular cone (Fig. 6.2) is filled with water. In what time will the water empty through an orifice O of cross-sectional area a at the vertex? Assume the velocity of exit is $v = k\sqrt{2gh}$, where h is the instantaneous height (*head*) of the water level above O, and k is the *discharge coefficient*.

▌ At time t the water level is at h. At time $t + dt$, $dt > 0$, the water level is at $h + dh$, where $dh < 0$. We have

Change in volume of water = amount of water leaving

$$-\pi r^2\,dh = av\,dt = ak\sqrt{2gh}\,dt$$

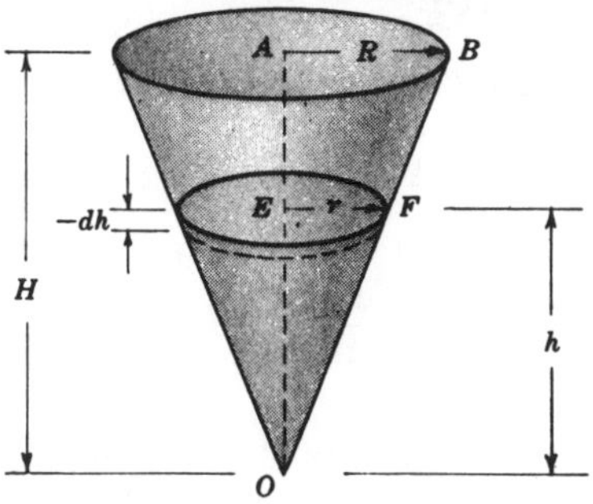

Fig. 6.2

From similar triangles OAB and OEF, $r = Rh/H$. Then the above equation becomes $-\dfrac{\pi R^2 h^2}{H^2}\,dh = ak\sqrt{2gh}\,dt$. Its solution, subject to the condition $h = H$ at $t = 0$, is $t = \dfrac{2\pi R^2}{5akH^2\sqrt{2g}}(H^{5/2} - h^{5/2})$. The time required for emptying is the time when $h = 0$, or $t = \dfrac{\pi R^2}{5ak}\sqrt{\dfrac{2H}{g}}$.

6.95 A hemispherical tank of radius R is initially filled with water. At the bottom of the tank, there is a hole of radius r through which the water drains under the influence of gravity. Find an expression for the depth of the water in the tank at any time t.

▌ Let the origin be chosen at the lowest point of the tank, let y be the instantaneous depth of the water, and let x be the instantaneous radius of the free surface of the water (Fig. 6.3). Then in an infinitesimal interval dt, the water level will fall by the amount dy, and the resultant decrease in the volume of water in the tank will be $dV = \pi x^2\,dy$. This, of course, must equal in magnitude the volume of water that leaves the orifice during the same interval dt. Now by *Torricelli's law*, the velocity with which a liquid issues from an orifice is $v = \sqrt{2gh}$, where g is the acceleration of gravity and h is the instantaneous height, or *head*, of the liquid above the orifice.

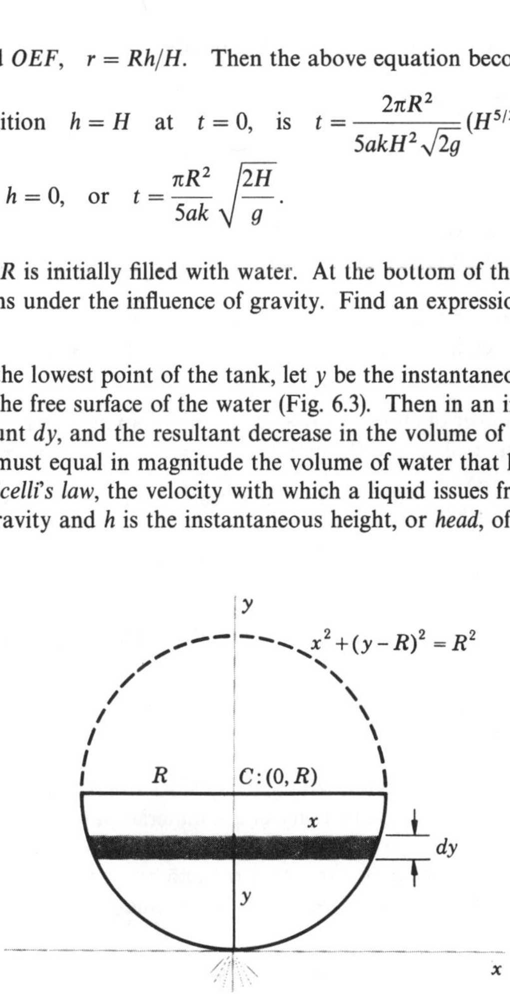

Fig. 6.3 Vertical plane section through the center of a hemispherical tank.

In the interval dt, then, a stream of water of length $v \times dt = \sqrt{2gy}\,dt$ and of cross-sectional area πr^2 will emerge from the outlet. The volume of this stream of water is $dV = \text{area} \times \text{length} = \pi r^2\sqrt{2gy}\,dt$. Now, equating the magnitudes of our two expressions for dV, we obtain the differential equation

$$-\pi x^2\,dy = \pi r^2\sqrt{2gy}\,dt \qquad (1)$$

The minus sign indicates that as t increases, the depth y decreases.

Before this equation can be solved, x must be expressed in terms of y. This is easily done through the use of the equation of the circle which describes a maximal vertical cross section of the tank: $x^2 + (y - R)^2 = R^2$, or $x^2 = 2yR - y^2$. With this relation, *(1)* can be written as $\pi(2yR - y^2)\,dy = -\pi r^2\sqrt{2gy}\,dt$. This is a simple separable equation that can be solved without difficulty. Separation yields $(2Ry^{1/2} - y^{3/2})\,dy = -r^2\sqrt{2g}\,dt$, and integration then gives $\frac{4}{3}Ry^{3/2} - \frac{2}{5}y^{5/2} = -r^2\sqrt{2g}\,t + c$. Since $y = R$ when $t = 0$, we find $\frac{14}{15}R^{5/2} = c$, and thus $\frac{4}{3}Ry^{3/2} - \frac{2}{5}y^{5/2} = -r^2\sqrt{2g}\,t + \frac{14}{15}R^{5/2}$.

6.96 Determine how long it will take the tank described in the previous problem to empty.

▮ We require t corresponding to $y = 0$. From the result of the previous problem, we have $0 = -r^2\sqrt{2g}\,t + \frac{14}{15}R^{5/2}$, from which $t = \dfrac{14}{15}\dfrac{R^{5/2}}{r^2\sqrt{2g}}$.

6.97 A 100-gal tank is filled with brine containing 60 lb of dissolved salt. Water runs into the tank at the rate of 2 gal/min and the mixture, kept uniform by stirring, runs out at the same rate. How much salt is in the tank after 1 h?

▮ Let s be the number of pounds of salt in the tank after t minutes, so that the concentration then is $s/100$ lb/gal. During the interval dt, $2\,dt$ gal of water flows into the tank, and $2\,dt$ gal of brine containing $\dfrac{2s}{100}\,dt = \dfrac{s}{50}\,dt$ lb of salt flows out. Thus, the change ds in the amount of salt in the tank is $ds = -\dfrac{s}{50}\,dt$. Integrating yields $s = ce^{-t/50}$.

At $t = 0$, $s = 60$; hence, $c = 60$ and the solution becomes $s = 60e^{-t/50}$. When $t = 60$ min, $s = 60e^{-6/5} = 60(0.301) = 18$ lb.

6.98 The air in a certain room with dimensions $150 \times 50 \times 12$ ft tested at 0.2 percent CO_2. Fresh air containing 0.05 percent CO_2 was then admitted by ventilators at the rate of 9000 ft³/min. Find the percentage of CO_2 after 20 min.

▮ Let x denote the number of cubic feet of CO_2 in the room at time t, so that the concentration of CO_2 then is $x/90{,}000$. During the interval dt, the amount of CO_2 entering the room is $9000(0.0005)\,dt$ ft³, and the amount leaving is $9000\dfrac{x}{90{,}000}\,dt$ ft³. Hence, the change dx in the interval dt is $dx = 9000\left(0.0005 - \dfrac{x}{90{,}000}\right)dt = -\dfrac{x-45}{10}\,dt$. Integrating yields $10\ln(x-45) = -t + \ln c_1$ or $x = 45 + ce^{-t/10}$.

At $t = 0$, $x = 0.002(90{,}000) = 180$. Then $c = 180 - 45 = 135$, and the solution becomes $x = 45 + 135e^{-t/10}$. When $t = 20$, $x = 45 + 135e^{-2} = 63$. The percentage of CO_2 is then $\dfrac{63}{90{,}000} = 0.0007 = 0.07$ percent.

6.99 Under certain conditions the constant quantity Q in calories/second of heat flowing through a wall is given by $Q = -kA\,dT/dx$, where k is the conductivity of the material, A (cm²) is the area of a face of the wall perpendicular to the direction of flow, and T is the temperature x cm from that face such that T decreases as x increases. Find the heat flow per hour through 1 m² of a refrigerator room wall 125 cm thick for which $k = 0.0025$, if the temperature of the inner face is -5°C and that of the outer face is 75°C. (See Fig. 6.4.)

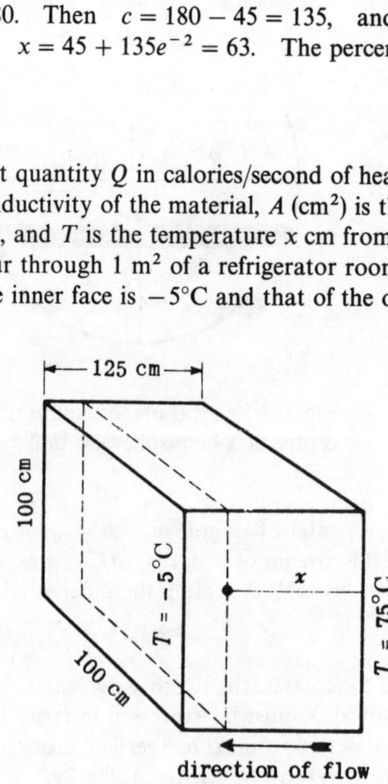

Fig. 6.4

Let x denote the distance of a point within the wall from the outer face. Integrating $dT = -\frac{Q}{kA}\,dx$ from $x = 0$, $T = 75$ to $x = 125$, $T = -5$, we get $\int_{75}^{-5} dT = -\frac{Q}{kA}\int_0^{125} dx$ or $80 = \frac{Q}{kA}(125)$, from which $Q = \frac{80kA}{125} = \frac{80(0.0025)(100)^2}{125} = 16$ cal/s. Thus, the flow of heat per hour is $3600Q = 57{,}600$ cal.

6.100 A steam pipe 20 cm in diameter is protected with a covering 6 cm thick for which $k = 0.0003$. Find the heat loss per hour through a meter length of the pipe if the surface of the pipe is at 200°C and the outer surface of the covering is at 30°C. (See Fig. 6.5.)

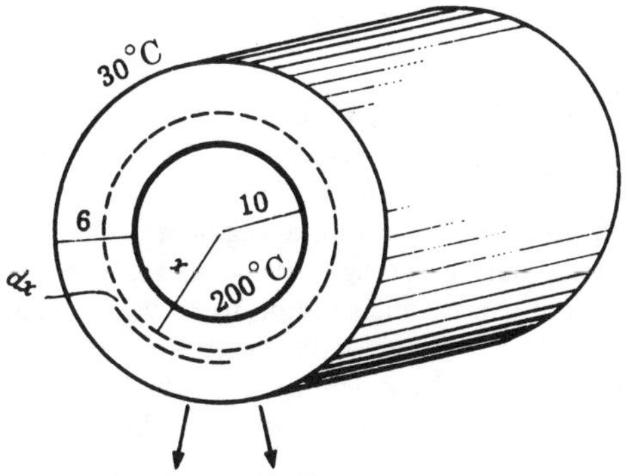

Fig. 6.5

At a distance $x > 10$ cm from the center of the pipe, heat is flowing across a cylindrical shell of surface area $2\pi x$ cm² per centimeter of length of pipe. From Problem 6.99,

$$Q = -kA\frac{dT}{dx} = -2\pi kx\frac{dT}{dx} \qquad \text{or} \qquad 2\pi k\,dT = -Q\frac{dx}{x}$$

Integrating between the limits $T = 30$, $x = 16$ and $T = 200$, $x = 10$, we get $2\pi k\int_{30}^{200} dT = -Q\int_{16}^{10}\frac{dx}{x}$, or $340\pi k = Q(\ln 16 - \ln 10) = Q\ln 1.6$. Then $Q = \frac{340\pi k}{\ln 1.6}$ cal/s, and the heat loss per hour through a meter length of pipe is $100(60)^2Q = 245{,}000$ cal.

6.101 Find the temperature at a distance $x > 10$ cm from the center of the pipe described in the previous problem.

Integrating $2\pi k\,dT = -\frac{340\pi k}{\ln 1.6}\frac{dx}{x}$ between the limits $T = 30$, $x = 16$ and $T = T$, $x = x$, we get $\int_{30}^{T} dT = -\frac{170}{\ln 1.6}\int_{16}^{x}\frac{dx}{x}$, or $T - 30 = -\frac{170}{\ln 1.6}\ln\frac{x}{16}$. Then $T = 30 + \frac{170}{\ln 1.6}\ln\frac{16}{x}$.

Check: When $x = 10$, $T = 30 + \frac{170}{\ln 1.6}\ln 1.6 = 200$°C. When $x = 16$, $T = 30 + 0 = 30$°C.

6.102 Find the time required for a cylindrical tank of radius 8 ft and height 10 ft to empty through a round hole of radius 1 in at the bottom of the tank, given that water will issue from such a hole with velocity approximately $v = 4.8\sqrt{h}$ ft/s, where h is the depth of the water in the tank.

The volume of water that runs out per second may be thought of as the volume a cylinder 1 in in radius and of height v. Hence, the volume which runs out in dt seconds is $\pi\left(\frac{1}{12}\right)^2(4.8\sqrt{h})\,dt = \frac{\pi}{144}(4.8\sqrt{h})\,dt$. Denoting by dh the corresponding drop in the water level in the tank, we note that the volume of water which runs out in time dt is also given by $64\pi\,dh$. Hence,

$$\frac{\pi}{144}(4.8\sqrt{h})\,dt = -64\pi\,dh \qquad \text{or} \qquad dt = -\frac{64(144)}{4.8}\frac{dh}{\sqrt{h}} = -1920\frac{dh}{\sqrt{h}}.$$

Integrating between $t = 0$, $h = 10$ and $t = t$, $h = 0$, we get $\int_0^t dt = -1920\int_{10}^{0}\frac{dh}{\sqrt{h}}$, from which $t = -3840\sqrt{h}\Big|_{10}^{0} = 3840\sqrt{10}\,\text{s} = 3\text{ h }22\text{ min}.$

6.103 As a possible model of a diffusion process in the bloodstream in the human body, consider a solution moving with constant velocity v through a cylindrical tube of length L and radius r. We suppose that as the solution moves through the tube, some of the solute which it contains diffuses through the wall of the tube into an ambient solution of the same solute of lower concentration, while some continues to be transported through the tube. As variables, we let x be a distance coordinate along the tube and $y(x)$ be the concentration of the solute at any point x, assumed uniform over the cross section of the tube. As boundary conditions, we assume that $y(0) = y_0$ and $y(L) = y_L(<y_0)$ are known. Find an expression for the concentration $y(x)$ at any point along the tube.

▌ As a principle to use in formulating this problem, we have *Frick's law: The time rate at which a solute diffuses through a thin membrane in a direction perpendicular to the membrane is proportional to the area of the membrane and to the difference between the concentrations of the solute on the two sides of the membrane.*

We begin by considering conditions in a typical segment of the tube between x and $x + \Delta x$ (Fig. 6.6). The concentration of the solution entering the segment is $y(x)$; the concentration of the solution leaving the segment is $y(x + \Delta x)$. In the time Δt that it takes the solution to move through the segment, an amount of solute equal to concentration × volume $= y(x)\pi r^2\,\Delta x$ enters the left end of the segment, and the amount $y(x + \Delta x)\pi r^2\,\Delta x$ leaves the right end of the segment. The difference, $[y(x) - y(x + \Delta x)]\pi r^2\,\Delta x$, must have left the segment by diffusion through the wall of the tube. The expression for this amount, as given by Frick's law, is

$$\text{Rate of diffusion} \times \text{time} = k(2\pi r\,\Delta x)[y(x + \theta\,\Delta x) - c]\,\Delta t$$

where $x + \theta\,\Delta x$, for $0 < \theta < 1$, is a typical point between x and $x + \Delta x$ at which to assume an "average" value of the concentration, and c, assumed constant, is the concentration of the solute in the fluid surrounding the tube. Equating the two expressions we have found for the loss of solute by diffusion, we have

$$[y(x) - y(x + \Delta x)]\pi r^2\,\Delta x = k(2\pi r\,\Delta x)[y(x + \theta\,\Delta x) - c]\,\Delta t$$

Since $\Delta x = v\,\Delta t$, this simplifies to $\dfrac{y(x) - y(x + \Delta x)}{\Delta x} = \dfrac{2k}{rv}[y(x + \theta\,\Delta x) - c]$ and, taking limits (as though the system were continuous), we get $-\dfrac{dy}{dx} = \dfrac{2k}{rv}[y(x) - c]$.

By hypothesis, $y > c$; hence $y(x) - c \neq 0$, and we can solve this equation by separating variables to obtain $\dfrac{dy}{y - c} = -\dfrac{2k}{rv}dx$. Integration then gives $\ln(y - c) = -\dfrac{2k}{rv}x + \ln B$. Putting $x = 0$ and $y = y_0$, we find that $\ln B = \ln(y_0 - c)$, and the solution becomes $\ln\dfrac{y - c}{y_0 - c} = -\dfrac{2k}{rv}x$, or $y = c + (y_0 - c)e^{-2kx/rv}$.

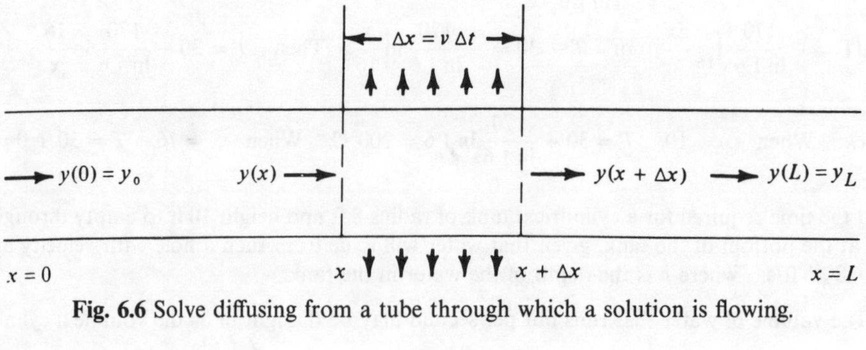

Fig. 6.6 Solve diffusing from a tube through which a solution is flowing.

ELECTRIC CIRCUIT PROBLEMS

6.104 An RL circuit has an emf of 5 V, a resistance of 50 Ω, an inductance of 1 H, and no initial current. Find the current in the circuit at any time t.

▌ Here $E = 5$, $R = 50$, and $L = 1$, so (1) of Problem 1.87 becomes $dI/dt + 50I = 5$. Its solution is $I = ce^{-50t} + \frac{1}{10}$ (see Problem 5.31). At $t = 0$, $I = 0$; thus, $0 = ce^{-50(0)} + \frac{1}{10}$, or $c = -\frac{1}{10}$. The current at any time t is then $I = -\frac{1}{10}e^{-50t} + \frac{1}{10}$.

The quantity $-\frac{1}{10}e^{-50t}$ in this result is called the *transient current*, since this quantity goes to zero ("dies out") as $t \to \infty$. The quantity $\frac{1}{10}$ is called the *steady-state current*. As $t \to \infty$, the current I approaches the value of the steady-state current.

6.105 An RL circuit has an emf given (in volts) by $3 \sin 2t$, a resistance of $10\,\Omega$, an inductance of 0.5 H, and an initial current of 6 A. Find the current in the circuit at any time t.

▌ Here $E = 3 \sin 2t$, $R = 10$, and $L = 0.5$, so (1) of Problem 1.87 becomes $dI/dt + 20I = 6 \sin 2t$. Its solution, from Problem 5.51, is $I = ce^{-20t} + \frac{30}{101} \sin 2t - \frac{3}{101} \cos 2t$. At $t = 0$, $I = 6$; hence, $6 = ce^{-20(0)} + \frac{30}{101} \sin 2(0) - \frac{3}{101} \cos 2(0)$ or $6 = c - \frac{3}{101}$, whence $c = \frac{609}{101}$. The current at any time t is then $I = \frac{609}{101}e^{-20t} + \frac{30}{101} \sin 2t - \frac{3}{101} \cos 2t$.

As in Problem 6.104, the current is the sum of a transient current, here $\frac{609}{101}e^{-20t}$, and a steady-state current, $\frac{30}{101} \sin 2t - \frac{3}{101} \cos 2t$.

6.106 Rewrite the steady-state current of Problem 6.105 in the form $A \sin(2t - \phi)$. The angle ϕ is called the *phase angle*.

▌ Since $A \sin(2t - \phi) = A(\sin 2t \cos \phi - \cos 2t \sin \phi)$, we require

$$I_s = \frac{30}{101} \sin 2t - \frac{3}{101} \cos 2t = A \cos \phi \sin 2t - A \sin \phi \cos 2t$$

Thus, we have $A \cos \phi = \frac{30}{101}$ and $A \sin \phi = \frac{3}{101}$. It now follows that

$$\left(\tfrac{30}{101}\right)^2 + \left(\tfrac{3}{101}\right)^2 = A^2 \cos^2 \phi + A^2 \sin^2 \phi = A^2(\cos^2 \phi + \sin^2 \phi) = A^2$$

and $\tan \phi = \dfrac{A \sin \phi}{A \cos \phi} = \dfrac{3/101}{30/101} = \dfrac{1}{10}$. Consequently, I_s has the required form if $A = \sqrt{\dfrac{909}{(101)^2}} = \dfrac{3}{\sqrt{101}}$ and $\phi = \arctan \dfrac{1}{10}$.

6.107 Determine the amplitude and frequency of the steady-state current in the previous problem.

▌ The amplitude is $A = 3/\sqrt{101}$, and the frequency is $f = \dfrac{2}{2\pi} = \dfrac{1}{\pi}$.

6.108 A resistor of $15\,\Omega$ and an inductance of 3 H are connected in series with a 60-Hz sinusoidal voltage source having amplitude 110 V. Find an expression for the steady-state current at any time t if initially there is no current in the system.

▌ Here $R = 15$, $L = 3$, and $E = 110 \sin 2\pi(60)t = 110 \sin 120\pi t$. Thus, (1) of Problem 1.87 becomes $dI/dt + 5I = (110/3) \sin 120\pi t$; its solution, from Problem 5.53, is

$$I = \frac{22}{3}\,\frac{\sin 120\pi t - 24\pi \cos 120\pi t}{1 + 576\pi^2} + ce^{-5t}$$

When $t = 0$, $I = 0$. Then $c = \dfrac{22(24\pi)}{3(1 + 576\pi^2)}$ and $I = \dfrac{22}{3}\,\dfrac{\sin 120\pi t - 24\pi \cos 120\pi t + 24\pi e^{-5t}}{1 + 576\pi^2}$.

As $t \to \infty$, $I \to \dfrac{22}{3}\,\dfrac{\sin 120\pi t - 24\pi \cos 120\pi t}{1 + 576\pi^2}$, which is the steady-state current.

6.109 Rewrite the steady-state current of the previous problem in the form $A \sin(120\pi t - \phi)$.

▌ Since $A \sin(120\pi t - \phi) = A \sin 120\pi t \cos \phi - A \cos 120\pi t \sin \phi$, we must have $A \cos \phi = \dfrac{22}{3(1 + 576\pi^2)}$ and $A \sin \phi = \dfrac{(22)(24\pi)}{3(1 + 576\pi^2)}$. It now follows that

$$\left[\frac{22}{3(1 + 576\pi^2)}\right]^2 + \left[\frac{(22)(24\pi)}{3(1 + 576\pi^2)}\right]^2 = A^2 \cos^2 \phi + A^2 \sin^2 \phi = A^2 \qquad \text{or} \qquad A = \frac{22}{3\sqrt{1 + 576\pi^2}}$$

and $$\tan\phi = \frac{A\sin\phi}{A\cos\phi} = \frac{22(24\pi)/3(1+576\pi^2)}{22/3(1+576\pi^2)} = 24\pi \quad \text{or} \quad \phi = \arctan 24\pi = 1.56 \text{ rad}$$

6.110 Determine the amplitude and frequency of the steady-state current of the previous problem.

The amplitude is $A = \dfrac{22}{3\sqrt{1+576\pi^2}} \approx 0.097$, while the frequency is $f = \dfrac{120\pi}{2\pi} = 60$.

6.111 Determine the period of the steady-state current in Problem 6.109.

The period is the reciprocal of the frequency. The results of the previous problem show that the period is 1/60

6.112 The steady-state current in a circuit is known to be $\frac{5}{17}\sin t - \frac{3}{17}\cos t$. Rewrite this current in the form $A\sin(t-\phi)$.

Since $A\sin(t-\phi) = A\sin t\cos\phi - A\cos t\sin\phi$, we must have $A\cos\phi = \frac{5}{17}$ and $A\sin\phi = \frac{3}{17}$. It then follows that $(\frac{5}{17})^2 + (\frac{3}{17})^2 = A^2\cos^2\phi + A^2\sin^2\phi = A^2$, so $A = \sqrt{2/17}$. Also, $\tan\phi = \dfrac{\sin\phi}{\cos\phi} = \dfrac{A\sin\phi}{A\cos\phi} = \dfrac{3/17}{5/17} = \dfrac{3}{5}$ or $\phi = \arctan\dfrac{3}{5} = 0.54$ rad. The current is $\sqrt{2/17}\sin(t-0.54)$.

6.113 Rewrite the steady-state current of the previous problem in the form $A\cos(t-\phi)$.

Since $A\cos(t-\phi) = A\cos t\cos\phi + A\sin t\sin\phi$, it follows that $A\cos\phi = -\frac{3}{17}$ and $A\sin\phi = \frac{5}{17}$. Then $\left(\dfrac{-3}{17}\right)^2 + \left(\dfrac{5}{17}\right)^2 = A^2\cos^2\phi + A^2\sin^2\phi = A^2$, so $A = \sqrt{2/17}$, as before. Now, however, $\tan\phi = \dfrac{\sin\phi}{\cos\phi} = \dfrac{A\sin\phi}{A\cos\phi} = \dfrac{5/17}{-3/17} = -\dfrac{5}{3}$ so $\phi = \arctan(-\frac{5}{3}) = -1.03$ rad. The current is $\sqrt{2/17}\cos(t+1.03)$.

6.114 Determine the amplitude and frequency of the steady-state current in the previous three problems.

The amplitude is $A = \sqrt{2/17}$, while the frequency is $f = \dfrac{1}{2\pi}$ (the numerator of f is the coefficient of t).

6.115 Determine the period of the steady-state current in Problem 6.112.

From the results of the previous problem, we have period $= 1/f = 2\pi$.

6.116 An RL circuit with no source of emf has an initial current given by I_0. Find the current at any time t.

In this case (1) of Problem 1.87 becomes $\dfrac{dI}{dt} + \dfrac{R}{L}I = 0$. Its solution (see Problem 5.8 with I replacing N and $k = -R/L$) is $I = ce^{-(R/L)t}$. At $t = 0$, $I = I_0$; hence $I_0 = ce^{-(R/L)(0)} = c$, and the current is, as a function of time, $I = I_0 e^{-(R/L)t}$.

6.117 Determine the steady-state current for the circuit described in the previous problem.

As $t \to \infty$, I tends to zero. Thus, when the steady state is reached, there is no current flowing through the circuit.

6.118 Determine the current in a simple series RL circuit having a resistance of 10 Ω, an inductance of 1.5 H, and an emf of 9 V if initially the current is 6 A.

Here $E(t) = 9$, $R = 10$, and $L = 1.5$. Then (1) of Problem 1.87 becomes $\dfrac{dI}{dt} + \dfrac{20}{3}I = 6$; its solution is $I = \frac{9}{10} + ce^{-(20/3)t}$ (see Problem 5.33). At $t = 0$, $I = 6$; hence $6 = \frac{9}{10} + ce^{-(20/3)(0)}$, so that $c = 6 - \frac{9}{10} = \frac{51}{10}$. Thus, $I = \frac{9}{10} + \frac{51}{10}e^{-(20/3)t}$.

6.119 Identify the transient component of the current found in the previous problem.

As $t \to \infty$, $I \to I_s = \frac{9}{10}$. The transient component is $I_t = I - I_s = \frac{9}{10} + \frac{51}{10}e^{-(20/3)t} - \frac{9}{10} = \frac{51}{10}e^{-(20/3)t}$.

6.120 An RC circuit has an emf (in volts) given by $400 \cos 2t$, a resistance of $100\ \Omega$, and a capacitance of 10^{-2} F. Initially there is no charge on the capacitor. Find the current in the circuit at any time t.

▌ We first find the charge q on the capacitor and then the current using the formula $I = dq/dt$. Here $E = 400 \cos 2t$, $R = 100$, and $C = 10^{-2}$. Then (1) of Problem 1.90 becomes $dq/dt + q = 4 \cos 2t$; its solution is $q = ce^{-t} + \frac{8}{5} \sin 2t + \frac{4}{5} \cos 2t$ (see Problem 5.52).

At $t = 0$, $q = 0$; hence, $0 = ce^{-(0)} + \frac{8}{5} \sin 2(0) + \frac{4}{5} \cos 2(0)$, so that $c = -\frac{4}{5}$. Thus

$$q = -\frac{4}{5} e^{-t} + \frac{8}{5} \sin 2t + \frac{4}{5} \cos 2t \quad \text{and} \quad I = \frac{dq}{dt} = \frac{4}{5} e^{-t} + \frac{16}{5} \cos 2t - \frac{8}{5} \sin 2t.$$

6.121 A resistor $R = 5\ \Omega$ and a condenser $C = 0.02$ F are connected in series with a battery $E = 100$ V. If at $t = 0$ the charge on the condenser is 5 C, find Q and the current I for $t > 0$.

▌ With $E = 100$, $R = 5$, and $C = 0.02$, (1) of Problem 1.90 becomes $dq/dt + 10q = 20$; its solution is $q = 2 + ce^{-10t}$ (see Problem 5.30). At $t = 0$, $q = 5$; hence, $5 = 2 + ce^{-10(0)}$ and $c = 3$. Thus $q = 2 + 3e^{-10t}$ and $I = dq/dt = -30e^{-10t}$.

6.122 Specify the steady-state and transient components of the current found in Problem 6.120.

▌ The current is $I = \frac{4}{5}e^{-t} + \frac{16}{5} \cos 2t - \frac{8}{5} \sin 2t$. As $t \to \infty$, the current approaches the steady-state value $I_s = \frac{16}{5} \cos 2t, - \frac{8}{5} \sin 2t$. The transient component is $I_t = I - I_s = \frac{4}{5}e^{-t}$.

6.123 Determine the amplitude, frequency, and period of the steady-state current of the previous problem.

▌ The amplitude is $A = \sqrt{\left(\frac{16}{5}\right)^2 + \left(-\frac{8}{5}\right)^2} = \frac{8}{\sqrt{5}}$; the frequency is $f = \frac{2}{2\pi} = \frac{1}{\pi}$, and the period is $\frac{1}{f} = \pi$.

6.124 Rewrite the steady-state current in Problem 6.122 in the form $A \cos (2t + \phi)$.

▌ Since $A \cos (2t + \phi) = A \cos 2t \cos \phi - A \sin 2t \sin \phi$, we have $A \cos \phi = \frac{16}{5}$ and $A \sin \phi = \frac{8}{5}$. From the previous problem, we know the amplitude A is $8/\sqrt{5}$. Also, $\tan \phi = \frac{A \sin \phi}{A \cos \phi} = \frac{8/5}{16/5} = \frac{1}{2}$, so $\phi = \arctan \frac{1}{2} = 0.4636$ rad. Thus $I_s = \frac{8}{\sqrt{5}} \cos (2t + 0.4636)$.

6.125 An RC circuit has an emf of 5 V, a resistance of $10\ \Omega$, a capacitance of 10^{-2} F, and initially a charge of 5 C on the capacitor. Find an expression for the charge on the capacitor at any time t.

▌ Here $E(t) = 5$, $R = 10$, and $C = 0.01$, so (1) of Problem 1.90 becomes $\dot{q} + 10q = \frac{1}{2}$; its solution is $q = \frac{1}{20} + ce^{-10t}$ (see Problem 5.34). At $t = 0$, $q = 5$; hence $5 = \frac{1}{20} + ce^{-10(0)}$, so that $c = \frac{99}{20}$. Thus, $q = \frac{1}{20}(1 + 99e^{-10t})$.

6.126 Determine the current flowing through the circuit described in the previous problem.

▌ $$I = \frac{dq}{dt} = -\frac{99}{2} e^{-10t}.$$

6.127 Find the charge on the capacitor in a simple RC circuit having a resistance of $10\ \Omega$, a capacitance of 0.001 F, and an emf of $100 \sin 120\pi t$ V, if there is no initial charge on the capacitor.

▌ Here $E(t) = 100 \sin 120\pi t$, $R = 10$, and $C = 0.001$. Then (1) of Problem 1.90 becomes $\dot{q} + 100q = 10 \sin 120\pi t$, and its solution is $q = \frac{10 \sin 120\pi t - 12\pi \cos 120\pi t}{100 + 144\pi^2} + Ae^{-100t}$ (see Problem 5.54).

At $t = 0$, $q = 0$; hence $0 = \frac{0 - 12\pi}{100 + 144\pi^2} + A$, or $A = \frac{3\pi}{25 + 36\pi^2}$. Then

$$q = \frac{10 \sin 120\pi t - 12\pi \cos 120\pi t}{100 + 144\pi^2} + \frac{3\pi}{25 + 36\pi^2} e^{-100t}.$$

6.128 Determine the steady-state current in the circuit described in the previous problem.

▌ The current is $I = \dfrac{dq}{dt} = 120\pi \dfrac{10\cos 120\pi t + 12\pi \sin 120\pi t}{100 + 144\pi^2} - \dfrac{300\pi}{25 + 36\pi^2} e^{-100t}$. As $t \to \infty$, $I \to I_s = 120\pi \dfrac{10\cos 120\pi t + 12\pi \sin 120\pi t}{100 + 144\pi^2}$.

6.129 Find the charge (as a function of time) on the capacitor in a simple RC circuit having no applied electromagnetic force if the initial charge is Q_0.

▌ With $E(t) = 0$, (1) of Problem 1.90 becomes $\dot{q} + \dfrac{1}{RC} q = 0$. Its solution is $q = ce^{-t/RC}$ (see Problem 5.8 with q replacing N and $k = -1/RC$). At $t = 0$, $q = Q_0$; so $Q_0 = ce^{-(0)/RC} = c$, and $q = Q_0 e^{-t/RC}$.

6.130 Find the current in the circuit described in the previous problem.

▌ By differentiating the result of that problem, we get $I = -\dfrac{Q_0}{RC} e^{-t/RC}$.

6.131 Determine the steady-state current in the previous problem.

▌ As $t \to \infty$, $I \to 0 = I_s$.

6.132 Find an expression for the charge on the capacitor of a simple series LC circuit (consisting of an inductor and a capacitor only) if the initial charge on the capacitor is Q_0 and there is no initial current in the circuit.

▌ Applying Kirchoff's loop law (see Problem 1.81), we have $L\dfrac{dI}{dt} = L\dfrac{d^2Q}{dt^2}$ for the potential drop across L and $\dfrac{Q}{C}$ for the potential drop across C. Then $L\dfrac{d^2Q}{dt^2} + \dfrac{Q}{C} = 0$.

Since $\dfrac{dQ}{dt} = I$, we have $\dfrac{d^2Q}{dt^2} = \dfrac{dI}{dt} = \dfrac{dI}{dQ}\dfrac{dQ}{dt} = I\dfrac{dI}{dQ}$ so that the last equation becomes $LI\dfrac{dI}{dQ} + \dfrac{Q}{C} = 0$ or $LI\,dI + \dfrac{Q}{C}\,dQ = 0$. Integration then yields $\frac{1}{2}LI^2 + \dfrac{Q^2}{2C} = C_1$.

Since $I = 0$ when $Q = Q_0$, we have $C_1 = Q_0^2/2C$. Substituting for C_1 and solving for I yield $I = \dfrac{dQ}{dt} = \pm\dfrac{1}{\sqrt{LC}}\sqrt{Q_0^2 - Q^2}$, which is separable. Integration then yields

$$\int \frac{dQ}{\sqrt{Q_0^2 - Q^2}} = \pm \int \frac{dt}{\sqrt{LC}} \qquad \text{or} \qquad \sin^{-1}\frac{Q}{Q_0} = \pm\frac{t}{\sqrt{LC}} + C_2$$

Since $Q = Q_0$ for $t = 0$, we find $C_2 = \pi/2$. Thus, we have $\sin^{-1}\dfrac{Q}{Q_0} = \dfrac{\pi}{2} \pm \dfrac{t}{\sqrt{LC}}$ or $Q = Q_0 \cos\dfrac{t}{\sqrt{LC}}$

6.133 Find the amplitude, period, and frequency of the charge in the previous problem.

▌ The amplitude is Q_0, the period is $2\pi\sqrt{LC}$, and the frequency is the reciprocal of the period, $1/2\pi\sqrt{LC}$.

6.134 Determine the current in the circuit described in Problem 6.132.

▌ Since $Q = Q_0 \cos\dfrac{t}{\sqrt{LC}}$, $I = \dfrac{dQ}{dt} = -\dfrac{Q_0}{\sqrt{LC}} \sin\dfrac{t}{\sqrt{LC}}$.

6.135 Determine the amplitude, frequency, and period of the current in the previous problem.

▌ The amplitude is $Q_0/\sqrt{LC}$, while the frequency and period are identical to those of the charge (see Problem 6.133

MECHANICS PROBLEMS

6.136 Derive a first-order differential equation governing the motion of a vertically falling body of mass m that is influenced only by gravity g and air resistance, which is proportional to the velocity of the body.

▌ Assume that both gravity and mass remain constant and, for convenience, choose the downward direction as the positive direction. Then by Newton's second law of motion, the net force acting on a body is equal to the time rate of change of the momentum of the body; or, for constant mass, $F = m\frac{dv}{dt}$, where F is the net force on the body and v is the velocity of the body, both at time t.

For the problem at hand, there are two forces acting on the body: (1) the force due to gravity given by the weight w of the body, which equals mg, and (2) the force due to air resistance given by $-kv$, where $k \geq 0$ is a constant of proportionality. The minus sign is required because this force opposes the velocity; that is, it acts in the upward, or negative, direction (see Fig. 6.7). The net force F on the body is, therefore, $F = mg - kv$, so that we have

$$mg - kv = m\frac{dv}{dt} \qquad \text{or} \qquad \frac{dv}{dt} + \frac{k}{m}v = g \tag{1}$$

as the equation of motion for the body. If air resistance is negligible, then $k = 0$ and (*1*) simplifies to $dv/dt = g$.

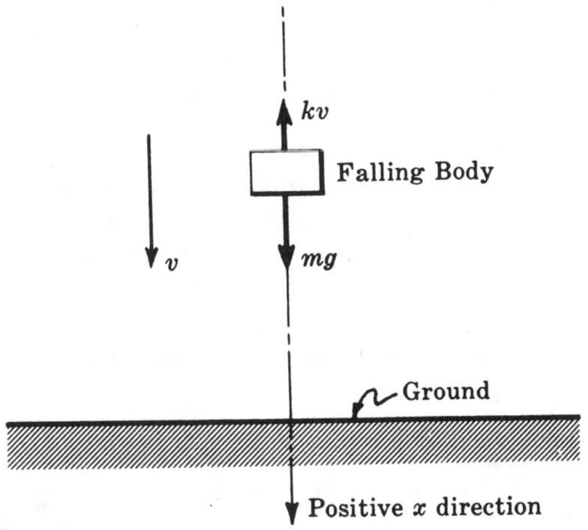

Fig. 6.7

6.137 A body of mass 5 slugs is dropped from a height of 100 ft with zero velocity. Assuming no air resistance, find an expression for the velocity of the body at any time t.

▌ Choose the coordinate system as in Fig. 6.8. Then, since there is no air resistance, (*1*) of Problem 6.136 becomes $dv/dt = g$; its solution is $v = gt + c$ (see Problem 5.41). When $t = 0$, $v = 0$ (initially the body has zero velocity); hence $0 = g(0) + c$, so that $c = 0$. Thus, $v = gt$ or, for $g = 32\ \text{ft/s}^2$, $v = 32t$.

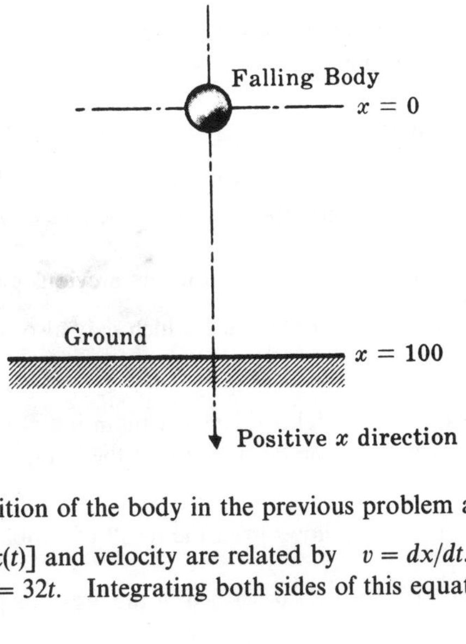

Fig. 6.8

6.138 Find an expression for the position of the body in the previous problem at any time t.

▌ Position [as measured by $x(t)$] and velocity are related by $v = dx/dt$. It then follows from the result of the previous problem that $dx/dt = 32t$. Integrating both sides of this equation with respect to time, we get

$x = 16t^2 + c$. But $x = 0$ at $t = 0$ (see Figure 6.8), so we have $0 = 16(0)^2 + c$ from which $c = 0$. Thus, the position of the body at any time t is $x = 16t^2$.

6.139 Determine the time required for the body described in Problem 6.137 to reach the ground.

▌ The position of the body is given by the result of the previous problem. We require t when $x = 100$; hence we have $100 = 16t^2$ or $t = 2.5$ s.

6.140 A body of mass 2 slugs is dropped from a height of 800 ft with zero velocity. Find an expression for the velocity of the body at any time t if the force due to air resistance is $-\frac{1}{2}v$ lb.

▌ Here $m = 2$ and $k = \frac{1}{2}$. With $g = 32$, (*1*) of Problem 6.136 becomes $\dfrac{dv}{dt} + \dfrac{1}{4}v = 32$. The solution to this differential equation is given in Problem 5.35 as $v = 128 + ce^{-t/4}$. At $t = 0$, $v = 0$; hence $0 = 128 + ce^0$, or $c = -128$. The velocity at any time t (before the body reaches the ground) is then $v = 128 - 128e^{-t/4}$.

6.141 Find an expression for the position of the body described in the previous problem.

▌ Since $v = dx/dt$, it follows from the result of the previous problem that $\dfrac{dx}{dt} = 128 - 128e^{-t/4}$. Integrating directly with respect to time, we obtain $x = 128t + 512e^{-t/4} + c$.

Since the positive x direction is downward (see Problem 6.136), we take the origin to be the point at which the body was released; then the ground is at $x = 800$. At $t = 0$, $x = 0$; so $0 = 128(0) + 512e^0 + c$, or $c = -512$. The position of the body at any time t (before it reaches the ground) is then $x = -512 + 128t + 512e^{-t/4}$.

6.142 Find an expression for the limiting (or terminal) velocity of a freely falling body satisfying the conditions of Problem 6.136.

▌ The limiting (or terminal) velocity is that velocity for which $dv/dt = 0$. Substituting this requirement into (*1*) of Problem 6.136, we find $0 + \dfrac{k}{m}v = g$ or $v_{ter} = mg/k$.

This equation is valid only when $k \neq 0$. If $k = 0$, then (*1*) of Problem 6.136 becomes $dv/dt = g$. In that case, the condition $dv/dt = 0$ cannot be satisfied; thus, there is no limiting velocity in the absence of air resistance.

6.143 Determine the limiting velocity of the body described in Problem 6.140.

▌ With $m = 2$, $k = \frac{1}{2}$, and $g = 32$, we have $v_{ter} = 2(32)/\frac{1}{2} = 128$ ft/s. (Note that as $t \to \infty$, the velocity derived in Problem 6.140 also tends to $v = 128$.)

6.144 A mass of 2 kg is dropped from a height of 200 m with a velocity of 3 m/s. Find an expression for the velocity of the object if the force due to air resistance is $-50v$ N.

▌ Here $m = 2$ and $k = 50$. With $g = 9.8$, (*1*) of Problem 6.136 becomes $\dfrac{dv}{dt} + 25v = 9.8$. Its solution is $v = 0.392 + ce^{-25t}$ (see Problem 5.36). At $t = 0$, $v = 3$; hence $3 = 0.392 + ce^0$ or $c = 2.608$. The velocity at any time t (before the mass reaches the ground) is then $v = 0.392 + 2.608e^{-25t}$.

6.145 Determine the limiting velocity for the object described in the previous problem.

▌ Here we find $v_{ter} = mg/k = 2(9.8)/50 = 0.392$ m/s, which may also be obtained by letting $t \to \infty$ in the result of Problem 6.144.

6.146 A body weighing 64 lb is dropped from a height of 100 ft with an initial velocity of 10 ft/s. It is known that air resistance is proportional to the velocity of the body and that the limiting velocity for this body is 128 ft/s. Find the constant of proportionality.

▌ Here $mg = 64$ and $v_{ter} = 128$. It follows from the result of Problem 6.142 that $128 = 64/k$, or $k = \frac{1}{2}$.

6.147 Find an expression for the velocity of the body described in the previous problem at any time t before that body reaches the ground.

With $mg = 64$, it follows that $m = 2$ slugs. We have from the previous problem that $k = \frac{1}{2}$, so (1) of Problem 6.136 becomes $\frac{dv}{dt} + \frac{1}{4}v = 32$, which has as its solution $v = 128 + ce^{-t/4}$ (see Problem 5.35). At $t = 0$, $v = 10$; hence $10 = 128 + ce^0$, or $c = -118$. The velocity is then $v = 128 - 118e^{-t/4}$.

6.148 A body of mass 10 slugs is dropped from a height of 1000 ft with no initial velocity. The body encounters air resistance proportional to its velocity. If the limiting velocity is known to be 320 ft/s, find the constant of proportionality.

With $m = 2$ and $v_{ter} = 320$, the result of Problem 6.142 becomes $320 = 2(32)/k$ or $k = \frac{1}{5}$.

6.149 Find an expression for the velocity of the body described in the previous problem.

With $m = 2$, $k = \frac{1}{5}$, and $g = 32$, (1) of Problem 6.136 becomes $\frac{dv}{dt} + \frac{1}{10}v = 32$; its solution is $v = 320 + ce^{-t/10}$ (see Problem 5.37). At $t = 0$, $v = 0$; hence $0 = 320 + ce^0$, so $c = -320$. The velocity at any time t (before the body reaches the ground) is $v = 320(1 - e^{-t/10})$.

6.150 Find an expression for the position of the body described in Problem 6.148.

Using the result of the previous problem along with $v = dx/dt$, we may write $\frac{dx}{dt} = 320 - 320e^{-t/10}$. Integrating directly with respect to time, we obtain $x = 320t + 3200e^{-t/10} + c$. Since the positive direction is assumed to be downward (see Problem 6.136), we take the origin to be the point where the body was released (and the ground as $x = 1000$). Then $x = 0$ at $t = 0$, and $0 = 320(0) + 3200e^0 + c$ so that $c = -3200$. The position of the body is then $x = 320t - 3200 + 320e^{-t/10}$.

6.151 Determine the time required for the body described in Problem 6.148 to attain a speed of 160 ft/s.

Substituting $v = 160$ into the result of Problem 6.149 gives us $160 = 320(1 - e^{-t/10})$, from which $e^{-t/10} = \frac{1}{2}$. Then $t = -10 \ln \frac{1}{2} = 6.93$ s.

6.152 A body of mass m is thrown vertically into the air with an initial velocity v_0. If the body encounters air resistance proportional to its velocity, find the equation for its motion in the coordinate system of Fig. 6.9.

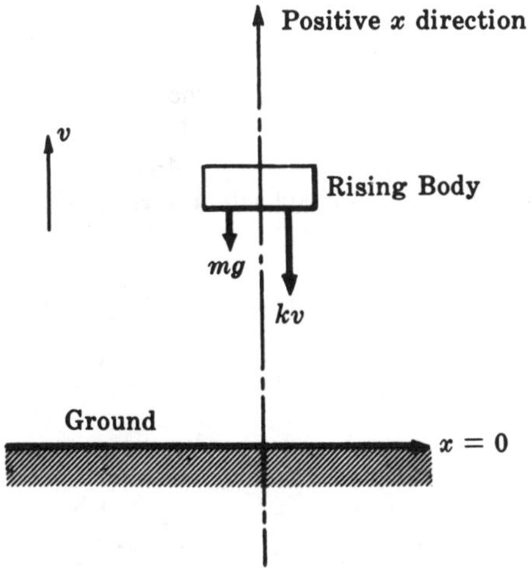

Fig. 6.9

There are two forces on the body: (1) the force due to gravity given by mg and (2) the force due to air resistance given by kv, which impedes the motion of the body. Since both of these forces act in the downward or negative direction, the net force on the body is $-mg - kv$. Then from Newton's second law of motion, we have, as the equation of motion, $m\frac{dv}{dt} = -mg - kv$ or

$$\frac{dv}{dt} + \frac{k}{m}v = -g \tag{1}$$

6.153 Find an expression for the velocity of the body described in the previous problem.

▮ The solution to (*1*) of the previous problem is given in Problem 5.37 as $v = ce^{-(k/m)t} - mg/k$. At $t = 0$, $v = v_0$; hence $v_0 = ce^{-(k/m)0} - (mg/k)$, or $c = v_0 + (mg/k)$. The velocity of the body at any time t is then $v = \left(v_0 + \frac{mg}{k}\right)e^{-(k/m)t} - \frac{mg}{k}$.

6.154 Find the time at which the body described in Problem 6.152 reaches its maximum height.

▮ The body reaches its maximum height when $v = 0$. Substituting this value into the result of the previous problem, we obtain

$$0 = \left(v_0 + \frac{mg}{k}\right)e^{-(k/m)t} - \frac{mg}{k} \qquad \text{or} \qquad e^{-(k/m)t} = \frac{1}{1 + v_0k/mg}$$

Taking the logarithms of both sides gives us $-\frac{k}{m}t = \ln\frac{1}{1 + v_0k/mg}$, from which we find $t = \frac{m}{k}\ln\left(1 + \frac{v_0k}{mg}\right)$.

6.155 An object is thrown vertically upward from the ground with initial velocity 1960 cm/s. Neglecting air resistance, find its velocity at any time t.

▮ With $k = 0$, (*1*) of Problem 6.152 becomes $dv/dt = -g$, which may be integrated directly to yield $v = -gt + c$. At $t = 0$, $v = 1960$; hence we have $1960 = -g(0) + c = c$. With this value of c and $g = 980\ \text{cm/s}^2$, the velocity becomes $v = 1960 - 980t$.

6.156 Determine the total time required for the object described in Problem 6.155 to return to the starting point.

▮ We require t when $x = 0$. Since $v = dx/dt$, it follows from the result of the previous problem that $dx/dt = 1960 - 980t$. Integrating this equation with respect to t, we get $x = 1960t - 490t^2 + c$. At $t = 0$, $x = 0$; hence we have $0 = 1960(0) - 490(0)^2 + c = c$, and the position of the object is given by

$$x = 1960t - 490t^2 \tag{1}$$

Setting $x = 0$ in (*1*) and solving for t, we obtain $0 = 1960t - 490t^2 = 490t(4 - t)$, so that $t = 0$ or 4. The time needed for the object to return to the ground is $t = 4$ s.

6.157 Determine the maximum height attained by the body described in Problem 6.155.

▮ The maximum height occurs when $v = 0$. Substituting this value into the result of Problem 6.155, we get $0 = 1960 - 980t$, or $t = 2$ s. For that value of t, (*1*) of the previous problem yields $x = 1960(2) - 490(2)^2 = 1960$ cm.

6.158 A body of mass 2 slugs is dropped with no initial velocity and encounters an air resistance that is proportional to the square of its velocity. Find an expression for the velocity of the body at any time t.

▮ The force due to air resistance is $-kv^2$, so that Newton's second law of motion becomes $m\frac{dv}{dt} = mg - kv^2$ or $2\frac{dv}{dt} = 64 - kv^2$. Rewriting this equation in differential form, we have $\frac{2}{64 - kv^2}dv - dt = 0$, which is separable. By partial fractions,

$$\frac{2}{64 - kv^2} = \frac{2}{(8 - \sqrt{k}v)(8 + \sqrt{k}v)} = \frac{1/8}{8 - \sqrt{k}v} + \frac{1/8}{8 + \sqrt{k}v}$$

so our differential equation can be rewritten as $\frac{1}{8}\left(\frac{1}{8 - \sqrt{k}v} + \frac{1}{8 + \sqrt{k}v}\right)dv - dt = 0$. Integration gives

$$\frac{1}{8}\left[-\frac{1}{\sqrt{k}}\ln|8 - \sqrt{k}v| + \frac{1}{\sqrt{k}}\ln|8 + \sqrt{k}v|\right] - t = c$$

which can be rewritten as $\ln\left|\frac{8 + \sqrt{k}v}{8 - \sqrt{k}v}\right| = 8\sqrt{k}t + 8\sqrt{k}c$ or as $\frac{8 + \sqrt{k}v}{8 - \sqrt{k}v} = c_1e^{8\sqrt{k}t}$, where $c_1 = \pm e^{8\sqrt{k}c}$.

At $t = 0$, we are given $v = 0$. This implies $c_1 = 1$, and the velocity is given by $\dfrac{8 + \sqrt{kv}}{8 - \sqrt{kv}} = e^{8\sqrt{kt}}$ or $v = \dfrac{8}{\sqrt{k}} \tanh 4\sqrt{kt}$. Without additional information, we cannot obtain a numerical value for the constant k.

6.159 A 192-lb object falls from rest at time $t = 0$ in a medium offering a resistance of $3v^2$ (in pounds). Find an expression for the velocity of this object.

▌ From Newton's second law, $m\,dv/dt = mg - 3v^2$, so that $\dfrac{192}{g}\dfrac{dv}{dt} = 192 - 3v^2$, or $2\dfrac{dv}{dt} = 64 - v^2$. Then, separating variables and integrating, we get

$$\int \frac{dv}{64 - v^2} = \int \frac{dt}{2} \qquad \text{or} \qquad \frac{1}{16}\ln\frac{8 + v}{8 - v} = \frac{t}{2} + c$$

Since $v = 0$ at $t = 0$, we find that $c = 0$. Then $\dfrac{1}{16}\ln\dfrac{8 + v}{8 - v} = \dfrac{t}{2}$, and $v = 8\dfrac{e^{4t} - e^{-4t}}{e^{4t} + e^{-4t}}$.

6.160 Find an expression for the position of the object described in the previous problem.

▌ Since $v = dx/dt$, we integrate the result of Problem 6.159 subject to $x = 0$ at $t = 0$, to find, for the distance traveled, $x = 2\ln\dfrac{e^{4t} + e^{-4t}}{2} = 2\ln\cosh 4t$.

6.161 Determine the limiting velocity for the object described in Problem 6.159.

▌ The limiting velocity is $\lim\limits_{t\to\infty} 8\dfrac{e^{4t} - e^{-4t}}{e^{4t} + e^{-4t}} = \lim\limits_{t\to\infty} 8\dfrac{1 - e^{-8t}}{1 + e^{-8t}} = 8$ ft/s, which can also be obtained by setting $\dfrac{dv}{dt} = 32 - \dfrac{v^2}{2} = 0$.

Observe that the result of Problem 6.142 is not valid here, because the resistance of the medium is not proportional to the velocity of the object. It is proportional instead to the square of the velocity.

6.162 A boat of mass m is traveling with velocity v_0. At $t = 0$ the power is shut off. Assuming water resistance proportional to v^n, where n is a constant and v is the instantaneous velocity, find v as a function of the distance traveled.

▌ Let x be the distance traveled after time $t > 0$. The only force acting on the boat is the water resistance, so we have $m\dfrac{dv}{dt} = -kv^n$, where k is a constant of proportionality. Then we have $m\dfrac{dv}{dt} = m\dfrac{dv}{dx}\dfrac{dx}{dt} = mv\dfrac{dv}{dx} = -kv^n$, which we write as $mv^{1-n}\,dv = -k\,dx$.

Case 1, $n \neq 2$: With $v = v_0$ at $x = 0$, integration gives $v^{2-n} = v_0^{2-n} - \dfrac{k}{m}(2 - n)x$.

Case 2, $n = 2$: Again with $v = v_0$ at $x = 0$, integration now yields $v = v_0 e^{-kx/m}$.

6.163 A ship weighing 48,000 tons starts from rest under the force of a constant propeller thrust of 200,000 lb. Find its velocity as a function of time t, given that the water resistance in pounds is $10{,}000v$, with v the velocity measured in feet/second. Also find the terminal velocity in miles per hour.

▌ Since mass (slugs) × acceleration (ft/s²) = net force (lb) = propeller thrust − resistance, we have $\dfrac{48{,}000(2000)}{32}\dfrac{dv}{dt} = 200{,}000 - 10{,}000v$, from which $\dfrac{dv}{dt} + \dfrac{v}{300} = \dfrac{20}{300}$. Integrating gives $ve^{t/300} = \dfrac{20}{300}\int e^{t/300}\,dt = 20e^{t/300} + C$. Because $v = 0$ when $t = 0$, we have $C = -20$, so that $v = 20 - 20e^{-t/300} = 20(1 - e^{-t/300})$.

As $t \to \infty$, $v \to 20$; the terminal velocity thus is 20 ft/s = 13.6 mi/h. This may also be obtained from the differential equation with $dv/dt \to 0$.

6.164 A boat is being towed at the rate of 12 mi/h. At the instant $(t = 0)$ that the towing line is cast off, a man in the boat begins to row in the direction of motion, exerting a force of 20 lb. If the combined weight of the man and boat is 480 lb and the resistance (in pounds) is equal to $1.75v$, where v is measured in feet/second, find the speed of the boat after $\frac{1}{2}$ min.

▌ Since mass (slugs) × acceleration (ft/s²) = net force (lb) = forward force − resistance, we have $\frac{480}{32}\frac{dv}{dt} = 20 - 1.75v$ from which $\frac{dv}{dt} + \frac{7}{60}v = \frac{4}{3}$. Integrating gives $ve^{7t/60} = \frac{4}{3}\int e^{7t/60}\,dt = \frac{80}{7}e^{7t/60} + C$. When $t = 0$, $v = \frac{12(5280)}{(60)^2} = \frac{88}{5}$, so that $C = \frac{216}{35}$. Then $v = \frac{80}{7} + \frac{216}{35}e^{-7t/60}$.
Now when $t = 30$, $v = \frac{80}{7} + \frac{216}{35}e^{-3.5} = 11.6$ ft/s.

6.165 A mass is being pulled across ice on a sled, the total weight including the sled being 80 lb. The resistance offered by the ice to the runners is negligible, and the air offers a resistance in pounds equal to five times the velocity (v ft/s) of the sled. Find the constant force (in pounds) that must be exerted on the sled to give it a terminal velocity of 10 mi/h, and the velocity and distance traveled at the end of 48 s.

▌ Since mass (slugs) × acceleration (ft/s²) = net force (lb) = forward force − resistance, we have $\frac{80}{32}\frac{dv}{dt} = F - 5v$, or $\frac{dv}{dt} + 2v = \frac{2}{5}F$, where F (in pounds) is the forward force. Integrating then yields $v = \frac{F}{5} ce^{-2t}$. When $t = 0$, $v = 0$ so that $c = -\frac{F}{5}$ and $v = \frac{F}{5}(1 - e^{-2t})$. As $t \to \infty$, $\frac{F}{5} \to v_{ter} = \frac{10(5280)}{(60)^2} = \frac{44}{3}$. The required force thus is $F = \frac{220}{3}$ lb.
Substituting this value for F gives us $v = \frac{44}{3}(1 - e^{-2t})$. So, when $t = 48$, we have $v = \frac{44}{3}(1 - e^{-96}) = \frac{44}{3}$ ft/s. The distance traveled is $s = \int_0^{48} v\,dt = \frac{44}{3}\int_0^{48}(1 - e^{-2t})\,dt = 697$ ft.

6.166 A spring of negligible weight hangs vertically. A mass of m slugs is attached to the other end. If the mass is moving with velocity v_0 ft/s when the spring is unstretched, find the velocity v as a function of the stretch x in feet.

▌ According to Hooke's law, the spring force (the force opposing the stretch) is proportional to the stretch. Thus, we have $m\frac{dv}{dt} = mg - kx$, which we can write as $m\frac{dv}{dx}\frac{dx}{dt} = mv\frac{dv}{dx} = mg - kx$, since $\frac{dx}{dt} = v$. Integrating then gives $mv^2 = 2mgx - kx^2 + C$. Now $v = v_0$ when $x = 0$ so that $C = mv_0^2$, and $mv^2 = 2mgx - kx^2 + mv_0^2$.

6.167 A parachutist is falling with speed 176 ft/s when his parachute opens. If the air resistance is $Wv^2/256$ lb, where W is the total weight of the man and parachute, find his speed as a function of the time t after the parachute opens.

▌ Since net force on system = weight of system − air resistance, we have $\frac{W}{g}\frac{dv}{dt} = W - \frac{Wv^2}{256}$, from which $\frac{dv}{v^2 - 256} = -\frac{dt}{8}$. Integrating between the limits $t = 0$, $v = 176$ and $t = t$, $v = v$ gives

$$\int_{176}^{v}\frac{dv}{v^2 - 256} = -\frac{1}{8}\int_0^t dt \qquad \text{or} \qquad \frac{1}{32}\ln\frac{v - 16}{v + 16}\bigg|_{176}^{v} = -\frac{t}{8}\bigg|_0^t$$

from which we get $\ln\frac{v - 16}{v + 16} - \ln\frac{5}{6} = -4t$. Exponentiation then gives $\frac{v - 16}{v + 16} = \frac{5}{6}e^{-4t}$, or $v = 16\frac{6 + 5e^{-4t}}{6 - 5e^{-4t}}$.
Note that the parachutist quickly attains an approximately constant speed—the terminal speed of 16 ft/sec.

6.168 A body of mass m slugs falls from rest in a medium for which the resistance (in pounds) is proportional to the square of the velocity (in feet per second). If the terminal velocity is 150 ft/s, find the velocity at the end of 2 s and the time required for the velocity to become 100 ft/s.

▌ Let v denote the velocity of the body at time t. Then we have

net force on body = weight of body − resistance, and the equation of motion is $m\frac{dv}{dt} = mg - Kv^2$. Some

simplification is possible if we choose to write $K = 2mk^2$. Then the equation of motion reduces to
$\frac{dv}{dt} = 2(16 - k^2v^2)$ or $\frac{dv}{k^2v^2 - 16} = -2\,dt$.

Integrating now gives $\ln\frac{kv-4}{kv+4} = -16kt + \ln c$, from which $\frac{kv-4}{kv+4} = ce^{-16kt}$. Since $v = 0$ when $t = 0$, we find $c = -1$ and $\frac{kv-4}{kv+4} = -e^{-16kt}$. Also, $v = 150$ for $t \to \infty$, so $k = \frac{2}{75}$ and our solution becomes $\frac{v-150}{v+150} = -e^{-0.43t}$.

When $t = 2$, $\frac{v-150}{v+150} = -e^{-0.86} = -0.423$ and $v = 61$ ft/s.
When $v = 100$, $e^{-0.43t} = 0.2 = e^{-1.6}$, so $t = 3.7$ s.

.169 A body of mass m falls from rest in a medium for which the resistance (in pounds) is proportional to the velocity (in feet per second). If the specific gravity of the medium is one-fourth that of the body and if the terminal velocity is 24 ft/s, find the velocity at the end of 3 s and the distance traveled in 3 s.

▌ Let v denote the velocity of the body at time t. In addition to the two forces acting as in Problem 6.136, there is a third force which results from the difference in specific gravities. This force is equal in magnitude to the weight of the medium which the body displaces, and it opposes gravity. Thus, we have
net force on body = weight of body − buoyant force − resistance, and the equation of motion is
$m\frac{dv}{dt} = mg - \frac{1}{4}mg - Kv = \frac{3}{4}mg - Kv$. With $g = 32$ ft/s^2 and K taken as $3mk$, the equation becomes
$\frac{dv}{dt} = 3(8 - kv)$ or $\frac{dv}{8 - kv} = 3\,dt$. Integrating from $t = 0$, $v = 0$ to $t = t$, $v = v$ gives
$-\frac{1}{k}\ln(8 - kv)\Big|_0^v = 3t\Big|_0^t$, from which $-\ln(8 - kv) + \ln 8 = 3kt$, so that $kv = 8(1 - e^{-3kt})$. When $t \to \infty$, $v = 24$, so $k = 1/3$ and $v = 24(1 - e^{-t})$. Thus, when $t = 3$, $v = 24(1 - e^{-3}) = 22.8$ ft/s.

Since $v = \frac{dx}{dt} = 24(1 - e^{-t})$, we integrate between $t = 0$, $x = 0$ and $t = 3$, $x = x$ to find
$x\Big|_0^x = 24(t + e^{-t})\Big|_0^3$ or $x = 24(2 + e^{-3}) = 49.2$ ft as the distance traveled in 3 s.

170 The gravitational pull on a mass m at a distance s feet from the center of the earth is proportional to m and inversely proportional to s^2. (*a*) Find the velocity attained by the mass in falling from rest at a distance $5R$ from the center to the earth's surface, where $R = 4000$ mi is the radius of the earth. (*b*) What velocity would correspond to a fall from an infinite distance; that is, with what velocity must the mass be propelled vertically upward to escape the earth's gravitational pull? (All other forces, including friction, are to be neglected.)

▌ The gravitational force at a distance s from the earth's center is km/s^2. To determine k, we note that the force is mg when $s = R$; thus $mg = km/R^2$ and $k = gR^2$. The equation of motion is then
$m\frac{dv}{dt} = m\frac{ds}{dt}\frac{dv}{ds} = mv\frac{dv}{ds} = -\frac{mgR^2}{s^2}$, or $v\,dv = -gR^2\frac{ds}{s^2}$, the minus sign indicating that v increases as s decreases.

(*a*) Integrating from $v = 0$, $s = 5R$ to $v = v$, $s = R$, we get $\int_0^v v\,dv = -gR^2\int_{5R}^{R}\frac{ds}{s^2}$, from which
$\frac{1}{2}v^2 = gR^2\left(\frac{1}{R} - \frac{1}{5R}\right) = \frac{4}{5}gR$, so that $v^2 = \frac{8}{5}(32)(4000)(5280)$. Then $v = 2560\sqrt{165}$ ft/s or approximately 6 mi/s.

(*b*) Integrating now from $v = 0$, $s \to \infty$ to $v = v$, $s = R$, we get $\int_0^v v\,dv = -gR^2\int_\infty^R \frac{ds}{s^2}$, from which $v^2 = 2gR$. Then $v = 6400\sqrt{33}$ ft/s or approximately 7 mi/s.

171 A uniform chain of length a is placed on a horizontal frictionless table so that a length b of the chain dangles over the side. How long will it take for the chain to slide off the table?

▌ Suppose that at time t a length x of the chain is dangling over the side (Fig. 6.10). Assume that the density (mass per unit length) of the chain is σ. Then the net force acting on the chain is $\sigma g x$, and we have $\sigma g x = \sigma a\frac{dv}{dt}$.

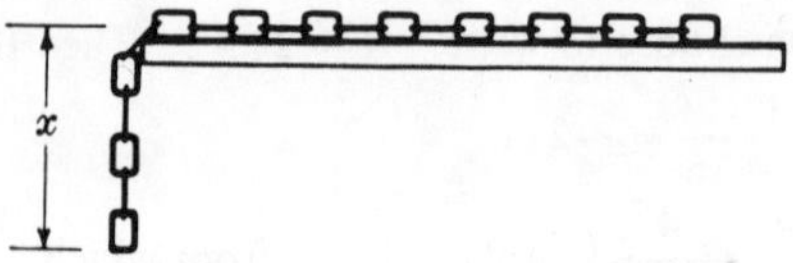

Fig. 6.10

Now, since $\dfrac{dv}{dt} = \dfrac{dv}{dx}\dfrac{dx}{dt} = v\dfrac{dv}{dx}$, this becomes $v\dfrac{dv}{dx} = \dfrac{gx}{a}$. Integrating and using the fact that $x = b$ when $v = 0$, we get $v = \dfrac{dx}{dt} = \sqrt{\dfrac{g}{a}}\sqrt{x^2 - b^2}$. Separating the variables, integrating again, and using $x = b$ when $t = 0$, we get, finally, $\ln \dfrac{x + \sqrt{x^2 - b^2}}{b} = \sqrt{\dfrac{g}{a}}\,t$. Since the chain slides off when $x = a$, the time taken is $T = \sqrt{\dfrac{a}{g}} \ln \dfrac{a + \sqrt{a^2 - b^2}}{b}$.

GEOMETRICAL PROBLEMS

6.172 Find the orthogonal trajectories of the family of curves $y = cx^2$.

▮ It follows from Problem 1.95 that the orthogonal trajectories satisfy the differential equation $\dfrac{dy}{dx} = \dfrac{-x}{2y}$. This equation has the differential form $x\,dx + 2y\,dy = 0$, which is separable. Its solution is $\int x\,dx + \int 2y\,dy = c$, or $\frac{1}{2}x^2 + y^2 = c$, which is the family of orthogonal trajectories. These orthogonal trajectories are ellipses. Some members of this family, along with some members of the original family of parabolas, are shown in Fig. 6.11. Note that each ellipse intersects each parabola at right angles.

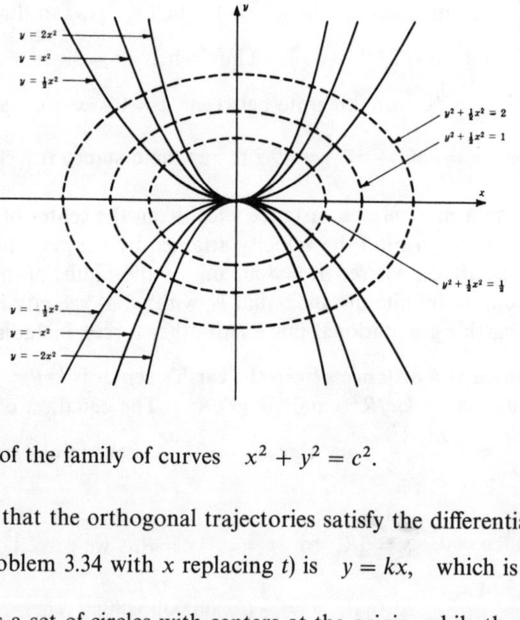

Fig. 6.11

6.173 Find the orthogonal trajectories of the family of curves $x^2 + y^2 = c^2$.

▮ It follows from Problem 1.97 that the orthogonal trajectories satisfy the differential equation $\dfrac{dy}{dx} = \dfrac{y}{x}$. Its solution (see Problem 4.71 or Problem 3.34 with x replacing t) is $y = kx$, which is the family of orthogonal trajectories.

The original family of curves is a set of circles with centers at the origin, while the orthogonal trajectories

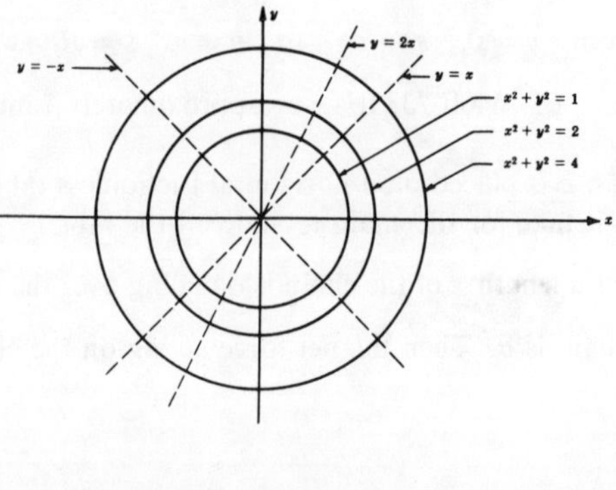

Fig. 6.12

are straight lines through the origin. Some members of each family are shown in Fig. 6.12. Observe that each straight line intersects each circle at right angles.

6.174 Find the orthogonal trajectories of the family of curves $xy = C$.

It follows from Problem 1.99 that the orthogonal trajectories satisfy the differential equation $y' = x/y$ or, in differential form, $x\,dx - y\,dy = 0$. This equation is separable; integrating term by term, we get $\frac{1}{2}x^2 - \frac{1}{2}y^2 = k$ or $x^2 - y^2 = C$, where $C = 2k$. Both the original family of curves and its orthogonal trajectories are hyperbolas, as shown in Fig. 6.13.

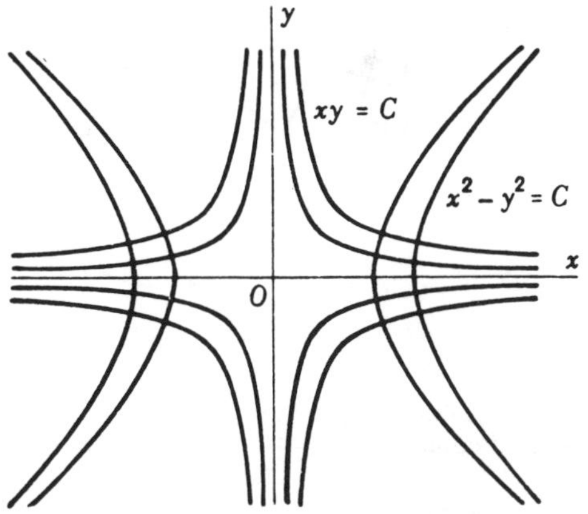

Fig. 6.13

6.175 Find the orthogonal trajectories of the family of curves $x^2 + y^2 = cx$.

It follows from Problem 1.96 that the orthogonal trajectories satisfy the differential equation $\dfrac{dy}{dx} = \dfrac{2xy}{x^2 - y^2}$, whose solution is given in Problem 3.127 as $x^2 + y^2 = ky$. Both the original family of curves and its orthogonal trajectories are circles.

6.176 Find the orthogonal trajectories of the family of cardioids $\rho = C(1 + \sin\theta)$.

It follows from Problem 1.100 that the orthogonal trajectories satisfy the differential equation $d\rho/\rho + (\sec\theta + \tan\theta)\,d\theta = 0$. This equation is separable; integrating term by term we obtain the equation for the orthogonal trajectories as

$$\ln\rho + \ln(\sec\theta + \tan\theta) - \ln\cos\theta = \ln C \qquad \text{or} \qquad \rho = \frac{C\cos\theta}{\sec\theta + \tan\theta} = C(1 - \sin\theta)$$

6.177 Find the orthogonal trajectories of the family of curves $y = ce^x$.

It follows from Problem 1.98 that the orthogonal trajectories satisfy the differential equation $dy/dx = -1/y$ or, in differential form, $y\,dy + 1\,dx = 0$. This equation is separable; integrating term by term, we obtain the equation for the orthogonal trajectories as $\frac{1}{2}y^2 + x = c$.

6.178 Find the orthogonal trajectories of the family of curves $y^2 = 4cx$.

Differentiating the given equation with respect to x, we obtain $2yy' = 4c$ or $\dfrac{dy}{dx} = \dfrac{2c}{y}$. Since $y^2 = 4cx$, it follows that $c = y^2/4x$. Substituting this result into the last equation, we obtain $\dfrac{dy}{dx} = \dfrac{y}{2x}$ as the differential equation for every member of the given family of curves. The differential equation for its orthogonal trajectories (see Problem 1.94) is then $\dfrac{dy}{dx} = -\dfrac{2x}{y}$ or, in differential form, $y\,dy + 2x\,dx = 0$. This equation is separable; integrating term by term, we find $\frac{1}{2}y^2 + x^2 = k^2$, where the integration constant has been written as a square to emphasize the fact that it cannot be negative since it is equal to the sum of two squares. Typical curves of the two families are shown in Fig. 6.14.

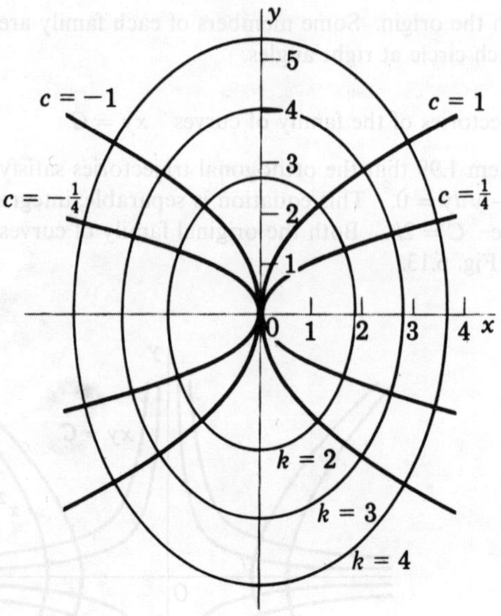

Fig. 6.14

6.179 Show that the family of confocal conics $\dfrac{x^2}{C} + \dfrac{y^2}{C - \lambda} = 1$, where C is an arbitrary constant, is self-orthogonal.

▮ Differentiating the equation of the family with respect to x yields $\dfrac{x}{C} + \dfrac{yp}{C - \lambda} = 0$, where $p = \dfrac{dy}{dx}$. Solving this for C, we find $C = \dfrac{\lambda x}{x + yp}$, so that $C - \lambda = \dfrac{-\lambda py}{x + yp}$. When these replacements are made in the equation of the family, the differential equation of the family is found to be $(x + yp)(px - y) - \lambda p = 0$.

Since this equation is unchanged when p is replaced by $-1/p$, it is also the differential equation of the orthogonal trajectories of the given family. The graphs of several members of this family are shown in Fig. 6.15. If $C > \lambda$, then the graph is an ellipse; if $C < \lambda$, it is a hyperbola.

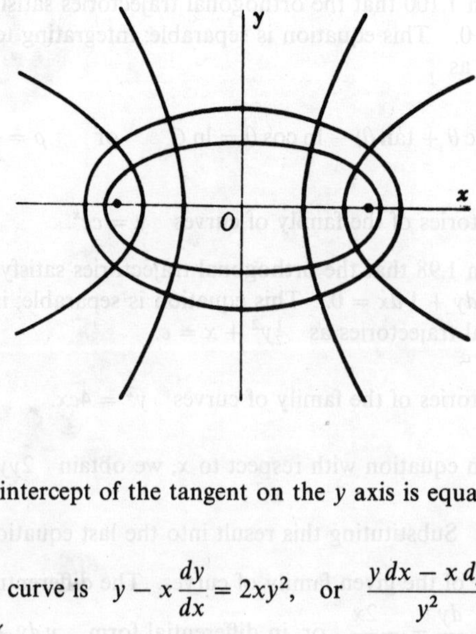

Fig. 6.15

6.180 At each point (x, y) of a curve the intercept of the tangent on the y axis is equal to $2xy^2$ (see Fig. 6.16). Find the curve.

▮ The differential equation of the curve is $y - x\dfrac{dy}{dx} = 2xy^2$, or $\dfrac{y\,dx - x\,dy}{y^2} = 2x\,dx$. Integrating yields $\dfrac{x}{y} = x^2 + c$ or $x - x^2y = cy$. $\left(\text{The differential equation may also be obtained directly from the figure as } \dfrac{dy}{dx} = \dfrac{y - 2xy^2}{x}.\right)$

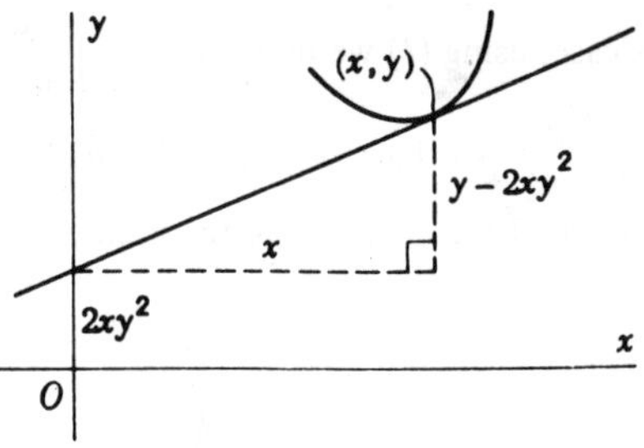

Fig. 6.16

6.181 At each point (x, y) of a curve the subtangent is proportional to the square of the abscissa. Find the curve if it also passes through the point $(1, e)$.

The differential equation of the curve is $y\dfrac{dx}{dy} = kx^2$ or $\dfrac{dx}{x^2} = k\dfrac{dy}{y}$, where k is the proportionality factor. Integrating yields $k \ln y = -\dfrac{1}{x} + C$. When $x = 1$, $y = e$; thus $k = -1 + C$ or $C = k + 1$, and the curve has the equation $k \ln y = -1/x + k + 1$.

6.182 Find the family of curves for which the length of the part of the tangent between the point of contact (x, y) and the y axis is equal to the y intercept of the tangent.

We have $x\sqrt{1 + \left(\dfrac{dy}{dx}\right)^2} = y - x\dfrac{dy}{dx}$, or $x^2 = y^2 - 2xy\dfrac{dy}{dx}$. The transformation $y = vx$ reduces the latter equation to $(1 + v^2)\,dx + 2vx\,dv = 0$, which we write as $\dfrac{dx}{x} + \dfrac{2v\,dv}{1 + v^2} = 0$. Integrating then gives $\ln x + \ln(1 + v^2) = \ln C$. Since $v = y/x$, we have $x\left(1 + \dfrac{y^2}{x^2}\right) = C$ or $x^2 + y^2 = Cx$ as the equation of the family of curves.

6.183 Determine a curve such that the length of its tangent included between the x and y axes is a constant $a > 0$.

Let (x, y) be any point P on the required curve and (X, Y) any point Q on the tangent line AB (Fig. 6.17). The equation of line AB, passing through (x, y) with slope y', is $Y - y = y'(X - x)$. We set $X = 0$ and $Y = 0$ in turn to obtain the y and x intercepts $\overline{OA} = y - xy'$ and $\overline{OB} = x - y/y' = -(y - xy')/y'$. Then the length of AB, apart from sign, is $\sqrt{\overline{OA}^2 + \overline{OB}^2} = (y - xy')\sqrt{1 + y'^2}/y'$. Since this must equal $\pm a$, we have on solving for y,

$$y = xy' \pm \frac{ay'}{\sqrt{1 + y'^2}} = xp \pm \frac{ap}{\sqrt{1 + p^2}} \qquad \text{where} \qquad y' = p \tag{1}$$

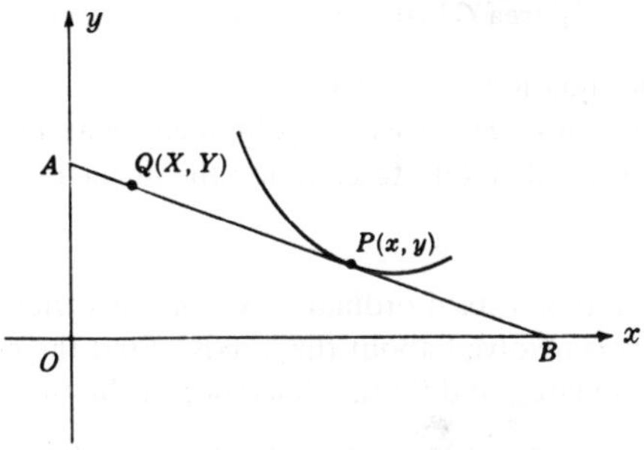

Fig. 6.17

To solve (*1*), we differentiate both sides with respect to x to get

$$y' = p = x\frac{dp}{dx} + p \pm \frac{a}{(1 + p^2)^{3/2}}\frac{dp}{dx} \qquad \text{or} \qquad \frac{dp}{dx}\left[x \pm \frac{a}{(1 + p^2)^{3/2}}\right] = 0$$

Case 1, $dp/dx = 0$: In this case $p = c$ and the general solution is $y = cx \pm \dfrac{ac}{\sqrt{1 + c^2}}$.

Case 2, $dp/dx \neq 0$: In this case, using *(1)* we find $x = \mp\dfrac{a}{(1+p^2)^{3/2}}$ and $y = \pm\dfrac{ap^3}{(1+p^2)^{3/2}}$, from which $x^{2/3} = \dfrac{a^{2/3}}{1+p^2}$ and $y^{2/3} = \dfrac{a^{2/3}p^2}{1+p^2}$ so that $x^{2/3} + y^{2/3} = a^{2/3}$. This is a singular solution and is the equation of a *hypocycloid* (Fig. 6.18), which is the envelope of the family of lines found in Case 1 and is the required curve.

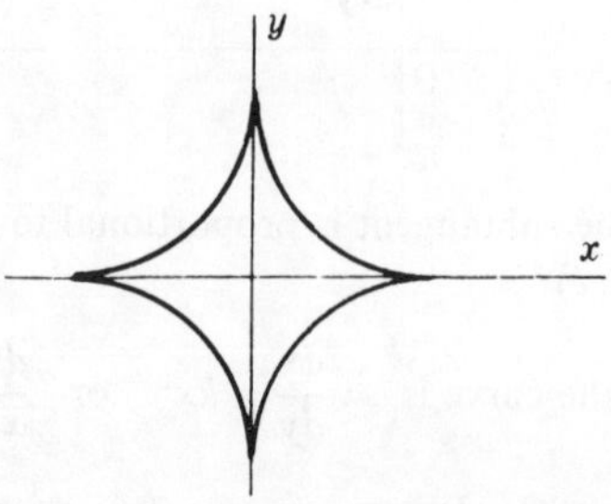

Fig. 6.18

6.184 Through any point (x, y) of a certain curve which passes through the origin, lines are drawn parallel to the coordinate axes. Find the curve given that it divides the rectangle formed by the two lines and the axes into two areas, one of which is three times the other.

▮ There are two cases, illustrated in Figs. 6.19 and 6.20.

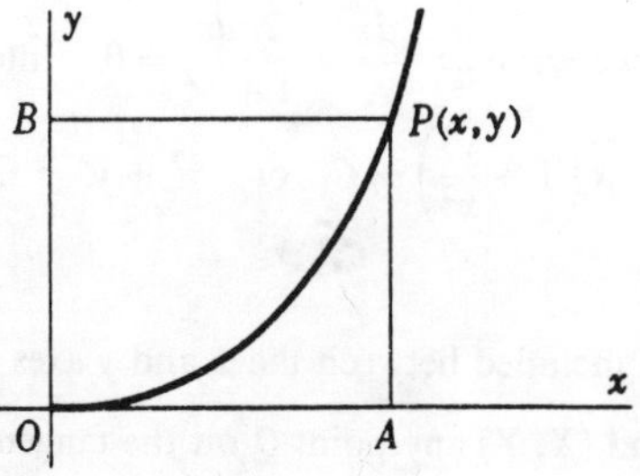

Fig. 6.19

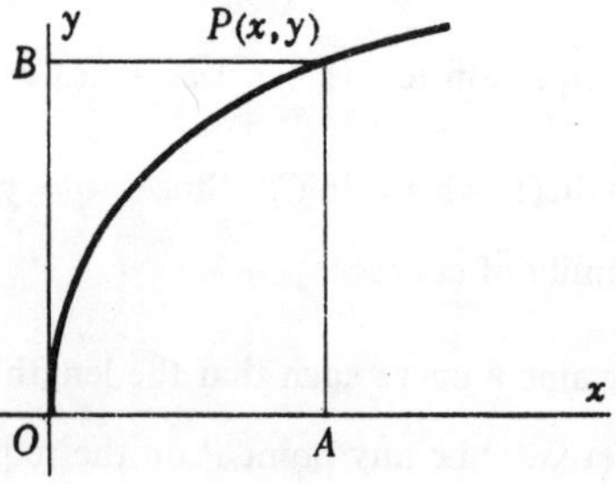

Fig. 6.20

Case 1: Here (3)(area OAP) = area OPB. Then $3\int_0^x y\,dx = xy - \int_0^x y\,dx$, or $4\int_0^x y\,dx = xy$. To obtain the differential equation, we differentiate with respect to x, obtaining $4y = y + x\dfrac{dy}{dx}$ or $\dfrac{dy}{dx} = \dfrac{3y}{x}$. An integration yields the family of curves $y = Cx^3$.

Case 2: Here area OAP = (3)(area OPB) and $4\int_0^x y\,dx = 3xy$. The differential equation is $\dfrac{dy}{dx} = \dfrac{y}{3x}$, and the family of curves has the equation $y^3 = Cx$.

Since the differential equation in each case was obtained by a differentiation, extraneous solutions may have been introduced. It is necessary therefore to compute the areas as a check. In each of the above cases, the curves satisfy the conditions.

6.185 The areas bounded by the x axis, a fixed ordinate $x = a$, a variable ordinate, and the part of a certain curve intercepted by the ordinates is revolved about the x axis. Find the curve if the volume generated is proportional to *(a)* the sum of the two ordinates, and *(b)* the difference of the two ordinates.

▮ *(a)* Let A be the length of the fixed ordinate. The differential equation obtained by differentiating $\pi\int_a^x y^2\,dx = k(y + A)$ is $\pi y^2 = k\,dy/dx$. Integrating then yields $y(C - \pi x) = k$, from which $y = \dfrac{k}{C - \pi x}$.

Then

$$\pi\int_a^x y^2\,dx = \pi\int_a^x \frac{k^2\,dx}{(C-\pi x)^2} = \frac{k^2}{C-\pi x} - \frac{k^2}{C-\pi a} = k(y - A) \neq k(y + A)$$

Thus, the solution is extraneous and no curve exists having the property *(a)*.

(*b*) Repeating the above procedure with $\pi \int_0^x y^2\,dx = k(y - A)$, we obtain the differential equation $\pi y^2 = k\,dy/dx$, whose solution is $y(C - \pi x) = k$. It can be shown (as we tried to do in part *a*) that this equation satisfies the condition and thus represents the family of curves with the required property.

6.186 Find the curve such that, at any point on it, the angle between the radius vector and the tangent is equal to one-third the angle of inclination of the tangent.

▮ Let θ denote the angle of inclination of the radius vector, τ the angle of inclination of the tangent, and ψ the angle between the radius vector and the tangent. Since $\psi = \tau/3 = (\psi + \theta)/3$, we have $\psi = \frac{1}{2}\theta$ and $\tan \psi = \tan \frac{1}{2}\theta$. Now

$$\tan \psi = \rho \frac{d\theta}{d\rho} = \tan \frac{\theta}{2} \qquad \text{so that} \qquad \frac{d\rho}{\rho} = \cot \frac{\theta}{2}\,d\theta$$

Integrating then yields $\ln \rho = 2 \ln \sin \frac{1}{2}\theta + \ln C_1$, or $\rho = C_1 \sin^2 \frac{1}{2}\theta = C(1 - \cos \theta)$.

6.187 The area of the sector formed by an arc of a curve and the radii vectors to the end points is one-half the length of the arc. Find the curve.

▮ Let the radii vectors be given by $\theta = \theta_1$ and $\theta = \theta$. Then $\frac{1}{2}\int_{\theta_1}^{\theta} \rho^2\,d\theta = \frac{1}{2}\int_{\theta_1}^{\theta} \sqrt{\left(\frac{d\rho}{d\theta}\right)^2 + \rho^2}\,d\theta.$

Differentiating with respect to θ yields the differential equation

$$\rho^2 = \sqrt{\left(\frac{d\rho}{d\theta}\right)^2 + \rho^2} \quad \text{or} \quad d\rho = \pm\rho\sqrt{\rho^2 - 1}\,d\theta.$$

If $\rho^2 = 1$, this latter equation reduces to $d\rho = 0$. It is easily verified that $\rho = 1$ satisfies the condition of the problem. If $\rho^2 \neq 1$, we write the equation in the form $\dfrac{d\rho}{\rho\sqrt{\rho^2 - 1}} = \pm d\theta$ and obtain the solution $\rho = \sec(C \pm \theta)$. Thus, the conditions are satisfied by the circle $\rho = 1$ and the family of curves $\rho = \sec(C + \theta)$. Note that the families $\rho = \sec(C + \theta)$ and $\rho = \sec(C - \theta)$ are the same.

6.188 Find the curve for which the portion of the tangent between the point of contact and the foot of the perpendicular through the pole to the tangent is one-third the radius vector to the point of contact.

▮ In Fig. 6.21, $\rho = 3a = 3\rho \cos(\pi - \psi) = -3\rho \cos \psi$, so that $\cos \psi = -\frac{1}{3}$ and $\tan \psi = -2\sqrt{2}$. In Fig. 6.22, $\rho = 3a = 3\rho \cos \psi$ and $\tan \psi = 2\sqrt{2}$. Combining the two cases, we get $\tan \psi = \rho \dfrac{d\theta}{d\rho} = \pm 2\sqrt{2}$, from which $\dfrac{d\rho}{\rho} = \pm \dfrac{d\theta}{2\sqrt{2}}$. The required curves are the families $\rho = Ce^{\theta/2\sqrt{2}}$ and $\rho = Ce^{-\theta/2\sqrt{2}}$.

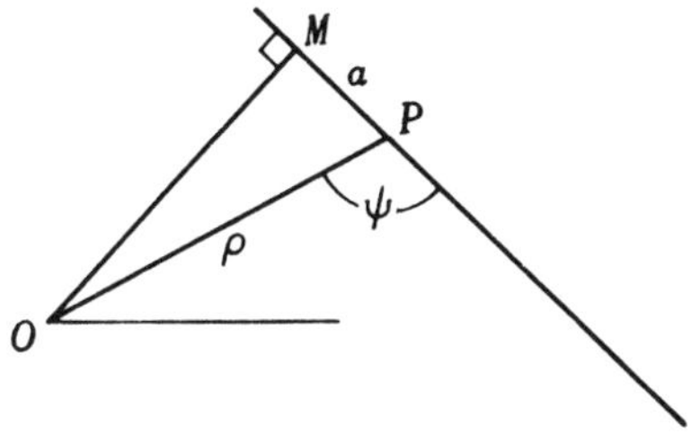

Fig. 6.21

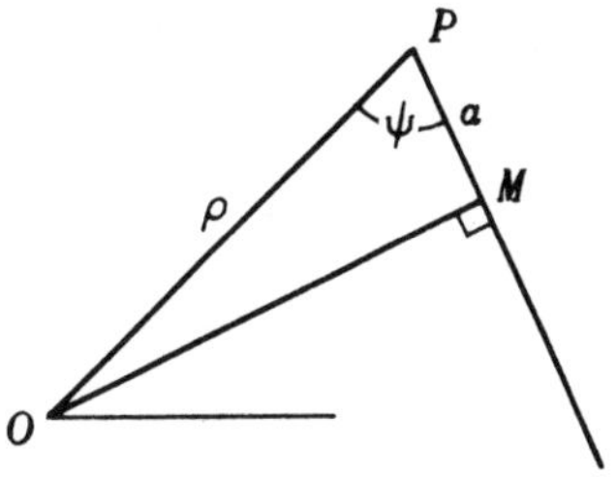

Fig. 6.22

6.189 Find the shape assumed by a flexible chain suspended between two points and hanging under its own weight.

▮ Let the y axis pass through the lowest point of the chain (Fig. 6.23), let s be the arc length from this point to a variable point (x, y), and let $w(s)$ be the linear density of the chain. We obtain the equation of the curve from the fact that the portion of the chain between the lowest point and (x, y) is in equilibrium under the action of three forces: the horizontal tension T_0 at the lowest point; the variable tension T at (x, y), which acts along the tangent because of the flexibility of the chain; and a downward force equal to the weight of the chain between these two points. Equating the horizontal component of T to T_0 and the vertical component of T to the weight of the chain gives $T \cos \theta = T_0$ and $T \sin \theta = \int_0^s w(s)\,ds$. It follows from the first of these equations that $T \sin \theta = T_0 \tan \theta = T_0 \dfrac{dy}{dx}$, so $T_0 y' = \int_0^s w(s)\,ds$. We eliminate the integral here by

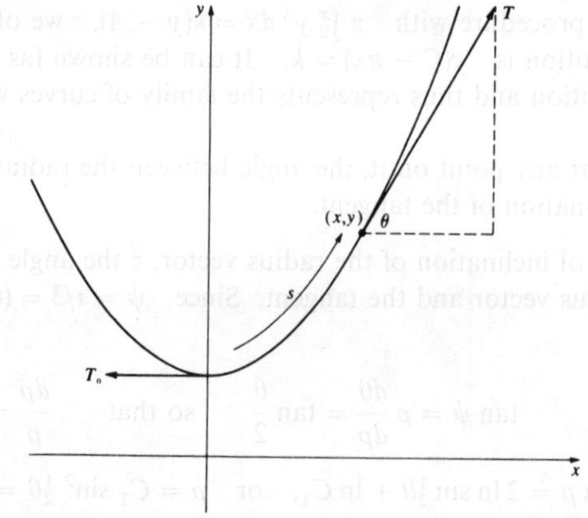

Fig. 6.23

differentiating with respect to x:

$$T_0 y'' = \frac{d}{dx}\int_0^s w(s)\,ds = \frac{d}{ds}\int_0^s w(s)\,ds\,\frac{ds}{dx} = w(s)\sqrt{1+(y')^2}$$

Thus $T_0 y'' = w(s)\sqrt{1+(y')^2}$ is the differential equation of the desired curve, and the curve itself is found by solving this equation. To proceed further, we must have definite information about the function $w(s)$.

We shall assume that $w(s)$ is a constant w_0, so that $y'' = a\sqrt{1+(y')^2}$, where $a = w_0/T_0$. Substituting $y' = p$ and $y'' = dp/dx$ then yields $\dfrac{dp}{\sqrt{1+p^2}} = a\,dx$. Integration and use of the fact that $p = 0$ when $x = 0$ now give $\log(p + \sqrt{1+p^2}) = ax$. Solving for p then yields $p = \dfrac{dy}{dx} = \dfrac{1}{2}(e^{ax} - e^{-ax})$. If we place the x axis at the proper height, so that $y = 1/a$ when $x = 0$, we get

$$y = \frac{1}{2a}(e^{ax} + e^{-ax}) = \frac{1}{a}\cosh ax$$

as the equation of the curve assumed by a uniform flexible chain hanging under its own weight. This curve is called a *catenary*, from the Latin word for chain, *catena*.

6.190 A point P is dragged along the xy plane by a string PT of length a. If T starts at the origin and moves along the positive y axis, and if P starts at $(a, 0)$, what is the path of P?

▮ It is easy to see from Fig. 6.24 that the differential equation of the path is $\dfrac{dy}{dx} = -\dfrac{\sqrt{a^2 - x^2}}{x}$. On separating variables and integrating (and using the fact that $y = 0$ when $x = a$), we find that $y = a\ln\left(\dfrac{a + \sqrt{a^2 - x^2}}{x}\right) - \sqrt{a^2 - x^2}$. This is the equation of a *tractrix*, from the Latin word *tractum*, meaning drag.

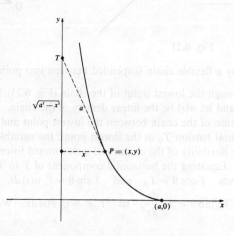

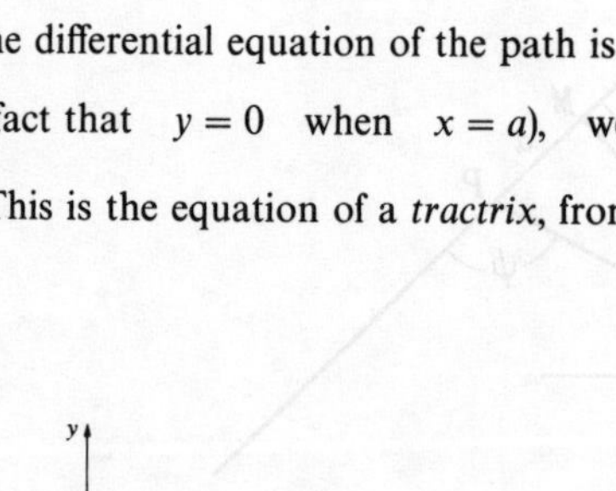

Fig. 6.24

6.191 A rabbit starts at the origin and runs up the y axis with speed a. At the same time a dog, running with speed b, starts at the point $(c, 0)$ and pursues the rabbit. Find a differential equation describing the path of the dog.

▌ At time t, measured from the instant both start, the rabbit will be at the point $R = (0, at)$ and the dog at $D = (x, y)$ (Fig. 6.25). Since the line DR is tangent to the path, we have $\dfrac{dy}{dx} = \dfrac{y - at}{x}$ or $xy' - y = -at$. To eliminate t, we begin by differentiating this last equation with respect to x, which gives $xy'' = -a\dfrac{dt}{dx}$. Since $\dfrac{ds}{dt} = b$, we have $\dfrac{dt}{dx} = \dfrac{dt}{ds}\dfrac{ds}{dx} = -\dfrac{1}{b}\sqrt{1 + (y')^2}$, where the minus sign appears because s increases as x decreases. When these two equations are combined, we obtain the differential equation of the path:

$$xy'' = k\sqrt{1 + (y')^2} \qquad k = \frac{a}{b}$$

The substitution $y' = p$ and $y'' = dp/dx$ reduces this to $\dfrac{dp}{\sqrt{1 + p^2}} = k\dfrac{dx}{x}$, and on integrating and using the initial condition $p = 0$ when $x = c$, we find that $\ln(p + \sqrt{1 + p^2}) = \ln\left(\dfrac{x}{c}\right)^k$. Then $p = \dfrac{dy}{dx} = \dfrac{1}{2}\left[\left(\dfrac{x}{c}\right)^k - \left(\dfrac{c}{x}\right)^k\right]$.

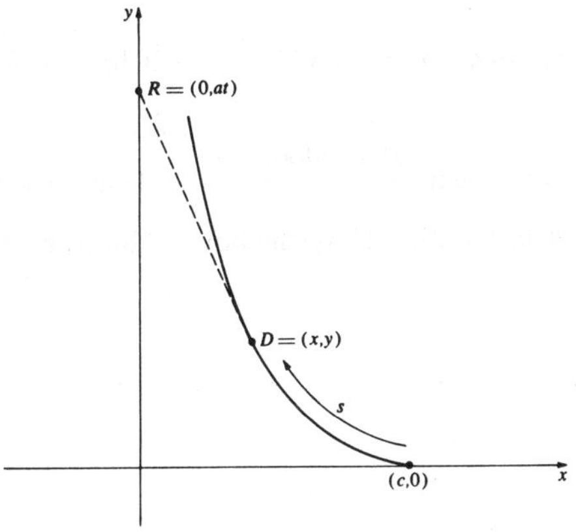

Fig. 6.25

6.192 The y axis and the line $x = c$ are the banks of a river whose current has uniform speed a in the negative y direction. A boat enters the river at the point $(c, 0)$ and heads directly toward the origin with speed b relative to the water. What is the path of the boat?

▌ The components of the boat's velocity (Fig. 6.26) are $\dfrac{dx}{dt} = -b\cos\theta$ and $\dfrac{dy}{dt} = -a + b\sin\theta$, so

$$\frac{dy}{dx} = \frac{-a + b\sin\theta}{-b\cos\theta} = \frac{-a + b(-y/\sqrt{x^2 + y^2})}{-b(x/\sqrt{x^2 + y^2})} = \frac{a\sqrt{x^2 + y^2} + by}{bx}$$

This equation is homogeneous, and its solution is $c^k(y + \sqrt{x^2 + y^2}) = x^{k+1}$, where $k = a/b$. It is clear that the fate of the boat depends on the relation between a and b.

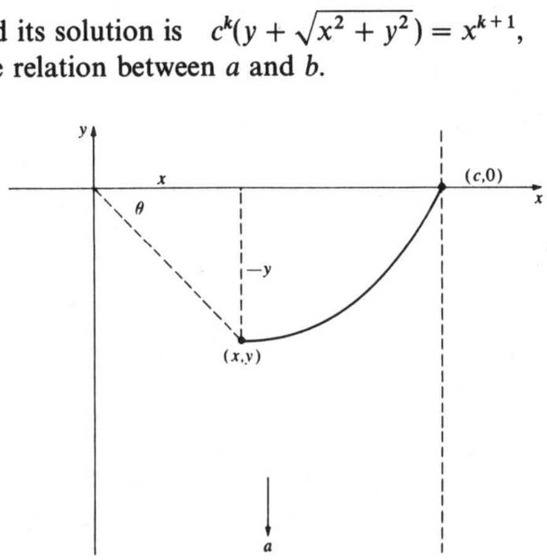

Fig. 6.26

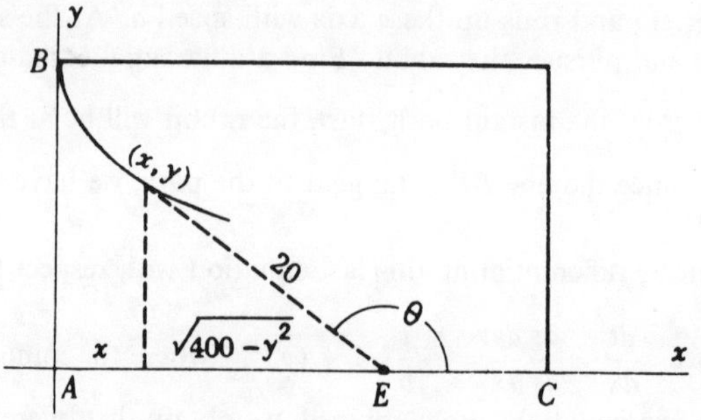

Fig. 6.27

6.193 A boy, standing in corner A of a rectangular pool (Fig. 6.27), has a boat in the adjacent corner B on the end of a string 20 ft long. He walks along the side of the pool toward C, keeping the string taut. Locate the boy and boat when the latter is 12 ft from AC.

▌ We choose the coordinate system so that AC is along the x axis, and AB is along the y axis. Let (x, y) be the position of the boat when the boy has reached E, and let θ denote the angle of inclination of the string. Then $\tan\theta = \dfrac{dy}{dx} = \dfrac{-y}{\sqrt{400-y^2}}$ or $dx = -\dfrac{\sqrt{400-y^2}}{y}\,dy$. Integrating gives $x = -\sqrt{400-y^2} + 20\ln\dfrac{20+\sqrt{400-y^2}}{y} + C$ To find C, we note that when the boat is at B, we have $x = 0$ and $y = 20$. Then $C = 0$, and $x = -\sqrt{400-y^2} + 20\ln\dfrac{20+\sqrt{400-y^2}}{y}$ is the equation of the boat's path.

Now $AE = x + \sqrt{400-y^2} = 20\ln\dfrac{20+\sqrt{400-y^2}}{y}$. Hence, when the boat is 12 ft from AC (that is, when $y = 12$), $x + 16 = 20\ln 3 = 22$. Thus, the boy is 22 ft from A, and the boat is 6 ft from AB.

CHAPTER 7

Linear Differential Equations—Theory of Solutions

VRONSKIAN

.1 Define the *Wronskian* of the set of functions $\{z_1(x), z_2(x), \dots, z_n(x)\}$ on the interval $a \le x \le b$, where each function possesses $n-1$ derivatives.

I

The Wronskian is the determinant $$W(z_1, z_2, \dots, z_n) = \begin{vmatrix} z_1 & z_2 & \cdots & z_n \\ z_1' & z_2' & \cdots & z_n' \\ z_1'' & z_2'' & \cdots & z_n'' \\ \cdots & \cdots & \cdots & \cdots \\ z_1^{(n-1)} & z_2^{(n-1)} & \cdots & z_n^{(n-1)} \end{vmatrix}$$

.2 Find the Wronskian of $\{\sin 2x, \cos 2x\}$.

I

$$W(\sin 2x, \cos 2x) = \begin{vmatrix} \sin 2x & \cos 2x \\ \dfrac{d(\sin 2x)}{dx} & \dfrac{d(\cos 2x)}{dx} \end{vmatrix} = \begin{vmatrix} \sin 2x & \cos 2x \\ 2\cos 2x & -2\sin 2x \end{vmatrix} = -2$$

.3 Find the Wronskian of $\{3\sin 2x, 4\sin 2x\}$.

I

$$W(3\sin 2x, 4\sin 2x) = \begin{vmatrix} 3\sin 2x & 4\sin 2x \\ \dfrac{d(3\sin 2x)}{dx} & \dfrac{d(4\sin 2x)}{dx} \end{vmatrix} = \begin{vmatrix} 3\sin 2x & 4\sin 2x \\ 6\cos 2x & 8\cos 2x \end{vmatrix}$$
$$= (3\sin 2x)(8\cos 2x) - (4\sin 2x)(6\cos 2x) = 0$$

.4 Find the Wronskian of $\{\sin 3x, \cos 3x\}$.

I

$$W(\sin 3x, \cos 3x) = \begin{vmatrix} \sin 3x & \cos 3x \\ \dfrac{d(\sin 3x)}{dx} & \dfrac{d(\cos 3x)}{dx} \end{vmatrix} = \begin{vmatrix} \sin 3x & \cos 3x \\ 3\cos 3x & -3\sin 3x \end{vmatrix}$$
$$= (\sin 3x)(-3\sin 3x) - (\cos 3x)(3\sin 3x) = -3\sin^2 3x - 3\cos^2 3x = -3$$

.5 Find the Wronskian of $\{1, x\}$.

I

$$W(1, x) = \begin{vmatrix} 1 & x \\ \dfrac{d(1)}{dx} & \dfrac{d(x)}{dx} \end{vmatrix} = \begin{vmatrix} 1 & x \\ 0 & 1 \end{vmatrix} = 1(1) - x(0) = 1$$

.6 Find the Wronskian of $\{3x, 5x\}$.

I

$$W(3x, 5x) = \begin{vmatrix} 3x & 5x \\ \dfrac{d(3x)}{dx} & \dfrac{d(5x)}{dx} \end{vmatrix} = \begin{vmatrix} 3x & 5x \\ 3 & 5 \end{vmatrix} = 3x(5) - (5x)(3) = 0$$

.7 Find the Wronskian of $\{t, t^2\}$.

I

$$W(t, t^2) = \begin{vmatrix} t & t^2 \\ \dfrac{d(t)}{dt} & \dfrac{d(t^2)}{dt} \end{vmatrix} = \begin{vmatrix} t & t^2 \\ 1 & 2t \end{vmatrix} = t(2t) - t^2(1) = t^2$$

.8 Find the Wronskian of $\{t, t^3\}$.

I

$$W(t, t^3) = \begin{vmatrix} t & t^3 \\ \dfrac{d(t)}{dt} & \dfrac{d(t^3)}{dt} \end{vmatrix} = \begin{vmatrix} t & t^3 \\ 1 & 3t^2 \end{vmatrix} = t(3t^2) - t^3(1) = 2t^3$$

7.9 Find the Wronskian of $\{t^2, t^3\}$.

▌ $$W(t^2, t^3) = \begin{vmatrix} t^2 & t^3 \\ 2t & 3t^2 \end{vmatrix} = t^2(3t^2) - t^3(2t) = t^4$$

7.10 Find the Wronskian of $\{3t^2, 2t^3\}$.

▌ $$W(3t^2, 2t^3) = \begin{vmatrix} 3t^2 & 2t^3 \\ 6t & 6t^2 \end{vmatrix} = 3t^2(6t^2) - 2t^3(6t) = 6t^4$$

7.11 Find the Wronskian of $\{3t^2, 2t^2\}$.

▌ $$W(3t^2, 2t^2) = \begin{vmatrix} 3t^2 & 2t^2 \\ 6t & 4t \end{vmatrix} = 3t^2(4t) - 2t^2(6t) = 0$$

7.12 Find the Wronskian of $\{t^3, 5t^3\}$.

▌ $$W(t^3, 5t^3) = \begin{vmatrix} t^3 & 5t^3 \\ 3t^2 & 15t^2 \end{vmatrix} = t^3(15t^2) - 5t^3(3t^2) = 0$$

7.13 Find the Wronskian of $\{t^2, \frac{1}{2}t^6\}$.

▌ $$W(t^2, \tfrac{1}{2}t^6) = \begin{vmatrix} t^2 & \frac{1}{2}t^6 \\ 2t & 3t^5 \end{vmatrix} = t^2(3t^5) - \tfrac{1}{2}t^6(2t) = 2t^7$$

7.14 Find the Wronskian of $\{2t^3, 3t^7\}$.

▌ $$W(2t^3, 3t^7) = \begin{vmatrix} 2t^3 & 3t^7 \\ 6t^2 & 21t^6 \end{vmatrix} = 2t^3(21t^6) - 3t^7(6t^2) = 24t^9$$

7.15 Find the Wronskian of $\{e^x, e^{-x}\}$.

▌ $$W(e^x, e^{-x}) = \begin{vmatrix} e^x & e^{-x} \\ e^x & -e^{-x} \end{vmatrix} = e^x(-e^{-x}) - e^{-x}e^x = -2$$

7.16 Find the Wronskian of $\{5e^x, 7e^{-x}\}$.

▌ $$W(5e^x, 7e^{-x}) = \begin{vmatrix} 5e^x & 7e^{-x} \\ 5e^x & -7e^{-x} \end{vmatrix} = 5e^x(-7e^{-x}) - 7e^{-x}(5e^x) = -70$$

7.17 Find the Wronskian of $\{5e^{2x}, 7e^{3x}\}$.

▌ $$W(5e^{2x}, 7e^{3x}) = \begin{vmatrix} 5e^{2x} & 7e^{3x} \\ 10e^{2x} & 21e^{3x} \end{vmatrix} = 5e^{2x}(21e^{3x}) - 7e^{3x}(10e^{2x}) = 35e^{5x}$$

7.18 Find the Wronskian of $\{7e^{-3x}, 4e^{-3x}\}$.

▌ $$W(7e^{-3x}, 4e^{-3x}) = \begin{vmatrix} 7e^{-3x} & 4e^{-3x} \\ -21e^{-3x} & -12e^{-3x} \end{vmatrix} = 7e^{-3x}(-12e^{-3x}) - 4e^{-3x}(-21e^{-3x}) = 0$$

7.19 Find the Wronskian of $\{e^x, xe^x\}$.

▌ $$W(e^x, xe^x) = \begin{vmatrix} e^x & xe^x \\ e^x & e^x + xe^x \end{vmatrix} = e^x(e^x + xe^x) - xe^x(e^x) = e^{2x}$$

7.20 Find the Wronskian of $\{x^3, |x^3|\}$ on $[-1, 1]$.

▌ We have $$|x^3| = \begin{cases} x^3 & \text{if } x \ge 0 \\ -x^3 & \text{if } x < 0 \end{cases} \quad \text{so} \quad \frac{d(|x^3|)}{dx} = \begin{cases} 3x^2 & \text{if } x > 0 \\ 0 & \text{if } x = 0 \\ -3x^2 & \text{if } x < 0 \end{cases}$$

Then, for $x > 0$, $$W(x^3, |x^3|) = \begin{vmatrix} x^3 & x^3 \\ 3x^2 & 3x^2 \end{vmatrix} \equiv 0$$

For $x < 0$, $$W(x^3, |x^3|) = \begin{vmatrix} x^3 & -x^3 \\ 3x^2 & -3x^2 \end{vmatrix} \equiv 0$$

For $x = 0$, $$W(x^3, |x^3|) = \begin{vmatrix} 0 & 0 \\ 0 & 0 \end{vmatrix} = 0$$

Thus, $W(x^3, |x^3|) \equiv 0$ on $[-1, 1]$.

21 Find the Wronskian of $\{1, x, x^2\}$.

▮ $$W(1, x, x^2) = \begin{vmatrix} 1 & x & x^2 \\ 0 & 1 & 2x \\ 0 & 0 & 2 \end{vmatrix} = 1\begin{vmatrix} 1 & 2x \\ 0 & 2 \end{vmatrix} - 0\begin{vmatrix} x & x^2 \\ 0 & 2 \end{vmatrix} + 0\begin{vmatrix} x & x^2 \\ 1 & 2x \end{vmatrix}$$
$$= 1[1(2) - 2x(0)] - 0 + 0 = 2$$

22 Find the Wronskian of $\{x, 2x^2, -x^3\}$.

▮ $$W(x, 2x^2, -x^3) = \begin{vmatrix} x & 2x^2 & -x^3 \\ 1 & 4x & -3x^2 \\ 0 & 4 & -6x \end{vmatrix} = x\begin{vmatrix} 4x & -3x^2 \\ 4 & -6x \end{vmatrix} - 1\begin{vmatrix} 2x^2 & -x^3 \\ 4 & -6x \end{vmatrix} + 0\begin{vmatrix} 2x^2 & -x^3 \\ 4x & -3x^2 \end{vmatrix}$$
$$= x(-12x^2) - 1(-8x^3) + 0 = -4x^3$$

23 Find the Wronskian of $\{x^2, x^3, x^4\}$.

▮ $$W(x^2, x^3, x^4) = \begin{vmatrix} x^2 & x^3 & x^4 \\ 2x & 3x^2 & 4x^3 \\ 2 & 6x & 12x^2 \end{vmatrix} = x^2\begin{vmatrix} 3x^2 & 4x^3 \\ 6x & 12x^2 \end{vmatrix} - x^3\begin{vmatrix} 2x & 4x^3 \\ 2 & 12x^2 \end{vmatrix} + x^4\begin{vmatrix} 2x & 3x^2 \\ 2 & 6x \end{vmatrix}$$
$$= x^2(12x^4) - x^3(16x^3) + x^4(6x^2) = 2x^6$$

24 Find the Wronskian of $\{x^2, -2x^2, 3x^3\}$.

▮ $$W(x^2, -2x^2, 3x^3) = \begin{vmatrix} x^2 & -2x^2 & 3x^3 \\ 2x & -4x & 9x^2 \\ 2 & -4 & 18x \end{vmatrix} = x^2\begin{vmatrix} -4x & 9x^2 \\ -4 & 18x \end{vmatrix} - (-2x^2)\begin{vmatrix} 2x & 9x^2 \\ 2 & 18x \end{vmatrix} + 3x^3\begin{vmatrix} 2x & -4x \\ 2 & -4 \end{vmatrix}$$
$$= x^2(-36x^2) - (-2x^2)(18x^2) + 3x^3(0) = 0$$

25 Find the Wronskian of $\{x^{-2}, x^2, 2 - 3x\}$.

▮ $$W(x^{-2}, x^2, 2-3x) = \begin{vmatrix} x^{-2} & x^2 & 2-3x \\ \dfrac{d(x^{-2})}{dx} & \dfrac{d(x^2)}{dx} & \dfrac{d(2-3x)}{dx} \\ \dfrac{d^2(x^{-2})}{dx^2} & \dfrac{d^2(x^2)}{dx^2} & \dfrac{d^2(2-3x)}{dx^2} \end{vmatrix} = \begin{vmatrix} x^{-2} & x^2 & 2-3x \\ -2x^{-3} & 2x & -3 \\ 6x^{-4} & 2 & 0 \end{vmatrix} = 36x^{-2} - 32x^{-3}$$

26 Find the Wronskian of $\{e^t, e^{-t}, e^{2t}\}$.

▮ $$W(e^t, e^{-t}, e^{2t}) = \begin{vmatrix} e^t & e^{-t} & e^{2t} \\ \dfrac{d(e^t)}{dt} & \dfrac{d(e^{-t})}{dt} & \dfrac{d(e^{2t})}{dt} \\ \dfrac{d^2(e^t)}{dt^2} & \dfrac{d^2(e^{-t})}{dt^2} & \dfrac{d^2(e^{2t})}{dt^2} \end{vmatrix} = \begin{vmatrix} e^t & e^{-t} & e^{2t} \\ e^t & -e^{-t} & 2e^{2t} \\ e^t & e^{-t} & 4e^{2t} \end{vmatrix} = -6e^{2t}$$

27 Find the Wronskian of $\{1, \sin 2t, \cos 2t\}$.

▮ $$W(1, \sin 2t, \cos 2t) = \begin{vmatrix} 1 & \sin 2t & \cos 2t \\ 0 & 2\cos 2t & -2\sin 2t \\ 0 & -4\sin 2t & -4\cos 2t \end{vmatrix} = -8\cos^2 2t - 8\sin^2 2t = -8$$

7.28 Find the Wronskian of $\{t, t-3, 2t+5\}$.

▮

$$W = \begin{vmatrix} t & t-3 & 2t+5 \\ 1 & 1 & 2 \\ 0 & 0 & 0 \end{vmatrix} = 0$$

7.29 Find the Wronskian of $\{t^3, t^3+t, 2t^3-7t\}$.

▮

$$W = \begin{vmatrix} t^3 & t^3+t & 2t^3-7t \\ 3t^2 & 3t^2+1 & 6t^2-7 \\ 6t & 6t & 12t \end{vmatrix} = 0$$

7.30 Find the Wronskian of $\{t^3, t^3+t, 2t^3-7\}$.

▮

$$W = \begin{vmatrix} t^3 & t^3+t & 2t^3-7 \\ 3t^2 & 3t^2+1 & 6t^2 \\ 6t & 6t & 12t \end{vmatrix} = 42t$$

7.31 Find the Wronskian of $\{\sin t, \cos t, 2\sin t - \cos t\}$.

▮

$$W = \begin{vmatrix} \sin t & \cos t & 2\sin t - \cos t \\ \cos t & -\sin t & 2\cos t + \sin t \\ -\sin t & -\cos t & -2\sin t + \cos t \end{vmatrix} = 0$$

7.32 Find the Wronskian of $\{t^3, t^2, t, 1\}$.

▮

$$W = \begin{vmatrix} t^3 & t^2 & t & 1 \\ 3t^2 & 2t & 1 & 0 \\ 6t & 2 & 0 & 0 \\ 6 & 0 & 0 & 0 \end{vmatrix} = 12$$

7.33 Find the Wronskian of $\{e^{m_1t}, e^{m_2t}, e^{m_3t}, e^{m_4t}\}$, where m_1, m_2, m_3, and m_4 are constants.

▮

$$W = \begin{vmatrix} e^{m_1t} & e^{m_2t} & e^{m_3t} & e^{m_4t} \\ m_1e^{m_1t} & m_2e^{m_2t} & m_3e^{m_3t} & m_4e^{m_4t} \\ m_1^2e^{m_1t} & m_2^2e^{m_2t} & m_3^2e^{m_3t} & m_4^2e^{m_4t} \\ m_1^3e^{m_1t} & m_2^3e^{m_1t} & m_3^3e^{m_3t} & m_4^3e^{m_4t} \end{vmatrix} = e^{m_1t}e^{m_2t}e^{m_3t}e^{m_4t} \begin{vmatrix} 1 & 1 & 1 & 1 \\ m_1 & m_2 & m_3 & m_4 \\ m_1^2 & m_2^2 & m_3^2 & m_4^2 \\ m_1^3 & m_2^3 & m_3^3 & m_4^3 \end{vmatrix}$$

This last determinant is a *Vondermonde determinant* and is equal to $(m_4 - m_1)(m_4 - m_2)(m_4 - m_3)(m_3 - m_1)(m_3 - m_2)(m_2 - m_1)$. Thus, $W = e^{(m_1+m_2+m_3+m_4)t}(m_4 - m_1)(m_4 - m_2)(m_4 - m_3)(m_3 - m_1)(m_3 - m_2)(m_2 - m_1)$.

LINEAR INDEPENDENCE

7.34 Determine whether the set $\{e^x, e^{-x}\}$ is linearly dependent on $(-\infty, \infty)$.

▮ Consider the equation

$$c_1e^x + c_2e^{-x} \equiv 0 \quad (1)$$

We must determine whether there exist values of c_1 and c_2, *not both zero*, that will satisfy (*1*). Rewriting (*1*), we have $c_2e^{-x} \equiv -c_1e^x$ or $c_2 \equiv -c_1e^{2x}$. For any nonzero value of c_1, the left side of this equation is a constant, whereas the right side is not; hence the equality is not valid. It follows that the *only* solution to this latter equation, and therefore to (*1*), is $c_1 = c_2 = 0$. Thus, the set is not linearly dependent; rather it is linear independent.

7.35 Rework the previous problem using differentiation.

▮ We begin again with the equation $c_1e^x + c_2e^{-x} \equiv 0$. Differentiating it now, we obtain $c_1e^x - c_2e^{-x} \equiv 0$. These two equations are a set of simultaneous linear equations for the unknowns c_1 and c_2. Solving them, we find that the only solution is $c_1 = c_2 = 0$, so the functions are linearly independent.

7.36 Rework Problem 7.34 using the Wronskian.

▌ The Wronskian of $\{e^x, e^{-x}\}$ is found in Problem 7.15 to be -2. Since it is nonzero for at least one point in the interval of interest (it is, in fact, nonzero everywhere), the functions are linearly independent on that interval.

7.37 Is the set $\{x^2, x, 1\}$ linearly dependent on $(-\infty, \infty)$?

▌ Consider the equation $c_1x^2 + c_2x + c_3 \equiv 0$. Since this equation is valid for all x only if $c_1 = c_2 = c_3 = 0$, the given set is linearly independent. Note that if any of the c's were not zero, then the equation could hold for at most two values of x, the roots of the equation, *and not for all x*.

7.38 Rework the previous problem using differentiation.

▌ We begin again with the equation $c_1x^2 + c_2x + c_3 \equiv 0$. Differentiating this equation, we get $2c_1x + c_2 \equiv 0$. Differentiating once again, we get $2c_1 \equiv 0$. These three equations are a set of simultaneous linear equations for the unknowns c_1, c_2, and c_3. Solving them, we find that the only solution is $c_1 = c_2 = c_3 = 0$, which implies that the functions are linearly independent.

7.39 Rework Problem 7.37 using the Wronskian.

▌ The Wronskian of $\{x^2, x, 1\}$ is found in Problem 7.21 to be 2. Since it is nonzero for at least one point in the interval of interest (it is, in fact, nonzero everywhere), the functions are linearly independent on that interval.

7.40 Determine whether the set $\{\sin 2x, \cos 2x\}$ is linearly independent on $(-\infty, \infty)$.

▌ Consider the equation $c_1 \sin 2x + c_2 \cos 2x \equiv 0$. Differentiating, we obtain $2c_1 \cos 2x - 2c_2 \sin 2x \equiv 0$. These two equations are a set of simultaneous linear equations for the unknowns c_1 and c_2. The easiest way to solve them is to multiply the first by $2 \sin 2x$, multiply the second by $\cos 2x$, and then add the resulting equations together. Doing so, we find that the only solution is $c_1 = c_2 = 0$, which implies that the functions are linearly independent.

7.41 Rework the previous problem using the Wronskian.

▌ The Wronskian of $\{\sin 2x, \cos 2x\}$ is found in Problem 7.2 to be -2. Since it is nonzero for at least one point in the interval of interest (it is, in fact, nonzero everywhere), the functions are linearly independent on that interval.

7.42 Determine whether the set $\{\sin 3x, \cos 3x\}$ is linearly independent on $(-\infty, \infty)$.

▌ Consider the equation $c_1 \sin 3x + c_2 \cos 3x \equiv 0$. Differentiating, we obtain $3c_1 \cos 3x - 3c_2 \sin 3x \equiv 0$. To solve these two equations simultaneously for the unknowns c_1 and c_2, we multiply the first by $3 \sin 3x$, multiply the second by $\cos 3x$, and then add. The only solution is $c_1 = c_2 = 0$, which implies that the functions are linearly independent.

7.43 Rework the previous problem using the Wronskian.

▌ The Wronskian of $\{\sin 3x, \cos 3x\}$ is found in Problem 7.4 to be -3. Since it is nonzero for at least one point in the interval of interest (it is, in fact, nonzero everywhere), the functions are linearly independent on that interval.

7.44 Is the set $\{t^2, t\}$ linearly independent on $(-\infty, \infty)$?

▌ Consider the equation $c_1t^2 + c_2t \equiv 0$. Since a polynomial equals zero for all values of t in $(-\infty, \infty)$ if and only if the coefficients are zero, it follows that $c_1 = c_2 = 0$, from which we conclude that the original functions are linearly independent.

7.45 Rework the previous problem using differentiation.

▌ We begin again with the equation $c_1t^2 + c_2t \equiv 0$. Differentiating gives $2c_1t + c_2 \equiv 0$. These two equations are a system of linear equations which may be solved simultaneously for c_1 and c_2. Instead, we may differentiate the second equation to obtain $2c_1 \equiv 0$. It follows from this that $c_1 = 0$, and then from the second equation that $c_2 = 0$. Thus, the only solution is $c_1 = c_2 = 0$, and the functions are linearly independent.

7.46 Rework Problem 7.44 using the Wronskian.

The Wronskian of $\{t^2, t\}$ is found in Problem 7.7 to be t^2. Since it is nonzero for at least one point in the interval of interest (for example, at $t = 2$, we have $W = 4 \neq 0$), the functions are linearly independent on that interval.

7.47 Is the set $\{t^2, t^3\}$ linearly independent on $(-\infty, \infty)$?

Consider the equation $c_1t^2 + c_2t^3 \equiv 0$. This is a third-degree polynomial in t. Since a polynomial is zero for all values of t in $(-\infty, \infty)$ if and only if all of its coefficients are zero, it follows that $c_1 = c_2 = 0$, from which we conclude that the functions are linearly independent.

7.48 Rework the previous problem using differentiation.

We begin again with the equation $c_1t^2 + c_2t^3 \equiv 0$. Differentiating, we obtain $2c_1t + 3c_2t^2 \equiv 0$. These two equations are a set of linear equations which may be solved simultaneously for c_1 and c_2. Instead, however, we may differentiate twice more, obtaining successively $2c_1 + 6c_2t \equiv 0$ and $6c_2 \equiv 0$. It follows from these equations first that $c_2 = 0$, and then that $c_1 = 0$. Thus, the only solution to the first equation is $c_1 = c_2 = 0$, and the functions are linearly independent.

7.49 Rework Problem 7.47 using the Wronskian.

The Wronskian of $\{t^2, t^3\}$ is found in Problem 7.9 to be t^4. Since it is nonzero for at least one point in the interval of interest (for example, at $t = 1$, $W = 1 \neq 0$), the functions are linearly independent on that interval.

7.50 Determine whether $\{e^x, xe^x\}$ is linearly independent on $(-\infty, \infty)$.

Consider the equation $c_1e^x + c_2xe^x \equiv 0$. Differentiating, we obtain $c_1e^x + c_2(e^x + xe^x) \equiv 0$. These two equations may be solved simultaneously for c_1 and c_2. We begin by subtracting the first from the second, and then recall that e^x is never zero. Thus we find that the only solution is $c_1 = c_2 = 0$. It follows that the functions are linearly independent.

7.51 Rework the previous problem using the Wronskian.

The Wronskian of $\{e^x, xe^x\}$ is found in Problem 7.19 to be e^{2x}. Since it is nonzero for at least one point in the interval of interest (it is, in fact, nonzero everywhere), it follows that the functions are linearly independent on that interval.

7.52 Determine whether $\{3 \sin 2x, 4 \sin 2x\}$ is linearly independent on $(-\infty, \infty)$.

Consider the equation $c_1(3 \sin 2x) + c_2(4 \sin 2x) \equiv 0$. By inspection, we see that there exist constants c_1 and c_2, not both zero (in particular, $c_1 = 4$ and $c_2 = -3$), that satisfy this equation for all values of x in $(-\infty, \infty)$; thus, the functions are linearly dependent.

7.53 Can the Wronskian be used to determine whether the functions $3 \sin 2x$ and $4 \sin 2x$ are linearly independent on $(-\infty, \infty)$?

It is shown in Problem 7.3 that the Wronskian of these two functions is identically zero, so no conclusions can be drawn about linear independence.

7.54 Redo the previous problem if in addition it is known that the two functions are solutions of the same linear homogeneous differential equation.

If the Wronskian of a set of functions is zero, and if those functions are all solutions to the same linear homogeneous differential equation, then the functions are linearly dependent. Thus, it now follows from Problem 7.3 that $3 \sin 2x$ and $4 \sin 2x$ are linearly dependent.

7.55 Determine whether $\{3x, 5x\}$ is linearly independent on $(-\infty, \infty)$.

Consider the equation $c_1(3x) + c_2(5x) \equiv 0$. By inspection, we see that there exist constants c_1 and c_2 not both zero (in particular, $c_1 = 5$ and $c_2 = 3$) that satisfy this equation for all values of x in $(-\infty, \infty)$. Thus, the functions are linearly dependent.

7.56 Can the Wronskian be used to determine whether the functions $3x$ and $5x$ are linearly independent on $(-\infty, \infty)$?

▌ It is shown in Problem 7.6 that the Wronskian of these two functions is identically zero, so no conclusions can be drawn about linear independence.

7.57 Redo the previous problem knowing that $y_1 = 3x$ and $y_2 = 5x$ are both solutions of $y'' = 0$.

▌ Since both functions are solutions to the same linear homogeneous differential equation, and since their Wronskian is identically zero, the two functions are linearly dependent.

7.58 Determine whether $\{t^3, 5t^3\}$ is linearly independent on $(-\infty, \infty)$.

▌ Consider the equation $c_1(t^3) + c_2(5t^3) \equiv 0$. By inspection, we see that there exist constants c_1 and c_2 not both zero ($c_1 = -5$, $c_2 = 1$ is one pair; $c_1 = 10$, $c_2 = -2$ is another) that satisfy this equation for all values of t in $(-\infty, \infty)$; therefore the functions are linearly dependent.

7.59 What conclusion can one draw about the linear independence of the functions t^3 and $5t^3$ on $(-\infty, \infty)$ by computing their Wronskian?

▌ Since their Wronskian is identically zero (see Problem 7.12), no conclusion can be drawn about linear independence.

7.60 Redo the previous problem knowing that $y_1 = t^3$ and $y_2 = 5t^3$ are both solutions of $d^4y/dt^4 = 0$.

▌ Since both functions are solutions to the same linear homogeneous differential equation, and since their Wronskian is identically zero, the two functions are linearly dependent.

7.61 Determine whether $\{7e^{-3x}, 4e^{-3x}\}$ is linearly dependent on $(-\infty, \infty)$.

▌ Consider the equation $c_1(7e^{-3x}) + c_2(4e^{-3x}) \equiv 0$. By inspection, we see that there exist constants c_1 and c_2 not both zero ($c_1 = 4$ and $c_2 = -7$ is one pair; $c_1 = 1$ and $c_2 = -7/4$ is another) that satisfy this equation for all values of x in $(-\infty, \infty)$; hence the functions are linearly dependent.

7.62 What conclusion can one draw about the linear independence of the functions $7e^{-3x}$ and $4e^{-3x}$ on $(-\infty, \infty)$ by computing their Wronskian?

▌ Since their Wronskian is identically zero (see Problem 7.18), no conclusion can be drawn about linear independence.

7.63 Redo the previous problem knowing that both functions are solutions to $y' + 3y = 0$.

▌ Since both functions are solutions to the same linear homogeneous differential equation, and since their Wronskian is identically zero, the two functions are linearly dependent.

7.64 What conclusions can one draw about the linear independence of the functions $5e^{2x}$ and $7e^{3x}$ on $(-\infty, \infty)$ by computing their Wronskian?

▌ The Wronskian of these two functions is $35e^{5x}$ (see Problem 7.17). Since it is nonzero for at least one point in the interval of interest (it is, in fact, nonzero everywhere), the functions are linearly independent on that interval.

7.65 What conclusions can one draw about the linear independence of the functions t^2 and $\frac{1}{2}t^6$ on $(0, 5)$ by computing their Wronskian?

▌ The Wronskian of these two functions is $2t^7$ (see Problem 7.13). Since it is nonzero for at least one point in the interval of interest (for example, at $t = 1$, $W = 2 \neq 0$), the functions are linearly independent on $(0, 5)$.

7.66 What conclusions can one draw about the linear independence of the functions $3t^2$ and $2t^2$ on $(0, 8)$ by computing their Wronskian?

▌ Since their Wronskian is identically zero (see Problem 7.11), no conclusion can be drawn about linear independence.

7.67 Redo the previous problem knowing that both functions are solutions to $d^3y/dt^3 = 0$.

▌ Since both functions are solutions to the same linear homogeneous differential equation, and since their Wronskian is identically zero, the two functions are linearly dependent on $(0, 8)$.

7.68 What conclusions can one draw about the linear independence of the functions x^3 and $|x^3|$ on $(-1, 1)$ by computing their Wronskian?

▮ Since the Wronskian is identically zero (see Problem 7.20), no conclusion can be drawn about linear independence.

7.69 Determine whether the set $\{x^3, |x^3|\}$ is linearly dependent on $[-1, 1]$.

▮ Consider the equation $c_1x^3 + c_2|x^3| \equiv 0$. Recall that $|x^3| = x^3$ if $x \ge 0$, and $|x^3| = -x^3$ if $x < 0$. Thus, our equation becomes

$$c_1x^3 + c_2x^3 \equiv 0 \qquad \text{for } x \ge 0$$
$$c_1x^3 - c_2x^3 \equiv 0 \qquad \text{for } x \le 0$$

Solving these two equations simultaneously for c_1 and c_2, we find that the *only* solution is $c_1 = c_2 = 0$. The given set is, therefore, linearly independent.

7.70 Can both x^3 and $|x^3|$ be solutions of the same linear homogeneous differential equation?

▮ No, for if they were, then we would have two solutions of the same linear homogeneous differential equation having an identically zero Wronskian, which would imply that the two functions are linearly dependent. We know, however, from the previous problem that the two functions are linearly *independent* on $(-1, 1)$.

7.71 Determine whether x^3 and $|x^3|$ are linearly dependent on $[-1, 0]$.

▮ It follows from Problem 7.69 that, for linear dependence, we must satisfy $c_1x^3 - c_2x^3 \equiv 0$. Observe that this is the only condition that is operable here, because we do not include any positive values of the independent variable x. By inspection, we see that there exist constants c_1 and c_2 not both zero (for example, $c_1 = c_2 = 7$) that satisfy this equation for all values of x in the interval of interest; therefore the functions are linearly dependent there.

7.72 Determine whether x^3 and $|x^3|$ are linearly dependent on $[0, 1]$.

▮ It follows from Problem 7.69 that we must now satisfy $c_1x^3 + c_2x^3 \equiv 0$. Observe that this is the only operable condition because we do not include negative values of x. By inspection, we see that there exist constants c_1 and c_2 not both zero (for example, $c_1 = -c_2 = 3$) that satisfy this equation for all values of x in the interval of interest; hence the functions are linearly dependent there.

7.73 Determine whether the set $\{1 - x, 1 + x, 1 - 3x\}$ is linearly dependent on $(-\infty, \infty)$.

▮ Consider the equation $c_1(1 - x) + c_2(1 + x) + c_3(1 - 3x) \equiv 0$, which can be rewritten as $(-c_1 + c_2 - 3c_3)x + (c_1 + c_2 + c_3) \equiv 0$. This linear equation can be satisfied for all x only if both coefficients are zero. Thus, we require

$$-c_1 + c_2 - 3c_3 = 0 \qquad \text{and} \qquad c_1 + c_2 + c_3 = 0$$

Solving these equations simultaneously, we find that $c_1 = -2c_3$ and $c_2 = c_3$, with c_3 arbitrary. Choosing $c_3 = 1$ (any other nonzero number would do), we obtain $c_1 = -2$, $c_2 = 1$, and $c_3 = 1$ as a set of constants, not all zero, that satisfy the original equation. Thus, the given set of functions is linearly dependent.

7.74 Determine whether the set $\{5, 3 - 2x, 2 + x - \frac{1}{2}x^2\}$ is linearly dependent on $(-\infty, \infty)$.

▮ Consider the equation $c_1(5) + c_2(3 - 2x) + c_3(2 + x - \frac{1}{2}x^2) \equiv 0$, which we may rewrite as $(-\frac{1}{2}c_3)x^2 + (-2c_2 + c_3)x + (5c_1 + 3c_2 + 2c_3) \equiv 0$. The left side is a second-degree polynomial in x. Since a polynomial is zero for all values of x in $(-\infty, \infty)$ if and only if all of its coefficients are zero, it follows that

$$-\tfrac{1}{2}c_3 = 0 \qquad \text{and} \qquad -2c_2 + c_3 = 0 \qquad \text{and} \qquad 5c_1 + 3c_2 + 2c_3 = 0$$

Solving this set of equations simultaneously, we find that the only solution to it, and therefore to the original equation, is $c_1 = c_2 = c_3 = 0$. Hence the functions are linearly independent.

7.75 Rework the previous problem using differentiation.

▮ We begin again with the rewritten equation $(-\frac{1}{2}c_3)x^2 + (-2c_2 + c_3)x + (5c_1 + 3c_2 + 2c_3) \equiv 0$. Differentiating this equation twice, we obtain successively $-c_3x + (-2c_2 + c_3) \equiv 0$ and $-c_3 \equiv 0$. Solving this set of equations one at a time in reverse order, we find that $c_3 = c_2 = c_1 = 0$ is the only solution; therefore the functions are linearly independent.

7.76 Rework Problem 7.74 using the Wronskian.

▌ Here,

$$W = \begin{vmatrix} 5 & 3-2x & 2+x-\frac{1}{2}x^2 \\ 0 & -2 & 1-x \\ 0 & 0 & -1 \end{vmatrix} = -10$$

which is nonzero for at least one point in the interval of interest (it is, in fact, nonzero everywhere); hence the functions are linearly independent.

7.77 Determine whether $\{x^2, -2x^2, x^3\}$ is linearly dependent on $(-\infty, \infty)$.

▌ Consider the equation $c_1x^2 + c_2(-2x^2) + c_3x^3 \equiv 0$. By inspection, we see that there exist constants c_1, c_2, and c_3 not all zero (for example, $c_1 = 2$, $c_2 = 1$, and $c_3 = 0$) that satisfy this equation for all values of x. Therefore, the functions are linearly dependent.

Alternatively, we can rearrange our equation to $(c_1 - 2c_2)x^2 + c_3x^3 \equiv 0$. The left side is a third-degree polynomial; it is zero for all values of x in $(-\infty, \infty)$ if and only if the coefficients are zero, that is, if and only if $c_1 - 2c_2 = 0$ and $c_3 = 0$. Solving these last two equations simultaneously, we find that $c_3 = 0$ and $c_1 = 2c_2$, with c_2 arbitrary. Choosing $c_2 = 1$ (any other nonzero number would do equally well), we arrive at the same conclusion as before—namely that the functions are linearly dependent.

7.78 What conclusions can one draw about the linear dependence of $\{x^2, -2x^2, x^3\}$ on $\{-\infty, \infty\}$ by computing their Wronskian?

▌ The Wronskian of this set is shown in Problem 7.24 to be zero, so no conclusion about linear dependence may be drawn.

7.79 Redo the previous problem if in addition it is known that all three functions are solutions of $d^4y/dx^4 = 0$.

▌ Since all three functions are solutions of the same linear homogeneous differential equation, and since their Wronskian is identically zero, the three functions are linearly dependent.

7.80 Determine whether the functions e^t, e^{-t}, and e^{2t} are linearly dependent on $(-\infty, \infty)$.

▌ Their Wronskian is $-6e^{2t}$ (see Problem 7.26), which is nonzero for at least one point in the interval of interest (it is, in fact, nonzero everywhere); thus the functions are linearly independent.

7.81 The functions $\sin t$, $\cos t$, and $2\sin t - \cos t$ are all solutions of the differential equation $y'' + y = 0$. Are these functions linearly independent on $(-\infty, \infty)$?

▌ The Wronskian of these functions is found in Problem 7.31 to be zero. Since the functions are all solutions of the same linear homogeneous differential equation, they are linearly dependent.

7.82 The functions t^3, $t^3 + t$, and $2t^3 - 7$ are all solutions of the differential equation $d^4y/dt^4 = 0$. Are these functions linearly independent on $(-\infty, \infty)$?

▌ The Wronskian of these functions is $42t$ (see Problem 7.30). Since it is nonzero for at least one point in the interval of interest (for example, at $t = 1$, $W = 42$), the functions are linearly independent.

7.83 The functions t, $t - 3$, and $2t + 5$ are all solutions of the differential equation $d^2y/dt^2 = 0$. Are these functions linearly dependent on $(-\infty, \infty)$?

▌ The Wronskian of these functions is identically zero (see Problem 7.28). Since the functions are all solutions of the same linear homogeneous differential equation, they are linearly dependent.

7.84 Determine whether the functions e^{2t}, e^{3t}, e^{-t}, and e^{-5t} are linearly dependent on $(-\infty, \infty)$.

▌ Using the result of Problem 7.33 with $m_1 = 2$, $m_2 = 3$, $m_3 = -1$, and $m_4 = -5$, we find that the Wronskian of these four functions is never zero. Thus, the functions are linearly independent.

GENERAL SOLUTIONS OF HOMOGENEOUS EQUATIONS

7.85 Show that the equation $\frac{d^2y}{dx^2} - \frac{dy}{dx} - 2y = 0$ has two distinct solutions of the form $y = e^{ax}$.

▌ If $y = e^{ax}$ is a solution for some value of a, then the given equation is satisfied when we replace y with e^{ax}, $\frac{dy}{dx}$ with ae^{ax}, and $\frac{d^2y}{dx^2}$ with a^2e^{ax}. Doing so, we obtain $\frac{d^2y}{dx^2} - \frac{dy}{dx} - 2y = e^{ax}(a^2 - a - 2) = 0$, which is satisfied when $a = -1$ or 2. Thus $y = e^{-x}$ and $y = e^{2x}$ are solutions.

7.86 Show that $y = c_1e^{-x} + c_2e^{2x}$ is a solution of the differential equation of the previous problem for any values of the arbitrary constants c_1 and c_2.

▌ Since both e^{-x} and e^{2x} are solutions to $y'' - y' - 2y = 0$, and since this differential equation is linear and homogeneous, the result follows immediately from the principle of superposition.

7.87 Find the general solution of $y'' - y' - 2y = 0$.

▌ This is a second-order linear homogeneous differential equation with continuous coefficients on $(-\infty, \infty)$ having the property that the coefficient of the highest derivative is nonzero on this interval. This equation possesses two linearly independent solutions. Two solutions, e^{-x} and e^{2x}, were produced in Problem 7.85; they are linearly independent because their Wronskian is $W = \begin{vmatrix} e^{-x} & e^{2x} \\ -e^{-x} & 2e^{2x} \end{vmatrix} = 3e^x \neq 0$. Hence the general solution is $y = c_1e^{-x} + c_2e^{2x}$.

7.88 Show that the differential equation $x^3\frac{d^3y}{dx^3} - 6x\frac{dy}{dx} + 12y = 0$ has three linearly independent solutions of the form $y = x^r$.

▌ By making the replacements

$$y = x^r \qquad \frac{dy}{dx} = rx^{r-1} \qquad \frac{d^2y}{dx^2} = r(r-1)x^{r-2} \qquad \frac{d^3y}{dx^3} = r(r-1)(r-2)x^{r-3}$$

in the left member of the given equation, we obtain $x^r(r^3 - 3r^2 - 4r + 12) = 0$, which is satisfied when $r = 2$, 3, or -2. The corresponding solutions $y = x^2$, $y = x^3$, and $y = x^{-2}$ are linearly independent because

$$W = \begin{vmatrix} x^2 & x^3 & x^{-2} \\ 2x & 3x^2 & -2x^{-3} \\ 2 & 6x & 6x^{-4} \end{vmatrix} = 20 \neq 0.$$

7.89 Show that $y = c_1x^2 + c_2x^3 + c_3x^{-2}$ is a solution of the differential equation of the previous problem for any values of the arbitrary constants c_1, c_2, and c_3.

▌ Since x^2, x^3, and x^{-2} are all solutions to $x^3y''' - 6xy' + 12y = 0$ (see Problem 7.88), and since this differential equation is linear and homogeneous, the result follows immediately from the principle of superposition.

7.90 Is the solution given in the previous problem the general solution to $x^3y''' - 6xy' + 12y = 0$ on $(1, 5)$?

▌ Yes. The differential equation is linear, of order 3, and homogeneous; it has continuous coefficients on $(1, 5)$ with the property that the coefficient of the highest derivative is not zero *on this interval.* Thus, this equation possesses three linearly independent solutions, which we found in Problem 7.88. The general solution is the superposition of these three linearly independent solutions.

7.91 Two solutions of $y'' - 2y' + y = 0$ are e^x and $5e^x$. Show that $y = c_1e^x + 5c_2e^x$ is also a solution.

▌ Since the differential equation is linear and homogeneous, the result is immediate from the principle of superposition.

7.92 Determine whether $y = c_1e^x + 5c_2e^x$ is the general solution of the differential equation in the previous problem.

▌ The differential equation is linear, of order 2, and homogeneous; it has continuous coefficients with the property that the coefficient of the highest derivative is nonzero everywhere. It follows that the general solution is the superposition of two linearly independent solutions. However, because $W(e^x, 5e^x) = \begin{vmatrix} e^x & 5e^x \\ e^x & 5e^x \end{vmatrix} \equiv 0$, the functions are not linearly independent, and their superposition does not comprise the general solution.

.93 Show that xe^x is a solution of the differential equation in Problem 7.91.

Substituting $y = xe^x$, $y' = e^x + xe^x$, and $y'' = 2e^x + xe^x$ into the left side of the differential equation, we obtain $y'' - 2y' + y = 2e^x + xe^x - 2(e^x + xe^x) + xe^x = 0$. Thus, xe^x satisfies the differential equation for all values of x and is a solution on $(-\infty, \infty)$.

.94 Determine whether $y = c_1 e^x + c_2 xe^x$ is the general solution of the differential equation in Problem 7.91.

It follows from Problem 7.50 that e^x and xe^x are linearly independent. Since both functions are solutions of $y'' - 2y' + y = 0$, and since this is a second-order linear homogeneous differential equation with continuous coefficients having the property that the coefficient of the highest derivative is nonzero everywhere, the superposition of these two linearly independent solutions does comprise the general solution.

.95 Determine whether $y = c_1 e^x + c_2 e^{-x}$ is the general solution of $y'' - 2y' + y = 0$.

Since y as given is not a solution (that is, it does not satisfy the differential equation when substituted into the left side), it cannot be the general solution.

.96 Determine whether $y = c_1 e^x + c_2 e^{-x}$ is the general solution of $y''' - y' = 0$.

It is not. The general solution of a third-order linear homogeneous differential equation with constant coefficients must be formed from the superposition of three linearly independent solutions. Although e^x and e^{-x} are solutions and are linearly independent (see Problem 7.34), they constitute only two functions; they are one short of the number needed to form the solution of a third-order differential equation.

97 Determine whether $y = c_1 \sin 2x$ is a solution of $y'' + 4y = 0$ for any value of the arbitrary constant c_1 if it is known that $\sin 2x$ is a solution.

Since the differential equation is linear and homogeneous, the result follows immediately from the principle of superposition.

98 Determine whether $y = c_1 \sin 2x$ is the general solution of $y'' + 4y = 0$.

It is not. The general solution of a second-order linear homogeneous differential equation with constant coefficients must be formed from the superposition of two linearly independent solutions of that equation; here we have only one such function, namely $\sin 2x$.

99 Show that $y_1 = \sin 2x$ and $y_2 = 1$ are linearly independent on $(-\infty, \infty)$.

The Wronskian of these two functions is $W = \begin{vmatrix} \sin 2x & 1 \\ 2\cos 2x & 0 \end{vmatrix} = -2\cos 2x$, which is nonzero for at least one point in $(-\infty, \infty)$. In particular, at $x = 0$, $W = -2 \neq 0$. Therefore the functions are linearly independent.

100 Determine whether $y = c_1 \sin 2x + c_2$ is the general solution of $y'' + 4y = 0$.

Although $y_1 = \sin 2x$ and $y_2 = 1$ are linearly independent (see the previous problem), their superposition is *not* the general solution because one of the functions, namely $y_2 = 1$, is not a solution to the differential equation.

01 Determine whether $y = c_1(3 \sin 2x) + c_2(4 \sin 2x)$ is a solution of $y'' + 4y = 0$ if it is known that both $3 \sin 2x$ and $4 \sin 2x$ are solutions.

It is. The result follows immediately from the principle of superposition.

02 Determine whether $y = c_1(3 \sin 2x) + c_2(4 \sin 2x)$ is the general solution of $y'' + 4y = 0$.

It is not. Although both functions are solutions to the differential equation, they are not linearly independent (see Problem 7.52). Hence the proposed solution is not the superposition of two linearly independent solutions and is not the general solution to the differential equation.

03 Determine whether $y = c_1 \sin 2x + c_2 \cos 2x$ is a solution of $y'' + 4y = 0$ if it is known that both $\sin 2x$ and $\cos 2x$ are solutions.

It is. The result follows immediately from the principle of superposition. See also Problem 2.20.

7.104 Determine whether $y = c_1 \sin 2x + c_2 \cos 2x$ is the general solution of $y'' + 4y = 0$.

■ It is. Since the two solutions are linearly independent (see Problem 7.40), their superposition is the general solution of this second-order linear homogeneous differential equation with constant coefficients.

7.105 Determine whether $y = c_1 + c_2 x$ is a solution of $y''' = 0$ for any values of the arbitrary constants c_1 and c_2 if it is known that $y_1 = 1$ and $y_2 = x$ are solutions.

■ It is. Since the differential equation is linear and homogeneous, the result follows immediately from the principle of superposition.

7.106 Determine whether $y = c_1 + c_2 x$ is the general solution of $y''' = 0$.

■ It is not. The general solution of a third-order linear homogeneous differential equation with constant coefficients must be formed from the superposition of three linearly independent solutions. Although $y_1 = 1$ and $y_2 = x$ are linearly independent (see Problem 7.5), they are only two in number and therefore one short of the required number of solutions.

7.107 Determine whether $y = c_1(1 - x) + c_2(1 + x) + c_3(1 - 3x)$ is a solution of $y''' = 0$ for any values of the arbitrary constants c_1, c_2, and c_3 if it is known that $y_1 = 1 - x$, $y_2 = 1 + x$, and $y_3 = 1 - 3x$ are solutions.

■ It is. The result follows immediately from the principle of superposition.

7.108 Determine whether $y = c_1(1 - x) + c_2(1 + x) + c_3(1 - 3x)$ is the general solution of $y''' = 0$.

■ It is not. The general solution is the superposition of three linearly independent solutions, but the functions $1 - x$, $1 + x$, and $1 - 3x$ are linearly dependent (see Problem 7.73).

7.109 Determine whether $y = c_1 + c_2 x + c_3 x^2$ is a solution of $y''' = 0$ if it is known that 1, x, and x^2 are all solutions.

■ It is. The result follows immediately from the principle of superposition.

7.110 Determine whether $y = c_1 + c_2 x + c_3 x^2$ is the general solution of $y''' = 0$.

■ It is, because the three functions 1, x, and x^2 are linearly independent (see Problem 7.37) and their number (three) is the same as the order of the differential equation.

7.111 Determine whether $y = d_1(5) + d_2(3 - 2x) + d_3(2 + x - \frac{1}{2}x^2)$ is the general solution of $y''' = 0$.

■ It is. The three functions 5, $3 - 2x$, and $2 + x - \frac{1}{2}x^2$ are all solutions of $y''' = 0$ (as may be verified by direct substitution), and they are linearly independent (see Problems 7.74 through 7.76). Since there are three such functions, their superposition is the general solution.

7.112 Problems 7.110 and 7.111 identify two general solutions to the same differential equation. How is this possible?

■ The two solutions must be algebraically equivalent. We can rewrite the solution given in Problem 7.111 as $y = (5d_1 + 3d_2 + 2d_3) + (-2d_2 + d_3)x + (-\frac{1}{2}d_3)x^2$. Then with $c_1 = 5d_1 + 3d_2 + 2d_3$, $c_2 = -2d_2 + d_3$, and $c_3 = -\frac{1}{2}d_3$, that solution is identical to the one given in Problem 7.110.

7.113 Show that $\dfrac{d^4y}{dx^4} - \dfrac{d^3y}{dx^3} - 3\dfrac{d^2y}{dx^2} + 5\dfrac{dy}{dx} - 2y = 0$ has only two linearly independent solutions of the form $y = e^{ax}$.

■ Substituting for y and its derivatives in the given equation, we get $e^{ax}(a^4 - a^3 - 3a^2 + 5a - 2) = 0$, which is satisfied when $a = 1, 1, 1, -2$. Since

$$\begin{vmatrix} e^x & e^{-2x} \\ e^x & -2e^{-2x} \end{vmatrix} \neq 0 \quad \text{but} \quad \begin{vmatrix} e^x & e^x & e^x & e^{-2x} \\ e^x & e^x & e^x & -2e^{-2x} \\ e^x & e^x & e^x & 4e^{-2x} \\ e^x & e^x & e^x & -8e^{-2x} \end{vmatrix} = 0$$

the linearly independent solutions are $y = e^x$ and $y = e^{-2x}$.

7.114 Verify that $y = e^x$, $y = xe^x$, $y = x^2e^x$, and $y = e^{-2x}$ are four linearly independent solutions of the equation of Problem 7.113, and write the primitive.

▌ By Problem 7.113, $y = e^x$ and $y = e^{-2x}$ are solutions. By direct substitution in the given equation it is found that the others are solutions. And, since

$$W = \begin{vmatrix} e^x & xe^x & x^2e^x & e^{-2x} \\ e^x & xe^x + e^x & x^2e^x + 2xe^x & -2e^{-2x} \\ e^x & xe^x + 2e^x & x^2e^x + 4xe^x + 2e^x & 4e^{-2x} \\ e^x & xe^x + 3e^x & x^2e^x + 6xe^x + 6e^x & -8e^{-2x} \end{vmatrix} = e^x \begin{vmatrix} 1 & 0 & 0 & 1 \\ 1 & 1 & 0 & -2 \\ 1 & 2 & 2 & 4 \\ 1 & 3 & 6 & -8 \end{vmatrix} = -54e^x \neq 0$$

these solutions are linearly independent. The primitive is $y = c_1e^x + c_2xe^x + c_3x^2e^x + c_4e^{-2x}$.

7.115 Verify that $y = e^{-2x}\cos 3x$ and $y = e^{-2x}\sin 3x$ are solutions of $\dfrac{d^2y}{dx^2} + 4\dfrac{dy}{dx} + 13y = 0$, and write the primitive.

▌ Substituting for y and its derivatives, we find that the equation is satisfied by the proposed solutions. Since $W = 3e^{-4x} \neq 0$, these solutions are linearly independent. The primitive is $y = e^{-2x}(c_1\cos 3x + c_2 \sin 3x)$.

7.116 Show that the differential equation $\dfrac{d^3y}{dt^3} - 2\dfrac{d^2y}{dt^2} - \dfrac{dy}{dt} + 2y = 0$ has solutions of the form $y = e^{mt}$, where m denotes a constant.

▌ Substituting $y = e^{mt}$, $\dfrac{dy}{dt} = me^{mt}$, $\dfrac{d^2y}{dt^2} = m^2e^{mt}$, and $\dfrac{d^3y}{dt^3} = m^3e^{mt}$ into the left side of the differential equation, we get

$$m^3e^{mt} - 2m^2e^{mt} - me^{mt} + 2e^{mt} = e^{mt}(m^3 - 2m^2 - m + 2) = e^{mt}(m-1)(m+1)(m-2)$$

This is equal to zero when $m = \pm 1$ or 2. Thus, $y_1 = e^t$, $y_2 = e^{-t}$, and $y_3 = e^{2t}$ are solutions of the differential equation.

7.117 Are the three distinct solutions found in the previous problem linearly independent?

▌ The Wronskian of e^t, e^{-t}, and e^{2t} was determined in Problem 7.26 to be $-6e^{2t}$. Since it is nonzero for at least one point in every interval (it is, in fact, nonzero everywhere), the functions are linearly independent.

7.118 What is the general solution of the differential equation in Problem 7.116?

▌ Since the differential equation is linear, homogeneous, and of order 3 with constant coefficients (which implies continuous coefficients having the property that the coefficient of the highest derivative is nonzero everywhere), it follows that the general solution is the superposition of any three linearly independent solutions. From the results of the previous two problems, it follows that the general solution is $y = c_1e^t + c_2e^{-t} + c_3e^{2t}$.

7.119 Three solutions of $\dfrac{d^3y}{dt^3} - 2\dfrac{d^2y}{dt^2} + \dfrac{dy}{dt} - 2y = 0$ are known to be $\sin t$, $\cos t$, and $2\sin t - \cos t$. Is the general solution $y = c_1\cos t + c_2 \sin t + c_3(2\sin t - \cos t)$?

▌ No. The general solution is the superposition of three linearly independent solutions of the differential equation. The three solutions given here have a zero Wronskian (see Problem 7.31) and are, therefore, linearly dependent. Their superposition cannot be the general solution.

7.120 Three solutions of $\dfrac{d^3y}{dt^3} + 4\dfrac{dy}{dt} = 0$ are known to be 1, $\sin 2t$, and $\cos 2t$. Is the general solution $y = c_1 + c_2\sin 2t + c_3\cos 2t$?

▌ Yes. The three functions have a nonzero Wronskian (see Problem 7.27) and are, therefore, linearly independent. Their superposition is the general solution.

7.121 Show that the differential equation $\dfrac{d^4x}{dt^4} - 5\dfrac{d^2x}{dt^2} + 4x = 0$ has solutions of the form $x = e^{rt}$, where r denotes a real constant.

▌ Substituting $x = e^{rt}$, $\frac{dx}{dt} = re^{rt}$, $\frac{d^2x}{dt^2} = r^2e^{rt}$, $\frac{d^3x}{dt^3} = r^3e^{rt}$, and $\frac{d^4x}{dt^4} = r^4e^{rt}$ into the left side of the differential equation, we get $r^4e^{rt} - 5r^2e^{rt} + 4e^{rt} = e^{rt}(r^4 - 5r^2 + 4) = e^{rt}(r^2 - 4)(r^2 - 1)$, which is equal to zero when $r = \pm 1$ or ± 2. Thus, $y_1 = e^t$, $y_2 = e^{-t}$, $y_3 = e^{2t}$, and $y_4 = e^{-2t}$ are solutions of the differential equation.

7.122 Is $y = c_1e^t + c_2e^{-t} + c_3e^{2t} + c_3e^{-2t}$ a solution of the differential equation of the previous problem?

▌ Yes, by the principle of superposition.

7.123 Is the solution given in the previous problem the general solution of the differential equation of Problem 7.121?

▌ Yes. The Wronskian of the four solutions is nonzero (see Problem 7.33 with $m_1 = 1$, $m_2 = -1$, $m_3 = 2$, and $m_4 = -2$). Thus they are linearly independent and their superposition is the general solution.

7.124 Show that the differential equation $\frac{d^2y}{dx^2} + 4y = 0$ has solutions of the form $e^{\alpha x}$ if α may be complex.

▌ Substituting $y = e^{\alpha x}$, $y' = \alpha e^{\alpha x}$, and $y'' = \alpha^2 e^{\alpha x}$ into the left side of the differential equation, we get $\alpha^2e^{\alpha x} + 4e^{\alpha x} = e^{\alpha x}(\alpha^2 + 4)$, which is equal to zero only when $\alpha = \pm i2$, where $i = \sqrt{-1}$. Thus, $y_1 = e^{i2x}$ and $y_2 = e^{-i2x}$ are solutions.

7.125 Is $y = d_1e^{i2x} + d_2e^{-i2x}$ a solution of the differential equation in the previous problem?

▌ Yes. The result follows immediately from the previous problem and the principle of superposition.

7.126 Rewrite the solution given in the previous problem as the superposition of real-valued functions.

▌ Using Euler's relations, we have

$$y = d_1e^{i2x} + d_2e^{-i2x} = d_1(\cos 2x + i \sin 2x) + d_2(\cos 2x - i \sin 2x) = (d_1 + d_2)\cos 2x + (id_1 - id_2)\sin 2x$$

or $y = c_1 \cos 2x + c_2 \sin 2x$, where $c_1 = d_1 + d_2$ and $c_2 = id_1 - id_2$.

7.127 Show that the differential equation $\frac{d^2y}{dx^2} - 6\frac{dy}{dx} + 25y = 0$ has solutions of the form $e^{\alpha x}$ if α may be complex.

▌ Substituting $y = e^{\alpha x}$ and its derivatives into the left side of the differential equation, we get $\alpha^2e^{\alpha x} - 6\alpha e^{\alpha x} + 25e^{\alpha x} = e^{\alpha x}(\alpha^2 - 6\alpha + 25)$, which is equal to zero only when $\alpha^2 - 6\alpha + 25 = 0$ or when $\alpha = 3 \pm i4$. Thus, $y_1 = e^{(3+i4)x}$ and $y_2 = e^{(3-i4)x}$ are solutions.

7.128 Is $y = d_1e^{(3+i4)x} + d_2e^{(3-i4)x}$ a solution of the differential equation in the previous problem?

▌ Yes. The result follows immediately from the solution to the previous problem and the principle of superposition.

7.129 Rewrite the solution given in the previous problem as the superposition of real-valued functions.

▌ Using Euler's relations, we have

$$\begin{aligned} y = d_1e^{(3+i4)x} + d_2e^{(3-i4)x} &= d_1e^{3x}e^{i4x} + d_2e^{3x}e^{-i4x} = e^{3x}[d_1(\cos 4x + i \sin 4x) + d_2(\cos 4x + i \sin 4x)] \\ &= e^{3x}[(d_1 + d_2)\cos 4x + (id_1 - id_2)\sin 4x] \end{aligned}$$

or $y = c_1e^{3x}\cos 4x + c_2e^{3x}\sin 4x$, where $c_1 = d_1 + d_2$ and $c_2 = id_1 - id_2$.

7.130 Show that the differential equation $\frac{d^2x}{dt^2} + 4\frac{dx}{dt} + 11x = 0$ has solutions of the form $x = e^{\alpha t}$, where α may be complex.

▌ Substituting $x = e^{\alpha t}$ and its derivatives into the left side of the differential equation, we get $\alpha^2e^{\alpha t} + 4\alpha e^{\alpha t} + 11e^{\alpha t} = e^{\alpha t}(\alpha^2 + 4\alpha + 11)$, which is equal to zero only when $\alpha^2 + 4\alpha + 11 = 0$, or when $\alpha = -2 \pm i\sqrt{7}$. Thus, $x_1 = e^{(-2+i\sqrt{7})t}$ and $x_2 = e^{(-2-i\sqrt{7})t}$ are solutions.

7.131 Is $x = d_1e^{(-2+i\sqrt{7})t} + d_2e^{(-2-i\sqrt{7})t}$ a solution of the differential equation in the previous problem?

Yes. The result follows immediately from the solution to the previous problem and the principle of superposition.

7.132 Rewrite the solution given in the previous problem as the superposition of real-valued functions.

Using Euler's relations, we have

$$x = d_1 e^{(-2+i\sqrt{7})t} + d_2 e^{(-2-i\sqrt{7})t} = d_1 e^{-2t} e^{i\sqrt{7}t} + d_2 e^{-2t} e^{-i\sqrt{7}t} = e^{-2t}(d_1 e^{i\sqrt{7}t} + d_2 e^{-i\sqrt{7}t})$$
$$= e^{-2t}[d_1(\cos\sqrt{7}t + i\sin\sqrt{7}t) + d_2(\cos\sqrt{7}t - i\sin\sqrt{7}t)] = e^{-2t}[(d_1 + d_2)\cos\sqrt{7}t + (id_1 - id_2)\sin\sqrt{7}t]$$

or $x = c_1 e^{-2t}\cos\sqrt{7}t + c_2 e^{-2t}\sin\sqrt{7}t$, where $c_1 = d_1 + d_2$ and $c_2 = id_1 - id_2$.

GENERAL SOLUTIONS OF NONHOMOGENEOUS EQUATIONS

7.133 Find the general solution $y'' - y' - 2y = 2x + 1$ if one solution is known to be $y = -x$.

The general solution to the associated homogeneous equation, $y'' - y' - 2y = 0$, is found in Problem 7.87 to be $y = c_1 e^{-x} + c_2 e^{2x}$. The general solution to the nonhomogeneous differential equation is $y = c_1 e^{-x} + c_2 e^{2x} - x$.

7.134 Find the general solution of $y'' - y' - 2y = \cos x + 3\sin x$ if one solution is known to be $y = -\sin x$.

As in the previous problem, the general solution to the associated homogeneous equation is $y = c_1 e^{-x} + c_2 e^{2x}$. Therefore, the general solution to the given differential equation is $y = c_1 e^{-x} + c_2 e^{2x} - \sin x$.

7.135 Find the general solution of $y'' - 2y' + y = x^2$ if one solution is known to be $y = x^2 + 4x + 6$.

The general solution to the associated homogeneous equation, $y'' - 2y' + y = 0$, is found in Problem 7.94 to be $y = c_1 e^x + c_2 x e^x$. Therefore, the general solution to the given nonhomogeneous differential equation is $y = c_1 e^x + c_2 x e^x + x^2 + 4x + 6$.

7.136 Find the general solution of $y'' - 2y' + y = 2e^{3x}$ if one solution is known to be $y = \frac{1}{2}e^{3x}$.

As in the previous problem, the general solution to the associated homogeneous equation is $y = c_1 e^x + c_2 x e^x$. Therefore, the general solution to the given nonhomogeneous equation is $y = c_1 e^x + c_2 x e^x + \frac{1}{2}e^{3x}$.

7.137 Find the general solution of $x^3 \dfrac{d^3y}{dx^3} - 6x \dfrac{dy}{dx} + 12y = 12\ln x - 4$ if one solution is known to be $y = \ln x$.

The general solution to the associated homogeneous differential equation is found in Problem 7.89 to be $y = c_1 x^2 + c_2 x^3 + c_3 x^{-2}$. Therefore, the solution to the given nonhomogeneous differential equation is $y = c_1 x^2 + c_2 x^3 + c_3 x^{-2} + \ln x$.

7.138 Find the general solution of $y'' + 4y = e^{3x}$ if one solution is known to be $y = \frac{1}{13}e^{3x}$.

The general solution to the associated homogeneous equation, $y'' + 4y = 0$, is found in Problem 7.104 to be $y = c_1 \sin 2x + c_2 \cos 2x$. Therefore, the general solution to the given nonhomogeneous differential equation is $y = c_1 \sin 2x + c_2 \cos 2x + \frac{1}{13}e^{3x}$.

7.139 Find the general solution of $y'' + 4y = 8x$ if one solution is known to be $y = 2x$.

As in the previous problem, the general solution to the associated homogeneous differential equation is $y = c_1 \sin 2x + c_2 \cos 2x$, so the general solution to the given differential equation is $y = c_1 \sin 2x + c_2 \cos 2x + 2x$.

7.140 Use the information of the previous two problems to ascertain a particular solution of $y'' + 4y = e^{3x} + 8x$.

Since the differential equation is linear, a particular solution is $y = \frac{1}{13}e^{3x} + 2x$.

7.141 A particular solution of $y''' = \sin 2x$ is $y = \frac{1}{8}\cos 2x$, while a particular solution of $y''' = \cos 2x$ is $y = -\frac{1}{8}\sin 2x$. Determine a particular solution of $y''' = 3\sin 2x + 5\cos 2x$.

Since the differential equation is linear, a particular solution is $y = 3(\frac{1}{8}\cos 2x) + 5(-\frac{1}{8}\sin 2x)$.

7.142 Find the general solution of $y''' = 3\sin 2x + 5\cos 2x$.

▌ The general solution of the associated homogeneous differential equation is shown in Problem 7.110 to be $y = c_1 + c_2x + c_3x^2$. Using the result of the previous problem, we conclude that the general solution of the given equation is $y = c_1 + c_2x + c_3x^2 + \frac{3}{8}\cos 2x - \frac{5}{8}\sin 2x$.

7.143 A particular solution of $y''' - 2y'' - y' + 2y = x$ is $y = \frac{1}{2}x + \frac{1}{4}$, while a particular solution of $y''' - 2y'' - y' + 2y = \sin x$ is $y = \frac{2}{10}\sin x + \frac{1}{10}\cos x$. Determine a solution of $y''' - 2y'' - y' + 2y = 7x - 3\sin x$.

▌ Since the differential equation is linear, a particular solution is $y = 7(\frac{1}{2}x + \frac{1}{4}) - 3(\frac{2}{10}\sin x + \frac{1}{10}\cos x)$.

7.144 Use the information of the previous problem to obtain a solution of $y''' - 2y'' - y' + 2y = -3x$.

▌ The solution is $y = -3(\frac{1}{2}x + \frac{1}{4}) = -\frac{3}{2}x - \frac{3}{4}$.

7.145 Use the information of Problem 7.143 to obtain a solution of $y''' - 2y'' - y' + 2y = \frac{5}{2}\sin x$.

▌ The solution is $y = \frac{5}{2}(\frac{2}{10}\sin x + \frac{1}{10}\cos x) = \frac{1}{2}\sin x + \frac{1}{4}\cos x$.

7.146 A particular solution of $\dfrac{d^4x}{dt^4} - 5\dfrac{d^2x}{dt^2} + 4x = 20e^{3t}$ is $x = \frac{1}{2}e^{3t}$, while a particular solution of $\dfrac{d^4x}{dt^4} - 5\dfrac{d^2x}{dt^2} + 4x = 8t$ is $x = 2t$. Determine a particular solution of $\dfrac{d^4x}{dt^4} - 5\dfrac{d^2x}{dt^2} + 4x = 10e^{3t} - 20t$.

▌ Since $10e^{3t} - 20t = \frac{1}{2}(20e^{3t}) - \frac{5}{2}(8t)$, a particular solution is $y = \frac{1}{2}(\frac{1}{2}e^{3t}) - \frac{5}{2}(2t) = \frac{1}{4}e^{3t} - 5t$.

7.147 Use the information of the previous problem to determine a particular solution of $\dfrac{d^4x}{dt^4} - 5\dfrac{d^2x}{dt^2} + 4x = 60e^{3t}$.

▌ Since $60e^{3t} = 3(20e^{3t})$, a particular solution of this differential equation is $y = 3(\frac{1}{2}e^{3t}) = \frac{3}{2}e^{3t}$.

7.148 Find the general solution of $y'' + y = x^2$, if one solution is $y = x^2 - 2$, and if two solutions of $y'' + y = 0$ are $\sin x$ and $\cos x$.

▌ The Wronskian of $\{\sin x, \cos x\}$ is $W = \begin{vmatrix} \sin x & \cos x \\ \cos x & -\sin x \end{vmatrix} = -\sin^2 x - \cos^2 x = -1 \neq 0$, so these two functions are linearly independent. The general solution to the associated homogeneous differential equation, $y'' + y = 0$, is then $y = c_1 \sin x + c_2 \cos x$. Therefore, the general solution to the nonhomogeneous equation is $y = c_1 \sin x + c_2 \cos x + x^2 - 2$.

7.149 Find the general solution of $y'' + y = x^2$ if a particular solution is known to be $y = x^2 + 3\sin x - 2$.

▌ As in the previous problem, the general solution of the associated homogeneous equation is $y = C_1 \sin x + c_2 \cos x$; then a general solution to the nonhomogeneous equation is $y = C_1 \sin x + C_2 \cos x + x^2 + 3\sin x - 2$.

7.150 Explain why the results of the previous two problems, in which we have generated two different general solutions to the same differential equation, are not contradictory.

▌ The two general solutions must be algebraically equivalent to one another. In particular, the solution given in Problem 7.149 can be rewritten as $y = (C_1 + 3)\sin x + C_2 \cos x + x^2 - 2$, which is identical in form to the solution given in Problem 7.148 if we define $c_1 = C_1 + 3$ and $C_2 = c_2$.

7.151 Find the general solution of $y''' = 12$ if one solution is $y = 2x^3$ and three solutions of $y''' = 0$ are 1, x, and x^2.

▌ The general solution of the associated homogeneous equation, $y''' = 0$, is shown in Problems 7.109 and 7.110 to be $y = c_1 + c_2x + c_3x^2$. Then the general solution to the nonhomogeneous equation is $y = c_1 + c_2x + c_3x^2 + 2x^3$.

7.152 Rework the previous problem if instead of $y = 2x^3$ a particular solution is known to be $y = 3x - 4x^2 + 2x^3$.

▮ As before, the solution to the associated homogeneous equation is $y = C_1 + C_2x + C_3x^2$; thus the general solution to the nonhomogeneous equation is $y = C_1 + C_2x + C_3x^2 + 3x - 4x^2 + 2x^3$.

7.153 Explain why the results of the previous two problems, in which we generated two different general solutions to the same differential equation, are not contradictory.

▮ The two general solutions must be algebraically equivalent to each other. In particular, the solution given in Problem 7.152 can be rewritten as $y = C_1 + (C_2 + 3)x + (C_3 - 4)x^2 + 2x^3$, which is identical in form to the solution given in Problem 7.151 if we define $c_2 = C_2 + 3$ and $c_3 = C_3 - 4$.

7.154 Find the general solution of $y'' + 4y = e^{2x}$ if one solution is $y = \frac{1}{6}e^{2x}$ and two solutions of $y'' + 4y = 0$ are $3 \sin 2x$ and $4 \sin 2x$.

▮ Since we do not have enough information to write the general solution of the associated homogeneous equation (see Problem 7.102), we cannot write the general solution of the nonhomogeneous equation.

CHAPTER 8

Linear Homogeneous Differential Equations with Constant Coefficients

DISTINCT REAL CHARACTERISTIC ROOTS

8.1 Solve $y'' - y' - 2y = 0$.

The characteristic (or auxiliary) equation is $\lambda^2 - \lambda - 2 = 0$, which can be factored into $(\lambda + 1)(\lambda - 2) = 0$. Since the roots $\lambda_1 = -1$ and $\lambda_2 = 2$ are real and distinct, the solution is $y = c_1 e^{-x} + c_2 e^{2x}$.

8.2 Solve $y'' - 7y' = 0$.

The characteristic (or auxiliary) equation is $\lambda^2 - 7\lambda = 0$, which can be factored into $(\lambda - 0)(\lambda - 7) = 0$. Since the roots $\lambda_1 = 0$ and $\lambda_2 = 7$ are real and distinct, the solution is $y = c_1 e^{0x} + c_2 e^{7x} = c_1 + c_2 e^{7x}$.

8.3 Solve $y'' - 5y = 0$.

The characteristic equation is $\lambda^2 - 5 = 0$, which can be factored into $(\lambda - \sqrt{5})(\lambda + \sqrt{5}) = 0$. Since the roots $\lambda_1 = \sqrt{5}$ and $\lambda_2 = -\sqrt{5}$ are real and distinct, the solution is $y = c_1 e^{\sqrt{5}x} + c_2 e^{-\sqrt{5}x}$.

8.4 Write the solution to the previous problem in terms of hyperbolic functions.

In succeeding steps we may write

$$\begin{aligned} y &= c_1(\tfrac{1}{2}e^{\sqrt{5}x} + \tfrac{1}{2}e^{-\sqrt{5}x}) + c_1(\tfrac{1}{2}e^{\sqrt{5}x} - \tfrac{1}{2}e^{-\sqrt{5}x}) + c_2(\tfrac{1}{2}e^{\sqrt{5}x} + \tfrac{1}{2}e^{-\sqrt{5}x}) - c_2(\tfrac{1}{2}e^{\sqrt{5}x} - \tfrac{1}{2}e^{-\sqrt{5}x}) \\ &= c_1 \cosh \sqrt{5}x + c_1 \sinh \sqrt{5}x + c_2 \cosh \sqrt{5}x - c_2 \sinh \sqrt{5}x \\ &= (c_1 + c_2) \cosh \sqrt{5}x + (c_1 - c_2) \sinh \sqrt{5}x = k_1 \cosh \sqrt{5}x + k_2 \sinh \sqrt{5}x \end{aligned}$$

where $k_1 = c_1 + c_2$ and $k_2 = c_1 - c_2$.

8.5 Solve $d^2x/dt^2 - 16x = 0$.

The characteristic equation is $\lambda^2 - 16 = 0$, which has the roots $\lambda = \pm 4$. Since these are real and distinct, the solution is $x(t) = c_1 e^{4t} + c_2 e^{-4t}$.

8.6 Write the solution to the previous problem in terms of hyperbolic functions.

In succeeding steps we may write $x(t)$ as

$$\begin{aligned} x(t) &= c_1(\tfrac{1}{2}e^{4t} + \tfrac{1}{2}e^{-4t}) + c_1(\tfrac{1}{2}e^{4t} - \tfrac{1}{2}e^{-4t}) + c_2(\tfrac{1}{2}e^{4t} + \tfrac{1}{2}e^{-4t}) - c_2(\tfrac{1}{2}e^{4t} - \tfrac{1}{2}e^{-4t}) \\ &= c_1 \cosh 4t + c_1 \sinh 4t + c_2 \cosh 4t - c_2 \sinh 4t \\ &= (c_1 + c_2) \cosh 4t + (c_1 - c_2) \sinh 4t = k_1 \cosh 4t + k_2 \sinh 4t \end{aligned}$$

8.7 Solve $d^2r/dt^2 - \omega^2 r = 0$, where ω denotes a positive constant.

The characteristic equation is $\lambda^2 - \omega^2 = 0$, which has the real and distinct roots $\lambda = \pm\omega$. The solution is thus $r(t) = Ae^{\omega t} + Be^{-\omega t}$, where A and B denote arbitrary constants.

8.8 Write the solution to the previous problem in terms of hyperbolic functions.

Since $r(t) = Ae^{\omega t} + Be^{-\omega t}$, we have

$$\begin{aligned} r(t) &= A(\tfrac{1}{2}e^{\omega t} + \tfrac{1}{2}e^{-\omega t}) + A(\tfrac{1}{2}e^{\omega t} - \tfrac{1}{2}e^{-\omega t}) + B(\tfrac{1}{2}e^{\omega t} + \tfrac{1}{2}e^{-\omega t}) - B(\tfrac{1}{2}e^{\omega t} - \tfrac{1}{2}e^{-\omega t}) \\ &= A \cosh \omega t + A \sinh \omega t + B \cosh \omega t - B \sinh \omega t \\ &= (A + B) \cosh \omega t + (A - B) \sinh \omega t = C \cosh \omega t + D \sinh \omega t \end{aligned}$$

where $C = A + B$ and $D = A - B$.

8.9 Solve $\dfrac{d^2y}{dt^2} - 4\dfrac{dy}{dt} + y = 0$.

▌ The characteristic equation is $\lambda^2 - 4\lambda + 1 = 0$, which, by the quadratic formula, has roots $\lambda = \dfrac{4 \pm \sqrt{16-4}}{2} = 2 \pm \sqrt{3}$. Since these roots are real and distinct, the solution is $y = C_1 e^{(2+\sqrt{3})t} + C_2 e^{(2-\sqrt{3})t} = C_1 e^{3.732t} + C_2 e^{0.2679t}$, where C_1 and C_2 denote arbitrary constants.

8.10 Solve $\dfrac{d^2I}{dt^2} + 60\dfrac{dI}{dt} + 500I = 0$.

▌ The characteristic equation is $\lambda^2 + 60\lambda + 500 = 0$, which can be factored into $(\lambda + 50)(\lambda + 10) = 0$. Since the roots $\lambda_1 = -50$ and $\lambda_2 = -10$ are real and distinct, the solution is $I(t) = c_1 e^{-50t} + c_2 e^{-10t}$.

8.11 Solve $\ddot{x} + 20\dot{x} + 64x = 0$.

▌ The characteristic (or auxiliary) equation is $\lambda^2 + 20\lambda + 64 = 0$, which can be factored into $(\lambda + 4)(\lambda + 16) = 0$. Since the roots $\lambda_1 = -4$ and $\lambda_2 = -16$ are real and distinct, the solution is $x = c_1 e^{-4t} + c_2 e^{-16t}$.

8.12 Solve $\ddot{x} + 128\dot{x} + 96x = 0$.

▌ The characteristic equation is $\lambda^2 + 128\lambda + 96 = 0$, which by the quadratic formula has roots $\lambda = \dfrac{-128 \pm \sqrt{(128)^2 - 4(96)}}{2} = -64 \pm 20\sqrt{10}$. Since these roots are real and distinct, the solution is $x = C_1 e^{(-64+20\sqrt{10})t} + C_2 e^{(-64-20\sqrt{10})t} = C_1 e^{-0.7544t} + C_2 e^{-127.2t}$.

8.13 Solve $\dfrac{d^2y}{dx^2} + \dfrac{dy}{dx} - 6y = 0$.

▌ We write the equation as $(D^2 + D - 6)y = (D - 2)(D + 3)y = 0$. Then the characteristic roots are 2 and -3; since they are real and distinct, the primitive is $y = C_1 e^{2x} + C_2 e^{-3x}$.

8.14 Solve $\dfrac{d^3y}{dx^3} - \dfrac{d^2y}{dx^2} - 12\dfrac{dy}{dx} = 0$.

▌ We write the equation as $(D^3 - D^2 - 12D)y = D(D - 4)(D + 3)y = 0$. Then the characteristic roots are 0, 4, and -3. They are real and distinct, and the primitive is $y = C_1 + C_2 e^{4x} + C_3 e^{-3x}$.

8.15 Solve $\dfrac{d^3y}{dx^3} + 2\dfrac{d^2y}{dx^2} - 5\dfrac{dy}{dx} - 6y = 0$.

▌ We write the equation as $(D^3 + 2D^2 - 5D - 6)y = (D - 2)(D + 1)(D + 3)y = 0$. Then the characteristic roots are 2, -1, and -3. They are real and distinct, so the primitive is $y = C_1 e^{2x} + C_2 e^{-x} + C_3 e^{-3x}$.

8.16 Solve $2y'' - 5y' + 2y = 0$.

▌ The auxiliary equation is $2m^2 - 5m + 2 = 0$ or $(2m - 1)(m - 2) = 0$, so the real and distinct characteristic roots are $m = 1/2$ and $m = 2$. Then the general solution is $y = c_1 e^{x/2} + c_2 e^{2x}$.

8.17 Solve $\dfrac{d^2x}{dt^2} + 9\dfrac{dx}{dt} + 14x = 0$.

▌ The characteristic equation is $\lambda^2 + 9\lambda + 14 = 0$, which can be factored into $(\lambda + 2)(\lambda + 7) = 0$. Since the roots $\lambda_1 = -2$ and $\lambda_2 = -7$ are real and distinct, the solution is $x = c_1 e^{-2t} + c_2 e^{-7t}$.

8.18 Solve $\ddot{q} + 1000\dot{q} + 50{,}000q = 0$.

▌ The characteristic equation is $\lambda^2 + 1000\lambda + 50{,}000 = 0$, which, by the quadratic formula, has roots $\lambda = \dfrac{-1000 \pm \sqrt{(1000)^2 - 4(50{,}000)}}{2} = -52.79$ and -947.2. Since these roots are real and distinct, the solution is $q = c_1 e^{-52.79t} + c_2 e^{-947.2t}$.

8.19 Solve $\dfrac{d^2Q}{dt^2} + 1000\dfrac{dQ}{dt} + 160{,}000Q = 0$.

▌ The characteristic equation is $\lambda^2 + 1000\lambda + 160{,}000 = 0$, or $(\lambda + 200)(\lambda + 800) = 0$. Since the roots $\lambda_1 = -200$ and $\lambda_2 = -800$ are real and distinct, the solution is $Q = c_1 e^{-200t} + c_2 e^{-800t}$.

8.20 Solve $\dfrac{d^2x}{dt^2} + k\dfrac{dx}{dt} = 0$, where k denotes a real constant.

▌ The characteristic equation is $m^2 + km = 0$, which can be factored into $m(m + k) = 0$. The roots $m_1 = 0$ and $m_2 = -k$ are real and distinct, so the solution is $x = C_1 + C_2 e^{-kt}$.

8.21 Solve $\dfrac{d^2x}{dt^2} - \dfrac{g}{10}x = 0$, where g denotes a positive constant.

▌ The characteristic equation is $m^2 - g/10 = 0$, which has the roots $m = \pm\sqrt{g/10}$. The solution is then $x = C_1 e^{\sqrt{g/10}t} + C_2 e^{-\sqrt{g/10}t}$.

8.22 Solve $\dfrac{d^2x}{dt^2} - \dfrac{k}{m}x = 0$, where both k and m denote positive constants.

▌ The characteristic equation is $\lambda^2 - k/m = 0$, which has as its roots $\lambda = \pm\sqrt{k/m}$. The solution is thus $x = c_1 e^{\sqrt{k/m}t} + c_2 e^{-\sqrt{k/m}t}$.

8.23 Solve $y'' - \frac{9}{2}y' + 2y = 0$.

▌ The auxiliary equation is $m^2 - \frac{9}{2}m + 2 = 0$, which may be written as $2m^2 - 9m + 4 = 0$. This last equation can be factored into $(2m - 1)(m - 4) = 0$. The roots are $m_1 = \frac{1}{2}$ and $m_2 = 4$; the solution is $y = Ae^{x/2} + Be^{4x}$.

8.24 Solve $y'' - \frac{1}{4}y' - \frac{1}{8}y = 0$.

▌ The auxiliary equation is $\lambda^2 - \frac{1}{4}\lambda - \frac{1}{8} = 0$, or $8\lambda^2 - 2\lambda - 1 = 0$. This last equation can be factored into $(2\lambda - 1)(4\lambda + 1) = 0$. The roots are $\lambda_1 = \frac{1}{2}$ and $\lambda_2 = -\frac{1}{4}$; the solution is $y = c_1 e^{x/2} + c_2 e^{-x/4}$.

8.25 Solve $\ddot{y} - 2\dot{y} + \frac{1}{2}y = 0$.

▌ The characteristic equation is $m^2 - 2m + \frac{1}{2} = 0$, or $2m^2 - 4m + 1 = 0$. Using the quadratic formula, we obtain the roots to this last equation as $m = \dfrac{4 \pm \sqrt{(-4)^2 - 4(2)(1)}}{2(2)} = 1 \pm \sqrt{2}/2$. The solution is $y = Ae^{(1+\sqrt{2}/2)t} + Be^{(1-\sqrt{2}/2)t}$.

8.26 Solve $\dfrac{d^3y}{dx^3} - 3\dfrac{d^2y}{dx^2} + 2\dfrac{dy}{dx} = 0$.

▌ The characteristic equation is $\lambda^3 - 3\lambda^2 + 2\lambda = 0$, which may be factored into $\lambda(\lambda - 1)(\lambda - 2) = 0$. The roots, $\lambda_1 = 0$, $\lambda_2 = 1$, and $\lambda_3 = 2$, are real and distinct. The solution is $y = c_1 + c_2 e^x + c_3 e^{2x}$.

8.27 Solve $\dfrac{d^3y}{dx^3} - \dfrac{dy}{dx} = 0$.

▌ The characteristic equation is $\lambda^3 - \lambda = 0$, which may be factored into $\lambda(\lambda - 1)(\lambda + 1) = 0$. The roots, 0 and ± 1, are real and distinct, so the solution is $y = c_1 + c_2 e^x + c_3 e^{-x}$.

8.28 Solve $y''' - 6y'' + 11y' - 6y = 0$.

▌ The characteristic equation is $\lambda^3 - 6\lambda^2 + 11\lambda - 6 = 0$, which can be factored into $(\lambda - 1)(\lambda - 2)(\lambda - 3) = 0$. The roots are $\lambda_1 = 1$, $\lambda_2 = 2$, and $\lambda_3 = 3$; hence the solution is $y = c_1 e^x + c_2 e^{2x} + c_3 e^{3x}$.

8.29 Solve $y^{(4)} - 9y'' + 20y = 0$.

▌ The characteristic equation is $\lambda^4 - 9\lambda^2 + 20 = 0$, which can be factored into $(\lambda - 2)(\lambda + 2)(\lambda - \sqrt{5})(\lambda + \sqrt{5}) = 0$. The roots are $\lambda_1 = 2$, $\lambda_2 = -2$, $\lambda_3 = \sqrt{5}$, and $\lambda_4 = -\sqrt{5}$; hence the solution is

$$y = c_1 e^{2x} + c_2 e^{-2x} + c_3 e^{\sqrt{5}x} + c_4 e^{-\sqrt{5}x} = k_1 \cosh 2x + k_2 \sinh 2x + k_3 \cosh \sqrt{5}x + k_4 \sinh \sqrt{5}x$$

8.30 Solve $\dfrac{d^4x}{dt^4} - 13\dfrac{d^2x}{dt^2} + 36x = 0$.

▌ The auxiliary equation is $\lambda^4 - 13\lambda^2 + 36 = 0$, which may be factored first into $(\lambda^2 - 4)(\lambda^2 - 9) = 0$ and then into $(\lambda - 2)(\lambda + 2)(\lambda - 3)(\lambda + 3) = 0$. The roots, ± 2 and ± 3, are real and distinct. The solution is

$$x = c_1e^{2t} + c_2e^{-2t} + c_3e^{3t} + c_4e^{-3t} = k_1 \cosh 2t + k_2 \sinh 2t + k_3 \cosh 3t + k_4 \sinh 3t$$

8.31 Solve $\dfrac{d^4x}{dt^4} - 10\dfrac{d^3x}{dt^3} + 35\dfrac{d^2x}{dt^2} - 50\dfrac{dx}{dt} + 24x = 0$.

▌ The characteristic equation is $m^4 - 10m^3 + 35m^2 - 50m + 24 = 0$, which can be factored into $(m-1)(m-2)(m-3)(m-4) = 0$. The roots are 1, 2, 3, and 4, so the solution is $x = C_1e^t + C_2e^{2t} + C_3e^{3t} + C_4e^{4t}$.

8.32 Solve $\dfrac{d^5x}{dt^5} - 10\dfrac{d^4x}{dt^4} + 35\dfrac{d^3x}{dt^3} - 50\dfrac{d^2x}{dt^2} + 24\dfrac{dx}{dt} = 0$.

▌ The characteristic equation is $m^5 - 10m^4 + 35m^3 - 50m^2 + 24m = 0$, which may be factored into $m(m-1)(m-2)(m-3)(m-4) = 0$. The roots are $m_1 = 0$, $m_2 = 1$, $m_3 = 2$, $m_4 = 3$, and $m_5 = 4$; the solution is $x = c_1 + c_2e^t + c_3e^{2t} + c_4e^{3t} + c_5e^{4t}$.

8.33 Solve $x^{(5)} - 3x^{(4)} - 3x^{(3)} + 9\ddot{x} + 2\dot{x} - 6x = 0$.

▌ The characteristic equation is $\lambda^5 - 3\lambda^4 - 3\lambda^3 + 9\lambda^2 + 2\lambda - 6 = 0$, which we factor into $(\lambda^2 - 1)(\lambda^2 - 2)(\lambda - 3) = 0$ or $(\lambda - 1)(\lambda + 1)(\lambda - \sqrt{2})(\lambda + \sqrt{2})(\lambda - 3) = 0$. The roots are ± 1, $\pm\sqrt{2}$, and 3, which are real and distinct; thus the solution is $x = Ae^t + Be^{-t} + Ce^{\sqrt{2}t} + De^{-\sqrt{2}t} + Ee^{3t}$, where A, B, C, D, and E are arbitrary constants.

8.34 Solve $y' - 5y = 0$.

▌ The characteristic equation is $\lambda - 5 = 0$, which has the single root $\lambda_1 = 5$. The solution is then $y = c_1e^{5x}$.

8.35 Solve $(2D^3 - D^2 - 5D - 2)y = 0$.

▌ The auxiliary equation is $2m^3 - m^2 - 5m - 2 = 0$ or $(2m + 1)(m + 1)(m - 2) = 0$. The roots are $m = -1/2$, -1, and 2. The general solution is then $y = c_1e^{-x/2} + c_2e^{-x} + c_3e^{2x}$.

8.36 Solve $(D^3 + D^2 - 2D)y = 0$.

▌ The auxiliary equation is $m^3 + m^2 - 2m = 0$ or $m(m-1)(m+2) = 0$, so that $m = 0$, 1, and -2. The solution is then $y = C_1 + C_2e^x + C_3e^{-2x}$.

8.37 Solve $(D + 6)y = 0$.

▌ The characteristic equation is $m + 6 = 0$, which has the single root $m = -6$. The general solution is $y = Ae^{-6x}$.

8.38 Solve $(2D - 5)y = 0$.

▌ The characteristic equation is $2\lambda - 5 = 0$, which has the single root $\lambda = 5/2$. The general solution is $y = Ae^{5x/2}$.

8.39 Solve $(D^3 - 2D^2 - D + 2)y = 0$.

▌ The characteristic equation is $\lambda^3 - 2\lambda^2 - \lambda + 2 = 0$ or $(\lambda^2 - 1)(\lambda - 2) = 0$. The roots are ± 1 and 2, and the solution is $y = c_1e^x + c_2e^{-x} + c_3e^{2x}$.

8.40 Solve $(D^4 - 8D^2 + 15)x = 0$.

▌ The characteristic equation is $\lambda^4 - 8\lambda^2 + 15 = 0$ or $(\lambda^2 - 3)(\lambda^2 - 5) = 0$. The roots are $\pm\sqrt{3}$ and $\pm\sqrt{5}$, and the solution is $x = C_1e^{\sqrt{3}t} + C_2e^{-\sqrt{3}t} + C_3e^{\sqrt{5}t} + C_4e^{-\sqrt{5}t}$

8.41 A second-order linear homogeneous differential equation for $y(x)$ with constant coefficients has $\lambda_1 = 2$ and $\lambda_2 = 4$ as the roots of its characteristic equation. What is the differential equation?

▌ The characteristic equation is $(\lambda - 2)(\lambda - 4) = 0$ or $\lambda^2 - 6\lambda + 8 = 0$. The associated differential equation is $y'' - 6y' + 8y = 0$.

8.42 A second-order linear homogeneous differential equation for $y(x)$ with constant coefficients has $\lambda = \pm\sqrt{17}$ as the roots of its characteristic equation. What is the differential equation?

▌ The characteristic equation is $(\lambda - \sqrt{17})[\lambda - (-\sqrt{17})] = 0$ or $\lambda^2 - 17 = 0$. The associated differential equation is $y'' - 17y = 0$.

8.43 Find a second-order linear homogeneous differential equation in $x(t)$ with constant coefficients that has, for the roots of its characteristic equation, $2 \pm \sqrt{5}$.

▌ The characteristic equation is $[\lambda - (2 + \sqrt{5})][\lambda - (2 - \sqrt{5})] = 0$ or $\lambda^2 - 4\lambda - 1 = 0$. The associated differential equation is $\ddot{x} - 4\dot{x} - x = 0$.

8.44 Find a third-order linear homogeneous differential equation in $x(t)$ with constant coefficients that has, for the roots of its characteristic equation, -1 and $\pm\sqrt{3}$.

▌ The characteristic equation is $[m - (-1)](m - \sqrt{3})[m - (-\sqrt{3})] = 0$ or $m^3 + m^2 - 3m - 3 = 0$. The associated differential equation is $\dfrac{d^3x}{dt^3} + \dfrac{d^2x}{dt^2} - 3\dfrac{dx}{dt} - 3x = 0$.

8.45 A fourth-order linear homogeneous differential equation in $y(t)$ with constant coefficients has, as the roots of its characteristic equation, $\frac{1}{2} \pm \sqrt{5}$ and $-1 \pm \sqrt{8}$. What is the differential equation?

▌ The characteristic equation is $[m - (\frac{1}{2} + \sqrt{5})][m - (\frac{1}{2} - \sqrt{5})][m - (-1 + \sqrt{8})][m - (-1 - \sqrt{8})] = 0$, which we can simplify to $(m^2 - m - \frac{19}{4})(m^2 + 2m - 7) = 0$ or $m^4 + m^3 - \frac{55}{4}m^2 - \frac{5}{2}m + \frac{133}{4} = 0$. The associated differential equation is

$$\frac{d^4y}{dt^4} + \frac{d^3y}{dt^3} - \frac{55}{4}\frac{d^2y}{dt^2} - \frac{5}{2}\frac{dy}{dt} + \frac{133}{4}y = 0 \quad \text{or} \quad (4D^4 + 4D^3 - 55D^2 - 10D + 133)y = 0$$

8.46 Show that $(D - a)(D - b)(D - c)y = (D - b)(D - c)(D - a)y$.

▌ We expand both sides of this equation and show they are equal:

$$(D - a)(D - b)(D - c)y = (D - a)(D - b)\left(\frac{dy}{dx} - cy\right) = (D - a)\left[\frac{d^2y}{dx^2} - (b + c)\frac{dy}{dx} + bcy\right]$$

$$= \frac{d^3y}{dx^3} - (a + b + c)\frac{d^2y}{dx^2} + (ab + bc + ac)\frac{dy}{dx} - abcy$$

$$(D - b)(D - c)(D - a)y = (D - b)(D - c)\left(\frac{dy}{dx} - ay\right) = (D - b)\left[\frac{d^2y}{dx^2} - (a + c)\frac{dy}{dx} + acy\right]$$

$$= \frac{d^3y}{dx^3} - (a + b + c)\frac{d^2y}{dx^2} + (ab + ac + bc)\frac{dy}{dx} - abcy$$

8.47 Verify that $y = C_1e^{ax} + C_2e^{bx} + C_3e^{cx}$ satisfies the differential equation $(D - a)(D - b)(D - c)y = 0$.

▌ We need to show that $(D - a)(D - b)(D - c)(C_1e^{ax} + C_2e^{bx} + C_3e^{cx}) = 0$. For the first term on the left, we have

$$(D - a)(D - b)(D - c)C_1e^{ax} = (D - b)(D - c)(D - a)C_1e^{ax} = (D - b)(D - c)0 = 0$$

and similarly for the other two terms.

DISTINCT COMPLEX CHARACTERISTIC ROOTS

8.48 Show that if the characteristic equation of a second-order linear homogeneous differential equation with *real* constant coefficients has complex roots, then these roots must be complex conjugates.

▌ Denote the two roots as $a + ib$ and $c + id$, where $i = \sqrt{-1}$. Then the characteristic equation is $[\lambda - (a + ib)][\lambda - (c + id)] = 0$ or $\lambda^2 - \lambda[(a + ib) + (c + id)] + [(a + ib)(c + id)] = 0$. The associated differential equation (with y as the dependent variable and x as the independent variable) is

$$\frac{d^2y}{dx^2} - [(a + ib) + (c + id)]\frac{dy}{dx} + [(a + ib)(c + id)]y = 0.$$

If the coefficients of this equation are real, then $(a + ib) + (c + id) = (a + c) + i(b + d)$ must be real, which requires that $d = -b$. Then the coefficient of y becomes $(a + ib)(c - ib) = (ac + b^2) + ib(c - a)$. This coefficient is real only if $b = 0$ or $c = a$. We discard $b = 0$ as a possibility, because if it is so then the roots are not complex as hypothesized. Thus, for the differential equation to have complex roots $a + ib$ and $c + id$, we must have $d = -b$ and $c = a$, which implies that the roots are complex conjugates.

8.49 Derive a real-valued solution for a second-order linear differential equation with real constant coefficients if the roots of its characteristic equation are complex.

▌ Assume the unknown function is $y(x)$. By the previous problem, the roots of the differential equation must be complex conjugates; denote them as $\lambda_1 = a + ib$ and $\lambda_2 = a - ib$. Then two linearly independent solutions are $e^{(a+ib)x}$ and $e^{(a-ib)x}$, and the general complex solution is $y(x) = d_1 e^{(a+ib)x} + d_2 e^{(a-ib)x}$. Using Euler's relations, $e^{ibx} = \cos bx + i \sin bx$ and $e^{-ibx} = \cos bx - i \sin bx$, we can rewrite this solution as

$$\begin{aligned} y &= d_1 e^{ax} e^{ibx} + d_2 e^{ax} e^{-ibx} = e^{ax}(d_1 e^{ibx} + d_2 e^{-ibx}) \\ &= e^{ax}[d_1(\cos bx + i \sin bx) + d_2(\cos bx - i \sin bx)] \\ &= e^{ax}[(d_1 + d_2) \cos bx + i(d_1 - d_2) \sin bx] \end{aligned}$$

If we define $c_1 = d_1 + d_2$ and $c_2 = i(d_1 - d_2)$ as two new arbitrary constants, we can write the general solution as $y = c_1 e^{ax} \cos bx + c_2 e^{ax} \sin bx$. This equation is real if and only if c_1 and c_2 are both real, which occurs if and only if d_1 and d_2 are complex conjugates. Since we are interested in the general *real* solution, we must restrict d_1 and d_2 to be a conjugate pair.

8.50 Solve $\dfrac{d^2y}{dx^2} - 6\dfrac{dy}{dx} + 25y = 0.$

▌ The characteristic equation is $\lambda^2 - 6\lambda + 25 = 0$. Using the quadratic formula, we find its roots to be $\lambda = \dfrac{-(-6) \pm \sqrt{(-6)^2 - 4(25)}}{2} = \dfrac{6 \pm \sqrt{-64}}{2} = 3 \pm i4$. Since these roots are complex conjugates, the solution is (from the result of Problem 8.49 with $a = 3$ and $b = 4$) $y = e^{3x}(c_1 \cos 4x + c_2 \sin 4x)$.

8.51 Solve $\dfrac{d^2y}{dx^2} - 10\dfrac{dy}{dx} + 29y = 0.$

▌ The characteristic equation is $\lambda^2 - 10\lambda + 29 = 0$. Using the quadratic formula, we find its roots to be $\lambda = \dfrac{-(-10) \pm \sqrt{(-10)^2 - 4(29)}}{2} = \dfrac{10 \pm \sqrt{-16}}{2} = 5 \pm i2$. Since these roots are complex conjugates, the solution is (from the result of Problem 8.49 with $a = 5$ and $b = 2$) $y = e^{5x}(c_1 \cos 2x + c_2 \sin 2x)$.

8.52 Solve $\dfrac{d^2y}{dx^2} + 9y = 0.$

▌ The characteristic equation is $\lambda^2 + 9 = 0$, which has as its roots $\lambda = \pm i3 = 0 \pm i3$. Since these roots are complex conjugates, the solution is (from Problem 8.49 with $a = 0$ and $b = 3$) $y = e^{0x}(c_1 \cos 3x + c_2 \sin 3x) = c_1 \cos 3x + c_2 \sin 3x$.

8.53 Solve $\dfrac{d^2x}{dt^2} + 8\dfrac{dx}{dt} + 25x = 0.$

▌ The characteristic equation is $\lambda^2 + 8\lambda + 25 = 0$. Using the quadratic formula, we find its roots to be $\lambda = \dfrac{-8 \pm \sqrt{(8)^2 - 4(25)}}{2} = \dfrac{-8 \pm \sqrt{-36}}{2} = -4 \pm i3$. Since these roots are complex conjugates, the solution is $x = e^{-4t}(c_1 \cos 3t + c_2 \sin 3t)$.

8.54 Solve $\dfrac{d^2x}{dt^2} + 4\dfrac{dx}{dt} + 8x = 0.$

▌ The characteristic equation is $\lambda^2 + 4\lambda + 8 = 0$. Using the quadratic formula, we find its roots to be $\lambda = \dfrac{-4 \pm \sqrt{(4)^2 - 4(8)}}{2} = \dfrac{-4 \pm \sqrt{-16}}{2} = -2 \pm i2$. Since these roots are complex conjugates, the solution is $x = e^{-2t}(c_1 \cos 2t + c_2 \sin 2t)$.

8.55 Solve $\dfrac{d^2Q}{dt^2} + 8\dfrac{dQ}{dt} + 52Q = 0.$

The characteristic equation is $\lambda^2 + 8\lambda + 52 = 0$. Using the quadratic formula, we find its roots to be $\lambda = \dfrac{-8 \pm \sqrt{(8)^2 - 4(52)}}{2} = \dfrac{-8 \pm \sqrt{-144}}{2} = -4 \pm i6$. Since these roots are complex conjugates, the solution is $Q = e^{-4t}(c_1 \cos 6t + c_2 \sin 6t)$.

8.56 Solve $\dfrac{d^2I}{dt^2} + 100\dfrac{dI}{dt} + 50{,}000I = 0.$

The characteristic equation is $m^2 + 100m + 50{,}000 = 0$, which has as its roots $m = \dfrac{-100 \pm \sqrt{(100)^2 - 4(50{,}000)}}{2} = -50 \pm i50\sqrt{19}$. Since these roots are complex conjugates, the solution is $I = e^{-50t}(c_1 \cos 50\sqrt{19}t + c_2 \sin 50\sqrt{19}t)$.

8.57 Solve $\ddot{x} + 16x = 0.$

The characteristic equation is $m^2 + 16 = 0$, which has as its roots $m = \pm i4 = 0 \pm i4$. Since these roots are complex conjugates, the solution is $x = c_1 \cos 4t + c_2 \sin 4t$.

8.58 Solve $\ddot{x} + 64x = 0.$

The characteristic equation is $m^2 + 64 = 0$, which has as its roots $m = \pm i8 = 0 \pm i8$. Since these roots are complex conjugates, the solution is $x = c_1 \cos 8t + c_2 \sin 8t$.

8.59 Solve $y'' + 4y = 0.$

The characteristic equation is $\lambda^2 + 4 = 0$, which has roots $\lambda_1 = 2i$ and $\lambda_2 = -2i$. Since these roots are complex conjugates with real part equal to zero, the solution is $y = c_1 \cos 2x + c_2 \sin 2x$.

8.60 Solve $\ddot{y} + 50y = 0.$

The characteristic equation is $\lambda^2 + 50 = 0$, which has roots $\lambda = \pm i\sqrt{50}$. The solution is $y = c_1 \cos \sqrt{50}t + c_2 \sin \sqrt{50}t$.

8.61 Solve $\ddot{x} + 96x = 0.$

The characteristic equation is $\lambda^2 + 96 = 0$, which has roots $\lambda_1 = i\sqrt{96}$ and $\lambda_2 = -i\sqrt{96}$. The solution is $x = C_2 \cos \sqrt{96}t + C_1 \sin \sqrt{96}t$, where C_2 and C_1 denote arbitrary constants.

8.62 Solve $\ddot{x} + 12.8\dot{x} + 64x = 0.$

The characteristic equation is $m^2 + 12.8m + 64 = 0$, which has roots $m = \dfrac{-12.8 \pm \sqrt{(12.8)^2 - 4(64)}}{2} = -6.4 \pm i4.8$. The solution is $x = e^{-6.4t}(c_1 \cos 4.8t + c_2 \sin 4.8t)$.

8.63 Solve $\ddot{x} + \frac{1}{32}\dot{x} + 96x = 0.$

The characteristic equation is $m^2 + \frac{1}{32}m + 96 = 0$, which has roots $m = \dfrac{-(1/32) \pm \sqrt{(1/32)^2 - 4(96)}}{2} = -0.015625 \pm i9.7979$. The solution is $x = e^{-0.015625t}(C_1 \cos 9.7979t + C_2 \sin 9.7979t)$.

8.64 Solve $\dfrac{d^2I}{dt^2} + 20\dfrac{dI}{dt} + 200I = 0.$

The characteristic equation is $\lambda^2 + 20\lambda + 200 = 0$, which has roots $\lambda_1 = -10 + i10$ and $\lambda_2 = -10 - i10$. The solution is $I = e^{-10t}(c_1 \cos 10t + c_2 \sin 10t)$.

8.65 Solve $y'' - 3y' + 4y = 0.$

▮ The characteristic equation is $\lambda^2 - 3\lambda + 4 = 0$, which has roots $\lambda_1 = \frac{3}{2} + i\frac{\sqrt{7}}{2}$ and $\lambda_2 = \frac{3}{2} - i\frac{\sqrt{7}}{2}$. The solution is $y = c_1 e^{(3/2)x} \cos\frac{\sqrt{7}}{2}x + c_2 e^{(3/2)x} \sin\frac{\sqrt{7}}{2}x$.

.66 Solve $y'' + 4y' + 5y = 0$.

▮ The characteristic equation is $\lambda^2 + 4\lambda + 5 = 0$, which has roots $\lambda_1 = -2 + i$ and $\lambda_2 = -2 - i$. The solution is $y = c_1 e^{-2x} \cos x + c_2 e^{-2x} \sin x$.

.67 Solve $(D^2 + D + 2)y = 0$.

▮ The characteristic equation is $m^2 + m + 2 = 0$, which has roots $m = \frac{-1 \pm \sqrt{1-8}}{2} = -\frac{1}{2} \pm i\frac{\sqrt{7}}{2}$. The solution is $y = e^{-x/2}\left(A \cos\frac{\sqrt{7}x}{2} + B \sin\frac{\sqrt{7}x}{2}\right)$, where A and B denote arbitrary constants.

.68 Solve $(D^2 - 2D + 10)y = 0$.

▮ The characteristic roots are $1 \pm 3i$, so that the primitive is $y = e^x(C_1 \cos 3x + C_2 \sin 3x)$. This solution may also be written as $C_3 e^x \sin(3x + C_4)$ or $C_3 e^x \cos(3x + C_5)$.

.69 Solve $\frac{d^2q}{dt^2} + 200\frac{dq}{dt} + 400{,}000q = 0$.

▮ The characteristic equation is $\lambda^2 + 200\lambda + 400{,}000 = 0$, which has roots $\lambda = \frac{-200 \pm \sqrt{40{,}000 - 1{,}600{,}000}}{2} = -100 \pm i100\sqrt{39}$. The solution is $q = e^{-100t}(c_1 \cos 100\sqrt{39}t + c_2 \sin 100\sqrt{39}t)$.

70 Solve $\frac{d^2q}{dt^2} + 400\frac{dq}{dt} + 200{,}000q = 0$.

▮ The characteristic equation is $\lambda^2 + 400\lambda + 200{,}000 = 0$, which has roots $\lambda = -200 \pm i400$. The solution is $q = e^{-200t}(A \cos 400t + B \sin 400t)$.

71 Solve $\frac{d^2I}{dt^2} + 40\frac{dI}{dt} + 800\bar{I} = 0$.

▮ The characteristic equation is $\lambda^2 + 40\lambda + 800 = 0$, which has the roots $\lambda_1 = -20 + i20$ and $\lambda_2 = -20 - i20$. The solution is $I = e^{-20t}(c_1 \cos 20t + c_2 \sin 20t)$.

72 Solve $\ddot{x} + 25x = 0$.

▮ The characteristic equation is $m^2 + 25 = 0$, which has roots $m_1 = i5$ and $m_2 = -i5$. The solution is $x = C_1 \cos 5t + C_2 \sin 5t$.

73 Solve $\ddot{x} + 128x = 0$.

▮ The characteristic equation is $m^2 + 128 = 0$, which has roots $m = \pm i\sqrt{128}$. The solution is $x = C_1 \cos\sqrt{128}t + C_2 \sin\sqrt{128}t$.

74 Solve $\ddot{x} + 3gx = 0$ where g denotes a positive constant.

▮ The characteristic equation is $m^2 + 3g = 0$, which has roots $m = \pm i\sqrt{3g}$. The solution is $x = C_1 \cos\sqrt{3g}t + C_2 \sin\sqrt{3g}t$.

75 Solve $(D^2 + 488)y = 0$.

▮ The characteristic equation is $m^2 + 488 = 0$, which has roots $m = \pm i\sqrt{488} = \pm i22.09$. The solution is $y = C_1 \cos 22.09x + C_2 \sin 22.09x$.

8.76 Determine the characteristic equation of a second-order linear homogeneous differential equation with real coefficients if one solution is $e^{-t}\cos 5t$.

▮ This particular solution corresponds to the roots $-1 \pm i5$. Thus, the characteristic equation is $[\lambda - (-1 + i5)][\lambda - (-1 - i5)] = 0$, or $\lambda^2 + 2\lambda + 26 = 0$.

8.77 Solve the previous problem if one solution is $e^{2t}\sin 3t$.

▮ This particular solution corresponds to the roots $2 \pm i3$. Thus, the characteristic equation is $[\lambda - (2 + i3)][\lambda - (2 - i3)] = 0$, or $\lambda^2 - 4\lambda + 13$.

8.78 Solve Problem 8.76 if one solution is $\cos\sqrt{3}t$.

▮ This particular solution corresponds to the roots $0 \pm i\sqrt{3}$. The characteristic equation is $[\lambda - i\sqrt{3}][\lambda - (-i\sqrt{3})] = 0$ or $\lambda^2 + 3 = 0$.

8.79 Find the general solution to a second-order linear homogeneous differential equation for $x(t)$ with real coefficients if one root of the characteristic equation is $3 + i7$.

▮ Since the roots of the characteristic equation must be a conjugate pair (see Problem 8.48), the second root is $3 - i7$. The general solution is then $x = e^{3t}(c_1 \cos 7t + c_2 \sin 7t)$.

8.80 Solve the previous problem if, instead, one root of the characteristic equation is $-i8$.

▮ It follows from Problem 8.48 that the second root is $+i8$. The general solution is then $x = c_1 \cos 8t + c_2 \sin 8t$.

8.81 Find the general solution to a second-order linear homogeneous differential equation for $x(t)$ with real coefficients if a particular solution is $e^{2t}\cos 5t$.

▮ A second linearly independent solution is $e^{2t}\sin 5t$, so the general solution is $x = e^{2t}(c_1 \cos 5t + c_2 \sin 5t)$. (The given particular solution is obtained by taking $c_1 = 1$ and $c_2 = 0$.)

8.82 Solve the previous problem if, instead, a particular solution is $3e^{-t}\sin 4t$.

▮ Such a particular solution can occur only if the roots of the characteristic equation are $-1 \pm i4$, which implies that the general solution is $x = e^{-t}(c_1 \cos 4t + c_2 \sin 4t)$. The given particular solution is the special case $c_1 = 0$, $c_2 = 3$.

8.83 Solve Problem 8.81 if, instead, a particular solution is $8 \cos 3t$.

▮ Such a particular solution can occur only if the roots of the characteristic equation are $0 \pm i3$, which implies that the general solution is $x = c_1 \cos 3t + c_2 \sin 3t$. The given particular solution is the special case $c_1 = 8$, $c_2 = 0$.

8.84 Solve $\dfrac{d^4y}{dx^4} + 10\dfrac{d^2y}{dx^2} + 9y = 0$.

▮ The characteristic equation is $m^4 + 10m^2 + 9 = 0$, or $(m^2 + 1)(m^2 + 9) = 0$; it has roots $m_1 = i$, $m_2 = -i$, $m_3 = i3$, and $m_4 = -i3$. Since the roots are distinct, the solution is $y = d_1e^{ix} + d_2e^{-ix} + d_3e^{i3x} + d_4e^{-i3x}$.

Using Euler's relations (see Problem 8.49), we can combine the first two terms and then the last two terms, rewriting this solution as $y = c_1 \cos x + c_2 \sin x + c_3 \cos 3x + c_4 \sin 3x$.

8.85 Solve $\dfrac{d^4y}{dx^4} + \dfrac{d^3y}{dx^3} + \dfrac{d^2y}{dx^2} + 2y = 0$.

▮ The characteristic equation is $m^4 + m^3 + m^2 + 2 = 0$, which we factor into $(m^2 - m + 1)(m^2 + 2m + 2) = 0$; its roots are $1/2 \pm i\sqrt{3}/2$ and $-1 \pm i$. Since the roots are distinct, the solution is $y = d_1e^{(1/2 + i\sqrt{3}/2)x} + d_2e^{(1/2 - i\sqrt{3}/2)x} + d_3e^{(-1+i)x} + d_4e^{(-1-i)x}$.

Using Euler's relations (see Problem 8.49), we can combine the first two terms and then the last two terms, rewriting this solution as $y = e^{x/2}(c_1 \cos \sqrt{3}x/2 + c_2 \sin \sqrt{3}x/2) + e^{-x}(c_3 \cos x + c_4 \sin x)$.

8.86 Solve $y^{(4)} + 4y'' - y' + 6y = 0$.

The auxiliary equation is $m^4 + 4m^2 - m + 6 = 0$, which we factor into $(m^2 + m + 3)(m^2 - m + 2) = 0$; its roots are $-1/2 \pm i\sqrt{11}/2$ and $1/2 \pm i\sqrt{7}$. The solution is $y = d_1 e^{(-1/2 + i\sqrt{11}/2)x} + d_2 e^{(-1/2 - i\sqrt{11}/2)x} + d_3 e^{(1/2 + i\sqrt{7})x} + d_4 e^{(1/2 - i\sqrt{7})x}$.

Using Euler's relations, we can combine the first two terms and then the last two terms, rewriting this solution as $y = e^{-x/2}(c_1 \cos \sqrt{11}x/2 + c_2 \sin \sqrt{11}x/2) + e^{x/2}(c_3 \cos \sqrt{7}x + c_4 \sin \sqrt{7}x)$.

8.87 Solve $y^{(4)} - 4y''' + 7y'' - 4y' + 6y = 0$.

The charateristic equation, $\lambda^4 - 4\lambda^3 + 7\lambda^2 - 4\lambda + 6 = 0$, has roots $\lambda_1 = 2 + i\sqrt{2}$, $\lambda_2 = 2 - i\sqrt{2}$, $\lambda_3 = i$, and $\lambda_4 = -i$. The solution is $y = d_1 e^{(2+i\sqrt{2})x} + d_2 e^{(2-i\sqrt{2})x} + d_3 e^{ix} + d_4 e^{-ix}$. Using Euler's relations, we can combine the first two terms and then the last two terms, to rewrite the solution as $y = c_1 e^{2x} \cos \sqrt{2}x + c_2 e^{2x} \sin \sqrt{2}x + c_3 \cos x + c_4 \sin x$.

8.88 Find the solution of a fourth-order linear homogeneous differential equation for $x(t)$ with real constant coefficients if the roots of its characteristic equation are $2 \pm i3$ and $-5 \pm i8$.

The solution is $x = e^{2t}(c_1 \cos 3t + c_2 \sin 3t) + e^{-5t}(c_3 \cos 8t + c_4 \sin 8t)$.

8.89 Solve Problem 8.88 if the roots are $-15 \pm i11$ and $-7 \pm i83$.

The solution is $x = e^{-15t}(c_1 \cos 11t + c_4 \sin 11t) + e^{-7t}(c_3 \cos 83t + c_4 \sin 83t)$.

8.90 Solve Problem 8.88 if the roots are $\frac{1}{2} \pm i\frac{1}{2}$ and $\frac{1}{2} \pm i\frac{1}{4}$.

The solution is $x = e^{t/2}(c_1 \cos \frac{1}{2}t + c_2 \sin \frac{1}{2}t + c_3 \cos \frac{1}{4}t + c_4 \sin \frac{1}{4}t)$.

8.91 Solve Problem 8.88 if the roots are $-7 \pm i3$ and $\pm i3$.

The solution is

$$x = e^{-7t}(c_1 \cos 3t + c_2 \sin 3t) + c_3 \cos 3t + c_4 \sin 3t = (c_1 e^{-7t} + c_3) \cos 3t + (c_2 e^{-7t} + c_4) \sin 3t$$

8.92 Solve Problem 8.88 if the roots are $2 \pm i$ and $\pm i$.

The solution is $x = e^{2t}(c_1 \cos t + c_2 \sin t) + c_3 \cos t + c_4 \sin t$.

8.93 Determine a differential equation associated with the previous problem.

The characteristic equation is

$$\begin{aligned} 0 &= [\lambda - (2+i)][\lambda - (2-i)][\lambda - i][\lambda - (-i)] = (\lambda^2 - 4\lambda + 5)(\lambda^2 + 1) \\ &= \lambda^4 - 4\lambda^3 + 6\lambda^2 - 4\lambda + 5 \end{aligned}$$

The associated differential equation is $\dfrac{d^4x}{dt^4} - 4\dfrac{d^3x}{dt^3} + 6\dfrac{d^2x}{dt^2} - 4\dfrac{dx}{dt} + 5x = 0$.

8.94 Determine a differential equation associated with Problem 8.90.

The characteristic equation is

$$\begin{aligned} 0 &= \left[\lambda - \left(\frac{1}{2} + i\frac{1}{2}\right)\right]\left[\lambda - \left(\frac{1}{2} - i\frac{1}{2}\right)\right]\left[\lambda - \left(\frac{1}{2} + i\frac{1}{4}\right)\right]\left[\lambda - \left(\frac{1}{2} - i\frac{1}{4}\right)\right] = \left(\lambda^2 - \lambda + \frac{1}{2}\right)\left(\lambda^2 - \lambda + \frac{5}{16}\right) \\ &= \lambda^4 - 2\lambda^3 + \frac{29}{16}\lambda^2 - \frac{13}{16}\lambda + \frac{5}{32} \end{aligned}$$

An associated differential equation is

$$\frac{d^4x}{dt^4} - 2\frac{d^3x}{dt^3} + \frac{29}{16}\frac{d^2x}{dt^2} - \frac{13}{16}\frac{dx}{dt} + \frac{5}{32}x = 0 \quad \text{or} \quad 32\frac{d^4x}{dt^4} - 64\frac{d^3x}{dt^3} + 58\frac{d^2x}{dt^2} - 26\frac{dx}{dt} + 5x = 0$$

which may be written as $(32D^4 - 64D^3 + 58D^2 - 26D + 5)x = 0$.

8.95 Find the solution of a sixth-order linear homogeneous differential equation for $x(t)$ with real coefficients if the roots of its characteristic equation are $2 \pm i4$, $-3 \pm i5$, and $-1 \pm i2$.

Since the roots are distinct, the solution is

$$x(t) = e^{2t}(c_1 \cos 4t + c_2 \sin 4t) + e^{-3t}(c_3 \cos 5t + c_4 \sin 5t) + e^{-t}(c_5 \cos 2t + c_6 \sin 2t)$$

8.96 Solve the previous problem if the roots are, instead, $-16 \pm i108$, $-14 \pm i53$, and $-2 \pm i64$.

Since the roots are distinct, the solution is

$$x(t) = e^{-16t}(c_1 \cos 108t + c_2 \sin 108t) + e^{-14t}(c_3 \cos 53t + c_4 \sin 53t) + e^{-2t}(c_5 \cos 64t + c_6 \sin 64t)$$

8.97 Solve Problem 8.95 if the roots are, instead, $-\frac{1}{2} \pm i2$, $2 \pm i\frac{1}{2}$, and $\pm i4$.

Since the roots are distinct, the solution is

$$x(t) = e^{-t/2}(c_1 \cos 2t + c_2 \sin 2t) + e^{2t}(c_3 \cos \tfrac{1}{2}t + c_4 \sin \tfrac{1}{2}t) + c_5 \cos 4t + c_6 \sin 4t$$

8.98 Find the solution of an eighth-order linear homogeneous differential equation for $y(x)$ with real coefficients if the roots of its characteristic equation are $3 \pm i$, $-3 \pm i$, $1 \pm i3$, and $-1 \pm i3$.

Since the roots are distinct, the general solution is

$$y(x) = e^{3x}(c_1 \cos x + c_2 \sin x) + e^{-3x}(c_3 \cos x + c_4 \sin x) + e^{x}(c_5 \cos 3x + c_6 \sin 3x) + e^{-x}(c_7 \cos 3x + c_8 \sin 3x)$$

8.99 Solve the previous problem if the roots are, instead, $7 \pm i8$, $8 \pm i9$, $\frac{1}{2} \pm i4$, and $-\frac{1}{2} \pm i\frac{1}{2}$.

Since the roots are distinct, the general solution is

$$y(x) = e^{7x}(c_1 \cos 8x + c_2 \sin 8x) + e^{8x}(c_3 \cos 9x + c_4 \sin 9x) + e^{x/2}(c_5 \cos 4x + c_6 \sin 4x) + e^{-x/2}(c_7 \cos \tfrac{1}{2}x + c_8 \sin \tfrac{1}{2}x)$$

8.100 Find the general solution to a fourth-order linear homogeneous differential equation for $x(t)$ with real coefficients if two roots of the characteristic equation are $2 + i3$ and $-2 - i4$.

Since the roots must be in conjugate pairs, the other two roots are $2 - i3$ and $-2 + i4$. The general solution is then $x = e^{2t}(c_1 \cos 3t + c_2 \sin 3t) + e^{-2t}(c_3 \cos 4t + c_4 \sin 4t)$.

8.101 Solve $\dfrac{d^4x}{dt^4} - 6\dfrac{d^3x}{dt^3} + 15\dfrac{d^2x}{dt^2} - 18\dfrac{dx}{dt} + 10x = 0$ if a particular solution is $5e^{2t}\cos t$.

This particular solution corresponds to the complex roots $2 \pm i$ of the characteristic equation $\lambda^4 - 6\lambda^3 + 15\lambda^2 - 18\lambda + 10 = 0$. Thus, $[\lambda - (2 + i)][\lambda - (2 - i)] = \lambda^2 - 4\lambda + 5$ is a factor of the characteristic polynomial (the left side of the characteristic equation). Dividing by this factor, we find that $\lambda^2 - 2\lambda + 2$ is also a factor. Thus, two additional roots are $-1 \pm i$, and the general solution is $x = e^{2t}(c_1 \cos t + c_2 \sin t) + e^{-t}(c_3 \cos t + c_4 \sin t)$. The given particular solution is the special case $c_1 = 5$, $c_2 = c_3 = c_4 = 0$.

8.102 Solve $\dfrac{d^4y}{dx^4} + 4\dfrac{d^3y}{dx^3} + 9\dfrac{d^2y}{dx^2} + 16\dfrac{dy}{dx} + 20y = 0$ if a particular solution is $\sin 2x$.

This particular solution corresponds to the roots $\pm i2$ of the characteristic equation $\lambda^4 + 4\lambda^3 + 9\lambda^2 + 16\lambda + 20 = 0$. Thus $[\lambda - i2][\lambda - (-i2)] = \lambda^2 + 4$ is a factor of the characteristic polynomial (the left side of the characteristic equation). Dividing by this factor, we find that $\lambda^2 + 4\lambda + 5$ is also a factor. Thus, two additional roots are $-2 \pm i$. The general solution is then $y = c_1 \cos 2x + c_2 \sin 2x + c_3 e^{-2x} \cos x + c_4 e^{-2x} \sin x$.

8.103 Solve $\dfrac{d^6y}{dx^6} - 4\dfrac{d^5y}{dx^5} + 16\dfrac{d^4y}{dx^4} - 12\dfrac{d^3y}{dx^3} + 41\dfrac{d^2y}{dx^2} - 8\dfrac{dy}{dx} + 26y = 0$ if two solutions are $\sin x$ and $e^{2x} \sin 3x$.

The particular solution $\sin x$ corresponds to the characteristic roots $\pm i$, while $e^{2x} \sin 3x$ corresponds to the characteristic roots $2 \pm i3$, so both $[\lambda - i][\lambda - (-i)] = \lambda^2 + 1$ and $[\lambda - (2 + i3)][\lambda - (2 - i3)] = \lambda^2 - 4\lambda + 13$ are factors of the characteristic polynomial. Dividing the characteristic polynomial $\lambda^6 - 4\lambda^5 + 16\lambda^4 - 12\lambda^3 + 41\lambda^2 - 8\lambda + 26$ by both factors successively, we find that $\lambda^2 + 2$ is also a factor. Thus, two additional roots are $\pm i\sqrt{2}$. The general solution is

$$y = c_1 \cos x + c_2 \sin x + c_3 e^{2x} \cos 3x + c_4 e^{2x} \sin 3x + c_5 \cos \sqrt{2}x + c_6 \sin \sqrt{2}x$$

DISTINCT REAL AND COMPLEX CHARACTERISTIC ROOTS

8.104 Solve $y''' + 2y'' + 5y' - 26y = 0$.

The characteristic equation is $\lambda^3 + 2\lambda^2 + 5\lambda - 26 = 0$, which we factor into $(\lambda - 2)(\lambda^2 + 4\lambda + 13) = 0$; its roots are $\lambda_1 = 2$, $\lambda_2 = -2 + i3$, and $\lambda_3 = -2 - i3$. Since they are all distinct, the solution is $y = c_1 e^{2x} + d_2 e^{(-2+i3)x} + d_3 e^{(-2-i3)x}$.

Using the result of Problem 8.49, we can rewrite this solution as $y = c_1 e^{2x} + e^{-2x}(c_2 \cos 3x + c_3 \sin 3x)$, which is real.

8.105 Solve $y''' + 5y'' + 7y' + 13y = 0$.

The characteristic equation is $\lambda^3 + 5\lambda^2 + 7\lambda - 13 = 0$, which can be factored into $(\lambda - 1)(\lambda^2 + 6\lambda + 13) = 0$. Its roots are $\lambda_1 = 1$, $\lambda_2 = -3 + i2$, and $\lambda_3 = -3 - i2$. Since they are all distinct, the solution is $y = c_1 e^{x} + d_2 e^{(-3+i2)x} + d_3 e^{(-3-i2)x}$.

Using the result of Problem 8.49, we can rewrite this solution as $y = c_1 e^{x} + e^{-3x}(c_2 \cos 2x + c_3 \sin 2x)$, which is real.

8.106 Solve $y''' + 8y'' + 37y' + 50y = 0$.

The characteristic equation is $\lambda^3 + 8\lambda^2 + 37\lambda + 50 = 0$, which we factor into $(\lambda + 2)(\lambda^2 + 6\lambda + 25) = 0$. Its roots are $\lambda_1 = -2$, $\lambda_2 = 3 + i4$, and $\lambda_3 = 3 - i4$. Since they are all distinct, the solution is $y = c_1 e^{-2x} + d_2 e^{(3+i4)x} + d_3 e^{(3-i4)x}$. Using the result of Problem 8.49, we can rewrite this solution as $y = c_1 e^{-2x} + e^{3x}(c_2 \cos 4x + c_3 \sin 4x)$.

8.107 Solve $\dfrac{d^3y}{dx^3} - 7\dfrac{d^2y}{dx^2} - \dfrac{dy}{dx} + 87y = 0$.

The characteristic equation is $\lambda^3 - 7\lambda^2 - \lambda + 87 = 0$, which has the roots $\lambda_1 = -3$, $\lambda_2 = 5 + i2$, and $\lambda_3 = 5 - i2$. Since they are all distinct, the solution is $y = c_1 e^{-3x} + d_2 e^{(5+i2)x} + d_3 e^{(5-i2)x}$. This can be rewritten, using Euler's relations (Problem 8.49), as $y = c_1 e^{-3x} + e^{5x}(c_2 \cos 2x + c_3 \sin 2x)$.

8.108 Solve $y''' + y' = 0$.

The characteristic equation is $m^3 + m = 0$, with roots $m_1 = 0$, $m_2 = i$, and $m_3 = -i$. Since the roots are all distinct, the solution is $y = c_1 e^{0x} + d_2 e^{ix} + d_3 e^{-ix}$. By the result of Problem 8.49, this can be rewritten as $y = c_1 + c_2 \cos x + c_3 \sin x$.

8.109 Solve $y''' + 4y' = 0$.

The characteristic equation is $m^3 + 4m = 0$ or $m(m^2 + 4) = 0$, which has roots $m_1 = 0$, $m_2 = i2$, and $m_3 = -i2$. Since these roots are all distinct, the solution is $y = c_1 e^{0x} + d_2 e^{i2x} + d_3 e^{-i2x}$. By the result of Problem 8.49, this can be rewritten as $y = c_1 + c_2 \cos 2x + c_3 \sin 2x$.

8.110 Solve $y''' - 6y'' + 2y' + 36y = 0$.

The characteristic equation is $\lambda^3 - 6\lambda^2 + 2\lambda + 36 = 0$, with roots $\lambda_1 = -2$, $\lambda_2 = 4 + i\sqrt{2}$, and $\lambda_3 = 4 - i\sqrt{2}$. The solution is $y = c_1 e^{-2x} + d_2 e^{(4+i\sqrt{2})x} + d_3 e^{(4-i\sqrt{2})x}$. This can be rewritten, using Euler's relations (Problem 8.49), as $y = c_1 e^{-2x} + c_2 e^{4x} \cos \sqrt{2}x + c_3 e^{4x} \sin \sqrt{2}x$.

8.111 Solve $\dfrac{d^3x}{dt^3} - \dfrac{d^2x}{dt^2} - 12\dfrac{dx}{dt} - 40x = 0$.

The characteristic equation is $\lambda^3 - \lambda^2 - 12\lambda - 40 = 0$, whose roots are $\lambda_1 = 5$, $\lambda_2 = -2 + i2$, and $\lambda_3 = -2 - i2$. The solution is thus $x = c_1 e^{5t} + d_2 e^{(-2+i2)t} + d_3 e^{(-2-i2)t}$. This can be rewritten (see Problem 8.49) as $x = c_1 e^{5t} + e^{-2t}(c_2 \cos 2t + c_3 \sin 2t)$.

8.112 Solve $\dfrac{d^3x}{dt^3} + 5\dfrac{d^2x}{dt^2} + 26\dfrac{dx}{dt} - 150x = 0$.

The characteristic equation is $m^3 + 5m^2 + 26m - 150 = 0$, which has roots $m_1 = 3$, $m_2 = -4 + i\sqrt{34}$, and $m_3 = -4 - i\sqrt{34}$. The solution is $x = c_1 e^{3t} + e^{-4t}(c_2 \cos \sqrt{34}t + c_3 \sin \sqrt{34}t)$.

8.113 Solve $\dfrac{d^3Q}{dt^3} - 5\dfrac{d^2Q}{dt^2} + 25\dfrac{dQ}{dt} - 125Q = 0.$

▌ The characteristic equation is $m^3 - 5m^2 + 25m - 125 = 0$, which we factor into $(m-5)(m^2+25) = 0$; it has as roots $m_1 = 5$, $m_2 = i5$, and $m_3 = -i5$. The solution is then $Q = c_1e^{5t} + c_2\cos 5t + c_3\sin 5t$.

8.114 Solve $\dfrac{d^3I}{dr^3} - \dfrac{d^2I}{dr^2} + 2\dfrac{dI}{dr} - 2I = 0.$

▌ The characteristic equation is $m^3 - m^2 + 2m - 2 = 0$; its roots are $m_1 = 1$, $m_2 = i\sqrt{2}$, and $m_3 = -i\sqrt{2}$. The solution is $I = c_1e^r + c_2\cos\sqrt{2}r + c_3\sin\sqrt{2}r$.

8.115 Solve $\dfrac{d^3r}{d\theta^3} + 12.8\dfrac{d^2r}{d\theta^2} + 64\dfrac{dr}{d\theta}.$

▌ The characteristic equation, $m^3 + 12.8m^2 + 64m = 0$, can be factored into $m(m^2 + 12.8m + 64) = 0$ and has as its roots $m_1 = 0$, $m_2 = -6.4 + i4.8$, and $m_3 = -6.4 - i4.8$. The solution is thus

$$r = c_1e^{0\theta} + e^{-6.4\theta}(c_2\cos 4.8\theta + c_3\sin 4.8\theta) = c_1 + c_2e^{-6.4\theta}\cos 4.8\theta + c_3e^{-6.4\theta}\sin 4.8\theta$$

8.116 Solve $\dddot{y} + 64\dot{y} = 0.$

▌ The characteristic equation, $\lambda^3 + 64\lambda = 0$, may be factored into $\lambda(\lambda^2 + 64) = 0$ and has as roots $\lambda_1 = 0$, $\lambda_2 = i8$, and $\lambda_3 = -i8$. The solution is $y = c_1 + c_2\cos 8t + c_3\sin 8t$.

8.117 Solve $\dfrac{d^4y}{dx^4} - 81y = 0.$

▌ The characteristic equation is $\lambda^4 - 81 = 0$, which we factor into $(\lambda^2 - 9)(\lambda^2 + 9) = 0$; its roots are $\lambda_1 = 3$, $\lambda_2 = -3$, $\lambda_3 = i3$, and $\lambda_4 = -i3$. Since they are all distinct, the solution is $y = c_1e^{3x} + c_2e^{-3x} + d_1e^{i3x} + d_2e^{-i3x}$.

Using the result of Problem 8.49 on the last two terms, we may rewrite this solution as $y = c_1e^{3x} + c_2e^{-3x} + c_3\cos 3x + c_4\sin 3x$.

8.118 Solve $\dfrac{d^4y}{dx^4} - y = 0.$

▌ The characteristic equation, $\lambda^4 - 1 = 0$, can be factored into $(\lambda^2 - 1)(\lambda^2 + 1) = 0$: its roots are $\lambda_1 = 1$, $\lambda_2 = -1$, $\lambda_3 = i$, and $\lambda_4 = -i$. Since they are all distinct, the solution is $y = c_1e^x + c_2e^{-x} + d_1e^{ix} + d_2e^{-ix}$.

Using the result of Problem 8.49 on the last two terms, we may rewrite this solution as $y = c_1e^x + c_2e^{-x} + c_3\cos x + c_4\sin x$.

8.119 Solve $\dfrac{d^4y}{dx^4} + \dfrac{d^2y}{dx^2} - 20y = 0.$

▌ The characteristic equation is $\lambda^4 + \lambda^2 - 20 = 0$, which we factor into $(\lambda^2 - 4)(\lambda^2 + 5) = 0$. Its roots are $\lambda_1 = 2$, $\lambda_2 = -2$, $\lambda_3 = i\sqrt{5}$, and $\lambda_4 = -i\sqrt{5}$. Thus, the solution is $y = c_1e^{2x} + c_2e^{-2x} + d_1e^{i\sqrt{5}x} + d_2e^{-i\sqrt{5}x}$, which may be rewritten as $y = c_1e^{2x} + c_2e^{-2x} + c_3\cos\sqrt{5}x + c_4\sin\sqrt{5}x$.

8.120 Solve $\dfrac{d^4y}{dx^4} + 2\dfrac{d^3y}{dx^3} + 5\dfrac{d^2y}{dx^2} - 26\dfrac{dy}{dx} = 0.$

▌ The characteristic equation is $\lambda^4 + 2\lambda^3 + 5\lambda^2 - 26\lambda = 0$, which we factor into $\lambda(\lambda - 2)(\lambda^2 + 4\lambda + 13) = 0$. Its roots are $\lambda_1 = 0$, $\lambda_2 = 2$, $\lambda_3 = -2 + i3$, and $\lambda_4 = -2 - i3$. Since they are all distinct, the solution is $y = c_1 + c_2e^{2x} + d_1e^{(-2+i3)x} + d_2e^{(-2-i3)x}$.

Using the result of Problem 8.49 on the last two terms, we may rewrite this solution as $y = c_1 + c_2e^{2x} + e^{-2x}(c_3\cos 3x + c_4\sin 3x)$.

8.121 Solve $y^{(4)} + 5y^{(3)} + 7y'' + 13y' = 0.$

▌ The characteristic equation is $m^4 + 5m^3 + 7m^2 + 13m = 0$, which we can factor into $m(m-1)(m^2 + 6m + 13)$; its roots are $m_1 = 0$, $m_2 = 1$, $m_3 = -3 + i2$, and $m_4 = -3 - i2$. The solution is thus $y = c_1 + c_2e^x + d_1e^{(-3+i2)x} + d_2e^{(-3-i2)x}$, which may be rewritten (see Problem 8.49) as $y = c_1 + c_2e^x + e^{-3x}(c_3 \cos 2x + c_4 \sin 2x)$.

8.122 Solve $y^{(4)} + y^{(3)} - y' - y = 0$.

▌ The characteristic equation is $m^4 + m^3 - m - 1 = 0$, which we can factor into $(m^2 - 1)(m^2 + m + 1) = 0$; its roots are $m_1 = 1$, $m_2 = -1$, $m_3 = -1/2 + i\sqrt{3}/2$, and $m_4 = -1/2 - i\sqrt{3}/2$. The solution is $y = c_1e^x + c_2e^{-x} + d_1e^{(-1/2+i\sqrt{3}/2)x} + d_2e^{(-1/2-i\sqrt{3}/2)x}$. This may be rewritten (see Problem 8.49) as $y = c_1e^x + c_2e^{-x} + e^{-x/2}\left(c_3 \cos \frac{\sqrt{3}}{2}x + c_4 \sin \frac{\sqrt{3}}{2}x\right)$.

8.123 Solve $y^{(4)} + y^{(3)} - 4y' - 16y = 0$.

▌ The characteristic equation is $m^4 + m^3 - 4m - 16 = 0$, which we factor into $(m - 2)(m + 2)(m^2 + m + 4) = 0$. Its roots are $m_1 = 2$, $m_2 = -2$, $m_3 = -1/2 + i\sqrt{15}/2$, and $m_4 = -1/2 - i\sqrt{15}/2$. The solution is $y = c_1e^{2x} + c_2e^{-2x} + e^{-x/2}\left(c_3 \cos \frac{\sqrt{15}}{2}x + c_4 \sin \frac{\sqrt{15}}{2}x\right)$.

8.124 Solve $y^{(4)} - 6y^{(3)} + 16y'' + 54y' - 225y = 0$.

▌ The characteristic equation is $m^4 - 6m^3 + 16m^2 + 54m - 225 = 0$, which we factor into $(m - 3)(m + 3)(m^2 - 6m + 25) = 0$. Its roots are $m_1 = 3$, $m_2 = -3$, $m_3 = 3 + i4$, and $m_4 = 3 - i4$. The solution is then $y = c_1e^{3x} + c_2e^{-3x} + e^{3x}(c_3 \cos 4x + c_4 \sin 4x)$.

8.125 Solve $y^{(4)} + y^{(3)} - 2y'' - 6y' - 4y = 0$.

▌ The characteristic equation is $m^4 + m^3 - 2m^2 - 6m - 4$, with the roots $m_1 = -1$, $m_2 = 2$, $m_3 = -1 + i$, and $m_4 = -1 - i$. The solution is $y = c_1e^{-x} + c_2e^{2x} + e^{-x}(c_3 \cos x + c_4 \sin x)$.

8.126 Solve $(D^3 + 4D)y = 0$.

▌ This equation may be factored into $D(D^2 + 4)y = 0$, with characteristic roots 0 and $\pm 2i$. The solution is $y = C_1 + C_2 \cos 2x + C_3 \sin 2x$.

8.127 Solve $(D^4 + 5D^2 - 36)y = 0$.

▌ This equation may be factored into $(D^2 - 4)(D^2 + 9)y = 0$. The characteristic roots are ± 2 and $\pm 3i$, and the primitive is $y = Ae^{2x} + Be^{-2x} + C_3 \cos 3x + C_4 \sin 3x$. This may be written as $y = C_1 \cosh 2x + C_2 \sinh 2x + C_3 \cos 3x + C_4 \sin 3x$, since $\cosh 2x = \frac{1}{2}(e^{2x} + e^{-2x})$ and $\sinh 2x = \frac{1}{2}(e^{2x} - e^{-2x})$.

8.128 Solve $(D^4 - 16)y = 0$.

▌ The auxiliary equation is $m^4 - 16 = 0$ or $(m^2 + 4)(m^2 - 4) = 0$, with roots $\pm 2i$ and ± 2. Then the general solution is $y = c_1 \cos 2x + c_2 \sin 2x + c_3e^{2x} + c_4e^{-2x}$.

8.129 Find the solution of a fourth-order linear homogeneous differential equation for $x(t)$ with real coefficients if the roots of its characteristic equation are $2 \pm i$ and ± 1.

▌ The solution is $x = e^{2t}(c_1 \cos t + c_2 \sin t) + c_3e^t + c_4e^{-t}$.

8.130 Determine a differential equation associated with the previous problem.

▌ The characteristic equation is

$$\begin{aligned} 0 &= [\lambda - (2 + i)][\lambda - (2 - i)](\lambda - 1)[\lambda - (-1)] = (\lambda^2 - 4\lambda + 5)(\lambda^2 - 1) \\ &= \lambda^4 - 4\lambda^3 + 4\lambda^2 + 4\lambda - 5 \end{aligned}$$

An associated differential equation is $\frac{d^4x}{dt^4} - 4\frac{d^3x}{dt^3} + 4\frac{d^2x}{dt^2} + 4\frac{dx}{dt} - 5x = 0$.

8.131 Find the solution of a sixth-order linear homogeneous differential equation for $x(t)$ with real coefficients if the roots of its characteristic equation are $3 \pm i\pi$, $\pm i2\pi$, and ± 5.

▮ Since the roots are distinct, the solution is

$$x = e^{3t}(c_1 \cos \pi t + c_2 \sin \pi t) + c_3 \cos 2\pi t + c_4 \sin 2\pi t + c_5 e^{5t} + c_6 e^{-5t}.$$

8.132 Solve Problem 8.131 if the roots are, instead, $-3 \pm i3$, 4, 5, and ± 6.

▮ Since the roots are distinct, the solution is

$$x = e^{-3t}(c_1 \cos 3t + c_2 \sin 3t) + c_3 e^{4t} + c_4 e^{5t} + c_5 e^{6t} + c_6 e^{-6t}.$$

8.133 Solve Problem 8.131 if the roots are $\pm i\frac{1}{2}$, 1, 2, 3, and 0.

▮ The solution is $x = c_1 \cos \frac{1}{2}t + c_2 \sin \frac{1}{2}t + c_3 e^{t} + c_4 e^{2t} + c_5 e^{3t} + c_6$.

8.134 Determine a differential equation associated with the previous problem.

▮ The auxiliary equation is

$$0 = (m - i\tfrac{1}{2})(m + i\tfrac{1}{2})(m-1)(m-2)(m-3)(m-0) = m^6 - 6m^5 + \tfrac{45}{4}m^4 - \tfrac{30}{4}m^3 + \tfrac{11}{4}m^2 - \tfrac{6}{4}m$$

or $4m^6 - 24m^5 + 45m^4 - 30m^3 + 11m^2 - 6m = 0$. An associated differential equation is $(4D^6 - 24D^5 + 45D^4 - 30D^3 + 11D^2 - 6D)x = 0$.

8.135 Find the solution of a twelfth-order linear homogeneous differential equation for $x(t)$ with real coefficients if the roots of its auxiliary equation are $-2 \pm i3$, $2 \pm i3$, $\pm i5$, $\pm i19$, ± 13, 3, and 0.

▮ Since all the roots are distinct, the solution is

$$x = e^{-2t}(c_1 \cos 3t + c_2 \sin 3t) + e^{2t}(c_3 \cos 3t + c_4 \sin 3t) + c_5 \cos 5t + c_6 \sin 5t + c_7 \cos 19t + c_8 \sin 19t$$
$$+ c_9 e^{13t} + c_{10} e^{-13t} + c_{11} e^{3t} + c_{12}.$$

8.136 Find the general solution to a fifth-order linear homogeneous differential equation for $x(t)$ with real coefficients if three solutions are $\cos 2t$, $e^{-t}\sin 3t$, and e^{2t}.

▮ If $\cos 2t$ and $e^{-t}\sin 3t$ are solutions, then so too are $\sin 2t$ and $e^{-t}\cos 3t$. Since these are the remaining two linearly independent solutions, the general solution is $x = c_1 \cos 2t + c_2 \sin 2t + e^{-t}(c_3 \cos 3t + c_4 \sin 3t) + c_5 e^{2t}$.

8.137 Determine the characteristic equation of a third-order linear homogeneous differential equation with real coefficients if two solutions are $\cos 3t$ and e^{3t}.

▮ To generate $\cos 3t$, two roots of the characteristic equation must be $\pm i3$. To generate e^{3t}, another root must be 3. Thus, the characteristic equation is $0 = (\lambda - i3)(\lambda + i3)(\lambda - 3) = \lambda^3 - 3\lambda^2 + 9\lambda - 27$.

8.138 Solve $\dddot{x} + 7\ddot{x} + \dot{x} + 7x = 0$ if a particular solution is $\sin t$.

▮ To generate $\sin t$, two roots of the characteristic equation must be $\pm i$, which implies that $(\lambda - i)(\lambda + i) = \lambda^2 + 1$ is a factor of the characteristic polynomial, $\lambda^3 + 7\lambda^2 + \lambda + 7$. Dividing by this factor, we find that $\lambda + 7$ is also a factor. Therefore, the roots are -7 and $\pm i$, and the general solution is $x = c_1 e^{-7t} + c_2 \cos t + c_3 \sin t$.

8.139 Solve $\dfrac{d^5x}{dt^5} + 4\dfrac{d^4x}{dt^4} + 33\dfrac{d^3x}{dt^3} + 100\dfrac{d^2x}{dt^2} + 200\dfrac{dx}{dt} = 0$ if two solutions are 8 and $\frac{1}{2}\sin 5t$.

▮ The particular solution 8 corresponds to the root $\lambda = 0$ of the characteristic equation, while the solution $\frac{1}{2}\sin 5t$ corresponds to the roots $\pm i5$. Thus, $(\lambda - 0)(\lambda - i5)[\lambda - (-i5)] = \lambda^3 + 25\lambda$ is a factor of the characteristic polynomial, $\lambda^5 + 4\lambda^4 + 33\lambda^3 + 100\lambda^2 + 200\lambda$. Dividing both sides of the characteristic equation by $\lambda^3 + 25\lambda$, we obtain $\lambda^2 + 4\lambda + 8 = 0$, which implies that $-2 \pm i2$ are two other roots. The general solution is then $x = c_1 + c_2 \cos 5t + c_3 \sin 5t + e^{-2t}(c_4 \cos 2t + c_5 \sin 2t)$.

8.140 Solve $(32D^5 - 40D^4 - 20D^3 + 50D^2 - 7D - 5)x = 0$ if two solutions are $-3e^t \sin \frac{1}{2}t$ and e^{-t}.

▮ The particular solution $-3e^t \sin \frac{1}{2}t$ corresponds to the roots $1 \pm i\frac{1}{2}$, while the solution e^{-t} corresponds to the root -1. Thus, $[\lambda - (-1)][\lambda - (1 + i\frac{1}{2})][\lambda - (1 - i\frac{1}{2})] = \lambda^3 - \lambda^2 - \frac{3}{4}\lambda + \frac{5}{4}$ is a factor of the characteristic polynomial, $32\lambda^5 - 40\lambda^4 - 20\lambda^3 + 50\lambda^2 - 7\lambda - 5$, as is $4\lambda^3 - 4\lambda^2 - 3\lambda + 5$. Dividing by this last factor yields $8\lambda^2 - 2\lambda - 1 = 0$, which implies that $\lambda = \frac{1}{2}$ and $\lambda = -\frac{1}{4}$ are two additional roots. The general solution is then $x = e^t(c_1 \cos \frac{1}{2}t + c_2 \sin \frac{1}{2}t) + c_3 e^{-t} + c_4 e^{t/2} + c_5 e^{-t/4}$.

REPEATED CHARACTERISTIC ROOTS

8.141 Solve $y'' + 4y' + 4y = 0$.

▌ The characteristic equation is $\lambda^2 + 4\lambda + 4 = 0$, which has the roots $\lambda_1 = \lambda_2 = -2$. The solution is $y = c_1e^{-2x} + c_2xe^{-2x}$.

8.142 Solve $y'' + 6y' + 9y = 0$.

▌ The characteristic equation is $\lambda^2 + 6\lambda + 9 = 0$ or $(\lambda + 3)^2 = 0$, which has the roots $\lambda_1 = \lambda_2 = -3$. The solution is $y = c_1e^{-3x} + c_2xe^{-3x}$.

8.143 Solve $y'' - 2y' + y = 0$.

▌ The characteristic equation is $\lambda^2 - 2\lambda + 1 = 0$ or $(\lambda - 1)^2 = 0$, which has the roots $\lambda_1 = \lambda_2 = 1$. The solution is $y = c_1e^x + c_2xe^x$.

8.144 Solve $y'' + 2y' + y = 0$.

▌ The characteristic equation is $\lambda^2 + 2\lambda + 1 = 0$ or $(\lambda + 1)^2 = 0$, which has the roots $\lambda_1 = \lambda_2 = -1$. The solution is $y = C_1e^{-x} + C_2xe^{-x}$.

8.145 Solve $\ddot{x} - 8\dot{x} + 16x = 0$.

▌ The characteristic equation is $\lambda^2 - 8\lambda + 16 = 0$ or $(\lambda - 4)^2 = 0$, which has the roots $\lambda_1 = \lambda_2 = 4$. The solution is $x = C_1e^{4t} + C_2te^{4t} = e^{4t}(C_1 + C_2t)$.

8.146 Solve $\ddot{x} + 10\dot{x} + 25x = 0$.

▌ The characteristic equation is $m^2 + 10m + 25 = 0$ or $(m + 5)^2 = 0$, which has the roots $m_1 = m_2 = -5$. The solution is $x = C_1e^{-5t} + C_2te^{-5t}$.

8.147 Solve $\dfrac{d^2Q}{dt^2} + 1000\dfrac{dQ}{dt} + 250{,}000Q = 0$.

▌ The characteristic equation is $m^2 + 1000m + 250{,}000 = 0$ or $(m + 500)^2 = 0$, which has the repeated roots $m_1 = m_2 = -500$. The solution is $Q = c_1e^{-500t} + c_2te^{-500t}$.

8.148 Solve $\ddot{x} + 16\dot{x} + 64 = 0$.

▌ The characteristic equation is $m^2 + 16m + 64 = 0$ or $(m + 8)^2 = 0$, which has the repeated roots $m_1 = m_2 = -8$. The solution is $x = c_1e^{-8t} + c_2te^{-8t}$.

8.149 Solve $\dfrac{d^2I}{dt^2} - 60\dfrac{dI}{dt} + 900I = 0$.

▌ The characteristic equation is $\lambda^2 - 60\lambda + 900 = 0$ or $(\lambda - 30)^2 = 0$, which has $\lambda = 30$ as a double root. The solution is $I = Ae^{30t} + Bte^{30t} = (A + Bt)e^{30t}$.

8.150 Solve $(4D^2 - 4D + 1)y = 0$.

▌ The characteristic equation is $4\lambda^2 - 4\lambda + 1 = 0$ or $(2\lambda - 1)^2 = 0$, which has $\lambda = \frac{1}{2}$ as a double root. The solution is $y = Ae^{x/2} + Bxe^{x/2}$.

8.151 Solve $(16D^2 + 8D + 1)y = 0$.

▌ The characteristic equation is $16m^2 + 8m + 1 = 0$ or $(4m + 1)^2 = 0$, which has $m = -\frac{1}{4}$ as a double root. The solution is $y = Ae^{-x/4} + Bxe^{-x/4}$.

8.152 Solve $(D^2 + 30D + 225)y = 0$.

▌ The characteristic equation is $m^2 + 30m + 225 = 0$ or $(m + 15)^2 = 0$, which has $m = -15$ as a double root. The solution is $y = c_1e^{-15x} + c_2xe^{-15x}$.

8.153 Solve $y'' = 0$.

▌ The characteristic equation is $\lambda^2 = 0$, which has the roots $\lambda_1 = \lambda_2 = 0$. The solution is then $y = c_1e^{0x} + c_2xe^{0x} = c_1 + c_2x$.

8.154 Solve $y''' = 0$.

▌ The characteristic equation is $\lambda^3 = 0$, which has zero as a triple root. The solution is then $y = c_1e^{0x} + c_2xe^{0x} + c_3x^2e^{0x} = c_1 + c_2x + c_3x^2$.

8.155 Solve $\dfrac{d^4y}{dx^4} = 0$.

▌ The characteristic equation is $\lambda^4 = 0$, which has zero as a quadruple root. The solution is then $y = c_1 + c_2x + c_3x^2 + c_4x^3$.

8.156 Solve $\dfrac{d^3y}{dx^3} + 6\dfrac{d^2y}{dx^2} + 12\dfrac{dy}{dx} + 8y = 0$.

▌ The characteristic equation is $\lambda^3 + 6\lambda^2 + 12\lambda + 8 = 0$ or $(\lambda + 2)^3 = 0$, which has $\lambda = -2$ as a triple root. The solution is $y = c_1e^{-2x} + c_2xe^{-2x} + c_3x^2e^{-2x} = e^{-2x}(c_1 + c_2x + c_3x^2)$.

8.157 Solve $y^{(4)} + 8y''' + 24y'' + 32y' + 16y = 0$.

▌ The characteristic equation is $\lambda^4 + 8\lambda^3 + 24\lambda^2 + 32\lambda + 16 = 0$ or $(\lambda + 2)^4 = 0$. Since $\lambda_1 = -2$ is a root of multiplicity four, the solution is $y = c_1e^{-2x} + c_2xe^{-2x} + c_3x^2e^{-2x} + c_4x^3e^{-2x}$.

8.158 Solve $\dfrac{d^3Q}{dt^3} + 3\dfrac{d^2Q}{dt^2} + 3\dfrac{dQ}{dt} + Q = 0$.

▌ The characteristic equation is $\lambda^3 + 3\lambda^2 + 3\lambda + 1 = 0$ or $(\lambda + 1)^3 = 0$, which has $\lambda = -1$ as a triple root. The solution is $Q = C_1e^{-t} + C_2te^{-t} + C_3t^2e^{-t}$.

8.159 Solve $Q^{(4)} + 4Q^{(3)} + 6\ddot{Q} + 4\dot{Q} + Q = 0$.

▌ The characteristic equation is $\lambda^4 + 4\lambda^3 + 6\lambda^2 + 4\lambda + 1 = 0$ or $(\lambda + 1)^4 = 0$, which has $\lambda = -1$ as a root of multiplicity four. The solution is $Q = C_1e^{-t} + C_2te^{-t} + C_3t^2e^{-t} + C_4t^3e^{-t}$ or $Q = (C_1 + C_2t + C_3t^2 + C_4t^3)e^{-t}$.

8.160 Solve $Q^{(5)} + 5Q^{(4)} + 10Q^{(3)} + 10\ddot{Q} + 5\dot{Q} + Q = 0$.

▌ The characteristic equation is $\lambda^5 + 5\lambda^4 + 10\lambda^3 + 10\lambda^2 + 5\lambda + 1 = 0$ or $(\lambda + 1)^5 = 0$, which has $\lambda = -1$ as a root of multiplicity five. The solution is $Q = C_1e^{-t} + C_2te^{-t} + C_3t^2e^{-t} + C_4t^3e^{-t} + C_5t^4e^{-t}$.

8.161 Solve $\dfrac{d^4r}{d\theta^4} - 12\dfrac{d^3r}{d\theta^3} + 54\dfrac{d^2r}{d\theta^2} - 108\dfrac{dr}{d\theta} + 81r = 0$.

▌ The characteristic equation is $\lambda^4 - 12\lambda^3 + 54\lambda^2 - 108\lambda + 81 = 0$ or $(\lambda - 3)^4 = 0$, which has $\lambda = 3$ as a root of multiplicity four. The solution is $r = Ae^{3\theta} + B\theta e^{3\theta} + C\theta^2e^{3\theta} + D\theta^3e^{3\theta}$.

8.162 Find the solution of a third-order linear homogeneous differential equation for $x(t)$ with real coefficients if its characteristic equation has $\lambda = \frac{1}{2}$ as a triple root.

▌ The solution is $x = c_1e^{t/2} + c_2te^{t/2} + c_3t^2e^{t/2}$.

8.163 Find the solution of a fourth-order linear homogeneous differential equation for $x(t)$ with real coefficients if its characteristic equation has $\lambda = \frac{1}{2}$ as a quadruple root.

▌ The solution is $x = e^{t/2}(c_1 + c_2t + c_3t^2 + c_4t^3)$.

8.164 Determine a differential equation associated with the previous problem.

▌ The characteristic equation is

$$0 = (\lambda - \tfrac{1}{2})^5 = \lambda^5 - \tfrac{5}{2}\lambda^4 + \tfrac{5}{2}\lambda^3 - \tfrac{5}{4}\lambda^2 + \tfrac{5}{16}\lambda - \tfrac{1}{32} \qquad \text{or} \qquad 32\lambda^5 - 80\lambda^4 + 80\lambda^3 - 40\lambda^2 + 10\lambda - 1 = 0$$

An associated differential equation is $(32D^5 - 80D^4 + 80D^3 - 40D^2 + 10D - 1)x = 0$.

8.165 Find the solution of a sixth-order linear homogeneous differential equation for $x(t)$ with real coefficients if its characteristic equation has $\lambda = -8$ as a root of multiplicity six.

▌ The solution is $x = e^{-8t}(c_1 + c_2t + c_3t^2 + c_4t^3 + c_5t^4 + c_6t^5)$.

8.166 Solve the previous problem if the differential equation has order 10.

▌ A tenth-order differential equation must have 10 roots to its characteristic equation, and since we are given only six of them we cannot solve the problem. We can, however, say that the solution to the previous problem will be *part* of the solution to this problem.

8.167 Determine a differential equation associated with Problem 8.165.

▌ The characteristic equation is

$$0 = (\lambda + 8)^6 = \lambda^6 + 48\lambda^5 + 960\lambda^4 + 10{,}240\lambda^3 + 61{,}440\lambda^2 + 196{,}608\lambda + 262{,}144$$

An associated differential equation is

$$x^{(6)} + 48x^{(5)} + 960x^{(4)} + 10{,}240x^{(3)} + 61{,}440\ddot{x} + 196{,}608\dot{x} + 262{,}144x = 0$$

8.168 Find the solution of a ninth-order linear homogeneous differential equation for $x(t)$ with real coefficients if its characteristic equation has a single root of multiplicity nine.

▌ Denote the root by λ. The solution is $x = e^{\lambda t}(c_1 + c_2t + c_3t^2 + c_4t^3 + c_5t^4 + c_6t^5 + c_7t^6 + c_8t^7 + c_9t^8)$.

CHARACTERISTIC ROOTS OF VARIOUS TYPES

8.169 Solve $\dfrac{d^3y}{dx^3} - 3\dfrac{d^2y}{dx^2} + 4y = 0$.

▌ The characteristic equation is $m^3 - 3m^2 + 4 = 0$; its roots are $m_1 = m_2 = 2$ and $m_3 = -1$. Since the first two are repeated, $c_1e^{2x} + c_2xe^{2x}$ is part of the solution. The last root, which is distinct from the others, gives c_3e^{-x} as part of the solution. The complete solution is then $y = c_1e^{2x} + c_2xe^{2x} + c_3e^{-x}$.

8.170 Solve $\dfrac{d^3y}{dx^3} + 4\dfrac{d^2y}{dx^2} - 3\dfrac{dy}{dx} - 18y = 0$.

▌ The characteristic equation is $m^3 + 4m^2 - 3m - 18 = 0$, with roots -3, -3, and 2. Since the first two are repeated, $c_1e^{-3x} + c_2xe^{-3x}$ is part of the general solution. The last root, which is distinct from the others, gives c_3e^{2x} as part of the solution. The complete solution is then $y = c_1e^{-3x} + c_2xe^{-3x} + c_3e^{2x}$.

8.171 Solve $(4D^3 - 28D^2 - 31D - 8)y = 0$.

▌ The characteristic equation is $4m^3 - 28m^2 - 31m - 8 = 0$, with roots $-\frac{1}{2}$, $-\frac{1}{2}$, and 8. The general solution is $y = c_1e^{-x/2} + c_2xe^{-x/2} + c_3e^{8x}$.

8.172 Find the general solution to a fourth-order linear homogeneous differential equation for $y(x)$ with real coefficients if the roots of its characteristic equation are -1, -1, -1, and 2.

▌ Since the first three roots are repeated, $e^{-x}(c_1 + c_2x + c_3x^2)$ is part of the general solution. The last root, which is distinct from the others, adds c_4e^{2x} to the general solution. The primitive is then $y = e^{-x}(c_1 + c_2x + c_3x^2) + c_4e^{2x}$.

8.173 Solve Problem 8.172 if the roots are -1, -1, 2, and 2.

▌ The first two roots are repeated and contribute $e^{-x}(c_1 + c_2x)$ to the general solution. The last two roots, which are different from the first two and are also repeated, contribute $e^{2x}(c_3 + c_4x)$ to the general solution. The complete solution is $y = e^{-x}(c_1 + c_2x) + e^{2x}(c_3 + c_4x)$.

8.174 Solve Problem 8.172 if the roots are $\frac{1}{2}$, $\frac{1}{2}$, 3, and 3.

▌ The general solution is $y = e^{x/2}(c_1 + c_2x) + e^{3x}(c_3 + c_4x)$.

8.175 Solve Problem 8.172 if the roots are $\pm i2$ and $\pm i2$.

▌ The root $i2$ has multiplicity two, so it contributes $d_1e^{i2x} + d_2xe^{i2x}$ to the general solution. Similarly, the double root $-i2$ contributes $d_3e^{-i2x} + d_4xe^{-i2x}$ to the general solution. The complete solution is

$$y = d_1e^{i2x} + d_2xe^{i2x} + d_3e^{-i2x} + d_4xe^{-i2x} = (d_1e^{i2x} + d_3e^{-i2x}) + x(d_2e^{i2x} + d_3e^{-i2x})$$

Using Euler's relations (see Problem 8.49) on each set of terms in parentheses, we can rewrite the latter solution as $y = (c_1 \cos 2x + c_2 \sin 2x) + x(c_3 \cos 2x + c_4 \sin 2x)$.

8.176 Solve Problem 8.172 if the roots are $3 \pm i5$ and $3 \pm i5$.

▌ The root $3 + i5$ has multiplicity two, so it contributes $d_1e^{(3+i5)x} + d_2xe^{(3+i5)x}$ to the general solution. Similarly, the double root $3 - i5$ contributes $d_3e^{(3-i5)x} + d_4xe^{(3-i5)x}$ to the general solution. The complete solution is

$$\begin{aligned} y &= d_1e^{(3+i5)x} + d_2xe^{(3+i5)x} + d_3e^{(3-i5)x} + d_4xe^{(3-i5)x} \\ &= e^{3x}[(d_1e^{i5x} + d_3e^{-i5x}) + x(d_2e^{i5x} + d_4e^{-i5x})] \\ &= e^{3x}[c_1 \cos 5x + c_2 \sin 5x + x(c_3 \cos 5x + c_4 \sin 5x)] \end{aligned}$$

8.177 Solve Problem 8.172 if the roots are $\frac{1}{2} \pm i3$ and $\frac{1}{2} \pm i3$.

▌ The general solution is $y = e^{x/2}[c_1 \cos 3x + c_2 \sin 3x + x(c_3 \cos 3x + c_4 \sin 3x)]$.

8.178 Solve Problem 8.172 if the roots are $-1 \pm i$ and $-1 \pm i$.

▌ The general solution is $y = e^{-x}[c_1 \cos x + c_2 \sin x + x(c_3 \cos x + c_4 \sin x)]$.

8.179 Solve Problem 8.172 if the roots are $7 \pm i23$ and $7 \pm i23$.

▌ The general solution is $y = e^{7x}(c_1 \cos 23x + c_2 \sin 23x) + xe^{7x}(c_3 \cos 23x + c_4 \sin 23x)$.

8.180 Solve Problem 8.172 if the roots are $\pm i\frac{1}{4}$ and $\pm i\frac{1}{4}$.

▌ The general solution is $y = c_1 \cos \frac{1}{4}x + c_2 \sin \frac{1}{4}x + x(c_3 \cos \frac{1}{4}x + c_4 \sin \frac{1}{4}x)$.

8.181 Solve Problem 8.172 if the roots are ± 4 and ± 4.

▌ There are two real roots of multiplicity two, so the general solution is $y = e^{4x}(c_1 + c_2x) + e^{-4x}(c_3 + xc_4)$.

8.182 Solve Problem 8.172 if the roots are -6, -6, and $2 \pm i4$.

▌ The double real root -6 contributes $e^{-6x}(c_1 + c_2x)$ to the general solution; the distinct complex roots contribute $e^{2x}(c_3 \cos 4x + c_4 \sin 4x)$. The complete solution is $y = e^{-6x}(c_1 + c_2x) + e^{2x}(c_3 \cos 4x + c_4 \sin 4x)$.

8.183 Solve Problem 8.172 if the roots are 2, 2, and $\pm i2$.

▌ The general solution is $y = e^{2x}(c_1 + xc_2) + c_3 \cos 2x + c_4 \sin 2x$.

8.184 Solve $(D^4 + 6D^3 + 5D^2 - 24D - 36)y = 0$.

▌ This may be rewritten as $(D - 2)(D + 2)(D + 3)^2y = 0$, which has characteristic roots 2, -2, -3, and -3. The primitive is $y = C_1e^{2x} + C_2e^{-2x} + C_3e^{-3x} + C_4xe^{-3x}$.

8.185 Solve $(D^4 - D^3 - 9D^2 - 11D - 4)y = 0$.

▌ This may be rewritten as $(D + 1)^3(D - 4)y = 0$, which has characteristic roots -1, -1, -1, and 4. The primitive is $y = e^{-x}(C_1 + C_2x + C_3x^2) + C_4e^{4x}$.

8.186 Solve $(D^4 + 4D^2)y = 0$.

▌ This may be rewritten as $D^2(D^2 + 4)y = 0$, which has the characteristic roots 0, 0, and $\pm i2$. The general solution is $y = c_1 + c_2x + c_3 \cos 2x + c_4 \sin 2x$.

8.187 Solve $(D^4 - 6D^3 + 13D^2 - 12D + 4)y = 0$.

▌ This differential equation may be rewritten as $(D - 1)^2(D - 2)^2y = 0$, which has the characteristic roots 1, 1, 2, and 2. The primitive is $y = e^x(c_1 + c_2x) + e^{2x}(c_3 + c_4x)$.

.188 Solve $y^{(4)} - 8y''' + 32y'' - 64y' + 64y = 0$.

▌ The characteristic equation $m^4 - 8m^3 + 32m^2 - 64m + 64 = 0$ has roots $2 \pm i2$ and $2 \pm i2$; hence $\lambda_1 = 2 + i2$ and $\lambda_2 = 2 - i2$ are both roots of multiplicity two. The solution is

$$y = e^{2x}(c_1 \cos 2x + c_3 \sin 2x) + xe^{2x}(c_2 \cos 2x + c_4 \sin 2x) = (c_1 + c_2 x)e^{2x} \cos 2x + (c_3 + c_4 x)e^{2x} \sin 2x$$

.189 Solve $\dfrac{d^4y}{dx^4} - 12\dfrac{d^3y}{dx^3} + 56\dfrac{d^2y}{dx^3} - 120\dfrac{dy}{dx} + 100y = 0$.

▌ The characteristic equation has roots $3 \pm i$ and $3 \pm i$, so both $3 + i$ and $3 - i$ are roots of multiplicity two. The solution is $y = (c_1 + c_2 x)e^{3x} \cos x + (c_3 + c_4 x)e^{3x} \sin x$.

.190 Solve $\dfrac{d^4y}{dx^4} + 4\dfrac{d^3y}{dx^3} + 14\dfrac{d^2y}{dx^2} + 20\dfrac{dy}{dx} + 25y = 0$.

▌ The characteristic equation has roots $-1 \pm i2$ and $-1 \pm i2$, so both $-1 + i2$ and $-1 - i2$ are roots of multiplicity two. The solution is $y = (c_1 + c_2 x)e^{-x} \cos 2x + (c_3 + c_4 x)e^{-x} \sin 2x$.

191 Solve $y^{(4)} + 8y'' + 16y = 0$.

▌ The characteristic equation has roots $\pm i2$ and $\pm i2$. The solution is $y = (c_1 + c_2 x) \cos 2x + (c_3 + c_4 x) \sin 2x$.

.192 Solve $y^{(4)} + 14y'' + 49y = 0$.

▌ The characteristic equation has roots $m_1 = i7$ and $m_2 = -i7$, both of multiplicity two. The solution is $y = (c_1 + c_2 x) \cos 7x + (c_3 + c_4 x) \sin 7x$.

.193 Find the general solution of a sixth-order linear homogeneous differential equation for $y(x)$ with real coefficients if its characteristic equation has roots $i5$ and $-i5$, each with multiplicity three.

▌ The solution is

$$\begin{aligned} y &= e^{i5x}(d_1 + d_2 x + d_3 x^2) + e^{-i5x}(d_4 + d_5 x + d_6 x^2) \\ &= (d_1 e^{i5x} + d_4 e^{-i5x}) + x(d_2 e^{i5x} + d_5 e^{-i5x}) + x^2(d_3 e^{i5x} + d_6 e^{-i5x}) \\ &= (c_1 \cos 5x + c_4 \sin 5x) + x(c_2 \cos 5x + c_5 \sin 5x) + x^2(c_3 \cos 5x + c_6 \sin 5x) \\ &= (c_1 + c_2 x + c_3 x^2) \cos 5x + (c_4 + c_5 x + c_6 x^2) \sin 5x \end{aligned}$$

194 Solve the previous problem if the roots are $i3$ and $-i3$, each of multiplicity three.

▌ The solution is $y = (c_1 + c_2 x + c_3 x^2) \cos 3x + (c_4 + c_5 x + c_6 x^2) \sin 3x$.

195 Solve Problem 8.193 if the roots are $-3 + i5$ and $-3 - i5$, each of multiplicity three.

▌ The solution is $y = (c_1 + c_2 x + c_3 x^2)e^{-3x} \cos 5x + (c_4 + c_5 x + c_6 x^2)e^{-3x} \sin 5x$.

196 Solve Problem 8.193 if the roots are $0.2 + i0.7$ and $0.2 - i0.7$, each of multiplicity three.

▌ The solution is $y = (c_1 + c_2 x + c_3 x^2)e^{0.2x} \cos 0.7x + (c_4 + c_5 x + c_6 x^2)e^{0.2x} \sin 0.7x$.

197 Find the general solution of an eighth-order linear homogeneous differential equation for $y(x)$ with real coefficients if its characteristic equation has roots $\frac{1}{2} + i3$ and $\frac{1}{2} - i3$, each with multiplicity four.

▌ The solution is $y = (c_1 + c_2 x + c_3 x^2 + c_4 x^3)e^{x/2} \cos 3x + (c_5 + c_6 x + c_7 x^2 + c_8 x^3)e^{x/2} \sin 3x$.

198 Solve the previous problem if the roots are $-3 + i\frac{1}{4}$ and $-3 - i\frac{1}{4}$, each of multiplicity four.

▌ The solution is $y = (c_1 + c_2 x + c_3 x^2 + c_4 x^3)e^{-3x} \cos \frac{1}{4}x + (c_5 + c_6 x + c_7 x^2 + c_8 x^3)e^{-3x} \sin \frac{1}{4}x$.

199 Solve Problem 8.197 if the roots are $i6$ and $-i6$, each of multiplicity four.

▌ The solution is $y = (c_1 + c_2 x + c_3 x^2 + c_4 x^3) \cos 6x + (c_5 + c_6 x + c_7 x^2 + c_8 x^3) \sin 6x$.

200 Solve Problem 8.197 if the roots are $2 \pm i2$ and $-2 \pm i2$, each of multiplicity two.

▮ Since $2+i2$ and $2-i2$ are both roots of multiplicity two, they contribute $(c_1+c_2x)e^{2x}\cos 2x+(c_3+c_4x)e^{2x}\sin 2x$ to the general solution (see Problem 8.188). Similarly, since both $-2+i2$ and $-2-i2$ are roots of multiplicity two, they contribute $(c_5+c_6x)e^{-2x}\cos 2x+(c_7+c_8x)e^{-2x}\sin 2x$ to the general solution. The complete solution is the sum of these two contributions, namely

$$y=(c_1+c_2x)e^{2x}\cos 2x+(c_3+c_4x)e^{2x}\sin 2x+(c_5+c_6x)e^{-2x}\cos 2x+(c_7+c_8x)e^{-2x}\sin 2x$$

8.201 Solve Problem 8.197 if the roots are $3\pm i2$ and $4\pm i5$, each of multiplicity two.

▮ The solution is $y=(c_1+c_2x)e^{3x}\cos 2x+(c_3+c_4x)e^{3x}\sin 2x+(c_5+c_6x)e^{4x}\cos 5x+(c_7+c_8x)e^{4x}\sin 5x$.

8.202 Solve Problem 8.197 if the roots are $-3\pm i5$, each of multiplicity three, and $-5\pm i6$, each of multiplicity one.

▮ Since $-3+i5$ and $-3-i5$ are both roots of multiplicity three, they contribute $(c_1+c_2x+c_3x^2)e^{-3x}\cos 5x+(c_4+c_5x+c_6x^2)e^{-3x}\sin 5x$ to the general solution (see Problem 8.195). In contrast, $-5\pm i6$ are both simple roots, so they contribute $c_7e^{-5x}\cos 6x+c_8e^{-5x}\sin 6x$ to the general solution. The complete solution is the sum of these two contributions, namely

$$y=(c_1+c_2x+c_3x^2)e^{-3x}\cos 5x+(c_4+c_5x+c_6x^2)e^{-3x}\sin 5x+c_7e^{-5x}\cos 6x+c_8e^{-5x}\sin 6x$$

8.203 Solve Problem 8.197 if the roots are $-16\pm i25$, each of multiplicity three, and $-\frac{1}{2}$ of multiplicity two.

▮ The solution is $y=(c_1+c_2x+c_3x^2)e^{-16x}\cos 25x+(c_4+c_5x+c_6x^2)e^{-16x}\sin 25x+(c_7+c_8x)e^{-x/2}$.

8.204 Solve Problem 8.197 if the roots are $-16\pm i25$, each of multiplicity two, and $-\frac{1}{2}$ of multiplicity four.

▮ The solution is $y=(c_1+c_2x)e^{-16x}\cos 25x+(c_3+c_4x)e^{-16x}\sin 25x+(c_5+c_6x+c_7x^2+c_8x^3)e^{-x/2}$.

8.205 Solve $y^{(5)}-y^{(4)}-2y'''+2y''+y'-y=0$.

▮ The characteristic equation can be factored into $(\lambda-1)^3(\lambda+1)^2=0$; hence, $\lambda_1=1$ is a root of multiplicity three and $\lambda_2=-1$ is a root of multiplicity two. The solution is $y=c_1e^x+c_2xe^x+c_3x^2e^x+c_4e^{-x}+c_5xe^{-x}$.

8.206 Find the general solution of a fifth-order linear homogeneous differential equation for $y(x)$ with real coefficients if its characteristic equation has roots 2, 2, -3, -3, and 4.

▮ The solution is $y=(c_1+c_2x)e^{2x}+(c_3+c_4x)e^{-3x}+c_5e^{4x}$.

8.207 Solve the previous problem if the roots are 2, -3, -3, -3, and 4.

▮ The solution is $y=c_1e^{2x}+c_2e^{4x}+(c_3+c_4x+c_5x^2)e^{-3x}$.

8.208 Solve Problem 8.206 if the roots are 2, 2, 2, 2, and 4.

▮ The solution is $y=(c_1+c_2x+c_3x^2+c_4x^3)e^{2x}+c_5e^{4x}$.

8.209 Solve Problem 8.206 if the roots are 2, 2, 2, and $3\pm i4$.

▮ The solution is $y=(c_1+c_2x+c_3x^2)e^{2x}+e^{3x}(c_4\cos 4x+c_5\sin 4x)$.

8.210 Solve Problem 8.206 if the roots are 2, $3\pm i4$, and $3\pm i4$.

▮ The solution is $y=c_1e^{2x}+(c_2+c_3x)e^{3x}\cos 4x+(c_4+c_5x)e^{3x}\sin 4x$.

8.211 Solve Problem 8.206 if the roots are 2, 2, 2, 2, and $3+i4$.

▮ This cannot be. Since the coefficients of the differential equation are real, the complex roots must occur in conjugate pairs. Thus, if $3+i4$ is a root, then $3-i4$ is also a root. We now have six characteristic roots for a fifth-order differential equation, which is impossible.

8.212 Find the general solution of a sixth-order linear homogeneous differential equation in $x(t)$ with real coefficients if its characteristic equation has roots 1, 2, 2, 3, 3, and 3.

▮ The solution is $x=c_1e^t+(c_2+c_3t)e^{2t}+(c_4+c_5t+c_6t^2)e^{3t}$.

8.213 Solve Problem 8.212 if the roots are 2, 2, 3, 3, 3, and 3.

▌ The solution is $x = (c_1 + c_2 t)e^{2t} + (c_3 + c_4 t + c_5 t^2 + c_6 t^3)e^{3t}$.

8.214 Solve Problem 8.212 if the roots are 2, 3, 3, 3, 3, and 3.

▌ The solution is $x = c_1 e^{2t} + (c_2 + c_3 t + c_4 t^2 + c_5 t^3 + c_6 t^4)e^{3t}$.

8.215 Solve Problem 8.212 if the roots are $-3, -3, -3, -3$, and $-3 \pm i\pi$.

▌ The solution is $x = (c_1 + c_2 t + c_3 t^2 + c_4 t^3)e^{-3t} + e^{-3t}(c_5 \cos \pi t + c_6 \sin \pi t)$.

8.216 Solve Problem 8.212 if the roots are $-3, -3, -3 \pm i\pi$, and $-3 \pm i\pi$.

▌ The solution is $x = (c_1 + c_2 t)e^{-3t} + (c_3 + c_4 t)e^{-3t} \cos \pi t + (c_5 + c_6 t)e^{-3t} \sin \pi t$.

8.217 Solve Problem 8.212 if the roots are $2, 3, -3 \pm i\pi$, and $-3 \pm i\pi$.

▌ The solution is $x = c_1 e^{2t} + c_2 e^{3t} + (c_3 + c_4 t)e^{-3t} \cos \pi t + (c_5 + c_6 t)e^{-3t} \sin \pi t$.

8.218 Solve Problem 8.212 if the roots are $-2 \pm i\pi, -3 \pm i\pi$, and $-3 \pm i\pi$.

▌ The solution is $x = c_1 e^{-2t} \cos \pi t + c_2 e^{-2t} \sin \pi t + (c_3 + c_4 t)e^{-3t} \cos \pi t + (c_5 + c_6 t)e^{-3t} \sin \pi t$.

8.219 Solve Problem 8.212 if the roots are $-3 \pm i\pi, -3 \pm i\pi$, and $-3 \pm i\pi$.

▌ The solution is $x = (c_1 + c_2 t + c_3 t^2)e^{-3t} \cos \pi t + (c_4 + c_5 t + c_6 t^2)e^{-3t} \sin \pi t$.

8.220 Solve $(D + 2)^3(D - 3)^4(D^2 + 2D + 5)y = 0$.

▌ The auxiliary equation $(m + 2)^3(m - 3)^4(m^2 + 2m + 5) = 0$ has roots $-2, -2, -2, 3, 3, 3, 3$, and $-1 \pm 2i$. The general solution is

$$y = (c_1 + c_2 x + c_3 x^2)e^{-2x} + (c_4 + c_5 x + c_6 x^2 + c_7 x^3)e^{3x} + e^{-x}(c_8 \cos 2x + c_9 \sin 2x)$$

8.221 Find the general solution to a fourth-order linear homogeneous differential equation for $x(t)$ with real coefficients if two particular solutions are $3e^{2t}$ and $6t^2 e^{-t}$.

▌ To have $3e^{2t}$ as a solution, $m = 2$ must be a characteristic root. To have $6t^2 e^{-t}$ as a solution, $m = -1$ must be a root *of multiplicity three.* Thus, we know four characteristic roots, which is the complete set for this differential equation. The general solution is $x = c_1 e^{2t} + (c_2 + c_3 t + c_4 t^2)e^{-t}$.

8.222 Solve the previous problem if two particular solutions are $\frac{1}{2}te^{-t}$ and $-8te^{2t}$.

▌ To generate these solutions both $m_1 = -1$ and $m_2 = 2$ must be roots *of multiplicity two.* Thus, we know four characteristic roots, which is the complete set here. The general solution is $x = (c_1 + c_2 t)e^{-t} + (c_3 + c_4 t)e^{2t}$.

8.223 Determine the differential equation associated with the previous problem.

▌ The characteristic equation is $[m - (-1)]^2(m - 2)^2 = 0$, so the differential equation is $(D + 1)^2(D - 2)^2 x = 0$. This may be expanded to $(D^4 - 2D^3 - 3D^2 + 4D + 4)x = 0$.

8.224 Determine the form of the general solution to a fifth-order linear homogeneous differential equation for $x(t)$ with real coefficients if a particular solution is $t^3 e^{3t}$.

▌ To have $t^3 e^{3t}$ as a solution, $m = 3$ must be a root of at least multiplicity four. Thus, four roots of the characteristic equation are 3, 3, 3, and 3. Since the differential equation is of order 5, there must be one additional real root. Denote it as k.

If $k = 3$, then 3 is a root of multiplicity five, and the general solution is $y = (c_1 + c_2 t + c_3 t^2 + c_4 t^3 + c_5 t^4)e^{3t}$. If $k \neq 3$, then the general solution is $y = (c_1 + c_2 t + c_3 t^2 + c_4 t^3)e^{3t} + c_5 e^{kt}$.

8.225 Find the general solution to a sixth-order linear homogeneous differential equation for $x(t)$ with real coefficients if one solution is $t^2 \sin t$.

▮ The t^2 portion of the given particular solution implies that the associated characteristic root has multiplicity three. Since $\sin t$ can be generated only from roots $\pm i$, which must occur in conjugate pairs, it follows that both $\pm i$ are roots of multiplicity three. Since this yields six roots, we have the complete set, and the general solution is $x = (c_1 + c_2t + c_3t^2)\cos t + (c_4 + c_5t + c_6t^2)\sin t$.

8.226 Solve the previous problem if two particular solutions are $\sin t$ and $4te^{-3t}\cos 2t$.

▮ The particular solution $\sin t$ can be generated only from the characteristic roots $\pm i$. Similarly, $e^{-3t}\cos 2t$ can be generated only from the characteristic roots $-3 \pm i2$. Since this function is multiplied by t, it follows that $-3 \pm i2$ are both roots of at least multiplicity two. Thus, we have identified as characteristic roots $\pm i$, $-3 \pm i2$, and $-3 \pm i2$. These six roots form a complete set of characteristic roots for a sixth-order differential equation, so the general solution is

$$x = c_1 \cos t + c_2 \sin t + (c_3 + c_4t)e^{-3t}\cos 2t + (c_5 + c_6t)e^{-3t}\sin 2t$$

8.227 Determine the differential equation associated with Problem 8.225.

▮ The characteristic equation is $(m - i)^3[m - (-i)]^3 = 0$, or $(m^2 + 1)^3 = 0$. The corresponding differential equation is $(D^2 + 1)^3x = 0$, which may be expanded to $(D^6 + 3D^4 + 3D^2 + 1)x = 0$.

8.228 Find the general solution of a twelfth-order linear homogeneous differential equation for $x(t)$ with real coefficients if its characteristic equation has roots 1, 2, 2, 3, 3, 3, $\pm i$, $2 \pm i3$, $2 \pm i3$.

▮ The solution is

$$x = c_1e^t + (c_2 + c_3t)e^{2t} + (c_4 + c_5t + c_6t^2)e^{3t} + c_7\cos t + c_8 \sin t + (c_9 + c_{10}t)e^{2t}\cos 3t + (c_{11} + c_{12}t)e^{2t}\sin 3t$$

8.229 Solve the previous problem if the roots are 0, 0, $\pm i$, $\pm i$, $2 \pm i3$, $2 \pm i3$, $2 \pm i3$.

▮ The solution is

$$x = c_1 + c_2t + (c_3 + c_4t)\cos t + (c_5 + c_6t)\sin t + (c_7 + c_8t + c_9t^2)e^{2t}\cos 3t + (c_{10} + c_{11}t + c_{12}t^2)e^{2t}\sin 3t$$

EULER'S EQUATION

8.230 Develop a method for obtaining nontrivial solutions to Euler's equation,

$$b_nx^ny^{(n)} + b_{n-1}x^{n-1}y^{(n-1)} + \cdots + b_2x^2y'' + b_1xy' + b_0y = 0$$

where b_j $(j = 0, 1, \ldots, n)$ is a constant.

▮ An Euler equation can always be transformed into a linear differential equation with constant coefficients through the change of variables $z = \ln x$ or $x = e^z$. With the notation $D = \dfrac{d}{dz}$, it follows from this equation and the chain rule that

$$\frac{dy}{dx} = \frac{dy}{dz}\frac{dz}{dx} = \frac{dy}{dz}\frac{1}{x} = \frac{1}{x}Dy$$

$$\frac{d^2y}{dx^2} = \frac{d}{dx}\left(\frac{1}{x}Dy\right) = -\frac{1}{x^2}Dy + \frac{1}{x}\frac{d}{dx}(Dy) = -\frac{1}{x^2}Dy + \frac{1}{x}\frac{d}{dz}(Dy)\frac{dz}{dx} = -\frac{1}{x^2}Dy + \frac{1}{x}D(Dy)\frac{1}{x} = \frac{1}{x^2}D(D-1)y$$

Similarly, $\dfrac{d^3y}{dx^3} = \dfrac{1}{x^3}D(D-1)(D-2)y$, and in general

$$\frac{d^ny}{dx^n} = \frac{1}{x^n}D(D-1)(D-2)(D-3)\cdots(D-n+1)y$$

By substituting these derivatives into an Euler equation, we obtain a linear differential equation without variable coefficients, which may be solved like the other problems in this chapter.

8.231 Solve $2x^2y'' + 11xy' + 4y = 0$.

▮ This is an Euler equation. If we set $x = e^z$, it follows from Problem 8.230 that $y' = \dfrac{1}{x}Dy$ and $y'' = \dfrac{1}{x^2}D(D-1)y$, and the given differential equation becomes $2D(D-1)y + 11Dy + 4y = 0$ or $(2D^2 + 9D + 4)y = 0$. Now all derivatives are taken with respect to z. From the result of Problem 8.23 (with

z replacing x), the solution is

$$y = Ae^{z/2} + Be^{4z} = A(e^z)^{1/2} + B(e^z)^4 = Ax^{1/2} + Bx^4$$

232 Solve $x^2y'' - 2y = 0$.

This is an Euler equation with $b_1 = 0$. If we set $x = e^z$, it follows from Problem 8.230 that $y' = \frac{1}{x}Dy$ and $y'' = \frac{1}{x^2}D(D-1)y$, and the given differential equation becomes $D(D-1)y - 2y = 0$ or $(D^2 - D - 2)y = 0$. Now all derivatives are taken with respect to z. From the result of Problem 8.1 (with z replacing x), the solution is

$$y = c_1e^{-z} + c_2e^{2z} = c_1(e^z)^{-1} + c_2(e^z)^2 = c_1x^{-1} + c_2x^2$$

233 Solve $x^2y'' - 6xy' = 0$.

This is an Euler equation with $b_0 = 0$. If we set $x = e^z$, it follows from Problem 8.230 that $y' = \frac{1}{x}Dy$ and $y'' = \frac{1}{x^2}D(D-1)y$, so that the given differential equation becomes $D(D-1)y - 6Dy = 0$ or $(D^2 - 7D)y = 0$. Now all derivatives are taken with respect to z. By Problem 8.2, the solution is $y = c_1 + c_2e^{7z} = c_1 + c_2(e^z)^7 = c_1 + c_2x^7$.

234 Solve $x^2y'' + xy' - 5y = 0$.

This is an Euler equation. If we set $x = e^z$, it follows that $y' = \frac{1}{x}Dy$ and $y'' = \frac{1}{x^2}D(D-1)y$, so that the given differential equation becomes $D(D-1)y + Dy - 5y = 0$ or $(D^2 - 5)y = 0$. The independent variable in this last equation is z, and by Problem 8.3 the solution is

$$y = c_1e^{\sqrt{5}z} + c_2e^{-\sqrt{5}z} = c_1(e^z)^{\sqrt{5}} + c_2(e^z)^{-\sqrt{5}} = c_1x^{\sqrt{5}} + c_2x^{-\sqrt{5}}$$

235 Solve $x^2y'' + 5xy' + 4y = 0$.

This is an Euler equation. Using the substitutions suggested in Problem 8.230, we obtain $D(D-1)y + 5Dy + 4y = 0$ or $(D^2 + 4D + 4)y = 0$. This equation is similar in form to that given in Problem 8.141, except now the independent variable is z; its solution is $y = c_1e^{-2z} + c_2ze^{-2z} = c_1(e^z)^{-2} + c_2z(e^z)^{-2}$. Since $x = e^z$ and $z = \ln x$, it follows that $y = c_1x^{-2} + c_2(\ln x)x^{-2} = (c_1 + c_2 \ln x)/x^2$.

236 Solve $x^2y'' - xy' + y = 0$.

This is an Euler equation. Using the substitutions suggested in Problem 8.230, we obtain $D(D-1)y - Dy + y = 0$ or $(D^2 - 2D + 1)y = 0$. This equation is similar in form to that given in Problem 8.143, except now the independent variable is z; its solution is $y = c_1e^z + c_2ze^z$. But $x = e^z$ and $z = \ln x$, so this solution may be rewritten as $y = c_1x + c_2x \ln x$.

237 Solve $x^2y'' + xy' + 4y = 0$.

This is an Euler equation. Using the substitutions suggested in Problem 8.230, we rewrite the differential equation as $D(D-1)y + Dy + 4y = 0$ or $(D^2 + 4)y = 0$. This equation is similar in form to that given in Problem 8.59, except now the independent variable is z; its solution is

$$y = c_1 \cos 2z + c_2 \sin 2z = c_1 \cos(2 \ln x) + c_2 \sin(2 \ln x) = c_1 \cos(\ln x^2) + c_2 \sin(\ln x^2)$$

238 Solve $x^2y'' + xy' + 50y = 0$.

This is an Euler equation. Using the substitutions suggested in Problem 8.230, we rewrite the differential equation as $D(D-1)y + Dy + 50y = 0$ or $(D^2 + 50)y = 0$. This equation is similar in form to that given in Problem 8.60, except now the independent variable is z; its solution is $y = c_1 \cos\sqrt{50}z + c_2 \sin\sqrt{50}x = c_1 \cos(\sqrt{50}\ln x) + c_2 \sin(\sqrt{50}\ln x)$.

239 Solve $x^2y'' - 5xy' + 25y = 0$.

This is an Euler equation. Using the substitutions suggested in Problem 8.230, we rewrite the differential equation as $D(D-1)y - 5Dy + 25y = 0$ or $(D^2 - 6D + 25)y = 0$. This equation is similar in form to that given in Problem 8.50, except now the independent variable is z; its solution is $y = e^{3z}(c_1 \cos 4z + c_2 \sin 4z) = (e^z)^3(c_1 \cos 4z + c_2 \sin 4z)$. But since $x = e^z$ and $z = \ln x$, we have $y = x^3[c_1 \cos(4 \ln x) + c_2 \sin(4 \ln x)] = x^3[c_1 \cos(\ln x^4) + c_2 \sin(\ln x^4)]$.

8.240 Solve $x^3y''' - 3x^2y'' + 6xy' - 6y = 0$.

This is an Euler equation. Using the substitutions suggested in Problem 8.230, we rewrite the differential equation as $D(D-1)(D-2)y - 3D(D-1)y + 6Dy - 6y = 0$, or $(D^3 - 6D^2 + 11D - 6)y = 0$. This equation is similar in form to that given in Problem 8.28, except now the independent variable is z; its solution is

$$y = c_1e^z + c_2e^{2z} + c_3e^{3z} = c_1e^z + c_2(e^z)^2 + c_3(e^z)^3 = c_1x + c_2x^2 + c_3x^3$$

8.241 Solve $x^3y''' - 2xy' + 4y = 0$.

This is an Euler equation with the coefficient of x^2y'' equal to zero. Using the substitutions suggested in Problem 8.230, we rewrite the differential equation as $D(D-1)(D-2)y - 2Dy + 4y = 0$ or $(D^3 - 3D^2 + 4)y = 0$. This equation is similar in form to that given in Problem 8.169, except now the independent variable is z; its solution is

$$y = c_1e^{2z} + c_2ze^{2z} + c_3e^{-z} = c_1(e^z)^2 + c_2z(e^z)^2 + c_3(e^z)^{-1} = c_1x^2 + c_2x^2 \ln x + c_3x^{-1}.$$

8.242 Solve $4x^3y''' - 16x^2y'' - 55xy' - 8y = 0$.

This is an Euler equation. Using the substitutions suggested in Problem 8.230, we rewrite the differential equation as $4D(D-1)(D-2)y - 16D(D-1) - 55Dy - 8y = 0$, or $(4D^3 - 28D^2 - 31D - 8)y = 0$. This equation is identical in form to that given in Problem 8.171, except now the independent variable is z; its solution is $y = c_1e^{-z/2} + c_2ze^{-z/2} + c_3e^{8z}$. Since $x = e^z$ and $z = \ln x$, this solution becomes $y = c_1x^{-1/2} + c_2x^{-1/2} \ln x + c_3x^8$.

8.243 Solve $x^3y''' + 2x^2y'' + 4xy' - 4y = 0$.

This is an Euler equation. Using the substitutions suggested in Problem 8.230, we rewrite the differential equation as $D(D-1)(D-2)y + 2D(D-1)y + 4Dy - 4y = 0$, or $(D^3 - D^2 + 4D - 4)y = 0$. Its characteristic equation is $\lambda^3 - \lambda^2 + 4\lambda - 4 = 0$, which has as its roots 1 and $\pm i2$. The solution is

$$y = c_1e^z + c_2 \cos 2z + c_3 \sin 2z = c_1x + c_2 \cos(2 \ln x) + c_3 \sin(2 \ln x) = c_1x + c_2 \cos(\ln x^2) + c_3 \sin(\ln x^2)$$

8.244 Solve $x^3y''' - 3x^2y'' - 3xy' + 36y = 0$.

This is an Euler equation. Using the substitutions suggested in Problem 8.230, we rewrite the differential equation as $D(D-1)(D-2)y - 3D(D-1)y - 3Dy + 36y = 0$, or $(D^3 - 6D^2 + 2D + 36)y = 0$. This equation is similar in form to that given in Problem 8.110, except now the independent variable is z; its solution is

$$y = c_1e^{-2z} + c_2e^{4z} \cos\sqrt{2}z + c_3e^{4z} \sin\sqrt{2}z = c_1x^{-2} + c_2x^4 \cos(\sqrt{2} \ln x) + c_3x^4 \sin(\sqrt{2} \ln x)$$

8.245 Solve $x^4y^{(4)} + 14x^3y^{(3)} + 55x^2y'' + 65xy' + 16y = 0$.

This is an Euler equation. Using the substitutions suggested in Problem 8.230, we rewrite the differential equation as

$$D(D-1)(D-2)(D-3)y + 14D(D-1)(D-2) + 55D(D-1) + 65Dy + 16y = 0$$

or $(D^4 + 8D^3 + 24D^2 + 32D + 16)y = 0$. This equation is identical in form to that given in Problem 8.157, except now the independent variable is z; its solution is

$$y = c_1e^{-2z} + c_2ze^{-2z} + c_3z^2e^{-2z} + c_4z^3e^{-2z} = c_1x^{-2} + c_2(\ln x)x^{-2} + c_3(\ln x)^2x^{-2} + c_4(\ln x)^3x^{-2}$$

8.246 Solve $x^4y^{(4)} + 6x^3y^{(3)} - 2x^2y'' - 8xy' + 20y = 0$.

This is an Euler equation. Using the substitutions suggested in Problem 8.230, we can rewrite the equation as

$$D(D-1)(D-2)(D-3)y + 6D(D-1)(D-2)y - 2D(D-1)y - 8Dy + 20y = 0$$

or $(D^4 - 9D^2 + 20)y = 0$. This equation is similar in form to that given in Problem 8.29, except now the independent variable is z; its solution is

$$y = c_1e^{2z} + c_2e^{-2z} + c_3e^{\sqrt{5}z} + c_4e^{-\sqrt{5}z} = c_1x^2 + c_2x^{-2} + c_3x^{\sqrt{5}} + c_4x^{-\sqrt{5}}$$

CHAPTER 9

The Method of Undetermined Coefficients

EQUATIONS WITH EXPONENTIAL RIGHT SIDE

.1 Solve $y' - 5y = e^{2x}$.

▌ We assume a particular solution of the form $y_p = A_0 e^{2x}$. The general solution to the associated homogeneous equation is shown in Problem 8.34 to be $y_h = c_1 e^{5x}$. Since y_p is not a linear combination of y_h, there is no need to modify it.

Substituting y_p into the given nonhomogeneous differential equation, we obtain $2A_0 e^{2x} - 5A_0 e^{2x} = e^{2x}$ or $-3A_0 e^{2x} = e^{2x}$, from which we find $A_0 = -\frac{1}{3}$. Then $y_p = -\frac{1}{3}e^{2x}$, and the general solution to the nonhomogeneous equation is $y = y_h + y_p = c_1 e^{5x} - \frac{1}{3}e^{2x}$.

.2 Solve $y' + 6y = e^{3x}$.

▌ We assume a particular solution of the form $y_p = A_0 e^{3x}$, where A_0 denotes an unknown constant which must be determined. The general solution to the associated homogeneous equation is found in Problem 8.37 to be $y_h = Ae^{-6x}$.

Since y_p and y_h have no terms in common except perhaps for a multiplicative constant, there is no need to modify y_p. Substituting it into the given differential equation, we obtain $3A_0 e^{3x} + 6A_0 e^{3x} = e^{3x}$, or $9A_0 e^{3x} = e^{3x}$, from which $A_0 = \frac{1}{9}$. Then $y_p = \frac{1}{9}e^{3x}$, and the general solution to the nonhomogeneous equation is $y = y_h + y_p = Ae^{-6x} + \frac{1}{9}e^{3x}$.

.3 Solve $y' + 6y = 18e^{3x}$.

▌ Both y_p and y_h of the previous problem are valid here. Substituting y_p into this differential equation, we obtain $3A_0 e^{3x} + 6A_0 e^{3x} = 18e^{3x}$, or $9A_0 e^{3x} = 18e^{3x}$. Thus, $A_0 = 2$ and $y_p = 2e^{3x}$. The general solution to the nonhomogeneous differential equation is then $y = y_h + y_p = Ae^{-6x} + 2e^{3x}$.

.4 Solvc $y' + 6y = 4e^{-5x}$.

▌ We assume a particular solution of the form $y_p = A_0 e^{-5x}$, with $y_h = Ae^{-6x}$ as in Problem 9.2.

Since y_p and y_h have no terms in common except perhaps for a multiplicative constant, there is no need to modify y_p. Substituting it into the nonhomogeneous differential equation, we get $-5A_0 e^{-5x} + 6A_0 e^{-5x} = 4e^{-5x}$, or $A_0 e^{-5x} = 4e^{-5x}$. Then $A_0 = 4$, and $y_p = 4e^{-5x}$. The general solution to the nonhomogeneous equation is thus $y = y_h + y_p = Ae^{-6x} + 4e^{-5x}$.

5 Solve $y' + 6y = 6e^{6x}$.

▌ We assume a particular solution of the form $y_p = A_0 e^{6x}$, with y_h as in the previous problem.

Since y_p and y_h are linearly independent, there is no need to modify y_p. Substituting it into the given differential equation, we obtain $6A_0 e^{6x} + 6A_0 e^{6x} = 6e^{6x}$, or $12A_0 e^{6x} = 6e^{6x}$. Thus $A_0 = \frac{1}{2}$, and $y_p = \frac{1}{2}e^{6x}$. The general solution to the nonhomogeneous equation is then $y = y_h + y_p = Ae^{-6x} + \frac{1}{2}e^{6x}$.

6 Solve $2y' - 5y = 6e^{6x}$.

▌ We assume y_p as in the previous problem, but now the general solution to the associated homogeneous differential equation, as found in Problem 8.38, is $y_h = Ae^{5x/2}$. Since y_p and y_h are linearly independent, no modification of y_p is necessary.

Substituting y_p into this differential equation, we obtain $2(6A_0 e^{6x}) - 5A_0 e^{6x} = 6e^{6x}$, or $7A_0 e^{6x} = 6e^{6x}$. Thus $A_0 = \frac{6}{7}$, and $y_p = \frac{6}{7}e^{6x}$. The general solution of the given differential equation is then $y = y_h + y_p = Ae^{5x/2} + \frac{6}{7}e^{6x}$.

7 Solve $y'' - 7y' = 6e^{6x}$.

▌ We assume a particular solution of the form $y_p = A_0 e^{6x}$. The general solution of the associated homogeneous differential equation is found in Problem 8.2 to be $y_h = c_1 + c_2 e^{7x}$. Since e^{6x} is not a linear combination of 1 and e^{7x}, there is no need to modify y_p.

Substituting y_p into the given nonhomogeneous differential equation, we obtain $36A_0e^{6x} - 7(6A_0e^{6x}) = 6e^{6x}$, or $-6A_0e^{6x} = 6e^{6x}$. Thus $A_0 = -1$, and $y_p = -e^{6x}$. The general solution to the given differential equation is then $y = y_h + y_p = c_1 + c_2e^{7x} - e^{6x}$.

9.8 Solve $y'' - 7y' = e^{8x}$.

▌ We assume a particular solution of the form $y_p = A_0e^{8x}$, with y_h as in the previous problem. Since y_h has no terms in common with y_p except perhaps for a multiplicative constant, there is no need to modify y_p.

Substituting y_p into the given differential equation, we get $64A_0e^{8x} - 7(8A_0e^{8x}) = e^{8x}$, or $8A_0e^{8x} = e^{8x}$. Thus $A_0 = \frac{1}{8}$, and $y_p = \frac{1}{8}e^{8x}$. The general solution to the nonhomogeneous differential equation is then $y = y_h + y_p = c_1 + c_2e^{7x} + \frac{1}{8}e^{8x}$.

9.9 Solve $y'' + 6y' + 9y = 100e^{2x}$.

▌ We assume a particular solution of the form $y_p = A_0e^{2x}$. The complementary solution is found in Problem 8.142 to be $y_c = c_1e^{-3x} + c_2xe^{-3x}$. Since y_p is not part of y_c, there is no need to modify y_p.

Differentiating y_p twice and substituting $y = y_p$ into the given differential equation, we obtain $4A_0e^{2x} + 6(2A_0e^{2x}) + 9A_0e^{2x} = 100e^{2x}$, or $25A_0e^{2x} = 100e^{2x}$. Thus $A_0 = 4$, and $y_p = 4e^{2x}$. The general solution is then $y = y_c + y_p = c_1e^{-3x} + c_2xe^{-3x} + 4e^{2x}$.

9.10 Solve $y'' - y' - 2y = e^{3x}$.

▌ We assume a particular solution of the form $y_p = A_0e^{3x}$. The general solution to the associated homogeneous equation is found in Problem 8.1 to be $y_h = c_1e^{-x} + c_2e^{2x}$, and since y_p is not part of y_h, no modification of y_p is required.

Substituting y_p into the differential equation, we obtain $9A_0e^{3x} - 3A_0e^{3x} - 2A_0e^{3x} = e^{3x}$, or $4A_0e^{3x} = e^{3x}$. Thus $A_0 = \frac{1}{4}$, and $y_p = \frac{1}{4}e^{3x}$. The general solution then is $y = c_1e^{-x} + c_2e^{2x} + \frac{1}{4}e^{3x}$.

9.11 Solve $\dfrac{d^2y}{dt^2} - 4\dfrac{dy}{dt} + y = 3e^{2t}$.

▌ We assume a particular solution of the form $y_p = A_0e^{2t}$. The general solution of the associated homogeneous differential equation is shown in Problem 8.9 to be $y_h = C_1e^{3.732t} + C_2e^{0.2679t}$; since e^{2t} cannot be written as a linear combination of $e^{3.732t}$ and $e^{0.2679t}$, there is no need to modify y_p.

Substituting y_p into the given nonhomogeneous differential equation, we obtain $4A_0e^{2t} - 4(2A_0e^{2t}) + A_0e^{2t} = 3e^{2t}$, or $-3A_0e^{2t} = 3e^{2t}$. Thus $A_0 = -1$, and $y_p = -e^{2t}$. The general solution to the nonhomogeneous equation is then $y = y_h + y_p = C_1e^{3.732t} + C_2e^{0.2679t} - e^{2t}$.

9.12 Solve $\dfrac{d^2y}{dt^2} - 4\dfrac{dy}{dt} + y = 2e^{3t}$.

▌ We assume a particular solution of the form $y_p = A_0e^{3t}$, and y_h of the previous problem is valid here as well. Since y_p and y_h have no terms in common except perhaps for a multiplicative constant, there is no need to modify y_p.

Substituting y_p into the given differential equation yields $9A_0e^{3t} - 4(3A_0e^{3t}) + A_0e^{3t} = 2e^{3t}$, or $-2A_0e^{3t} = 2e^{3t}$. Thus $A_0 = -1$, and $y_p = -e^{3t}$. The general solution to the given differential equation is then $y = y_h + y_p = C_1e^{3.732t} + C_2e^{0.2679t} - e^{3t}$.

9.13 Solve $\dfrac{d^2x}{dt^2} + 4\dfrac{dx}{dt} + 8x = e^{-2t}$.

▌ We assume a particular solution of the form $x_p = A_0e^{-2t}$. The general solution to the associated homogeneous problem is found in Problem 8.54 to be $x_h = c_1e^{-2t}\cos 2t + c_2e^{-2t}\sin 2t$. Since the functions e^{-2t}, $e^{-2t}\sin 2t$, and $e^{-2t}\cos 2t$ are linearly independent, there is no need to modify x_p.

Substituting x_p into the given nonhomogeneous differential equation, we get $4A_0e^{-2t} + 4(-2A_0e^{-2t}) + 8A_0e^{-2t} = e^{-2t}$, or $4A_0e^{-2t} = e^{-2t}$. Thus $A_0 = \frac{1}{4}$ and $x_p = \frac{1}{4}e^{-2t}$. The general solution to the nonhomogeneous equation is then $x = x_h + x_p = e^{-2t}(c_1\cos 2t + c_2\sin 2t + \frac{1}{4})$.

9.14 Solve $\dfrac{d^2I}{dt^2} - 60\dfrac{dI}{dt} + 900I = 5e^{10t}$.

▌ We assume a solution of the form $I_p = A_0e^{10t}$. The general solution of the associated homogeneous problem is found in Problem 8.149 to be $I_h = Ae^{30t} + Bte^{30t}$. Since I_p and I_h have no terms in common, except perhaps for a multiplicative constant, there is no need to modify I_p. Substituting it into the given nonhomogeneous differential equation, we obtain $100A_0e^{10t} - 60(10A_0e^{10t}) + 900A_0e^{10t} = 5e^{10t}$, or $400A_0e^{10t} = 5e^{10t}$. Thus $A_0 = 0.0125$, and $I_p = 0.0125e^{10t}$. The general solution to the nonhomogeneous equation is then $I = I_h + I_p = Ae^{30t} + Bte^{30t} + 0.0125e^{10t}$.

9.15 Solve $\dfrac{d^3x}{dt^3} + 5\dfrac{d^2x}{dt^2} + 26\dfrac{dx}{dt} - 150x = 20e^{-t}$.

▌ We assume a particular solution of the form $x_p = A_0e^{-t}$. The general solution to the associated homogeneous differential equation is found in Problem 8.112 to be $x_h = c_1e^{3t} + e^{-4t}(c_1 \cos\sqrt{34}t + c_3 \sin\sqrt{34}t)$. Since x_p cannot be obtained from x_h by any choice of the arbitrary constants c_1, c_2, and c_3, there is no need to modify it.

Substituting x_p into the given nonhomogeneous equation, we get $-A_0e^{-t} + 5A_0e^{-t} + 26(-A_0e^{-t}) - 150A_0e^{-t} = 20e^{-t}$, or $-172A_0e^{-t} = 20e^{-t}$. Thus $A_0 = -20/172$, and $x_p = -\frac{20}{172}e^{-t}$. The general solution to the nonhomogeneous equation is then

$$x = x_h + x_p = c_1e^{3t} + e^{-4t}(c_2 \cos\sqrt{34}t + c_3 \sin\sqrt{34}t) - \tfrac{5}{43}e^{-t}$$

9.16 Solve $\dfrac{d^3Q}{dt^3} - 5\dfrac{d^2Q}{dt^2} + 25\dfrac{dQ}{dt} - 125Q = -60e^{7t}$.

▌ We assume a particular solution of the form $Q_p = A_0e^{7t}$. The general solution to the associated homogeneous equation is shown in Problem 8.113 to be $Q_h = c_1e^{5t} + c_2 \cos 5t + c_3 \sin 5t$. Since Q_p cannot be obtained from Q_h by any choice of the constants c_1, c_2, and c_3, there is no need to modify Q_p.

Substituting Q_p into the given differential equation, we get $343A_0e^{7t} - 5(49A_0e^{7t}) + 25(7A_0e^{7t}) - 125A_0e^{7t} = -60e^{7t}$, or $148A_0e^{7t} = -60e^{7t}$. Therefore $A_0 = -\frac{15}{37}$, and $Q_p = -\frac{15}{37}e^{7t}$. The general solution to the nonhomogeneous differential equation is then $Q = Q_h + Q_p = c_1e^{5t} + c_2 \cos 5t + c_3 \sin 5t - \frac{15}{37}e^{7t}$.

9.17 Solve $y^{(4)} - 6y^{(3)} + 16y'' + 54y' - 225y = 100e^{-2x}$.

▌ We assume a particular solution of the form $y_p = A_0e^{-2x}$. The general solution to the associated homogeneous equation is shown in Problem 8.124 to be $y_h = c_1e^{3x} + c_2e^{-3x} + c_3e^{3x} \cos 4x + c_4e^{3x} \sin 4x$. Since y_p cannot be obtained from y_h by any choice of the constants c_1 through c_4, there is no need to modify it.

Substituting y_p into the given nonhomogeneous differential equation, we obtain $16A_0e^{-2x} - 6(-8A_0e^{-2x}) + 16(4A_0e^{-2x}) + 54(-2A_0e^{-2x}) - 225A_0e^{-2x} = 100e^{2x}$, or $-205A_0e^{-2x} = 100e^{-2x}$. Thus $A_0 = -100/205 = -\frac{20}{41}$ and $y_p = -\frac{20}{41}e^{-2x}$. The general solution to the nonhomogeneous equation is then

$$y = y_h + y_p = c_1e^{3x} + c_2e^{-3x} + c_3e^{3x} \cos 4x + c_4e^{3x} \sin 4x - \tfrac{20}{41}e^{-2x}.$$

EQUATIONS WITH CONSTANT RIGHT SIDE

9.18 Solve $y' - 5y = 8$.

▌ We assume a particular solution of the form $y_p = A_0$ where A_0 is a constant to be determined. The general solution of the associated homogeneous equation is found in Problem 8.34 to be $y_h = c_1e^{5x}$. Since any nonzero constant A_0 is linearly independent of e^{5x}, there is no need to modify y_p.

Substituting y_p into the nonhomogeneous equation and noting that $y_p' = 0$, we get $0 - 5A_0 = 8$, or $A_0 = -\frac{8}{5}$. Then $y_p = -\frac{8}{5}$ and the general solution to the nonhomogeneous equation is $y = y_h + y_p = c_1e^{5x} - \frac{8}{5}$.

9.19 Solve $y'' - y' - 2y = 7$.

▌ We assume a particular solution of the form $y_p = A_0$. The general solution to the associated homogeneous differential equation is shown in Problem 8.1 to be $y_h = c_1e^{-x} + c_2e^{2x}$. Since y_p cannot be obtained from y_h by any choice of c_1 and c_2, there is no need to modify it.

Substituting y_p and its derivatives (all of which are zero) into the nonhomogeneous differential equation, we get $0 - 0 - 2A_0 = 7$ or $A_0 = -\frac{7}{2}$. Then $y_p = -\frac{7}{2}$ and the general solution to the nonhomogeneous equation is $y = c_1e^{-x} + c_2e^{2x} - \frac{7}{2}$.

9.20 Solve $\dfrac{d^2Q}{dt^2} + 8\dfrac{dQ}{dt} + 52Q = 26$.

▌ We assume a particular solution of the form $Q_p = A_0$. The general solution to the associated homogeneous equation is shown in Problem 8.55 to be $Q_h = c_1e^{-4t}\cos 6t + c_2e^{-4t}\sin 6t$. Since Q_p is not part of Q_h, it requires no modification. Substituting Q_p into the nonhomogeneous differential equation, we obtain $0 + 8(0) + 52A_0 = 26$ or $A_0 = \frac{1}{2}$. Thus $Q_p = \frac{1}{2}$ and the general solution to the nonhomogeneous equation is $Q = c_1e^{-4t}\cos 6t + c_2e^{-4t}\sin 6t + \frac{1}{2}$.

9.21 Solve $\dfrac{d^2q}{dt^2} + 100\dfrac{dq}{dt} + 50{,}000q = 2200$.

▌ We assume a particular solution of the form $q_p = A_0$. The general solution to the associated homogeneous equation is found in Problem 8.56 (with q replacing I) to be $q_h = c_1e^{-50t}\cos 50\sqrt{19}t + c_2e^{-50t}\sin 50\sqrt{19}t$. Since q_p is not part of q_h, no modification to q_p is necessary.

Substituting q_p into the nonhomogeneous differential equation, we get $0 + 100(0) + 50{,}000A_0 = 2200$, or $A_0 = q_p = \frac{11}{250}$. The general solution to the nonhomogeneous equation is then $q = q_h + q_p = c_1e^{-50t}\cos 50\sqrt{19}t + c_2e^{-50t}\sin 50\sqrt{19}t + \frac{11}{250}$.

9.22 Solve $\ddot{q} + 20\dot{q} + 200q = 24$.

▌ We assume a particular solution of the form $q_p = A_0$. The general solution to the associated homogeneous equation is, from Problem 8.64 with q replacing I, $q_h = c_1e^{-10t}\cos 10t + c_2e^{-10t}\sin 10t$. Since an arbitrary constant A_0 cannot be expressed as a linear combination of $e^{-10t}\cos 10t$ and $e^{-10t}\sin 10t$, there is no need to modify q_p.

Substituting q_p into the nonhomogeneous differential equation, we obtain $200A_0 = 24$, or $A_0 = q_p = \frac{3}{25}$. The general solution to the nonhomogeneous equation is then $q = c_1e^{-10t}\cos 10t + c_2e^{-10t}\sin 10t + \frac{3}{25}$.

9.23 Solve $\ddot{q} + 400\dot{q} + 200{,}000q = 2000$.

▌ We assume a particular solution $q_p = A_0$. The general solution to the associated homogeneous equation is found in Problem 8.70 to be $q_h = e^{-200t}(A\cos 400t + B\sin 400t)$, which has no terms in common with q_p; thus, q_p needs no modification.

Substituting q_p into the given nonhomogeneous differential equation, we get $200{,}000A_0 = 2000$, or $A_0 = q_p = 0.01$. The general solution to the nonhomogeneous equation is then $q = e^{-200t}(A\cos 400t + B\sin 400t) + 0.01$.

9.24 Solve $\ddot{q} + 1000\dot{q} + 50{,}000q = 2200$.

▌ We assume a particular solution of the form $q_p = A_0$. The general solution to the associated homogeneous equation is shown in Problem 8.18 to be $q_h = c_1e^{-52.79t} + c_2e^{-947.2t}$. Since q_p and q_h have no common terms, there is no need to modify q_p.

Substituting q_p into the given nonhomogeneous differential equation, we obtain $50{,}000A_0 = 2200$, or $A_0 = q_p = 0.044$. The general solution to the nonhomogeneous equation is then $q = c_1e^{-52.79t} + c_2e^{-947.2t} + 0.044$.

9.25 Solve $\dfrac{d^2Q}{dt^2} + 1000\dfrac{dQ}{dt} + 250{,}000Q = 24$.

▌ We assume a particular solution of the form $Q_p = A_0$. The general solution to the associated homogeneous equation is shown in Problem 8.147 to be $Q_h = c_1e^{-500t} + c_2te^{-500t}$. Since Q_p is not a linear combination of e^{-500t} and te^{-500t}, no modification of Q_p is required.

Substituting Q_p into the given nonhomogeneous differential equation, we get $0 + 1000(0) + 250{,}000A_0 = 24$, or $A_0 = Q_p = 9.6 \times 10^{-5}$. The general solution to the nonhomogeneous equation is then $Q = c_1e^{-500t} + c_2te^{-500t} + 9.6 \times 10^{-5}$.

9.26 Solve $\dfrac{d^2x}{dt^2} - \dfrac{g}{10}x = \dfrac{g}{5}$, where g denotes a positive constant.

▌ The right side of this equation is a constant, so we assume a particular solution of the form $x_p = A_0$. The general solution to the associated homogeneous equation is found in Problem 8.21 to be $x_h = C_1e^{\sqrt{g/10}\,t} + C_2e^{-\sqrt{g/10}\,t}$. Since x_p cannot be obtained from x_h by suitably choosing C_1 and C_2, we do not need to modify x_p.

Substituting x_p into the nonhomogeneous differential equation, we get $0 - \dfrac{g}{10}A_0 = \dfrac{g}{5}$, or $A_0 = x_p = -2$. The general solution to the nonhomogeneous equation is then $x = C_1e^{\sqrt{g/10}\,t} + C_2e^{-\sqrt{g/10}\,t} - 2$.

9.27 Solve $\dfrac{d^2x}{dt^2} - \dfrac{g}{10}x = \dfrac{3g}{20}$, where g denotes a positive constant.

▮ Both x_p and x_h are as in the previous problem. Substituting x_p into this differential equation, we get $0 - \dfrac{g}{10}A_0 = \dfrac{3g}{20}$, or $A_0 = x_p = -\dfrac{3}{2}$. Thus, $x = x_h + x_p = C_1e^{\sqrt{g/10}t} + C_2e^{-\sqrt{g/10}t} - \frac{3}{2}$.

9.28 Solve $\dfrac{d^3x}{dt^3} + 5\dfrac{d^2x}{dt^2} + 26\dfrac{dx}{dt} - 150x = 30$.

▮ We assume a particular solution of the form $x_p = A_0$. The general solution to the associated homogeneous differential equation is found in Problem 8.112 to be $x_h = c_1e^{3t} + e^{-4t}(c_2\cos\sqrt{34}t + c_3\sin\sqrt{34}t)$. Since x_p cannot be obtained from x_h no matter how the arbitrary constants c_1 through c_3 are chosen, there is no need to modify x_p.

Substituting x_p into the given differential equation, we obtain $150A_0 = 30$, or $A_0 = x_p = -\frac{1}{5}$. The general solution is then $x = c_1e^{3t} + e^{-4t}(c_2\cos\sqrt{34}t + c_3\sin\sqrt{34}t) - \frac{1}{5}$.

9.29 Solve $\dfrac{d^3Q}{dt^3} - 5\dfrac{d^2Q}{dt^2} + 25\dfrac{dQ}{dt} - 125Q = 1000$.

▮ We assume a particular solution of the form $Q_p = A_0$. The complementary function is shown in Problem 8.113 to be $Q_c = c_1e^{5t} + c_2\cos 5t + c_3\sin 5t$. Since Q_p is not a linear combination of e^{5t}, $\cos 5t$, and $\sin 5t$, there is no need to modify Q_p.

Substituting Q_p into the given differential equation, we obtain $0 - 5(0) + 25(0) - 125A_0 = 1000$. Thus $A_0 = Q_p = -8$, and the general solution to the nonhomogeneous differential equation is $Q = Q_c + Q_p = c_1e^{5t} + c_2\cos 5t + c_3\sin 5t - 8$.

9.30 Solve $y^{(4)} - 6y^{(3)} + 16y'' + 54y' - 225y = -75$.

▮ We assume a particular solution of the form $y_p = A_0$. The complementary function is found in Problem 8.124 to be $y_c = c_1e^{3x} + c_2e^{-3x} + c_3e^{3x}\cos 4x + c_4e^{3x}\sin 4x$. Since y_p cannot be obtained from y_c by any choice of the constants c_1 through c_4, there is no need to modify y_p.

Substituting y_p into the given differential equation, we find $-225A_0 = -75$. Therefore, $A_0 = y_p = \frac{1}{3}$, and the general solution to the nonhomogeneous equation is $y = y_c + y_p = c_1e^{3x} + c_2e^{-3x} + c_3e^{3x}\cos 4x + c_4e^{3x}\sin 4x + \frac{1}{3}$.

EQUATIONS WITH POLYNOMIAL RIGHT SIDE

9.31 Solve $y' - 5y = 3x + 1$.

▮ Since the right side of the differential equation is a first-degree polynomial, we try a general first-degree polynomial as a particular solution. We assume $y_p = A_1x + A_0$, where the coefficients A_1 and A_0 must be determined. The solution to the associated homogeneous equation is shown in Problem 8.34 to be $y_h = c_1e^{5x}$. Since no part of y_p solves the homogeneous equation, there is no need to modify y_p.

Substituting y_p into the given nonhomogeneous equation and noting that $y_p' = A_1$, we get $A_1 - 5(A_1x + A_0) = 3x + 1$, or $(-5A_1)x + (A_1 - 5A_0) = 3x + 1$. Equating the coefficients of like powers of x, we obtain

$$\begin{aligned} -5A_1 &= 3 \qquad \text{or} \qquad A_1 = -\tfrac{3}{5} \\ A_1 - 5A_0 &= 1 \qquad \text{or} \qquad A_0 = -\tfrac{8}{25} \end{aligned}$$

Then $y_p = -\frac{3}{5}x - \frac{8}{25}$, and the general solution is $y = y_h + y_p = c_1e^{5x} - \frac{3}{5}x - \frac{8}{25}$.

9.32 Solve $y' - 5y = 8x$.

▮ The right side of this differential equation is a first-degree polynomial, so both y_p and y_h of the previous problem are valid here. Substituting y_p into the given differential equation, we get $A_1 - 5(A_1x + A_0) = 8x$, or $(-5A_1)x + (A_1 - 5A_0) = 8x + 0$. Equating the coefficients of like powers of x, we obtain

$$\begin{aligned} -5A_1 &= 8 \qquad \text{or} \qquad A_1 = -\tfrac{8}{5} \\ A_1 - 5A_0 &= 0 \qquad \text{or} \qquad A_0 = -\tfrac{8}{25} \end{aligned}$$

Thus $y_p = -\frac{8}{5}x - \frac{8}{25}$, and the general solution to the given nonhomogeneous equation is $y = y_h + y_p = c_1e^{5x} - \frac{8}{5}x - \frac{8}{25}$.

9.33 Solve $y' - 5y = 2x^2 - 5$.

▌ Since the right side is a second-order polynomial, we assume a general second-order polynomial for y_p, namel $y_p = A_2x^2 + A_1x + A_0$. The complementary solution remains as it was in the previous two problems: $y_h = c_1e^{5x}$. Since y_p and y_h have no terms in common except perhaps for a multiplicative constant, there is no need to modify y_p.

Substituting y_p along with $y_p' = 2A_2x + A_1$ into the given differential equation, we find that $2A_2x + A_1 - 5(A_2x^2 + A_1x + A_0) = 2x^2 - 5$, or $(-5A_2)x^2 + (2A_2 - 5A_1)x + (A_1 - 5A_0) = 2x^2 + 0x - 5$. Equating coefficients of like powers of x, we obtain

$$\begin{aligned} -5A_2 &= 2 \quad \text{or} \quad A_2 = -0.4 \\ 2A_2 - 5A_1 &= 0 \quad \text{or} \quad A_1 = -0.16 \\ A_1 - 5A_0 &= -5 \quad \text{or} \quad A_0 = 0.968 \end{aligned}$$

Then $y_p = -0.4x^2 - 0.16x + 0.968$, and the general solution to the given differential equation is $y = c_1e^{5x} - 0.4x^2 - 0.16x + 0.968$.

9.34 Solve $2y' - 5y = 2x^2 - 5$.

▌ The particular solution y_p of the previous problem is valid here, but the complementary solution is now $y_c = Ae^{5x/2}$ (see Problem 8.38). Substituting y_p into the differential equation, we get $2(2A_2x + A_1) - 5(A_2x^2 + A_1x + A_0) = 2x^2 - 5$, or $(-5A_2)x^2 + (4A_2 - 5A_1)x + (2A_1 - 5A_0) = 2x^2 + 0x - 5$. Equating coefficients of like powers of x, we obtain

$$\begin{aligned} -5A_2 &= 2 \quad \text{or} \quad A_2 = -0.4 \\ 4A_2 - 5A_1 &= 0 \quad \text{or} \quad A_1 = -0.32 \\ 2A_1 - 5A_0 &= -5 \quad \text{or} \quad A_0 = 0.872 \end{aligned}$$

The general solution is then $y = c_1e^{5x/2} - 0.4x^2 - 0.32x + 0.872$.

9.35 Solve $y'' - y' - 2y = 4x^2$.

▌ We assume a particular solution of the form $y_p = A_2x^2 + A_1x + A_0$. The general solution to the associate homogeneous differential equation is found in Problem 8.1 to be $y_h = c_1e^{-x} + c_2e^{2x}$. Since y_p and y_h have no terms in common except perhaps for a multiplicative constant, there is no need to modify y_p.

Substituting y_p into the given differential equation, we get $2A_2 - (2A_2x + A_1) - 2(A_2x^2 + A_1x + A_0) = 4x^2$ or, equivalently, $(-2A_2)x^2 + (-2A_2 - 2A_1)x + (2A_2 - A_1 - 2A_0) = 4x^2 + 0x + 0$. Equating the coefficients of like powers of x, we obtain

$$\begin{aligned} -2A_2 &= 4 \quad \text{or} \quad A_2 = -2 \\ -2A_2 - 2A_1 &= 0 \quad \text{or} \quad A_1 = 2 \\ 2A_2 - A_1 - 2A_0 &= 0 \quad \text{or} \quad A_0 = -3 \end{aligned}$$

Then $y_p = -2x^2 + 2x - 3$, and the general solution is $y = y_h + y_p = c_1e^{-x} + c_2e^{2x} - 2x^2 + 2x - 3$.

9.36 Solve $\dfrac{d^2y}{dt^2} - 4\dfrac{dy}{dt} + y = 3t - 4$.

▌ Since the right side is a first-order polynomial in t, we assume a general first-order polynomial in t as the for of a particular solution. We try $y_p = A_1t + A_0$. The complementary solution is found in Problem 8.9 to be $y_c = C_1e^{3.732t} + C_2e^{0.2679t}$. Since y_p and y_c have no terms in common, there is no need to modify y_p.

Substituting y_p, $y_p' = A_1$, and $y_p'' = 0$ into the given differential equation, we get $0 - 4A_1 + A_1t + A_0 = 3t - 4$, or $A_1t + (-4A_1 + A_0) = 3t - 4$. Equating coefficients of like powers of t yields

$$\begin{aligned} A_1 &= 3 \\ -4A_1 + A_0 &= -4 \quad \text{or} \quad A_0 = 8 \end{aligned}$$

Then $y_p = 3t + 8$, and the general solution is $y = y_c + y_p = C_1e^{3.732t} + C_2e^{0.2679t} + 3t + 8$.

9.37 Solve $\dfrac{d^2y}{dt^2} - 4\dfrac{dy}{dt} + y = t^2 - 2t + 3$.

▌ Since the right side of this differential equation is a second-degree polynomial, we try a second-degree polynomial (with undetermined coefficients) as the form of y_p, namely $y_p = A_2t^2 + A_1t + A_0$. The complementary solution of the previous problem is valid here, and since y_p and y_c have no terms in common, there is no need to modify y_p.

Substituting y_p into the given differential equation, we obtain $2A_2 - 4(2A_2t + A_1) + A_2t^2 + A_1t + A_0 = t^2 - 2t + 3$, or $A_2t^2 + (-8A_2 + A_1)t + (2A_2 - 4A_1 + A_0) = t^2 - 2t + 3$. Equating coefficients of like powers of t, we get

$$\begin{aligned} A_2 &= 1 \\ -8A_2 + A_1 &= -2 \quad \text{or} \quad A_1 = 6 \\ 2A_2 - 4A_1 + A_0 &= 3 \quad \text{or} \quad A_0 = 25 \end{aligned}$$

Then $y_p = t^2 + 6t + 25$, and the general solution to the nonhomogeneous equation is $y = C_1e^{3.732t} + C_2e^{0.2679t} + t^2 + 6t + 25$.

9.38 Solve $\dfrac{d^2y}{dt^2} - 4\dfrac{dy}{dt} + y = 2t^3 + 3t^2 - 1$.

▌ Since the right side of the differential equation is a third-degree polynomial, we assume a general third-degree polynomial as the form of a particular solution. We try $y_p = A_3t^3 + A_2t^2 + A_1t + A_0$. The complementary solution is again that of Problem 9.36.

Substituting y_p along with $y_p' = 3A_3t^2 + 2A_2t + A_1$ and $y_p'' = 6A_3t + 2A_2$ into the nonhomogeneous differential equation, we get

$$6A_3t + 2A_2 - 4(3A_3t^2 + 2A_2t + A_1) + A_3t^3 + A_2t^2 + A_1t + A_0 = 2t^3 + 3t^2 - 1$$

or

$$A_3t^3 + (-12A_3 + A_2)t^2 + (6A_3 - 8A_2 + A_1)t + (2A_2 - 4A_1 + A_0) = 2t^3 + 3t^2 + 0t - 1$$

Equating coefficients of like powers of t, we obtain

$$\begin{aligned} A_3 &= 2 \\ -12A_3 + A_2 &= 3 \quad \text{or} \quad A_2 = 27 \\ 6A_3 - 8A_2 + A_1 &= 0 \quad \text{or} \quad A_1 = 204 \\ 2A_2 - 4A_1 + A_0 &= -1 \quad \text{or} \quad A_0 = 761 \end{aligned}$$

Then $y_p = 2t^3 + 27t^2 + 204t + 761$, and the general solution to the nonhomogeneous equation is $y = C_1e^{3.732t} + C_2e^{0.2679t} + 2t^3 + 27t^2 + 204t + 761$.

9.39 Solve $\dfrac{d^2x}{dt^2} + 4\dfrac{dx}{dt} + 8x = -3t + 1$.

▌ We try a particular solution of the form $x_p = A_1t + A_0$. The complementary solution is shown in Problem 8.54 to be $x_c = c_1e^{-2t}\cos 2t + c_2e^{-2t}\sin 2t$. Substituting x_p along with its derivatives into the given differential equation, we obtain $0 + 4A_1 + 8(A_1t + A_0) = -3t + 1$, or $(8A_1)t + (4A_1 + 8A_0) = -3t + 1$. Equating coefficients of like powers of t, we find that

$$\begin{aligned} 8A_1 &= -3 \quad \text{or} \quad A_1 = -\tfrac{3}{8} \\ 4A_1 + 8A_0 &= 1 \quad \text{or} \quad A_0 = \tfrac{5}{16} \end{aligned}$$

Thus $y_p = -\frac{3}{8}t + \frac{5}{16}$, and the general solution to the nonhomogeneous equation is $x = c_1e^{-2t}\cos 2t + c_2e^{-2t}\sin 2t - \frac{3}{8}t + \frac{5}{16}$.

9.40 Solve $\dfrac{d^2x}{dt^2} + 4\dfrac{dx}{dt} + 8x = 8t^2 + 8t + 18$.

▌ Since the right side is a second-degree polynomial, we try $x_p = A_2t^2 + A_1t + A_0$. The complementary solution of the previous problem is valid here. Substituting x_p along with its derivatives into the given differential equation, we get $2A_2 + 4(2A_2t + A_1) + 8(A_2t^2 + A_1t + A_0) = 8t^2 + 8t + 18$, or $(8A_2)t^2 + (8A_2 + 8A_1)t + (2A_2 + 4A_1 + 8A_0) = 8t^2 + 8t + 18$. Equating coefficients of like powers of t, we obtain

$$\begin{aligned} 8A_2 &= 8 \quad \text{or} \quad A_2 = 1 \\ 8A_2 + 8A_1 &= 8 \quad \text{or} \quad A_1 = 0 \\ 2A_2 + 4A_1 + 8A_0 &= 18 \quad \text{or} \quad A_0 = 2 \end{aligned}$$

Thus $x_p = t^2 + 2$, and the general solution to the nonhomogeneous equation is $x = c_1e^{-2t}\cos 2t + c_2e^{-2t}\sin 2t + t^2 + 2$.

9.41 Solve $\dfrac{d^2x}{dt^2} + 4\dfrac{dx}{dt} + 8x = -t^2$.

Since the right side of the differential equation is a second-degree polynomial, y_p of the previous problem is appropriate here. The complementary solution is given by y_c of Problem 9.39, and since it has no terms in common with the particular solution, no modifications are necessary.

Substituting y_p and its derivatives into the given differential equation, we get $2A_2 + 4(2A_2t + A_1) + 8(A_2t^2 + A_1t + A_0) = -t^2$, or $(8A_2)t^2 + (8A_2 + 8A_1)t + (2A_2 + 4A_1 + 8A_0) = -t^2 + 0t + 0$. Equating coefficients of like powers of t, we obtain

$$\begin{aligned} 8A_2 &= 1 \quad &\text{or} \quad A_2 &= -\tfrac{1}{8} \\ 8A_2 + 8A_1 &= 0 \quad &\text{or} \quad A_1 &= \tfrac{1}{8} \\ 2A_2 + 4A_1 + 8A_0 &= 0 \quad &\text{or} \quad A_0 &= -\tfrac{1}{32} \end{aligned}$$

Then $x_p = -\frac{1}{8}t^2 + \frac{1}{8}t - \frac{1}{32}$, and the general solution to the nonhomogeneous equation is $x = c_1e^{-2t}\cos 2t + c_2e^{-2t}\sin 2t - \frac{1}{8}t^2 + \frac{1}{8}t - \frac{1}{32}$.

9.42 Solve $\dfrac{d^2x}{dt^2} + 4\dfrac{dx}{dt} + 8x = 16t^3 - 40t^2 - 60t + 4$.

The complementary solution is that of Problem 9.39. Since the right side of the given nonhomogeneous equation is a third-degree polynomial, we try a third-degree polynomial as the form of a particular solution. We assume $x_p = A_3t^3 + A_2t^2 + A_1t + A_0$. Then $x_p' = 3A_3t^2 + 2A_2t + A_1$ and $x_p'' = 6A_3t + 2A_2$. Substituting these quantities and x_p into the given differential equation, we obtain

$$6A_3t + 2A_2 + 4(3A_3t^2 + 2A_2t + A_1) + 8(A_3t^3 + A_2t^2 + A_1t + A_0) = 16t^3 - 40t^2 - 60t + 4$$

or $$(8A_3)t^3 + (12A_3 + 8A_2)t^2 + (6A_3 + 8A_2 + 8A_1)t + (2A_2 + 4A_1 + 8A_0) = 16t^3 - 40t^2 - 60t + 4$$

Equating coefficients of like powers of t, we get

$$\begin{aligned} 8A_3 &= 16 \quad &\text{or} \quad A_3 &= 2 \\ 12A_3 + 8A_2 &= -40 \quad &\text{or} \quad A_2 &= -8 \\ 6A_3 + 8A_2 + 8A_1 &= -60 \quad &\text{or} \quad A_1 &= -1 \\ 2A_2 + 4A_1 + 8A_0 &= 4 \quad &\text{or} \quad A_0 &= 3 \end{aligned}$$

Then $x_p = 2t^3 - 8t^2 - t + 3$, and the general solution to the nonhomogeneous equation is $x = c_1e^{-2t}\cos 2t + c_2e^{-2t}\sin 2t + 2t^3 - 8t^2 - t + 3$.

9.43 Solve $\dfrac{d^2x}{dt^2} + 4\dfrac{dx}{dt} + 8x = 8t^4 + 16t^3 - 12t^2 - 24t - 6$.

The complementary solution is, again, x_c of Problem 9.39. Since the right side of the given nonhomogeneous differential equation is a fourth-degree polynomial, we assume a general fourth-degree polynomial as the form of a particular solution. That is, we let $x_p = A_4t^4 + A_3t^3 + A_2t^2 + A_1t + A_0$. Since x_p has no term in common with the complementary solution except perhaps for a multiplicative constant, there is no need to modify x_p. Substituting x_p and its derivatives $x_p' = 4A_4t^3 + 3A_3t^2 + 2A_2t + A_1$ and $x_p'' = 12A_4t^2 + 6A_3t + 2A_2$ into the nonhomogeneous differential equation, we get

$$12A_4t^2 + 6A_3t + 2A_2 + 4(4A_4t^3 + 3A_3t^2 + 2A_2t + A_1) + 8(A_4t^4 + A_3t^3 + A_2t^2 + A_1t + A_0) = 8t^4 + 16t^3 - 12t^2 - 24t - 6$$

or $$(8A_4)t^4 + (16A_4 + 8A_3)t^3 + (12A_4 + 12A_3 + 8A_2)t^2 + (6A_3 + 8A_2 + 8A_1)t + (2A_2 + 4A_1 + 8A_0) = 8t^4 + 16t^3 - 12t^2 - 24t - 6$$

Equating coefficients of like powers of t yields

$$\begin{aligned} 8A_4 &= 8 \quad &\text{or} \quad A_4 &= 1 \\ 16A_4 + 8A_3 &= 16 \quad &\text{or} \quad A_3 &= 0 \\ 12A_4 + 12A_3 + 8A_2 &= -12 \quad &\text{or} \quad A_2 &= -3 \\ 6A_3 + 8A_2 + 8A_1 &= -24 \quad &\text{or} \quad A_1 &= 0 \\ 2A_2 + 4A_1 + 8A_0 &= -6 \quad &\text{or} \quad A_0 &= 0 \end{aligned}$$

Then $x_p = t^4 - 3t^2$, and the general solution to the nonhomogeneous equation is $x = c_1e^{-2t}\cos 2t + c_2e^{-2t}\sin 2t + t^4 - 3t^2$.

9.44 Solve $\dfrac{d^2I}{dt^2} - 60\dfrac{dI}{dt} + 900I = 1800t^3 - 300t$.

▌ We assume a particular solution of the form $I_p = A_3t^3 + A_2t^2 + A_1t + A_0$. The general solution of the associated homogeneous equation is found in Problem 8.149 to be $I_c = Ae^{30t} + Bte^{30t}$. Since I_p and I_c have no terms in common, there is no need to modify I_p.

Substituting I_p and its derivatives into the given nonhomogeneous differential equation, we obtain $6A_3t + 2A_2 - 60(3A_3t^2 + 2A_2t + A_1) + 900(A_3t^3 + A_2t^2 + A_1t + A_0) = 1800t^3 - 300t$, or

$$(900A_3)t^3 + (-180A_3 + 900A_2)t^2 + (6A_3 - 120A_2 + 900A_1)t + (2A_2 - 60A_1 + 900A_0) = 1800t^3 - 300t$$

Equating coefficients of like powers of t yields

$$\begin{aligned} 900A_3 &= 1800 &\quad\text{or}\quad A_3 &= 2 \\ -180A_3 + 900A_2 &= 0 &\quad\text{or}\quad A_2 &= 0.4 \\ 6A_3 - 120A_2 + 900A_1 &= -300 &\quad\text{or}\quad A_1 &= -0.2933 \\ 2A_2 - 60A_1 + 900A_0 &= 0 &\quad\text{or}\quad A_0 &= 0.0204 \end{aligned}$$

Then $I_p = 2t^3 + 0.4t^2 - 0.2933t + 0.0204$, and the general solution to the nonhomogeneous equation is $I = Ae^{30t} + Bte^{30t} + 2t^3 + 0.4t^2 - 0.2933t + 0.0204$.

9.45 Solve $\dfrac{d^2I}{dt^2} - 60\dfrac{dI}{dt} + 900I = 900t^4 + 1800t^3 - 3600t^2$.

▌ We assume a particular solution of the form $I_p = A_4t^4 + A_3t^3 + A_2t^2 + A_1t + A_0$. The complementary solution is I_c of the previous problem, and since it has no term in common with I_p, there is no need to modify I_p.

Substituting I_p and its first two derivatives into the given nonhomogeneous equation, we get

$$12A_4t^2 + 6A_3t + 2A_2 - 60(4A_4t^3 + 3A_3t^2 + 2A_2t + A_1) + 900(A_4t^4 + A_3t^3 + A_2t^2 + A_1t + A_0)$$
$$= 900t^4 + 1800t^3 - 3600t^2$$

or

$$(900A_4)t^4 + (-240A_4 + 900A_3)t^3 + (12A_4 - 180A_3 + 900A_2)t^2 + (6A_3 - 120A_2 + 900A_1)t$$
$$+ (2A_2 - 60A_1 + 900A_0) = 900t^4 + 1800t^3 - 3600t^2 + 0t + 0$$

Equating coefficients of like powers of t, we obtain

$$\begin{aligned} 900A_4 &= 900 &\quad\text{or}\quad A_4 &= 1 \\ -240A_4 + 900A_3 &= 1800 &\quad\text{or}\quad A_3 &= 2.267 \\ 12A_4 - 180A_3 + 900A_2 &= -3600 &\quad\text{or}\quad A_2 &= -3.560 \\ 6A_3 - 120A_2 + 900A_1 &= 0 &\quad\text{or}\quad A_1 &= -0.490 \\ 2A_2 - 60A_1 + 900A_0 &= 0 &\quad\text{or}\quad A_0 &= 0.025 \end{aligned}$$

Then $I_p = t^4 + 2.267t^3 - 3.56t^2 - 0.490t + 0.025$, and the general solution is $I = Ae^{30t} + Bte^{30t} + t^4 + 2.267t^3 - 3.56t^2 - 0.490t + 0.025$.

9.46 Solve $\dfrac{d^2I}{dt^2} - 60\dfrac{dI}{dt} + 900I = 4500t^5$.

▌ We assume a particular solution of the form $I_p = A_5t^5 + A_4t^4 + A_3t^3 + A_2t^2 + A_1t + A_0$, which is a general fifth-degree polynomial in t. The complementary solution is I_c of Problem 9.44. Substituting I_p and its first two derivatives, $I_p'' = 20A_5t^3 + 12A_4t^2 + 6A_3t + 2A_2$ and $I_p' = 5A_5t^4 + 4A_4t^3 + 3A_3t^2 + 2A_2t + A^1$, into the given differential equation and rearranging yield

$$(900A_5)t^5 + (-300A_5 + 900A_4)t^4 + (20A_5 - 240A_4 + 900A_3)t^3 + (12A_4 - 180A_3 + 900A_2)t^2$$
$$+ (6A_3 - 120A_2 + 900A_1)t + (2A_2 - 60A_1 + 900A_0) = 4500t^5 + 0t^4 + 0t^3 + 0t^2 + 0t + 0$$

Equating coefficients of like powers of t, we get

$$\begin{aligned} 900A_5 &= 4500 \\ -300A_5 + 900A_4 &= 0 \\ 20A_5 - 240A_4 + 900A_3 &= 0 \\ 12A_4 - 180A_3 + 900A_2 &= 0 \\ 6A_3 - 120A_2 + 900A_1 &= 0 \\ 2A_2 - 60A_1 + 900A_0 &= 0 \end{aligned}$$

The solution of this system is $A_5 = 5$, $A_4 = 1.6667$, $A_3 = 0.3333$, $A_2 = 0.0444$, $A_1 = 0.0037$, and $A_0 = 0.0001$. Thus, the general solution to the nonhomogeneous equation is
$I = I_c + I_p = Ae^{30t} + Bte^{30t} + 5t^5 + 1.667t^4 + 0.3333t^3 + 0.0444t^2 + 0.0037t + 0.0001$.

9.47 Solve $\dfrac{d^3x}{dt^3} + 5\dfrac{d^2x}{dt^2} + 26\dfrac{dx}{dt} - 150x = 600t$.

▌ We try a particular solution of the form $x_p = A_1t + A_0$. The general solution to the associated homogeneous equation is found in Problem 8.112 to be $x_h = c_1e^{3t} + c_2e^{-4t}\cos\sqrt{34}t + c_3e^{-4t}\sin\sqrt{34}t$. Since there is no need to modify x_p, we substitute it and its derivatives into the given differential equation, obtaining
$0 + 0 + 26A_1 - 150(A_1t + A_0) = 600t$, or $(-150A_1)t + (26A_1 - 150A_0) = 600t$. It follows that

$$\begin{aligned} -150A_1 &= 600 \quad \text{or} \quad A_1 = -4 \\ 26A_1 - 150A_0 &= 0 \quad \text{or} \quad A_0 = -0.693 \end{aligned}$$

Then $x_p = -4t - 0.693$, and the general solution is
$x = c_1e^{3t} + c_2e^{-4t}\cos\sqrt{34}t + c_3e^{-4t}\sin\sqrt{34}t - 4t - 0.693$.

9.48 Solve $\dfrac{d^3x}{dt^3} + 5\dfrac{d^2x}{dt^2} + 26\dfrac{dx}{dt} - 150x = 600t^2$.

▌ We assume a particular solution of the form $x_p = A_2t^2 + A_1t + A_0$, while the general solution x_h of the previous problem is valid here. Substituting x_p into the given differential equation yields
$0 + 5(2A_2) + 26(2A_2t + A_1) - 150(A_2t^2 + A_1t + A_0) = 600t^2$, or
$(-150A_2)t^2 + (52A_2 - 150A_1)t + (10A_2 + 26A_1 - 150A_0) = 600t^2 + 0t + 0$. Equating coefficients of like powers of t, we find

$$\begin{aligned} -150A_2 &= 600 \quad \text{or} \quad A_2 = -4 \\ 52A_2 - 150A_1 &= 0 \quad \text{or} \quad A_1 = -1.387 \\ 10A_2 + 26A_1 - 150A_0 &= 0 \quad \text{or} \quad A_0 = -0.507 \end{aligned}$$

Then $x = x_h + x_p = c_1e^{3t} + c_2e^{-4t}\cos\sqrt{34}t + c_3e^{-4t}\sin\sqrt{34}t - 4t^2 - 1.387t - 0.507$.

9.49 Solve $\dfrac{d^3x}{dt^3} + 5\dfrac{d^2x}{dt^2} + 26\dfrac{dx}{dt} - 150x = 600t^3$.

▌ We assume a particular solution of the form $x_p = A_3t^3 + A_2t^2 + A_1t + A_0$, while x_h of Problem 9.47 is valid here. Substituting x_p and its derivatives into the given differential equation and equating coefficients of like powers of t in the resulting equation, we obtain

$$\begin{aligned} -150A_3 &= 600 \quad \text{or} \quad A_3 = -4 \\ 78A_3 - 150A_2 &= 0 \quad \text{or} \quad A_2 = -2.08 \\ 30A_3 + 52A_2 - 150A_1 &= 0 \quad \text{or} \quad A_1 = -1.521 \\ 6A_3 + 10A_2 + 26A_1 - 150A_0 &= 0 \quad \text{or} \quad A_0 = -0.562 \end{aligned}$$

Then $x = x_h + x_p = c_1e^{3t} + c_2e^{-4t}\cos\sqrt{34}t + c_3e^{-4t}\sin\sqrt{34}t - 4t^3 - 2.08t^2 - 1.521t - 0.562$.

9.50 Solve $\dfrac{d^3x}{dt^3} + 5\dfrac{d^2x}{dt^2} + 26\dfrac{dx}{dt} - 150x = 600t^4$.

▌ We assume a particular solution of the form $x_p = A_4t^4 + A_3t^3 + A_2t^2 + A_1t + A_0$, while x_h of Problem 9.47 is valid here. Substituting x_p and its derivatives into the given differential equation and equating coefficients of like powers of t in the result, we get

$$\begin{aligned} -150A_4 &= 600 \quad \text{or} \quad A_4 = -4 \\ 104A_4 - 150A_3 &= 0 \quad \text{or} \quad A_3 = -2.773 \\ 60A_4 + 78A_3 - 150A_2 &= 0 \quad \text{or} \quad A_2 = -3.042 \\ 24A_4 + 30A_3 + 52A_2 - 150A_1 &= 0 \quad \text{or} \quad A_1 = -2.249 \\ 6A_3 + 10A_2 + 26A_1 - 150A_0 &= 0 \quad \text{or} \quad A_0 = -0.704 \end{aligned}$$

Then the general solution is
$x = x_h + x_p = c_1e^{3t} + c_2e^{-4t}\cos\sqrt{34}t + c_3e^{-4t}\sin\sqrt{34}t - 4t^4 - 2.773t^3 - 3.042t^2 - 2.249t - 0.704$.

.51 Solve $\dfrac{d^3Q}{dt^3} - 5\dfrac{d^2Q}{dt^2} + 25\dfrac{dQ}{dt} - 125Q = -625t^4 + 250t^3 - 150t^2 + 60t + 137.$

▌ We try a particular solution of the form $Q_p = A_4t^4 + A_3t^3 + A_2t^2 + A_1t + A_0$. The complementary solution is shown in Problem 8.113 to be $Q_c = c_1e^{5t} + c_2\cos 5t + c_3\sin 5t$. Since Q_p and Q_c have no terms in common, there is no need to modify Q_p. Substituting Q_p and its derivatives into the given differential equation and equating coefficients of like powers of t in the result, we obtain

$$\begin{aligned}
-125A_4 &= -625 &\text{or}\quad A_4 &= 5\\
100A_4 - 125A_3 &= 250 &\text{or}\quad A_3 &= 2\\
-60A_4 + 75A_3 - 125A_2 &= -150 &\text{or}\quad A_2 &= 0\\
24A_4 - 30A_3 + 50A_2 - 125A_1 &= 60 &\text{or}\quad A_1 &= 0\\
6A_3 - 10A_2 + 25A_1 - 125A_0 &= 137 &\text{or}\quad A_0 &= -1
\end{aligned}$$

Then the general solution is $Q = Q_c + Q_p = c_1e^{5t} + c_2\cos 5t + c_3\sin 5t + 5t^4 + 2t^3 - 1$.

52 Solve $\dfrac{d^3Q}{dt^3} - 5\dfrac{d^2Q}{dt^2} + 25\dfrac{dQ}{dt} - 125Q = 5000t^5 - 3000t^3.$

▌ We assume a particular solution of the form $Q_p = A_5t^5 + A_4t^4 + A_3t^3 + A_2t^2 + A_1t + A_0$, while Q_c remains as in the previous problem. Substituting Q_p and its derivatives into the given differential equation and equating coefficients of like powers of t in the result, we get

$$\begin{aligned}
-125A_5 &= 5000\\
125A_5 - 125A_4 &= 0\\
-100A_5 + 100A_4 - 125A_3 &= -3000\\
60A_5 - 60A_4 + 75A_3 - 125A_2 &= 0\\
24A_4 - 30A_3 + 50A_2 - 125A_1 &= 0\\
6A_3 - 10A_2 + 25A_1 - 125A_0 &= 0
\end{aligned}$$

This system may be solved to yield $A_5 = A_4 = -40$, $A_3 = 24$, $A_2 = 14.4$, $A_1 = -7.68$, and $A_0 = -1.536$. Then the general solution is
$Q = Q_p + Q_c = c_1e^{5t} + c_2\cos 5t + c_3\sin 5t - 40t^5 - 40t^4 + 24t^3 + 14.4t^2 - 7.68t - 1.536$.

53 Solve $(D^4 - 16)y = 80x^2$.

▌ We assume a particular solution of the form $y_p = A_2x^2 + A_1x + A_0$. The general solution to the associated homogeneous equation is shown in Problem 8.128 to be $y_h = c_1\cos 2x + c_2\sin 2x + c_3e^{2x} + c_4e^{-2x}$. Since y_p and y_h have no terms in common, there is no need to modify y_p.

Substituting y_p into the nonhomogeneous equation and noting that $D^4y_p = 0$, we find $0 - 16(A_2x^2 + A_1x + A_0) = 80x^2$, or $(-16A_2)x^2 + (-16A_1)x + (-16A_0) = 80x^2 + 0x + 0$. Equating coefficients of like powers of x then yields $-16A_2 = 80$, so that $A_2 = -5$, and $A_1 = A_0 = 0$. Thus $y_p = -5x^2$, and the general solution is $y = c_1\cos 2x + c_2\sin 2x + c_3e^{2x} + c_4e^{-2x} - 5x^2$.

54 Solve $(D^4 - 16)y = 80x^5 - 16$.

▌ We assume a particular solution of the form $y_p = A_5x^5 + A_4x^4 + A_3x^3 + A_2x^2 + A_1x + A_0$, while y_h of the previous problem is valid here as well. Substituting y_p and $D^4y_p = 120A_5x + 24A_4$ into the given differential equation, equating coefficients of like powers of x in the result, and solving for the coefficients of y_p, we find that $A_5 = -5$, $A_4 = A_3 = A_2 = 0$, $A_1 = -37.5$, and $A_0 = 1$. Then $y_p = -5x^5 - 37.5x + 1$, and the general solution is $y = c_1\cos 2x + c_2\sin 2x + c_3e^{2x} + c_4e^{-2x} - 5x^5 - 37.5x + 1$.

QUATIONS WHOSE RIGHT SIDE IS THE PRODUCT OF POLYNOMIAL AND AN EXPONENTIAL

55 Solve $y' - 5y = xe^{2x}$.

▌ Since the right side of this equation is the product of a first-degree polynomial and an exponential, we assume a particular solution of the same form—a general first-degree polynomial times an exponential. We try $y_p = (A_1x + A_0)e^{2x}$. The solution to the associated homogeneous differential equation is found in Problem 8.34 to be $y_h = c_1e^{5x}$. Since y_p and y_h have no terms in common, there is no need to modify y_p.

Substituting y_p into the given differential equation, while noting that $y_p' = A_1e^{2x} + 2(A_1x + A_0)e^{2x}$, we obtain $A_1e^{2x} + 2(A_1x + A_0)e^{2x} - 5(A_1x + A_0)e^{2x} = xe^{2x}$, which may be simplified to $(-3A_1)x + (A_1 - 3A_0) = x$. Equating coefficients of like powers of x, we obtain

$$-3A_1 = 1 \quad \text{or} \quad A_1 = -\tfrac{1}{3}$$
$$A_1 - 3A_0 = 0 \quad \text{or} \quad A_0 = -\tfrac{1}{9}$$

Then $y_p = (-\frac{1}{3}x - \frac{1}{9})e^{2x}$, and the general solution to the nonhomogeneous differential equation is $y = y_h + y_p = c_1e^{5x} - \frac{1}{3}xe^{2x} - \frac{1}{9}e^{2x}$.

9.56 Solve $y' - 5y = (2x - 1)e^{2x}$.

▌ The right side of this equation is again a first-degree polynomial times an exponential, and both y_p and y_h of the previous problem are valid here. Substituting y_p into the given differential equation and simplifying, we obtain $(-3A_1)x + (A_1 - 3A_0) = 2x - 1$. Equating coefficients of like powers of x yields

$$-3A_1 = 2 \quad \text{or} \quad A_1 = -\tfrac{2}{3}$$
$$A_1 - 3A_0 = -1 \quad \text{or} \quad A_0 = \tfrac{1}{9}$$

Then $y_p = (-\frac{2}{3}x + \frac{1}{9})e^{2x}$, and the general solution to the nonhomogeneous differential equation is $y = y_h + y_p = c_1e^{5x} + (-\frac{2}{3}x + \frac{1}{9})e^{2x}$.

9.57 Solve $y' - 5y = 3x^2e^{2x}$.

▌ Since the right side of this equation is a second-degree polynomial times an exponential, we assume a particular solution in the form of a general second-degree polynomial times an exponential: $y_p = (A_2x^2 + A_1x + A_0)e^{2x}$. From Problem 8.34 we have, as the general solution of the associated homogeneous differential equation, $y_h = c_1e^{5x}$. Because y_p and y_h have no terms in common except perhaps for a multiplicative constant, there is no need to modify y_p.

Substituting y_p and $y_p' = [2A_2x^2 + (2A_2 + 2A_1)x + A_1 + 2A_0]e^{2x}$ into the given differential equation yields, after simplification, $(-3A_2)x^2 + (2A_2 - 3A_1)x + (A_1 - 3A_0) = 3x^2 + 0x + 0$. Equating coefficients of like powers of x, we obtain

$$-3A_2 = 3 \quad \text{or} \quad A_2 = -1$$
$$2A_2 - 3A_1 = 0 \quad \text{or} \quad A_1 = -\tfrac{2}{3}$$
$$A_1 - 3A_0 = 0 \quad \text{or} \quad A_0 = -\tfrac{2}{9}$$

Then $y_p = (-\frac{2}{9}x^2 - \frac{2}{3}x - 1)e^{2x}$, and the general solution to the nonhomogeneous differential equation is $y = y_h + y_p = c_1e^{5x} + (-\frac{2}{9}x^2 - \frac{2}{3}x - 1)e^{2x}$.

9.58 Solve $y' - 5y = (-9x^2 + 6x)e^{2x}$.

▌ The right side of this equation is again a second-degree polynomial times an exponential, and y_p, y_p', and y_h are all as in the previous problem. Substituting the first two into the given differential equation yields

$$[2A_2x^2 + (2A_2 + 2A_1)x + A_1 + 2A_0]e^{2x} - 5(A_2x^2 + A_1x + A_0)e^{2x} = (-9x^2 + 6x)e^{2x}$$

After simplifying and equating the coefficients of like powers of x, we have

$$-3A_2 = -9 \quad \text{or} \quad A_2 = 3$$
$$2A_2 - 3A_1 = 6 \quad \text{or} \quad A_1 = 0$$
$$A_1 - 3A_0 = 0 \quad \text{or} \quad A_0 = 0$$

Then $y_p = 3x^2e^{2x}$, and the general solution to the given differential equation is $y = y_h + y_p = c_1e^{5x} + 3x^2e^{2x}$.

9.59 Solve $y' - 5y = (2x^3 - 5)e^{2x}$.

▌ Since the right side of this equation is the product of a third-degree polynomial and an exponential, we assume a particular solution in the form of a general third-degree polynomial times an exponential: $y_p = (A_3x^3 + A_2x^2 + A_1x + A_0)e^{2x}$. The complementary solution is found in Problem 8.34 to be $y_h = c_1e^{5x}$. Since y_p and y_h have no terms in common, there is no need to modify y_p.

Substituting y_p and $y_p' = (3A_3x^2 + 2A_2x + A_1)e^{2x} + 2(A_3x^3 + A_2x^2 + A_1x + A_0)e^{2x}$ into the given differential equation yields, after simplification,

$$(-3A_3)x^3 + (3A_3 - 3A_2)x^2 + (2A_2 - 3A_1)x + (A_1 - 3A_2) = 2x^3 + 0x^2 + 0x - 5$$

Equating coefficients of like powers of x and solving the resulting system of equations, we obtain $A_3 = A_2 = -\frac{2}{3}$, $A_1 = -\frac{4}{9}$, and $A_0 = \frac{41}{27}$. The general solution is then $y = y_h + y_p = c_1 e^{5x} + (-\frac{2}{3}x^3 - \frac{2}{3}x^2 - \frac{4}{9}x + \frac{41}{27})e^{2x}$.

9.60 Solve $y' + 6y = 18xe^{-3x}$.

▌ We assume a particular solution of the form $y_p = (A_1 x + A_0)e^{-3x}$, which is the product of a first-degree polynomial and an exponential. The complementary function is shown in Problem 8.37 to be $y_c = Ae^{-6x}$. Since y_p and y_c have no terms in common, there is no need to modify y_p.

Substituting y_p into the given differential equation yields, after simplification, $(3A_1)x + (A_1 + 3A_0) = 18x + 0$. Equating coefficients of like powers of x then yields $A_1 = 6$ and $A_0 = -2$, so $y_p = 6x - 2$ and the general solution to the nonhomogeneous equation is $y = Ae^{-6x} + (6x - 2)e^{-3x}$.

9.61 Solve $y' + 6y = 9x^3 e^x - 12x^2 e^x$.

▌ The right side of this equation is $(9x^3 - 12x^2)e^x$, which is the product of a third-degree polynomial and e^x, so we assume $y_p = (A_3 x^3 + A_2 x^2 + A_1 x + A_0)e^x$. Also, y_c of the previous problem is valid here; since it has no term in common with y_p, there is no need to modify y_p.

Substituting y_p into the given differential equation yields

$$(3A_3 x^2 + 2A_2 x + A_1)e^x + (A_3 x^3 + A_2 x^2 + A_1 x + A_0)e^x + 6(A_3 x^3 + A_2 x^2 + A_1 x + A_0)e^x = (9x^3 - 12x^2)e^x$$

After this equation is simplified and the coefficients of like powers of x are equated, we have

$$\begin{aligned} 7A_3 &= 9 &\quad\text{or}\quad A_3 &= 1.29 \\ 3A_3 + 7A_2 &= -12 &\quad\text{or}\quad A_2 &= -2.27 \\ 2A_2 + 7A_1 &= 0 &\quad\text{or}\quad A_1 &= 0.65 \\ A_1 + 7A_0 &= 0 &\quad\text{or}\quad A_0 &= -0.09 \end{aligned}$$

The general solution is then $y = y_h + y_p = Ae^{-6x} + (1.29x^3 - 2.27x^2 + 0.65x - 0.09)e^x$.

.62 Solve $y'' - 7y' = (3 - 36x)e^{4x}$.

▌ We try $y_p = (A_1 x + A_0)e^{4x}$, a first-degree polynomial times an exponential, as a particular solution. The complementary solution is shown in Problem 8.2 to be $y_c = c_1 + c_2 e^{7x}$. Since y_p and y_c have no term in common, there is no need to modify y_p.

Substituting $y_p' = (4A_1 x + A_1 + 4A_0)e^{4x}$, and $y_p'' = (16A_1 x + 8A_1 + 16A_0)e^{4x}$ into the given differential equation and simplifying, we obtain $(-12A_1)x + (A_1 - 12A_0) = -36x + 3$. Equating coefficients of like powers of x yields a system of two equations from which we find $A_1 = 3$ and $A_0 = 0$. The general solution is then $y = y_c + y_p = c_1 + c_2 e^{7x} + 3xe^{4x}$.

.63 Solve $y'' - 7y' = (-80x^2 - 108x + 38)e^{2x}$.

▌ We try as a particular solution $y_p = (A_2 x^2 + A_1 x + A_0)e^{2x}$, while y_c of the previous problem is valid here as well. Since y_c has no term in common with y_p, there is no need to modify y_p.

Substituting $y_p' = (2A_2 x^2 + 2A_2 x + 2A_1 x + A_1 + 2A_0)e^{2x}$ and $y_p'' = (4A_2 x^2 + 8A_2 x + 4A_1 x + 2A_2 + 4A_1 + 4A_0)e^{2x}$ into the given differential equation, simplifying the result, and then equating coefficients of like powers of x, we get

$$\begin{aligned} -10A_2 &= -80 &\quad\text{or}\quad A_2 &= 8 \\ -6A_2 - 10A_1 &= -108 &\quad\text{or}\quad A_1 &= 6 \\ 2A_2 - 3A_1 - 10A_0 &= 38 &\quad\text{or}\quad A_0 &= -4 \end{aligned}$$

The general solution to the given nonhomogeneous equation is then $y = c_1 + c_2 e^{7x} + (8x^2 + 6x - 4)e^{2x}$.

64 Solve $y'' - y' + 2y = (6x^2 + 8x + 7)e^x$.

▌ We try as a particular solution $y_p = (A_2 x^2 + A_1 x + A_0)e^x$. The complementary solution is found in Problem 8.1 to be $y_c = c_1 e^{-x} + c_2 e^{2x}$. Since these two solutions have no term in common, there is no need to modify y_p.

Substituting y_p, $y'_p = (A_2x^2 + 2A_2x + A_1x + A_1 + A_0)e^x$, and $y''_p = (A_2x^2 + 4A_2x + A_1x + 2A_2 + 2A_1 + A_0)e^x$ into the given differential equation, simplifying, and then equating coefficients of like powers of x, we get

$$\begin{aligned} 2A_2 &= 6 \quad \text{or} \quad A_2 = 3 \\ 2A_2 + 2A_1 &= 8 \quad \text{or} \quad A_1 = 1 \\ 2A_1 + A_1 + 2A_0 &= 7 \quad \text{or} \quad A_0 = 0 \end{aligned}$$

Thus, the general solution is $y = c_1e^{-x} + c_2e^{2x} + (3x^2 + x)e^x$.

9.65 Solve $y'' - y' + 2y = (x^2 - x + 4)e^x$.

▎ The expressions for y_p, y'_p, y''_p, and y_c of the previous problem are all valid here. Substituting the first three into the given differential equation and simplifying, we get $(2A_2)x^2 + (2A_2 + 2A_1)x + (2A_2 + A_1 + 2A_0) = x^2 - x + 4$. By equating coefficients of like powers of x and solving the resulting system of equations, we find that $A_2 = \frac{1}{2}$, $A_1 = -1$, and $A_0 = 2$. The general solution is then $y = y_c + y_p = c_1e^{-x} + c_2e^{2x} + (\frac{1}{2}x^2 - x + 2)e^x$.

9.66 Solve $y'' - y' + 2y = (x^2 - x + 4)e^{4x}$.

▎ The complementary solution is that of Problem 9.64: $y_c = c_1e^{-x} + c_2e^{2x}$, and we try a particular solution of the form $y_p = (A_2x^2 + A_1x + A_0)e^{4x}$. Since y_c and y_p have no terms in common, we need not modify y_p.

Substituting y_p, $y'_p = (4A_2x^2 + 2A_2x + 4A_1x + A_1 + 4A_0)e^{4x}$, and $y''_p = (16A_2x^2 + 16A_2x + 16A_1x + 2A_2 + 8A_1 + 16A_0)e^{4x}$ into the nonhomogeneous differential equation and simplifying yield

$$(14A_2)x^2 + (14A_2 + 14A_1)x + (2A_2 + 7A_1 + 14A_0) = x^2 - x + 4$$

Equating the coefficients of like powers of x, we obtain a system of three equations that yields $A_2 = 0.071$, $A_1 = -0.143$, and $A_0 = 0.347$. The general solution is then $y = y_c + y_p = c_1e^{-x} + c_2e^{2x} + (0.071x^2 - 0.143x + 0.347)e^{4x}$.

9.67 Solve $\dfrac{d^2x}{dt^2} + 4\dfrac{dx}{dt} + 8x = (20t^2 + 16t - 78)e^{2t}$.

▎ The complementary solution is shown in Problem 8.54 to be $x_c = c_1e^{-2t}\cos 2t + c_2e^{-2t}\sin 2t$. We try a particular solution of the form $x_p = (A_2t^2 + A_1t + A_0)e^{2t}$, and since this function has no term in common with x_c there is no need to modify it.

Substituting x_p, $x'_p = (2A_2t^2 + 2A_2t + 2A_1t + A_1 + 2A_0)e^{2t}$, and $x''_p = (4A_2t^2 + 8A_2t + 4A_1t + 2A_2 + 4A_1 + 4A_0)e^{2t}$ into the given differential equation, we get, after simplifying

$$(20A_2)t^2 + (16A_2 + 20A_1)t + (2A_2 + 8A_1 + 20A_0) = 20t^2 + 16t - 78$$

Then equating the coefficients of like powers of t yields

$$\begin{aligned} 20A_2 &= 20 \quad \text{or} \quad A_2 = 1 \\ 16A_2 + 20A_1 &= 16 \quad \text{or} \quad A_1 = 0 \\ 2A_2 + 8A_1 + 20A_0 &= -78 \quad \text{or} \quad A_0 = -4 \end{aligned}$$

Thus the general solution is $x = c_1e^{-2t}\cos 2t + c_2e^{-2t}\sin 2t + (t^2 - 4)e^{2t}$.

9.68 Solve $\dfrac{d^2x}{dt^2} + 4\dfrac{dx}{dt} + 8x = (5t^2 - 14t + 11)e^{-3t}$.

▎ The complementary solution x_c of the previous problem is valid here, but now we assume a particular solution of the form $x_p = (A_2t^2 + A_1t + A_0)e^{-3t}$. Substituting x_p, $x'_p = (-3A_2t^2 + 2A_2t - 3A_1t + A_1 - 3A_0)e^{-3t}$, and $x''_p = (9A_2t^2 - 12A_2t + 9A_1t + 2A_2 - 6A_1 + 9A_0)e^{-3t}$ into the given differential equation yields, after simplification,

$$(5A_2)t^2 + (-4A_2 + 5A_1)t + (2A_2 - 2A_1 + 5A_0) = 5t^2 - 14t + 11$$

By equating coefficients of like powers of t and solving the resulting system of equations, we find that $A_2 = 1$, $A_1 = -2$, and $A_0 = 1$. The general solution is then $x = c_1e^{-2t}\cos 2t + c_2e^{-2t}\sin 2t + (t^2 - 2t + 1)e^{-3t}$.

9.69 Solve $\dfrac{d^2x}{dt^2} + 4\dfrac{dx}{dt} + 8x = (29t^3 + 30t^2 - 52t - 20)e^{3t}$.

▌ The complementary solution of the preceding two problems is valid here, and we try a particular solution of the form $x_p = (A_3t^3 + A_2t^2 + A_1t + A_0)e^{3t}$. This trial solution needs no modification.

Substituting x_p, $x_p' = (3A_3t^3 + 3A_3t^2 + 3A_2t^2 + 2A_2t + 3A_1t + A_1 + 3A_0)e^{3t}$, and $x_p'' = (9A_3t^3 + 18A_3t^2 + 9A_2t^2 + 6A_3t + 12A_2t + 9A_1t + 2A_2 + 6A_1 + 9A_0)e^{3t}$ into the given differential equation yields, after simplification,

$$(29A_3)t^3 + (30A_3 + 29A_2)t^2 + (6A_3 + 20A_2 + 29A_1)t + (2A_2 + 10A_1 + 29A_0) = 29t^3 + 30t^2 - 52t - 20$$

Then equating the coefficients of like powers of t and solving the resulting system of equations, we find that $A_3 = 1$, $A_2 = 0$, $A_1 = -2$, and $A_0 = 0$. The general solution is thus $x = x_c + x_p = c_1e^{-2t}\cos 2t + c_2e^{-2t}\sin 2t + (t^3 - 2t)e^{3t}$.

9.70 Solve $\ddot{x} + 9\dot{x} + 14x = (12t^2 + 22t + 27)e^{-t}$.

▌ The complementary solution is shown in Problem 8.17 to be $x_c = c_1e^{-2t} + c_2e^{-7t}$. We try a particular solution of the form $x_p = (A_2t^2 + A_1t + A_0)e^{-t}$, which needs no modification.

Substituting x_p, $x_p' = (-A_2t^2 + 2A_2t - A_1t + A_1 - A_0)e^{-t}$, and $x_p'' = (A_2t^2 - 4A_2t + A_1t + 2A_2 - 2A_1 + A_0)e^{-t}$ into the given differential equation and simplifying the result, we get

$$(6A_2)t^2 + (14A_2 + 6A_1)t + (2A_2 + 7A_1 + 6A_0) = 12t^2 + 22t + 27$$

Then equating the coefficients of like powers of t, we obtain

$$\begin{aligned} 6A_2 \qquad\qquad &= 12 \quad \text{or} \quad A_2 = 2 \\ 14A_2 + 6A_1 \qquad &= 22 \quad \text{or} \quad A_1 = -1 \\ 2A_2 + 7A_1 + 6A_0 &= 27 \quad \text{or} \quad A_0 = 5 \end{aligned}$$

Thus, the general solution is $x = c_1e^{-2t} + c_2e^{-7t} + (2t^2 - t + 5)e^{-t}$.

9.71 Solve $\ddot{x} + 9\dot{x} + 14x = (144t^3 + 156t^2 + 24t)e^{2t}$.

▌ The complementary solution x_c of the previous problem is valid here as well. We try a particular solution of the form $x_p = (A_3t^3 + A_2t^2 + A_1t + A_0)e^{2t}$, which needs no modification.

Substituting x_p, $x_p' = (2A_3t^3 + 3A_3t^2 + 2A_2t^2 + 2A_2t + 2A_1t + A_1 + 2A_0)e^{2t}$, and $x_p'' = (4A_3t^3 + 12A_3t^2 + 4A_2t^2 + 6A_3t + 8A_2t + 4A_1t + 2A_2 + 4A_1 + 4A_0)e^{2t}$ into the given differential equation and simplifying yield

$$(36A_3)t^3 + (39A_3 + 36A_2)t^2 + (6A_3 + 26A_2 + 36A_1)t + (2A_2 + 13A_1 + 36A_0) = 144t^3 + 156t^2 + 24t$$

Equating the coefficients of like powers of t, we then have

$$\begin{aligned} 36A_3 \qquad\qquad\qquad &= 144 \quad \text{or} \quad A_3 = 4 \\ 39A_3 + 36A_2 \qquad\qquad &= 156 \quad \text{or} \quad A_2 = 0 \\ 6A_3 + 26A_2 + 36A_1 \qquad &= 24 \quad \text{or} \quad A_1 = 0 \\ 2A_2 + 13A_1 + 36A_0 &= 0 \quad \text{or} \quad A_0 = 0 \end{aligned}$$

The general solution is thus $x = x_c + x_p = c_1e^{-2t} + c_2e^{-7t} + 4t^3e^{2t}$.

9.72 Solve $\ddot{x} + 10\dot{x} + 25x = (2t - 10)e^{3t}$.

▌ The complementary solution is found in Problem 8.146 to be $x_c = C_1e^{-5t} + C_2te^{-5t}$. We assume a particular solution of the form $x_p = (A_1t + A_0)e^{3t}$, and since it has no terms in common with x_c, it requires no modification.

Substituting x_p, $x_p' = (3A_1t + A_1 + 3A_0)e^{3t}$, and $x_p'' = (9A_1t + 6A_1 + 9A_0)e^{3t}$ into the given differential equation, we have, after simplification, $(64A_1)t + (16A_1 + 64A_0) = 2t - 10$. Equating the coefficients of like powers of t, we get a pair of equations whose solution is $A_1 = \dfrac{1}{32}$ and $A_2 = \dfrac{-21}{128}$. Combining x_c and x_p, we then form the general solution $x = (C_1 + C_2t)e^{-5t} + (4t - 21)e^{3t}/128$.

9.73 Solve $\ddot{x} + 10\dot{x} + 25x = (320t^3 + 48t^2 - 66t + 122)e^{3t}$

▌ The complementary solution x_c of the previous problem is valid here, but we now assume a particular solution of the form $x_p = (A_3t^3 + A_2t^2 + A_1t + A_0)e^{3t}$. Substituting x_p and its first two derivatives (see Problem 9.69) into the given differential equation and simplifying, we get

$$(64A_3)t^3 + (48A_3 + 64A_2)t^2 + (6A_3 + 32A_2 + 64A_1)t + (2A_2 + 16A_1 + 64A_0) = 320t^3 + 48t^2 - 66t + 122$$

Then equating coefficients of like powers of t, we obtain

$$\begin{aligned} 64A_3 &= 320 \quad \text{or} \quad A_3 = 5 \\ 48A_3 + 64A_2 &= 48 \quad \text{or} \quad A_2 = -3 \\ 6A_3 + 32A_2 + 64A_1 &= -66 \quad \text{or} \quad A_1 = 0 \\ 2A_2 + 16A_1 + 64A_0 &= 122 \quad \text{or} \quad A_0 = 2 \end{aligned}$$

The general solution is then $x = (C_1 + C_2t)e^{-5t} + (5t^3 - 3t^2 + 2)e^{3t}$.

9.74 Solve $y''' - 6y'' + 11y' - 6y = 2xe^{-x}$.

▌ The complementary solution is found in Problem 8.28 to be $y_c = c_1e^x + c_2e^{2x} + c_3e^{3x}$. We try a particular solution of the form $y_p = (A_1x + A_0)e^{-x}$, which has no term in common with y_c and therefore requires no modification. Substituting

$$y_p' = -A_1xe^{-x} + A_1e^{-x} - A_0e^{-x}$$
$$y_p'' = A_1xe^{-x} - 2A_1e^{-x} + A_0e^{-x}$$

and

$$y_p''' = -A_1xe^{-x} + 3A_1e^{-x} - A_0e^{-x}$$

into the given differential equation and simplifying, we obtain $(-24A_1)x + (26A_1 - 24A_0) = 2x + 0$. Equating coefficients of like powers of x and solving the resulting system yield $A_1 = -\frac{1}{12}$ and $A_0 = -\frac{13}{144}$. Then $y_p = -\frac{1}{12}xe^{-x} - \frac{13}{144}e^{-x}$, and the general solution is $y = c_1e^x + c_2e^{2x} + c_3e^{3x} - \frac{1}{12}xe^{-x} - \frac{13}{144}e^{-x}$.

9.75 Solve $\dfrac{d^3Q}{dt^3} - 5\dfrac{d^2Q}{dt^2} + 25\dfrac{dQ}{dt} - 125Q = (-522t^2 + 465t - 387)e^{2t}$.

▌ The complementary solution is found in Problem 8.113 to be $Q_c = c_1e^{5t} + c_2\cos 5t + c_3\sin 5t$. We try a particular solution of the form $Q_p = (A_2t^2 + A_1t + A_0)e^{2t}$, which requires no modification because Q_p has no term (except perhaps for a multiplicative constant) in common with Q_c.

Substituting Q_p,

$$Q_p' = (2A_2t^2 + 2A_2t + 2A_1t + A_1 + 2A_0)e^{2t}$$
$$Q_p'' = (4A_2t^2 + 8A_2t + 4A_1t + 2A_2 + 4A_1 + 4A_0)e^{2t}$$

and

$$Q_p''' = (8A_2t^2 + 24A_2t + 8A_1t + 12A_2 + 12A_1 + 8A_0)e^{2t}$$

into the given differential equation and simplifying, we get

$$(-87A_2)t^2 + (34A_2 - 87A_1)t + (2A_2 + 17A_1 - 87A_0) = -522t^2 + 465t - 387$$

Equating the coefficients of like terms, we obtain

$$\begin{aligned} -87A_2 &= -522 \quad \text{or} \quad A_2 = 6 \\ 34A_2 - 87A_1 &= 465 \quad \text{or} \quad A_1 = -3 \\ 2A_2 + 17A_1 - 87A_0 &= -387 \quad \text{or} \quad A_0 = 4 \end{aligned}$$

The general solution is then $Q = c_1e^{5t} + c_2\cos 5t + c_3\sin 5t + (6t^2 - 3t + 4)e^{2t}$.

9.76 Solve $(D^4 + 4D^2)y = (3x - 1)e^{4x}$.

▌ The complementary solution is found in Problem 8.186 to be $y_c = c_1 + c_2x + c_3\cos 2x + c_4\sin 2x$. We assume a particular solution of the form $y_p = (A_1x + A_0)e^{4x}$, which needs no modification because y_p has no terms in common with y_c.

Substituting $y_p'' = (16A_1x + 8A_1 + 16A_0)e^{4x}$ and $y_p'''' = (256A_1x + 256A_1 + 256A_0)e^{4x}$ into the given differential equation and simplifying yield $(320A_1)x + (288A_1 + 320A_0) = 3x - 1$. Equating coefficients of like terms and solving the resulting system, we obtain $A_1 = 0.00938$ and $A_0 = -0.01156$. The general solution is then $y = y_c + y_p = c_1 + c_2x + c_3\cos 2x + c_4\sin 2x + (0.00938x - 0.01156)e^{4x}$.

EQUATIONS WHOSE RIGHT SIDE CONTAINS SINES AND COSINES

.77 Solve $y' - 5y = \sin x$.

The complementary solution is found in Problem 8.34 to be $y_c = c_1 e^{5x}$. We assume a particular solution of the form $y_p = A_0 \sin x + B_0 \cos x$, which needs no modification because y_p and y_c have no terms in common except perhaps for a multiplicative constant.

Substituting y_p and its derivative into the given differential equation yields $A_0 \cos x - B_0 \sin x - 5(A_0 \sin x + B_0 \cos x) = \sin x$, which we rearrange as

$$(-5A_0 - B_0)\sin x + (A_0 - 5B_0)\cos x = 1 \sin x + 0 \cos x$$

Equating the coefficients of like terms, we obtain the system

$$-5A_0 - B_0 = 1 \qquad A_0 - 5B_0 = 0$$

from which we find $A_0 = -\frac{5}{26}$ and $B_0 = -\frac{1}{26}$. Then the general solution is $y = y_c + y_p = c_1 e^{5x} - \frac{5}{26}\sin x - \frac{1}{26}\cos x$.

.78 Solve $y' + 6y = -2\cos 3x$.

The complementary solution is shown in Problem 8.37 to be $y_c = Ae^{-6x}$. We assume a particular solution of the form $y_p = A_0 \sin 3x + B_0 \cos 3x$, which need not be modified.

Substituting y_p and its derivative $y_p' = 3A_0 \cos 3x - 3B_0 \sin 3x$ into the differential equation and rearranging, we obtain

$$(6A_0 - 3B_0)\sin 3x + (3A_0 + 6B_0)\cos 3x = 0 \sin 3x + (-2)\cos 3x$$

Equating coefficients of like terms and solving the resulting system, we find $A_0 = -\frac{2}{15}$ and $B_0 = -\frac{4}{15}$. The general solution is then $y = Ae^{-6x} - \frac{2}{15}\sin 3x - \frac{4}{15}\cos 3x$.

.79 Solve $2y' - 5y = \sin 2x - 7\cos 2x$.

The complementary solution is shown in Problem 8.38 to be $y_c = Ae^{5x/2}$. We assume a particular solution of the form $y_p = A_0 \sin 2x + B_0 \cos 2x$, which needs no modification. Substituting y_p and its derivative into the differential equation and rearranging yield

$$(-5A_0 - 4B_0)\sin 2x + (4A_0 - 5B_0)\cos 2x = 1 \sin 2x + (-7)\cos 2x$$

Equating coefficients of like terms and solving the resulting system, we find $A_0 = -\frac{33}{41}$ and $B_0 = \frac{31}{41}$. The general solution is then $y = y_c + y_p = Ae^{5x/2} - \frac{33}{41}\sin 2x + \frac{31}{41}\cos 2x$.

.80 Solve $y'' - 7y' = 48\sin 4x + 84\cos 4x$.

The complementary solution is found in Problem 8.2 to be $y_c = c_1 + c_2 e^{7x}$. We assume a particular solution of the form $y_p = A_0 \sin 4x + B_0 \cos 4x$, which needs no modification.

Substituting $y_p' = 4A_0 \cos 4x - 4B_0 \sin 4x$ and $y_p'' = -16A_0 \sin 4x - 16B_0 \cos 4x$ into the given differential equation and rearranging give

$$(-16A_0 + 28B_0)\sin 4x + (-28A_0 - 16B_0)\cos 4x = 48\sin 4x + 84\cos 4x$$

Equating coefficients of like terms and solving the resulting system, we get $A_0 = -3$ and $B_0 = 0$. The general solution is then $y = c_1 + c_2 e^{7x} - 3\sin 4x$.

.81 Solve $y'' - 6y' + 25y = 48\sin 4x + 84\cos 4x$.

The complementary solution is shown in Problem 8.50 to be $y_c = c_1 e^{3x}\cos 4x + c_2 e^{3x}\sin 4x$, while y_p and its derivatives are as in the previous problem. There is no need to modify y_p because it is not a solution of the associated homogeneous differential equation for any choice of the arbitrary constants A_0 and B_0. (Observe that $e^{3x}\sin 4x$ is linearly independent of $\sin 4x$.)

Substituting y_p, y_p', and y_p'' into the given differential equation and rearranging yield

$$(9A_0 + 24B_0)\sin 4x + (-24A_0 + 9B_0)\cos 4x = 48\sin 4x + 84\cos 4x$$

Equating coefficients of like terms and solving the resulting system, we obtain $A_0 = -2.411$ and $B_0 = 2.904$. The general solution is then $y = c_1 e^{3x}\cos 4x + c_2 e^{3x}\sin 4x - 2.411\sin 4x + 2.904\cos 4x$.

9.82 Solve $y'' + 4y' + 5y = 2\cos x - 2\sin x$.

▮ The complementary solution is shown in Problem 8.66 to be $y_c = c_1e^{-2x}\cos x + c_2e^{-2x}\sin x$. We assume a particular solution of the form $y_p = A_0 \sin x + B_0 \cos x$, which needs no modification because y_p cannot be made equal to y_c by any nonzero choice of the arbitrary constants.

Substituting y_p and its derivatives into the differential equation and rearranging, we obtain

$$(4A_0 - 4B_0)\sin x + (4A_0 + 4B_0)\cos x = -2\sin x + 2\cos x$$

Equating coefficients of like terms and solving the resulting system yield $A_0 = 0$ and $B_0 = \frac{1}{2}$. The general solution is then $y = c_1e^{-2x}\cos x + c_2e^{-2x}\sin x + \frac{1}{2}\cos x$.

9.83 Solve $\ddot{x} + 4\dot{x} + 8x = 16\cos 4t$.

▮ The complementary solution is shown in Problem 8.54 to be $x_c = c_1e^{-2t}\cos 2t + c_2e^{-2t}\sin 2t$. We try a particular solution of the form $x_p = A_0 \sin 4t + B_0 \cos 4t$, which needs no modification.

Substituting x_p into the differential equation and rearranging, we get

$$(-8A_0 - 16B_0)\sin 4t + (16A_0 - 8B_0)\cos 4t = 0\sin 4t + 16\cos 4t$$

Equating coefficients of like terms yields the system

$$-8A_0 - 16B_0 = 0 \qquad 16A_0 - 8B_0 = 16$$

Solving, we find that $A_0 = \frac{4}{5}$ and $B_0 = -\frac{2}{5}$. The general solution is then $x = c_1e^{-2t}\cos 2t + c_2e^{-2t}\sin 2t + \frac{4}{5}\sin 4t - \frac{2}{5}\cos 4t$.

9.84 Solve $\ddot{x} + \frac{1}{32}\dot{x} + 96x = 96\cos 4t$.

▮ The complementary solution is found in Problem 8.63 to be $x_c = e^{-0.015625t}(C_1\cos 9.7979t + C_2\sin 9.7979t)$. We assume a particular solution of the form $x_p = A_0\sin 4t + B_0\cos 4t$ which, because it has no term in common with x_c, needs no modification.

Substituting x_p and its first two derivatives into the differential equation and rearranging, we obtain

$$(80A_0 - \tfrac{1}{8}B_0)\sin 4t + (\tfrac{1}{8}A_0 + 80B_0)\cos 4t = 96\cos 4t + 0\sin 4t$$

Equating coefficients of like terms and solving the resulting system, we find that $A_0 = 0.0019$ and $B_0 = 1.2000$. The general solution is then $x = e^{-0.015625t}(C_1\cos 9.7979t + C_2\sin 9.7979t) + 0.0019\sin 4t + 1.2000\cos 4t$.

9.85 Solve $\dfrac{d^2x}{dt^2} + 25x = 2\sin 2t$.

▮ The complementary solution is shown in Problem 8.72 to be $x_c = C_1\cos 5t + C_2\sin 5t$. We assume a particular solution of the form $x_p = A_0\sin 2t + B_0\cos 2t$, which requires no modification. Substituting x_p and its second derivative into the differential equation and rearranging, we get $(21A_0)\sin 2t + (21B_0)\cos 2t = 2\sin 2t + 0\cos 2t$. It follows that $A_0 = \frac{2}{21}$ and $B_0 = 0$. The general solution is then $x = x_c + x_p = C_1\cos 5t + C_2\sin 5t + \frac{2}{21}\sin 2t$.

9.86 Solve $\ddot{x} + 128x = 512(\sin 2t + \cos 2t)$.

▮ The complementary solution is $x_c = C_1\sin\sqrt{128}t + C_2\cos\sqrt{128}t$ (see Problem 8.73). We assume a particular solution of the form $x_p = A_0\sin 2t + B_0\cos 2t$, which requires no modification. Substituting x_p and its second derivative into the differential equation and rearranging, we get $(124A_0)\sin 2t + (124B_0)\cos 2t = 512\sin 2t + 512\cos 2t$. It follows that $A_0 = B_0 = \frac{512}{124} = 4.129$. The general solution is $x = C_1\sin\sqrt{128}t + C_2\cos\sqrt{128}t + 4.129(\sin 2t + \cos 2t)$.

9.87 Solve $\ddot{Q} + 8\dot{Q} + 52Q = 32\cos 2t$.

▮ The complementary solution is $Q_c = c_1e^{-4t}\cos 6t + c_2e^{-4t}\sin 6t$ (see Problem 8.55). We assume a particular solution of the form $Q_p = A_0\sin 2t + B_0\cos 2t$, which requires no modification. Substituting Q_p and its first two derivatives into the differential equation and rearranging, we obtain

$$(48A_0 - 16B_0)\sin 2t + (16A_0 + 48B_0)\cos 2t = 0\sin 2t + 32\cos 2t$$

By equating the coefficients of like terms and solving the resulting system, we find $A_0 = \frac{1}{5}$ and $B_0 = \frac{3}{5}$. The general solution is then $Q = c_1e^{-4t}\cos 6t + c_2e^{-4t}\sin 6t + \frac{1}{5}\sin 2t + \frac{3}{5}\cos 2t$.

9.88 Solve $\ddot{Q} + 8\dot{Q} + 25Q = 50 \sin 3t$.

▮ The complementary solution is shown in Problem 8.53 (with Q replacing x) to be $Q_c = e^{-4t}(c_1 \cos 3t + c_2 \sin 3t)$. We assume a particular solution of the form $Q_p = A_0 \sin 3t + B_0 \cos 3t$. This trial solution requires no modification because no part of it can be obtained from Q_c by any choice of the arbitrary constants c_1 and c_2. (Note the additional e^{-4t} term in Q_c.)

Substituting Q_p and its derivatives into the differential equation and rearranging, we get

$$(16A_0 - 24B_0) \sin 3t + (24A_0 + 16B_0) \cos 3t = 50 \sin 3t + 0 \cos 3t$$

Equating the coefficients of like terms and solving the resulting system, we obtain $A_0 = \frac{50}{52}$ and $B_0 = -\frac{75}{52}$. The general solution is $Q = e^{-4t}(c_1 \cos 3t + c_2 \sin 3t) + \frac{50}{52} \sin 3t - \frac{75}{52} \cos 3t$.

9.89 Solve $\ddot{q} + 9\dot{q} + 14q = \frac{1}{2} \sin t$.

▮ The complementary solution is found in Problem 8.17 (with q replacing x) to be $q_c = c_1 e^{-2t} + c_2 e^{-7t}$. We assume a particular solution of the form $q_p = A_0 \sin t + B_0 \cos t$, which requires no modification.

Substituting q_p and its derivatives into the differential equation and rearranging, we get $(13A_0 - 9B_0) \sin t + (9A_0 + 13B_0) \cos t = \frac{1}{2} \sin t + 0 \cos t$. Equating coefficients of like terms yields the system

$$13A_0 - 9B_0 = \tfrac{1}{2} \qquad 9A_0 + 13B_0 = 0$$

Thus $A_0 = \frac{13}{500}$ and $B_0 = -\frac{9}{500}$. The general solution is then $q = c_1 e^{-2t} + c_2 e^{-7t} + \frac{13}{500} \sin t - \frac{9}{500} \cos t$.

9.90 Solve $\ddot{I} + 100\dot{I} + 50{,}000I = -400{,}000 \sin 100t$.

▮ The complementary solution is shown in Problem 8.56 to be $I_c = e^{-50t}(c_1 \cos 50\sqrt{19}t + c_2 \sin 50\sqrt{19}t)$. We assume a particular solution of the form $I_p = A_0 \sin 100t + B_0 \cos 100t$, which needs no modification because it has no terms in common with I_c. Substituting I_p into the differential equation yields, after rearranging,

$$(40{,}000A_0 - 10{,}000B_0) \sin 100t + (10{,}000A_0 + 40{,}000B_0) \cos 100t = -400{,}000 \sin 100t + 0 \cos 100t$$

By equating coefficients of like terms, and solving the resulting system, we find

$$40{,}000A_0 - 10{,}000B_0 = -400{,}000 \qquad 10{,}000A_0 + 40{,}000B_0 = 0$$

Thus $A_0 = -\frac{160}{17}$ and $B_0 = \frac{40}{17}$, and the general solution is

$$I = e^{-50t}(c_1 \cos 50\sqrt{19}t + c_2 \sin 50\sqrt{19}t) - \tfrac{160}{17} \sin 100t + \tfrac{40}{17} \cos 100t$$

9.91 Solve $\ddot{q} + 100\dot{q} + 50{,}000q = 4000 \cos 100t$.

▮ This problem is very similar to the previous one. The associated homogeneous differential equations are identical, except that here q replaces I. Thus, q_c is identical in form to I_c. We assume the same form for the particular solution (with q replacing I), so most of the analysis of the previous problem remains valid. Only the right sides of the differential equations are different.

Substituting the particular solution into this differential equation, rearranging, and equating coefficients of like terms yield the system

$$40{,}000A_0 - 10{,}000B_0 = 0 \qquad 10{,}000A_0 + 40{,}000B_0 = 4000$$

Now $A_0 = \frac{4}{170}$ and $B_0 = \frac{16}{170}$, and the general solution is

$$q = e^{-50t}(c_1 \cos 50\sqrt{19}t + c_2 \sin 50\sqrt{19}t) + \tfrac{4}{170} \sin 100t + \tfrac{16}{170} \cos 100t$$

9.92 Solve $y'' - y' - 2y = \sin 2x$.

▮ The complementary solution is shown in Problem 8.1 to be $y_c = c_1 e^{-x} + c_2 e^{2x}$. We assume a particular solution of the form $y_p = A_0 \sin 2x + B_0 \cos 2x$, which needs no modification.

Substituting y_p, $y_p' = 2A_0 \cos 2x - 2B_0 \sin 2x$, and $y_p'' = -4A_0 \sin 2x - 4B_0 \cos 2x$ into the differential equation and rearranging yield

$$(-6A_0 + 2B_0) \sin 2x + (-6B_0 - 2A_0) \cos 2x = 1 \sin 2x + 0 \cos 2x$$

Equating coefficients of like terms and solving the resulting system, we find that $A_0 = -\frac{3}{20}$ and $B_0 = \frac{1}{20}$. Then $y_p = -\frac{3}{20} \sin 2x + \frac{1}{20} \cos 2x$, and the general solution is $y = c_1 e^{-x} + c_2 e^{2x} - \frac{3}{20} \sin 2x + \frac{1}{20} \cos 2x$.

9.93 Solve $\ddot{q} + 400\dot{q} + 200{,}000q = 2000 \cos 200t$.

The complementary solution is $q_c = e^{-200t}(A \cos 400t + B \sin 400t)$ (see Problem 8.70). We assume a particular solution of the form $q_p = A_0 \sin 200t + B_0 \cos 200t$, which has no term in common with q_c and therefore needs no modification.

Substituting q_c into the given differential equation and rearranging yield

$$(160{,}000A_0 - 80{,}000B_0) \sin 200t + (80{,}000A_0 + 160{,}000B_0) \cos 200t = 0 \sin 200t + 2000 \cos 200t$$

Equating coefficients of like terms and solving the resulting system, we get $A_0 = 0.005$ and $B_0 = 0.01$. The general solution is $q = e^{-200t}(A \cos 400t + B \sin 400t) + 0.005 \sin 200t + 0.01 \cos 200t$.

9.94 Solve $\ddot{q} + 1000\dot{q} + 50{,}000q = 4000 \cos 100t$.

The complementary solution is $q_c = c_1 e^{-52.79t} + c_2 e^{-947.2t}$ (see Problem 8.18). We try a particular solution of the form $q_p = A_0 \sin 100t + B_0 \cos 100t$, which needs no modification.

Substituting q_p into the given differential equation and rearranging, we get

$$(40{,}000A_0 - 100{,}000B_0) \sin 100t + (100{,}000A_0 + 40{,}000B_0) \cos 100t = 4000 \cos 100t$$

Equating coefficients of like terms yields

$$40{,}000A_0 - 100{,}000B_0 = 0 \qquad 100{,}000A_0 + 40{,}000B_0 = 4000$$

Solving this system, we find $A_0 = \frac{1}{29}$ and $B_0 = \frac{2}{145}$. The general solution is $q = c_1 e^{-52.79t} + c_2 e^{-947.2t} + \frac{1}{29} \sin 100t + \frac{2}{145} \cos 100t$.

9.95 Solve $\dfrac{d^2I}{dt^2} + 40\dfrac{dI}{dt} + 800I = 8 \cos t$.

The complementary solution is found in Problem 8.71 to be $I_c = c_1 e^{-20t} \cos 20t + c_2 e^{-20t} \sin 20t$. We assume a particular solution of the form $I_p = A_0 \sin t + B_0 \cos t$, which has no terms in common with I_c and so needs no modification.

Substituting I_p into the given differential equation and rearranging, we have $(799A_0 - 40B_0) \sin t + (40A_0 - 799B_0) \cos t = 8 \cos t$. Equating coefficients of like terms and solving the resulting system, we find $A_0 = 320/640{,}001$ and $B_0 = 6392/640{,}001$. The general solution is then $I = c_1 e^{-20t} \cos 20t + c_2 e^{-20t} \sin 20t + \dfrac{320}{640{,}001} \sin t + \dfrac{6392}{640{,}001} \cos t$.

9.96 Solve $\dfrac{d^2r}{dt^2} - \omega^2 r = -g \sin \omega t$ when both g and ω are positive constants.

The complementary solution is $r_c = Ae^{\omega t} + Be^{-\omega t}$ (see Problem 8.7). We assume a particular solution of the form $r_p = A_0 \sin \omega t + B_0 \cos \omega t$, which requires no modification.

Substituting r_p into the differential equation and rearranging lead to $(-2\omega^2 A_0) \sin \omega t + (-2\omega^2 B_0) \cos \omega t = -g \sin \omega t + 0 \cos \omega t$. By equating coefficients of like terms, we find that $A_0 = g/(2\omega^2)$ and $B_0 = 0$. The general solution is then $r = Ae^{\omega t} + Be^{-\omega t} + \dfrac{g}{2\omega^2} \sin \omega t$.

9.97 Solve $\dfrac{d^3Q}{dt^3} - 5\dfrac{d^2Q}{dt^2} + 25\dfrac{dQ}{dt} - 125Q = 504 \cos 2t - 651 \sin 2t$.

The complementary solution is $Q_c = c_1 e^{5t} + c_2 \cos 5t + c_3 \sin 3t$ (see Problem 8.113). We try a particular solution of the form $Q_p = A_0 \sin 2t + B_0 \cos 2t$, which needs no modification.

Substituting Q_p into the given differential equation and rearranging, we obtain

$$(-105A_0 - 42B_0) \sin 2t + (42A_0 - 105B_0) \cos 2t = -651 \sin 2t + 504 \cos 2t$$

Equating coefficients of like terms and solving the resulting system, we find $A_0 = 7$ and $B_0 = -2$, so the general solution is $Q = c_1 e^{5t} + c_2 \cos 5t + c_3 \sin 5t + 7 \sin 2t - 2 \cos 2t$.

9.98 Solve $y^{(4)} - 6y^{(3)} + 16y'' + 54y' - 225y = 1152 \cos 3x - 3924 \sin 3x$.

The complementary solution is, from Problem 8.124, $y_c = c_1 e^{3x} + c_2 e^{-3x} + c_3 e^{3x} \cos 4x + c_4 e^{3x} \sin 4x$. We assume a particular solution of the form $y_p = A_0 \sin 3x + B_0 \cos 3x$, which has no terms in common with y_c and needs no modification.

Substituting y_p into the given differential equation and rearranging yield

$$(-288A_0 - 324B_0)\sin 3x + (324A_0 - 288B_0)\cos 3x = -3924\sin 3x + 1152\cos 3x$$

By equating coefficients of like terms, we obtain the system

$$-288A_0 - 324B_0 = -3924 \qquad 324A_0 - 288B_0 = 1152$$

and find that $A_0 = 8$ and $B_0 = 5$. Then the general solution is $y = c_1e^{3x} + c_2e^{-3x} + c_3e^{3x}\cos 4x + c_4e^{3x}\sin 3x + 8\sin 3x + 5\cos 3x$.

EQUATIONS WHOSE RIGHT SIDE CONTAINS A PRODUCT INVOLVING SINES AND COSINES

9.99 Solve $y' + 6y = 3e^{2x}\sin 3x$.

▌ The complementary solution is $y_c = Ae^{-6x}$ (see Problem 8.37). Since the right side of the nonhomogeneous differential equation is the product of an exponential and a sine, we try a particular solution of the form $y_p = A_0e^{2x}\sin 3x + B_0e^{2x}\cos 3x$. Since y_p has no term in common with y_c (except perhaps for a multiplicative constant), there is no need to modify y_p.

Substituting y_p and $y_p' = 2A_0e^{2x}\sin 3x + 3A_0e^{2x}\cos 3x + 2B_0e^{2x}\cos 3x - 3B_0e^{2x}\sin 3x$ into the differential equation yields

$$2A_0e^{2x}\sin 3x + 3A_0e^{2x}\cos 3x + 2B_0e^{2x}\cos 3x - 3B_0e^{2x}\sin 3x + 6(A_0e^{2x}\sin 3x + B_0e^{2x}\cos 3x) = 3e^{2x}\sin 3x$$

which may be rearranged and simplified to $(8A_0 - 3B_0)\sin 3x + (3A_0 + 8B_0)\cos 3x = 3\sin 3x + 0\cos 3x$. Equating coefficients of like terms, we obtain the system

$$8A_0 - 3B_0 = 3 \qquad 3A_0 + 8B_0 = 0$$

from which we find that $A_0 = \frac{24}{73}$ and $B_0 = -\frac{9}{73}$. The general solution is then $y = Ae^{-6x} + \frac{24}{73}e^{2x}\sin 3x - \frac{9}{73}e^{2x}\cos 3x$.

9.100 Solve $y' + 6y = 2e^{-x}\cos 2x$.

▌ The complementary solution of the previous problem is valid here, and we assume a particular solution of the form $y_p = A_0e^{-x}\sin 2x + B_0e^{-x}\cos 2x$. Because it has no term in common with y_c, y_p needs no modification.

Substituting y_p into the given differential equation and rearranging, we obtain $(5A_0 - 2B_0)\sin 2x + (2A_0 + 5B_0)\cos 2x = 0\sin 2x + 2\cos 2x$. Equating coefficients of like terms and solving the resulting system of equations, we find $A_0 = \frac{4}{29}$ and $B_0 = \frac{10}{29}$. The general solution is then $y = Ae^{-6x} + \frac{4}{29}e^{-x}\sin 2x + \frac{10}{29}e^{-x}\cos 2x$.

9.101 Solve $y' + 6y = 11e^{x}\sin x + 23e^{x}\cos x$.

▌ The complementary solution again is that of Problem 9.99. We try a particular solution of the form $y_p = A_0e^{x}\sin x + B_0e^{x}\cos x$, which has no term in common with y_c and therefore requires no modification.

Substituting y_p into the differential and rearranging lead to $(7A_0 - B_0)\sin x + (A_0 + 7B_0)\cos x = 11\sin x + 23\cos x$. Equating coefficients of like terms and solving the resulting equations, we find that $A_0 = 2$ and $B_0 = 3$. The general solution is then $y = Ae^{-6x} + 2e^{x}\sin x + 3e^{x}\cos x$.

9.102 Solve $2y' - 5y = 41e^{-x}\cos x - 11e^{-x}\sin x$.

▌ The complementary solution is $y_c = Ae^{5x/2}$ (see Problem 8.38). We try a particular solution of the form $y_p = A_0e^{-x}\sin x + B_0e^{-x}\cos x$, which needs no modification.

Substituting y_p into the given differential equation and simplifying yield $(-7A_0 - 2B_0)\sin x + (2A_0 - 7B_0)\cos x = -11\sin x + 41\cos x$. Equating coefficients of like terms and solving the resulting system, we obtain $A_0 = 3$ and $B_0 = -5$. The general solution is $y = Ae^{5x/2} + 3e^{-x}\sin x - 5e^{-x}\cos x$.

9.103 Solve $y'' + 6y' + 9y = 16e^{-x}\cos 2x$.

▌ The complementary solution is $y_c = c_1e^{-3x} + c_2xe^{-3x}$ (see Problem 8.142). We try a particular solution of the form $y_p = A_0e^{-x}\sin 2x + B_0e^{-x}\cos 2x$, which needs no modification because it has no term in common with y_c.

Substituting y_p, $y_p' = (-A_0 - 2B_0)e^{-x}\sin 2x + (2A_0 - B_0)e^{-x}\cos 2x$, and $y_p'' = (-3A_0 + 4B_0)e^{-x}\sin 2x + (-4A_0 - 3B_0)e^{-x}\cos 2x$ into the given differential equation and rearranging yield $(-8B_0)\sin 2x + (8A_0)\cos 2x = 16\cos 2x$. By equating coefficients of like terms, we find $A_0 = 2$ and $B_0 = 0$. The general solution is then $y = c_1e^{-3x} + c_2xe^{-3x} + 2e^{-x}\sin 2x$.

9.104 Solve $\dfrac{d^2y}{dx^2} - 6\dfrac{dy}{dx} + 25y = 30e^{3x}\sin x$.

▌ The complementary solution is $y_c = e^{3x}(c_1\cos 4x + c_2\sin 4x)$ (see Problem 8.50). We try a particular solution of the form $y_p = A_0e^{3x}\sin x + B_0e^{3x}\cos x$. Since no part of y_p can be generated from y_c by any choices of the arbitrary constants, there is no need to modify y_p.

Substituting y_p, $y_p' = (3A_0 - B_0)e^{3x}\sin x + (A_0 + 3B_0)e^{3x}\cos x$, and $y_p'' = (8A_0 - 6B_0)e^{3x}\sin x + (6A_0 + 8B_0)e^{3x}\cos x$ into the given differential equation and simplifying, we get $(15A_0)\sin x + (15B_0)\cos x = 30\sin x$. Therefore, $A_0 = 2$ and $B_0 = 0$, and the general solution is $y = e^{3x}(c\cos 4x + c_2\sin 4x) + 2e^{3x}\sin x$.

9.105 Solve $(D^2 - 2D)y = e^x\sin x$.

▌ The complementary solution is $y = C_1 + C_2e^{2x}$. As a particular solution we try $y = Ae^x\sin x + Be^x\cos x$, which requires no modification.

Substituting y, $Dy = (A - B)e^x\sin x + (A + B)e^x\cos x$, and $D^2y = -2Be^x\sin x + 2Ae^x\cos x$ into the given equation yields $(-2A)\sin x + (-2B)\cos x = 1\sin x$. Equating coefficients of like terms, we find that $A = -\frac{1}{2}$ and $B = 0$. Hence, the primitive is $y = C_1 + C_2e^{2x} - \frac{1}{2}e^x\sin x$.

9.106 Solve $(D^4 - 16)y = 60e^x\sin 3x$.

▌ The complementary solution is, from Problem 8.128, $y_c = c_1\cos 2x + c_2\sin 2x + c_3e^{2x} + c_4e^{-2x}$. We try a particular solution of the form $y_p = A_0e^x\sin 3x + B_0e^x\cos 3x$, which requires no modification. Since

$$y_p' = (A_0 - 3B_0)e^x\sin 3x + (3A_0 + B_0)e^x\cos 3x$$

$$y_p'' = (-8A_0 - 6B_0)e^x\sin 3x + (6A_0 - 8B_0)e^x\cos 3x$$

$$y_p^{(3)} = (-26A_0 + 18B_0)e^x\sin 3x + (-18A_0 - 26B_0)e^x\cos 3x$$

and

$$y_p^{(4)} = (28A_0 + 96B_0)e^x\sin 3x + (-96A_0 + 28B_0)e^x\cos 3x$$

the given differential equation becomes, after substitution and simplification, $(12A_0 + 96B_0)\sin 3x + (-96A_0 + 12B_0)\cos 3x = 60\sin 3x$. Equating coefficients of like terms and solving the resulting system, we find that $A_0 = \frac{1}{13}$ and $B_0 = \frac{8}{13}$. The general solution is then $y = y_c + y_p = c_1\cos 2x + c_2\sin 2x + c_3e^{2x} + c_4e^{-2x} + \frac{1}{13}e^x\sin 3x + \frac{8}{13}e^x\cos 3x$.

9.107 Solve $\dfrac{d^3Q}{dt^3} - 5\dfrac{d^2Q}{dt^2} + 25\dfrac{dQ}{dt} - 125Q = 5000e^{-t}\cos 2t$.

▌ The complementary solution is $Q_c = c_1e^{5t} + c_2\cos 5t + c_3\sin 5t$ (see Problem 8.113). We try a particular solution of the form $Q_p = A_0e^{-t}\sin 2t + B_0e^{-t}\cos 2t$, which requires no modification. Since

$$Q_p' = (-A_0 - 2B_0)e^{-t}\sin 2t + (2A_0 - B_0)e^{-t}\cos 2t$$

$$Q_p'' = (-3A_0 + 4B_0)e^{-t}\sin 2t + (-4A_0 - 3B_0)e^{-t}\cos 2t$$

and

$$Q_p''' = (11A_0 + 2B_0)e^{-t}\sin 2t + (-2A_0 + 11B)e^{-t}\cos 2t$$

the given differential equation becomes, after substitution and simplification, $(-124A_0 - 68B_0)\sin 2t + (68A_0 - 124B_0)\cos 2t = 5000\cos 2t$. By equating coefficients of like terms and solving the resulting system, we find $A_0 = 17$ and $B_0 = -31$, and the general solution is $Q = c_1e^{5t} + c_2\cos 5t + c_3\sin 5t + 17e^{-t}\sin 2t - 31e^{-t}\cos 2t$.

9.108 Solve $y' + 6y = (20x + 3)\sin 2x$.

▌ The complementary solution is shown in Problem 8.37 to be $y_c = Ae^{-6x}$. Since the right side of the differentia equation is the product of a first-degree polynomial and $\sin 2x$, we try a particular solution of the form $y_p = (A_1x + A_0)\sin 2x + (B_1x + B_0)\cos 2x$. This trial solution has no terms in common with y_c, so it needs no modification.

Substituting y_p and $y_p' = (A_1 - 2B_1x - 2B_0)\sin 2x + (2A_1x + 2A_0 + B_1)\cos 2x$ into the given differential equation and simplifying, we get

$$(6A_1 - 2B_1)x\sin 2x + (A_1 + 6A_0 - 2B_0)\sin 2x + (2A_1 + 6B_1)x\cos 2x + (2A_0 + B_1 + 6B_0)\cos 2x$$
$$= 20x\sin 2x + 3\sin 2x + 0x\cos 2x + 0\cos 2x$$

Equating coefficients of like terms yields the system

$$\begin{aligned} 6A_1 \qquad - 2B_1 \qquad &= 20 \\ A_1 + 6A_0 \qquad - 2B_0 &= 3 \\ 2A_1 \qquad + 6B_1 \qquad &= 0 \\ 2A_0 + B_1 + 6B_0 &= 0 \end{aligned}$$

which we solve to find that $A_1 = 3$, $A_0 = 0.05$, $B_1 = -1$, and $B_0 = 0.15$. The general solution is then $y = Ae^{-6x} + (3x + 0.05)\sin 2x + (-x + 0.15)\cos 2x$.

109 Determine the form of a particular solution to $y' + 6y = (x^2 - 1)\sin 2x$.

▮ The complementary solution is $y_c = Ae^{-6x}$ (see Problem 8.37). Since the right side of the given differential equation is the product of a second-degree polynomial and sin 2x, we assume a particular solution of the more general form $y_p = (A_2x^2 + A_1x + A_0)\sin 2x + (B_2x^2 + B_1x + B_0)\cos 2x$. This form of y_p has no terms in common with the complementary function, and therefore it requires no modification.

110 Determine the form of a particular solution to $y' + 6y = (20x^2 - 10x + 7)\cos 2x$.

▮ The complementary solution is that of the previous problem. The right side of the given differential equation is the product of a second-degree polynomial and cos 2x, so y_p of the previous problem is again an appropriate trial form for the particular solution. Since it has no terms in common with y_c, it needs no modification.

111 Determine the form of a particular solution to $y' + 6y = 18x\cos 5x$.

▮ The complementary solution is $y_c = Ae^{-6x}$ (see Problem 8.37). Since the right side of the given differential equation is the product of a first-degree polynomial and cos 5x, we try a particular solution having the more general form $y_p = (A_1x + A_0)\sin 5x + (B_1x + B_0)\cos 5x$. This form of y_p has no terms in common with y_c and so requires no modification.

112 Determine the form of a particular solution to $2y' - 5y = -29x^2\cos x$.

▮ The complementary solution is $y_c = Ae^{5x/2}$ (see Problem 8.38). Since the right side of the differential equation is the product of a second-degree polynomial and cos x, we assume a particular solution having the more general form $y_p = (A_2x^2 + A_1x + A_0)\sin x + (B_2x^2 + B_1x + B_0)\cos x$. Since y_c and y_p have no terms in common, there is no need to modify y_p.

113 Solve the differential equation of the previous problem.

▮ Substituting y_p of that problem and

$$y_p' = (2A_2x + A_1 - B_2x^2 - B_1x - B_0)\sin x + (A_2x^2 + A_1x + A_0 + 2B_2x + B_1)\cos x$$

into the given differential equation and simplifying, we obtain

$$(-5A_2 - 2B_2)x^2\sin x + (4A_2 - 5A_1 - 2B_1)x\sin x + (2A_1 - 5A_0 - 2B_0)\sin x + (2A_2 - 5B_2)x^2\cos x$$
$$+ (2A_1 + 4B_2 - 5B_1)x\cos x + (2A_0 + 2B_1 - 5B_0)\cos x = -29x^2\cos x$$

By equating coefficients of like terms and solving the resulting system of six simultaneous equations, we find that $A_2 = -2$, $A_1 = -\frac{80}{29}$, $A_0 = -\frac{1136}{841}$, $B_2 = 5$, $B_1 = \frac{84}{29}$, and $B_0 = \frac{520}{841}$. Substituting these results into y_p and y_c of the previous problem, we generate the general solution $y = y_c + y_p = Ae^{5x/2} - (2x^2 + \frac{80}{29}x + \frac{1136}{841})\sin x + (5x^2 + \frac{84}{29}x + \frac{520}{841})\cos x$.

114 Determine the form of a particular solution for $y'' - 7y' = (2x - 1)\sin 2x$.

▮ The complementary solution is $y_c = c_1 + c_2e^{7x}$ (see Problem 8.2). Since the right side of the given differential equation is the product of a first-degree polynomial and sin 2x, we assume $y_p = (A_1x + A_0)\sin 2x + (B_1x + B_0)\cos 2x$. Since y_p and y_c have no terms in common, there is no need to modify y_p.

9.115 Solve the differential equation of the previous problem.

▮ Substituting y_p,

$$y_p' = (A_1 - 2B_1x - 2B_0)\sin 2x + (2A_1x + 2A_0 + B_1)\cos 2x$$

and

$$y_p'' = (-4A_1x - 4A_0 - 4B_1)\sin 2x + (4A_1 - 4B_1x - 4B_0)\cos 2x$$

into the given differential equation and simplifying yield

$$(-4A_1 + 14B_1)x\sin x + (-7A_1 - 4A_0 - 4B_1 + 14B_0)\sin x + (-14A_1 - 4B_1)x\cos x$$
$$+ (4A_1 - 14A_0 - 7B_1 - 4B_0)\cos x = 2x\sin 2x + (-1)\sin 2x$$

Equating coefficients of like terms leads to a system of four equations whose solution is $A_1 = -0.0377$, $A_0 = -0.0571$, $B_1 = 0.1321$, and $B_0 = -0.0689$. Then the general solution is $y = y_c + y_p = c_1 + c_2e^{7x} + (-0.0377x - 0.0571)\sin 2x + (0.1321x - 0.0689)\cos 2x$.

9.116 Determine the form of a particular solution for $y'' - 7y' = \frac{1}{2}x\cos 2x$.

▮ The form is identical to that of y_p in Problem 9.114.

9.117 Determine the form of a particular solution for $y'' - 7y = 3x\cos 3x$.

▮ As in Problem 9.114, the complementary solution is $y_c = c_1 + c_2e^{7x}$. Since the right side of the differential equation is the product of a first-degree polynomial and $\cos 3x$, we try $y_p = (A_1x + A_0)\sin 3x + (B_1x + B_0)\cos 3x$. This expression has no terms in common with y_c, so it need not be modified.

9.118 Determine the form of a particular solution for $y'' - 7y' = (3x^2 - 2x + 11)\cos 2\pi x$.

▮ We try a particular solution of the form $y_p = (A_2x^2 + A_1x + A_0)\sin 2\pi x + (B_2x^2 + B_1x + B_0)\cos 2\pi x$, which is a generalization of the form of the right side of the given equation. Since y_p has no terms in common with the complementary function $y_c = c_1 + c_2e^{7x}$, it needs no modification.

9.119 Determine the form of a particular solution for $y'' - 7y' = 6x^2\sin 2\pi x$.

▮ The form is identical to that of y_p in the previous problem.

9.120 Determine the form of a particular solution for $y'' - y' - 2y = x^3\sin 7x$.

▮ The complementary solution is $y_c = c_1e^{-x} + c_2e^{2x}$ (see Problem 8.1). We try a particular solution of the form

$$y_p = (A_3x^3 + A_2x^2 + A_1x + A_0)\sin 7x + (B_3x^3 + B_2x^2 + B_1x + B_0)\cos 7x$$

Since y_p and y_c have no terms in common, y_p need not be modified.

9.121 Determine the form of a particular solution for $\dfrac{d^3y}{dx^3} + 6\dfrac{d^2y}{dx^2} + 12\dfrac{dy}{dx} + 8y = 2x\sin 3x$.

▮ The complementary solution is $y_c = c_1e^{2x} + c_2xe^{2x} + c_3x^2e^{2x}$ (see Problem 8.156). We try a particular solution of the generalized form $y_p = (A_1x + A_0)\sin 3x + (B_1x + B_0)\cos 3x$. Since y_p and y_c have no terms in common, y_p needs no modification.

9.122 Determine the form of a particular solution for $y^{(4)} + y^{(3)} - 2y'' - 6y' - 4y = (x^2 - 5x)\cos x$.

▮ The general solution to the associated homogeneous differential equation is shown in Problem 8.125 to be $y_h = c_1e^{-x} + c_2e^{2x} + c_3e^{-x}\cos x + c_4e^{-x}\sin x$. We try a particular solution of the form $y_p = (A_2x^2 + A_1x + A_0)\sin x + (B_2x^2 + B_1x + B_0)\cos x$. Since no part of y_p can be obtained from y_h by any choice of the constants c_1 through c_4, there is no need to modify y_p; it has the proper form as written.

9.123 Determine the form of a particular solution for $y'' - \frac{9}{2}y' + 2y = (3x - 4)e^x\sin 2x$.

▮ The complementary solution is $y_c = Ae^{x/2} + Be^{4x}$ (see Problem 8.23). Since the right side of the differential equation is the product of a sine term, an exponential, and a first-degree polynomial, we try a particular solution of the same but more generalized form, noting that wherever a sine term occurs, the particular solution may have an identical component with a cosine term. We try, therefore,

$$y_p = (A_1x + A_0)e^x\sin 2x + (B_1x + B_0)e^x\cos x$$

Since y_c and y_p have no terms in common, there is no need to modify y_p.

9.124 Determine the form of a particular solution for $y'' - 10y' + 29y = x^2e^{-2x}\cos 3x$.

▌ The complementary solution is, from Problem 8.51, $y_c = c_1e^{5x}\cos 2x + c_2e^{5x}\sin 2x$. Since the right side of the differential equation is the product of a cosine term, an exponential, and a second-degree polynomial, we try a particular solution of the same but more generalized form, noting that wherever a cosine term occurs, a particular solution may have an identical component with a sine term. We try

$$y_p = (A_2x^2 + A_1x + A_0)e^{-2x}\sin 3x + (B_2x^2 + B_1x + B_0)e^{-2x}\cos 3x$$

Since y_c and y_p have no terms in common, there is no need to modify y_p.

9.125 Determine the form of a particular solution for $y'' - 10y' + 29y = (103x^2 - 27x + 11)e^{-2x}\cos 3x$.

▌ The form is identical to that of y_p of the previous problem.

9.126 Determine the form of a particular solution for $\dfrac{d^2I}{dt^2} + 60\dfrac{dI}{dt} + 500I = (2t - 50)e^{-10t}\sin 50t$.

▌ The complementary solution is $I_c = c_1e^{-50t} + c_2e^{-10t}$ (see Problem 8.10). We try a particular solution having the same general form as the right side of the given differential equation, coupled with an associated term involving cos $50t$:

$$I_p = (A_1t + A_0)e^{-10t}\sin 50t + (B_1t + B_0)e^{-10t}\cos 50t$$

Since no part of I_p can be obtained from I_c by any choice of the constants c_1 and c_2, it follows that I_p does not need modification.

9.127 Determine the form of a particular solution for $\dfrac{d^4x}{dt^4} - 13\dfrac{d^2x}{dt^2} + 36x = t(5 - 12t^2)e^{-2t}\cos 8t$.

▌ The complementary solution is, from Problem 8.30, $x_c = c_1e^{2t} + c_2e^{-2t} + c_3e^{3t} + c_4e^{-3t}$. The right side of the differential equation is actually the product of a *third*-degree polynomial, an exponential, and a cosine term. We therefore try

$$x_p = (A_3t^3 + A_2t^2 + A_1t + A_0)e^{-2t}\sin 8t + (B_3t^3 + B_2t^2 + B_1t + B_0)e^{-2t}\cos 8t$$

Since no part of x_p can be formed from x_c by any choice of c_1 through c_4, there is no need to modify x_p.

MODIFICATIONS OF TRIAL PARTICULAR SOLUTIONS

9.128 Determine the form of a particular solution to $y' - 5y = 2e^{5x}$.

▌ The complementary solution is $y_c = c_1e^{5x}$ (see Problem 8.34). Since the right side of the given nonhomogeneous equation is an exponential, we try a particular solution of the same form, namely A_0e^{5x}. However, this is exactly the same as y_c except for a multiplicative constant (both are a constant times e^{5x}). Therefore, we must modify it. We do so by multiplying it by x to get $y_p = A_0xe^{5x}$, which is distinct from y_c and is an appropriate candidate for a particular solution.

9.129 Solve the differential equation of the previous problem.

▌ Substituting y_p of the previous problem and its derivative $y_p' = A_0e^{5x} + 5A_0xe^{5x}$ into the differential equation, we get $(A_0e^{5x} + 5A_0xe^{5x}) - 5(A_0xe^{5x}) = 2e^{5x}$, or $A_0e^{5x} = 2e^{5x}$, so that $A_0 = 2$. Then $y_{p1} = 2xe^{5x}$, and the general solution is $y = y_c + y_p = c_1e^{5x} + 2xe^{5x} = (c_1 + 2x)e^{5x}$.

9.130 Determine the form of a particular solution to $y' + 6y = 3e^{-6x}$.

▌ The complementary solution is $y_c = Ae^{-6x}$ (see Problem 8.37). Since the right side of the given differential equation is an exponential, we try a particular solution having the same form, namely A_0e^{-6x}. However, since this is the same as y_c except for a multiplicative constant, it must be modified. Multiplying by x, we get $y_p = A_0xe^{-6x}$, which is different from y_c and the appropriate candidate for a particular solution.

9.131 Solve the differential equation of the previous problem.

▌ Substituting y_p of the previous problem and $y_p' = A_0e^{-6x} - 6A_0xe^{-6x}$ into the given differential equation and simplifying, we obtain $A_0e^{-6x} = 3e^{-6x}$, so that $A_0 = 3$. Then $y_p = 3xe^{-6x}$, and the general solution is $y = Ae^{-6x} + 3xe^{-6x}$.

9.132 Determine the form of a particular solution to $2y' - 5y = -4e^{5x/2}$.

▮ The complementary solution is $y_c = Ae^{5x/2}$ (see Problem 8.38). We try a particular solution having the same form as the right side of the given differential equation, namely $A_0e^{5x/2}$. Since this is identical to y_c, we modify it by multiplying by x. The new candidate is $y_p = A_0xe^{5x/2}$, which is distinct from y_c and therefore an appropriate trial solution.

9.133 Solve the differential equation of the previous problem.

▮ By substituting y_p of the previous problem and $y_p' = A_0e^{5x/2} + \frac{5}{2}A_0xe^{5x/2}$ into the given differential equation and simplifying, we find that $A_0 = -2$. Then $y_p = -2xe^{5x/2}$, and the general solution is $y = y_c + y_p = Ae^{5x/2} - 2xe^{5x/2}$.

9.134 Determine the form of a particular solution to $y'' - y' - 2y = e^{2x}$.

▮ The complementary solution is $y = c_1e^{-x} + c_2e^{2x}$ (see Problem 8.1). We try the particular solution A_0e^{2x}, which is similar in form to the right side of the given equation. However, it also has the form of part of the complementary function (let $c_2 = A_0$), so it must be modified. Multiplying by x, we get $y_p = A_0xe^{2x}$, which has no term in common with the complementary function and is, therefore, in proper form for a particular solution.

9.135 Solve the differential equation of the previous problem.

▮ Substituting y_p of the previous problem with its derivatives $y_p' = A_0e^{2x} + 2A_0xe^{2x}$ and $y_p'' = 4A_0e^{2x} + 4A_0xe^{2x}$ into the given differential equation and simplifying we find that $A_0 = \frac{1}{3}$. Then $y_p = \frac{1}{3}xe^{2x}$, and the general solution is $y = c_1e^{-x} + c_2e^{2x} + \frac{1}{3}xe^{2x}$.

9.136 Determine the form of a particular solution to $y'' - y' - 2y = 2e^{-x}$.

▮ The complementary solution is $y_c = c_1e^{-x} + c_2e^{2x}$ (see Problem 8.1). We try a particular solution of the form A_0e^{-x}, but since this has the same form as part of y_c (let $c_1 = A_0$), we must modify it. Multiplying the right side by x, we get $y_p = A_0xe^{-x}$; since this is distinct from y_c, it is in proper form for a particular solution.

9.137 Solve the differential equation of the previous problem.

▮ Substituting y_p of the previous problem and its derivatives into the differential equation, we obtain $-2A_0e^{-x} + A_0xe^{-x} - (A_0e^{-x} - A_0xe^{-x}) - 2(A_0xe^{-x}) = 2e^{-x}$, from which we find that $A_0 = -\frac{2}{3}$. Then a particular solution is $y_p = -\frac{2}{3}xe^{-x}$, and the general solution is $y = c_1e^{-x} + c_2e^{2x} - \frac{2}{3}xe^{-x}$.

9.138 Determine the form of a particular solution to $y'' + 4y' + 4y = e^{-2x}$.

▮ The complementary solution is found in Problem 8.141 to be $y_c = c_1e^{-2x} + c_2xe^{-2x}$. We try a particular solution of the form A_0e^{-2x}, similar to the right side of the given differential equation. However, this is also part of y_c (let $c_1 = A_0$), so we modify it by multiplying by x. This gives us A_0xe^{-2x}, which is also part of y_c (let $c_2 = A_0$) and must be modified. Multiplying again by x, we generate $y_p = A_0x^2e^{-2x}$. Since this result has no terms in common with y_c, it is the proper form for a particular solution.

9.139 Solve the differential equation of the previous problem.

▮ Substituting y_p of the previous problem and its derivatives $y_p' = -2A_0x^2e^{-2x} + 2A_0xe^{-2x}$ and $y_p'' = 4A_0x^2e^{-2x} - 8A_0xe^{-2x} + 2A_0e^{-2x}$ into the differential equation and simplifying, we find that $A_0 = \frac{1}{2}$. Then $y_p = \frac{1}{2}x^2e^{-2x}$, and the general solution is $y = y_c + y_p = c_1e^{-2x} + c_2xe^{-2x} + \frac{1}{2}x^2e^{-2x}$.

9.140 Determine the form of a particular solution to $y'' + 6y' + 9y = 12e^{-3x}$.

▮ The complementary solution is shown in problem 8.142 to be $y_c = c_1e^{-3x} + c_2xe^{-3x}$. We try the particular solution A_0e^{-3x}, which is a general form of the right side of the given equation. However, it has a term in common with y_c (take $c_1 = A_0$), so it must be modified. To do so, we multiply by x, to get A_0xe^{-3x}. But this also is part of y_c (let $c_2 = A_0$) and must be modified. Multiplying by x once more, we obtain $y_p = A_0x^2e^{-3x}$, which has no term in common with y_c and so is the proper form for a particular solution.

9.141 Solve the differential equation of the previous problem.

▮ Substituting y_p of the previous problem and its derivatives into the given differential equation yields

$$(9x^2 - 12x + 2)A_0e^{-3x} + 6(-3x^2 + 2x)A_0e^{-3x} + 9A_0x^2e^{-3x} = 12e^{-3x}$$

from which we find that $A_0 = 6$. Then $y_p = 6x^2e^{-3x}$, and the general solution is $y = (c_1 + c_2x + 6x^2)e^{-3x}$

9.142 Determine the form of a particular solution to $\ddot{x} + 20\dot{x} + 64x = 60e^{-4t}$.

The complementary solution is $x_c = c_1e^{-4t} + c_2e^{-16t}$ (see Problem 8.11). We try a particular solution of the form A_0e^{-4t}, but because it is part of x_c (let $c_1 = A_0$), it must be modified. We multiply it by t, the independent variable, and obtain $x_p = A_0te^{-4t}$. Since this has no term in common with x_c, it is the proper form for a particular solution.

9.143 Solve the differential equation of the previous problem.

Substituting x_p of the previous problem and its derivatives into the given differential equation, we get

$$16A_0te^{-4t} - 8A_0e^{-4t} + 20(-4A_0te^{-4t} + A_0e^{-4t}) + 64A_0te^{-4t} = 60e^{-4t}$$

from which $A_0 = 5$. Then a particular solution is $x_p = 5te^{-4t}$, and the general solution is $x = c_1e^{-4t} + c_2e^{-16t} + 5te^{-4t}$.

9.144 Determine the form of a particular solution to $\ddot{x} + 10\dot{x} + 25x = 20e^{-5t}$.

The complementary solution is $x_c = C_1e^{-5t} + C_2te^{-5t}$ (see Problem 8.146). We try the particular solution A_0e^{-5t}, which is similar in form to the right side of the given differential equation. But because it is part of x_c (let $C_1 = A_0$), we modify it by multiplying by t. The result, A_0te^{-5t}, is also part of x_c (let $C_2 = A_0$), so it too must be modified.

Multiplying again by the independent variable t, we obtain $x_p = A_0t^2e^{-5t}$. Since this has no term in common with x_c, it is the proper form for a particular solution.

9.145 Solve the differential equation of the previous problem.

Substituting x_p of the previous problem and its derivatives into the given differential equation, we obtain

$$25A_0t^2e^{-5t} - 20A_0te^{-5t} + 2A_0e^{-5t} + 10(-5A_0t^2e^{-5t} + 2A_0te^{-5t}) + 25A_0t^2e^{-5t} = 20e^{-5t}$$

which can be simplified to $A_0 = 10$. Then $x_p = 10t^2e^{-5t}$, and the general solution is $x = C_1e^{-5t} + C_2te^{-5t} + 10t^2e^{-5t}$.

9.146 Determine the form of a particular solution to $\dfrac{d^3y}{dx^3} + 6\dfrac{d^2y}{dx^2} + 12\dfrac{dy}{dx} + 8y = 12e^{-2x}$.

The complementary solution is $y_c = (c_1 + c_2x + c_3x^2)e^{-2x}$ (see Problem 8.145). We try the particular solution A_0e^{-2x}, which is similar in form to the right side of the differential equation. But it is part of y_c (let $c_1 = A_0$), so we modify it by multiplying by x. The result, A_0xe^{-2x}, is also part of y_c (let $c_2 = A_0$) and must also be modified.

Multiplying again by x, we get $A_0x^2e^{-2x}$, which again is part of y_c (let $c_3 = A_0$). Multiplying once more by x, we obtain $y_p = A_0x^3e^{-2x}$, which has no terms in common with y_c and thus is the proper form for a particular solution.

9.147 Solve the differential equation of the previous problem.

Substituting y_p of the previous problem and its derivatives

$$y_p' = -2A_0x^3e^{-2x} + 3A_0x^2e^{-2x}$$
$$y_p'' = 4A_0x^3e^{-2x} - 12A_0x^2e^{-2x} + 6A_0xe^{-2x}$$

and

$$y_p''' = -8A_0x^3e^{-2x} + 36A_0x^2e^{-2x} - 36A_0xe^{-2x} + 6A_0e^{-2x}$$

into the given differential equation and simplifying, we find that $A_0 = 2$. Then $y_p = 2x^3e^{-2x}$ and the general solution is $y = (c_1 + c_2x + c_3x^2)e^{-2x} + 2x^3e^{-2x}$.

9.148 Determine the form of a particular solution to $\dfrac{d^3Q}{dt^3} + 3\dfrac{d^2Q}{dt^2} + 3\dfrac{dQ}{dt} + Q = 5e^{-t}$.

The complementary solution is, from Problem 8.158, $Q_c = C_1e^{-t} + C_2te^{-t} + C_3t^2e^{-t}$. We try a particular solution similar in form to the right side of the differential equation, namely A_0e^{-t}. Since this is part of the complementary solution (let $C_1 = A_0$), it must be modified. If we multiply by t or t^2, the result will be of the form A_0te^{-t} or $A_0t^2e^{-t}$ and will also be part of the complementary function. However, if we multiply by t^3, we obtain $Q_p = A_0t^3e^{-t}$, which is not part of the complementary function and is, therefore, the proper form for a particular solution.

9.149 Determine the form of a particular solution to $Q^{(4)} + 4Q^{(3)} + 6\ddot{Q} + 4\dot{Q} + Q = -23e^{-t}$.

▮ The complementary solution is, from Problem 8.159, $Q_c = C_1e^{-t} + C_2te^{-t} + C_3t^2e^{-t} + C_4t^3e^{-t}$. We try a particular solution having the same form as the right side of the differential equation, namely A_0e^{-t}. Since this is part of Q_c (for $C_1 = A_0$), it must be modified. To do so, we multiply by the smallest positive integral power of t that eliminates any duplication between it and Q_c. This is the fourth power, and the result is the proper form for a particular solution, $Q_p = A_0t^4e^{-t}$.

9.150 Determine the form of a particular solution to $Q^{(5)} + 5Q^{(4)} + 10Q^{(3)} + 10\ddot{Q} + 5\dot{Q} + Q = -3e^{-t}$.

▮ The complementary solution is, from Problem 8.160, $Q_p = (C_1 + C_2t + C_3t^2 + C_4t^3 + C_5t^4)e^{-t}$. We first try a particular solution of the form A_0e^{-t}. Since this is part of Q_c, we modify it by multiplying by the smallest positive integral power of t that eliminates any duplication of Q_c. This is the fifth power, and the proper form for a particular solution is $Q_p = A_0t^5e^{-t}$.

9.151 Determine the form of a particular solution to $y^{(4)} + 8y^{(3)} + 24y'' + 32y' + 16y = 7e^{-2x}$.

▮ The complementary solution is found in Problem 8.157 to be $y_c = c_1e^{-2x} + c_2xe^{-2x} + c_3x^2e^{-2x} + c_4x^3e^{-2x}$. We first try a particular solution of the form A_0e^{-2x}. Then, since this is part of y_c, we modify it by multiplying by the smallest positive integral power of x that eliminates any duplication of y_c. This is the fourth power, and the proper form for a particular solution is $y_p = A_0x^4e^{-2x}$.

9.152 Determine the form of a particular solution to $(D^4 - 6D^3 + 13D^2 - 12D + 4)y = 6e^x$.

▮ The complementary solution is, from Problem 8.187, $y_c = e^x(c_1 + c_2x) + e^{2x}(c_3 + c_4x)$. We try a particular solution having the same form as the right side of the differential equation, namely A_0e^x. Then, since this is part of y_c, we modify it by multiplying by x^2, the smallest positive integral power of x that eliminates any duplication. Thus, the proper form is $y_p = A_0x^2e^x$.

9.153 Rework the previous problem if, instead, the right side of the differential equation is $-3e^{2x}$.

▮ The complementary solution y_c of that problem remains valid, and a first try for the particular solution is A_0e^{2x}, which has the same form as the new right side. It must be modified, however, because it is part of y_c (for $c_3 = A_0$). We multiply by x^2, the smallest positive integral power of x that eliminates any duplication of y_c. The result is $y_p = A_0x^2e^{2x}$.

9.154 Determine the form of a particular solution to $y'' - 7y' = -3$.

▮ The complementary solution is $y_c = c_1 + c_2e^{7x}$ (see Problem 8.2). We try the particular solution A_0, but since this is part of y_c (for $c_1 = A_0$), it must be modified. Multiplying by x, we get $y_p = A_0x$, which is not part of y_c and is, therefore, the proper form for a particular solution.

9.155 Solve the differential equation of the previous problem.

▮ Substituting y_p, $y_p' = A_0$, and $y_p'' = 0$ into the given differential equation yields $0 - 7A_0 = -3$, or $A_0 = \frac{3}{7}$. Then $y_p = \frac{3}{7}x$, and the general solution is $y = c_1 + c_2e^{7x} + \frac{3}{7}x$.

9.156 Determine the form of a particular solution to $y'' - 7y' = -3x$.

▮ The complementary solution y_c of Problem 9.154 remains valid. Since the right side of the differential equation is a first-degree polynomial, we try $A_1x + A_0$, which is a general first-degree polynomial. But part of this trial solution, namely A_0, is also part of y_c, so it must be modified. To do so, we multiply it by the smallest positive integral power of x that will eliminate duplication. This is the first power, which gives us $y_p = x(A_1x + A_0) = A_1x^2 + A_0x$.

9.157 Solve the differential equation of the previous problem.

▮ Substituting y_p of the previous problem, $y_p' = 2A_1x + A_0$, and $y_p'' = 2A_1$ into the given differential equation yields, after rearrangement, $(-14A_1)x + (2A_1 - 7A_0) = -3x + 0$. Equating coefficients of like powers of x, we find $A_1 = \frac{3}{14}$ and $A_0 = \frac{3}{49}$. Then $y_p = \frac{3}{14}x^2 + \frac{3}{49}x$, and the general solution is $y = c_1 + c_2e^{7x} + \frac{3}{14}x^2 + \frac{3}{49}x$.

9.158 Determine the form of a particular solution to $y'' - 7y' = -3x^2$.

▌ As in Problem 9.154, the complementary solution is $y_c = c_1 + c_2 e^{7x}$. We try as a particular solution $A_2x^2 + A_1x + A_0$, which has the same degree as the right side of the given differential equation. But part of this duplicates part of y_c (let $c_1 = A_0$), so it needs to be modified. We multiply it by x, which is the smallest positive integral power of x that eliminates duplication, obtaining $y_p = A_2x^3 + A_1x^2 + A_0x$. Since there is now no duplication, y_p is in proper form.

9.159 Determine the form of a particular solution to $(D^4 + 4D^2)y = 6$.

▌ The complementary function is, from Problem 8.186, $y_c = c_1 + c_2x + c_3 \cos 2x + c_4 \sin 2x$. We first try the particular solution A_0. Because this is part of y_c (for $c_1 = A_0$), we modify it by multiplying by the smallest positive integral power of x that eliminates duplication. The first power will not work, because A_0x is part of y_c (for $c_2 = A_0$). The second power does work, however, so a particular solution is $y_p = A_0x^2$.

9.160 Redo the previous problem if, instead, the right side of the differential equation is $6x$.

▌ y_c remains valid, but we now use a general first-degree polynomial as our initial try for y_p: $A_1x + A_0$. However, this is part of y_c (for $c_1 = A_0$, $c_2 = A_1$), so we modify it by multiplying by x. The result is $A_1x^2 + A_0x$. But part of this (namely A_0x, a constant times x) is also part of y_c, so we must modify it by multiplying again by x. The result, $y_p = A_1x^3 + A_2x^2$, does not duplicate any part of y_c and is therefore in proper form.

.161 Redo Problem 9.159 if the right side of the differential equation is $6x^2 - 2x + 5$.

▌ The complementary solution y_c of Problem 9.159 remains valid, but we now use a general second-degree polynomial as our initial try for y_p: $A_2x^2 + A_1x + A_0$. Because parts of this expression (namely A_1x and A_0) are also part of the complementary function, we must modify it. To do so, we multiply by x^2, the smallest positive integral power of x that eliminates duplication of y_c. The proper form is then $y_p = A_2x^4 + A_1x^3 + A_0x^2$.

.162 Determine the form of a particular solution to $y'' = 9x^2 + 2x + 1$.

▌ The complementary function is $y_c = c_1 + c_2x$ (see Problem 8.153). Since the right side of the differential equation is a second-degree polynomial, we try a general second-degree polynomial as the form of y_p: $A_2x^2 + A_1x + A_0$. However, this trial solution has a first-power term and a constant term in common with y_c. We must modify it by multiplying by the smallest positive integral power of x that eliminates duplication between y_p and y_c. This is x^2, which gives $y_p = A_2x^4 + A_1x^3 + A_0x^2$.

.163 Solve the differential equation of the previous problem.

▌ Differentiating y_p of the previous problem twice and then substituting into the differential equation, we obtain $12A_2x^2 + 6A_1x + 2A_0 = 9x^2 + 2x - 1$, from which we find $A_2 = \frac{3}{4}$, $A_1 = \frac{1}{3}$, and $A_0 = -\frac{1}{2}$. Then $y_p = \frac{3}{4}x^4 + \frac{1}{3}x^3 - \frac{1}{2}x^2$, and the general solution is $y = c_1x + c_0 + \frac{3}{4}x^4 + \frac{1}{3}x^3 - \frac{1}{2}x^2$.

The particular solution also can be obtained by twice integrating both sides of the differential equation with respect to x.

.164 Determine the form of a particular solution to $y''' = 3x^2$.

▌ The complementary function is $y_c = c_1 + c_2x + c_3x^2$ (see Problem 8.154). Since the right side of the differential equation is a second-degree polynomial, we try $A_0 = A_1x + A_2x^2$. However, since this is identical in form to y_c, it must be modified. To do so, we multiply by x^3, the smallest positive integral power of x that eliminates any duplication of y_c. The result is $y_p = A_2x^5 + A_1x^4 + A_0x^3$.

.165 Determine the form of a particular solution to $y''' = -2x^2 + 9x + 18$.

▌ The particular solution is identical to that of the previous problem.

.166 Determine the form of a particular solution to $\dfrac{d^4y}{dx^4} = 12x^2 - 60$.

▌ The complementary solution is, from Problem 8.155, $y_c = c_1 + c_2x + c_3x^2 + c_4x^3$.

Since the right side of the given differential equation is a second-degree polynomial, we try, as a particular solution, the general second-degree polynomial $A_0 + A_1x + A_2x^2$. But this is part of y_c for suitable choices of c_1 through c_4, and so must be modified. To do so, we multiply by x^4, the smallest positive integral power of x that eliminates any duplication of y_c. The result is a proper particular solution: $y_p = A_0x^4 + A_1x^5 + A_2x^6$.

9.167 Solve the differential equation of the previous problem.

▌ Differentiating y_p four times successively and substituting the result into the given differential equation, we get $24A_0 + 120A_1x + 360A_2x^2 = 12x^2 - 60$. Then, by equating the coefficients of like powers of x, we conclude that $A_0 = -\frac{5}{2}$, $A_1 = 0$, and $A_2 = \frac{1}{30}$. Thus $y_p = \frac{1}{30}x^6 - \frac{5}{2}x^4$, and the general solution is $y = c_1 + c_2x + c_3x^2 + c_4x^3 + \frac{1}{30}x^6 - \frac{5}{2}x^4$.

9.168 Determine the form of a particular solution to $\dfrac{d^4y}{dx^4} = 30x^5 - 2x^2 + 5$.

▌ The complementary solution is, again, $y_c = c_1 + c_2x + c_3x^2 + c_4x^3$. The right side of the differential equation is a fifth-degree polynomial, so we try the particular solution $A_5x^5 + A_4x^4 + A_3x^3 + A_2x^2 + A_1x + A_0$. But because this trial solution and y_c have terms in common, it must be modified. Again we multiply by x^4, the smallest positive integral power of x that eliminates any duplication of y_c. The result is a particular solution in proper form, $y_p = A_5x^9 + A_4x^8 + A_3x^7 + A_2x^6 + A_1x^5 + A_0x^4$.

9.169 Solve $\dfrac{d^2x}{dt^2} + k\dfrac{dx}{dt} - g$, where k and g are positive constants.

▌ The complementary solution is $x_c = C_1 + C_2e^{-kt}$ (see Problem 8.20). Since the right side of the differential equation is a constant, we first try the particular solution A_0. But this term is part of x_c (for $C_1 = A_0$), so we modify it by multiplying by t. The result is $x_p = A_0t$, which is distinct from the terms comprising x_c.

Substituting x_p into the given differential equation, we get $0 + kA_0 = -g$, from which $A_0 = -g/k$. The general solution is then $x = x_c + x_p = C_1 + C_2e^{-kt} - \dfrac{g}{k}t$.

9.170 Determine the form of a particular solution to $y'' + 4y = \cos 2x$.

▌ The complementary solution is $y_c = c_1 \cos 2x + c_2 \sin 2x$ (see Problem 8.59). Since the right side of the given differential equation is a cosine term, we try $A_0 \sin 2x + B_0 \cos 2x$ as a particular solution. But this trial solution is identical in form to y_c (with $c_1 = B_0$ and $c_2 = A_0$), so we must modify it. Multiplying by x, we get $y_p = A_0x \sin 2x + B_0x \cos 2x$, which is distinct from y_c and therefore in proper form.

9.171 Redo the previous problem if, instead, the right side of the differential equation is $-3 \sin 2x$.

▌ The particular solution is identical to y_p of the previous problem.

9.172 Determine the form of a particular solution to $\ddot{x} + 16x = 2 \sin 4t$.

▌ The complementary solution is $x_c = c_1 \cos 4t + c_2 \sin 4t$ (see Problem 8.57). Since the right side of the differential equation is a sine term, we try $A_0 \sin 4t + B_0 \cos 4t$ as a particular solution. But y_c and this trial solution are of identical form, so we must modify it. We do so by multiplying by t, getting $x_p = A_0t \sin 4t + B_0t \cos 4t$. Since all the terms of x_p are distinct from those of x_c, it is in proper form.

9.173 Solve the differential equation of the previous problem.

▌ By substituting x_p and $x_p'' = (-16A_0t - 8B_0) \sin 4t + (8A_0 - 16B_0t) \cos 4t$ into the given differential equation and simplifying, we get

$$(-8B_0) \sin 4t + (8A_0) \cos 4t = 2 \sin 4t + 0 \cos 4t$$

Then equating the coefficients of like terms yields $A_0 = 0$ and $B_0 = -\frac{1}{4}$. The general solution is thus $x = x_c + x_p = c_1 \cos 4t + c_2 \sin 4t - \frac{1}{4}t \cos 4t$.

9.174 Determine the form of a particular solution to $\ddot{x} + 64x = 64 \cos 8t$.

▌ The complementary solution is $x_c = x_1 \cos 8t + c_2 \sin 8t$ (see Problem 8.58). Since the right side of the differential equation is a cosine term, we try $A_0 \sin 8t + B_0 \cos 8t$ as the form of a particular solution. But it is identical to x_c when $c_1 = B_0$ and $c_2 = A_0$, so we must modify the trial solution. We do so by multiplying by t, obtaining $x_p = A_0t \sin 8t + B_0t \cos 8t$. This is in proper form, because no part of it can be formed from x_c by a suitable choice of c_1 or c_2.

9.175 Solve the differential equation of the previous problem.

▌ By substituting x_p and its second derivative into the given differential equation, we obtain

$$(-64A_0t - 16B_0)\sin 8t + (16A_0 - 64B_0t)\cos 8t + 64(A_0t\sin 8t + B_0t\cos 8t) = 64\cos 8t$$

which can be simplified to $(-16B_0)\sin 8t + (16A_0)\cos 8t = 0\sin 8t + 64\cos 8t$. Equating coefficients of like terms, we conclude that $A_0 = 4$ and $B_0 = 0$. Then the general solution is $x = c_1\cos 8t + c_2\sin 8t + 4t\sin 8t$.

9.176 Solve $\ddot{x} + 3gx = 3g\sin\sqrt{3g}t$, where g is a positive constant.

▌ The complementary solution is, from Problem 8.74, $x_c = C_1\cos\sqrt{3g}t + C_2\sin\sqrt{3g}t$. We try $A_0\sin\sqrt{3g}t + B_0\cos\sqrt{3g}t$ as a particular solution, and then modify it to $x_p = A_0t\sin\sqrt{3g}t + B_0t\cos\sqrt{3g}t$ by multiplying by t.

Substituting x_p into the given differential equation and simplifying, we get $(-2\sqrt{3g}B_0)\sin\sqrt{3g}t + (2\sqrt{3g}A_0)\cos\sqrt{3g}t = 3g\sin\sqrt{3g}t$. By equating coefficients of like terms, we find that $B_0 = -\frac{1}{2}\sqrt{3g}$ and $A_0 = 0$. The general solution is then $x = x_c + x_p = c_1\cos\sqrt{3g}t + c_2\sin\sqrt{3g}t - \frac{1}{2}\sqrt{3g}t\cos\sqrt{3g}t$.

9.177 Determine the form of a particular solution to $(D^2 + 4)y = x^2\sin 2x$.

▌ The complementary solution is $y_c = C_1\cos 2x + C_2\sin 2x$. Since the right side of the differential equation is a second-degree polynomial times a sine term, we try $(Bx^2 + Ex + G)\sin 2x + (Ax^2 + Cx + F)\cos 2x$ as a particular solution. But this trial solution has terms in common with y_c, namely a constant times $\sin 2x$ and a constant times $\cos 2x$, so we must modify it. We do so by multiplying by the smallest positive integral power of x that eliminates all commonality between y_c and y_p. This is the first power. As a result, we find

$$y_p = Ax^3\cos 2x + Bx^3\sin 2x + Cx^2\cos 2x + Ex^2\sin 2x + Fx\cos 2x + Gx\sin 2x$$

9.178 Solve the differential equation of the previous problem.

▌ Differentiating y_p twice yields

$$\begin{aligned}D^2y_p = &-4Ax^3\cos 2x - 4Bx^3\sin 2x + (12B - 4C)x^2\cos 2x + (-12A - 4E)x^2\sin 2x\\ &+ (6A + 8E - 4F)x\cos 2x + (6B - 8C - 4G)x\sin 2x + (2C + 4G)\cos 2x + (2E - 4F)\sin 2x\end{aligned}$$

Then substituting this and y_p into the given equation yields

$$\begin{aligned}12Bx^2\cos 2x - 12Ax^2\sin 2x + (6A + 8E)x\cos 2x + (6B - 8C)x\sin 2x + (2C + 4G)\cos 2x + (2E - 4F)\sin 2x\\ = x^2\sin 2x\end{aligned}$$

By equating coefficients of like terms, we find that $A = -\frac{1}{12}$, $B = C = G = 0$, $E = \frac{1}{16}$, and $F = \frac{1}{32}$. Then $y_p = -\frac{1}{12}x^3\cos 2x + \frac{1}{16}x^2\sin 2x + \frac{1}{32}x\cos 2x$, and the primitive is $y = C_1\cos 2x + C_2\sin 2x - \frac{1}{12}x^3\cos 2x + \frac{1}{16}x^2\sin 2x + \frac{1}{32}x\cos 2x$.

9.179 Solve $(D^2 + 4)y = 8\sin 2x$.

▌ The complementary solution is $y_c = c_1\cos 2x + c_2\sin 2x$. For a particular solution we would normally try $A\cos 2x + B\sin 2x$. However, since its terms appear in the complementary solution, we multiply by x to obtain the particular solution $y_p = x(A\cos 2x + B\sin 2x)$. Then, substituting in the given equation yields $-4A\cos 2x - 4B\sin 2x = 8\sin 2x$, so that $A = 0$ and $B = -2$. Then the required general solution is $y = c_1\cos 2x + c_2\sin 2x - 2x\sin 2x$.

9.180 Determine the form of a particular solution to $y'' + 4y = 8x\sin 2x$.

▌ As in the previous problem, the complementary solution is $y_c = c_1\cos 2x + c_2\sin 2x$. We try a particular solution of the form $(A_1x + A_0)\sin 2x + (B_1x + B_0)\cos 2x$ because the right side of the given differential equation is the product of a first-degree polynomial and $\sin 2x$. But two of the summands in this trial solution are identical in form with the summands in y_c, so it must be modified. Multiplying by x, we get $y_p = (A_1x^2 + A_0x)\sin 2x + (B_1x^2 + B_0x)\cos 2x$, which has no terms in common with y_c and is therefore in proper form.

9.181 Solve the differential equation of the previous problem.

▌ Differentiating y_p of the previous problem twice in succession yields

$$y_p'' = (-4A_1x^2 - 4A_0x + 2A_1 - 8B_1x - 4B_0)\sin 2x + (8A_1x + 4A_0 - 4B_1x^2 - 4B_0x + 2B_1)\cos 2x$$

Substituting this and y_p into the given differential equation and simplifying, we obtain

$$[(-8B_1)x + (2A_1 - 4B_0)] \sin 2x + [(8A_1)x + (4A_0 + 2B_1)] \cos 2x = 8x \sin 2x$$

By equating the coefficients of like terms and solving the resulting system of equations, we find that $A_1 = 0$, $A_0 = \frac{1}{2}$, $B_1 = -1$, and $B_0 = 0$. Combining these results with y_c and y_p of the previous problem, we form the general solution $y = c_1 \cos 2x + c_2 \sin 2x + \frac{1}{2}x \sin 2x - x^2 \cos 2x$.

9.182 Solve $y'' + 4y = (8 - 16x) \cos 2x$.

y_c and y_p of Problem 9.180 are valid here, as is y_p'' of the previous problem. Substituting y_p and y_p'' into the given differential equation and simplifying, we get

$$[(-8B_1)x + (2A_1 - 4B_0)] \sin 2x + [(8A_1)x + (4A_0 + 2B_1)] \cos 2x = -16x \cos 2x + 8 \cos 2x$$

By equating coefficients of like terms and solving the resulting system of equations, we find that $A_1 = -2$, $A_0 = 2$, $B_1 = 0$, and $B_0 = -1$. Combining these results with y_c and y_p of Problem 9.180, we form the general solution $y = (-2x^2 + 2x + c_2) \sin 2x + (-x + c_1) \cos 2x$.

9.183 Determine the form of a particular solution of $\ddot{x} + 16x = (80t - 16) \sin 4t$.

The complementary solution is $x_c = c_1 \cos 4t + c_2 \sin 4t$ (see Problem 8.57). We try a particular solution of the form $(A_1 t + A_0) \sin 4t + (B_1 t + B_0) \cos 4t$, but since it contains summands which are identical in form to the summands of x_c, it must be modified. Multiplying by t, we get $x_p = (A_1 t^2 + A_0 t) \sin 4t + (B_1 t^2 + B_0 t) \cos 4t$.

Since none of the summands of x_p is identical to a summand of x_c except perhaps for a multiplicative constant x_p is the proper form for a particular solution.

9.184 Solve the differential equation of the previous problem.

Differentiating x_p of the previous problem twice yields

$$\ddot{x}_p = (-16A_1 t^2 - 16A_0 t + 2A_1 - 16B_1 t - 8B_0) \sin 4t + (16A_1 t + 8A_0 - 16B_1 t^2 - 16B_0 t + 2B_1) \cos 4t$$

Then by substituting x_p and $\ddot{x}_p$ into the given differential equation and simplifying, we get

$$[(-16B_1)t + (2A_1 - 8B_0)] \sin 4t + [(16A_1)t + (8A_0 + 2B_1)] \cos 4t = (80t - 16) \sin 4t + (0t + 0) \cos 4t$$

Equating coefficients of like terms, we get a system of equations whose solution is $A_1 = 0$, $A_0 = 1.25$, $B_1 = -5$, and $B_0 = 2$. Combining these results with x_p and x_c of the previous problem, we form the general solution $x = c_1 \cos 4t + c_2 \sin 4t + 1.25t \sin 4t + (-5t^2 + 2t) \cos 4t$.

9.185 Determine the form of a particular solution to $\ddot{x} + 64x = 8t^2 \cos 8t$.

The complementary solution is $x_c = c_1 \cos 8t + c_2 \sin 8t$ (see Problem 8.58). Since the right side of the given differential equation is the product of a second-degree polynomial and a cosine term, we try a particular solution of the form $(A_2 t^2 + A_1 t + A_0) \sin 8t + (B_2 t^2 + B_1 t + B_0) \cos 8t$.

Two of the summands in this trial solution also appear in x_c for suitable choices of c_1 and c_2, so it must be modified. Multiplying by t, we get

$$x_p = (A_2 t^3 + A_1 t^2 + A_0 t) \sin 8t + (B_2 t^3 + B_1 t^2 + B_0 t) \cos 8t$$

Since the summands of x_p are linearly independent of the summands of x_c, we have the proper form for a particular solution.

9.186 Determine the form of a particular solution to $\ddot{x} + 96x = (t^3 + 3) \sin \sqrt{96}t$.

The complementary solution is, from Problem 8.61, $x_c = C_1 \sin \sqrt{96}t + C_2 \cos \sqrt{96}t$. We try a particular solution of the form $(A_3 t^3 + A_2 t^2 + A_1 t + A_0) \sin \sqrt{96}t + (B_3 t^3 + B_2 t^2 + B_1 t + B_0) \cos \sqrt{96}t$. But this trial solution has summands in common with x_c, so it must be modified. We multiply it by the smallest positive integral power of t that eliminates any duplication of terms of x_c—the first. Thus, we have

$$x_p = (A_3 t^4 + A_2 t^3 + A_1 t^2 + A_0 t) \sin \sqrt{96}t + (B_3 t^4 + B_2 t^3 + B_1 t^2 + B_0 t) \cos \sqrt{96}t$$

9.187 Determine the form of a particular solution to $(D^4 + 4D^2)y = x \cos 2x$.

The complementary solution is, from Problem 8.186, $x_c = c_1 + c_2 x + c_3 \cos 2x + c_4 \sin 2x$. We try a particular solution of the form $(A_1 x + A_0) \sin 2x + (B_1 x + B_0) \cos 2x$. But this trial solution and x_c have summands in common, so the trial solution must be modified. Multiplying it by x, we get $y_p = (A_1x^2 + A_0x)\sin 2x + (B_1x^2 + B_0x)\cos 2x$.

Since each summand of y_p is linearly independent of the summands of y_c, it is the proper form for a particular solution.

.188 Determine the form of a particular solution to $\dfrac{d^4y}{dx^4} + 8\dfrac{d^2y}{dx^2} + 16y = (1 - 8x)\sin 2x$.

▌ The characteristic equation of the associated homogeneous equation can be factored into $(\lambda^2 + 4)^2 = 0$, so the roots are $\pm i2$, each of multiplicity two. It follows from Problem 8.175 that the complementary function is $y_c = c_1\cos 2x + c_2\sin 2x + c_3x\cos 2x + c_4x\sin 2x$.

Since the right side of the given differential equation is a first-degree polynomial times $\sin 2x$, we try the particular solution $(A_1x + A_0)\sin 2x + (B_1x + B_0)\cos 2x$. But this is identical to y_c when $c_1 = B_0$, $c_2 = A_0$, $c_3 = B_1$, and $c_4 = A_1$. We must modify the trial solution, and we do so by multiplying it by x^2, the smallest positive integral power of x that eliminates any duplication. The result is $y_p = (A_1x^3 + A_0x^2)\sin 2x + (B_1x^3 + B_0x^2)\cos 2x$.

The summands of y_p are distinct from those of y_c, so it is the proper form for a particular solution.

.189 Determine the form of a particular solution to $(D^2 + 9)^3y = (2x^2 - 3x + 5)\cos 3x$.

▌ The roots of the characteristic equation of the associated homogeneous equation are $\pm i3$, each of multiplicity three. The complementary function is

$$y_c = (c_1 + c_2x + c_3x^2)\cos 3x + (c_4 + c_5x + c_6x^2)\sin 3x$$

We try a particular solution of the form $(A_2x^2 + A_1x + A_0)\sin 3x + (B_2x^2 + B_1x + B_0)\cos 3x$. However, this is identical in form to y_c and must be modified. To do so, we multiply by x^3, the smallest positive integral power of x that results in summands distinct from those of y_c. The result is $y_p = (A_2x^5 + A_1x^4 + A_0x^3)\sin 3x + (B_2x^5 + B_1x^4 + B_0x^3)\cos 3x$, which is the proper form for a particular solution.

.190 Determine the form of a particular solution for $\dfrac{d^2y}{dx^2} - 6\dfrac{dy}{dx} + 25y = 6e^{3x}\cos 4x$.

▌ The complementary solution is $y_c = e^{3x}(c_1\cos 4x + c_2\sin 4x)$ (see Problem 8.50). Since the right side of the differential equation is an exponential times a cosine term, we try the particular solution $A_0e^{3x}\sin 4x + B_0e^{3x}\cos 4x$. Because this has the same form as y_c, we modify it by multiplying by x. The result, $y_p = A_0xe^{3x}\sin 4x + B_0xe^{3x}\cos 4x$, consists of terms that are different from those of y_c, so it needs no further modification.

.191 Determine the form of a particular solution to $\dfrac{d^2y}{dx^2} - 10\dfrac{dy}{dx} + 29y = -8e^{5x}\sin 2x$.

▌ The complementary solution is $y_c = e^{5x}(c_1\cos 2x + c_2\sin 2x)$ (see Problem 8.51). The right side of the given differential equation is an exponential times a sine, so we try the particular solution $A_0e^{5x}\sin 2x + B_0e^{5x}\cos 2x$. Since this is identical in form to y_c, we modify it by multiplying by x. The result, $y_p = A_0xe^{5x}\sin 2x + B_0xe^{5x}\cos 2x$, consists of terms that cannot be obtained from y_c by any choice of the constants c_1 and c_2 and so needs no further modification.

.192 Determine the form of a particular solution to $y'' + 4y' + 5y = 60e^{-2x}\sin x$.

▌ The complementary solution is $y_c = c_1e^{-2x}\cos x + c_2e^{-2x}\sin x$ (see Problem 8.66). We try, as a particular solution, $A_0e^{-2x}\sin x + B_0e^{-2x}\cos x$; but because it is identical in form to y_c, it must be modified. We multiply it by x, getting $y_p = A_0xe^{-2x}\sin x + B_0xe^{-2x}\cos x$, which has no terms in common with y_c and so is in proper form.

193 Solve the differential equation of the previous problem.

▌ Differentiating y_p of the previous problem yields

$$y_p' = (-2A_0x + A_0 - B_0x)e^{-2x}\sin x + (A_0x - 2B_0x + B_0)e^{-2x}\cos x$$

and $$y_p'' = (3A_0x - 4A_0 + 4B_0x - 2B_0)e^{-2x}\sin x + (-4A_0x + 2A_0 + 3B_0x - 4B_0)e^{-2x}\cos x$$

Substituting y_p and its derivatives into the differential equation and simplifying, we get $-2B_0e^{-2x}\sin x + 2A_0e^{-2x}\cos x = 60e^{-2x}\sin x$; by equating coefficients of like terms, we find that $A_0 = 0$ and $B_0 = -30$. Then $y_p = -30xe^{-2x}\cos x$, and the general solution is $y = c_1e^{-2x}\cos x + c_2e^{-2x}\sin x - 30xe^{-2x}\cos x$.

9.194 Determine the form of a particular solution to $(D^2 - 2D + 10)y = 18e^x \cos 3x$.

▌ The complementary solution is, from Problem 8.68, $y_c = C_1e^x\cos 3x + C_2e^x\sin 3x$. We try the particular solution $A_0e^x\sin 3x + B_0e^x\cos 3x$, but since this is identical in form to y_c, it must be modified. We therefore multiply by x, obtaining $y_p = A_0xe^x\sin 3x + B_0xe^x\cos 3x$. Since there is no duplication between y_c and y_p, the latter is in proper form.

9.195 Solve the differential equation of the previous problem.

▌ By differentiating y_p of the previous problem twice, we get

$$y_p' = (A_0x + A_0 - 3B_0x)e^x\sin 3x + (3A_0x + B_0x + B_0)e^x\sin 3x$$

and

$$y_p'' = (-8A_0x + 2A_0 - 6B_0x - 6B_0)e^x\sin 3x + (6A_0x + 6A_0 - 8B_0x + 2B_0)e^x\cos 3x$$

Substituting y_p and its derivatives into the differential equation and simplifying then yield $(-6B_0)e^x\sin 3x + (6A_0)e^x\cos 3x = 18e^x\cos 3x$. By equating coefficients of like terms, we find that $A_0 = 3$ and $B_0 = 0$. Combining these results with y_c and y_p of the previous problem, we form the general solution $y = C_1e^x\cos 3x + C_2e^x\sin 3x + 3xe^x\sin 3x$.

9.196 Determine the form of a particular solution to $y^{(4)} - 8y^{(3)} + 32y'' - 64y' + 64y = 30e^{2x}\sin 2x$.

▌ The complementary solution is, from Problem 8.188, $y_c = (c_1 + c_2x)e^{2x}\cos 2x + (c_3 + c_4x)e^{2x}\sin 2x$. We try as a particular solution $A_0e^{2x}\sin 2x + B_0e^{2x}\cos 2x$, but since this is part of the complementary function (for $c_3 = A_0$, $c_1 = B_0$), it must be modified. Multiplying by x will also duplicate terms of y_c, so we multiply by x^2 to get $y_p = A_0x^2e^{2x}\sin 2x + B_0x^2e^{2x}\cos 2x$. Since the summands of y_p are distinct from those of y_c, y_p is the proper form for a particular solution.

9.197 Determine the form of a particular solution to $\dfrac{d^4y}{dx^4} + 4\dfrac{d^3y}{dx^3} + 8\dfrac{d^2y}{dx^2} + 8\dfrac{dy}{dx} + 4y = 2e^{-x}\cos x$.

▌ The characteristic equation of the associated homogeneous differential equation can be factored into $(m^2 + 2m + 2)^2 = 0$, so $-1 \pm i$ are both roots of multiplicity two, and the complementary solution is $y_c = (c_1 + c_2x)e^{-x}\cos x + (c_3 + c_4x)e^{-x}\sin x$.

We try as a particular solution $A_0e^{-x}\sin x + B_0e^{-x}\cos x$, but because this trial solution has terms in common with y_c, we must modify it. We thus multiply by x^2, the smallest positive integral power of x that eliminates any duplication, obtaining $y_p = A_0x^2e^{-x}\sin x + B_0x^2e^{-x}\cos x$.

9.198 Determine the form of a particular solution to $(D^2 + 2D + 2)^3y = 3e^{-x}\sin x$.

▌ The characteristic equation of the associated homogeneous differential equation is $(m^2 + 2m + 2)^3 = 0$, so $-1 \pm i$ are both roots of multiplicity three, and the complementary solution is $y_c = (c_1 + c_2x + c_3x^2)e^{-x}\cos x + (c_4 + c_5x + c_6x^2)e^{-x}\sin x$.

We try as a particular solution $A_0e^{-x}\sin x + B_0e^{-x}\cos x$, but because this trial solution is part of y_c, it must be modified. We multiply by x^3, the smallest positive integral power of x that eliminates any duplication of y_c. The result is $y_p = A_0x^3e^{-x}\sin x + B_0x^3e^{-x}\cos x$, which is the proper form of a particular solution.

9.199 Determine the form of a particular solution to $\dfrac{d^2y}{dx^2} - 6\dfrac{dy}{dx} + 25y = (2x - 1)e^{3x}\cos 4x$.

▌ The complementary solution is, from Problem 8.50, $y_c = c_1e^{3x}\cos 4x + c_2e^{3x}\sin 4x$. We try as a particular solution $(A_1x + A_0)e^{3x}\sin 4x + (B_1x + B_0)e^{3x}\cos 4x$. Two of the summands of this trial solution are identical in form to the summands of y_c, so it must be modified. Multiplying by x yields $y_p = (A_1x^2 + A_0x)e^{3x}\sin 4x + (B_1x^2 + B_0x)e^{3x}\cos 4x$, which has no summands in common with y_c.

9.200 Determine the form of a particular solution to $\dfrac{d^2y}{dx^2} + 10\dfrac{dy}{dx} + 29y = xe^{5x}\sin 2x$.

▌ The complementary solution is, from Problem 8.51, $y_c = c_1e^{5x}\cos 2x + c_2e^{5x}\sin 2x$. We try as a particular solution $(A_1x + A_0)e^{5x}\sin 2x + (B_1x + B_0)e^{5x}\cos 2x$. Since two of the summands of this trial solution are identical in form to the summands in y_c, we must modify it. We do so by multiplying by x, which results in $y_p = (A_1x^2 + A_0x)e^{5x}\sin 2x + (B_1x^2 + B_0x)e^{5x}\cos 2x$. Because y_p and y_c have no terms in common, this is of the proper form for a particular solution.

9.201 Determine the form of a particular solution to $y'' + 4y' + 5y = (x^2 + 5)e^{-2x}\sin x$.

The complementary solution is, from Problem 8.66, $y_c = c_1e^{-2x}\cos x + c_2e^{-2x}\sin x$. We try as a particular solution $(A_2x^2 + A_1x + A_0)e^{-2x}\sin x + (B_2x^2 + B_1x + B_0)e^{-2x}\cos x$. But since this trial solution has summands that are identical in form to those of y_c, it must be modified. Multiplying by x, we get

$$y_p = (A_2x^3 + A_1x^2 + A_0x)e^{-2x}\sin x + (B_2x^3 + B_1x^2 + B_0x)e^{-2x}\cos x$$

Since y_p does not duplicate any of the summands of y_c, it is of the proper form for a particular solution.

9.202 Determine the form of a particular solution to $y^{(4)} - 8y^{(3)} + 32y'' - 64y' + 64y = x^2e^{2x}\sin 2x$.

The complementary solution is, from Problem 8.177, $y_c = (c_1 + c_2x)e^{2x}\cos 2x + (c_3 + c_4x)e^{2x}\sin 2x$. We try as a particular solution $(A_2x^2 + A_1x + A_0)e^{2x}\sin 2x + (B_2x^2 + B_1x + B_0)e^{2x}\cos 2x$. Because this trial solution has terms in common with y_c, it must be modified. If we multiply by x, it will still have terms in common with y_c, so we multiply by x^2, getting

$$y_p = (A_2x^4 + A_1x^3 + A_0x^2)e^{2x}\sin 2x + (B_2x^4 + B_1x^3 + B_0x^2)e^{2x}\cos 2x$$

Because each of its summands is different from those in y_c, y_p is in proper form.

9.203 Determine the form of a particular solution to $\dfrac{d^4y}{dx^4} + 4\dfrac{d^3y}{dx^3} + 8\dfrac{d^2y}{dx^2} + 8\dfrac{dy}{dx} + 4y = (x - 4)e^{-x}\cos x$.

The complementary solution is, from Problem 9.197, $y_c = (c_1 + c_2x)e^{-x}\cos x + (c_3 + c_4x)e^{2x}\sin x$. We try as a particular solution $(A_1x + A_0)e^{-x}\sin x + (B_1x + B_0)e^{-x}\cos x$, but because it is identical in form to y_c, it must be modified. To eliminate any duplication of y_c, we multiply by x^2, obtaining

$$y_p = (A_1x^3 + A_0x^2)e^{-x}\sin x + (B_1x^3 + B_0x^2)e^{-x}\cos x$$

as the proper form for a particular solution.

9.204 Determine the form of a particular solution to $(D^2 + 2D + 2)^3 = xe^{-x}\sin x$.

The complementary solution is, from Problem 9.198,

$$y_c = (c_1 + c_2x + c_3x^2)e^{-x}\cos x + (c_4 + c_5x + c_6x^2)e^{-x}\sin x$$

We try as a particular solution $(A_1x + A_0)e^{-x}\sin x + (B_1x + B_0)e^{-x}\cos x$, because the right side of the differential equation is the product of a first-degree polynomial, e^{-x}, and a sine term. Since each summand in the trial solution also appears in y_c except for the arbitrary multiplicative constants, it must be modified. We multiply by x^3, the smallest positive integral power of x that eliminates any commonality with y_c. The result is

$$y_p = (A_1x^4 + A_0x^3)e^{-x}\sin x + (B_1x^4 + B_0x^3)e^{-x}\cos x$$

which is the proper form for a particular solution.

9.205 Redo the previous problem if the right side is replaced with $(5 - 3x)e^{-x}\cos x$.

The particular solution here is identical to y_p of the previous problem: Since the right side of the new differential equation is the product of a first-degree polynomial, e^{-x}, and a cosine term, the trial solution here is the same as that in the previous problem. And it must be modified in the same way to the same particular solution y_p.

EQUATIONS WHOSE RIGHT SIDE CONTAINS A COMBINATION OF TERMS

9.206 Solve $y' - 5y = e^{2x} + 8x$.

A particular solution corresponding to a right side of e^{2x} is found in Problem 9.1 to be $y_1 = -\frac{1}{3}e^{2x}$, and a particular solution corresponding to a right side of $8x$ is found in Problem 9.32 to be $y_2 = -\frac{8}{5}x - \frac{8}{25}$.

A particular solution to the given differential equation is then $y_1 + y_2 = -\frac{1}{3}e^{2x} - \frac{8}{5}x - \frac{8}{25}$. When combined with the complementary solution $y_c = c_1e^{5x}$ (see Problem 8.34), it yields the general solution $y = c_1e^{5x} - \frac{1}{3}e^{2x} - \frac{8}{5}x - \frac{8}{25}$.

9.207 Solve $y' - 5y = e^{2x} + \sin x$.

▌ A particular solution corresponding to a right side of e^{2x} remains y_1 of the previous problem, while a particular solution corresponding to $\sin x$ is found in Problem 9.77 to be $y_2 = -\frac{5}{26}\sin x - \frac{1}{26}\cos x$.

A particular solution to the given differential equation is then $y_1 + y_2 = -\frac{1}{3}e^{2x} - \frac{5}{26}\sin x - \frac{1}{26}\cos x$; when combined with the complementary solution, it yields the general solution $y = c_1e^{5x} - \frac{1}{3}e^{2x} - \frac{5}{26}\sin x - \frac{1}{26}\cos x$.

9.208 Solve $y' - 5y = 8x + \sin x$.

▌ A particular solution corresponding to $8x$ is found in Problem 9.32 to be $y_1 = -\frac{8}{5}x - \frac{8}{25}$, and a particular solution corresponding to $\sin x$ is found in Problem 9.77 to be $y_2 = -\frac{5}{26}\sin x - \frac{1}{26}\cos x$.

A particular solution to the given differential equation is then $y_1 + y_2 = -\frac{8}{5}x - \frac{8}{25} - \frac{5}{26}\sin x - \frac{1}{26}\cos x$, which, when combined with the complementary solution (see Problem 8.34) yields the general solution $y = c_1e^{5x} - \frac{8}{5}x - \frac{8}{25} - \frac{5}{26}\sin x - \frac{1}{26}\cos x$.

9.209 Solve $y' - 5y = 8x - \sin x$.

▌ The expressions for y_1 and y_2 of the previous problem are valid here, but now a particular solution is the difference of those two solutions, namely $y_1 - y_2 = -\frac{8}{5}x - \frac{8}{25} - (-\frac{5}{26}\sin x - \frac{1}{26}\cos x)$.

Combining this with the general solution to the associated homogeneous problem (see Problem 8.34), we obtain the general solution to the nonhomogeneous equation as $y = c_1e^{5x} - \frac{8}{5}x - \frac{8}{25} + \frac{5}{26}\sin x + \frac{1}{26}\cos x$.

9.210 Solve $y' - 5y = 8 - 2e^{5x}$.

▌ A particular solution corresponding to a right side of 8 is found in Problem 9.17 as $y_1 = -\frac{8}{5}$, and a particular solution corresponding to $2e^{5x}$ is found in Problems 9.128 and 9.129 to be $y_2 = 2xe^{5x}$.

A particular solution to the given differential equation is then $y_1 - y_2 = -\frac{8}{5} - 2xe^{5x}$. Combined with the complementary solution (see Problem 8.34), it yields the general solution $y = c_1e^{5x} - 2xe^{5x} - \frac{8}{5}$.

9.211 Solve $y' - 5y = xe^{2x} + 2x^2 - 5$.

▌ A particular solution corresponding to xe^{2x} is found in Problem 9.55 to be $y_1 = (-\frac{1}{3}x - \frac{1}{9})e^{2x}$, and a particular solution corresponding to $2x^2 - 5$ is found in Problem 9.33 to be $y_2 = -0.4x^2 - 0.16x + 0.968$.

A particular solution to the given differential equation is then $y_1 + y_2 = (-\frac{1}{3}x - \frac{1}{9})e^{2x} - 0.4x^2 - 0.16x + 0.968$ When combined with the complementary solution (see Problem 8.34), it yields the general solution $y = c_1e^{5x} + (-\frac{1}{3}x - \frac{1}{9})e^{2x} - 0.4x^2 - 0.16x + 0.968$.

9.212 Solve $y' - 5y = xe^{2x} + 8x + \sin x$.

▌ y_1 of the previous problem is valid here. In addition, a particular solution corresponding to $8x$ is, from Problem 9.32, $y_2 = -\frac{8}{5}x - \frac{8}{25}$; a particular solution corresponding to $\sin x$ is found in Problem 9.77 to be $y_3 = -\frac{5}{26}\sin x - \frac{1}{26}\cos x$. A particular solution to the given differential equation is then

$$y_1 + y_2 + y_3 = (-\tfrac{1}{3}x - \tfrac{1}{9})e^{2x} + (-\tfrac{8}{5}x - \tfrac{8}{25}) + (-\tfrac{5}{26}\sin x - \tfrac{1}{26}\cos x)$$

We combine this result with the solution to the associated homogeneous equation (see Problem 8.34) to obtain the general solution $y = c_1e^{5x} - \frac{1}{3}xe^{2x} - \frac{1}{9}e^{2x} - \frac{8}{5}x - \frac{8}{25} - \frac{5}{26}\sin x - \frac{1}{26}\cos x$.

9.213 Solve $y' + 6y = -2\cos 3x + 3e^{2x}\sin 3x$.

▌ A particular solution corresponding to $-2\cos 3x$ is found in Problem 9.78 to be $y_1 = -\frac{2}{15}\sin 3x - \frac{4}{15}\cos 3x$, and a particular solution corresponding to $3e^{2x}\sin 3x$ is found in Problem 9.99 to be $y_2 = \frac{24}{73}e^{2x}\sin 3x - \frac{9}{73}e^{2x}\cos 3x$. A particular solution to the given differential equation is then

$$y_1 + y_2 = -\tfrac{2}{15}\sin 3x - \tfrac{4}{15}\cos 3x + \tfrac{24}{73}e^{2x}\sin 3x - \tfrac{9}{73}e^{2x}\cos 3x$$

When combined with the complementary solution (see Problem 8.37), this yields the general solution

$$y = Ae^{-6x} - \tfrac{2}{15}\sin 3x - \tfrac{4}{15}\cos 3x + \tfrac{24}{73}e^{2x}\sin 3x - \tfrac{9}{73}e^{2x}\cos 3x$$

9.214 Rework the previous problem if the term e^{3x} is added to the right side of the differential equation.

▌ y_1 and y_2 of the previous problem are valid here. A particular solution corresponding to e^{3x} is found in Problem 9.2 as $y_3 = \frac{1}{9}e^{3x}$. A particular solution corresponding to the new right side is then $y_1 + y_2 + y_3$; when combined with the complementary solution (see Problem 8.37), it yields the general solution

$$y = Ae^{-6x} - \tfrac{2}{15}\sin 3x - \tfrac{4}{15}\cos 3x + \tfrac{24}{73}e^{2x}\sin 3x - \tfrac{9}{73}e^{2x}\cos 3x + \tfrac{1}{9}e^{3x}$$

9.215 Solve $y' + 6y = 4e^{-5x} - 6e^{6x} + 3e^{-6x}$.

▌ A particular solution corresponding to a right side of $4e^{-5x}$ is found in Problem 9.4 to be $y_1 = 4e^{-5x}$; a particular solution corresponding to $6e^{6x}$ is found in Problem 9.5 to be $y_2 = \frac{1}{2}e^{6x}$; and a particular solution corresponding to $3e^{-6x}$ is found in Problems 9.130 and 9.131 to be $y_3 = 3xe^{-6x}$.

A particular solution to the given differential equation is $y_1 - y_2 + y_3$. When combined with the complementary solution (see Problem 8.37), it yields the general solution $y = Ae^{-6x} + 4e^{-5x} - \frac{1}{2}e^{6x} + 3xe^{-6x}$.

9.216 Solve $y'' - 7y' = 6e^{6x} + e^{8x}$.

▌ A particular solution corresponding to $6e^{6x}$ is found in Problem 9.7 to be $y_1 = -e^{6x}$, and a particular solution corresponding to e^{8x} is found in Problem 9.8 to be $y_2 = \frac{1}{8}e^{8x}$.

A particular solution to the given differential equation is then $y_1 + y_2$. When combined with the complementary solution (see Problem 8.2), it yields the general solution $y = c_1 + c_2e^{7x} - e^{6x} + \frac{1}{8}e^{8x}$.

9.217 Solve $y'' - 7y' = -3x + 48\sin 4x + 84\cos 4x$.

▌ A particular solution corresponding to $-3x$ is found in Problems 9.156 and 9.157 to be $y_1 = \frac{3}{14}x^2 + \frac{3}{49}x$. A particular solution corresponding to $48\sin 4x + 84\cos 4x$ is found in Problem 9.80 to be $y_2 = -3\sin 4x$.

A particular solution to the given differential equation is then $y_1 + y_2$; when combined with the complementary solution (see Problem 8.2), it yields the general solution $y = c_1 + c_2e^{7x} + \frac{3}{14}x^2 + \frac{3}{49}x - 3\sin 4x$.

9.218 Solve $y'' - y' - 2y = 7 + e^{3x}$.

▌ A particular solution corresponding to a right side of 7 is found in Problem 9.19 to be $y_1 = -\frac{7}{2}$, and a particular solution corresponding to e^{2x} is found in Problem 9.10 to be $y_2 = \frac{1}{4}e^{3x}$.

A particular solution to the given differential equation is $y_1 + y_2$. When combined with the complementary solution (see Problem 8.1), it yields the general solution $y = c_1e^{-x} + c_2e^{2x} - \frac{7}{2} + \frac{1}{4}e^{3x}$.

9.219 Use the results of the previous problem to solve $y'' - y' - 2y = 14 - 3e^{3x}$.

▌ The right side of the given differential equation may be written as $2(7) - 3(e^{3x})$. Then we conclude that a particular solution to this differential equation is $2y_1 - 3y_2$, where y_1 and y_2 are as in the previous problem. When combined with the complementary solution, this yields the general solution $y = c_1e^{-x} + c_2e^{2x} - 7 - \frac{3}{4}e^{3x}$.

9.220 Solve $y'' - y' - 2y = e^{2x} + 2e^{-x}$.

▌ A particular solution corresponding to e^{2x} is found in Problems 9.134 and 9.135 to be $y_1 = \frac{1}{3}xe^{2x}$. A particular solution corresponding to $2e^{-x}$ is found in Problems 9.136 and 9.137 to be $y_2 = -\frac{2}{3}xe^{-x}$.

A particular solution to the given differential equation then is $y_1 + y_2$; when combined with the complementary solution (see Problem 8.1), it yields the general solution $y = c_1e^{-x} + c_2e^{2x} + \frac{1}{3}xe^{2x} - \frac{2}{3}xe^{-x}$.

9.221 Use the results of the previous problem to solve $y'' - y' - 2y = 3e^{2x} - 18e^{-x}$.

▌ The right side of this differential equation may be written as $3(e^{2x}) - 9(2e^{-x})$. It then follows from the previous problem that a particular solution to this differential equation is $3y_1 - 9y_2$. When combined with the complementary solution, this yields the general solution $y = c_1e^{-x} + c_2e^{2x} + xe^{2x} + 6xe^{-x}$.

9.222 Solve $y'' - y' - 2y = 7 + e^{3x} + e^{2x} + 2e^{-x}$.

▌ Combining y_1 and y_2 of both Problem 9.218 and Problem 9.220 with the complementary function, we obtain the general solution $y = c_1e^{-x} + c_2e^{2x} - \frac{7}{2} + \frac{1}{4}e^{3x} + \frac{1}{3}xe^{2x} - \frac{2}{3}xe^{-x}$.

9.223 Solve $y'' - y' - 2y = 4x^2 - \sin 2x$.

▌ A particular solution corresponding to a right side of $4x^2$ is found in Problem 9.35 to be $y_1 = -2x^2 + 2x - 3$. A particular solution corresponding to $\sin 2x$ is found in Problem 9.92 to be $y_2 = -\frac{3}{20}\sin 2x + \frac{1}{20}\cos 2x$.

A particular solution to the given differential equation is then $y_1 - y_2$; when combined with the complementary solution (see Problem 8.1), it yields the general solution $y = c_1e^{-x} + c_2e^{2x} - 2x^2 + 2x - 3 + \frac{3}{20}\sin 2x - \frac{1}{20}\cos 2x$.

9.224 Solve $\dfrac{d^2y}{dt^2} - 4\dfrac{dy}{dt} + y = 3e^{2t} + 3t - 4$.

▍ A particular solution corresponding to $3e^{2t}$ is found in Problem 9.11 to be $y_1 = -e^{2t}$, and a particular solution corresponding to $3t - 4$ is found in Problem 9.35 to be $y_2 = 3t + 8$.

A particular solution to the given differential equation then is $y_1 + y_2$. When combined with the complementary solution (see Problem 8.9), it yields the general solution $y = C_1e^{3.732t} + C_2e^{0.2679t} - e^{2t} + 3t + 8$.

9.225 Solve $\dfrac{d^2x}{dt^2} + 4\dfrac{dx}{dt} + 8x = e^{-2t} + (20t^2 + 16t - 78)e^{2t}$.

▍ A particular solution corresponding to e^{-2t} is found in Problem 9.13 to be $x_1 = \frac{1}{4}e^{-2t}$. A particular solution corresponding to $(20t^2 + 16t - 78)e^{2t}$ is found in Problem 9.67 to be $x_2 = (t^2 - 4)e^{2t}$. A particular solution to the given differential equation then is $x_1 + x_2$. When combined with the complementary solution (see Problem 8.54), it yields the general solution

$$x = c_1e^{-2t}\cos 2t + c_2e^{-2t}\sin 2t + \tfrac{1}{4}e^{-2t} + (t^2 - 4)e^{2t}$$

9.226 Solve $\dfrac{d^2x}{dt^2} + 4\dfrac{dx}{dt} + 8x = -t^2 + 5t^2e^{-3t} - 14te^{-3t} + 11e^{-3t}$.

▍ A particular solution corresponding to $-t^2$ is found in Problem 9.41 to be $x_1 = -\frac{1}{8}t^2 + \frac{1}{8}t - \frac{1}{32}$. A particular solution corresponding to $(5t^2 - 14t + 11)e^{-3t}$ is found in Problem 9.68 to be $x_2 = (t^2 - 2t + 1)e^{-3t}$. Then a particular solution to the given differential equation is $x_1 + x_2$; when combined with the complementary solution (see Problem 8.54), it yields the general solution

$$x = c_1e^{-2t}\cos 2t + c_2e^{-2t}\sin 2t - \tfrac{1}{8}t^2 + \tfrac{1}{8}t - \tfrac{1}{32} + (t^2 - 2t + 1)e^{-3t}$$

9.227 Solve $\ddot{q} + 400\dot{q} + 200{,}000q = 2000(1 + \cos 200t)$.

▍ A particular solution corresponding to 2000 is found in Problem 9.23 to be $q_1 = 0.01$. A particular solution corresponding to $2000\cos 200t$ is found in Problem 9.93 to be $q_2 = 0.005\sin 200t + 0.01\cos 200t$. Then a particular solution to the given differential equation is $q_1 + q_2$; when combined with the complementary solution (see Problem 8.70), it yields the general solution

$$q = e^{-200t}(A\cos 400t + B\sin 400t) + 0.01 + 0.005\sin 200t + 0.01\cos 200t$$

9.228 Solve $\ddot{x} + 10\dot{x} + 25x = 20e^{-5t} + 320t^3e^{3t} + 48t^2e^{3t} - 66te^{3t} + 122e^{3t}$.

▍ A particular solution corresponding to $20e^{-5t}$ is found in Problems 9.144 and 9.145 to be $x_1 = 10t^2e^{-5t}$; a particular solution corresponding to $(320t^3 + 48t^2 - 66t + 122)e^{3t}$ is found in Problem 9.73 to be $x_2 = (5t^3 - 3t^2 + 2)e^{3t}$.

A particular solution to the given differential equation then is $x_1 + x_2$. When combined with the complementary solution (see Problem 8.146), it yields the general solution $x = (C_1 + C_2t)e^{-5t} + 10t^2e^{-5t} + (5t^3 - 3t^2 + 2)e^{3t}$.

9.229 Solve $\dfrac{d^3Q}{dt^3} - 5\dfrac{d^2Q}{dt^2} + 25\dfrac{dQ}{dt} - 125Q = 1000(1 + 5e^{-t}\cos 2t)$.

▍ A particular solution corresponding to a right side of 1000 is found in Problem 9.29 to be $Q_1 = -8$. A particular solution corresponding to $5000e^{-5}\cos 2t$ is found in Problem 9.107 to be $Q_2 = 17e^{-t}\sin 2t - 31e^{-t}\cos 2t$. A particular solution to the given differential equation then is $Q_1 + Q_2$; when combined with the complementary solution (see Problem 8.113), it yields the general solution

$$Q = c_1e^{5t} + c_2\cos 5t + c_3\sin 5t - 8 + 17e^{-t}\sin 2t - 31e^{-t}\cos 2t$$

9.230 Rework the previous problem if the term $-60e^{7t}$ is added to the right side of the differential equation.

▍ Q_1 and Q_2 of the previous problem remain valid. In addition, a particular solution corresponding to $-60e^{7t}$ is found in Problem 9.16 to be $Q_3 = -\frac{15}{37}e^{7t}$. Then a particular solution to the new differential equation is $Q_1 + Q_2 + Q_3$, and the general solution is

$$Q = c_1e^{5t} + c_2\cos 5t + c_3\sin 5t - 8 + 17e^{-t}\sin 2t - 31e^{-t}\cos 2t - \tfrac{15}{37}e^{7t}$$

9.231 Solve $y^{(4)} - 6y^{(3)} + 16y'' + 54y' - 225y = 100e^{-2x} + 1152\cos 3x - 3924\sin 3x$.

A particular solution corresponding to $100e^{-2x}$ is found in Problem 9.17 to be $y_1 = -\frac{20}{41}e^{-2x}$; a particular solution corresponding to $1152\cos 3x - 3924\sin 3x$ is found in Problem 9.98 to be $y_2 = 8\sin 3x + 5\cos 3x$. Then a particular solution to the given differential equation is $y_1 + y_2$. When combined with the complementary solution (see Problem 8.124), it yields the general solution

$$y = c_1e^{3x} + c_2e^{-3x} + c_3e^{3x}\cos 4x + c_4e^{3x}\sin 4x - \tfrac{20}{41}e^{-2x} + 8\sin 3x + 5\cos 3x$$

9.232 Solve $(D^4 - 16)y = 80x^2 + 60e^x \sin 3x$.

A particular solution corresponding to $80x^2$ is found in Problem 9.53 to be $y_1 = -5x^2$. A particular solution corresponding to $60e^x \sin x$ is found in Problem 9.106 to be $y_2 = \frac{1}{13}e^x \sin 3x + \frac{8}{13}\cos 3x$. Then a particular solution to the given differential equation is $y_1 + y_2$; when combined with the complementary solution (see Problem 8.128), it yields the general solution

$$y = c_1\cos 2x + c_2\sin 2x + c_3e^{2x} + c_4e^{-2x} - 5x^2 + \tfrac{1}{13}e^x\sin 3x + \tfrac{8}{13}e^x\cos 3x$$

9.233 Solve $(D^2 + 2D + 4)y = 8x^2 + 12e^{-x}$.

The complementary solution is $e^{-x}(c_1\cos\sqrt{3}x + c_2\sin\sqrt{3}x)$. To obtain a particular solution, we may assume the trial solutions $ax^2 + bx + c$ and de^{-x} corresponding to $8x^2$ and $12e^{-x}$, respectively, since none of these terms is present in the complementary solution. Then substituting $y = ax^2 + bx + c + de^{-x}$ in the given equation, we find

$$4ax^2 + (4a + 4b)x + (2a + 2b + 4c) + 3de^{-x} = 8x^2 + 12e^{-x}$$

Equating corresponding coefficients on both sides of the equation and solving the resulting system yield $a = 2$, $b = -2$, $c = 0$, and $d = 4$. The particular solution is then $2x^2 - 2x + 4e^{-x}$. Thus the required general solution is $y = e^{-x}(c_1\cos\sqrt{3}x + c_2\sin\sqrt{3}x) + 2x^2 - 2x + 4e^{-x}$.

9.234 Solve Problem 9.233 if the term $10\sin 3x$ is added to the right side.

Corresponding to the additional term $10\sin 3x$ we assume the additional trial solution $h\cos 3x + k\sin 3x$, whose terms do not appear in the complementary solution. Substituting this into the equation $(D^2 + 2D + 4)y = 10\sin 3x$, we get $(6k - 5h)\cos 3x - (5k + 6h)\sin 3x = 10\sin 3x$, from which we find that $h = -\frac{12}{61}$ and $k = -\frac{10}{61}$. Then the required general solution is

$$y = e^{-x}(c_1\cos\sqrt{3}x + c_2\sin\sqrt{3}x) + 2x^2 - 2x + 4e^{-x} - \tfrac{12}{61}\cos 3x - \tfrac{10}{61}\sin 3x$$

9.235 Solve $(D^5 - 3D^4 + 3D^3 - D^2)y = x^2 + 2x + 3e^x$.

The auxiliary equation is $m^5 - 3m^4 + 3m^3 - m^2 = 0$ or $m^2(m-1)^3 = 0$. Thus $m = 0, 0, 1, 1,$ and 1, and the complementary solution is $c_1 + c_2x + (c_3 + c_4x + c_5x^2)e^x$.

Corresponding to the polynomial $x^2 + 2x$, we would normally assume the trial solution $ax^2 + bx + c$. However, some of its terms appear in the complementary solution. Multiplying by x^2 yields $ax^4 + bx^3 + cx^2$, which has no term that is in the complementary solution and so is the proper trial solution.

Similarly, corresponding to $3e^x$ we would normally assume the trial solution de^x. But since this term as well as dxe^x and dx^2e^x are in the complementary solution, we must use dx^3e^x. Thus our assumed trial solution is $ax^4 + bx^3 + cx^2 + dx^3e^x$. Substituting this in the given differential equation, we get

$$-12ax^2 + (72a - 6b)x + (18b - 72a - 2c) + 6de^x = x^2 + 2x + 3e^x$$

from which we find $a = -\frac{1}{12}$, $b = -\frac{4}{3}$, $c = -9$, and $d = \frac{1}{2}$. The general solution is then $y = c_1 + c_2x + (c_3 + c_4x + c_5x^2)e^x + \frac{1}{2}x^3e^x - \frac{1}{12}x^4 - \frac{4}{3}x^3 - 9x^2$.

9.236 Find a complete solution to the equation $y'' + 5y' + 6y = 3e^{-2x} + e^{3x}$.

The characteristic equation is $m^2 + 5m + 6 = 0$, and its roots are $m_1 = -2$ and $m_2 = -3$. Hence the complementary solution is $c_1e^{-2x} + c_2e^{-3x}$.

For a trial solution corresponding to the term $3e^{-2x}$ we would normally use Ae^{-2x}. However, e^{-2x} is a part of the complementary solution, so we must multiply it by x. For the term e^{3x} the normal choice for a trial solution, namely, Be^{3x}, is satisfactory as it stands. Hence we assume $y_p = Axe^{-2x} + Be^{3x}$.

Substituting this into the differential equation and simplifying yield $Ae^{-2x} + 30Be^{3x} = 3e^{-2x} + e^{3x}$, from which we find $A = 3$ and $B = \frac{1}{30}$. Hence $y_p = 3xe^{-2x} + \frac{1}{30}e^{3x}$, and a complete solution is $y = c_1e^{-2x} + c_2e^{3x} + 3xe^{-2x} + \frac{1}{30}e^{3x}$.

9.237 Find a particular solution to the equation $y'' + 3y' + 2y = 10e^{3x} + 4x^2$.

▮ If we wished, we could find y_p by beginning with the expression $Ae^{3x} + Bx^2 + Cx + D$, which means that we would handle the various terms all at the same time. On the other hand, we can also find y_p by first finding a particular integral corresponding to $10e^{3x}$, and then finding a particular integral corresponding to $4x^2$, and finally taking y_p to be their sum.

Using the second method, we assume $y_1 = Ae^{3x}$, substitute into the equation $y'' + 3y' + 2y = 10e^{3x}$, and find that $A = \frac{1}{2}$ and $y_1 = \frac{1}{2}e^{3x}$. Then we assume $y_2 = Bx^2 + Cx + D$, substitute into the equation $y'' + 3y' + 2y = 4x^2$, and find after equating coefficients of like terms that $B = 2$, $C = -6$, and $D = 7$. Hence $y_2 = 2x^2 - 6x + 7$ and, finally $y_p = y_1 + y_2 = \frac{1}{2}e^{3x} + 2x^2 - 6x + 7$.

9.238 Solve $(D^2 - 2D + 3)y = x^3 + \sin x$.

▮ The complementary solution is $y_c = e^x(C_1 \cos \sqrt{2}x + C_2 \sin \sqrt{2}x)$. As a particular solution try $y_p = Ax^3 + Bx^2 + Cx + E + F \sin x + G \cos x$. Substituting y_p and its derivatives into the given equation then yields

$$3Ax^3 + 3(B - 2A)x^2 + (3C - 4B + 6A)x + (3E - 2C + 2B) + 2(F + G)\sin x + 2(G - F)\cos x = x^3 \sin x.$$

Equating coefficients of like terms yields $A = \frac{1}{3}$, $B = \frac{2}{3}$, $C = \frac{2}{9}$, $E = -\frac{8}{27}$, and $F = G = \frac{1}{4}$. Thus, a particular solution of the given differential equation is $y_p = \frac{1}{3}x^3 + \frac{2}{3}x^2 + \frac{2}{9}x - \frac{8}{27} + \frac{1}{4}(\sin x + \cos x)$, and the primitive is

$$y = e^x(C_1 \cos \sqrt{2}x + C_2 \sin \sqrt{2}x) + \tfrac{1}{27}(9x^3 + 18x^2 + 6x - 8) + \tfrac{1}{4}(\sin x + \cos x)$$

9.239 Solve $(D^3 + 2D^2 - D - 2)y = e^x + x^2$.

▮ The complementary solution is $y_c = C_1e^x + C_2e^{-x} + C_3e^{-2x}$. We take as a particular solution $y_p = Ax^2 + Bx + C + Exe^x$. Substitution then gives $-2Ax^2 - 2(B + A)x + (4A - B - 2C) + 6Ee^x = e^x + x^2$.

By equating coefficients of like terms and solving, we find that $A = -\frac{1}{2}$, $B = \frac{1}{2}$, $C = -\frac{5}{4}$, and $E = \frac{1}{6}$. Hence $y_p = -\frac{1}{2}x^2 + \frac{1}{2}x - \frac{5}{4} + \frac{1}{6}xe^x$, and the general solution is $y = C_1e^x + C_2e^{-x} + C_3e^{-2x} - \frac{1}{2}x^2 + \frac{1}{2}x - \frac{5}{4} + \frac{1}{6}xe^x$.

9.240 Solve $y' - 5y = x^2e^x - xe^{5x}$.

▮ The complementary solution is $y_c = c_1e^{5x}$, and the right side of the differential equation is the difference of two terms, each in manageable form. For x^2e^x we assume a solution of the form $e^x(A_2x^2 + A_1x + A_0)$. For xe^{5x} we would try the solution

$$e^{5x}(B_1x + B_0) = B_1xe^{5x} + B_0e^{5x}$$

But this trial solution would have, disregarding multiplicative constants, the term e^{5x} in common with y_c. We therefore multiply by x to obtain $e^{5x}(B_1x^2 + B_0x)$. Now we take y_p to be the difference $y_p = e^x(A_2x^2 + A_1x + A_0) - e^{5x}(B_1x^2 + B_0x)$.

Substituting into the differential equation and simplifying, we get

$$e^x[(-4A_2)x^2 + (2A_2 - 4A_1)x + (A_1 - 4A_0)] + e^{5x}[(-2B_1)x - B_0] = e^x(x^2 + 0x + 0) + e^{5x}[(-1)x + 0]$$

Equating coefficients of like terms and solving the resulting system yields $A_2 = -\frac{1}{4}$, $A_1 = -\frac{1}{8}$, $A_0 = -\frac{1}{32}$, $B_1 = \frac{1}{2}$, and $B_0 = 0$. Then the general solution is $y = y_c + y_p = c_1e^{5x} + e^x(-\frac{1}{4}x^2 - \frac{1}{8}x - \frac{1}{32}) - \frac{1}{2}x^2e^{5x}$.

9.241 Determine the form of a particular solution for $y' - 5y = (x - 1)\sin x + (x + 1)\cos x$.

▮ The solution to the associated homogeneous equation is shown in Problem 8.34 to be $y_h = c_1e^{5x}$. An assumed solution corresponding to $(x - 1)\sin x$ is $(A_1x + A_0)\sin x + (B_1x + B_0)\cos x$, and no modification is required. An assumed solution corresponding to $(x + 1)\cos x$ is $(C_1x + C_0)\sin x + (D_1x + D_0)\cos x$. (Note that we have used constants C and D in the last expression, since the constants A and B already have been used.) We therefore take

$$y_p = (A_1x + A_0)\sin x + (B_1x + B_0)\cos x + (C_1x + C_0)\sin x + (D_1x + D_0)\cos x$$

Combining like terms we arrive at $y_p = (E_1x + E_0)\sin x + (F_1x + F_0)\cos x$ as the form of the particular solution.

9.242 Solve the differential equation of the previous problem.

▌ Substituting y_p of the previous problem and its derivative into the differential equation and simplifying, we obtain

$$(-5E_1 - F_1)x \sin x + (-5E_0 + E_1 - F_0) \sin x + (-5F_1 + E_1)x \cos x + (-5F_0 + E_0 + F_1) \cos x$$
$$= x \sin x - 1 \sin x + 1x \cos x + 1 \cos x$$

Equating coefficients of like terms and solving the resulting system of equations lead to $E_1 = -\frac{2}{13}$, $E_0 = \frac{71}{338}$, $F_1 = -\frac{3}{13}$, and $F_0 = -\frac{69}{338}$. By combining these results with y_c and y_p, we obtain the general solution

$$y = c_1 e^{5x} + (-\tfrac{2}{13} x + \tfrac{71}{338}) \sin x - (\tfrac{3}{13} x + \tfrac{69}{338}) \cos x$$

9.243 Solve $y' - 5y = 3e^x - 2x + 1$.

▌ y_h of Problem 9.241 is valid here. The right side of the given differential equation is the sum of two manageable functions: $3e^x$ and $-2x + 1$. For $3e^x$ we assume a particular solution of the form $A_0 e^{3x}$; for $-2x + 1$ we assume a solution of the form $B_1 x + B_0$. Thus, we try $y_p = A_0 e^x + B_1 x + B_0$.

Substituting y_p into the differential equation and simplifying, we obtain $(-4A_0)e^x + (-5B_1)x + (B_1 - 5B_0) = 3e^x + (-2)x + 1$. Then by equating coefficients of like terms, we find that $A_0 = -\frac{3}{4}$, $B_1 = \frac{2}{5}$, and $B_0 = -\frac{3}{25}$. Hence the general solution is $y = c_1 e^{5x} - \frac{3}{4}e^x + \frac{2}{5}x - \frac{3}{25}$.

CHAPTER 10

Variation of Parameters

FORMULAS

10.1 Discuss the solution of the nth-order linear differential equation

$$\frac{d^n y}{dx^n} + a_{n-1}(x)\frac{d^{n-1}y}{dx^{n-1}} + \cdots + a_2(x)\frac{d^2y}{dx^2} + a_1(x)\frac{dy}{dx} + a_0(x)y = \phi(x) \tag{1}$$

by the method of variation of parameters.

▌ Let $y_1(x), y_2(x), \ldots, y_n(x)$ be a set of n linearly independent solutions to the associated homogeneous differential equation—that is, (*1*) with $\phi(x) \equiv 0$. The complementary function is

$$y_c = c_1 y_1(x) + c_2 y_2(x) + \cdots + c_n y_n(x) \tag{2}$$

where $c_1, c_2, \ldots, c_n$ denote arbitrary constants. A particular solution to the nonhomogeneous differential equation is

$$y_p = v_1 y_1 + v_2 y_2 + \cdots + v_n y_n \tag{3}$$

where $y_i = y_i(x)$ for $i = 1, 2, \ldots, n$, and v_i $(i = 1, 2, \ldots, n)$ is an unknown function of x which still must be determined.

To find the v_i, we first solve the following linear equations simultaneously for the v_i':

$$\begin{aligned} v_1'y_1 + v_2'y_2 \quad + \cdots + \quad v_n'y_n &= 0 \\ v_1'y_1' + v_2'y_2' \quad + \cdots + \quad v_n'y_n' &= 0 \\ \cdots\cdots\cdots\cdots\cdots\cdots\cdots\cdots\cdots & \\ v_1'y_1^{(n-2)} + v_2'y_2^{(n-2)} + \cdots + v_n'y_n^{(n-2)} &= 0 \\ v_1'y_1^{(n-1)} + v_2'y_2^{(n-1)} + \cdots + v_n'y_n^{(n-1)} &= \phi(x) \end{aligned} \tag{4}$$

We then integrate each v_i' to obtain v_i, disregarding all constants of integration. (This is permissible because we are seeking only *one* particular solution.) Finally, we find the general solution as $y = y_c + y_p$.

10.2 Specialize (*4*) of the previous problem to a third-order differential equation.

▌ Equations (*4*) become

$$\begin{aligned} v_1'y_1 + v_2'y_2 + v_3'y_3 &= 0 \\ v_1'y_1' + v_2'y_2' + v_3'y_3' &= 0 \\ v_1'y_1'' + v_2'y_2'' + v_3'y_3'' &= \phi(x) \end{aligned}$$

10.3 Specialize (*4*) of Problem 10.1 to a second-order differential equation.

▌ Equations (*4*) become

$$v_1'y_1 + v_2'y_2 = 0 \qquad v_1'y_1' + v_2'y_2' = \phi(x)$$

10.4 Specialize (*4*) of Problem 10.1 to a first-order differential equation.

▌ Equations (*4*) become $v_1'y_1 = \phi(x)$.

10.5 Compare the method of variation of parameters to the method of undetermined coefficients.

▌ The method of variation of parameters can be applied to *all* linear differential equations. It is therefore more powerful than the method of undetermined coefficients, which generally is restricted to linear differential equations with constant coefficients and to particular forms of $\phi(x)$. Nonetheless, in those cases where both methods are applicable, the method of undetermined coefficients is often the more efficient and, hence, the preferable method, because no integration is involved. As a practical matter, the integration of $v_i'(x)$ may be impossible to perform.

FIRST-ORDER DIFFERENTIAL EQUATIONS

10.6 Solve $x\,dy/dx + y = \ln x$, for $x > 0$, if $y = 1/x$ is one solution of the associated homogeneous differential equation $x\,dy/dx + y = 0$.

▌ We first divide the nonhomogeneous differential equation by x, obtaining $\dfrac{dy}{dx} + \dfrac{1}{x}y = \dfrac{1}{x}\ln x$, which has the form of (*1*) in Problem 10.1; the lead coefficient is now unity. The complementary function is $y_c = c_1(1/x)$, so we assume a particular solution of the form $y_p = v_1(1/x)$.

With $\phi(x) = \dfrac{1}{x}\ln x$, it follows from Problem 10.4 that $v_1'\dfrac{1}{x} = \dfrac{1}{x}\ln x$, so that $v_1' = \ln x$. Then $v_1 = \int \ln x\,dx = x\ln x - x$, in which we have disregarded all constants of integration. Now we have $y_p = (x\ln x - x)(1/x) = \ln x - 1$, and the general solution to the given nonhomogeneous differential equation is $y = y_c + y_p = c_1(1/x) + \ln x - 1$.

10.7 Solve $y' + 2xy = 4x$.

▌ The solution to the associated homogeneous equation is found in Problem 5.9 to be $y_c = ce^{-x^2}$, so we assume a particular solution of the form $y_p = v_1e^{-x^2}$.

Since $y_1 = e^{-x^2}$ and $\phi(x) = 4x$, it follows from Problem 10.4 that $v_1'e^{-x^2} = 4x$, so that $v_1' = 4xe^{x^2}$. Then $v_1 = \int v_1'\,dx = \int 4xe^{x^2}\,dx = 2e^{x^2}$, in which we disregard constants of integration. Now we have $y_p = 2e^{x^2}e^{-x^2} = 2$, and the general solution is $y = y_c + y_p = ce^{-x^2} + 2$. (Compare this problem with Problem 5.43.)

10.8 Solve $y' - 5y = e^{2x}$.

▌ The complementary solution is found in Problem 8.34 to be $y_c = c_1e^{5x}$, so we assume $y_p = v_1e^{5x}$.

Here $y_1 = e^{5x}$ and $\phi(x) = e^{2x}$, and it follows from Problem 10.4 that $v_1'e^{5x} = e^{2x}$. Then $v_1' = e^{-3x}$ and $v_1 = \int v_1'\,dx = \int e^{-3x}\,dx = -\frac{1}{3}e^{-3x}$. Now we have $y_p = -\frac{1}{3}e^{-3x}e^{5x} = -\frac{1}{3}e^{2x}$, and the general solution is $y = y_c + y_p = c_1e^{5x} - \frac{1}{3}e^{2x}$. (Compare with Problem 9.1.)

10.9 Solve $y' - 5y = 8$.

▌ As in the previous problem, $y_c = c_1e^{5x}$, we assume $y_p = v_1e^{5x}$, and $y_1 = e^{5x}$. Here $\phi(x) = 8$, and it follows from Problem 10.4 that $v_1'e^{5x} = 8$. Thus, $v_1' = 8e^{-5x}$, and integration gives $v_1 = \int v_1'\,dx = \int 8e^{-5x}\,dx = -\frac{8}{5}e^{-5x}$. Then we have $y_p = (-\frac{8}{5}e^{-5x})e^{5x} = -\frac{8}{5}$, and the general solution is $y = y_c + y_p = c_1e^{5x} - \frac{8}{5}$. (Compare with Problem 9.18.)

10.10 Solve $y' - 5y = 3x + 1$.

▌ As in Problem 10.8, $y_c = c_1e^{5x}$, we assume $y_p = v_1e^{5x}$, and $y_1 = e^{5x}$. Here $\phi(x) = 3x + 1$, so it follows from Problem 10.4 that $v_1'(e^{5x}) = 3x + 1$. Then $v_1' = (3x + 1)e^{-5x}$, and integration gives

$$v_1 = \int v_1'\,dx = \int (3x+1)e^{-5x}\,dx = (-\tfrac{3}{5}x - \tfrac{8}{25})e^{-5x}$$

Now we have $y_p = (-\frac{3}{5}x - \frac{8}{25})e^{-5x}e^{5x} = -\frac{3}{5}x - \frac{8}{25}$, and the general solution is $y = c_1e^{5x} - \frac{3}{5}x - \frac{8}{25}$. (Compare with Problem 9.31.)

10.11 Solve $y' - 5y = \sin x$.

▌ As in Problem 10.8, $y_c = c_1e^{5x}$, we let $y_p = v_1e^{5x}$, and $y_1 = e^{5x}$. Here $\phi(x) = \sin x$. It follows from Problem 10.4 that $v_1'e^{5x} = \sin x$, from which we obtain $v_1' = e^{-5x}\sin x$ and (using integration by parts twice)

$$v_1 = \int e^{-5x}\sin x\,dx = (-\tfrac{5}{26}\sin x - \tfrac{1}{26}\cos x)e^{-5x}.$$

Now $y_p = (-\frac{5}{26}\sin x - \frac{1}{26}\cos x)e^{-5x}e^{5x} = -\frac{5}{26}\sin x - \frac{1}{26}\cos x$, so that the general solution is $y = c_1e^{5x} - \frac{5}{26}\sin x - \frac{1}{26}\cos x$. (Compare with Problem 9.77.)

10.12 Solve $y' - 5y = 2e^{5x}$.

▌ As in Problem 10.8, $y_c = c_1e^{5x}$, we let $y_p = v_1e^{5x}$, and $y_1 = e^{5x}$. Here $\phi(x) = 2e^{5x}$. It follows from Problem 10.4 that $v_1'e^{5x} = 2e^{5x}$, from which we obtain $v_1' = 2$ and $v_1 = \int 2\,dx = 2x$. Then $y_p = 2xe^{5x}$, and the general solution is $y = y_c + y_p = c_1e^{5x} + 2xe^{5x}$. (Compare with Problems 9.128 and 9.129.)

10.13 Solve $y' - 3x^2y = 12x^2$.

▌ The complementary solution is found in Problem 5.11 to be $y_c = ce^{x^3}$, so we assume $y_p = v_1e^{x^3}$. With $y_1 = e^{x^3}$ and $\phi(x) = 12x^2$, it follows from Problem 10.4 that $v_1'e^{x^3} = 12x^2$, or $v_1' = 12x^2e^{-x^3}$. Integration gives $v_1 = \int 12x^2e^{-x^3}\,dx = -4e^{-x^3}$. Then $y_p = -4e^{-x^3}e^{x^3} = -4$, and the general solution is $y = y_c + y_p = ce^{x^3} - 4$.

10.14 Solve $y' - 3x^2y = -12x^3 + 4$.

▌ As in the previous problem, $y_c = ce^{x^3}$, we assume $y_p = v_1e^{x^3}$, and $y_1 = e^{x^3}$. Here $\phi(x) = -12x^3 + 4$. It follows from Problem 10.4 that $v_1'e^{x^3} = -12x^3 + 4$ so that $v_1' = (-12x^3 + 4)e^{-x^3}$. Then

$$v_1 = \int(-12x^3 + 4)e^{-x^3}\,dx = \int -12x^3x^{-x^3}\,dx + \int 4e^{-x^3}\,dx$$

Integration by parts (with $u = 4x$ and $dv = -3x^2e^{-x^3}\,dx$) gives us $\int -12x^3e^{-x^3}\,dx = 4xe^{-x^3} - \int 4e^{-x^3}\,dx$, so that $v_1 = 4xe^{-x^3} - \int 4e^{-x^3}\,dx + \int 4e^{-x^3}\,dx = 4xe^{-x^3}$. Then $y_p = (4xe^{-x^3})e^{x^3} = 4x$, and the general solution is $y = y_c + y_p = ce^{x^3} + 4x$.

10.15 Solve $y' - 3xy = e^{3x^2/2}$.

▌ The complementary solution is found in Problem 5.10 to be $y_c = ce^{3x^2/2}$, so we assume a particular solution of the form $y_p = v_1e^{3x^2/2}$.

Here $y_1 = e^{3x^2/2} = \phi(x)$, and it follows from Problem 10.4 that $v_1'e^{3x^2/2} = e^{3x^2/2}$ or $v_1' = 1$. Thus $v_1 = x$, and $y_p = xe^{3x^2/2}$. The general solution is then $y = y_c + y_p = ce^{3x^2/2} + xe^{3x^2/2}$.

10.16 Solve $y' - 3xy = -6x$.

▌ As in the previous problem, $y_c = ce^{3x^2/2}$, we assume $y_p = v_1e^{3x^2/2}$, and $y_1 = e^{3x^2/2}$. Here $\phi(x) = -6x$ It follows from Problem 10.4 that $v_1'e^{3x^2/2} = -6x$, from which we conclude that $v_1' = -6xe^{-3x^2/2}$ and $v_1 = \int -6xe^{-3x^2/2}\,dx = 2e^{-3x^2/2}$. Then $y_p = 2e^{-3x^2/2}e^{3x^2/2} = 2$, and the general solution is $y = ce^{3x^2/2} + 2$.

10.17 Solve $y' - 3xy = -6xe^{-3x^2/2}$.

▌ As in Problem 10.15, $y_c = ce^{3x^2/2}$, we assume $y_p = v_1e^{3x^2/2}$, and $y_1 = e^{3x^2/2}$. Here $\phi(x) = -6xe^{3x^2/2}$. It follows from Problem 10.4 that $v_1'e^{3x^2/2} = -6xe^{3x^2/2}$ or $v_1' = -6x$. Then $v_1 = \int -6x\,dx = -3x^2$, and $y_p = -3x^2e^{3x^2/2}$. The general solution is then $y = y_c + y_p = ce^{3x^2/2} - 3x^2e^{3x^2/2} = (c - 3x^2)e^{-3x^2/2}$.

10.18 Solve $y' + 6y = 18e^{3x}$.

▌ The complementary solution is found in Problem 8.37 to be $y_c = Ae^{-6x}$, so we assume $y_p = v_1e^{-6x}$. Here $y_1 = e^{-6x}$ and $\phi(x) = 18e^{3x}$. It follows from Problem 10.4 that $v_1'e^{-6x} = 18e^{3x}$, from which we conclude that $v_1' = 18e^{9x}$ and so $v_1 = \int 18e^{9x}\,dx = 2e^{9x}$. Then $y_p = 2e^{9x}e^{-6x} = 2e^{3x}$, and the general solution is $y = y_c + y_p = Ae^{-6x} + 2e^{3x}$. (Compare with Problem 9.3.)

10.19 Solve $y' + 6y = -2\cos 3x$.

▌ As in the previous problem $y_c = Ae^{-6x}$, we assume $y_p = v_1e^{-6x}$, and $y_1 = e^{-6x}$. Here $\phi(x) = -2\cos 3x$.

It follows from Problem 10.4 that $v_1'e^{-6x} = -2\cos 3x$ or $v_1' = -2e^{6x}\cos 3x$. Applying integration by parts twice, we find that $v_1 = \int -2e^{6x}\cos 3x\,dx = (-\frac{2}{15}\sin 3x - \frac{4}{15}\cos 3x)e^{6x}$. Then

$$y_p = (-\tfrac{2}{15}\sin 3x - \tfrac{4}{15}\cos 3x)e^{6x}e^{-6x} = -\tfrac{2}{15}\sin 3x - \tfrac{4}{15}\cos 3x$$

and the general solution is $y = Ae^{-6x} - \frac{2}{15}\sin 3x - \frac{4}{15}\cos 3x$. (Compare with Problem 9.78.)

10.20 Solve $y' + 6y = 3e^{-6x}$.

▌ As in Problem 10.18, $y_c = Ae^{-6x}$, we let $y_p = v_1e^{-6x}$, and $y_1 = e^{-6x}$. Here $\phi(x) = 3e^{-6x}$. It follows from Problem 10.4 that $v_1'e^{-6x} = 3e^{-6x}$, from which we conclude that $v_1' = 3$ and $v_1 = \int 3\,dx = 3x$. Then $y_p = 3xe^{-6x}$, and the general solution is $y = Ae^{-6x} + 3xe^{-6x}$. (Compare with Problems 9.130 and 9.131.)

10.21 Solve $2y' - 5y = 2x^2 - 5$.

▌ The complementary solution is found in Problem 8.38 to be $y_c = Ae^{5x/2}$, so we assume $y_p = v_1e^{5x/2}$. He $y_1 = e^{5x/2}$, but before we can determine $\phi(x)$, we must write the differential equation in the form of (1) of Proble

10.1; that is, the coefficient of the highest derivative must be unity. Dividing the differential equation by 2, we get $y' - \frac{5}{2}y = x^2 - \frac{5}{2}$, so that $\phi(x) = x^2 - \frac{5}{2}$.

It now follows from Problem 10.4 that $v_1'e^{5x/2} = x^2 - \frac{5}{2}$ or $v_1' = (x^2 - \frac{5}{2})e^{-5x/2}$. Using integration by parts twice, we find $v_1 = \int (x^2 - \frac{5}{2})e^{-5x/2}\,dx = (-\frac{2}{5}x^2 - \frac{8}{25}x + \frac{109}{125})e^{-5x/2}$. Then

$$y_p = (-\tfrac{2}{5}x^2 - \tfrac{8}{25}x + \tfrac{109}{125})e^{-5x/2}e^{5x/2} = -\tfrac{2}{5}x^2 - \tfrac{8}{25}x + \tfrac{109}{125}$$

The general solution is thus $y = Ae^{5x/2} - \frac{2}{5}x^2 - \frac{8}{25}x + \frac{109}{125}$. (Compare with Problem 9.33.)

10.22 Solve $(3D - 1)y = 6e^{3x}$.

▮ The associated homogeneous equation is $(3D - 1)y = 0$, which has as its characteristic equation $3m - 1 = 0$ and as its characteristic value $m = 1/3$. The complementary function is $y_c = c_1e^{x/3}$, so we assume a particular solution of the form $y_p = v_1e^{x/3}$. Thus, $y_1 = e^{x/3}$.

To apply variation of parameters, we must have the coefficient of the highest derivative in the differential equation equal to unity. Dividing the given differential equation by 3, we get $(D - \frac{1}{3})y = 2e^{3x}$, so that $\phi(x) = 2e^{3x}$. It follows from Problem 10.4 that $v_1'e^{x/3} = 2e^{3x}$ or $v_1' = 2e^{8x/3}$. Then $v_1 = \int 2e^{8x/3}\,dx = \frac{3}{4}e^{8x/3}$, and $y_p = \frac{3}{4}e^{8x/3}e^{x/3} = \frac{3}{4}e^{3x}$. The general solution is $y = y_c + y_p = c_1e^{x/3} + \frac{3}{4}e^{3x}$.

10.23 Solve $(2D - 1)y = t^3e^{t/2}$.

▮ We first divide the differential equation by 2, obtaining $dy/dt - \frac{1}{2}y = \frac{1}{2}t^3e^{t/2}$, which is in the form of (*1*) of Problem 10.1 with t as the independent variable. The complementary function is found in Problem 5.5 to be $y_c = ce^{t/2}$; hence we assume $y_p = v_1e^{t/2}$.

Here $y_1 = e^{t/2}$ and $\phi(t) = \frac{1}{2}t^3e^{t/2}$. It follows from Problem 10.4 that $v_1'e^{t/2} = \frac{1}{2}t^3e^{t/2}$, from which we conclude that $v_1' = \frac{1}{2}t^3$ and $v_1 = \frac{1}{8}t^4$. Then $y_p = \frac{1}{8}t^4e^{t/2}$ and the general solution is $y = y_c + y_p = ce^{t/2} + \frac{1}{8}t^4e^{t/2}$.

10.24 Solve $dy/dt + e^ty = e^t$.

▮ The complementary solution is found in Problem 5.14 to be $y_c = ce^{-e^t}$, so we assume $y_p = v_1e^{-e^t}$.

Here $y_1 = e^{-e^t}$ and $\phi(t) = e^t$. It follows from Problem 10.4 that $v_1'e^{-e^t} = e^t$, from which we conclude that $v_1' = e^te^{e^t}$ and that $v_1 = \int e^te^{e^t}\,dt = e^{e^t}$. Then $y_p = e^{e^t}e^{-e^t} = 1$, and the general solution is $y = ce^{-e^t} + 1$.

10.25 Solve $\dfrac{dx}{dt} + \dfrac{1}{t}x = t^2$.

▮ The complementary solution is found in Problem 5.16 to be $x_c = c/t$, so we assume a particular solution of the form $x_p = v_1/t$.

Here $x_1 = 1/t$ and $\phi(t) = t^2$. It follows from Problem 10.4 that $v_1'(1/t) = t^2$, from which we conclude that $v_1' = t^3$ and $v_1 = t^4/4$. Then $x_p = (t^4/4)/t = \frac{1}{4}t^3$, and the general solution is $x = c/t + \frac{1}{4}t^3$.

10.26 Solve $\dfrac{dx}{dt} + \dfrac{1}{t}x = \sin 7t$.

▮ As in the previous problem, $x_c = c/t$, we assume $x_p = v_1/t$, and $x_1 = 1/t$. Here $\phi(t) = \sin 7t$. It then follows from Problem 10.4 that $v_1'(1/t) = \sin 7t$ or $v_1' = t\sin 7t$. Integration gives $v_1 = \int t\sin 7t\,dt = \dfrac{1}{49}\sin 7t - \dfrac{t}{7}\cos 7t$, so that $x_p = \dfrac{1}{49t}\sin 7t - \dfrac{1}{7}\cos 7t$. The general solution is $x = \dfrac{c}{t} + \dfrac{1}{49t}\sin 7t - \dfrac{1}{7}\cos 7t$.

SECOND-ORDER DIFFERENTIAL EQUATIONS

10.27 Solve $y'' - 2y' + y = e^x/x$.

▮ The complementary solution is found in Problem 8.143 to be $y_c = c_1e^x + c_2xe^x$; hence we assume $y_p = v_1e^x + v_2xe^x$.

Since $y_1 = e^x$, $y_2 = xe^x$, and $\phi(x) = e^x/x$, it follows from Problem 10.3 that

$$v_1'e^x + v_2'xe^x = 0 \qquad v_1'e^x + v_2'(e^x + xe^x) = \frac{e^x}{x}$$

Solving this set of equations simultaneously, we obtain $v_1' = -1$ and $v_2' = 1/x$. Thus,

$$v_1 = \int -1\,dx = -x \qquad v_2 = \int \frac{1}{x}\,dx = \ln|x|$$

and $y_p = -xe^x + xe^x \ln|x|$. The general solution is therefore $y = y_c + y_p = c_1e^x + (c_2 - 1)xe^x + xe^x \ln|x|$.

10.28 Solve $y'' - 2y' + y = e^x/x^2$.

▮ The general solution of the associated homogeneous differential equation is found in Problem 8.143 to be $y_c = c_1e^x + c_2xe^x$; hence we assume $y_p = v_1e^x + v_2xe^x$.

Since $y_1 = e^x$, $y_2 = xe^x$, and $\phi(x) = e^x/x^2$, it follows from Problem 10.3 that

$$v_1'e^x + v_2'xe^x = 0 \qquad v_1'e^x + v_2'(e^x + xe^x) = \frac{e^x}{x^2}$$

Solving this set of equations simultaneously, we obtain $v_1' = -1/x$ and $v_2' = 1/x^2$. Then

$$v_1 = \int v_1'\,dx = \int -\frac{1}{x}\,dx = -\ln|x| \qquad \text{and} \qquad v_2 = \int v_2'\,dx = \int \frac{1}{x^2}\,dx = -\frac{1}{x}$$

so that $y_p = -\ln|x|e^x - \frac{1}{x}xe^x = -e^x\ln|x| - e^x$. The general solution is then $y = y_c + y_p = (c_1 - 1)e^x + c_2xe^x - e^x\ln|x|$.

10.29 Solve $y'' - 2y' + y = e^x/x^3$.

▮ y_c, y_1, and y_2 of the previous problem remain valid, and we assume $y_p = v_1e^x + v_2xe^x$. Here $\phi(x) = e^x/x^3$. It follows from Problem 10.3 that

$$v_1'e^x + v_2'xe^x = 0 \qquad v_1'e^x + v_2'(e^x + xe^x) = \frac{e^x}{x^3}$$

Solving this set of equations simultaneously, we find that $v_1' = -1/x^2$ and $v_2' = 1/x^3$. Then $v_1 = \int -\frac{1}{x^2}\,dx = 1/x$ and $v_2 = \int \frac{1}{x^3}\,dx = -\frac{1}{2x^2}$. Thus, $y_p = \frac{1}{x}e^x + \left(-\frac{1}{2x^2}\right)xe^x = \frac{1}{2x}e^x$, and the general solution is

$$y = y_c + y_p = c_1e^x + c_2xe^x + \frac{1}{2x}e^x.$$

10.30 Solve $y'' - 2y' + y = e^{2x}$.

▮ The complementary solution is found in Problem 8.143 to be $y_c = c_1e^x + c_2xe^x$, so we assume $y_p = v_1e^x + v_2xe^x$.

Here $y_1 = e^x$, $y_2 = xe^x$, and $\phi(x) = e^{2x}$. It follows from Problem 10.3 that

$$v_1'e^x + v_2'xe^x = 0 \qquad v_1'e^x + v_2'(e^x + xe^x) = e^{2x}$$

Solving this set of equations, we find that $v_1' = -xe^x$ and $v_2' = e^x$. Then $v_1 = \int -xe^x\,dx = -xe^x + e^x$ and $v_2 = \int e^x\,dx = e^x$. Thus, $y_p = (-xe^x + e^x)e^x + e^x xe^x = e^{2x}$, and the general solution is $y = y_c + y_p = c_1e^x + c_2xe^x + e^2x$.

10.31 Solve $y'' + 6y' + 9y = e^{-3x}/x^5$.

▮ The complementary solution is found in Problem 8.142 to be $y_c = c_1e^{-3x} + c_2xe^{-3x}$; hence we assume $y_p = v_1e^{-3x} + v_2xe^{-3x}$.

Here $y_1 = e^{-3x}$, $y_2 = xe^{-3x}$, and $\phi(x) = x^{-5}e^{-3x}$. It follows from Problem 10.3 that

$$v_1'e^{-3x} + v_2'xe^{-3x} = 0 \qquad v_1'(-3e^{-3x}) + v_2'(e^{-3x} - 3xe^{-3x}) = x^{-5}e^{-3x}$$

Solving this set of equations, we get $v_1' = -x^{-4}$ and $v_2' = x^{-5}$, from which

$$v_1 = \int -x^{-4}\,dx = \tfrac{1}{3}x^{-3} \qquad \text{and} \qquad v_2 = \int x^{-5}\,dx = -\tfrac{1}{4}x^{-4}$$

Then $y_p = \frac{1}{3}x^{-3}e^{-3x} + (-\frac{1}{4}x^{-4})xe^{-3x} = \frac{1}{12}x^{-3}e^{-3x}$, and the general solution is $y = y_c + y_p = c_1e^{-3x} + c_2xe^{-3x} + \frac{1}{12}x^{-3}e^{-3x}$.

10.32 Solve $y'' + 6y' + 9y = 100e^{2x}$.

▌ The complementary solution is found in Problem 8.142 to be $y_c = c_1e^{-3x} + c_2xe^{-3x}$, so we assume $y_p = v_1e^{-3x} + v_2xe^{-3x}$.

Here $y_1 = e^{-3x}$, $y_2 = xe^{-3x}$, and $\phi(x) = 100e^{2x}$, so the results of Problem 10.3 become

$$v_1'e^{-3x} + v_2'xe^{-3x} = 0 \qquad v_1'(-3e^{-3x}) + v_2'(e^{-3x} - 3xe^{-3x}) = 100e^{2x}$$

Solving this set of equations simultaneously, we obtain $v_1' = -100xe^{5x}$ and $v_2' = 100e^{5x}$, from which

$$v_1 = \int -100xe^{5x}\,dx = -20xe^{5x} + 4e^{5x} \qquad \text{and} \qquad v_2 = \int 100e^{5x}\,dx = 20e^{5x}$$

Then $y_p = (-20xe^{5x} + 4e^{5x})e^{-3x} + 20e^{5x}xe^{-3x} = 4e^{2x}$. The general solution is $y = y_c + y_p = c_1e^{-3x} + c_2xe^{-3x} + 4e^{2x}$. (Compare this with the result of Problem 9.9.)

10.33 Solve $y'' + 6y' + 9y = 12e^{-3x}$.

▌ The complementary solution is found in Problem 8.142 to be $y_c = c_1e^{-3x} + c_2xe^{-3x}$, so we assume $y_p = v_1e^{-3x} + v_2xe^{-3x}$.

Here $y_1 = e^{-3x}$, $y_2 = xe^{-3x}$, and $\phi(x) = 12e^{-3x}$. It follows from Problem 10.3 that

$$v_1'e^{-3x} + v_2'xe^{-3x} = 0 \qquad v_1'(-3e^{-3x}) + v_2'(e^{-3x} - 3xe^{-3x}) = 12e^{-3x}$$

Solving this set of equations simultaneously, we obtain $v_1' = -12x$ and $v_2' = 12$. Then $v_1 = \int -12x\,dx = -6x^2$ and $v_2 = \int 12\,dx = 12x$. Thus, $y_p = -6x^2e^{-3x} + 12x^2e^{-3x} = 6x^2e^{-3x}$, and the general solution is $y = y_c + y_p = c_1e^{-3x} + c_2xe^{-3x} + 6x^2e^{-3x}$. (Compare with Problems 9.140 and 9.141.)

10.34 Solve $(D^2 - 6D + 9)y = e^{3x}/x^2$.

▌ The complementary solution is $y_c = c_1e^{3x} + c_2xe^{3x}$, so we assume $y_p = v_1e^{3x} + v_2xe^{3x}$. It follows from Problem 10.3 [with $y_1 = e^{3x}$, $y_2 = xe^{3x}$, and $\phi(x) = e^{3x}/x^2$] that

$$v_1'e^{3x} + v_2'xe^{3x} = 0 \qquad v_1'(3e^{3x}) + v_2'(e^{3x} + 3xe^{3x}) = \frac{e^{3x}}{x^2}$$

Solving this set of equations, we obtain $v_1' = -\dfrac{1}{x}$ and $v_2' = \dfrac{1}{x^2}$, so that $v_1 = \int -\dfrac{1}{x}\,dx = -\ln|x|$ and $v_2 = \int \dfrac{1}{x^2}\,dx = -\dfrac{1}{x}$. Then $y_p = (-\ln|x|)e^{3x} + \left(-\dfrac{1}{x}\right)xe^{3x} = -e^{3x}\ln|x| - e^{3x}$, and the general solution is $y = y_c + y_p = (c_1 - 1)e^{3x} + c_2xe^{3x} - e^{3x}\ln|x|$.

10.35 Solve $y'' - 7y' = 6e^{6x}$.

▌ The complementary solution is found in Problem 8.2 to be $y_c = c_1 + c_2e^{7x}$; hence we assume $y_p = v_1 + v_2e^{7x}$. Here $y_1 = 1$, $y_2 = e^{7x}$, and $\phi(x) = 6e^{6x}$. It follows from Problem 10.3 that

$$v_1'(1) + v_2'e^{7x} = 0 \qquad v_1'(0) + v_2'(7e^{7x}) = 6e^{6x}$$

Solving these equations, we obtain $v_2' = \frac{6}{7}e^{-x}$ and $v_1' = -\frac{6}{7}e^{6x}$. Then

$$v_1 = \int v_1'\,dx = \int -\tfrac{6}{7}e^{6x}\,dx = -\tfrac{1}{7}e^{6x} \qquad \text{and} \qquad v_2 = \int v_2'\,dx = \int \tfrac{6}{7}e^{-x}\,dx = -\tfrac{6}{7}e^{-x}$$

Thus $y_p = -\frac{1}{7}e^{6x} + (-\frac{6}{7}e^{-x})e^{7x} = -e^{6x}$, and the general solution is $y = y_c + y_p = c_1 + c_2e^{7x} - e^{6x}$. (Compare with Problem 9.7.)

10.36 Solve $y'' - 7y' = -3$.

▌ y_c, y_1, and y_2 are as in the previous problem, and again we assume $y_p = v_1 + v_2e^{7x}$. Here, however, $\phi(x) = -3$. It follows from Problem 10.3 that

$$v_1'(1) + v_2'e^{7x} = 0 \qquad v_1'(0) + v_2'(7e^{7x}) = -3$$

The solution to this set of equations is $v_1' = \frac{3}{7}$ and $v_2' = -\frac{3}{7}e^{-7x}$, so that $v_1 = \int \frac{3}{7}\,dx = \frac{3}{7}x$ and $v_2 = \int -\frac{3}{7}e^{-7x}\,dx = \frac{3}{49}e^{-7x}$. Thus $y_p = \frac{3}{7}x + \frac{3}{49}e^{-7x}e^{7x} = \frac{3}{7}x + \frac{3}{49}$. The general solution is then $y = y_c + y_p = c_1 + \frac{3}{49} + c_2e^{7x} + \frac{3}{7}x$. (Compare with Problems 9.154 and 9.155.)

10.37 Solve $y'' - 7y' = -3x$.

▌ y_c, y_1, and y_2 are as in Problem 10.35, and again we assume $y_p = v_1 + v_2e^{7x}$. Here, however, $\phi(x) = -3x$. It follows from Problem 10.3 that

$$v_1'(1) + v_2'e^{7x} = 0 \qquad v_1'(0) + v_2'(7e^{7x}) = -3x$$

The solution to this set of equations is $v_1' = \frac{3}{7}x$ and $v_2' = -\frac{3}{7}xe^{-7x}$. Then $v_1 = \int \frac{3}{7}x\,dx = \frac{3}{14}x^2$ and $v_2 = \int -\frac{3}{7}xe^{-7x}\,dx = \frac{3}{49}xe^{-7x} + \frac{3}{343}e^{-7x}$. Thus

$$y_p = \frac{3}{14}x^2 + (\frac{3}{49}xe^{-7x} + \frac{3}{343}e^{-7x})e^{7x} = \frac{3}{14}x^2 + \frac{3}{49}x + \frac{3}{343}$$

and the general solution is $y = y_c + y_p = c_1 + \frac{3}{343} + c_2e^{7x} + \frac{3}{14}x^2 + \frac{3}{49}x$. (Compare with Problems 9.156 and 9.157.)

10.38 Solve $y'' - y' - 2y = e^{3x}$.

▌ The complementary solution is found in Problem 8.1 to be $y_c = c_1e^{-x} + c_2e^{2x}$, so we assume that $y_p = v_1e^{-x} + v_2e^{2x}$.

Here $y_1 = e^{-x}$, $y_2 = e^{2x}$, and $\phi(x) = e^{3x}$. It follows from Problem 10.3 that

$$v_1'e^{-x} + v_2'e^{2x} = 0 \qquad v_1'(-e^{-x}) + v_2'(2e^{2x}) = e^{3x}$$

The solution to this set of equations is $v_1' = -\frac{1}{3}e^{4x}$ and $v_2' = \frac{1}{3}e^{x}$, so that

$$v_1 = \int -\frac{1}{3}e^{4x}\,dx = -\frac{1}{12}e^{4x} \qquad \text{and} \qquad v_2 = \int \frac{1}{3}e^{x}\,dx = \frac{1}{3}e^{x}$$

Then $y_p = -\frac{1}{12}e^{4x}e^{-x} + \frac{1}{3}e^{x}e^{2x} = \frac{1}{4}e^{3x}$, and the general solution is $y = y_c + y_p = c_1e^{-x} + c_2e^{2x} + \frac{1}{4}e^{3x}$. (Compare with Problem 9.10.)

10.39 Solve $y'' - y' - 2y = 4x^2$.

▌ y_c, y_1, and y_2 are as in the previous problem, and we let $y_p = v_1e^{-x} + v_2e^{2x}$. Here $\phi(x) = 4x^2$. It follows from Problem 10.3 that

$$v_1'e^{-x} + v_2'e^{2x} = 0 \qquad v_1'(-e^{-x}) + v_2'(2e^{2x}) = 4x^2$$

The solution to this set of equations is $v_1' = -\frac{4}{3}x^2e^{x}$ and $v_2' = \frac{4}{3}x^2e^{-2x}$. Then, using integration by parts twice on each successive integral, we calculate

$$v_1 = \int -\frac{4}{3}x^2e^{x}\,dx = -\frac{4}{3}(x^2 - 2x + 2)e^{x} \qquad \text{and} \qquad v_2 = \int \frac{4}{3}x^2e^{-2x}\,dx = -\frac{1}{3}(2x^2 + 2x + 1)e^{-2x}$$

Thus $y_p = -\frac{4}{3}(x^2 - 2x + 2)e^{x}e^{-x} - \frac{1}{3}(2x^2 + 2x + 1)e^{-2x}e^{2x} = -2x^2 + 2x - 3$, and the general solution is $y = y_c + y_p = c_1e^{-x} + c_2e^{2x} - 2x^2 + 2x - 3$. (Compare with Problem 9.35.)

10.40 Solve $y'' - y' - 2y = \sin 2x$.

▌ y_c, y_1, and y_2 are as in Problem 10.38, and we let $y_p = v_1e^{-x} + v_2e^{2x}$. Here $\phi(x) = \sin 2x$. It follows from Problem 10.3 that

$$v_1'e^{-x} + v_2'e^{2x} = 0 \qquad v_1'(-e^{-x}) + v_2'(2e^{2x}) = \sin 2x$$

Solving this set of equations simultaneously yields $v_1' = -\frac{1}{3}e^{x}\sin 2x$ and $v_2' = \frac{1}{3}e^{-2x}\sin 2x$. Then, using integration by parts twice on each successive integral, we obtain

$$v_1 = \int -\frac{1}{3}e^{x}\sin 2x\,dx = -\frac{1}{15}e^{x}(\sin 2x - 2\cos 2x) \qquad \text{and} \qquad v_2 = \int \frac{1}{3}e^{-2x}\sin 2x\,dx = -\frac{1}{12}e^{-2x}(\sin 2x + \cos 2x)$$

Thus $y_p = -\frac{1}{15}e^{x}(\sin 2x - 2\cos 2x)e^{-x} - \frac{1}{12}e^{-2x}(\sin 2x + \cos 2x)e^{2x} = -\frac{3}{20}\sin 2x + \frac{1}{20}\cos 2x$, and the general solution is $y = y_c + y_p = c_1e^{-x} + c_2e^{2x} - \frac{3}{20}\sin 2x + \frac{1}{20}\cos 2x$. (Compare with Problem 9.92.)

10.41 Solve $y'' - y' - 2y = e^{2x}$.

▌ y_c, y_1, and y_2 are as in Problem 10.38, and we assume $y_p = v_1e^{-x} + v_2e^{2x}$. Here $\phi(x) = e^{2x}$. It follows from Problem 10.3 that

$$v_1'e^{-x} + v_2'e^{2x} = 0 \qquad v_1'(-e^{-x}) + v_2'(2e^{2x}) = e^{2x}$$

The solution to this set of equations is $v_1' = -\frac{1}{3}e^{3x}$ and $v_2' = \frac{1}{3}$; integrating directly gives $v_1 = -\frac{1}{9}e^{3x}$ and $v_2 = \frac{1}{3}x$. Then $y_p = -\frac{1}{9}e^{3x}e^{-x} + \frac{1}{3}xe^{2x} = (\frac{1}{3}x - \frac{1}{9})e^{2x}$, and the general solution is $y = y_c + y_p = c_1e^{-x} + (c_2 - \frac{1}{9})e^{2x} + \frac{1}{3}xe^{2x}$. (Compare with Problems 9.134 and 9.135.)

10.42 Solve $\dfrac{d^2y}{dt^2} - 4\dfrac{dy}{dt} + y = 3e^{2t}$.

The complementary solution is found in Problem 8.9 to be $y_c = C_1e^{3.732t} + C_2e^{0.268t}$, so we assume that $y_p = v_1e^{3.732t} + v_2e^{0.268t}$.

Here $y_1 = e^{3.732t}$, $y_2 = e^{0.268t}$, and $\phi(t) = 3e^{2t}$. It follows from Problem 10.3 that

$$v_1'e^{3.732t} + v_2'e^{0.268t} = 0 \qquad v_1'(3.732e^{3.732t}) + v_2'(0.268e^{0.268t}) = 3e^{2t}$$

The solution to this set of equations is $v_1' = 0.866e^{-1.732t}$ and $v_2' = -0.866e^{1.732t}$. Then

$$v_1 = \int 0.866e^{-1.732t}\,dt = -0.5e^{-1.732t} \qquad v_2 = \int -0.866e^{1.732t}\,dt = -0.5e^{1.732t}$$

so that $y_p = -0.5e^{-1.732t}e^{3.732t} - 0.5e^{1.732t}e^{0.268t} = -e^{2t}$. The general solution is then $y = C_1e^{3.732t} + C_2e^{0.268t} - e^{2t}$. (Compare with Problem 9.11.)

0.43 Solve $\dfrac{d^2y}{dt^2} - 4\dfrac{dy}{dt} + y = 3t - 4$.

y_c, y_1, and y_2 are as in the previous problem, and we assume $y_p = v_1e^{3.732t} + v_2e^{0.268t}$. Here $\phi(t) = 3t - 4$. It follows from Problem 10.3 that

$$v_1'e^{3.732t} + v_2'e^{0.268t} = 0 \qquad v_1'(3.732e^{3.732t}) + v_2'(0.268e^{0.268t}) = 3t - 4$$

The solution to this set of equations is $v_1' = (0.866t - 1.155)e^{-3.732t}$ and $v_2' = (-0.866t + 1.155)e^{-0.268t}$. Then

$$v_1 = \int (0.866t - 1.155)e^{-3.732t}\,dt = (-0.232t + 0.248)e^{-3.732t}$$

$$v_2 = \int (-0.866t + 1.155)e^{-0.268t}\,dt = (3.231t + 7.748)e^{-0.268t}$$

and $$y_p = (-0.232t + 0.248)e^{-3.732t}e^{3.732t} + (3.231t + 7.748)e^{-0.268t}e^{0.268t} = 2.999t + 7.996$$

Thus, the general solution is $y = y_c + y_p = C_1e^{3.732t} + C_2e^{0.268t} + 2.999t + 7.996$. (Compare this result with that of Problem 9.36; the differences are due solely to roundoff.)

0.44 Solve $\dfrac{d^2x}{dt^2} + 4\dfrac{dx}{dt} + 8x = e^{-2t}$.

The complementary solution is found in Problem 8.54 to be $x_c = c_1e^{-2t}\cos 2t + c_2e^{-2t}\sin 2t$, so we assume a particular solution of the form $x_p = v_1e^{-2t}\cos 2t + v_2e^{-2t}\sin 2t$.

Here $x_1 = e^{-2t}\cos 2t$, $x_2 = e^{-2t}\sin 2t$, and $\phi(t) = e^{-2t}$. It follows from Problem 10.3 (with x replacing y) that

$$v_1'e^{-2t}\cos 2t + v_2'e^{-2t}\sin 2t = 0$$

$$v_1'(-2e^{-2t}\cos 2t - 2e^{-2t}\sin 2t) + v_2'(-2e^{-2t}\sin 2t + 2e^{-2t}\cos 2t) = e^{-2t}$$

The solution to this set of equations is $v_1' = -\frac{1}{2}\sin 2t$ and $v_2' = \frac{1}{2}\cos 2t$, and integration yields $v_1 = \frac{1}{4}\cos 2t$ and $v_2 = \frac{1}{4}\sin 2t$. Then

$$x_p = (\tfrac{1}{4}\cos 2t)e^{-2t}\cos 2t + (\tfrac{1}{4}\sin 2t)(e^{-2t}\sin 2t) = \tfrac{1}{4}e^{-2t}(\cos^2 2t + \sin^2 2t) = \tfrac{1}{4}e^{-2t}$$

The general solution is then $x = c_1e^{-2t}\cos 2t + c_2e^{-2t}\sin 2t + \frac{1}{4}e^{-2t}$. (Compare with Problem 9.13.)

.45 Solve $\dfrac{d^2x}{dt^2} + 4\dfrac{dx}{dt} + 8x = 16\cos 4t$.

x_c, x_1, and x_2 are as in the previous problem, and again we let $x_p = v_1e^{-2t}\cos 2t + v_2e^{-2t}\sin 2t$. Here $\phi(t) = 16\cos 4t$. It follows from Problem 10.3 (with y replaced by x) that

$$v_1'e^{-2t}\cos 2t + v_2'e^{-2t}\sin 2t = 0$$

$$v_1'(-2e^{-2t}\cos 2t - 2e^{-2t}\sin 2t) + v_2'(-2e^{-2t}\sin 2t + 2e^{-2t}\cos 2t) = 16\cos 4t$$

The solution to this set of equations is $v_1' = -8e^{2t}\cos 4t\sin 2t$ and $v_2' = 8e^{2t}\cos 4t\cos 2t$. Integrating yields

$$v_1 = e^{2t}(\sin 2t - \cos 2t - \tfrac{1}{5}\sin 6t + \tfrac{3}{5}\cos 6t) \qquad v_2 = e^{2t}(\sin 2t + \cos 2t + \tfrac{3}{5}\sin 6t + \tfrac{1}{5}\cos 6t)$$

Then $x_p = (\sin 2t - \cos 2t - \frac{1}{5}\sin 6t + \frac{3}{5}\cos 6t)\cos 2t + (\sin 2t + \cos 2t + \frac{3}{5}\sin 6t + \frac{1}{5}\cos 6t)\sin 2t$

$$= 2\sin 2t\cos 2t - (\cos^2 2t - \sin^2 2t) - \tfrac{1}{5}(\sin 6t\cos 2t - \cos 6t\sin 2t) + \tfrac{3}{5}(\cos 6t\cos 2t + \sin 6t\sin 2t)$$
$$= \sin 2(2t) - \cos 2(2t) - \tfrac{1}{5}\sin(6t - 2t) + \tfrac{3}{5}\cos(6t - 2t)$$
$$= \tfrac{4}{5}\sin 4t - \tfrac{2}{5}\cos 4t$$

and the general solution is $x = c_1e^{-2t}\cos 2t + c_2e^{-2t}\sin 2t + \frac{4}{5}\sin 4t - \frac{2}{5}\cos 4t$. (Compare with Problem 9.83.)

10.46 Solve $\ddot{x} + 25x = 5$.

▮ The complementary solution is found in Problem 8.72 to be $x_c = C_1\cos 5t + C_2\sin 5t$, so we assume a particular solution of the form $x_p = v_1\cos 5t + v_2\sin 5t$.

Here $x_1 = \cos 5t$, $x_2 = \sin 5t$, and $\phi(t) = 5$. It follows from Problem 10.3 [with $x(t)$ replacing $y(x)$] that

$$v_1'\cos 5t + v_2'\sin 5t = 0 \qquad v_1'(-5\sin 5t) + v_2'(5\cos 5t) = 5$$

The solution to this set of equations is $v_1' = -\sin 5t$ and $v_2' = \cos 5t$. Then integration yields $v_1 = \frac{1}{5}\cos 5t$ and $v_2 = \frac{1}{5}\sin 5t$, so that

$$x_p = \tfrac{1}{5}\cos 5t\cos 5t + \tfrac{1}{5}\sin 5t\sin 5t = \tfrac{1}{5}(\cos^2 5t + \sin^2 5t) = \tfrac{1}{5}$$

The general solution is then $x = x_c + x_p = C_1\cos 5t + C_2\sin 5t + \frac{1}{5}$.

10.47 Solve $\ddot{x} + 25x = 2\sin 2t$.

▮ x_c, x_1, and x_2 of the previous problem are valid here, and again we assume $x_p = v_1\cos 5t + v_2\sin 5t$. Also, $\phi(t) = 2\sin 2t$. It follows from Problem 10.3 (with y replaced by x) that

$$v_1'\cos 5t + v_2'\sin 5t = 0 \qquad v_1'(-5\sin 5t) + v_2'(5\cos 5t) = 2\sin 2t$$

The solution to this set of equations is $v_1' = -\frac{2}{5}\sin 2t\sin 5t$ and $v_2' = \frac{2}{5}\sin 2t\cos 5t$. Then

$$v_1 = \int -\tfrac{2}{5}\sin 2t\sin 5t\,dt = -\tfrac{1}{15}\sin 3t + \tfrac{1}{35}\sin 7t \qquad \text{and} \qquad v_2 = \int \tfrac{2}{5}\sin 2t\cos 5t\,dt = \tfrac{1}{15}\cos 3t - \tfrac{1}{35}\cos 7t$$

and
$$x_p = (-\tfrac{1}{15}\sin 3t + \tfrac{1}{35}\sin 7t)\cos 5t + (\tfrac{1}{15}\cos 3t - \tfrac{1}{35}\cos 7t)\sin 5t$$
$$= \tfrac{1}{15}(\sin 5t\cos 3t - \sin 3t\cos 5t) + \tfrac{1}{35}(\sin 7t\cos 5t - \cos 7t\sin 5t)$$
$$= \tfrac{1}{15}\sin(5t - 3t) + \tfrac{1}{35}\sin(7t - 5t) = \tfrac{2}{21}\sin 2t$$

The general solution is $x = C_1\cos 5t + C_2\sin 5t + \frac{2}{21}\sin 2t$. (Compare with Problem 9.85.)

10.48 Solve $\ddot{x} + 16x = 80$.

▮ The complementary solution is found in Problem 8.57 to be $x_c = c_1\cos 4t + c_2\sin 4t$, so we assume a particular solution of the form $x_p = v_1\cos 4t + v_2\sin 4t$.

Here $x_1 = \cos 4t$, $x_2 = \sin 4t$, and $\phi(t) = 80$. It then follows from Problem 10.3 (with x replacing y) that

$$v_1'\cos 4t + v_2'\sin 4t = 0 \qquad v_1'(-4\sin 4t) + v_2'(4\cos 4t) = 80$$

The solution to this set of equations is $v_1' = -20\sin 4t$ and $v_2' = 20\cos 4t$. Integration yields $v_1 = 5\cos 4t$ and $v_2 = 5\sin 4t$, so that $x_p = (5\cos 4t)\cos 4t + (5\sin 4t)\sin 4t = 5(\cos^2 4t + \sin^2 4t) = 5$. The general solution is then $x = x_c + x_p = c_1\cos 4t + c_2\sin 4t + 5$.

10.49 Solve $\ddot{x} + 16x = 2\sin 4t$.

▮ x_c, x_1, and x_2 of the previous problem are valid here, and again we let $x_p = v_1\cos 4t + v_2\sin 4t$. Also, $\phi(t) = 2\sin 4t$. It follows from Problem 10.3 (with x replacing y) that

$$v_1'\cos 4t + v_2'\sin 4t = 0 \qquad v_1'(-4\sin 4t) + v_2'(4\cos 4t) = 2\sin 4t$$

The solution to this set of equations is $v_2' = -\frac{1}{2}\sin^2 4t$ and $v_2' = \frac{1}{2}\sin 4t\cos 4t$. Then

$$v_1 = \int -\tfrac{1}{2}\sin^2 4t\,dt = -\tfrac{1}{4}t + \tfrac{1}{32}\sin 8t \qquad \text{and} \qquad v_2 = \int \tfrac{1}{2}\sin 4t\cos 4t\,dt = \tfrac{1}{16}\sin^2 4t$$

and $x_p = (-\frac{1}{4}t + \frac{1}{32}\sin 8t)\cos 4t + \frac{1}{16}\sin^2 4t\sin 4t$. But $\sin 8t = \sin 2(4t) = 2\sin 4t\cos 4t$, so that, after simplification, $x_p = -\frac{1}{4}t\cos 4t + \frac{1}{16}\sin 4t$. The general solution is then $x = c_1\cos 4t + (c_2 + \frac{1}{16})\sin 4t - \frac{1}{4}t\cos 4t$. (Compare with Problems 172 and 173.)

10.50 Solve $y'' + y = \sec x$.

▌ The characteristic equation of the associated homogeneous differential equation is $\lambda^2 + 1 = 0$, which admits the roots $\lambda = \pm i$. The complementary function is $y_c = c_1 \cos x + c_2 \sin x$, and we assume a particular solution of the form $y_p = v_1 \cos x + v_2 \sin x$.

Here $y_1 = \cos x$, $y_2 = \sin x$, and $\phi(x) = \sec x$. Then it follows from Problem 10.3 that

$$v_1' \cos x + v_2' \sin x = 0 \qquad v_1'(-\sin x) + v_2' \cos x = \sec x$$

The solution to this set of equations is $v_1' = -\tan x$ and $v_2' = 1$. Then $v_2 = x$ and $v_1 = \int -\tan x\, dx = \ln|\cos x|$, so that $y_p = (\ln|\cos x|)\cos x + x \sin x$. The general solution is then $y = c_1 \cos x + c_2 \sin x + (\ln|\cos x|)\cos x + x\sin x$.

10.51 Solve $y'' + 4y = \sin^2 2x$.

▌ The complementary solution is found in Problem 8.59 to be $y_c = c_1 \cos 2x + c_2 \sin 2x$, so we assume $y_p = v_1 \cos 2x + v_2 \sin 2x$.

With $y_1 = \cos 2x$, $y_2 = \sin 2x$, and $\phi(x) = \sin^2 2x$, it follows from Problem 10.3 that

$$v_1' \cos 2x + v_2' \sin 2x = 0 \qquad v_1'(-2\sin 2x) + v_2'(2\cos 2x) = \sin^2 2x$$

The solution to this set of equations is $v_1' = -\frac{1}{2}\sin^3 2x$ and $v_2' = \frac{1}{2}\sin^2 2x \cos 2x$. Then

$$v_1 = \int -\tfrac{1}{2}\sin^3 2x\, dx = \tfrac{1}{4}\cos 2x - \tfrac{1}{12}\cos^3 2x \qquad \text{and} \qquad v_2 = \int \tfrac{1}{2}\sin^2 2x \cos 2x\, dx = \tfrac{1}{12}\sin^3 2x$$

so that $y_p = \frac{1}{4}\cos^2 2x - \frac{1}{12}(\cos^4 2x - \sin^4 2x)$. But $\cos^4 2x - \sin^4 2x = \cos^2 2x - \sin^2 2x$, so $y_p = \frac{1}{6}\cos^2 2x + \frac{1}{12}\sin^2 2x$. Then the general solution is $y = c_1 \cos 2x + c_2 \sin 2x + \frac{1}{6}\cos^2 2x + \frac{1}{12}\sin^2 2x$.

10.52 Solve $y'' + 4y = \csc 2x$.

▌ y_c and the form of y_p assumed in the previous problem are valid here. Also, $y_1 = \cos 2x$, $y_2 = \sin 2x$, and $\phi(x) = \csc 2x$. It then follows from Problem 10.3 that

$$v_1' \cos 2x + v_2' \sin 2x = 0 \qquad v_2'(-2\sin 2x) + v_2'(2\cos 2x) = \csc 2x$$

The solution to this set of equations is $v_1' = -\frac{1}{2}$ and $v_2' = \frac{1}{2}\cot 2x$. Then $v_1 = -\frac{1}{2}x$ and $v_2 = \int \frac{1}{2}\cot 2x\, dx = \frac{1}{4}\ln|\sin 2x|$, so that $y_p = -\frac{1}{2}x\cos 2x + \frac{1}{4}\ln|\sin 2x|\sin 2x$. The general solution is thus $y = y_c + y_p = c_1\cos 2x + c_2 \sin 2x - \frac{1}{2}x\cos 2x + \frac{1}{4}\ln|\sin 2x|\sin 2x$.

10.53 Solve $(D^2 + 4)y = 4\sec^2 2x$.

▌ y_c, y_1, and y_2 of Problem 10.51 are valid here, and again we let $y_p = v_1 \cos 2x + v_2 \sin 2x$. Also, $\phi(x) = \sec^2 2x$. Then it follows from Problem 10.3 that

$$v_1' \cos 2x + v_2' \sin 2x = 0 \qquad v_1'(-2\sin 2x) + v_2'(2\cos 2x) = 4\sec^2 2x$$

Solving this set of equations, we find

$$v_1' = -2\frac{\sin 2x}{\cos^2 2x} \qquad \text{and} \qquad v_2' = 2\sec 2x$$

Then integration yields

$$v_1 = \int -2\frac{\sin 2x}{\cos^2 2x}\, dx = -\cos^{-1} 2x = -\sec 2x$$

$$v_2 = \int 2\sec 2x\, dx = \int 2\frac{\sec 2x(\sec 2x + \tan 2x)}{\sec 2x + \tan 2x}\, dx = \ln|\sec 2x + \tan 2x|$$

and $\qquad y_p = -\sec 2x \cos 2x + \ln|\sec 2x + \tan 2x|\sin 2x = -1 + \ln|\sec 2x + \tan 2x|\sin 2x.$

The general solution is then $y = c_1 \cos 2x + c_2 \sin 2x - 1 + \ln|\sec 2x + \tan 2x|\sin 2x$.

10.54 Solve $\ddot{x} + 64x = \sec 8t$.

▌ The complementary solution is found in Problem 8.58 to be $x_c = c_1 \cos 8t + c_2 \sin 8t$, so we assume $x_p = v_1 \cos 8t + v_2 \sin 8t$.

Here, also, $x_1 = \cos 8t$, $x_2 = \sin 8t$, and $\phi(t) = \sec 8t$. It follows from Problem 10.3 (with x replacing y) that

$$v_1' \cos 8t + v_2' \sin 8t = 0 \qquad v_1'(-8 \sin 8t) + v_2'(8 \cos 8t) = \sec 8t$$

The solution to this set of equations is $v_1' = -\frac{1}{8} \sec 8t \sin 8t = -\frac{1}{8} \tan 8t$ and $v_2' = \frac{1}{8}$. Then $v_1 = \int -\frac{1}{8} \tan 8t\, dt = \frac{1}{64} \ln |\cos 8t|$ and $v_2 = \frac{1}{8}t$. Thus $x_p = \frac{1}{64} \ln |\cos 8t| \cos 8t + \frac{1}{8}t \sin 8t$, and the general solution is $x = x_c + x_p = c_1 \cos 8t + c_2 \sin 8t + \frac{1}{64} \ln |\cos 8t| \cos 8t + \frac{1}{8}t \sin 8t$.

10.55 Solve $\ddot{x} + 64x = 64 \cos 8t$.

▌ x_c, x_1, and x_2 of the previous problem are valid here, and we assume x_p as in that problem. With $\phi(t) = 64 \cos 8t$, we have

$$v_1' \cos 8t + v_2' \sin 8t = 0 \qquad v_1'(-8 \sin 8t) + v_2'(8 \cos 8t) = 64 \cos 8t$$

The solution to this set of equations is $v_1' = -8 \sin 8t \cos 8t$ and $v_2' = 8 \cos^2 8t$. Integrating yields

$$v_1 = \int -8 \sin 8t \cos 8t\, dt = \frac{1}{2} \cos^2 8t \qquad v_2 = \int 8 \cos^2 8t\, dt = 4t + \frac{1}{4} \sin 16t = 4t + \frac{1}{2} \sin 8t \cos 8t$$

Then $$x_p = \frac{1}{2} \cos^2 8t \cos 8t + (4t + \frac{1}{2} \sin 8t \cos 8t) \sin 8t = 4t \sin 8t + \frac{1}{2} \cos 8t$$

and the general solution is $x = (c_1 + \frac{1}{2}) \cos 8t + c_2 \sin 8t + 4t \sin 8t$. (Compare with Problems 9.174 and 9.175.)

10.56 Rework the previous problem, taking a different antiderivative for v_1.

▌ Integrating v_1' differently, we obtain $v_1 = \int -8 \sin 8t \cos 8t\, dt = -\frac{1}{2} \sin^2 8t$. With v_2 in its original form, we then have

$$x_p = -\frac{1}{2} \sin^2 8t \cos 8t + (4t + \frac{1}{2} \sin 8t \cos 8t) \sin 8t = 4t \sin 8t$$

The general solution now is $x = c_1 \cos 8t + c_2 \sin 8t + 4t \sin 8t$, which is identical in form to the previous answer because c_1 denotes an arbitrary real number.

10.57 Solve $\ddot{y} - 4\dot{y} + 3y = (1 + e^{-t})^{-1}$.

▌ The associated homogeneous differential equation is $\ddot{y} - 4\dot{y} + 3y = 0$, which has as its general solution $y_c = c_1 e^t + c_2 e^{3t}$. We assume as a particular solution to the nonhomogeneous equation $y_p = v_1 e^t + v_2 e^{3t}$. Now, with $y_1 = e^t$, $y_2 = e^{3t}$, and $\phi(t) = (1 + e^{-t})^{-1}$, it follows from Problem 10.3 that

$$v_1' e^t + v_2' e^{3t} = 0 \qquad v_1' e^t + v_2'(3e^{3t}) = (1 + e^{-t})^{-1}$$

The solution to this set of equations is $v_1' = -\frac{1}{2} \frac{e^{-t}}{1 + e^{-t}}$ and $v_2' = \frac{1}{2} \frac{e^{-3t}}{1 + e^{-t}}$. Setting $u = 1 + e^{-t}$ yields

$$v_1 = \int -\frac{1}{2} \frac{e^{-t}}{1 + e^{-t}}\, dt = \frac{1}{2} \int \frac{1}{u}\, du = \frac{1}{2} \ln u = \frac{1}{2} \ln (1 + e^{-t})$$

And setting $u = e^{-t}$ yields

$$v_2 = \int \frac{1}{2} \frac{e^{-3t}}{1 + e^{-t}}\, dt = \frac{1}{2} \int \frac{e^{-t} e^{-2t}}{1 + e^{-t}}\, dt = -\frac{1}{2} \int \frac{u^2}{1 + u}\, du$$

$$= -\frac{1}{2} \int \left(u - 1 + \frac{1}{1 + u}\right) du = -\frac{1}{2} \left[\frac{1}{2} u^2 - u + \ln (1 + u)\right] = -\frac{1}{4} e^{-2t} + \frac{1}{2} e^{-t} - \frac{1}{2} \ln (1 + e^{-t})$$

Then $$y_p = [\tfrac{1}{2} \ln (1 + e^{-t})] e^t + [-\tfrac{1}{4} e^{-2t} + \tfrac{1}{2} e^{-t} - \tfrac{1}{2} \ln (1 + e^{-t})] e^{3t} = \tfrac{1}{2}(e^t - e^{3t}) \ln (1 + e^{-t}) - \tfrac{1}{4} e^t + \tfrac{1}{2} e^{2t}$$

and the general solution is $y = (c_1 - \frac{1}{4}) e^t + c_2 e^{3t} + \frac{1}{2}(e^t - e^{3t}) \ln (1 + e^{-t}) + \frac{1}{2} e^{2t}$.

10.58 Solve $d^2y/dt^2 - y = (1 + e^{-t})^{-2}$.

▌ The associated homogeneous differential equation, $d^2y/dt^2 - y = 0$, has as its characteristic equation $m^2 - 1 = 0$, which we may factor into $(m - 1)(m + 1) = 0$. The characteristic roots are ± 1, so the complementary function is $y_c = c_1 e^t + c_2 e^{-t}$. We assume a particular solution of the form $y_p = v_1 e^t + v_2 e^{-t}$. It follows from Problem 10.3 that

$$v_1' e^t + v_2' e^{-t} = 0 \qquad v_1' e^t + v_2'(-e^{-t}) = (1 + e^{-t})^{-2}$$

The solution to this set of equations is $v_1' = \frac{1}{2}\frac{e^{-t}}{(1+e^{-t})^2}$ and $v_2' = -\frac{1}{2}\frac{e^t}{(1+e^{-t})^2}$. Setting $u = 1 + e^{-t}$ yields

$$v_1 = \int \frac{1}{2}\frac{e^{-t}}{(1+e^{-t})^2}\,dt = -\frac{1}{2}\int \frac{du}{u^2} = \frac{1}{2}u^{-1} = \frac{1}{2}\frac{1}{1+e^{-t}}$$

Setting $u = e^{-t}$, noting that $e^t = e^{-t}e^{2t} = e^{-t}(e^{-t})^{-2}$, and using partial fractions, we obtain

$$v_2 = \int -\frac{1}{2}\frac{e^t}{(1+e^{-t})^2}\,dt = \frac{1}{2}\int \frac{-e^{-t}(e^{-t})^{-2}}{(1+e^{-t})^2}\,dt = \frac{1}{2}\int \frac{1}{u^2(1+u)^2}\,du = \frac{1}{2}\int\left[\frac{1}{u^2} - \frac{2}{u} + \frac{1}{(1+u)^2} + \frac{2}{1+u}\right]du$$

$$= \frac{1}{2}\left[-\frac{1}{u} - 2\ln u - \frac{1}{1+u} + 2\ln(1+u)\right] = -\frac{1}{2}e^t - \ln e^{-t} - \frac{1/2}{1+e^{-t}} + \ln(1+e^{-t})$$

$$= -\frac{1}{2}e^t + 1 - \frac{1/2}{1+e^{-t}} + \ln(1+e^{-t})$$

and $$y_p = \frac{1}{2}\frac{1}{1+e^{-t}}e^t + \left[-\frac{1}{2}e^t + 1 - \frac{1/2}{1+e^{-t}} + \ln(1+e^{-t})\right]e^{-t}$$

The latter equation may be simplified to $y_p = \frac{1}{2}e^t - 1 + e^{-t} + e^{-t}\ln(1+e^{-t})$, so that the general solution is $y = y_c + y_p = (c_1 + \frac{1}{2})e^t + (c_2 + 1)e^{-t} - 1 + e^{-t}\ln(1+e^{-t})$.

10.59 Solve $d^2y/dt^2 - y = e^{-t}\sin e^{-t} + \cos e^{-t}$.

▌ y_c and the form of y_p assumed in the previous problem are valid here. It follows from Problem 10.3 that

$$v_1'e^t + v_2'e^{-t} = 0 \qquad v_1'e^t + v_2'(-e^{-t}) = e^{-t}\sin e^{-t} + \cos e^{-t}$$

The solution to this set of equations is $v_1' = \frac{1}{2}(e^{-2t}\sin e^{-t} + e^{-t}\cos e^{-t})$ and $v_2' = -\frac{1}{2}(\sin e^{-t} + e^t\cos e^{-t})$. Setting $u = e^{-t}$, we obtain

$$v_1 = \int \tfrac{1}{2}(e^{-2t}\sin e^{-t} + e^{-t}\cos e^{-t})\,dt = \tfrac{1}{2}\int(-u\sin u - \cos u)\,du = -\tfrac{1}{2}\int u\sin u\,du - \tfrac{1}{2}\int\cos u\,du$$

$$= -\tfrac{1}{2}(\sin u - u\cos u) - \tfrac{1}{2}\sin u = -\sin u + \tfrac{1}{2}u\cos u = -\sin e^{-t} + \tfrac{1}{2}e^{-t}\cos e^{-t}$$

For v_2, integration yields

$$v_2 = \int -\tfrac{1}{2}(\sin e^{-t} + e^t\cos e^{-t})\,dt = -\tfrac{1}{2}\int d(e^t\cos e^{-t}) = -\tfrac{1}{2}e^t\cos e^{-t}$$

Then $y_p = (-\sin e^{-t} + \frac{1}{2}e^{-t}\cos e^{-t})e^t + (-\frac{1}{2}e^t\cos e^{-t})e^{-t} = -e^t\sin e^{-t}$ and the general solution is $y = y_c + y_p = c_1e^t + c_2e^{-t} - e^t\sin e^{-t}$.

10.60 Solve $y'' + \frac{1}{t}y' - \frac{1}{t^2}y = \ln t$, for $t > 0$, if it is known that two linearly independent solutions to the associated homogeneous differential equation are $y_1 = t$ and $y_2 = 1/t$.

▌ The complementary solution is $y_c = c_1t + c_2(1/t)$, so we assume $y_p = v_1t + v_2(1/t)$. With $\phi(t) = \ln t$, it follows from Problem 10.3 that

$$v_1't + \frac{v_2'}{t} = 0 \qquad v_1' + v_2'\left(-\frac{1}{t^2}\right) = \ln t$$

The solution to this set of equations is $v_1' = \frac{1}{2}\ln t$ and $v_2' = -\frac{1}{2}t^2\ln t$. Then integration yields

$$v_1 = \int \tfrac{1}{2}\ln t\,dt = \tfrac{1}{2}t\ln t - \tfrac{1}{2}t \qquad \text{and} \qquad v_2 = \int -\tfrac{1}{2}t^2\ln t\,dt = -\tfrac{1}{6}t^3\ln t + \tfrac{1}{18}t^3$$

so that $y_p = (\frac{1}{2}t\ln t - \frac{1}{2}t)t + (-\frac{1}{6}t^3\ln t + \frac{1}{18}t^3)(1/t) = \frac{1}{3}t^2\ln t - \frac{4}{9}t^2$. The general solution is then $y = y_c + y_p = c_1t + c_2(1/t) + \frac{1}{3}t^2\ln t - \frac{4}{9}t^2$.

10.61 Solve $t^2\ddot{y} - 2t\dot{y} + 2y = t\ln t$, for $t > 0$, if it is known that two linearly independent solutions of the associated homogeneous differential equation are $y_1 = t$ and $y_2 = t^2$.

▌ We first divide the differential equation by t^2, obtaining $\ddot{y} - 2t^{-1}\dot{y} + 2t^{-2}y = t^{-1}\ln t$, which has the form of (*1*) in Problem 10.1. Now $\phi(t) = t^{-1}\ln t$. The complementary solution is the same for either form of the associated homogeneous differential equation, so $y_c = c_1t + c_2t^2$ and we assume $y_p = v_1t + v_2t^2$.

It follows from Problem 10.3 that

$$v_1't + v_2't^2 = 0 \qquad v_1' + v_2'(2t) = t^{-1}\ln t$$

The solution to this set of equations is $v_1' = -t^{-1}\ln t$ and $v_2' = t^{-2}\ln t$. Then

$$v_1 = \int -t^{-1}\ln t\,dt = -\tfrac{1}{2}(\ln t)^2 \qquad \text{and} \qquad v_2 = \int t^{-2}\ln t\,dt = -t^{-1}\ln t - t^{-1}$$

so that $y_p = -\tfrac{1}{2}(\ln t)^2 t + (-t^{-1}\ln t - t^{-1})t^2 = -\tfrac{1}{2}t(\ln t)^2 - t\ln t - t$. The general solution is $y = y_c + y_p = (c_1 - 1)t + c_2 t^2 - \tfrac{1}{2}t(\ln t)^2 - t\ln t$.

10.62 Solve $t^2\ddot{y} - t\dot{y} = t^3 e^t$ if it is known that two linearly independent solutions to the associated homogeneous differential equation are $y_1 = 1$ and $y_2 = t^2$.

▮ We divide the differential equation by t^2, obtaining $\ddot{y} - \dfrac{1}{t}\dot{y} = te^t$, so that the coefficient of the highest derivative is unity. The complementary solution of the differential equation in either form is $y_c = c_1 + c_2 t^2$, so we assume $y_p = v_1 + v_2 t^2$. Since $\phi(t) = te^t$, it follows from Problem 10.3 that

$$v_1' + v_2' t^2 = 0 \qquad v_1'(0) + v_2'(2t) = te^t$$

The solution to this set of equations is $v_1' = -\tfrac{1}{2}t^2 e^t$ and $v_2' = \tfrac{1}{2}e^t$. Therefore,

$$v_1 = \int -\tfrac{1}{2}t^2 e^t\,dt = -\tfrac{1}{2}t^2 e^t + te^t - e^t \qquad \text{and} \qquad v_2 = \int \tfrac{1}{2}e^t\,dt = \tfrac{1}{2}e^t$$

Then $y_p = -\tfrac{1}{2}t^2 e^t + te^t - e^t + \tfrac{1}{2}e^t t^2 = te^t - e^t$, and the general solution is $y = y_c + y_p = c_1 + c_2 t^2 + te^t - e^t$.

10.63 Solve $(x^2 - 1)z'' - 2xz' + 2z = (x^2 - 1)^2$ if it is known that two linearly independent solutions to the associated homogeneous differential equation are $z_1 = x$ and $z_2 = x^2 + 1$.

▮ We divide the differential equation by $x^2 - 1$ so that the coefficient of the highest derivative is unity, obtaining $z'' - \dfrac{2x}{x^2+1}z' + \dfrac{2}{x^2+1}z = x^2 - 1$. The complementary solution remains $z_c = c_1 x + c_2(x^2 + 1)$, so we assume $z_p = v_1 x + v_2(x^2 + 1)$.

Since $\phi(x) = x^2 - 1$, it follows from Problem 10.3 (with z replacing y) that

$$v_1' x + v_2'(x^2 + 1) = 0 \qquad v_1' + v_2'(2x) = x^2 - 1$$

The solution to this set of equations is $v_1' = -x^2 - 1$ and $v_2' = x$. Integration yields $v_1 = -\tfrac{1}{3}x^3 - x$ and $v_2 = \tfrac{1}{2}x^2$, so that $z_p = (-\tfrac{1}{3}x^3 - x)(x) + (\tfrac{1}{2}x^2)(x^2 + 1) = \tfrac{1}{6}x^4 - \tfrac{1}{2}x^2$. Then the general solution is $z = z_c + z_p = c_1 x + c_2(x^2 + 1) + \tfrac{1}{6}x^4 - \tfrac{1}{2}x^2$.

10.64 Solve $(x^2 + x)z'' + (2 - x^2)z' - (2 + x)z = x(x + 1)^2$ if the complementary function is known to be $c_1 e^x + c_2 x^{-1}$.

▮ We divide the differential equation by $x^2 + x$ so that the coefficient of the highest derivative is unity, obtaining $z'' + \dfrac{2 - x^2}{x^2 + x}z' - \dfrac{2 - x^2}{x^2 + x}z = x + 1$. The complementary solution remains $z_c = c_1 e^x + c_2 x^{-1}$, so we assume $z_p = v_1 e^x + v_2 x^{-1}$. It follows from Problem 10.3 (with y replaced by z) that

$$v_1' e^x + v_2' x^{-1} = 0 \qquad v_1' e^x + v_2'(-x^{-2}) = x + 1$$

The solution to this set of equations is $v_1' = xe^{-x}$ and $v_2' = -x^2$. Integration yields $v_1 = -xe^{-x} - e^{-x}$ and $v_2 = -\tfrac{1}{3}x^3$, so that $z_p = (-xe^{-x} - e^{-x})(e^x) + (-\tfrac{1}{3}x^3)(x^{-1}) = -\tfrac{1}{3}x^2 - x - 1$. Then the general solution is $z = z_c + z_p = c_1 e^x + c_2 x^{-1} - \tfrac{1}{3}x^2 - x - 1$.

10.65 Solve $\dfrac{d^2 I}{dt^2} - 60\dfrac{dI}{dt} + 900I = 5e^{10t}$.

▮ The complementary solution is found in Problem 8.149 to be $I_c = Ae^{30t} + Bte^{30t}$, so we assume a particular solution of the form $I_p = v_1 e^{30t} + v_2 te^{30t}$.

Here $I_1 = e^{30t}$, $I_2 = te^{30t}$, and $\phi(t) = 5e^{10t}$. It follows from Problem 10.3 (with I replacing y) that

$$v_1' e^{30t} + v_2' te^{30t} = 0 \qquad v_1'(30e^{30t}) + v_2'(e^{30t} + 30te^{30t}) = 5e^{10t}$$

The solution to this set of equations is $v_1' = -5te^{-20t}$ and $v_2' = 5e^{-20t}$. Integration then gives

$$v_1 = \int -5te^{-20t}\,dt = (\tfrac{1}{4}t + \tfrac{1}{80})e^{-20t} \qquad \text{and} \qquad v_2 = \int 5e^{-20t}\,dt = -\tfrac{1}{4}e^{-20t}$$

so that $I_p = (\tfrac{1}{4}t + \tfrac{1}{80})e^{-20t}e^{30t} - \tfrac{1}{4}e^{-20t}te^{30t} = \tfrac{1}{80}e^{10t}$. The general solution is thus $I = Ae^{30t} + Bte^{30t} + \tfrac{1}{80}e^{10t}$. (Compare with Problem 9.14.)

10.66 Solve $\dfrac{d^2I}{dt^2} - 60\dfrac{dI}{dt} + 900I = 4500t^5$.

▌ I_c and the assumed form of I_p in the previous problem are valid here. Now, with $\phi(t) = 4500t^5$, we have

$$v_1'e^{30t} + v_2'te^{30t} = 0 \qquad v_1'(30e^{30t}) + v_2'(e^{30t} + 30te^{30t}) = 4500t^5$$

The solution to this set of equations is $v_1' = -4500t^6e^{-30t}$ and $v_2' = 4500t^5e^{-30t}$. Integration yields

$$v_1 = (150t^6 + 30t^5 + 5t^4 + \tfrac{2}{3}t^3 + \tfrac{1}{15}t^2 + \tfrac{1}{225}t + \tfrac{1}{6750})e^{-30t}$$

and $$v_2 = (-150t^5 - 25t^4 - \tfrac{10}{3}t^3 - \tfrac{1}{3}t^2 - \tfrac{1}{45}t - \tfrac{1}{1350})e^{-30t}$$

so that $$I_p = (150 - 150)t^6 + (30 - 25)t^5 + (5 - \tfrac{10}{3})t^4 + (\tfrac{2}{3} - \tfrac{1}{3})t^3 + (\tfrac{1}{15} - \tfrac{1}{45})t^2 + (\tfrac{1}{225} - \tfrac{1}{1350})t + \tfrac{1}{6750}$$

Then the general solution is $I = I_c + I_p = Ae^{30t} + Bte^{30t} + 5t^5 + \frac{5}{3}t^4 + \frac{1}{3}t^3 + \frac{2}{45}t^2 + \frac{1}{270}t + \frac{1}{6750}$. (Compare with Problem 9.46.)

10.67 Solve $\ddot{I} + 40\dot{I} + 800I = 8\cos t$.

▌ The complementary solution is found in Problem 8.71 to be $I_c = c_1e^{-20t}\cos 20t + c_2e^{-20t}\sin 20t$, so we assume $I_p = v_1e^{-20t}\cos 20t + v_2e^{-20t}\sin 20t$. Then it follows from Problem 10.3 (with I replacing y) that

$$v_1'e^{-20t}\cos 20t + v_2'e^{-20t}\sin 20t = 0$$
$$v_1'(-20e^{-20t}\cos 20t - 20e^{-20t}\sin 20t) + v_2'(-20e^{-20t}\sin 20t + 20e^{-20t}\cos 20t) = 8\cos t$$

The solution to this set of two equations is $v_1' = -\frac{2}{5}e^{20t}\sin 20t\cos t$ and $v_2' = \frac{2}{5}e^{20t}\cos 20t\cos t$, from which we find

$$v_1 = \left(-\frac{4}{761}\sin 19t + \frac{19}{3805}\cos 19t - \frac{4}{841}\sin 21t + \frac{21}{4205}\cos 21t\right)e^{20t}$$

$$v_2 = \left(\frac{19}{3805}\sin 19t + \frac{4}{761}\cos 19t + \frac{21}{4205}\sin 21t + \frac{4}{841}\cos 21t\right)e^{20t}$$

and
$$\begin{aligned}
I_p &= \frac{4}{761}(\sin 20t\cos 19t - \sin 19t\cos 20t) + \frac{19}{3805}(\cos 20t\cos 19t + \sin 20t\sin 19t)\\
&\quad - \frac{4}{841}(\sin 21t\cos 20t - \sin 20t\cos 21t) + \frac{21}{4205}(\cos 21t\cos 20t + \sin 21t\sin 20t)\\
&= \frac{4}{761}\sin(20t - 19t) + \frac{19}{3805}\cos(20t - 19t) - \frac{4}{841}\sin(21t - 19t) + \frac{21}{4205}\cos(21t - 20t)\\
&= \left(\frac{4}{761} - \frac{4}{841}\right)\sin t + \left(\frac{19}{3805} + \frac{21}{4205}\right)\cos t = \frac{320}{640{,}001}\sin t + \frac{6392}{640{,}001}\cos t
\end{aligned}$$

The general solution is then $I = I_c + I_p = c_1e^{-20t}\cos 20t + c_2e^{-20t}\sin 20t + \dfrac{320}{640{,}001}\sin t + \dfrac{6392}{640{,}001}\cos t$.

(Compare with Problem 9.95.)

10.68 Solve $\dfrac{d^2Q}{dt^2} + 8\dfrac{dQ}{dt} + 52Q = 26$.

▌ The complementary solution is found in Problem 8.55 to be $Q_c = c_1e^{-4t}\cos 6t + c_2e^{-4t}\sin 6t$, so we assume $Q_p = v_1e^{-4t}\cos 6t + c_2e^{-4t}\sin 6t$.

Here $Q_1 = e^{-4t}\cos 6t$, $Q_2 = e^{-4t}\sin 6t$, and $\phi(t) = 26$. It follows from Problem 10.3 (with Q replacing y) that

$$v_1'e^{-4t}\cos 6t + v_2'e^{-4t}\sin 6t = 0$$
$$v_1'(-4e^{-4t}\cos 6t - 6e^{-4t}\sin 6t) + v_2'(-4e^{-4t}\sin 6t + 6e^{-4t}\cos 6t) = 26$$

The solution to this set of equations is $v_1' = -\frac{13}{3}e^{4t}\sin 6t$ and $v_2' = \frac{13}{3}e^{4t}\cos 6t$. Integration yields $v_1 = \frac{1}{2}e^{4t}\cos 6t - \frac{1}{3}e^{4t}\sin 6t$ and $v_2 = \frac{1}{3}e^{4t}\cos 6t + \frac{1}{2}e^{4t}\sin 6t$, so that

$$Q_p = (\tfrac{1}{2}e^{4t}\cos 6t - \tfrac{1}{3}e^{4t}\sin 6t)e^{-4t}\cos 6t + (\tfrac{1}{3}e^{4t}\cos 6t + \tfrac{1}{2}e^{4t}\sin 6t)e^{-4t}\sin 6t = \tfrac{1}{2}\cos^2 6t + \tfrac{1}{2}\sin^2 6t = \tfrac{1}{2}$$

The general solution is then $Q = Q_c + Q_p = c_1e^{-4t}\cos 6t + c_2e^{-4t}\sin 6t + \frac{1}{2}$. (Compare with Problem 9.20.)

10.69 Solve $\dfrac{d^2Q}{dt^2} + 8\dfrac{dQ}{dt} + 52Q = 32\cos 2t$.

▌ Q_c and the form assumed for Q_p in the previous problem are valid here. With $\phi(t) = 32\cos 2t$, we have

$$v_1'e^{-4t}\cos 6t + v_2'e^{-4t}\sin 6t = 0$$
$$v_1'(-4e^{-4t}\cos 6t - 6e^{-4t}\sin 6t) + v_2'(-4e^{-4t}\sin 6t + 6e^{-4t}\cos 6t) = 32\cos 2t$$

with solution $v_1' = -\frac{16}{3}e^{4t}\sin 6t\cos 2t$ and $v_2' = \frac{16}{3}\cos 6t\cos 2t$. Integration yields $v_1 = (-\frac{1}{3}\sin 4t + \frac{1}{3}\cos 4t - \frac{2}{15}\sin 8t + \frac{4}{15}\cos 8t)e^{4t}$ and $v_2 = (\frac{1}{3}\sin 4t + \frac{1}{3}\cos 4t + \frac{4}{15}\sin 8t + \frac{2}{15}\cos 8t)e^{4t}$, so that

$$\begin{aligned} Q_p &= \tfrac{1}{3}(\cos 6t\cos 4t + \sin 6t\sin 4t) + \tfrac{1}{3}(\sin 6t\cos 4t - \sin 4t\cos 6t) \\ &\quad - \tfrac{2}{15}(\sin 8t\cos 6t - \sin 6t\cos 8t) + \tfrac{4}{15}(\cos 8t\cos 6t + \sin 8t\sin 6t) \\ &= \tfrac{1}{3}\cos(6t-4t) + \tfrac{1}{3}\sin(6t-4t) - \tfrac{2}{15}\sin(8t-6t) + \tfrac{4}{15}\cos(8t-6t) = \tfrac{3}{5}\cos 2t + \tfrac{1}{5}\sin 2t. \end{aligned}$$

The general solution is $Q = c_1e^{-4t}\cos 6t + e^{-4t}\sin 6t + \frac{3}{5}\cos 2t + \frac{1}{5}\sin 2t$. (Compare with Problem 9.87.)

10.70 Solve $\ddot{Q} + 8\dot{Q} + 25Q = 50\sin 3t$.

▌ The complementary solution (from Problem 8.53 with Q replacing x) is $Q_c = c_1e^{-4t}\cos 3t + c_2e^{-4t}\sin 3t$, so we assume $Q_p = v_1(e^{-4t}\cos 3t) + v_2(e^{-4t}\sin 3t)$. Then with $\phi(t) = 50\sin 3t$, we have

$$v_1'e^{-4t}\cos 3t + v_2'e^{-4t}\sin 3t = 0$$
$$v_1'(-4e^{-4t}\cos 3t - 3e^{-4t}\sin 3t) + v_2'(-4e^{-4t}\sin 3t + 3e^{-4t}\cos 3t) = 50\sin 3t$$

from which $v_1' = -\frac{50}{3}e^{4t}\sin^2 3t$ and $v_2' = \frac{50}{3}e^{4t}\sin 3t\cos 3t$. Integration yields $v_1 = (-\frac{25}{12} + \frac{25}{26}\sin 6t + \frac{25}{39}\cos 6t)e^{4t}$ and $v_2 = (\frac{25}{39}\sin 6t - \frac{25}{39}\cos 6t)e^{4t}$, so that

$$\begin{aligned} Q_p &= -\tfrac{25}{12}\cos 3t + \tfrac{25}{26}(\sin 6t\cos 3t - \sin 3t\cos 6t) + \tfrac{25}{39}(\cos 6t\cos 3t + \sin 6t\sin 3t) \\ &= -\tfrac{25}{12}\cos 3t + \tfrac{25}{26}\sin(6t-3t) + \tfrac{25}{39}\cos(6t-3t) = -\tfrac{75}{52}\cos 3t + \tfrac{25}{26}\sin 3t \end{aligned}$$

The general solution is then $Q = Q_c + Q_p = c_1e^{-4t}\cos 3t + c_2e^{-4t}\sin 3t - \frac{75}{52}\cos 3t + \frac{25}{26}\sin 3t$. (Compare with Problem 9.88.)

10.71 Solve $\ddot{Q} + 8\dot{Q} + 25Q = 90e^{-4t}\cos 3t$.

▌ Q_c and the form of Q_p assumed in the previous problem are valid here. In addition, we have

$$v_1'e^{-4t}\cos 3t + v_2'e^{-4t}\sin 3t = 0$$
$$v_1'(-4e^{-4t}\cos 3t - 3e^{-4t}\sin 3t) + v_2'(-4e^{-4t}\sin 3t + 3e^{-4t}\cos 3t) = 90e^{-4t}\cos 3t$$

with solution $v_1' = -30\cos 3t\sin 3t$ and $v_2' = 30\cos^2 3t$. Integration yields $v_1 = -5\sin^2 3t$ and $v_2 = 15t + \frac{5}{2}\sin 6t = 15t + \frac{5}{2}\sin 2(3t) = 15t + 5\sin 3t\cos 3t$. Then

$$Q_p = -5\sin^2 3t(e^{-4t}\cos 3t) + (15t + 5\sin 3t\cos 3t)(e^{-4t}\sin 3t) = 15te^{-4t}\sin 3t$$

and the general solution is $Q = c_1e^{-4t}\cos 3t + c_2e^{-4t}\sin 3t + 15te^{-4t}\sin 3t$.

10.72 Rework the previous problem, integrating v_1' differently.

▌ With v_2 as in the previous problem, but with $v_1 = \int -30\cos 3t\sin 3t\,dt = 5\cos^2 3t$, we have

$$\begin{aligned} Q_p &= (5\cos^2 3t)(e^{-4t}\cos 3t) + (15t + 5\sin 3t\cos 3t)(e^{-4t}\sin 3t) \\ &= 5e^{-4t}\cos 3t(\cos^2 3t + \sin^2 3t) + 15te^{-4t}\sin 3t = 5e^{-4t}\cos 3t + 15te^{-4t}\sin 3t \end{aligned}$$

The general solution is then

$$Q = Q_c + Q_p = (c_1 + 5)e^{-4t}\cos 3t + c_2e^{-4t}\sin 3t + 15te^{-4t}\sin 3t$$

10.73 Solve $\dfrac{d^2Q}{dt^2} + 1000\dfrac{dQ}{dt} + 250{,}000 = 24$.

▌ The complementary solution is found in Problem 8.147 to be $Q_c = c_1e^{-500t} + c_2te^{-500t}$, so we assume $Q_p = v_1e^{500t} + v_2te^{-500t}$. Then we have

$$v_1'e^{-500t} + v_2'te^{-500t} = 0$$
$$v_1'(-500e^{-500t}) + v_2'(e^{-500t} - 500te^{-500t}) = 24$$

from which we find $v_1' = -24te^{500t}$ and $v_2' = 24e^{500t}$. Then integration yields

$$v_1 = \int -24te^{500t}\,dt = \left[-\frac{24}{500}t + \frac{24}{(500)^2}\right]e^{500t} \qquad v_2 = \int 24e^{500t}\,dt = \frac{24}{500}e^{500t}$$

so that $Q_p = \dfrac{24}{(500)^2} = \dfrac{3}{31{,}250}$. The general solution is then $Q = c_1e^{-500t} + c_2te^{-500t} + \dfrac{3}{31{,}250}$. (Compare with Problem 9.25.)

10.74 Solve $\dfrac{d^2Q}{dt^2} + 1000\dfrac{dQ}{dt} + 250{,}000Q = 24te^{-500t}$.

▮ Q_c and the form of Q_p assumed in the previous problem are valid here. Then we have

$$v_1'e^{-500t} + v_2'te^{-500t} = 0$$
$$v_1'(-500e^{-500t}) + v_2'(e^{-500t} - 500te^{-500t}) = 24te^{-500t}$$

Solving this set of equations simultaneously yields $v_1' = -24t^2$ and $v_2' = 24t$, from which we find $v_1 = -8t^3$ and $v_2 = 12t^2$. Then $Q_p = -8t^3e^{-500t} + 12t^3e^{-500t} = 4t^3e^{-500t}$, and the general solution is $Q = c_1e^{-500t} + c_2te^{-500t} + 4t^3e^{-500t} = (c_1 + c_2t + 4t^3)e^{-500t}$.

10.75 Solve $\dfrac{d^2x}{dt^2} - \dfrac{g}{10}x = \dfrac{g}{5}$, where g denotes a positive constant.

▮ The general solution to the associated homogeneous equation is found in Problem 8.21 to be $x_c = C_1e^{\sqrt{g/10}\,t} + C_2e^{-\sqrt{g/10}\,t}$, so we assume a particular solution of the form $x_p = v_1e^{\sqrt{g/10}\,t} + v_2e^{-\sqrt{g/10}\,t}$. Then we have

$$v_1'e^{\sqrt{g/10}\,t} + v_2'e^{-\sqrt{g/10}\,t} = 0$$
$$v_1'\sqrt{\frac{g}{10}}\,e^{\sqrt{g/10}\,t} + v_2'\left(-\sqrt{\frac{g}{10}}\,e^{-\sqrt{g/10}\,t}\right) = \frac{g}{5}$$

Solving this set of equations yields $v_1' = \sqrt{g/10}e^{-\sqrt{g/10}\,t}$ and $v_2' = -\sqrt{g/10}e^{\sqrt{g/10}\,t}$, from which we find $v_1 = -e^{-\sqrt{g/10}\,t}$ and $v_2 = -e^{\sqrt{g/10}\,t}$. Thus $x_p = -e^{-\sqrt{g/10}\,t}e^{\sqrt{g/10}\,t} - e^{\sqrt{g/10}\,t}e^{-\sqrt{g/10}\,t} = -2$. The general solution is $x = x_c + x_p = C_1e^{\sqrt{g/10}\,t} + C_2e^{-\sqrt{g/10}\,t} - 2$. (Compare with Problem 9.26.)

10.76 Solve $(D^2 - 2D)y = e^x \sin x$.

▮ The complementary solution is $y_c = C_1 + C_2e^{2x}$, so we assume a particular solution of the form $y_p = v_1 + v_2e^{2x}$. It follows from Problem 10.3 that

$$v_1' + v_2'e^{2x} = 0 \qquad v_1'(0) + v_2'(2e^{2x}) = e^x \sin x$$

Then $v_1' = -\frac{1}{2}e^x \sin x$ and $v_2' = \frac{1}{2}e^{-x}\sin x$, and integration yields $v_1 = -\frac{1}{4}e^x(\sin x - \cos x)$ and $v_2 = -\frac{1}{4}e^{-x}(\sin x + \cos x)$. Thus $y_p = -\frac{1}{4}e^x(\sin x - \cos x) - \frac{1}{4}e^{-x}(\sin x + \cos x)e^{2x} = -\frac{1}{2}e^x \sin x$, and the general solution is $y = y_c + y_p = C_1 + C_2e^{2x} - \frac{1}{2}e^x \sin x$. (Compare with Problem 9.105.)

10.77 Solve $\ddot{q} + 20\dot{q} + 200q = 24$.

▮ The complementary solution is found in Problem 8.64 (with q here replacing I) to be $q_c = c_1e^{-10t}\cos 10t + c_2e^{-10t}\sin 10t$, so we assume $q_p = v_1e^{-10t}\cos 10t + c_2e^{-10t}\sin 10t$.

Here we have $q_1 = e^{-10t}\cos 10t$, $q_2 = e^{-10t}\sin 10t$, and $\phi(t) = 24$. It follows from Problem 10.3 (with q replacing y) that

$$v_1'e^{-10t}\cos 10t + v_2'e^{-10t}\sin 10t = 0$$
$$v_1'(-10e^{-10t}\cos 10t - 10e^{-10t}\sin 10t) + v_2'(-10e^{-10t}\sin 10t + 10e^{-10t}\cos 10t) = 24$$

The solution to this set of equations is $v_1' = -\frac{24}{10}e^{10t}\sin 10t$ and $v_2' = \frac{24}{10}e^{10t}\cos 10t$. Integration then yields $v_1 = (\frac{3}{25}\cos 10t - \frac{3}{25}\sin 10t)e^{10t}$ and $v_2 = (\frac{3}{25}\cos 10t + \frac{3}{25}\sin 10t)e^{10t}$, so that

$$q_p = \tfrac{3}{25}(\cos 10t - \sin 10t)e^{10t}(e^{-10t}\cos 10t) + \tfrac{3}{25}(\cos 10t + \sin 10t)e^{10t}(e^{-10t}\sin 10t) = \tfrac{3}{25}(\cos^2 10t + \sin^2 10t) = \tfrac{3}{25}$$

Thus the general solution is $q = q_c + q_p = c_1e^{-10t}\cos 10t + c_2e^{-10t}\sin 10t + \frac{3}{25}$. (Compare with Problem 9.22.)

10.78 Solve $\ddot{q} + 20\dot{q} + 200q = 24e^{-10t}\sin 10t$.

▮ q_c and the form assumed for q_p in the previous problem are valid here. Thus

$$v_1'e^{-10t}\cos 10t + v_2'e^{-10t}\sin 10t = 0$$
$$v_1'(-10e^{-10t}\cos 10t - 10e^{-10t}\sin 10t) + v_2'(-10e^{-10t}\sin 10t + 10e^{-10t}\cos 10t) = 24e^{-10t}\sin 10t$$

from which we find $v_1' = -2.4\sin^2 10t$ and $v_2' = 2.4\sin 10t\cos 10t$. Then integration gives $v_1 = -1.2t + 0.12\sin 10t\cos 10t$ and $v_2 = -0.12\cos^2 10t$, so that

$$q_p = (-1.2t + 0.12\sin 10t\cos 10t)e^{-10t}\cos 10t + (-0.12\cos^2 10t)e^{-10t}\sin 10t = -1.2te^{-10t}\cos 10t$$

and the general solution is $q = c_1e^{-10t}\cos 10t + c_2e^{-10t}\sin 10t - 1.2te^{-10t}\cos 10t$.

10.79 Redo the previous problem, integrating v_2' differently.

▮ With v_1 as in the previous problem, but with $v_2 = \int 2.4\sin 10t\cos 10t\,dt = 0.12\sin^2 10t$, we have

$$\begin{aligned} q_p &= (-1.2t + 0.12\sin 10t\cos 10t)e^{-10t}\cos 10t + (0.12\sin^2 10t)e^{-10t}\sin 10t \\ &= -1.2te^{-10t}\cos 10t + 0.12e^{-10t}\sin 10t(\cos^2 10t + \sin^2 10t) = -1.2te^{-10t}\cos 10t + 0.12e^{-10t}\sin 10t \end{aligned}$$

and the general solution becomes $q = c_1e^{-10t}\cos 10t + (c_2 + 0.12)e^{-10t}\sin 10t - 1.2te^{-10t}\cos 10t$.

10.80 Solve $\ddot{q} + 400\dot{q} + 200{,}000q = 2000$.

▮ The complementary solution is found in Problem 8.70 to be $q_c = e^{-200t}(A\cos 400t + B\sin 400t)$, so we assume a particular solution of the form $q_p = v_1e^{-200t}\cos 400t + v_2e^{-200t}\sin 400t$.

Here $q_1 = e^{-200t}\cos 400t$, $q_2 = e^{-200t}\sin 400t$, and $\phi(t) = 2000$. It follows from Problem 10.3 (with q replacing y) that

$$v_1'e^{-200t}\cos 400t + v_2'e^{-200t}\sin 400t = 0$$
$$v_1'(-200e^{-200t}\cos 400t - 400e^{-200t}\sin 400t) + v_2'(-200e^{-200t}\sin 400t + 400e^{-200t}\cos 400t) = 2000$$

The solution to this set of equations is $v_1' = -5e^{200t}\sin 400t$ and $v_2' = 5e^{200t}\cos 400t$. Integration gives

$$v_1 = 0.01e^{200t}\cos 400t - 0.005e^{200t}\sin 400t \qquad v_2 = 0.005e^{200t}\cos 400t + 0.01e^{200t}\sin 400t$$

so that $q_p = 0.01\cos^2 400t + 0.01\sin^2 400t = 0.01$, and the general solution is $q = q_c + q_p = e^{-200t}(A\cos 400t + B\sin 400t) + 0.01$. (Compare with Problem 9.23.)

10.81 Solve $\ddot{q} + 400\dot{q} + 200{,}000q = 2000\cos 200t$.

▮ q_c and the form assumed for q_p in the previous problem are valid here. Thus, it follows that

$$v_1'e^{-200t}\cos 400t + v_2'e^{-200t}\sin 400t = 0$$
$$v_1'(-200e^{-200t}\cos 400t - 400e^{-200t}\sin 400t) + v_2'(-200e^{-200t}\sin 400t + 400e^{-200t}\cos 400t) = 2000\cos 200t$$

The solution to this set of equations is $v_1' = -5e^{200t}\cos 200t\sin 400t$ and $v_2' = 5e^{200t}\cos 200t\cos 400t$, and integration yields

$$v_1 = (-\tfrac{1}{160}\sin 200t + \tfrac{1}{160}\cos 200t - \tfrac{1}{800}\sin 600t + \tfrac{3}{800}\cos 600t)e^{200t}$$

and

$$v_2 = (\tfrac{1}{160}\sin 200t + \tfrac{1}{160}\cos 200t + \tfrac{3}{800}\sin 600t + \tfrac{1}{800}\cos 600t)e^{200t}$$

After some simplification, these values for v_1 and v_2 give

$$\begin{aligned} q_p &= \tfrac{1}{160}\sin(400t - 200t) + \tfrac{1}{160}\cos(400t - 200t) - \tfrac{1}{800}\sin(600t - 400t) + \tfrac{3}{800}\cos(600t - 400t) \\ &= \tfrac{1}{200}\sin 200t + \tfrac{1}{100}\cos 200t \end{aligned}$$

The general solution is then $q = e^{-200t}(A\cos 400t + B\sin 400t) + \frac{1}{200}\sin 200t + \frac{1}{100}\cos 200t$. (Compare with Problem 9.93.)

10.82 Solve $\ddot{q} + 9\dot{q} + 14q = \frac{1}{2}\sin t$.

▌ The complementary solution is found in Problem 8.17 (with q here replacing x) to be $q_c = c_1e^{-2t} + c_2e^{-7t}$, so we assume $q_p = v_1e^{-2t} + v_2e^{-7t}$. Then

$$v_1'e^{-2t} + v_2'e^{-7t} = 0 \qquad v_1'(-2e^{-2t}) + v_2'(-7e^{-7t}) = \tfrac{1}{2}\sin t$$

so that $v_1' = \frac{1}{10}e^{2t}\sin t$ and $v_2' = -\frac{1}{10}e^{7t}\sin 5t$. Integration yields $v_1 = (\frac{1}{25}\sin t - \frac{1}{50}\cos t)e^{2t}$ and $v_2 = (-\frac{7}{500}\sin t + \frac{1}{500}\cos t)e^{7t}$, and thus

$$q_p = (\tfrac{1}{25} - \tfrac{7}{500})\sin t + (-\tfrac{1}{50} + \tfrac{1}{500})\cos t = \tfrac{13}{500}\sin t - \tfrac{9}{500}\cos t$$

The general solution is then $q = q_c + q_p = c_1e^{-2t} + c_2e^{-7t} + \frac{13}{500}\sin t - \frac{9}{500}\cos t$. (Compare with Problem 9.89.)

HIGHER-ORDER DIFFERENTIAL EQUATIONS

10.83 Solve $x^3y''' + x^2y'' - 2xy' + 2y = x\ln x$, for $x > 0$, if the complementary solution is $y_c = c_1x^{-1} + c_2x + c_3x^2$.

▌ We first divide the differential equation by x^3 so that the ccoefficient of the highest derivative is unity, as in (*1*) of Problem 10.1. The result is $y''' + x^{-1}y'' - 2x^{-2}y' + 2x^{-3}y = x^{-2}\ln x$, for which $\phi(x) = x^{-2}\ln x$. We assume a particular solution of the form $y_p = v_1x^{-1} + v_2x + v_3x^2$. It follows from Problem 10.2 that

$$v_1'x^{-1} + v_2'x + v_3'x^2 = 0$$
$$v_1'(-x^{-2}) + v_2' + v_3'(2x) = 0$$
$$v_1'(2x^{-3}) + v_2'(0) + v_3'(2) = x^{-2}\ln x$$

The solution to this set of simultaneous equations is $v_1' = \frac{1}{6}x\ln x$, $v_2' = -\frac{1}{2}x^{-1}\ln x$, and $v_3' = \frac{1}{3}x^{-2}\ln x$. Integration yields

$$v_1 = \tfrac{1}{12}x^2\ln x - \tfrac{1}{24}x^2 \qquad v_2 = -\tfrac{1}{4}(\ln x)^2 \qquad v_3 = -\tfrac{1}{3}x^{-1}\ln x - \tfrac{1}{3}x^{-1}$$

and, after substitution and simplification, we have $y_p = -\frac{1}{4}x[(\ln x)^2 + \ln x] - \frac{3}{8}x$. Then the general solution is $y = y_c + y_p = c_1x^{-1} + (c_2 - \frac{3}{8})x + c_3x^2 - \frac{1}{4}x[(\ln x)^2 + \ln x]$.

10.84 Solve $y''' + y' = \sec x$.

▌ The complementary solution is found in Problem 8.108 to be $y_c = c_1 + c_2\cos x + c_3\sin x$, so we assume a particular solution of the form $y_p = v_1 + v_2\cos x + v_3\sin x$.

Here $y_1 = 1$, $y_2 = \cos x$, $y_3 = \sin x$, and $\phi(x) = \sec x$. It follows from Problem 10.2 that

$$v_1' + v_2'\cos x + v_3'\sin x = 0$$
$$v_1'(0) + v_2'(-\sin x) + v_3'\cos x = 0$$
$$v_1'(0) + v_2'(-\cos x) + v_3'(-\sin x) = \sec x$$

Solving this set of equations simultaneously, we obtain $v_1' = \sec x$, $v_2' = -1$, and $v_3' = -\tan x$. Thus, $v_1 = \int \sec x\,dx = \ln|\sec x + \tan x|$, $v_2 = -\int dx = -x$, and $v_3 = \int -\tan x\,dx = \ln|\cos x|$. Substitution then yields $y_p = \ln|\sec x + \tan x| - x\cos x + (\sin x)\ln|\cos x|$. The general solution is therefore

$$y = y_c + y_p = c_1 + c_2\cos x + c_3\sin x + \ln|\sec x + \tan x| - x\cos x + (\sin x)\ln|\cos x|$$

10.85 Solve $(D^3 + D)y = \csc x$.

▌ y_c and the form assumed for y_p in the previous problem are valid here. Now, however, $\phi(x) = \csc x$, and it follows from Problem 10.2 that

$$v_1' + v_2'\cos x + v_3'\sin x = 0$$
$$v_1'(0) + v_2'(-\sin x) + v_3'\cos x = 0$$
$$v_1'(0) + v_2'(-\cos x) + v_3'(-\sin x) = \csc x$$

The solution to this set of equations is $v_1' = \csc x$, $v_2' = -(\cos x)/\sin x$, and $v_3' = -1$, from which we find that $v_1 = -\ln|\csc x + \cot x|$, $v_2 = -\ln|\sin x|$, and $v_3 = -x$. Then substitution gives $y_p = -\ln|\csc x + \cot x| - (\ln|\sin x|)\cos x - x\sin x$. The general solution is thus

$$y = c_1 + c_2\cos x + c_3\sin x - \ln|\csc x + \cot x| - (\ln|\sin x|)\cos x - x\sin x$$

10.86 Solve $y''' + 4y' = 4\cot 2x$.

The complementary solution is found in Problem 8.109 to be $y_c = c_1 + c_2\cos 2x + c_3\sin 2x$, so we assume $y_p = v_1 + v_2\cos 2x + v_3\sin 2x$. It follows from Problem 10.2 that

$$v_1' + v_2'\cos 2x + v_3'\sin 2x = 0$$
$$v_1'(0) + v_2'(-2\sin 2x) + v_3'(2\cos 2x) = 0$$
$$v_1'(0) + v_2'(-4\cos 2x) + v_3'(-4\sin 2x) = 4\cot 2x$$

The solution to this set of equations is $v_1'(\cos 2x)/\sin 2x$, $v_2' = -(\cos^2 2x)/\sin 2x$, and $v_3' = -\cos 2x$. Integration then gives

$$v_1 = \tfrac{1}{2}\ln|\sin 2x| \qquad v_2 = -\tfrac{1}{2}\ln|\csc 2x - \cot 2x| - \tfrac{1}{2}\cos 2x \qquad v_3 = -\tfrac{1}{2}\sin 2x$$

Substitution into the expression for y_p and combination with y_c finally lead to

$$y = c_1 + c_2\cos 2x + c_3\sin 2x + \tfrac{1}{2}\ln|\sin 2x| - \tfrac{1}{2}\ln|\csc 2x - \cot 2x|\cos 2x - \tfrac{1}{2}(\cos^2 2x + \sin^2 2x)$$
$$= c_1 - \tfrac{1}{2} + c_2\cos 2x + c_3\sin 2x + \tfrac{1}{2}\ln|\sin 2x| - \tfrac{1}{2}\ln|\csc 2x - \cot 2x|\cos 2x$$

10.87 Solve $y''' - 3y'' + 3y' - y = e^x/x$.

The complementary solution is $y_c = c_1e^x + c_2xe^x + c_3x^2e^x$, so we assume $y_p = v_1e^x + v_2xe^x + v_3x^2e^x$. It follows from Problem 10.2 that

$$v_1'e^x + v_2'xe^x + v_3'x^2e^x = 0$$
$$v_1'e^x + v_2'(e^x + xe^x) + v_3'(2xe^x + x^2e^x) = 0$$
$$v_1'e^x + v_2'(2e^x + xe^x) + v_3'(2e^x + 4xe^x + x^2e^x) = \frac{e^x}{x}$$

The solution to this set of equations is $v_1' = \tfrac{1}{2}x$, $v_2' = -1$, and $v_3' = \tfrac{1}{2}x^{-1}$, so that $v_1 = x^2$, $v_2 = -x$, and $v_3 = \tfrac{1}{2}\ln|x|$. Thus $y_p = \tfrac{1}{2}x^2e^x\ln|x|$, and the general solution is $y = y_c + y_p = e^x(c_1 + c_2x + c_3x^2 + \tfrac{1}{2}x^2\ln|x|)$.

10.88 Solve $\dfrac{d^3y}{dx^3} + 6\dfrac{d^2y}{dx^2} + 12\dfrac{dy}{dx} + 8y = 12e^{-2x}$.

The complementary solution is found in Problem 8.156 to be $y_c = (c_1 + c_2x + c_3x^2)e^{-2x}$, so we assume $y_p = v_1e^{-2x} + v_2xe^{-2x} + v_3x^2e^{-2x}$. Then we have

$$v_1'e^{-2x} + v_2'xe^{-2x} + v_3'x^2e^{-2x} = 0$$
$$v_1'(-2e^{-2x}) + v_2'(e^{-2x} - 2xe^{-2x}) + v_3'(2xe^{-2x} - 2x^2e^{-2x}) = 0$$
$$v_1'(4e^{-2x}) + v_2'(-4e^{-2x} + 4xe^{-2x}) + v_3'(2e^{-2x} - 8xe^{-2x} + 4x^2e^{-2x}) = 12e^{-2x}$$

The solution to this set of equations is $v_1' = 6x^2$, $v_2' = -12x$, and $v_3' = 6$, so that $v_1 = 2x^3$, $v_2 = -6x^2$, and $v_3 = 6x$. Then $y_p = 2x^3e^{-2x} - 6x^3e^{-2x} + 6x^3e^{-2x} = 2x^3e^{-2x}$. The general solution is thus $y = y_c + y_p = (c_1 + c_2x + c_3x^2 + 2x^3)e^{-2x}$. (Compare with Problems 9.146 and 9.147.)

10.89 Solve $y''' - 6y'' + 11y' - 6y = 2xe^{-x}$.

The complementary solution is found in Problem 8.28 to be $y_c = c_1e^x + c_2e^{2x} + c_3e^{3x}$, so we assume that $y_p = v_1e^x + v_2e^{2x} + v_3e^{3x}$. Then we have

$$v_1'e^x + v_2'e^{2x} + v_3'e^{3x} = 0$$
$$v_1'e^x + v_2'(2e^{2x}) + v_3'(3e^{3x}) = 0$$
$$v_1'e^x + v_2'(4e^{2x}) + v_3'(9e^{3x}) = 2xe^{-x}$$

The solution to this set of equations is $v_1' = xe^{-2x}$, $v_2' = -2xe^{-3x}$, and $v_3' = xe^{-4x}$, so that integration yields

$$v_1 = e^{-2x}(-\tfrac{1}{2}x - \tfrac{1}{4}) \qquad v_2 = e^{-3x}(\tfrac{2}{3}x + \tfrac{2}{9}) \qquad v_3 = e^{-4x}(-\tfrac{1}{4}x - \tfrac{1}{16})$$

Substitution and simplification then give $y_p = e^{-x}(-\tfrac{1}{12}x - \tfrac{13}{144})$, and the general solution is $y = c_1e^x + c_2e^{2x} + c_3e^{3x} + e^{-x}(-\tfrac{1}{12}x - \tfrac{13}{144})$. (Compare with Problem 9.74.)

10.90 Solve $y''' - 3y'' + 2y' = \dfrac{e^{2t}}{1 + e^t}$.

▮ The complementary solution is found in Problem 8.26 (with t here replacing x) to be $y_c = c_1 + c_2e^t + c_3e^{2t}$. Thus, we assume $y_p = v_1 + v_2e^t + v_3e^{2t}$, and it follows from Problem 10.2 that

$$v_1' + v_2'e^t + v_3'e^{2t} = 0$$
$$v_1'(0) + v_2'e^t + v_3'(2e^{2t}) = 0$$
$$v_1'(0) + v_2'e^t + v_3(4e^{2t}) = \frac{e^{2t}}{1+e^t}$$

The solution to this set of equations is $v_1' = \frac{1}{2}\frac{e^{2t}}{1+e^t}$, $v_2' = \frac{-e^t}{1+e^t}$, and $v_3' = \frac{1}{2}\frac{1}{1+e^t}$. Using the substitution $u = 1 + e^t$, we find

$$v_1 = \frac{1}{2}\int \frac{e^t}{1+e^t}e^t\,dt = \frac{1}{2}\int \frac{u-1}{u}\,du = \frac{1}{2}(1+e^t) - \frac{1}{2}\ln(1+e^t)$$
$$v_2 = \int \frac{-1}{1+e^t}e^t\,dt = -\ln(1+e^t)$$

and
$$v_3 = \frac{1}{2}\int \frac{1}{1+e^t}\,dt = \frac{1}{2}\int \frac{1}{e^{-t}+1}e^{-t}\,dt = -\frac{1}{2}\ln(e^{-t}+1)$$

and it follows that $y_p = \frac{1}{2}(1+e^t) - \frac{1}{2}\ln(1+e^t) - e^t\ln(1+e^t) - \frac{1}{2}e^{2t}\ln(e^{-t}+1)$. Then the general solution is $y = c_1 + c_2e^t + c_3e^{2t} - (\frac{1}{2} + e^t)\ln(1+e^t) - \frac{1}{2}e^{2t}\ln(e^{-t}+1)$.

10.91 Find a particular solution to $y''' - 3y'' + 2y' = \dfrac{e^{3t}}{1+e^t}$.

▮ y_c and the form assumed for y_p in the previous problem are valid here. It follows from Problem 10.2 that

$$v_1' + v_2'e^t + v_3'e^{2t} = 0$$
$$v_1'(0) + v_2'e^t + v_3'(2e^{2t}) = 0$$
$$v_1'(0) + v_2'e^t + v_3'(4e^{2t}) = \frac{e^{3t}}{1+e^t}$$

The solution to this set of equations is $v_1' = \frac{1}{2}\frac{e^{3t}}{1+e^t}$, $v_2' = \frac{-e^{2t}}{1+e^t}$, and $v_3' = \frac{1}{2}\frac{e^t}{1+e^t}$. Then integration yields

$$v_1 = \tfrac{1}{4}(1+e^t)^2 - (1+e^t) + \tfrac{1}{2}\ln(1+e^t) \qquad v_2 = -(1+e^t) + \ln(1+e^t) \qquad v_3 = \tfrac{1}{2}\ln(1+e^t)$$

so that
$$y_p = \tfrac{1}{4}(1+e^t)^2 - (1+e^t) + \tfrac{1}{2}\ln(1+e^t) + [-(1+e^t) + \ln(1+e^t)]e^t + \tfrac{1}{2}e^{2t}\ln(1+e^t)$$

10.92 Solve $t^3y''' + 3t^2y'' = 1$, for $t > 0$, if three linearly independent solutions to the associated homogeneous differential equation are $y_1 = t^{-1}$, $y_2 = 1$, and $y_3 = t$.

▮ The complementary solution is $y_c = c_1t^{-1} + c_2 + c_3t$, so we assume a particular solution of the form $y_p = v_1t^{-1} + v_2 + v_3t$. We divide the nonhomogeneous differential equation by t^3 so that the coefficient of the highest derivative is unity, as in (*1*) of Problem 10.1. The result is $y''' + 3t^{-1}y'' = t^{-3}$; therefore, $\phi(t) = t^{-3}$. It follows from Problem 10.2 that

$$v_1't^{-1} + v_2' + v_3't = 0 \qquad v_1'(-t^{-2}) + v_2'(0) + v_3 = 0 \qquad v_1'(2t^{-3}) + v_2'(0) + v_3'(0) = t^{-3}$$

from which we find $v_1' = \frac{1}{2}$, $v_2' = -t^{-1}$, and $v_3' = \frac{1}{2}t^{-2}$. Then integration yields $v_1 = \frac{1}{2}t$, $v_2 = -\ln t$, and $v_3 = -\frac{1}{2}t^{-1}$, so that $y_p = \frac{1}{2}t^{-1}t - \ln t - \frac{1}{2}t^{-1}t = -\ln t$. The general solution is then $y = y_c + y_p = c_1t^{-1} + c_2 + c_3t - \ln t$.

10.93 Solve $\dfrac{d^3Q}{dt^3} - 5\dfrac{d^2Q}{dt^2} + 25\dfrac{dQ}{dt} - 125Q = -60e^{7t}$.

▮ The complementary solution is found in Problem 8.113 to be $Q_c = c_1e^{5t} + c_2\cos 5t + c_3\sin 5t$, so we assume $Q_p = v_1e^{5t} + v_2\cos 5t + v_3\sin 5t$. Then we have

$$v_1'e^{5t} + v_2'\cos 5t + v_3'\sin 5t = 0$$
$$v_1'(5e^{5t}) + v_2'(-5\sin 5t) + v_3'(5\cos 5t) = 0$$
$$v_1'(25e^{5t}) + v_2'(-25\cos 5t) + v_3'(-25\sin 5t) = -60e^{7t}$$

The solution to this set of equations is $v_1' = -\frac{6}{5}e^{2t}$, $v_2' = \frac{6}{5}e^{7t}(-\sin 5t + \cos 5t)$, and $v_3' = \frac{6}{5}e^{7t}(\sin 5t + \cos 5t)$, and integration gives

$$v_1 = -\tfrac{3}{5}e^{2t} \qquad v_2 = \tfrac{6}{185}e^{7t}(6\cos 5t - \sin 5t) \qquad v_3 = \tfrac{6}{185}e^{7t}(\cos 5t + 6\sin 5t)$$

Then $$Q_p = -\tfrac{3}{5}e^{2t}e^{5t} + \tfrac{6}{185}e^{7t}(6\cos^2 5t + 6\sin^2 5t) = (-\tfrac{3}{5} + \tfrac{36}{185})e^{7t} = -\tfrac{15}{37}e^{7t}$$

and the general solution is $Q = Q_c + Q_p = c_1e^{5t} + c_2\cos 5t + c_3\sin 5t - \frac{15}{37}e^{7t}$. (Compare with Problem 9.16.)

10.94 Find a particular solution to $\dfrac{d^3Q}{dt^3} - 5\dfrac{d^2Q}{dt^2} + 25\dfrac{dQ}{dt} - 125Q = 1000.$

▮ Q_c and the form assumed for Q_p in the previous problem are valid here. Then we have

$$v_1'e^{5t} + v_2'\cos 5t + v_3'\sin 5t = 0$$
$$v_1'(5e^{5t}) + v_2'(-5\sin 5t) + v_3'(5\cos 5t) = 0$$
$$v_1'(25e^{5t}) + v_2'(-25\cos 5t) + v_3'(-25\sin 5t) = 1000$$

The solution to this set of equations is $v_1' = 20e^{-5t}$, $v_2' = 20\sin 5t - 20\cos 5t$, and $v_3' = -20\sin 5t - 20\cos 5t$; integration yields $v_1 = -4e^{-5t}$, $v_2 = -4\cos 5t - 4\sin 5t$, and $v_3 = 4\cos 5t - 4\sin 5t$. Thus

$$Q_p = (-4e^{-5t})e^{5t} + (-4\cos 5t - 4\sin 5t)\cos 5t + (4\cos 5t - 4\sin 5t)\sin 5t = -8$$

(Compare with Problem 9.29.)

10.95 Find a particular solution to $\dfrac{d^3Q}{dt^3} - 5\dfrac{d^2Q}{dt^2} + 25\dfrac{dQ}{dt} - 125Q = 5000e^{-t}\cos 2t.$

▮ Q_c and the form assumed for Q_p in Problem 10.93 are valid here. Then we have

$$v_1'e^{5t} + v_2'\cos 5t + v_3'\sin 5t = 0$$
$$v_1'(5e^{5t}) + v_2'(-5\sin 5t) + v_3'(5\cos 5t) = 0$$
$$v_1'(25e^{5t}) + v_2'(-25\cos 5t) + v_3'(-25\sin 5t) = 5000e^{-t}\cos 2t$$

The solution to this set of equations is $v_1' = 100e^{-6t}\cos 2t$; $v_2' = 100e^{-t}(-\cos 5t\cos 2t + \sin 5t\cos 2t)$; and $v_3' = 100e^{-t}(-\sin 5t\cos 2t - \cos 5t\cos 2t)$. Integration yields

$$v_1 = e^{-6t}(-15\cos 2t + 5\sin 2t)$$
$$v_2 = e^{-t}(-20\sin 3t - 10\cos 3t - 8\sin 7t - 6\cos 7t)$$
$$v_3 = e^{-t}(-10\sin 3t + 20\cos 3t - 6\sin 7t + 8\cos 7t)$$

Substitution and simplification then give the particular solution

$$\begin{aligned} Q_p &= e^{-t}[-15\cos 2t + 5\sin 2t + 20(\sin 5t\cos 3t - \sin 3t\cos 5t) - 10(\cos 5t\cos 3t + \sin 5t\sin 3t) \\ &\quad - 8(\sin 7t\cos 5t - \sin 5t\cos 7t) - 6(\cos 7t\cos 5t + \sin 7t\sin 5t)] \\ &= e^{-t}[-15\cos 2t + 5\sin 2t + 20\sin(5t - 3t) - 10\cos(5t - 3t) - 8\sin(7t - 5t) - 6\cos(7t - 5t)] \\ &= e^{-t}(-31\cos 2t + 17\sin 2t) \end{aligned}$$

(Compare with Problem 9.107.)

10.96 Solve $\dfrac{d^4y}{dx^4} - 9\dfrac{d^2y}{dx^2} = 54x^2.$

▮ The complementary solution is $y_c = c_1 + c_2x + c_3e^{3x} + c_4e^{-3x}$, so we assume $y_p = v_1 + v_2x + v_3e^{3x} + v_4e^{-3x}$. It follows from Problem 10.1 (with $n = 4$) that

$$v_1' + v_2'x + v_3'e^{3x} + v_4'e^{-3x} = 0$$
$$v_1'(0) + v_2' + v_3'(3e^{3x}) + v_4'(-3e^{-3x}) = 0$$
$$v_1'(0) + v_2'(0) + v_3'(9e^{3x}) + v_4'(9e^{-3x}) = 0$$
$$v_1'(0) + v_2'(0) + v_3'(27e^{3x}) + v_4'(-27e^{-3x}) = 54x^2$$

The solution to this set of equations is $v_1' = 6x^3$, $v_2' = -6x^2$, $v_3' = x^2e^{-3x}$, and $v_4' = -x^2e^{3x}$, and integration yields

$$v_1 = \tfrac{3}{2}x^4 \qquad v_2 = -2x^3 \qquad v_3 = e^{-3x}(-\tfrac{1}{3}x^2 - \tfrac{2}{9}x - \tfrac{2}{27}) \qquad v_4 = e^{3x}(-\tfrac{1}{3}x^2 + \tfrac{2}{9}x - \tfrac{2}{27})$$

By substituting these results into the expression for y_p and simplifying, we obtain $y_p = -\tfrac{1}{2}x^4 - \tfrac{2}{3}x^2 - \tfrac{4}{27}$. The general solution is then $y = y_c + y_p = c_1 - \tfrac{4}{27} + c_2 x + c_3 e^{3x} + c_4 e^{-3x} - \tfrac{1}{2}x^4 - \tfrac{2}{3}x^2$.

.97 Solve $\dfrac{d^4 y}{dx^4} = 5x$.

▌ The complementary solution is found in Problem 8.155 to be $y_c = c_1 + c_2 x + c_3 x^2 + c_4 x^3$, so we assume a particular solution of the form $y_p = v_1 + v_2 x + v_3 x^2 + v_4 x^3$. Here $y_1 = 1$, $y_2 = x$, $y_3 = x^2$, $y_4 = x^3$, and $\phi(x) = 5x$. It follows from Problem 10.1 (with $n = 4$) that

$$\begin{aligned} v_1' + v_2' x + v_3' x^2 + v_4' x^3 &= 0 \\ v_1'(0) + v_2' + v_3'(2x) + v_4'(3x^2) &= 0 \\ v_1'(0) + v_2'(0) + v_3'(2) + v_4'(6x) &= 0 \\ v_1'(0) + v_2'(0) + v_3'(0) + v_4'(6) &= 5x \end{aligned}$$

Solving this set of equations simultaneously, we obtain $v_1' = -\tfrac{5}{6}x^4$, $v_2' = \tfrac{5}{2}x^3$, $v_3' = -\tfrac{5}{2}x^2$, and $v_4' = \tfrac{5}{6}x$. Then

$$v_1 = -\tfrac{1}{6}x^5 \qquad v_2 = \tfrac{5}{8}x^4 \qquad v_3 = -\tfrac{5}{6}x^3 \qquad v_4 = \tfrac{5}{12}x^2$$

and $y_p = -\tfrac{1}{6}x^5 + \tfrac{5}{8}x^4(x) - \tfrac{5}{6}x^3(x^2) + \tfrac{5}{12}x^2(x^3) = \tfrac{1}{24}x^5$. Thus, the general solution is $y = c_1 + c_2 x + c_3 x^2 + c_4 x^3 + \tfrac{1}{24}x^5$. This solution also can be obtained simply by integrating both sides of the differential equation four times with respect to x.

.98 Solve $y^{(4)} + 8y^{(3)} + 24y'' + 32y' + 16y = 120e^{-2x}/x^2$.

▌ The complementary solution is found in Problem 8.157 to be $y = c_1 e^{-2x} + c_2 x e^{-2x} + c_3 x^2 e^{-2x} + c_4 x^3 e^{-2x}$, so we assume $y_p = v_1 e^{-2x} + v_2 x e^{-2x} + v_3 x^2 e^{-2x} + v_4 x^3 e^{-2x}$. It then follows from Problem 10.1 (with $n = 4$) that

$$\begin{aligned} v_1' e^{-2x} + v_2' x e^{-2x} + v_3' x^2 e^{-2x} + v_4' x^3 e^{-2x} &= 0 \\ v_1'(-2e^{-2x}) + v_2'(e^{-2x} - 2xe^{-2x}) + v_3'(2xe^{-2x} - 2x^2 e^{-2x}) + v_4'(3x^2 e^{-2x} - 2x^3 e^{-2x}) &= 0 \\ v_1'(4e^{-2x}) + v_2'(-4e^{-2x} + 4xe^{-2x}) + v_3'(2e^{-2x} - 8xe^{-2x} + 4x^2 e^{-2x}) & \\ + v_4'(6xe^{-2x} - 12x^2 e^{-2x} + 4x^3 e^{-2x}) &= 0 \\ v_1'(-8e^{-2x}) + v_2'(12e^{-2x} - 8xe^{-2x}) + v_3'(-12e^{-2x} + 24xe^{-2x} - 8x^2 e^{-2x}) & \\ + v_4'(6e^{-2x} - 36xe^{-2x} + 36x^2 e^{-2x} - 8x^3 e^{-2x}) &= \frac{120e^{-2x}}{x^2} \end{aligned}$$

The solution to this set of equations is $v_1' = -20x$, $v_2' = 60$, $v_3' = -60x^{-1}$, and $v_4' = 20x^{-2}$. Integration then yields $v_1 = -10x^2$, $v_2 = 60x$, $v_3 = -60 \ln|x|$, and $v_4 = -20x^{-1}$, so that

$$y_p = -10x^2 e^{-2x} + (60x)(xe^{-2x}) - 60(\ln|x|)\,x^2 e^{-2x} + (-20x^{-1})(x^3 e^{-2x}) = 30x^2 e^{-2x} - 60x^2 e^{-2x} \ln|x|$$

The general solution is then $y = [c_1 + c_2 x + (c_3 + 30)x^2 + c_4 x^3 - 60x^2 \ln|x|]e^{-2x}$.

99 Find an expression for a particular solution to $(D^4 + D^2)y = f(x)$.

▌ The complementary solution is $y_c = c_1 + c_2 x + c_3 \cos x + c_4 \sin x$, so we assume $y_p = v_1 + v_2 x + v_3 \cos x + v_4 \sin x$. It follows from Problem 10.1 with $n = 4$ and $\phi(x) = f(x)$ that

$$\begin{aligned} v_1' + v_2' x + v_3' \cos x + v_4' \sin x &= 0 \\ v_2' + v_3'(-\sin x) + v_4' \cos x &= 0 \\ v_3'(-\cos x) + v_4'(-\sin x) &= 0 \\ v_3' \sin x + v_4'(-\cos x) &= f(x) \end{aligned}$$

The solution to this set of equations is $v_1' = -xf(x)$, $v_2' = f(x)$, $v_3' = f(x)\sin x$, and $v_4' = -f(x)\cos x$, so that

$$v_1 = -\int xf(x)\,dx \qquad v_2 = \int f(x)\,dx \qquad v_3 = \int f(x)\sin x\,dx \qquad v_4 = -\int f(x)\cos x\,dx$$

and $$y_p = -\int xf(x)\,dx + x\int f(x)\,dx + \cos x\int f(x)\sin x\,dx - \sin x\int f(x)\cos x\,dx$$

10.100 Solve $(D^5 - 4D^3)y = 32e^{2x}$.

▮ The complementary function is $y_c = c_1 + c_2x + c_3x^2 + c_4e^{2x} + c_5e^{-2x}$, so we assume that $y_p = v_1 + v_2x + v_3x^2 + v_4e^{2x} + v_5e^{-2x}$. It follows from Problem 10.1 (with $n = 5$) that

$$v_1' + v_2'(x) + v_3'x^2 + v_4'e^{2x} + v_5'e^{-2x} = 0$$
$$v_1'(0) + v_2' + v_3'(2x) + v_4'(2e^{2x}) + v_5'(-2e^{-2x}) = 0$$
$$v_1'(0) + v_2'(0) + v_3'(2) + v_4'(4e^{2x}) + v_5'(4e^{-2x}) = 0$$
$$v_1'(0) + v_2'(0) + v_3'(0) + v_4'(8e^{2x}) + v_5'(-8e^{-2x}) = 0$$
$$v_1'(0) + v_2'(0) + v_3'(0) + v_4'(16e^{2x}) + v_5'(16e^{-2x}) = 32e^{2x}$$

The solution to this set of equations is

$$v_1' = -4x^2e^{2x} - 2e^{2x} \qquad v_2' = 8xe^{2x} \qquad v_3' = -4e^{2x} \qquad v_4' = 1 \qquad v_5' = e^{4x}$$

so that

$$v_1 = (-2x^2 + 2x - 2)e^{2x} \qquad v_2 = (4x - 2)e^{2x} \qquad v_3 = -2e^{2x} \qquad v_4 = x \qquad v_5 = \tfrac{1}{4}e^{4x}$$

Substituting these quantities gives, after simplification, $y_p = (x - \frac{7}{4})e^{2x}$, and so $y = y_c + y_p = c_1 + c_2x + c_3x^2 + (c_4 - \frac{7}{4})e^{2x} + c_5e^{-2x} + xe^{2x}$.

CHAPTER 11

Applications of Second-Order Linear Differential Equations

SPRING PROBLEMS

1.1 A steel ball weighing 128 lb is suspended from a spring, whereupon the spring stretches 2 ft from its natural length. The ball is started in motion with no initial velocity by displacing it 6 in above the equilibrium position. Assuming no air resistance, find the position of the ball at $t = \pi/12$ s.

▌ This is free, undamped motion. The differential equation governing the vibrations of the system is shown in Problem 1.75 to be $\ddot{x} + 16x = 0$. Its solution (see Problem 8.57) is $x = c_1 \cos 4t + c_2 \sin 4t$, from which we find $v = \dot{x} = -4c_1 \sin 4t + 4c_2 \cos 4t$.

The initial conditions for this motion are $x(0) = -\frac{1}{2}$ ft (the minus sign is required since the ball is initially displaced *above* the equilibrium position, which is the *negative* direction) and $v(0) = 0$. Applying these conditions to the equations for x and v, we obtain $-\frac{1}{2} = x(0) = c_1$ and $0 = v(0) = 4c_2$, respectively. Therefore $c_1 = -\frac{1}{2}$, $c_2 = 0$, and $x = -\frac{1}{2}\cos 4t$. At $t = \pi/12$, we have $x(\pi/12) = -\frac{1}{2}\cos(4\pi/12) = -\frac{1}{4}$ ft.

1.2 A spring for which $k = 48$ lb/ft hangs vertically with its upper end fixed. A mass weighing 16 lb is attached to the lower end. From rest, the mass is pulled down 2 in and released. Find the equation for the resulting motion of the mass, neglecting air resistance.

▌ With $m = 16/32 = 0.5$ slug and $k = 48$ lb/ft, (*1*) of Problem 1.69 becomes $\ddot{x} + 96x = 0$, which has as its solution $x = C_1 \sin\sqrt{96}t + C_2 \cos\sqrt{96}t$ (see Problem 8.61).

Differentiating with respect to t yields $v = dx/dt = \sqrt{96}\,(C_1 \cos\sqrt{96}t - C_2 \sin\sqrt{96}t)$. When $t = 0$ we have the initial conditions $x = \frac{1}{6}$ and $v = 0$. Then $C_2 = \frac{1}{6}$, $C_1 = 0$, and $x = \frac{1}{6}\cos\sqrt{96}t$.

1.3 A 20-lb weight suspended from the end of a vertical spring stretches it 6 in. Assuming no external forces and no air resistance, find the position of the weight at any time if initially the weight is pulled down 2 in from its rest position and released.

▌ From Hooke's law (see Problems 1.69 through 1.74), the spring constant k satisfies the equation $k(\frac{1}{2}) = 20$ or $k = 40$ lb/ft. With $m = 20/32$ slug, (*1*) of Problem 1.69 becomes $\ddot{x} + 64x = 0$, which has as its solution $x = c_1 \cos 8t + c_2 \sin 8t$ (see Problem 8.58).

Since $x(0) = \frac{1}{6}$ and $\dot{x}(0) = 0$, we have $c_1 = \frac{1}{6}$ and $c_2 = 0$. Thus we have $x = \frac{1}{6}\cos 8t$.

1.4 Solve the previous problem if the weight is initially pulled down 3 in and given an initial velocity of 2 ft/s downward.

▌ As in the previous problem $x = c_1 \cos 8t + c_2 \sin 8t$. Now, however, $x(0) = \frac{1}{4}$ ft and $\dot{x}(0) = 2$ ft/s. Applying these initial conditions, we find that

$$\tfrac{1}{4} = x(0) = c_1 \cos 0 + c_2 \sin 0 = c_1 \qquad \text{and} \qquad 2 = \dot{x}(0) = -8c_1 \sin 0 + 8c_2 \cos 0 = 8c_2$$

Thus $c_1 = c_2 = \frac{1}{4}$ and $x = \frac{1}{4}\cos 8t + \frac{1}{4}\sin 8t$.

1.5 Determine the motion of a mass m attached to a spring suspended from a fixed mounting if the vibrations are free and undamped.

▌ The differential equation governing such a system is found in Problem 1.70 to be $\ddot{x} + \dfrac{k}{m}x = 0$, where k is the spring constant. The roots of its characteristic equation are $\lambda_1 = \sqrt{-k/m}$ and $\lambda_2 = -\sqrt{-k/m}$, or, since both k and m are positive, $\lambda_1 = i\sqrt{k/m}$ and $\lambda_2 = -i\sqrt{k/m}$. Its solution is $x = c_1 \cos\sqrt{k/m}t + c_2 \sin\sqrt{k/m}t$.

Applying the initial conditions $x(0) = x_0$ and $v(0) = v_0$, we obtain $c_1 = x_0$ and $c_2 = v_0\sqrt{m/k}$. Thus the solution becomes

$$x = x_0 \cos\sqrt{k/m}t + v_0\sqrt{m/k}\sin\sqrt{k/m}t \tag{1}$$

1.6 Rewrite the displacement x found in the previous problem in the form $x = A\cos(\omega t - \phi)$.

Since $A\cos(\omega t - \phi) = A\cos\omega t\cos\phi + A\sin\omega t\sin\phi$, we require

$$A\cos\omega t\cos\phi + A\sin\omega t\sin\phi = x_0\cos\sqrt{k/m}t + v_0\sqrt{m/k}\sin\sqrt{k/m}t$$

For this equality to hold, we must have $\omega = \sqrt{k/m}$, $A\cos\phi = x_0$, and $A\sin\phi = v_0\sqrt{m/k}$. Now, since

$$A^2 = A^2(1) = A^2(\cos^2\phi + \sin^2\phi) = (A\cos\phi)^2 + (A\sin\phi)^2 = (x_0)^2 + (v_0\sqrt{k/m})^2$$

we have $A = \sqrt{x_0^2 + v_0^2(m/k)}$ and the phase angle ϕ is given implicitly by $\cos\phi = \dfrac{x_0}{A}$ and $\sin\phi = \dfrac{v_0\sqrt{m/k}}{A}$. To find ϕ explicitly, we write $\tan\phi = \dfrac{\sin\phi}{\cos\phi} = \dfrac{v_0\sqrt{m/k}}{x_0}$, so that $\phi = \arctan\dfrac{v_0\sqrt{m/k}}{x_0}$.

11.7 Determine the circular frequency, natural frequency, and period for the motion described in Problem 11.5.

The motion described by (*1*) of Problem 11.5 is called *simple harmonic motion*. The *circular frequency* of such motion is given by $\omega = \sqrt{k/m}$.

The *natural frequency*, or number of complete oscillations per second, is $f = \omega/2\pi = (1/2\pi)\sqrt{k/m}$.

The *period* of the motion, or the time required to complete one oscillation, is $T = 1/f = 2\pi\sqrt{m/k}$.

11.8 Determine the circular frequency, natural frequency, and period for the vibrations described in Problem 11.1.

In that problem, $k = 64$ lb/ft and $m = 4$ slugs. Using the formulas of the previous problem, we have

Circular frequency: $\omega = \sqrt{64/4} = 4$ cycles per second $= 4$ Hz
Natural frequency: $f = 4/2\pi = 2/\pi$ Hz
Period: $T = 1/(2/\pi) = \pi/2$ s

11.9 Determine the circular frequency, natural frequency, and period for the vibrations described in Problem 11.2.

In that problem $k = 48$ lb/ft and $m = 0.5$ slug. Using the formulas of Problem 11.7, we have

Circular frequency: $\omega = \sqrt{48/0.5} = 9.80$ Hz
Natural frequency: $f = 9.80/2\pi = 1.56$ Hz
Period: $T = 1/1.56 = 0.64$ s

11.10 Determine the circular frequency, natural frequency, and period for the vibrations described in Problem 11.3.

In that problem $k = 40$ lb/ft and $m = 0.625$ slug. Then $\omega = \sqrt{40/0.625} = 8$ Hz; $f = 8/2\pi = 4/\pi$ Hz; and $T = 1/(4/\pi) = \pi/4$ s.

11.11 Write the displacement found in Problem 11.4 in the form $x = A\cos(\omega t - \phi)$.

We have $k = 40$ lb/ft, $m = 0.625$ slug, $x_0 = \frac{1}{4}$ ft, $v_0 = 2$ ft/s, and $\omega = 8$ Hz (from Problem 11.10). Substituting these values into the formulas derived in Problem 11.6 yields $A = \sqrt{(\frac{1}{4})^2 + (2)^2(0.625/40)} = 0.35$ ft and $\phi = \arctan\dfrac{2\sqrt{0.625/40}}{1/4} = \arctan 1 = \pi/4$. Thus $x = 0.35\cos(8t - \pi/4)$.

11.12 Write the displacement found in Problem 11.4 in the form $x = A\sin(\omega t + \phi)$.

Since $A\sin(\omega t + \phi) = A\sin\omega t\cos\phi + A\sin\phi\cos\omega t$ and the displacement is $x = \frac{1}{4}\cos 8t + \frac{1}{4}\sin 8t$, we must have $\omega = 8$, $A\cos\phi = \frac{1}{4}$, and $A\sin\phi = \frac{1}{4}$. Squaring and adding give

$$(\tfrac{1}{4})^2 + (\tfrac{1}{4})^2 = A^2\cos^2\phi + A^2\sin^2\phi = A^2 \qquad \text{or} \qquad A = \sqrt{\tfrac{1}{8}} = 0.35$$

Also, because $\dfrac{A\sin\phi}{A\cos\phi} = \tan\phi = \dfrac{1/4}{1/4} = 1$, we have $\phi = \arctan 1 = \pi/4$. Thus $x = 0.35\sin(8t + \pi/4)$.

11.13 A 20-g mass suspended from the end of a vertical spring stretches the spring 4 cm from its natural length. Assuming no external forces on the mass and no air resistance, find the position of the mass at any time t if it is pulled 1 cm below its equilibrium position and set into motion with an initial velocity of 0.5 cm/s in the upward direction.

▌ By Hooke's law, the spring constant k satisfies the equation $20(980) = k(4)$, so that $k = 4900$ dynes/cm. In addition, $m = 20\ g$, $x_0 = 1$ cm, and $v_0 = -0.5$ cm/s (the minus sign is required because the initial velocity is in the upward or negative direction). Then from Problem 1.70 we have $\ddot{x} + 245x = 0$. The position of the mass at any time t is given by (*1*) of Problem 11.5 as $x = \cos\sqrt{245}t - 0.5\sqrt{20/4900}\sin\sqrt{245}t = \cos\sqrt{245}t - 0.03194\sin\sqrt{245}t$.

11.14 Determine the circular frequency, natural frequency, and period for the vibrations described in the previous problem.

▌ Using the formulas of Problem 11.7, we have $\omega = \sqrt{4900/20} = 15.65$ Hz; $f = 15.65/2\pi = 2.49$ Hz; and $T = 1/2.49 = 0.40$ s.

11.15 A 10-kg mass attached to a spring stretches it 0.7 m from its natural length. Assuming no external forces on the mass and no air resistance, find the position of the mass as a function of time if it is pushed up 0.05 m from its equilibrium position and set into motion with an initial velocity of 0.1 m/s in the upward direction.

▌ It follows from Problem 1.74 that $k = 140$ N/m. Then with $m = 10$ kg, $x_0 = -0.05$ m, and $v_0 = -0.1$ m/s, the motion of the mass is given by (*1*) of Problem 11.5 as

$$x = -0.05\cos\sqrt{140/10}t - 0.1\sqrt{10/140}\sin\sqrt{140/10}t = -0.05\cos\sqrt{14}t - 0.0267\sin\sqrt{14}t$$

11.16 Find the amplitude and period of the vibrations described in the previous problem.

▌ It follows from Problem 11.6 that the amplitude is $A = \sqrt{x_0^2 + v_0^2(m/k)} = \sqrt{(-0.05)^2 + (-0.1)^2(10/140)} = 0.0567$ m.
It follows from Problem 11.7 that the period is $T = 2\pi\sqrt{m/k} = 2\pi\sqrt{10/140} = 1.68$ s.

11.17 A 10-kg mass attached to a spring stretches it 0.7 m from its natural length. The mass is started in motion from the equilibrium position with an initial velocity of 1 m/s in the upward direction. Find the subsequent motion, if the force due to air resistance is $-90\dot{x}$ N.

▌ The differential equation governing the vibrations of this system is given in Problem 1.78 as $\ddot{x} + 9\dot{x} + 14x = 0$, and we have the initial conditions $x(0) = 0$ and $\dot{x}(0) = -1$. The solution to the differential equation is found in Problem 8.17 to be $x(t) = c_1e^{-2t} + c_2e^{-7t}$, and differentiation yields $\dot{x}(t) = -2c_1e^{-2t} - 7c_2e^{-7t}$.
Applying the initial conditions to these last two equations, we get

$$0 = x(0) = c_1 + c_2 \qquad \text{and} \qquad -1 = \dot{x}(0) = -2c_1 - 7c_2$$

Solving this set of equations simultaneously yields $c_1 = -\frac{1}{5}$ and $c_2 = \frac{1}{5}$, so that $x(t) = \frac{1}{5}(e^{-7t} - e^{-2t})$.

11.18 Classify the motion described in the previous problem.

▌ The vibrations are free and damped. The roots of the characteristic equation are real (see Problem 8.17), so the system is overdamped. Since $x \to 0$ as $t \to \infty$, the motion is transient.

11.19 A mass of 1/4 slug is attached to a spring, whereupon the spring is stretched 1.28 ft from its natural length. The mass is started in motion from the equilibrium position with an initial velocity of 4 ft/s in the downward direction. Find the subsequent motion of the mass if the force due to air resistance is $-2\dot{x}$ lb.

▌ The differential equation governing the vibrations of this system is given in Problem 1.77 as $\ddot{x} + 8\dot{x} + 25x = 0$, and we have the initial conditions $x(0) = 0$ and $\dot{x}(0) = 4$. The solution to the differential equation is given in Problem 8.53 as $x(t) = c_1e^{-4t}\cos 3t + c_2e^{-4t}\sin 3t$.
Differentiation of $x(t)$ yields $\dot{x}(t) = c_1(-4e^{-4t}\cos 3t - 3e^{-4t}\sin 3t) + c_2(-4e^{-4t}\sin 3t + 3e^{-4t}\cos 3t)$.
Applying the initial conditions, we obtain

$$0 = x(0) = c_1 \qquad \text{and} \qquad 4 = \dot{x}(0) = -4c_1 + 3c_2$$

from which we find $c_1 = 0$ and $c_2 = \frac{4}{3}$. Then $x(t) = \frac{4}{3}e^{-4t}\sin 3t$.

11.20 Classify the motion described in the previous problem.

▌ The vibrations are free and damped. Since the roots of the characteristic equation are complex conjugates (see Problem 8.53), the system is underdamped. Furthermore, $x \to 0$ as $t \to \infty$, so the motion is all transient.

11.21 A mass of $\frac{1}{4}$ slug is attached to a spring having a spring constant of 1 lb/ft. The mass is started in motion by displacing it 2 ft in the downward direction and giving it an initial velocity of 2 ft/s in the upward direction. Find the subsequent motion of the mass if the force due to air resistance is $-1\dot{x}$ lb.

▌ Here $m = \frac{1}{4}$, $a = 1$, $k = 1$, and the external force $F(t) = 0$, so (*1*) of Problem 1.69 becomes $\ddot{x} + 4\dot{x} + 4x = 0$. The solution to this differential equation is $x(t) = c_1 e^{-2t} + c_2 te^{-2t}$ [see Problem 8.141 with $x(t)$ replacing $y(x)$].

Differentiation yields $\dot{x}(t) = -2c_1 e^{-2t} + c_2(e^{-2t} - 2te^{-2t})$. Application of the initial conditions to the last two equations gives

$$2 = x(0) = c_1 \qquad \text{and} \qquad -2 = \dot{x}(0) = 2c_1 + c_2$$

Thus, $c_1 = 2$ and $c_2 = 2$, so that $x(t) = 2(1 + t)e^{-2t}$.

11.22 Classify the motion described in the previous problem.

▌ The vibrations are free and damped. Since the roots of the characteristic equation are real and equal (see Problem 8.141), the system is critically damped. Furthermore, $x \to 0$ as $t \to \infty$, so the motion is all transient.

11.23 Show that free damped motion is completely determined by the quantity $a^2 - 4\,km$, where a is the constant of proportionality for the air resistance (which is assumed proportional to the velocity of the mass), k is the spring constant, and m is the mass.

▌ For free damped motion, $F(t) \equiv 0$ and (*1*) of Problem 1.69 becomes $\ddot{x} + \dfrac{a}{m}\dot{x} + \dfrac{k}{m}x = 0$. The roots of the associated characteristic equation are then $\lambda_1 = \dfrac{-a + \sqrt{a^2 - 4km}}{2m}$ and $\lambda_2 = \dfrac{-a - \sqrt{a^2 - 4km}}{2m}$.

If $a^2 - 4km > 0$, the roots are real and distinct; if $a^2 - 4km = 0$, the roots are equal; if $a^2 - 4km < 0$, the roots are complex conjugates. The corresponding motions are, respectively, overdamped, critically damped, and oscillatory damped. Since the real parts of both roots are always negative, the resulting motion in all three cases is transient. (That the real parts are always negative follows from the fact that for overdamped motion we must have $\sqrt{a^2 - 4km} < a$, whereas in the other two cases the real parts are both $-a/2m$.)

11.24 A 20-lb weight suspended from the end of a vertical spring stretches the spring 6 in from its natural length. Assume that the only external force is a damping force given in pounds by av, where v is the instantaneous velocity in feet per second. Find an expression for the vibrations of the system if $a = 8$ slugs/s and the mass is set into motion by displacing it 2 in below its equilibrium position.

▌ With $mg = 20$, it follows that $m = 20/32 = 0.625$ slug; then from Hooke's law $20 = k(\frac{1}{2})$ or $k = 40$ lb/ft. With $a = 8$ and the external force $F(t) \equiv 0$, (*1*) of Problem 1.69 becomes $\ddot{x} + 12.8\dot{x} + 64x = 0$. Its solution is found in Problem 8.62 to be $x(t) = c_1 e^{-6.4t}\cos 4.8t + c_2 e^{-6.4t}\sin 4.8t$.

Applying the initial conditions $x(0) = \frac{1}{6}$ ft and $\dot{x}(0) = v(0) = 0$, we find that $c_1 = \frac{3}{18}$ and $c_2 = \frac{4}{18}$. Thus, $x(t) = \frac{1}{18}e^{-6.4t}(3\cos 4.8t + 4\sin 4.8t)$.

11.25 Classify the motion described in the previous problem.

▌ The vibrations are free and damped. Since the roots of the characteristic equation are complex conjugates (see Problem 8.62), the system is underdamped.

11.26 Solve Problem 11.24 if $a = 12.5$.

▌ Here $a/m = 12.5/0.625 = 20$, so (*1*) of Problem 1.69 becomes $\ddot{x} + 20\dot{x} + 64x = 0$. Its solution is found in Problem 8.11 to be $x(t) = c_1 e^{-4t} + c_2 e^{-16t}$.

Applying the initial conditions $x(0) = \frac{1}{6}$ and $\dot{x}(0) = 0$, we find $c_1 = \frac{4}{18}$ and $c_2 = -\frac{1}{18}$. Thus, $x(t) = \frac{1}{18}(4e^{-4t} - e^{-16t})$.

11.27 Classify the motion described in the previous problem.

▌ The vibrations are free and damped. Since the roots of the characteristic equation are real and unequal (see Problem 8.11), the system is overdamped.

11.28 Solve Problem 11.24 if $a = 10$.

▌ Here $a/m = 10/0.625 = 16$, so (1) of Problem 1.69 becomes $\ddot{x} + 16\dot{x} + 64x = 0$. Its solution is found in Problem 8.148 to be $x(t) = c_1e^{-8t} + c_2te^{-8t}$.

Applying the initial conditions $x(0) = \frac{1}{6}$ and $\dot{x}(0) = 0$, we find that $c_1 = \frac{1}{6}$ and $c_2 = \frac{8}{6}$. Thus, $x(t) = \frac{1}{6}(1 + 8t)e^{-8t}$.

11.29 Classify the motion described in the previous problem.

▌ The vibrations are free and damped. Since the roots of the characteristic equation are real and equal (see Problem 8.148), the system is critically damped: a smaller value of a would result in an underdamped system; a larger value, in an overdamped system.

11.30 Solve Problem 11.2 if the system is surrounded by a medium offering a resistance in pounds equal to $v/64$, where the velocity v is measured in feet per second.

▌ As in Problem 11.2, $m = 0.5$ slug, $k = 48$ lb/ft, $x(0) = \frac{1}{6}$, and $\dot{x}(0) = 0$. In addition, we now have $a = \frac{1}{64}$ slug/s, so (1) of Problem 1.69 becomes $\ddot{x} + \frac{1}{32}\dot{x} + 96x = 0$. Its solution is found in Problem 8.63 to be

$$x(t) = C_1e^{-0.015625t}\cos 9.7979t + C_2e^{-0.015625t}\sin 9.7979t.$$

Differentiating yields

$$\dot{x}(t) = e^{-0.015625t}[(9.7979C_2 - 0.015625C_1)\cos 9.7979t - (9.7979C_1 + 0.015625C_2)\sin 9.7979t]$$

Applying the initial conditions, we obtain

$$\tfrac{1}{6} = x(0) = C_1 \qquad \text{and} \qquad 0 = \dot{x}(0) = -0.015625C_1 + 9.7979C_2$$

from which we find $C_1 = \frac{1}{6}$ and $C_2 = 0.0002657$. Then $x(t) = e^{-0.015625t}(\frac{1}{6}\cos 9.7979t + 0.0002657\sin 9.7979t)$.

11.31 Find the amplitude and frequency of the motion of the previous problem.

▌ The natural frequency is $f = 9.7979/2\pi = 1.56$ Hz; it remains constant throughout the motion. In contrast, the amplitude is $A = \sqrt{(\frac{1}{6}e^{-0.015625t})^2 + (0.0002657e^{-0.015625t})^2} = 0.167e^{-0.015625t}$. Thus the amplitude decreases with each oscillation, owing to the effect of the damping factor $e^{-0.015625t}$.

At $t = 0$, there is no damping, and the amplitude is at its maximum of 0.167. The amplitude reaches two-thirds of its maximum when $e^{-0.015625t} = \frac{2}{3}$, or when $t = 26$ s. It is one-third of its maximum when $e^{-0.015625t} = \frac{1}{3}$, or when $t = 70$ s.

1.32 Find the amplitude and frequency of the motion described in Problem 11.24.

▌ The natural frequency is $f = 4.8/2\pi = 0.76$ Hz. The amplitude is $A = e^{-6.4t}\sqrt{(\frac{3}{18})^2 + (\frac{4}{18})^2} = \frac{5}{18}e^{-6.4t}$.

1.33 Solve Problem 11.2 if the system is surrounded by a medium offering a resistance in pounds equal to $64v$, where the velocity v is measured in feet per second.

▌ With $m = 0.5$, $k = 48$, $F(t) \equiv 0$, $x(0) = \frac{1}{6}$, $\dot{x}(0) = 0$, and now $a = 64$, we have, from (1) of Problem 1.69, $\ddot{x} + 128\dot{x} + 96x = 0$. Its solution is given in Problem 8.12 as $x = C_1e^{-0.7544t} + C_2e^{-127.2t}$. Differentiating once with respect to t yields $v = -0.7544C_1e^{-0.7544t} - 127.2C_2e^{-127.2t}$.

When $t = 0$, we have $x = \frac{1}{6}$ and $v = 0$. Thus

$$C_1 + C_2 = \tfrac{1}{6} \qquad \text{and} \qquad -0.7544C_1 - 127.2C_2 = 0$$

so that $C_1 = 0.1677$ and $C_2 = -0.0001$. Then $x = 0.1677e^{-0.7544t} - 0.0001e^{-127.2t}$.

1.34 Describe the motion of the system of the previous problem.

▌ The motion is not vibratory but overdamped. After the initial displacement, the mass moves slowly toward the position of equilibrium as t increases. The motion is completely transient.

1.35 Assume the system described in Problem 11.3 is surrounded by a medium that offers a resistance in pounds equal to ax, where a is a constant. Determine the value of a that generates critically damped motion.

▌ Here $m = 0.625$ and $k = 40$. It follows from Problem 11.23 that critically damped motion will occur when $0 = a^2 - 4km = a^2 - 4(40)(0.625)$, or when $a = 10$ slugs/s.

11.36 Assume the system described in Problem 11.1 is surrounded by a medium that offers a resistance in pounds equa to ax, where a is a constant. Determine the value of a that generates critically damped motion.

▌ Here $m = 128/32 = 4$ slugs, and $k = 64$ lb/ft (see Problem 1.71). It follows from Problem 11.23 that critically damped motion will occur when $0 = a^2 - 4km = a^2 - 4(64)(4)$ or when $a = 32$ slugs/s.

11.37 Assume the system described in Problem 11.13 is surrounded by a medium that offers a resistance in pounds equal to ax, where a is a constant. Determine the value of a that generates critically damped motion.

▌ Here $m = 20$ g and $k = 4900$ dynes/cm. It follows from Problem 11.23 that critically damped motion will occur when $0 = a^2 - 4km = a^2 - 4(4900)(20)$ or when $a = 280\sqrt{5}$ g/s.

11.38 A 10-kg mass is attached to a spring having a spring constant of 140 N/m. The mass is started in motion from the equilibrium position with an initial velocity of 1 m/s in the upward direction and with an applied external for $F(t) = 5 \sin t$. Find the subsequent motion of the mass if the force due to air resistance is $-90\dot{x}$ N.

▌ Here $m = 10$, $k = 140$, $a = 90$, and $F(t) = 5 \sin t$. The equation of motion, (*1*) of Problem 1.69, becomes $\ddot{x} + 9\dot{x} + 14x = \frac{1}{2}\sin t$. Its solution is found in Problem 9.89 (with q replaced by x) to be $x(t) = c_1e^{-2t} + c_2e^{-7t} + \frac{13}{500}\sin t - \frac{9}{500}\cos t$.

Applying the initial conditions $x(0) = 0$ and $\dot{x}(0) = -1$, we obtain $x = \frac{1}{500}(-90e^{-2t} + 99e^{-7t} + 13\sin t - 9\cos t)$.

11.39 Identify the transient and steady-state portions of the motion of the previous problem.

▌ The exponential terms that comprise the homogeneous (or complementary) solution represent an associated free overdamped motion. These terms quickly die out, and they represent the transient part of the motion. The terms that are part of the particular solution (see Problem 9.89) do not die out as $t \to \infty$, so they comprise th steady-state portion of the motion. Observe that the steady-state portion has the same frequency as the forcing function.

11.40 A 1-slug mass is attached to a spring having a spring constant of 8 lb/ft. The mass is set into motion from the equilibrium position with no initial velocity by applying an external force $F(t) = 16\cos 4t$. Find the subseque motion of the mass, if the force due to air resistance is $-4\dot{x}$ lb.

▌ Here $m = 1$ slug, $k = 8$ lb/ft, $a = 4$ slugs/s, and $F(t) = 16\cos 4t$. Then (*1*) of Problem 1.69 becomes $\ddot{x} + 4\dot{x} + 8x = 16\cos 4t$. Its solution is given in Problem 9.83 as $x(t) = c_1e^{-2t}\cos 2t + c_2e^{-2t}\sin 2t + \frac{4}{5}\sin 4t - \frac{2}{5}\cos 4t$, and differentiation yields

$$\dot{x}(t) = (-2c_1 + 2c_2)e^{-2t}\cos 2t + (-2c_1 - 2c_2)e^{-2t}\sin 2t + \tfrac{16}{5}\cos 4t + \tfrac{8}{5}\sin 4t$$

Applying the initial conditions, we obtain

$$0 = x(0) = c_1 - \tfrac{2}{5} \qquad \text{and} \qquad 0 = \dot{x}(0) = -2c_1 + 2c_2 + \tfrac{16}{5}$$

so that $c_1 = \frac{2}{5}$ and $c_2 = -\frac{6}{5}$. Then $x = e^{-2t}(\frac{2}{5}\cos 2t - \frac{6}{5}\sin 2t) + \frac{4}{5}\sin 4t - \frac{2}{5}\cos 4t$.

11.41 Describe the motion of the system of the previous problem.

▌ The motion consists of overdamped transient vibrations which are due to the homogeneous (or complementary) function, namely $e^{-2t}(\frac{2}{5}\cos 2t - \frac{6}{5}\sin 2t)$, along with a harmonic component that does not tend to zero as $t \to \infty$. The latter, namely $\frac{4}{5}\sin 4t - \frac{2}{5}\cos 4t$, is the steady-state part of the solution. The steady-state oscillations have a period and frequency equal to those of the forcing function, $F(t) = 16\cos 4t$. The natural frequency is $f = 4/2\pi = 0.637$ Hz, while the amplitude of the steady-state vibrations is $A = \sqrt{(4/5)^2 + (-2/5)^2} = \sqrt{20}/5$.

11.42 Derive the differential equation governing the motion of the mass in Problem 11.30 if, in addition, the "fixed" e of the spring undergoes a motion $y = \cos 4t$ ft.

▌ Take the origin as the equilibrium position of the spring with the mounting fixed $(y = 0)$, and let x den the distance of the mass from the origin (see Fig. 11.1). The restoring force on the spring is $-k(x - y) = -48(x - \cos 4t)$.

The force due to air resistance is $-\frac{1}{64}\dot{x}$ lb, so by Newton's second law of motion we have $-48(x - \cos 4t) - \frac{1}{64}\dot{x} = m\ddot{x}$. Since $m = 0.5$ slug, this equation may be written as $\ddot{x} + \frac{1}{32}\dot{x} + 96x = 96\cos$ Note that at $t = 0$, $x(0) = y(0) + \frac{1}{6} = 1 + \frac{1}{6} = \frac{7}{6}$, while $\dot{x}(0) = 0$.

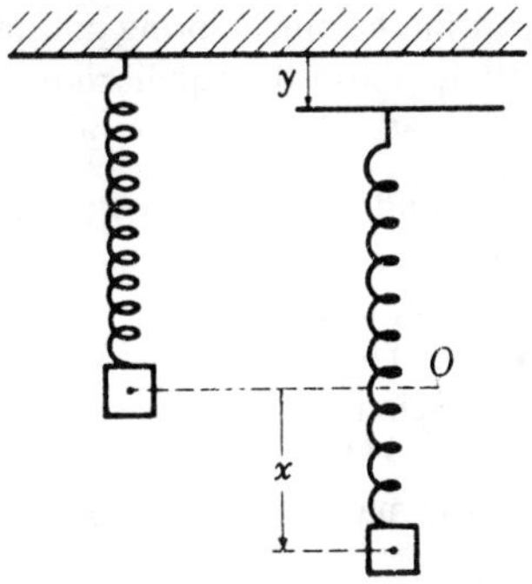

Fig. 11.1

11.43 Find an expression for the motion of the mass in the previous problem.

The solution to the differential equation of the previous problem is found in Problem 9.84 to be $x = e^{-0.0156t}(C_1 \cos 9.8t + C_2 \sin 9.8t) + 0.0019 \sin 4t + 1.2 \cos 4t$. Differentiating once with respect to t yields

$$v = e^{-0.0156t}[(9.8C_2 - 0.0156C_1)\cos 9.8t - (9.8C_1 + 0.0156C_2)\sin 9.8t] + 0.0076 \cos 4t - 4.8 \sin 4t$$

With $v(0) = 0$ and $x(0) = 7/6$, we find that $C_1 = -1/30$, $C_2 = -0.0008$, and $x = e^{-0.0156t}(-0.0333 \cos 9.8t - 0.0008 \sin 9.8t) + 0.0019 \sin 4t + 1.2 \cos 4t$.

11.44 Describe the motion of the system of the previous problem.

The motion consists of a damped harmonic component which gradually dies away (a transient component) and a harmonic (steady-state) component which remains. The steady-state oscillations have a period and a frequency equal to those of the forcing function $y = \cos 4t$, namely, a period of $2\pi/4 = 1.57$ s and a frequency of $4/2\pi = 0.637$ Hz. The steady-state amplitude is $\sqrt{(0.0019)^2 + (1.2)^2} = 1.2$ ft.

11.45 A mass of 20 lb is suspended from a spring which is thereby stretched 3 in. The upper end of the spring is then given a motion $y = 4(\sin 2t + \cos 2t)$ ft. Find the equation of the motion, neglecting air resistance.

Take the origin at the center of the mass when it is at rest. Let x represent the change in position of the mass at time t. The change in the length of the spring is $x - y$, the spring constant is $20/0.25 = 80$ lb/ft, and the net spring force is $-80(x - y)$. Thus $d^2x/dt^2 + 128x = 512(\sin 2t + \cos 2t)$.

Assuming that the spring starts from rest without any additional displacement, we have the initial conditions $x(0) = 0 + 4(\sin 0 + \cos 0) = 4$ and $\dot{x}(0) = 0$.

11.46 Find an expression for the motion of the mass in the previous problem.

The solution to the differential equation of the previous problem is found in Problem 9.86 to be $x = C_1 \cos \sqrt{128}t + C_2 \sin \sqrt{128}t + 4.129(\sin 2t + \cos 2t)$. Differentiating once with respect to t yields $v = -\sqrt{128}C_1 \sin \sqrt{128}t + \sqrt{128}C_2 \cos \sqrt{128}t + 8.258(-\sin 2t + \cos 2t)$.

Since $x(0) = 4$ and $v(0) = 0$, we have

$$4 = C_1 + 4.129 \qquad \text{and} \qquad \sqrt{128}C_2 + 8.258 = 0$$

from which $C_1 = 0.129$ and $C_2 = 0.730$. Then $x = -0.13 \cos \sqrt{128}t - 0.73 \sin \sqrt{128}t + 4.13(\sin 2t + \cos 2t)$.

11.47 A mass of 64 lb is attached to a spring for which $k = 50$ lb/ft and brought to rest. Find the position of the mass at time t if a force equal to $4 \sin 2t$ is applied to it.

Take the origin at the center of the mass when it is at rest. The equation of motion is then $\frac{64}{32}\frac{d^2x}{dt^2} + 50x = 4 \sin 2t$ or $\frac{d^2x}{dt^2} + 25x = 2 \sin 2t$. Its solution is found in Problem 9.84 to be $x = C_1 \cos 5t + C_2 \sin 5t + \frac{2}{21} \sin 2t$.

Differentiating once with respect to t yields $v = -5C_1 \sin 5t + 5C_2 \cos 5t + \frac{4}{21} \cos 2t$. Then, from the initial conditions $x = 0$ and $v = 0$ when $t = 0$, we find $C_1 = 0$, $C_2 = -\frac{4}{105}$, and $x = -0.038 \sin 5t + 0.095 \sin 2t$. The displacement here is the algebraic sum of two harmonic diplacements with *different* periods.

11.48 A 128-lb weight is attached to a spring having a spring constant of 64 lb/ft. The weight is started in motion with no initial velocity of displacing it 6 in above the equilibrium position and by simultaneously applying to the weight an external force $F(t) = 8 \sin 4t$. Assuming no air resistance, find the equation of motion of the weight.

▮ Here $m = 4$, $k = 64$, $a = 0$, and $F(t) = 8 \sin 4t$; hence, (*1*) of Problem 1.69 becomes $\ddot{x} + 16x = 2 \sin 4t$. Its solution is found in Problem 9.173 to be $x = c_1 \cos 4t + c_2 \sin 4t - \frac{1}{4}t \cos 4t$. Applying the initial conditions $x(0) = -\frac{1}{2}$ and $\dot{x}(0) = 0$, we obtain, finally, $x = -\frac{1}{2}\cos 4t + \frac{1}{16}\sin 4t - \frac{1}{4}t \cos 4t$.

Note that $|x| \to \infty$ as $t \to \infty$. This phenomenon is called *pure resonance*. It is due to the forcing function $F(t)$ having the same frequency as the circular frequency (see Problem 11.7) of the associated free undamped system.

11.49 Solve Problem 11.3 if, in addition, the mass is subjected to an externally applied force $F(t) = 40 \cos 8t$.

▮ With $m = 0.625$, $k = 40$, and $a = 0$, (*1*) of Problem 1.69 becomes $\ddot{x} + 64x = 64 \cos 8t$. Its solution is found in Problem 9.175 to be $x(t) = c_1 \cos 8t + c_2 \sin 8t + 4t \sin 8t$.

Applying the initial conditions $x(0) = \frac{1}{6}$ ft and $\dot{x}(0) = 0$, we find that $c_1 = \frac{1}{6}$ and $c_2 = 0$. Then $x(t) = \frac{1}{6}\cos 8t + 4t \sin 8t$.

11.50 Describe (physically) the motion of the previous problem as t increases.

▮ As t increases, the term $4t \sin 8t$ increases numerically without bound so that the amplitude of the motion increases without bound. The spring will ultimately break. This illustrates the phenomenon of *resonance* and shows what can happen when the frequency of the applied force is equal to the natural frequency of the system.

11.51 A mass of 16 lb is attached to a spring for which $k = 48$ lb/ft. Find the motion of the mass if, from rest, the support of the spring is given a motion $y = \sin\sqrt{3g}t$ ft.

▮ We take the origin at the center of the mass when it is at rest, and let x represent the change in position of the mass at time t. The stretch in the spring is then $x - y$, and the spring force is $-48(x - y)$. Thus, the equation of motion is $\frac{16}{g}\frac{d^2x}{dt^2} = -48(x - \sin\sqrt{3g}t)$ or $\frac{d^2x}{dt^2} + 3gx = 3g \sin\sqrt{3g}t$. Its solution is found in Problem 9.176 to be $x = C_1 \cos\sqrt{3g}t + C_2 \sin\sqrt{3g}t - \frac{1}{2}\sqrt{3g}t \cos\sqrt{3g}t$.

Differentiation gives $v = -C_1\sqrt{3g}\sin\sqrt{3g}t + C_2\sqrt{3g}\cos\sqrt{3g}t - \frac{1}{2}\sqrt{3g}\cos\sqrt{3g}t + \frac{3g}{2}t \sin\sqrt{3g}t$. Then, using the initial conditions $x = 0$ and $v = 0$ when $t = 0$, we find that $C_1 = 0$ and $C_2 = \frac{1}{2}$; thus, $x = \frac{1}{2}\sin\sqrt{3g}t - \frac{1}{2}\sqrt{3g}t \cos\sqrt{3g}t$.

The first term of this solution represents a simple harmonic motion, while the second represents a vibratory motion with increasing amplitude (resonance). As t increases, the amplitude of the oscillation increases until there is a mechanical breakdown.

MECHANICS PROBLEMS

11.52 A particle P of mass 2 g moves on the x axis toward the origin O acted upon by a force numerically equal to $8x$. Determine the differential equation governing the motion of the particle.

▮ Choose the positive direction to the right (Fig. 11.2). When $x > 0$, the net force is to the left (i.e., negative) and so is $-8x$. When $x < 0$, the net force is to the right (i.e., positive) and so is also $-8x$. Thus by Newton's law, $2\frac{d^2x}{dt^2} = -8x$ or $\frac{d^2x}{dt^2} + 4x = 0$.

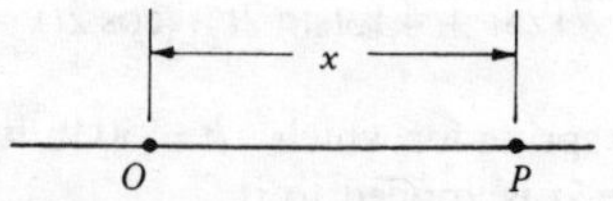

Fig. 11.2

11.53 Find an expression for the position of particle P of Problem 11.52 as a function of time if the particle is initially at rest at $x = 10$ cm.

▮ The solution to the differential equation of the previous problem is found in Problem 8.59 [with $x(t)$ here replacing $y(x)$] to be $x = c_1 \cos 2t + c_2 \sin 2t$. Then $v = \frac{dx}{dt} = -2c_1 \sin 2t + 2c_2 \cos 2t$.

Applying the initial conditions, we get $10 = x(0) = c_1$ and $0 = v(0) = 2c_2$ or $c_2 = 0$. Then the solution becomes $x = 10 \cos 2t$.

11.54 Describe the motion of the particle of the previous problem.

The graph of the motion is shown in Fig. 11.3. It is simple harmonic motion with an amplitude of 10 cm, a period of π s, and a natural frequency of $1/\pi$ Hz. The particle starts out at $x = 10$ cm at time zero and begins moving toward the origin, picking up speed as it moves. Its velocity is greatest in absolute value (i.e., its speed is greatest) at time $t = \pi/4$, when the particle reaches the origin. The velocity is negative at that time, so the particle continues through the origin; it begins slowing as its acceleration changes sign. The velocity reaches zero at time $t = \pi/2$, when $x = -10$. (Thus, at time $t = \pi/2$, the particle is at rest 10 cm to the left of the origin.) The particle then begins picking up velocity as it is accelerated toward the origin. It reaches the origin $(x = 0)$ at $t = 3\pi/4$ with maximum speed. Once it is through the origin, its velocity decreases until it again comes to rest at time $t = \pi$, now at $x = 10$. This completes one cycle of the motion; the particle will continue to repeat that cycle in the absence of other forces.

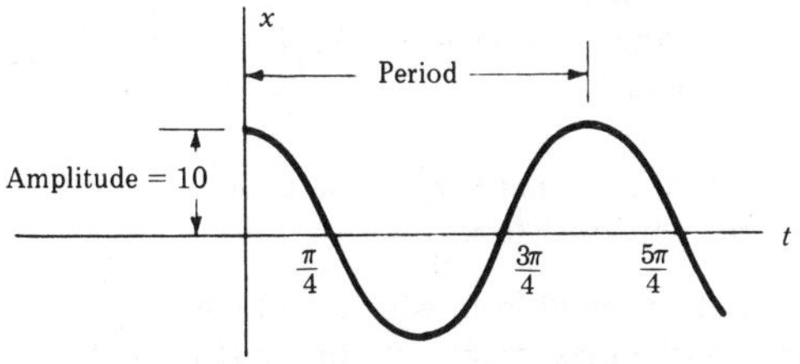

Fig. 11.3

11.55 Solve Problem 11.53 if the particle is also subject to a damping force (or resistance) that is numerically equal to eight times the instantaneous velocity.

The damping force is given by $-8\,dx/dt$, regardless of where the particle is. Thus, for example, if $x < 0$ and $dx/dt > 0$, then the particle is to the left of O and moving to the right, so the damping force must be acting to the left (i.e., negative). Thus by Newton's law, $2\dfrac{d^2x}{dt^2} = -8x - 8\dfrac{dx}{dt}$ or $\dfrac{d^2x}{dt^2} + 4\dfrac{dx}{dt} + 4x = 0$.

The solution to this equation is found in Problem 8.141 [with $x(t)$ herc replacing $y(x)$] to be $x = e^{-2t}(c_1 + c_2 t)$. Since $x = 10$ and $dx/dt = 0$ when $t = 0$, we have $c_1 = 10$ and $c_2 = 20$. Then $x = 10e^{-2t}(1 + 2t)$.

11.56 Describe the motion of the particle in the previous problem.

For all $t \geq 0$, $x = 10e^{-2t}(1 + 2t)$ is positive; furthermore, x tends to zero as $t \to \infty$. Thus particle P approaches the origin but never reaches it. In addition, the velocity of the particle, $v = dx/dt = -40te^{-2t}$ is negative for all $t \geq 0$, indicating that the particle is always heading in the same direction (the negative x direction); thus the motion is nonoscillatory.

11.57 A particle of mass m is repelled from the origin O along a straight line with a force equal to $k > 0$ times the distance from O. Determine the differential equation governing the motion of the mass.

Denote the line on which the particle moves as the x axis, taking the positive direction to be to the right of the origin. When $x > 0$, the net repellent force is to the right (i.e., positive) and is kx. When $x < 0$, the net repellent force is to the left (i.e., negative) and is also kx. Since the repellent force is kx in both cases, by Newton's second law of motion we have $m\dfrac{d^2x}{dt^2} = kx$ or $\dfrac{d^2x}{dt^2} - \dfrac{k}{m}x = 0$.

11.58 Find an expression for the position of the particle of the previous problem if it starts from rest at some initial position x_0.

The solution to the differential equation of the previous problem is found in Problem 8.22 to be $x(t) = c_1 e^{\sqrt{k/m}t} + c_2 e^{-\sqrt{k/m}t}$. Differentiating this equation yields

$$x'(t) = c_1\sqrt{k/m}\,e^{\sqrt{k/m}t} - c_2\sqrt{k/m}\,e^{-\sqrt{k/m}t}.$$

Applying the initial conditions, we obtain

$$x_0 = x(0) = c_1 + c_2 \qquad \text{and} \qquad 0 = x'(0) = c_1\sqrt{k/m} - c_2\sqrt{k/m}$$

Solving these two equations simultaneously, we find that $c_1 = c_2 = \frac{1}{2}x_0$, so that $x(t) = \frac{1}{2}x_0(e^{\sqrt{k/m}t} + e^{-\sqrt{k/m}t}) = x_0 \cosh \sqrt{k/m}\,t$.

11.59 Find the position and velocity of the mass described in Problem 11.58 after 2 s, if numerically $k = m$ and initially the particle starts from rest 12 ft to the right of the origin.

▌ With $k = m$ and $x_0 = 12$, the result of the previous problem becomes $x(t) = 6(e^t + e^{-t})$. Therefore, the velocity is $v(t) = dx/dt = 6(e^t - e^{-t})$. At $t = 2$, these equations become $x(2) = 6(e^2 + e^{-2}) = 45.15$ ft and $v(2) = 6(e^2 - e^{-2}) = 43.5$ ft/s.

11.60 Determine when the particle described in the previous problem will be 18 ft from the origin, and find its velocity at that time.

▌ Setting $x(t) = 6(e^t + e^{-t}) = 18$ and solving for t, we obtain $e^t + e^{-t} - 3 = 0$. Multiplying this last equation by e^t and rearranging then yield $(e^t)^2 - 3e^t + 1 = 0$, which may be solved for e^t with the quadratic formula. Thus, we find $e^t = \dfrac{3 \pm \sqrt{9-4}}{2}$ and $t = \ln \dfrac{3 \pm \sqrt{5}}{2}$. Since time is positive in this problem, we discard the negative choice and take $t = 0.962$ s. At that time, the velocity is $v(0.962) = 6(e^{0.962} - e^{-0.986}) = 13.4$ ft/s.

11.61 Determine the equation of motion for a mass m that is projected vertically upward from a point on the ground if the resistance of the air is proportional to its velocity.

▌ We designate the point on the ground from which the flight began as the origin O. We take upward as positive, and let x denote the distance of the mass from O at time t (Fig. 11.4). The mass is acted upon by two forces, a gravitational force of magnitude mg and a resistance of magnitude $Kv = K\dfrac{dx}{dt}$, both directed downward. Hence, $m\dfrac{d^2x}{dt^2} = -mg - K\dfrac{dx}{dt}$ or $\dfrac{d^2x}{dt^2} + k\dfrac{dx}{dt} = -g$, where $K = mk$.

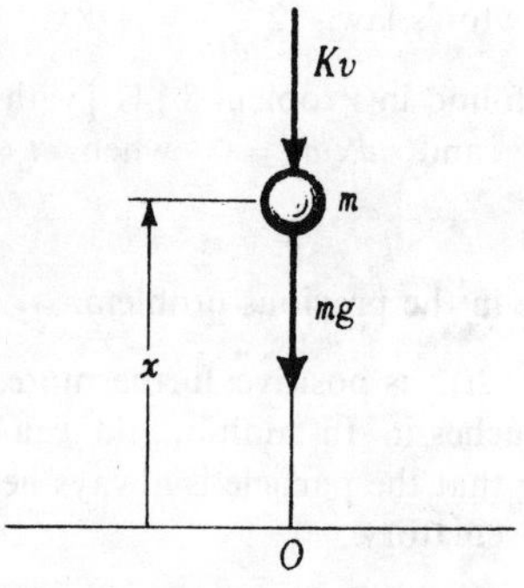

Fig. 11.4

11.62 Find the maximum height attained by the mass in the previous problem if it is projected with initial velocity v_0.

▌ The solution to the differential equation of the previous problem is found in Problem 9.169 to be $x = C_1 + C_2e^{-kt} - \dfrac{g}{k}t$. Differentiating yields $v = \dfrac{dx}{dt} = -kC_2e^{-kt} - \dfrac{g}{k}$.

When $t = 0$, $x = 0$ and $v = v_0$. Then $C_1 = -C_2$ and $-C_2 = \dfrac{v_0}{k} + \dfrac{g}{k^2}$. Making these replacements, we get $x = \dfrac{1}{k^2}(g + kv_0)(1 - e^{-kt}) - \dfrac{g}{k}t$. The maximum height is reached when $v = 0$. This occurs when $e^{-kt} = \dfrac{-g}{k^2C_2} = \dfrac{g}{g + kv_0}$ or $t = \dfrac{1}{k}\ln\dfrac{g + kv_0}{g}$. Then the maximum height is

$$x = \frac{1}{k^2}(g + kv_0)\left(1 - \frac{g}{g + kv_0}\right) - \frac{g}{k}\left(\frac{1}{k}\ln\frac{g + kv_0}{g}\right) = \frac{1}{k}\left(v_0 - \frac{g}{k}\ln\frac{g + kv_0}{g}\right).$$

11.63 A perfectly flexible cable hangs over a frictionless peg with 8 ft of cable on one side of the peg and 12 ft on the other side (see Fig. 11.5). Find the differential equation for the motion of the sliding cable.

▌ We denote the total mass of the cable by m, and the length (in feet) of cable that has moved over the peg at time t by x. At time t there are $8 - x$ ft of cable on one side and $12 + x$ ft on the other. The excess of

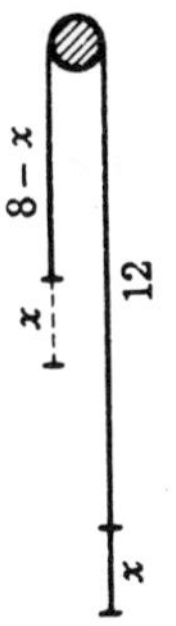

Fig. 11.5

$4 + 2x$ ft on one side produces an unbalanced force of $(4 + 2x)mg/20$ lb. Thus, $m\dfrac{d^2x}{dt^2} = (4 + 2x)\dfrac{mg}{20}$ or $\dfrac{d^2x}{dt^2} - \dfrac{g}{10}x = \dfrac{g}{5}$. Observe that the motion is not influenced by the mass of the cable.

.64 Find the time required for the cable of the previous problem to slide off the peg, starting from rest.

▮ The solution to the differential equation of the previous problem is found in Problem 9.26 to be $x = C_1e^{\sqrt{g/10}t} + C_2e^{-\sqrt{g/10}t} - 2$. Differentiating once with respect to t yields $v = \sqrt{g/10}(C_1e^{\sqrt{g/10}t} - C_2e^{-\sqrt{g/10}t})$. When $t = 0$, $x = 0$ and $v = 0$. Then $C_1 = C_2 = 1$ and $x = e^{\sqrt{g/10}t} + e^{-\sqrt{g/10}t} - 2 = 2\cosh\sqrt{g/10}t - 2$. Hence $t = \sqrt{10/g}\cosh^{-1}\frac{1}{2}(x + 2) = \sqrt{10/g}\ln\dfrac{x + 2 + \sqrt{x^2 + 4x}}{2}$. When $x = 8$ ft of cable has moved over the peg, $t = \sqrt{10/g}\ln(5 + 2\sqrt{6})$ s.

.65 Determine the differential equation for the motion of the cable described in Problem 11.63 if, in addition, the force of friction over the peg is equal to the weight of 1 ft of the cable.

▮ The force of friction retards the motion of the cable (so it is negative) and in absolute value is equal to $mg/20$. According to the analysis developed in Problem 11.63, it follows that the net force on the cable is $(4 + 2x)\dfrac{mg}{20} - \dfrac{mg}{20} = (3 + 2x)\dfrac{mg}{20}$. Then Newton's second law of motion gives $m\dfrac{d^2x}{dt^2} = (3 + 2x)\dfrac{mg}{20}$ or $\dfrac{d^2x}{dt^2} - \dfrac{g}{10}x = \dfrac{3g}{20}$.

66 Find the time required for the cable in the previous problem to slide off the peg, starting from rest.

▮ The solution to the differential equation of the previous problem is found in Problem 9.27 to be $x = C_1e^{\sqrt{g/10}t} + C_2e^{-\sqrt{g/10}t} - \frac{3}{2}$. Applying the initial conditions $x(0) = 0$ and $v(0) = 0$, we find $C_1 = C_2 = \frac{3}{4}$, so that $x = \frac{3}{4}e^{\sqrt{g/10}t} + \frac{3}{4}e^{-\sqrt{g/10}t} - \frac{3}{2} = \frac{3}{2}(\cosh\sqrt{g/10}t - 1)$.

We seek the value of t for which $x = 8$. Thus, we write $8 = \frac{3}{2}(\cosh\sqrt{g/10}t - 1)$, or $\cosh\sqrt{g/10}t = \frac{19}{3}$. Then $\sqrt{g/10}t = \cosh^{-1}\frac{19}{3} = \ln(\frac{19}{3} + \sqrt{(19/3)^2 - 1}) = 2.53268$, and $t = \sqrt{10/g}(2.53268) = 1.42$ s, where we have taken $g = 32$ ft/s^2.

67 Show directly that for $v > 0$, $\cosh^{-1} v = \ln(v \pm \sqrt{v^2 - 1})$.

▮ We set $y = \cosh^{-1} v$, so that $v = \cosh y = \frac{1}{2}(e^y + e^{-y})$; then we rewrite this as $e^y + e^{-y} - 2v = 0$. Multiplying by e^y and rearranging then yield $(e^y)^2 - 2ve^y + 1 = 0$, and the quadratic formula gives $e^y = \dfrac{2v \pm \sqrt{4v^2 - 4}}{2} = v \pm \sqrt{v^2 - 1}$. Thus $y = \ln(v \pm \sqrt{v^2 - 1})$. Since $y = \cosh^{-1} v$, the identity follows.

68 Show that if $\cosh^{-1} v$ is known to be positive (from physical considerations), then $\cosh^{-1} v = \ln(v + \sqrt{v^2 - 1})$ for $v > 0$.

▮ We have, from the previous problem, that $\cosh^{-1} v = \ln(v \pm \sqrt{v^2 - 1})$, so all that remains is to eliminate the minus sign in front of the square root as a possibility. If we set $y = \cosh^{-1} v$, it follows that $v = \cosh y$ and that $v > 1$. It also follows that $2v > 2$ and that $2v - 1 > 1$.

Then multiplying by -1 reverses the sense of the inequality, and adding v^2 gives us $v^2 - 2v + 1 < v^2 - 1$. Factoring yields $(v-1)^2 < v^2 - 1$ or $v - 1 < \sqrt{v^2 - 1}$, which we write as $v - \sqrt{v^2 - 1} < 1$. If this is so, then it must be that $\ln(v - \sqrt{v^2 - 1})$ is negative. But this is a contradiction if $\cosh^{-1} v$ is known to be positive, so the minus sign is impossible.

HORIZONTAL-BEAM PROBLEMS

11.69 Derive the differential equation for the deflection (bending) of a uniform (in material and shape) beam under specified loadings.

▌ It is convenient to think of the beam as consisting of fibers running lengthwise. In the bent beam shown in Fig. 11.6, the fibers of the upper half are compressed and those of the lower half are stretched, the two halves being separated by a neutral surface whose fibers are neither compressed nor stretched. The fiber which originally coincided with the horizontal axis of the beam now lies in the neutral surface along a curve called the elastic curve (or curve of deflection). We seek the equation of this curve.

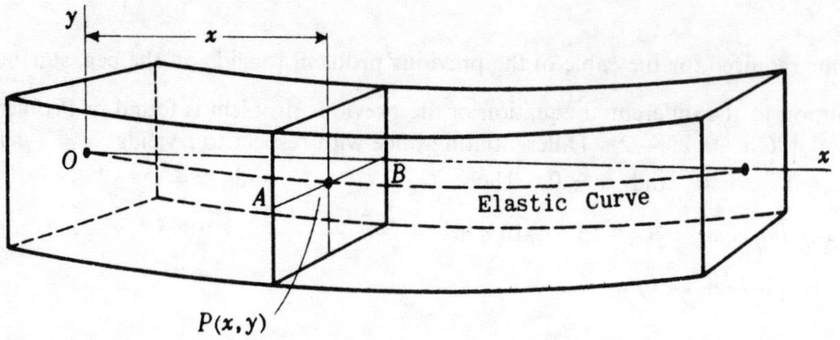

Fig. 11.6

Consider a cross section of the beam at a distance x from one end. Let AB be its intersection with the neutral surface, and P its intersection with the elastic curve. It is shown in mechanics that the moment M with respect to AB of all external forces acting on either of the two segments into which the beam is separated by the cross section is independent of the segment considered and is given by $M = \frac{EI}{R}$, where E is the modulus of elasticity of the beam, I is the moment of inertia of the cross section with respect to AB, and R is the radius of curvative of the elastic curve at P.

For convenience, we think of the beam as being replaced by its elastic curve, and the cross section by point P. We take the origin at the left end of the beam, with the x axis horizontal, and let P have coordinates (x, y). Since the slope dy/dx of the elastic curve is numerically small at all points, we have, approximately,

$$R = \frac{[1 + (dy/dx)^2]^{3/2}}{d^2y/dx^2} = \frac{1}{d^2y/dx^2} \quad \text{so that} \quad M = EI\frac{d^2y}{dx^2}$$

The bending moment M at the cross section (or at point P of the elastic curve) is the algebraic sum of the moments of the external forces acting on the segment of the beam (or segment of the elastic curve) about line AB in the cross section (or about point P of the elastic curve). We assume that upward forces give positive moments and downward forces give negative moments.

11.70 Find the bending moment at a distance x from the left end of a 30-ft beam resting on two vertical supports (see Fig. 11.7) if the beam carries a uniform load of 200 lb per foot of length and a load of 2000 lb at its middle.

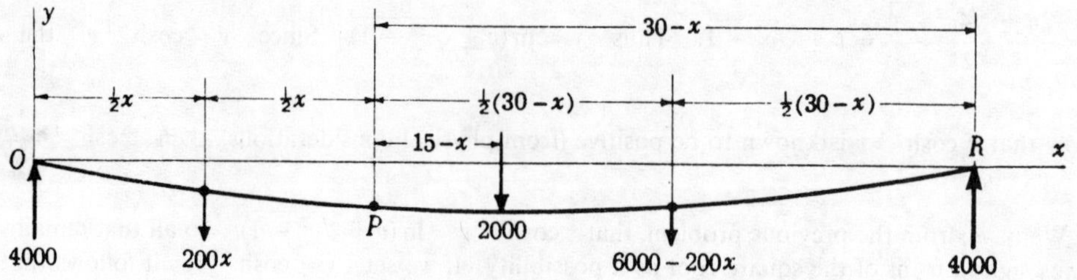

Fig. 11.7

▮ The external forces acting on OP in Fig. 11.7 are an upward force at O, x feet from P, equal to one-half the total load, or $\frac{1}{2}[2000 + 30(200)] = 4000$ lb, and a downward force of $200x$ lb, which we assume is concentrated at the middle of OP and thus $\frac{1}{2}x$ ft from P. The bending moment at P is then $M = 4000x - 200x(\frac{1}{2}x) = 4000x - 100x^2$.

11.71 Show that the bending moment in the previous problem is independent of the segment used to compute it.

▮ Consider the forces acting on segment PR. They are (1) an upward force of 4000 lb at R, $30 - x$ ft from P; (2) the load of 2000 lb acting downward at the middle of the beam, $15 - x$ ft from P; and (3) a force of $200(30 - x)$ lb downward, assumed to be concentrated at the middle of PR, $\frac{1}{2}(30 - x)$ ft from P. Then the total moment is, again,

$$M = 4000(30 - x) - 2000(15 - x) - 200(30 - x)\cdot\tfrac{1}{2}(30 - x) = 4000x - 100x^2$$

11.72 A horizontal beam of length $2l$ ft is freely supported at both ends. Find the equation of its elastic curve and its maximum deflection when the load is w lb per foot of length.

▮ We take the origin at the left end of the beam, with the x axis horizontal as in Fig. 11.8. Let P be any point on the elastic curve, with coordinates (x, y), and consider segment OP of the beam. There is an upward force of wl lb at O, x ft from P; there is also a load of wx lb at the midpoint of OP, $\frac{1}{2}x$ ft from P. Then, since $EI\,d^2y/dx^2 = M$, we have $EI\dfrac{d^2y}{dx^2} = wlx - wx(\frac{1}{2}x) = wlx - \frac{1}{2}wx^2$.

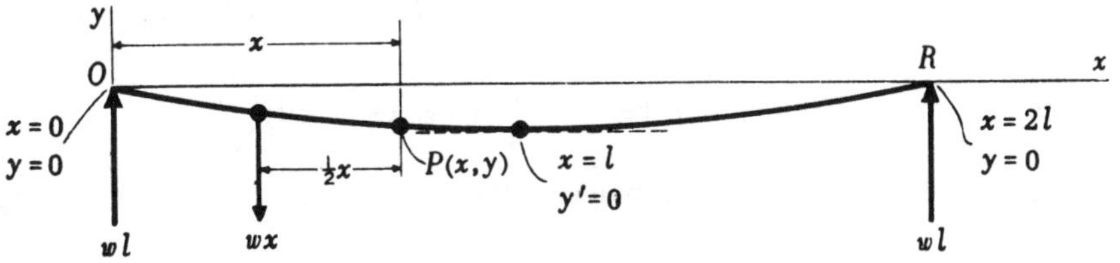

Fig. 11.8

Integrating once yields $EI\,dy/dx = \frac{1}{2}wlx^2 - \frac{1}{6}wx^3 + C_1$. At the middle of the beam, $x = l$ and $dy/dx = 0$. Applying these conditions then yields $C_1 = -\frac{1}{3}wl^3$, so that $EI\,dy/dx = \frac{1}{2}wlx^2 - \frac{1}{6}wx^3 - \frac{1}{3}wl^3$. A second integration now gives $EIy = \frac{1}{6}wlx^3 - \frac{1}{24}wx^4 - \frac{1}{3}wl^3x + C_2$. At point O, $x = y = 0$. Thus, $C_2 = 0$ and $y = \dfrac{w}{24EI}(4lx^3 - x^4 - 8l^3x)$.

11.73 Determine the maximum deflection of the beam in the previous problem.

▮ The deflection of the beam at any distance x from O is given by $-y$. The maximum deflection occurs at the middle of the beam $(x = l)$ and is $-y_{\max} = -\dfrac{w}{24EI}(4l^4 - l^4 - 8l^4) = \dfrac{5wl^4}{24EI}$.

11.74 Solve Problem 11.72 if there is, in addition, a load of W lb at the middle of the beam.

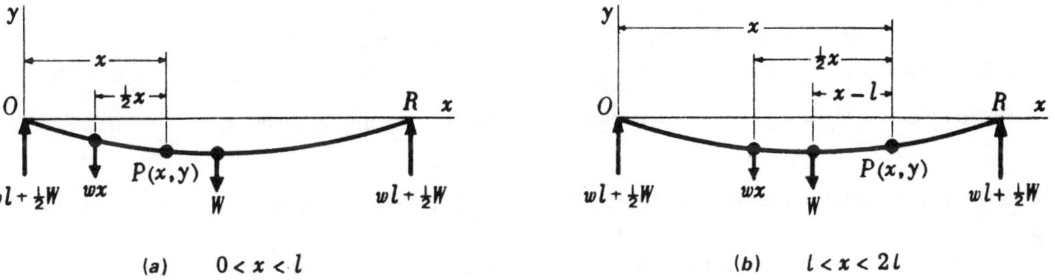

Fig. 11.9

▮ We choose the same coordinate system as in Problem 11.72. Since the forces acting on a segment OP of the beam differ according to whether P lies to the left or right of the midpoint, two cases must be considered.

When $0 < x < l$ [Fig. 11.9(*a*)], the forces acting on OP are an upward force of $wl + \frac{1}{2}W$ lb at O, x ft from P, and the load wx acting downward at the midpoint of OP, $\frac{1}{2}x$ ft from P. The bending moment is then

$$M = (wl + \tfrac{1}{2}W)x - wx(\tfrac{1}{2}x) = wlx + \tfrac{1}{2}Wx - \tfrac{1}{2}wx^2 \tag{1}$$

When $l < x < 2l$ [Fig. 11.9(*b*)], there is an additional force: the load of W lb at the midpoint of the beam, $x - l$ ft from P. The bending moment is then

$$M = (wl + \tfrac{1}{2}W)x - wx(\tfrac{1}{2}x) - W(x - l) = wlx + \tfrac{1}{2}Wx - \tfrac{1}{2}wx^2 - W(x - l) \tag{2}$$

Both (*1*) and (*2*) yield the bending moment $M = \frac{1}{2}wl^2 + \frac{1}{2}Wl$ when $x = l$. The two cases may be treated at the same time by noting that

$$wlx + \tfrac{1}{2}Wx - \tfrac{1}{2}wx^2 = wlx - \tfrac{1}{2}wx^2 - \tfrac{1}{2}W(l - x) + \tfrac{1}{2}Wl$$

and

$$wlx + \tfrac{1}{2}Wx - \tfrac{1}{2}wx^2 - W(x - l) = wlx - \tfrac{1}{2}wx^2 + \tfrac{1}{2}W(l - x) + \tfrac{1}{2}Wl$$

Then we may write $EI\,d^2y/dx^2 = wlx - \frac{1}{2}wx^2 \mp \frac{1}{2}W(l - x) + \frac{1}{2}Wl$ with the understanding that the upper sign holds for $0 < x < l$, and the lower for $l < x < 2l$.

Integrating this last equation twice yields $EIy = \frac{1}{6}wlx^3 - \frac{1}{24}wx^4 \mp \frac{1}{12}W(l - x)^3 + \frac{1}{4}Wlx^2 + C_1x + C_2$. Using the boundary conditions $x = y = 0$ at O and $x = 2l$ and $y = 0$ at R, we obtain $C_2 = \frac{1}{12}Wl^3$ and $C_1 = -\frac{1}{3}wl^3 - \frac{1}{2}Wl^2$. Then

$$\begin{aligned} EIy &= \tfrac{1}{6}wlx^3 - \tfrac{1}{24}wx^4 - \tfrac{1}{3}wl^3x \mp \tfrac{1}{12}W(l - x)^3 + \tfrac{1}{4}Wlx^2 - \tfrac{1}{2}Wl^2x + \tfrac{1}{12}Wl^3 \\ &= \tfrac{1}{6}wlx^3 - \tfrac{1}{24}wx^4 - \tfrac{1}{3}wl^3x - \tfrac{1}{12}W|l - x|^3 + \tfrac{1}{4}Wlx^2 - \tfrac{1}{2}Wl^2x + \tfrac{1}{12}Wl^3 \end{aligned}$$

and

$$y = \frac{w}{24EI}(4lx^3 - x^4 - 8l^3x) + \frac{W}{12EI}(3lx^2 - |l - x|^3 - 6l^2x + l^3)$$

11.75 Determine the maximum deflection of the beam in the previous problem.

▌ The maximum deflection occurs at the middle of the beam where $x = l$, and is $-y_{\max} = \dfrac{5wl^4}{24EI} + \dfrac{Wl^3}{6EI}$.

11.76 A horizontal beam of length l ft is fixed at one end but otherwise unsupported. Find the equation of its elastic curve when it carries a uniform load of w lb per foot of length.

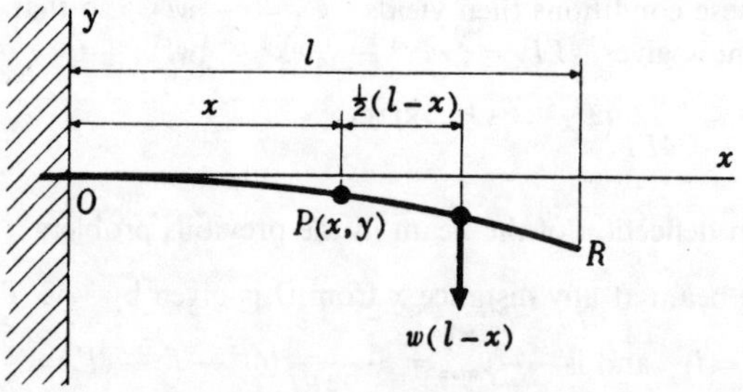

Fig. 11.10

▌ We take the origin at the fixed end and let P have coordinates (x, y). Consider the segment PR in Fig. 11.10. The only force is the weight $w(l - x)$ lb at the midpoint of PR, $\frac{1}{2}(l - x)$ ft from P. Then $EI\,d^2y/dx^2 = -w(l - x)[\frac{1}{2}(l - x)] = -\frac{1}{2}w(l - x)^2$, and integrating once yields $EI\,dy/dx = \frac{1}{6}w(l - x)^3 + C_1$. At O, $x = 0$ and $dy/dx = 0$; thus $C_1 = -\frac{1}{6}wl^3$ and we have $EI\,dy/dx = \frac{1}{6}w(l - x)^3 - \frac{1}{6}wl^3$.

Integrating once again gives us $EIy = -\frac{1}{24}w(l - x)^4 - \frac{1}{6}wl^3x + C_2$. Applying the conditions $x = y = 0$ at O then yields $C_2 = \frac{1}{24}wl^4$. Substitution and simpification finally give $y = \dfrac{w}{24EI}(4lx^3 - 6l^2x^2 - x^4)$.

11.77 Determine the maximum deflection of the beam in the previous problems.

▌ The maximum deflection, occurring at point R (where $x = l$), is $-y_{\max} = \dfrac{1}{8}\dfrac{wl^4}{EI}$.

11.78 A horizontal beam of length $3l$ ft is fixed at one end but otherwise unsupported. It carries a uniform load of w lb/ft and two loads of W lb each at distances l and $2l$ ft from the fixed end (see Fig. 11.11). Find the equation of its elastic curve.

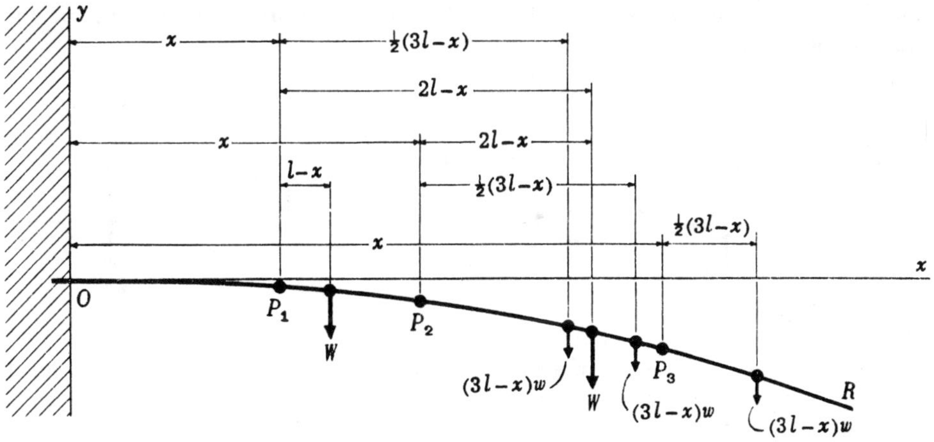

Fig. 11.11

We take the origin to be at the fixed end, and let point P have coordinates (x, y). There are three cases to be considered, according to whether P is on the interval $0 < x < l$, $l < x < 2l$, or $2l < x < 3l$. In each case, use will be made of the right-hand segment of the beam in computing the three bending moments.

When $0 < x < l$ ($P = P_1$ in Fig. 11.11), there are three forces acting on P_1R: the weight $(3l - x)w$ lb assumed to act at the midpoint of P_1R, $\frac{1}{2}(3l - x)$ ft from P_1; the load W lb, $(l - x)$ ft from P_1; and the load W lb, $(2l - x)$ ft from P_1. The bending moment about P_1 is then

$$M_1 = EI\frac{d^2y}{dx^2} = -(3l - x)w\left[\frac{1}{2}(3l - x)\right] - W(l - x) - W(2l - x) = -\frac{1}{2}w(3l - x)^2 - W(l - x) - W(2l - x)$$

Integrating yields $EI\,dy/dx = \frac{1}{6}w(3l - x)^3 + \frac{1}{2}W(l - x)^2 + \frac{1}{2}W(2l - x)^2 + C_1$. At point O, $x = 0$ and $dy/dx = 0$; these conditions give $C_1 = -\frac{9}{2}wl^3 - \frac{5}{2}Wl^2$, so that $EI\,dy/dx = \frac{1}{6}w(3l - x)^3 + \frac{1}{2}W(l - x)^2 + \frac{1}{2}W(2l - x)^2 - \frac{9}{2}wl^3 - \frac{5}{2}Wl^2$. A second integration yields $EIy = -\frac{1}{24}w(3l - x)^4 - \frac{1}{6}W(l - x)^3 - \frac{1}{6}W(2l - x)^3 - \frac{9}{2}wl^3x - \frac{5}{2}Wl^2x + C_2$. Since $x = y = 0$ at O, we have $C_2 = \frac{27}{8}wl^4 + \frac{3}{2}Wl^3$ and

$$EIy = -\tfrac{1}{24}w(3l - x)^4 - \tfrac{1}{6}W(l - x)^3 - \tfrac{1}{6}W(2l - x)^3 - \tfrac{9}{2}wl^3x - \tfrac{5}{2}Wl^2x + \tfrac{27}{8}wl^4 + \tfrac{3}{2}Wl^3 \qquad (1)$$

When $l < x < 2l$ ($P = P_2$ in Fig. 11.11), the bending moment about P_2 is $M_2 = EI\,d^2y/dx^2 = -\frac{1}{2}w(3l - x)^2 - W(2l - x)$. Integrating twice, we obtain $EIy = -\frac{1}{24}w(3l - x)^4 - \frac{1}{6}W(2l - x)^3 + C_3x + C_4$. Now we note that when $x = l$ this equation and (1) must agree in deflection and slope, so that $C_3 = C_1$ and $C_4 = C_2$. Thus, we have

$$EIy = -\tfrac{1}{24}w(3l - x)^4 - \tfrac{1}{6}W(2l - x)^3 - \tfrac{9}{2}wl^3x - \tfrac{5}{2}Wl^2x + \tfrac{27}{8}wl^4 + \tfrac{3}{2}Wl^3 \qquad (2)$$

When $2l < x < 3l$ ($P = P_3$ in Fig. 11.11), the bending moment about P_3 is $M_3 = EI\,d^2y/dx^2 = -\frac{1}{2}w(3l - x)^2$. Then, integrating twice and noting that the result must agree with (2) in deflection and slope, we obtain

$$EIy = -\tfrac{1}{24}w(3l - x)^4 + C_5x + C_6 = -\tfrac{1}{24}w(3l - x)^4 - \tfrac{9}{2}wl^3x - \tfrac{5}{2}Wl^2x - \tfrac{27}{8}Wl^4 + \tfrac{3}{2}Wl^3 \qquad (3)$$

Finally, we combine (1), (2), and (3) as follows:

$$y = \begin{cases} \dfrac{w}{24EI}(12lx^3 - 54l^2x^2 - x^4) + \dfrac{W}{6EI}(2x^3 - 9lx^2) & 0 \le x \le 1 \\[2ex] \dfrac{w}{24EI}(12lx^3 - 54l^2x^2 - x^4) + \dfrac{W}{6EI}(x^3 - 6lx^2 - 3l^2x + l^3) & l \le x \le 2l \\[2ex] \dfrac{w}{24EI}(12lx^3 - 54l^2x^2 - x^4) + \dfrac{W}{2EI}(3l^3 - 5l^2x) & 2l \le x \le 3l \end{cases}$$

9 Determine the maximum deflection of the beam in the previous problem.

The maximum deflection, occurring at point R (where $x = 3l$), is $-y_{max} = \dfrac{1}{8EI}(81wl^4 + 48Wl^3)$.

11.80 A horizontal beam of length l ft is fixed at both ends. Find the equation of its elastic curve if it carries a uniform load of w lb/ft.

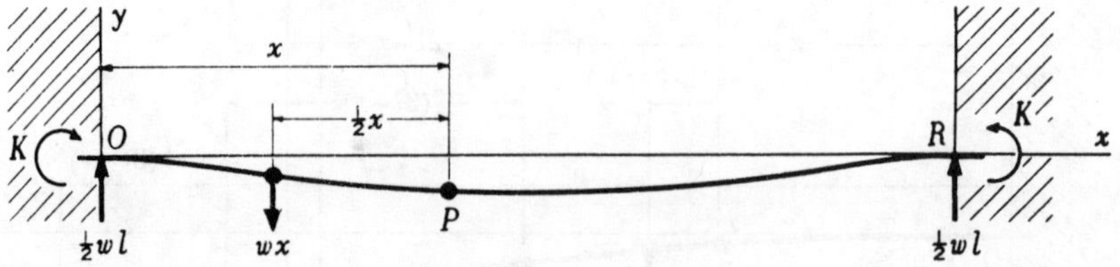

Fig. 11.12

▮ We take the origin at the left end of the beam and let P have coordinates (x, y), as in Fig. 11.12. The external forces acting on segment OP are a *couple* of unknown moment K exerted by the wall to keep the beam horizontal at O; an upward force of $\frac{1}{2}wl$ lb at O, x ft from P; and the load wx lb acting downward at the midpoint of OP, $\frac{1}{2}x$ ft from P. Thus, $EI\,d^2y/dx^2 = K + \frac{1}{2}wlx - \frac{1}{2}wx^2$. Integrating once and using the conditions $x = 0$ and $dy/dx = 0$ at O, we obtain $EI\,dy/dx = Kx + \frac{1}{4}wlx^2 - \frac{1}{6}wx^3$.

At point R, $x = l$ and $dy/dx = 0$ (since the beam is fixed there). Substituting these values into the last equation yields $0 = Kl + \frac{1}{4}wl^3 - \frac{1}{6}wl^3$, from which $K = -\frac{1}{12}wl^2$. Substitution for K, integration, and the use of $x = y = 0$ at O finally yield

$$EIy = -\frac{1}{24}wl^2x^2 + \frac{1}{12}wlx^3 - \frac{1}{24}wx^4 \quad \text{or} \quad y = \frac{wx^2}{24EI}(2lx - l^2 - x^2)$$

11.81 Determine the maximum deflection of the beam in the previous problem.

▮ The maximum deflection, occurring at the middle of the beam (where $x = \frac{1}{2}l$), is $-y_{\max} = \dfrac{wl^4}{384EI}$.

11.82 Solve Problem 11.80 if, in addition, there is a weight of W lb at the middle of the beam.

▮ We use the coordinate system of Problem 11.80. Figure 11.13 shows that two cases must be considered: x between 0 and $\frac{1}{2}l$, and x between $\frac{1}{2}l$ and l. When $0 < x < \frac{1}{2}l$, the external forces on the segment to the left of $P_1(x, y)$ are a couple of unknown moment K at O; an upward force of $\frac{1}{2}(wl + W)$ lb at O, x ft from P_1; and the load wx lb, $\frac{1}{2}x$ ft from P_1. Thus, $EI\,d^2y/dx^2 = K + \frac{1}{2}(wl + W)x - \frac{1}{2}wx^2 = K + \frac{1}{2}wlx - \frac{1}{2}wx^2 + \frac{1}{2}Wx$. Integrating once and using the conditions $x = 0$ and $dy/dx = 0$ at O, we obtain $EI\dfrac{dy}{dx} = Kx + \frac{1}{4}wlx^2 - \frac{1}{6}wx^3 + \frac{1}{4}Wx^2$. Integrating once again and using $x = y = 0$ at O, we get

$$EIy = \tfrac{1}{2}Kx^2 + \tfrac{1}{12}wlx^3 - \tfrac{1}{24}wx^4 + \tfrac{1}{12}Wx^3 \quad (1)$$

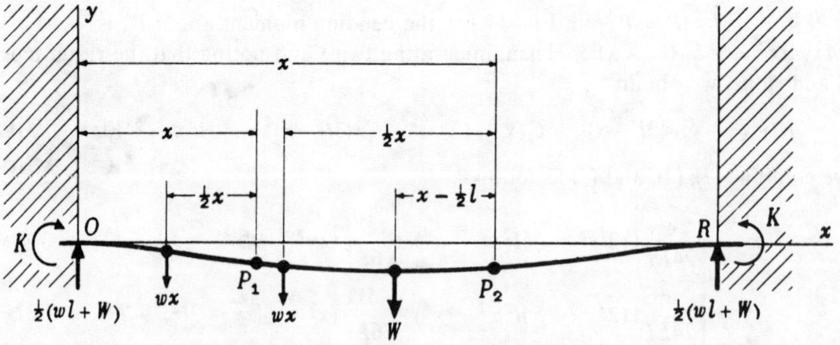

Fig. 11.13

When $\frac{1}{2}l < x < l$, there is in addition the weight W lb at the middle of the beam, $x - \frac{1}{2}l$ ft from P_2. Thus, $EI\,d^2y/dx^2 = K + \frac{1}{2}wlx - \frac{1}{2}wx^2 + \frac{1}{2}Wx - W(x - \frac{1}{2}l)$, Integrating twice yields $EIy = \frac{1}{2}Kx^2 + \frac{1}{12}wlx^3 - \frac{1}{24}wx^4 + \frac{1}{12}Wx^3 - \frac{1}{6}W(x - \frac{1}{2}l)^3 + C_1x + C_2$. When $x = \frac{1}{2}l$, the values of y and dy/dx here must agree with those for (1). Thus, $C_1 = C_2 = O$, and

$$EIy = \tfrac{1}{2}Kx^2 + \tfrac{1}{12}wlx^3 - \tfrac{1}{24}wx^4 + \tfrac{1}{12}Wx^3 - \tfrac{1}{6}W(x - \tfrac{1}{2}l)^3 \quad (2)$$

To determine K, we let $x = \frac{1}{2}l$ and $dy/dx = 0$ in the equation for $EI\,dy/dx$ above. This yields $K = -\frac{1}{12}wl^2 - \frac{1}{8}Wl$, so that (1) and (2) become

$$y = \begin{cases} \dfrac{w}{24EI}(2lx^3 - l^2x^2 - x^4) + \dfrac{W}{48EI}(4x^3 - 3lx^2) & 0 \le x \le \frac{1}{2}l \\ \dfrac{w}{24EI}(2lx^3 - l^2x^2 - x^4) + \dfrac{W}{48EI}(l^3 - 6l^2x + 9lx^2 - 4x^3) & \frac{1}{2}l \le x \le l \end{cases}$$

11.83 Find the maximum deflection of the beam in the previous problem.

▌ The maximum deflection, occurring at the middle of the beam, is $-y_{max} = \dfrac{1}{384EI}(wl^4 + 2Wl^3)$.

11.84 A horizontal beam of length l ft is fixed at one end and freely supported at the other end. Find the equation of the elastic curve if the beam carries a uniform load of w lb/ft and a weight W lb at the middle.

▌ We take the origin at the fixed end (Fig. 11.14) and let P have coordinates (x, y). There are two cases to be considered.

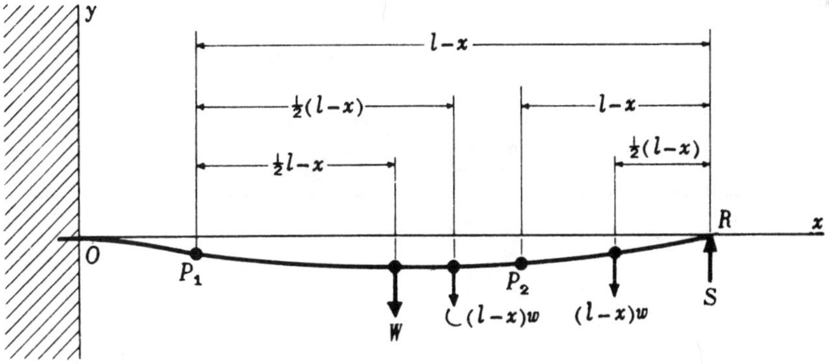

Fig. 11.14

When $0 < x < \frac{1}{2}l$, the external forces acting on the segment P_1R are an unknown upward force of S lb at R, $l - x$ ft from P_1; the load $w(l-x)$ lb at the midpoint of P_1R, $\frac{1}{2}(l-x)$ ft from P_1; and W lb, $\frac{1}{2}l - x$ ft from P_1. Thus, we have $EI\,d^2y/dx^2 = S(l-x) - w(l-x)[\frac{1}{2}(l-x)] - W(\frac{1}{2}l - x) = S(l-x) - \frac{1}{2}w(l-x)^2 - W(\frac{1}{2}l - x)$. Integrating once and using the conditions $x = 0$ and $dy/dx = 0$ at O, we obtain $EI\,dy/dx = -\frac{1}{2}S(l-x)^2 + \frac{1}{6}w(l-x)^3 + \frac{1}{2}W(\frac{1}{2}l - x)^2 + \frac{1}{2}Sl^2 - \frac{1}{6}wl^3 - \frac{1}{8}Wl^2$. Integrating again and using the conditions $x = y = 0$ at O yield

$$EIy = \tfrac{1}{6}S(l-x)^3 - \tfrac{1}{24}w(l-x)^4 - \tfrac{1}{6}W(\tfrac{1}{2}l - x)^3 + (\tfrac{1}{2}Sl^2 - \tfrac{1}{6}wl^3 - \tfrac{1}{8}Wl^2)x - \tfrac{1}{6}Sl^3 + \tfrac{1}{24}wl^4 + \tfrac{1}{48}Wl^3 \qquad (1)$$

When $\frac{1}{2}l < x < l$, the forces acting on P_2R are the unknown upward thrust S at R, $(l-x)$ ft from P_2, and the load $w(l-x)$ lb, $\frac{1}{2}(l-x)$ ft from P_2. Thus, $EI\,d^2y/dx^2 = S(l-x) - \frac{1}{2}w(l-x)^2$, from which we get $EIy = \frac{1}{6}S(l-x)^3 - \frac{1}{24}w(l-x)^4 + C_1x + C_2$. When $x = \frac{1}{2}l$, the values of EIy and $EI\,dy/dx$ here and in (1) must agree. Hence, C_1 and C_2 must have the values of the constants of integration in (1), so that

$$EIy = \tfrac{1}{6}S(l-x)^3 - \tfrac{1}{24}w(l-x)^4 + (\tfrac{1}{2}Sl^2 - \tfrac{1}{6}wl^3 - \tfrac{1}{8}Wl^2)x - \tfrac{1}{6}Sl^3 + \tfrac{1}{24}wl^4 + \tfrac{1}{28}Wl^3 \qquad (2)$$

To determine S, we note that $y = 0$ when $x = l$, so from (2), $S = \frac{3}{8}wl + \frac{5}{16}W$. This yields

$$y = \begin{cases} \dfrac{w}{48EI}(5lx^3 - 3l^2x^2 - 2x^4) + \dfrac{W}{96EI}(11x^3 - 9lx^2) & 0 \le x \le \frac{1}{2}l \\ \dfrac{w}{48EI}(5lx^3 - 3l^2x^2 - 2x^4) + \dfrac{W}{96EI}(2l^3 - 12l^2x + 15lx^2 - 5x^3) & \frac{1}{2}l \le x \le l \end{cases}$$

1.85 Locate the point of maximum deflection of the beam in the previous problem when $l = 10$ and $W = 10w$.

▌ It is clear from the result of Problem 11.84 that the maximum deflection occurs to the right of the midpoint of the beam. Thus we substitute $l = 10$ and $W = 10w$ in the second part of the solution, obtaining $y = \dfrac{w}{48EI}(-2x^4 + 25x^3 + 450x^2 - 6000x + 10{,}000)$. Since $dy/dx = 0$ at the point of maximum deflection, we solve $8x^3 - 75x^2 - 900x + 6000 = 0$, finding that the real root is $x = 5.6$ (approximately). Thus, the maximum deflection occurs approximately 5.6 ft from the fixed end.

BUOYANCY PROBLEMS

11.86 Determine a differential equation describing the vertical motion of a cylinder partially submerged in a liquid of density ρ, under the assumption that the motion is not damped.

▮ Denote the radius of the cylinder as R, its height as H, and its weight as W. Archimedes' principle states tha an object that is submerged (either partially or totally) in a liquid is acted on by an upward force equal to the weight of the liquid displaced.

Equilibrium occurs when the buoyant force upward on the cylinder is precisely equal to the weight of the cylinder. Assume that the axis of the cylinder is vertical, denote it by y, and take the upward direction to be the positive direction. At equilibrium, with the cylinder partially submerged, $\pi R^2 h\rho = W$ or $h = W/\pi R^2\rho$, where h is the number of units of the cylinder's height that are submerged at equilibrium (see Fig. 11.15). Let $y(t)$ denote the distance from the surface of the water to the equilibrium position as if it were marked on the cylinder. We adopt the convention that $y(t) > 0$ if the equilibrium position on the cylinder is above the surface, and $y(t) < 0$ if it is below the surface.

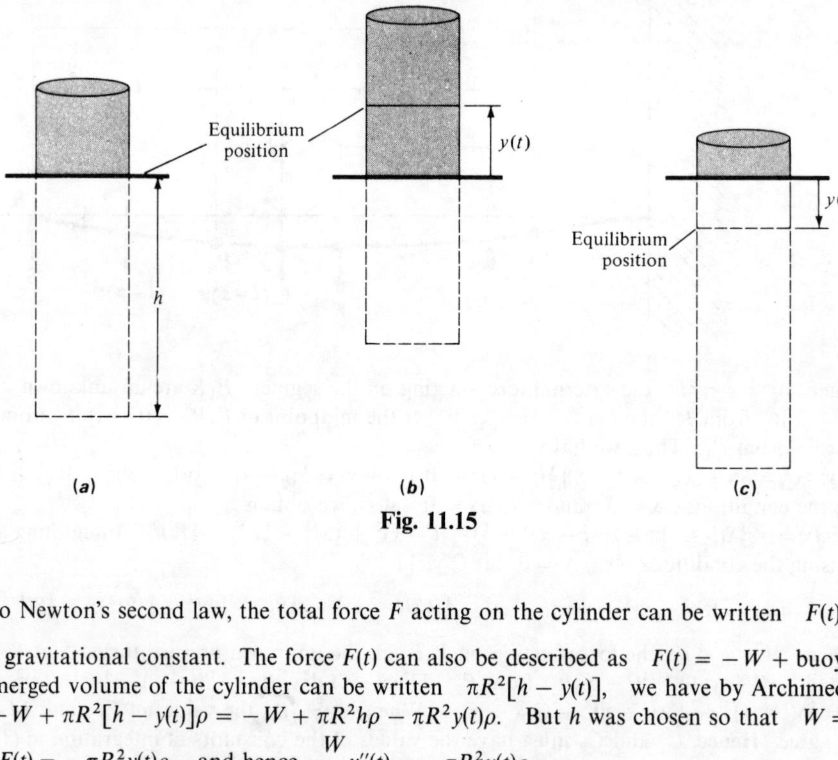

Fig. 11.15

According to Newton's second law, the total force F acting on the cylinder can be written $F(t) = \dfrac{W}{g}y''(t)$ where g is the gravitational constant. The force $F(t)$ can also be described as $F(t) = -W +$ buoyant force. Since the submerged volume of the cylinder can be written $\pi R^2[h - y(t)]$, we have by Archimedes' principle that $F(t) = -W + \pi R^2[h - y(t)]\rho = -W + \pi R^2 h\rho - \pi R^2 y(t)\rho$. But h was chosen so that $W = \pi R^2 h\rho$, so follows that $F(t) = -\pi R^2 y(t)\rho$, and hence $\dfrac{W}{g}y''(t) = -\pi R^2 y(t)\rho$.

Therefore, the initial-value problem describing the motion of the cylinder is

$$y'' + \frac{\pi R^2 \rho g}{W}y = 0 \qquad y(0) = y_0 \qquad y'(0) = v_0$$

where y_0 is the initial position of the cylinder, and v_0 its initial velocity.

11.87 In the notation of the previous problem, discuss what happens when $W > \pi R^2 H\rho$.

▮ If the weight W of the cylinder is greater than the buoyant force generated when the entire cylinder is submerged, then the cylinder must sink to the bottom of the liquid.

11.88 A cylinder with radius 3 in and weight $5\pi\,(\sim 15.71)$ lb is floating with its axis vertical in a pool of water (densit $\rho = 62.5\ \text{lb/ft}^3$). Determine a differential equation that describes its position $y(t)$ relative to equilibrium if it is raised 1 in above equilibrium and pushed downward with an initial velocity of 4 in/s.

▌ Here $R = \frac{1}{4}$ ft, $\rho = 62.5$ lb/ft^3, and $W = 5$ lb, so the result of Problem 11.86 becomes $y'' + \dfrac{\pi(1/4)^2(62.5)(32)}{5\pi} y = 0$ or $y'' + 25y = 0$. Its solution is found in Problem 8.72 (with y here replacing x) to be $y = c_1 \cos 5t + c_2 \sin 5t$.

With $y(0) = \frac{1}{12}$ ft and $y'(0) = -\frac{1}{3}$ ft/s as initial conditions, it follows that $c_1 = 1/12$ and $c_2 = -1/15$. Then $y = \frac{1}{60}(5 \cos 5t - 4 \sin 5t)$.

11.89 Solve the previous problem if, instead of water, the cylinder is floating in a liquid having density 125 lb/ft^3.

▌ With the data as given, the result of Problem 11.86 becomes $y'' + \dfrac{\pi(1/4)^2(125)(32)}{5\pi} y = 0$ or $y'' + 50y = 0$.

Its solution is found in Problem 8.60 to be $y = c_1 \cos \sqrt{50}t + c_2 \sin \sqrt{50}t$. Applying the initial conditions of the previous problem, we find $c_1 = 5/60$ and $c_2 = -2\sqrt{2}/60$, so that $y = \frac{1}{60}(5 \cos \sqrt{50}t - 2\sqrt{2} \sin \sqrt{50}t)$.

11.90 Find the period and the amplitude of the motion described in Problem 11.86.

▌ The motion is defined by the initial-value problem that is the result of Problem 11.86. The differential equation is of the second order and linear, with constant coefficients, having as the roots of its characteristic equation $\lambda = \pm iR\sqrt{\pi\rho g}/\sqrt{W}$. Hence the general solution to the differential equation is $y(t) = c_1 \cos(R\sqrt{\pi\rho g/W}t) + c_2 \sin(R\sqrt{\pi\rho g/W}t)$. The initial conditions imply $c_1 = y_0$ and $c_2 = \sqrt{W}v_0/R\sqrt{\pi\rho g}$, so that

$$y(t) = y_0 \cos\left(R\sqrt{\frac{\pi\rho g}{W}}\,t\right) + \frac{\sqrt{W}v_0}{R\sqrt{\pi\rho g}} \sin\left(R\sqrt{\frac{\pi\rho g}{W}}\,t\right)$$

Therefore, the motion of the cylinder is periodic, and the period T and amplitude A are given by $T = \dfrac{2\sqrt{\pi W}}{R\sqrt{\rho g}}$ and $A = \sqrt{y_0^2 + \dfrac{Wv_0^2}{R^2\pi\rho g}}$. The amplitude depends on the initial position and velocity, but the period is independent of the initial conditions. Notice also that the motion is independent of the height H (as long as the amplitude of the motion is smaller than $H/2$).

11.91 A cylindrical can, partially submerged in water (density 62.5 lb/ft^3) with its axis vertical, oscillates up and down with a period of $\frac{1}{2}$ s. Determine the weight of the can if its radius is 2 in.

▌ Taking $g = 32$ ft/s^2, $R = \frac{1}{6}$ ft, and $T = \frac{1}{2}$ Hz, we have, from the previous problem, $\dfrac{1}{2} = \dfrac{2\sqrt{\pi W}}{(1/6)\sqrt{(62.5)(32)}} = \dfrac{12\sqrt{\pi W}}{\sqrt{2000}}$. Therefore, $\pi W = 2000/(24)^2 \approx 3.4722$ and $W \approx 1.105$ lb.

11.92 What is the minimum height of the can in the previous problem?

▌ Since the can does not sink, it follows from Problem 11.87 that $W < \pi R^2 H\rho$; thus $H > \dfrac{W}{\pi R^2 \rho} = \dfrac{1.105}{\pi(1/6)^2(62.5)} = 0.203$ ft $= 2.43$ in.

11.93 A cylindrical buoy 2 ft in diameter floats in water (density 62.5 lb/ft^3) with its axis vertical. When depressed slightly and released, it vibrates with a period of 2 s. Find the weight of the cylinder.

▌ With $R = 1$, $g = 32$, and $T = 2$, it follows from Problem 11.90 that $2 = \dfrac{2\sqrt{\pi W}}{1\sqrt{(62.5)(32)}} = 2\sqrt{\dfrac{\pi W}{2000}}$.

Therefore, $W = 2000/\pi = 636.6$ lb.

11.94 Solve the previous problem by beginning with the differential equation of motion for the cylinder.

▌ With the numerical values given in the previous problem, the result of Problem 11.86 becomes $y'' + \dfrac{\pi(1)^2(62.5)(32)}{W} y = 0$ or $y'' + \dfrac{2000\pi}{W} y = 0$, which has as its solution $y = c_1 \cos \sqrt{2000\pi/W}t + c_2 \sin \sqrt{2000\pi/W}t$.

The period of these oscillations is $\dfrac{2\pi}{\sqrt{2000\pi/W}}$. Since the period is known to be 2 Hz, it follows that $2 = \dfrac{2\pi}{\sqrt{2000/W}}$, from which we find that $W = 2000/\pi = 636.6$ lb as before.

11.95 What is the minimum height of the buoy in Problem 11.93?

▌ Since the buoy does not sink, it follows from Problem 11.87 that $W < \pi R^2 H\rho$; thus $H > \dfrac{W}{\pi R^2 \rho} = \dfrac{636.6}{\pi(1)^2(62.5)} = 3.24$ ft.

11.96 Suppose a cylinder oscillates with its axis vertical in a liquid of density ρ_1. What is ρ_1 if the period of the oscillation is twice the period of oscillation in water?

▌ Let ρ denote the density of water (62.5 lb/ft^3). Then it follows from the result of Problem 11.90 that the period in water is $T_w = 2\sqrt{\pi W}/R\sqrt{\rho g}$, whereas the period in the liquid of unknown density is $T_l = 2\sqrt{\pi W}/R\sqrt{\rho_1 g}$. If $T_l = 2T_w$, then $\dfrac{2\sqrt{\pi W}}{R\sqrt{\rho_1 g}} = 2\dfrac{2\sqrt{\pi W}}{R\sqrt{\rho g}}$, or $\dfrac{1}{\sqrt{\rho_1}} = \dfrac{2}{\sqrt{\rho}}$. Thus $\rho_1 = \rho/4 = 62.5/4 = 15.625$ lb/ft^3.

11.97 Solve the previous problem if the period of oscillation in the liquid of density ρ_1 is three times the period of oscillation in water.

▌ Using the notation of the previous problem, we have $T_l = 3T_w$; hence $\dfrac{2\sqrt{\pi W}}{R\sqrt{\rho_1 g}} = 3\dfrac{2\sqrt{\pi W}}{R\sqrt{\rho g}}$, or $\dfrac{1}{\sqrt{\rho_1}} = \dfrac{3}{\sqrt{\rho}}$. Thus $\rho_1 = \rho/9 = 62.5/9 = 6.944$ lb/ft^3.

11.98 Suppose a rectangular box of width S_1, length S_2, and height H is floating in a liquid of density ρ, as indicated in Fig. 11.16. As for the cylinder of Problem 11.86, let $y(t)$ denote the position of the box relative to equilibrium, and suppose that W is the weight of the box. How large should H be so that the box will oscillate?

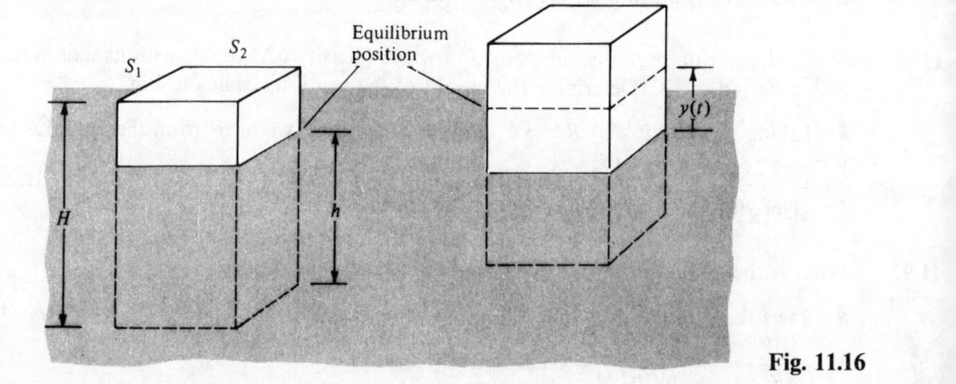

Fig. 11.16

▌ The volume of the box is S_1S_2H, and it will displace $S_1S_2H\rho$ lb of liquid when the box is completely submerged. If the box is to oscillate and not sink, then it must be that $W < S_1S_2H\rho$. Thus, $H > W/S_1S_2\rho$.

11.99 Find a differential equation for the position $y(t)$ of the box described in the previous problem.

▌ At equilibrium, h units of height of the box are submerged, such that $S_1S_2h\rho = W$ (see Fig. 11.16). (We adopt the sign conventions and notations of Problem 11.86.) The buoyant force acting on the box when $y(t)$ units of height are displaced from the equilibrium position is $S_1S_2\rho[h - y(t)]$. The net force on the box then is $-W + S_1S_2\rho[h - y(t)]$, so by Newton's second law of motion, we have $m\dfrac{d^2y}{dt^2} = -W + S_1S_2\rho[h - y(t)]$. Setting $m = W/g$ and noting that $W - S_1S_2\rho h = 0$, we simplify this differential equation to $y'' + \dfrac{S_1S_2\rho g}{W}y = 0$.

11.100 Suppose the box described in Problem 11.98 is displaced from the equilibrium position by y_0 units and given an initial velocity v_0. Find an equation that describes its position y as a function of time.

▮ Set $\omega = \sqrt{S_1 S_2 \rho g/W}$. Then the result of the previous problem becomes $y'' + \omega^2 y = 0$, which has as the roots of its characteristic equation $\lambda_1 = i\omega$ and $\lambda_2 = -i\omega$. Since the differential equation is of the second order, linear, and homogeneous with constant coefficients, its general solution is $y = c_1 \cos \omega t + c_2 \sin \omega t$.

Since $y' = -c_1\omega \sin \omega t + c_2 \omega \cos \omega t$, the initial conditions imply $y_0 = y(0) = c_1$ and $v_0 = y'(0) = c_2\omega$; that is, $c_1 = y_0$ and $c_2 = v_0/\omega$. Then

$$y = y_0 \cos \omega t + \frac{v_0}{\omega} \sin \omega t = y_0 \cos \sqrt{\frac{S_1 S_2 \rho g}{W}}\, t + \frac{\sqrt{W} v_0}{\sqrt{S_1 S_2 \rho g}} \sin \sqrt{\frac{S_1 S_2 \rho g}{W}}\, t$$

11.101 Determine the period and amplitude of the oscillations described in the previous problem.

▮ The frequency of the oscillations is $f = \dfrac{\omega}{2\pi} = \dfrac{\sqrt{S_1 S_2 \rho g}}{2\pi\sqrt{W}}$, so the period is $T = \dfrac{1}{f} = \dfrac{2\pi\sqrt{W}}{\sqrt{S_1 S_2 \rho g}}$.

The amplitude is $A = \sqrt{(y_0)^2 + \left(\dfrac{\sqrt{W} v_0}{\sqrt{S_1 S_2 \rho g}}\right)^2} = \sqrt{y_0^2 + \dfrac{W v_0^2}{S_1 S_2 \rho g}}$.

11.102 How does the period of the oscillations change if S_1 is doubled in the previous problem?

▮ If S_1 is doubled, then the period becomes $T = \dfrac{2\pi\sqrt{W}}{\sqrt{(2S_1)S_2 \rho g}} = \dfrac{1}{\sqrt{2}} \dfrac{2\pi\sqrt{W}}{\sqrt{S_1 S_2 \rho g}}$. Thus the period is reduced to about 71% of its original value.

11.103 How does the period of the oscillations in Problem 11.101 change if both S_1 and S_2 are tripled?

▮ If both S_1 and S_2 are tripled, the period becomes $T = \dfrac{2\pi\sqrt{W}}{\sqrt{(3S_1)(3S_2)\rho g}} = \dfrac{1}{3} \dfrac{2\pi\sqrt{W}}{\sqrt{S_1 S_2 \rho g}}$. Thus the period is reduced by a factor of three.

11.104 A prism whose cross section is an equilateral triangle with sides of length L is floating in a liquid of density ρ, with its height H parallel to the vertical axis. How large should H be so that the prism will oscillate?

▮ The volume of the prism is $\sqrt{3}L^2H/4$, so it will displace $\sqrt{3}\rho L^2 H/4$ lb of liquid when it is completely submerged. If the prism is to oscillate and not sink, then it must be that $W < \sqrt{3}\rho L^2 H/4$ or $H > 4W/\sqrt{3}\rho L^2$, where W is the weight of the prism.

11.105 Find a differential equation for the position $y(t)$ of the prism described in the previous problem,

▮ We adopt the conventions and notations of Problem 11.86. At equilibrium, h units of height of the prism are submerged, such that $\sqrt{3}\rho L^2 h/4 = W$. The buoyant force acting on the prism when $y(t)$ units of height are displaced from the equilibrium position is $\sqrt{3}\rho L^2[h - y(t)]/4$. The net force on the box then is $-W + \sqrt{3}\rho L^2[h - y(t)]/4$, so by Newton's second law of motion we have $m\dfrac{d^2y}{dt^2} = -W + \sqrt{3}\rho L^2[h - y(t)]/4$.

Setting $m = W/g$ and noting that $W - \sqrt{3}\rho L^2 h/4 = 0$ we simplify this equation to $y'' + \dfrac{\sqrt{3}\rho g L^2}{4W} y = 0$.

11.106 The prism described in Problem 11.104 is displaced y_0 units from its equilibrium position and given an initial velocity v_0. Find its position y as a function of time.

▮ Set $\omega^2 = \sqrt{3}\rho g L^2/4W$. Then the result of the previous problem becomes $y'' + \omega^2 y = 0$. Solving this equation with the initial conditions $y(0) = y_0$ and $y'(0) = v_0$ (see Problem 11.100), we obtain

$$y = y_0 \cos \omega t + \frac{v_0}{\omega} \sin \omega t = y_0 \cos \frac{\sqrt[4]{3}\sqrt{\rho g}Lt}{2\sqrt{W}} + \frac{2\sqrt{W} v_0}{\sqrt[4]{3}\sqrt{\rho g}L} \sin \frac{\sqrt[4]{3}\sqrt{\rho g}Lt}{2\sqrt{W}}$$

11.107 Determine the period of the oscillations obtained in the previous problem.

▮ The period is $T = \dfrac{1}{f} = \dfrac{2\pi}{\omega} = \dfrac{4\pi\sqrt{W}}{\sqrt[4]{3}\sqrt{\rho g}L}$.

11.108 How does the period of oscillations change if L is doubled in the previous problem?

▌ It follows from result of the previous problem that if L is doubled, then the period is halved.

ELECTRIC CIRCUIT PROBLEMS

11.109 Describe how to obtain two initial conditions for the current in a simple series RCL circuit with known emf, if initial conditions for the current and the charge on the capacitor are given at $t = 0$.

▌ Denote the current in the circuit and the charge on the capacitor at time t by $I(t)$ and $q(t)$, respectively. We are given $I(0)$, which is one initial condition for the current. From Kirchoff's loop law [see (*1*) of Problem 1.81], we have $RI + L\frac{dI}{dt} + \frac{1}{C}q - E(t) = 0$. Solving this equation for dI/dt and then setting $t = 0$, we obtain $\left.\frac{dI}{dt}\right|_{t=0} = \frac{1}{L}E(0) - \frac{R}{L}I(0) - \frac{1}{LC}q(0)$ as the second initial condition.

11.110 Find two initial conditions for the charge on the capacitor in the circuit of the previous problem.

▌ We are given $q(0)$ and $I(0)$; the first of these quantities provides one initial condition for $q(t)$. Since $dq/dt = I$, it follows that $\left.\frac{dq}{dt}\right|_{t=0} = I(0)$ is the second initial condition.

11.111 A series RCL circuit has $R = 10\ \Omega$, $C = 10^{-2}$ F, $L = \frac{1}{2}$ H, and an applied voltage $E = 12$ V. Assuming no initial current and no initial charge at $t = 0$ when the voltage is first applied, find the subsequent current in the system.

▌ Kirchoff's loop law (see Problem 1.85) gives $\frac{d^2I}{dt^2} + 20\frac{dI}{dt} + 200I = 0$, which has as its solution $I = e^{-10t}(c_1 \cos 10t + c_2 \sin 10t)$ (see Problem 8.64).

The initial conditions are $I(0) = 0$ and, from Problem 11.109, $\left.\frac{dI}{dt}\right|_{t=0} = \frac{12}{1/2} - \frac{10}{1/2}(0) - \frac{1}{(1/2)(10^{-2})}(0) = 24.$

These conditions yield $c_1 = 0$ and $c_2 = \frac{12}{5}$; thus, $I = \frac{12}{5}e^{-10t} \sin 10t$.

11.112 Solve the previous problem by first finding the charge on the capacitor.

▌ The differential equation for the charge on the capacitor is given in Problem 1.84 as $\ddot{q} + 20\dot{q} + 200q = 24$. Its solution is found in Problem 9.22 to be $q = e^{-10t}(c_1 \cos 10t + c_2 \sin 10t) + \frac{3}{25}$.

Initial conditions for the charge are $q(0) = 0$ and $\dot{q}(0) = 0$; applying them, we obtain $c_1 = c_2 = -3/25$. Therefore, $q = -e^{-10t}(\frac{3}{25} \cos 10t + \frac{3}{25} \sin 10t) + \frac{3}{25}$, and $I = \frac{dq}{dt} = \frac{12}{5}e^{-10t} \sin 10t$ as before.

11.113 A series RCL circuit with $R = 6\ \Omega$, $C = 0.02$ F, and $L = 0.1$ H has no applied voltage. Find the subsequent current in the circuit if the initial charge on the capacitor is $\frac{1}{10}$ C and the initial current is zero.

▌ Using Kirchoff's loop law (see Problem 1.86), we get $\frac{d^2I}{dt^2} + 60\frac{dI}{dt} + 500I = 0$, which has as its solution $I = c_1e^{-50t} + c_2e^{-10t}$ (see Problem 8.10). We are given the initial condition $I(0) = 0$, so that $c_1 + c_2 = 0$.

Differentiation of our expression for I yields $\frac{dI}{dt} = -50c_1e^{-50t} - 10c_2e^{-10t}$, from which we conclude that $\left.\frac{dI}{dt}\right|_{t=0} = -50c_1 - 10c_2$. Moreover, from the result of Problem 11.109 we have

$$\left.\frac{dI}{dt}\right|_{t=0} = \frac{1}{0.1}(0) - \frac{6}{0.1}(0) - \frac{1}{(0.1)(0.02)}\frac{1}{10} = -50$$

which implies that $-50 = -50c_1 - 10c_2$. This is a second equation in c_1 and c_2. Solving the two simultaneously, we obtain $c_1 = \frac{5}{4}$ and $c_2 = -\frac{5}{4}$, so that $I = \frac{5}{4}(e^{-50t} - e^{-10t})$.

11.114 Solve the previous problem by first finding the charge on the capacitor.

▌ With $R = 6$, $C = 0.02$, $L = 0.1$, and $E = 0$, (*1*) of Problem 1.82 becomes $\frac{d^2q}{dt^2} + 60\frac{dq}{dt} + 500q = 0$, which is the same differential equation as in the previous problem (with I replaced by q). Its solution is $q = c_1e^{-50t} + c_2e^{-10t}$.

We are given the initial condition $q(0) = \frac{1}{10}$, so we have $\frac{1}{10} = c_1 + c_2$. Also, we have $I(0) = \dot{q}(0) = 0$, so that $-50c_1 - 10c_2 = 0$. Solving these two equations simultaneously yields $c_1 = -\frac{1}{40}$ and $c_2 = \frac{5}{40}$, so that $q = -\frac{1}{40}e^{-50t} + \frac{5}{40}e^{-10t}$. Then $I = \dfrac{dq}{dt} = \dfrac{5}{4}e^{-50t} - \dfrac{5}{4}e^{-10t}$ as before.

11.115 A series RCL circuit consists of an inductance of 0.05 H, a resistance of 5 Ω, a capacitance of 4×10^{-4} F, and a constant emf of 110 V. Find the current flowing through the circuit as a function of time if initially there is no current in the circuit and no charge on the capacitor.

▌ With $L = 0.05$, $R = 5$, $C = 4 \times 10^{-4}$, and $E = 110$, (2) of Problem 1.81 becomes $\ddot{I} + 100\dot{I} + 50{,}000I = 0$. The solution to this equation was found in Problem 8.56 to be $I = c_1e^{-50t}\cos 50\sqrt{19}t + c_2e^{-50t}\sin 50\sqrt{19}t$.

Applying the initial condition $I(0) = 0$, we find that $c_1 = 0$, so this last equation becomes $I = c_2e^{-50t}\sin 50\sqrt{19}t$, which has as its derivative

$$\frac{dI}{dt} = c_2(-50e^{-50t}\sin 50\sqrt{19}t + 50\sqrt{19}e^{-50t}\cos 50\sqrt{19}t) \tag{1}$$

With $q(0) = I(0) = 0$, it follows from the result of Problem 11.109 that $\dot{I}(0) = \dfrac{1}{L}E(0) = \dfrac{110}{0.05} = 2200$. Then we make use of (1) to find that $2200 = \dot{I}(0) = c_2(50\sqrt{19})$; thus, $c_2 = 44/\sqrt{19}$ and $I = \dfrac{44\sqrt{19}}{19}e^{-50t}\sin 50\sqrt{19}t$.

11.116 Solve the previous problem by first finding the capacitance.

▌ We make use of (1) of Problem 1.82, which becomes $\ddot{q} + 100\dot{q} + 50{,}000q = 2200$. The solution to this equation is found in Problem 9.21 to be $q = c_1e^{-50t}\cos 50\sqrt{19}t + c_2e^{-50t}\sin 50\sqrt{19}t + \frac{11}{250}$. Since $q = 0$ at $t = 0$, this gives us $0 = c_1 + \frac{11}{250}$, or $c_1 = -\frac{11}{250}$.

We also have $I(0) = \dot{q}(0) = 0$. Since differentiation yields

$$\dot{q} = c_1(-50e^{-50t}\cos 50\sqrt{19}t - 50\sqrt{19}e^{-50t}\sin 50\sqrt{19}t) + c_2(-50e^{-50t}\sin 50\sqrt{19}t + 50\sqrt{19}e^{-50t}\cos 50\sqrt{19}t)$$

we have $0 = \dot{q}(0) = -50c_1 + 50\sqrt{19}c_2$ or $c_2 = -11\sqrt{19}/4750$. Then

$q = e^{-50t}\left(-\dfrac{11}{250}\cos 50\sqrt{19}t - \dfrac{11\sqrt{19}}{4750}\sin 50\sqrt{19}t\right) + \dfrac{11}{250}$ and $I = \dfrac{dq}{dt} = \dfrac{44\sqrt{19}}{19}e^{-50t}\sin 50\sqrt{19}t$ as before.

11.117 Solve Problem 11.115 if instead of the constant emf there is an alternating emf of $200\cos 100t$.

▌ Now we have $L = 0.05$, $R = 5$, $C = 4(10)^{-4}$, and $E = 200\cos 100t$, so that (2) of Problem 1.81 becomes $\ddot{I} + 100\dot{I} + 50{,}000I = -400{,}000\sin 100t$. Its solution is found in Problem 9.90 to be

$$I = c_1e^{-50t}\cos 50\sqrt{19}t + c_2e^{-50t}\sin 50\sqrt{19}t + \tfrac{40}{17}\cos 100t - \tfrac{160}{17}\sin 100t$$

Applying the initial condition $I(0) = 0$ yields $0 = c_1 + \frac{40}{17}$ or $c_1 = -\frac{40}{17}$.

With $I(0) = q(0) = 0$, we obtain the second initial condition by using the result of Problem 11.109: $\dot{I}(0) = \dfrac{1}{L}E(0) = \dfrac{1}{0.05}(200) = 4000$. Then differentiation gives

$$\dot{I} = c_1(-50e^{-50t}\cos 50\sqrt{19}t - 50\sqrt{19}\sin 50\sqrt{19}t) + c_2(-50e^{-50t}\sin 50\sqrt{19}t + 50\sqrt{19}\cos 50\sqrt{19}t)$$
$$-\frac{4000}{17}\sin 100t - \frac{16{,}000}{17}\cos 100t$$

so we have $4000 = I(0) = -50c_1 + 50\sqrt{19}c_2 - \dfrac{16{,}000}{17}$, or $c_2 = \dfrac{1640\sqrt{19}}{323}$. Substitution of the values of c_1 and c_2 yields $I = e^{-50t}\left(-\dfrac{40}{17}\cos 50\sqrt{19}t + \dfrac{1640\sqrt{19}}{323}\sin 50\sqrt{19}t\right) + \dfrac{40}{17}(\cos 100t - 4\sin 100t)$.

11.118 Solve the previous problem by first finding the charge on the condenser.

▌ With $L = 0.05$, $R = 5$, $C = 4 \times 10^{4}$, and $E = 200\cos 100t$, (1) of Problem 1.82 becomes $\ddot{q} + 100\dot{q} + 50{,}000q = 4000\cos 100t$. Its solution is found in Problem 9.91 to be

$q = c_1e^{-50t}\cos 50\sqrt{19}t + c_2e^{-50t}\sin 50\sqrt{19}t + \frac{16}{170}\cos 100t + \frac{4}{170}\sin 100t$.

Then differentiation yields

$$\dot{q} = c_1(-50e^{-50t}\cos 50\sqrt{19}t - 50\sqrt{19}e^{-50t}\sin 50\sqrt{19}t) + c_2(-50e^{-50t}\sin 50\sqrt{19}t + 50\sqrt{19}e^{-50t}\cos 50\sqrt{19}t)$$
$$-\tfrac{160}{17}\sin 100t + \tfrac{40}{17}\cos 100t$$

Applying the initial conditions, we obtain $0 = q(0) = c_1 + \frac{16}{170}$, or $c_1 = -\frac{16}{170}$, and $0 = I(0) = \dot{q}(0) = -50c_1 + 50\sqrt{19}c_2 + \frac{40}{17}$, from which we find $c_2 = -12\sqrt{19}/1615$. These values for c_1 and c_2 give $q = e^{-50t}\left(-\dfrac{16}{170}\cos 50\sqrt{19}t - \dfrac{12\sqrt{19}}{1615}\sin 50\sqrt{19}t\right) + \dfrac{4}{170}(4\cos 100t + \sin 100t)$. Differentiation yields the same expression for $I = dq/dt$ as before.

11.119 An RCL circuit has $R = 180\,\Omega$, $C = 1/280$ F, $L = 20$ H, and an applied voltage $E(t) = 10\sin t$. Assuming no initial charge on the capacitor, but an initial current of 1 A at $t = 0$ when the voltage is first applied, find the subsequent charge on the capacitor.

▌ The differential equation governing this system is formulated in Problem 1.82 as $\ddot{q} + 9\dot{q} + 14q = \frac{1}{2}\sin t$, where q denotes the charge on the capacitor. The solution to this equation is found in Problem 9.89 to be $q = c_1e^{-2t} + c_2e^{-7t} + \frac{13}{500}\sin t - \frac{9}{500}\cos t$.
Applying the initial conditions $q(0) = 0$ and $\dot{q}(0) = 1$, we obtain $c_1 = 110/500$ and $c_2 = -101/500$. Hence, $q = \frac{1}{500}(110e^{-2t} - 101e^{-7t} + 13\sin t - 9\cos t)$.

11.120 Determine the transient and steady-state components of the charge in the previous problem.

▌ Since the homogeneous (complementary) function, $\frac{110}{500}e^{-2t} - \frac{101}{500}e^{-7t}$, tends to zero as $t \to \infty$, it is the transient component. The steady-state component is the remaining part of the charge, namely $q_s = \frac{1}{500}(13\sin t - 9\cos t)$.

11.121 Determine the amplitude, period, and frequency of the steady-state charge of the previous problem.

▌ The amplitude is $A = \frac{1}{500}\sqrt{(13)^2 + (-9)^2} = 0.0316$. The natural frequency is $f = 2\pi$, so the period is $T = 1/f = 1/2\pi$.

11.122 A series circuit contains the components $L = 1$ H, $R = 1000\,\Omega$, and $C = 4 \times 10^{-6}$ F. At $t = 0$, while the circuit is completely passive (that is, while $Q = I = 0$), a battery supplying a constant voltage of $E = 24$ V is suddenly switched into the circuit. Find the charge on the capacitor as a function of time. (Here Q denotes the charge on the capacitor.)

▌ Substituting the numerical values for R, L, C, and E into (*1*) of Problem 1.82 yields $\dfrac{d^2Q}{dt^2} + 1000\dfrac{dQ}{dt} + \dfrac{Q}{4 \times 10^{-6}} = 24$. The solution to this equation is found in Problem 9.25 to be $Q = c_1e^{-500t} + c_2te^{-500t} + 9.6 \times 10^{-5}$.

Differentiation now yields $\dfrac{dQ}{dt} = I = -500c_1e^{-500t} + c_2(e^{-500t} - 500te^{-500t})$. Substituting $Q = 0$ at $t = 0$, we find $c_1 = -9.6 \times 10^{-5}$. Substituting $I = 0$ at $t = 0$, we find $0 = -500c_1 + c_2$, so that $c_2 = -4.8 \times 10^{-2}$. Hence $Q = -9.6 \times 10^{-5}e^{-500t} - 4.8 \times 10^{-2}te^{-500t} + 9.6 \times 10^{-5}$.

11.123 Determine the current as a function of time in the circuit of the previous problem.

▌ $I = \dfrac{dQ}{dt} = 24te^{-500t}$, in amperes.

11.124 A series RLC circuit has $R = 4\,\Omega$, $L = \frac{1}{2}$ H, $C = \frac{1}{26}$ F, and a constant emf of 13 V. Find the charge on the capacitor as a function of time if initially the circuit is completely passive.

▌ Substituting the given values for R, L, C, and the emf into (*1*) of Problem 1.82, we obtain $\ddot{Q} + 8\dot{Q} + 52Q = 26$, where Q denotes the charge on the capacitor. The solution to this equation is found in Problem 9.20 to be $Q = c_1e^{-4t}\cos 6t + c_2e^{-4t}\sin 6t + \frac{1}{2}$. The current I is then

$$I = \frac{dQ}{dt} = c_1(-4e^{-4t}\cos 6t - 6e^{-4t}\sin 6t) + c_2(-4e^{-4t}\sin 6t + 6e^{-4t}\cos 6t)$$

Applying the initial conditions, we obtain $0 = Q(0) = c_1 + \frac{1}{2}$, or $c_1 = -\frac{1}{2}$, and $0 = I(0) = -4c_1 + 6c_2$, from which $c_2 = -\frac{1}{3}$. Substitution of these values then yields $Q(t) = -e^{-4t}\dfrac{\cos 6t}{2} - e^{-4t}\dfrac{\sin 6t}{3} + \dfrac{1}{2}$.

11.125 Find the steady-state current in the circuit of the previous problem.

▮ Differentiation of the result of the previous problem yields $I = \dfrac{dQ}{dt} = \dfrac{13}{3}e^{-4t}\sin 6t$. Since this tends to zero as $t \to \infty$, the steady-state current is zero.

11.126 Find the steady-state charge on the capacitor in the circuit of Problem 11.124.

▮ The steady-state charge may be computed as $\lim\limits_{t\to\infty} Q = \lim\limits_{t\to\infty}\left(-e^{-4t}\dfrac{\cos 6t}{2} - e^{-4t}\dfrac{\sin 6t}{3} + \dfrac{1}{2}\right) = \dfrac{1}{2}$.

11.127 Solve Problem 11.124 if, instead of being constant, the emf is $E = 16\cos 2t$.

▮ The differential equation becomes $\ddot{Q} + 8\dot{Q} + 52Q = 32\cos 2t$, which has as its solution $Q = c_1e^{-4t}\cos 6t + c_2e^{-4t}\sin 6t + \frac{3}{5}\cos 2t + \frac{1}{5}\sin 2t$ (see Problem 9.87). Thus the current is now

$$I = \frac{dQ}{dt} = c_1(-4e^{-4t}\cos 6t - 6e^{-4t}\sin 6t) + c_2(-4e^{-4t}\sin 6t + 6e^{-4t}\cos 6t) - \tfrac{6}{5}\sin 2t + \tfrac{2}{5}\cos 2t$$

Applying the initial conditions, we obtain $0 = Q(0) = c_1 + \frac{3}{5}$, or $c_1 = -\frac{3}{5}$, and $0 = I(0) = -4c_1 + 6c_2 + \frac{2}{5}$, from which we find $c_2 = -\frac{7}{15}$. Substitution of these values yields

$$Q = -3e^{-4t}\frac{\cos 6t}{5} - 7e^{-4t}\frac{\sin 6t}{15} + \frac{3\cos 2t}{5} + \frac{\sin 2t}{5}$$

11.128 Find the steady-state charge on the capacitor in the previous problem.

▮ Since the homogeneous (complementary) function, $-\frac{3}{5}e^{-4t}\cos 6t - \frac{7}{15}e^{-4t}\sin 6t$, tends to zero as $t \to \infty$, the particular solution is the steady-state component. The steady-state charge is thus $Q_s = \frac{3}{5}\cos 2t + \frac{1}{5}\sin 2t$.

11.129 Express the steady-state charge of the previous problem in the form $A\sin(2t + \phi)$.

▮ Since $A\sin(2t+\phi) = A\cos 2t\sin\phi + A\sin 2t\cos\phi = \frac{3}{5}\cos 2t + \frac{1}{5}\sin 2t$, we require $A\sin\phi = \frac{3}{5}$ and $A\cos\phi = \frac{1}{5}$.

It now follows that $(\frac{3}{5})^2 + (\frac{1}{5})^2 = A^2\sin^2\phi + A^2\cos^2\phi = A^2$, so that $A = \sqrt{10}/5$. Moreover, $\tan\phi = \dfrac{A\sin\phi}{A\cos\phi} = \dfrac{3/5}{1/5} = 3$, so $\phi = \arctan 3 \approx 4\pi/10$. Thus $Q_s \approx (\sqrt{10}/5)\sin(2t + 4\pi/10)$.

11.130 An inductance of 2 H, a resistance of 16 Ω, and a capacitance of 0.02 F are connected in series with an emf $E = 100\sin 3t$. At $t = 0$ the charge on the capacitor and the current in the circuit are zero. Find the charge at $t > 0$.

▮ Letting Q and I be the instantaneous charge and current at time t, we find by Kirchhoff's laws (see Problem 1.82) that $\dfrac{d^2Q}{dt^2} + 8\dfrac{dQ}{dt} + 25Q = 50\sin 3t$. The solution to this equation is found in Problem 9.88 to be $Q = c_1e^{-4t}\cos 3t + c_2e^{-4t}\sin 3t + \frac{50}{52}\sin 3t - \frac{75}{52}\cos 3t$. The current is then

$$I = c_1(-4e^{-4t}\cos 3t - 3e^{-4t}\sin 3t) + c_2(-4e^{-4t}\sin 3t + 3e^{-4t}\cos 3t) + \tfrac{150}{52}\cos 3t - \tfrac{225}{52}\sin 3t$$

Applying the initial conditions, we obtain $0 = Q(0) = c_1 - \frac{75}{52}$, or $c_1 = \frac{75}{52}$, and $0 = I(0) = -4c_1 + 3c_2 + \frac{150}{52}$, so that $c_2 = \frac{50}{52}$. These values yield $Q = \frac{25}{52}(2\sin 3t - 3\cos 3t) + \frac{25}{52}e^{-4t}(3\cos 3t + 2\sin 3t)$.

1.131 Determine the current in the circuit of the previous problem at $t > 0$.

▮ $I = \dfrac{dQ}{dt} = \dfrac{75}{52}(2\cos 3t + 3\sin 3t) - \dfrac{25}{52}e^{-4t}(17\sin 3t + 6\cos 3t)$. The first term is the *steady-state* current and the second, which becomes negligible as time increases, is the *transient* current.

11.132 An electric circuit consists of an inductance of 0.1 H, a resistance of 20 Ω, and a capacitance of 25 μF = 25×10^{-6} F. Find the charge q and the current i at time t, given the initial conditions $q = 0.05$ C and $i = dq/dt = 0$ when $t = 0$.

▌ Since $L = 0.1$, $R = 20$, $c = 25 \times 10^{-6}$, and $E(t) = 0$, (1) of Problem 1.82 reduces to $\frac{d^2q}{dt^2} + 200\frac{dq}{dt} + 400{,}000q = 0$. The solution to this differential equation is found in Problem 8.69 to be $q = c_1e^{-100t}\cos 100\sqrt{39}t + c_2e^{-100t}\sin 100\sqrt{39}t$. Differentiation yields

$$\dot{q} = c_1(-100e^{-100t}\cos 100\sqrt{39}t - 100\sqrt{39}e^{-100t}\sin 100\sqrt{39}t)$$
$$+ c_2(-100e^{-100t}\sin 100\sqrt{39}t + 100\sqrt{39}e^{-100t}\cos 100\sqrt{39}t)$$

Applying the initial conditions, we obtain $0.05 = q(0) = c_1$ and $0 = \dot{q}(0) = -100c_1 + 100\sqrt{39}c_2$, from which $c_2 = 0.05/\sqrt{39} = 0.008$. Substitution then gives $q = e^{-100t}(0.05\cos 624.5t + 0.008\sin 624.5t)$.

11.133 Find the steady-state current for the circuit of the previous problem.

▌ Differentiating the result of the previous problem, we find $I = \frac{dq}{dt} = -32e^{-100t}\sin 624.5t$. Since this quantity tends to zero as $t \to \infty$, the current is all transient and there is no steady-state current.

11.134 Solve Problem 11.132 if there is an initial current of -0.2 A in the circuit.

▌ This change affects only the initial conditions for the problem, from which we now obtain $0.05 = q(0) = c_1$ as before, and $-0.2 = \dot{q}(0) = -100c_1 + 100\sqrt{39}c_2$. This latter equation yields $c_2 = 0.0077$, so the result becomes $q = e^{-100t}(0.05\cos 624.5t + 0.0077\sin 624.5t)$.

11.135 A circuit consists of an inductance of 0.05 H, a resistance of 20 Ω, a capacitance of 100 μF, and an emf $E = 100$ V. Find i and q, given the initial conditions $q = 0$ and $i = 0$ when $t = 0$.

▌ Here (1) of Problem 1.82 becomes $\frac{d^2q}{dt^2} + 400\frac{dq}{dt} + 200{,}000q = 2000$. The solution to this differential equation is found in Problem 9.23 to be $q = e^{-200t}(A\cos 400t + B\sin 400t) + 0.01$. Differentiating with respect to t yields $i = \frac{dq}{dt} = 200e^{-200t}[(-A + 2B)\cos 400t + (-B - 2A)\sin 400t]$.

Use of the initial conditions yields $A = -0.01$ and $B = -0.005$. Then substitution yields $q = e^{-200t}(-0.01\cos 400t - 0.005\sin 400t) + 0.01$ and $i = 5e^{-200t}\sin 400t$.

11.136 Solve the previous problem if the constant emf is replaced with a variable emf $E(t) = 100\cos 200t$.

▌ The differential equation now becomes $\frac{d^2q}{dt^2} + 400\frac{dq}{dt} + 200{,}000q = 2000\cos 200t$, which has as its solution $q = e^{-200t}(A\cos 400t + B\sin 400t) + 0.01\cos 200t + 0.005\sin 200t$ (see Problem 9.93). Therefore, $i = e^{-200t}[(-200A + 400B)\cos 400t + (-200B - 400A)\sin 400t] - 2\sin 200t + \cos 200t$.

Use of the initial conditions yields $A = -0.01$ and $B = -0.0075$. Then

$$q = e^{-200t}(-0.01\cos 400t - 0.0075\sin 400t) + 0.01\cos 200t + 0.005\sin 200t$$

and
$$i = e^{-200t}(-\cos 400t + 5.5\sin 400t) - 2\sin 200t + \cos 200t$$

11.137 A series circuit contains an inductance $L = 1$ H, a resistance $R = 1000$ Ω, and a capacitance $C = 6.25 \times 10^{-6}$ F. At $t = 0$, with the capacitor bearing a charge of 1.5×10^{-3} C, a switch is closed so that the capacitor discharges through the (now) closed circuit. Find Q and i as functions of t.

▌ With $E(t) = 0$, (1) of Problem 1.82 becomes $\frac{d^2Q}{dt^2} + 1000\frac{dQ}{dt} + 160{,}000Q = 0$. Its solution is found in Problem 8.19 to be $Q = c_1e^{-200t} + c_2e^{-800t}$, and differentiating yields $i = -200c_1e^{-200t} - 800c_2e^{-800t}$.

Substitution of the initial data into the equations for Q and i yields

$$1.5 \times 10^{-3} = c_1 + c_2 \qquad 0 = -200c_1 - 800c_2$$

Solving these two equations simultaneously, we obtain $c_1 = 2 \times 10^{-3}$ and $c_2 = -5 \times 10^{-4}$. Hence $Q = 2 \times 10^{-3}e^{-200t} - 5 \times 10^{-4}e^{-800t}$ and $i = -0.4e^{-200t} + 0.4e^{-800t}$.

38 When is the absolute value of the current a maximum for the circuit of the previous problem?

The extreme value of the current occurs when $di/dt = 0$. Hence we must solve the equation $\frac{di}{dt} = 80e^{-200t} - 320e^{-800t} = 0$. Multiplying by e^{200t} and dividing by 320, we obtain $e^{-600t} = \frac{1}{4}$, from which $600t = \ln 4$ and $t = 0.00231$ s.

39 What is the extreme value of the current in the circuit of Problem 11.137?

We use the result of the previous problem, noting that the extreme value of the current occurs at $t = 0.00231$. Then $i = -0.4e^{-0.462} + 0.4e^{-1.848} = -0.189$ A.

40 A series RCL circuit with $R = 5\,\Omega$, $C = 10^{-2}$ F, and $L = \frac{1}{8}$ H has an applied voltage $E(t) = \sin t$. Find the steady-state current in the circuit.

We use (2) of Problem 1.81, which becomes $\frac{d^2I}{dt^2} + 40\frac{dI}{dt} + 800I = 8\cos t$. Its solution is found in Problem 9.95 to be $I = e^{-20t}(c_1 \cos 20t + c_2 \sin 20t) + \frac{6392}{640{,}001}\cos t + \frac{320}{640{,}001}\sin t$. The complementary function, composed of the first term of this equation, tends to zero rapidly, leaving as the steady-state solution

$$I_s = \frac{1}{640{,}001}(6392\cos t + 320\sin t).$$

41 Solve the previous problem if there is no applied voltage.

With no external emf, the current in the circuit must tend to zero. If, initially, the system is passive, then the current is always zero. In any event, $I_s = 0$.

42 Find the amplitude and frequency of the steady-state current in Problem 11.140.

The amplitude is $A = \sqrt{\left(\frac{6392}{640{,}001}\right)^2 + \left(\frac{320}{640{,}001}\right)^2} = 0.001$. The period of both $\sin t$ and $\cos t$ is 2π, so the frequency is $1/2\pi$.

43 For a series circuit consisting of an inductance L, a resistance R, a capacitance C, and an emf $E(t) = E_0 \sin \omega t$, derive the formula for the steady-state current $i = \frac{E_0}{Z}\left(\frac{R}{Z}\sin\omega t - \frac{X}{Z}\cos\omega t\right) = \frac{E_0}{Z}\sin(\omega t - \theta)$, where $X = L\omega - 1/C\omega$, $Z = \sqrt{X^2 + R^2}$, and θ is determined from $\sin\theta = X/Z$ and $\cos\theta = R/Z$.

By differentiating $L\frac{d^2q}{dt^2} + R\frac{dq}{dt} + \frac{q}{C} = E_0 \sin\omega t$ and using $i = \frac{dq}{dt}$, we obtain $L\frac{d^2i}{dt^2} + R\frac{di}{dt} + \frac{i}{C} = \left(LD^2 + RD + \frac{1}{C}\right)i = \omega E_0 \cos\omega t$. The required steady-state solution is the particular integral of this equation:

$$i = \frac{\omega E_0}{LD^2 + RD + 1/C}\cos\omega t = \frac{\omega E_0}{RD - \left(L\omega - \frac{1}{C\omega}\right)\omega}\cos\omega t$$

$$= \frac{\omega E_0(RD + X\omega)}{R^2D^2 - X^2\omega^2}\cos\omega t = \frac{E_0}{R^2 + X^2}(R\sin\omega t - X\cos\omega t)$$

$$= \frac{E_0}{Z}\left(\frac{R}{Z}\sin\omega t - \frac{X}{Z}\cos\omega t\right) = \frac{E_0}{Z}\sin(\omega t - \theta)$$

44 A series circuit consisting of an inductance L, a capacitance C, and an emf E is known as a harmonic oscillator. Find q and i when $E = E_0 \cos\omega t$ and the initial conditions are $q = q_0$ and $i = i_0$ when $t = 0$. Assume that $\omega \neq 1\sqrt{CL}$.

With $R = 0$, (1) of Problem 1.82 becomes $\frac{d^2q}{dt^2} + \frac{q}{CL} = \frac{E_0}{L}\cos\omega t$. If $\omega \neq 1/\sqrt{CL}$, the solution to this differential equation is

$$q = A\cos\frac{1}{\sqrt{CL}}t + B\sin\frac{1}{\sqrt{CL}}t + \frac{E_0}{L}\frac{1}{D^2 + 1/CL}\cos\omega t = A\cos\frac{1}{\sqrt{CL}}t + B\sin\frac{1}{\sqrt{CL}}t + \frac{E_0 C}{1-\omega^2 CL}\cos\omega t$$

and $$i = \frac{1}{\sqrt{CL}}\left(-A\sin\frac{1}{\sqrt{CL}}t + B\cos\frac{1}{\sqrt{CL}}t\right) - \frac{E_0 C\omega}{1-\omega^2 CL}\sin\omega t$$

Use of the initial conditions then yields $A = q_0 - \dfrac{E_0 C}{1-\omega^2 CL}$ and $B = \sqrt{CL}i_0$. Then

$$q = \left(q_0 - \frac{E_0 C}{1-\omega^2 CL}\right)\cos\frac{1}{\sqrt{CL}}t + \sqrt{CL}i_0\sin\frac{1}{\sqrt{CL}}t + \frac{E_0 C}{1-\omega^2 CL}\cos\omega t$$

and $$i = i_0\cos\frac{1}{\sqrt{CL}}t - \frac{1}{\sqrt{CL}}\left(q_0 - \frac{E_0 C}{1-\omega^2 CL}\right)\sin\frac{1}{\sqrt{CL}}t - \frac{E_0 C\omega}{1-\omega^2 CL}\sin\omega t$$

11.145 Solve the previous problem under the condition that $\omega = 1/\sqrt{CL}$.

▮ Here (*1*) of Problem 1.82 becomes $\dfrac{d^2q}{dt^2} + \omega^2 q = \dfrac{E_0}{L}\cos\omega t$. Then $q = A\cos\omega t + B\sin\omega t + \dfrac{E_0}{2L\omega}t\sin\omega t$, and $i = \omega(-A\sin\omega t + B\cos\omega t) + \dfrac{E_0}{2L}\left(\dfrac{1}{\omega}\sin\omega t + t\cos\omega t\right)$.

Use of the initial conditions now yields $A = q_0$ and $B = i_0/\omega$. Then

$$q = q_0\cos\omega t + \frac{i_0}{\omega}\sin\omega t + \frac{E_0}{2L\omega}t\sin\omega t, \quad \text{and} \quad i = i_0\cos\omega t - q_0\omega\sin\omega t + \frac{E_0}{2L}\left(\frac{1}{\omega}\sin\omega t + t\cos\omega t\right).$$

Note that here the frequency of the impressed emf is the natural frequency of the oscillator, that is, the frequency when there is no impressed emf. The circuit is in resonance, since the reactance $X = L\omega - 1/C\omega$ is zero when $\omega = 1/\sqrt{CL}$. The presence of the term $(E_0 t/2L)\cos\omega t$, whose amplitude increases with t, indicates that eventually such a circuit will destroy itself.

CHAPTER 12

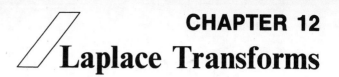

Laplace Transforms

TRANSFORMS OF ELEMENTARY FUNCTIONS

12.1 Find the Laplace transform of $f(x) \equiv 1$.

We have $\mathscr{L}\{1\} = \int_0^\infty e^{-sx}(1)\,dx$. For $s = 0$,

$$\int_0^\infty e^{-sx}\,dx = \int_0^\infty e^{-(0)(x)}\,dx = \lim_{R\to\infty} \int_0^R 1\,dx = \lim_{R\to\infty} x\Big|_0^R = \lim_{R\to\infty} R = \infty$$

Hence the integral diverges. For $s \neq 0$,

$$\int_0^\infty e^{-sx}\,dx = \lim_{R\to\infty} \int_0^R e^{-sx}\,dx = \lim_{R\to\infty} \left(-\frac{1}{s}e^{-sx}\right)\Bigg|_{x=0}^{x=R} = \lim_{R\to\infty} \left(\frac{-1}{s}e^{-sR} + \frac{1}{s}\right)$$

When $s < 0$, $-sR > 0$; hence the limit is ∞ and the integral diverges. When $s > 0$, $-sR < 0$; hence, the limit is $1/s$ and the integral converges. Thus, $\mathscr{L}\{1\} = \dfrac{1}{s}$, for $s > 0$.

12.2 Find $\mathscr{L}\{e^{ax}\}$.

$$\mathscr{L}\{e^{ax}\} = \int_0^\infty e^{-sx}e^{ax}\,dx = \lim_{R\to\infty} \int_0^R e^{(a-s)x}\,dx = \lim_{R\to\infty} \left(\frac{e^{(a-s)x}}{a-s}\right)\Bigg|_{x=0}^{x=R}$$

$$= \lim_{R\to\infty} \left(\frac{e^{(a-s)R} - 1}{a-s}\right) = \frac{1}{s-a} \qquad \text{for } s > a$$

Note that when $s \le a$, the improper integral diverges.

12.3 Find the Laplace transform of $f(t) = t$.

$$\mathscr{L}(t) = \int_0^\infty e^{-st}(t)\,dt = \lim_{P\to\infty} \int_0^P te^{-st}\,dt = \lim_{P\to\infty} \left(t\frac{e^{-st}}{-s} - \frac{e^{-st}}{s^2}\right)\Bigg|_0^P$$

$$= \lim_{P\to\infty} \left(\frac{1}{s^2} - \frac{e^{-sP}}{s^2} - \frac{Pe^{-sP}}{s}\right) = \frac{1}{s^2} \qquad \text{for } s > 0$$

where we have used integration by parts.

12.4 Find the Laplace transform of $f(x) = x^2$.

Using integration by parts twice, we find that

$$\mathscr{L}\{x^2\} = \int_0^\infty e^{-sx}x^2\,dx = \lim_{R\to\infty} \int_0^R x^2e^{-sx}\,dx = \lim_{R\to\infty} \left(-\frac{x^2}{s}e^{-sx} - \frac{2x}{s^2}e^{-sx} - \frac{2}{s^3}e^{-sx}\right)\Bigg|_{x=0}^{x=R}$$

$$= \lim_{R\to\infty} \left(-\frac{R^2}{s}e^{-sR} - \frac{2R}{s^2}e^{-sR} - \frac{2}{s^3}e^{-sR} + \frac{2}{s^3}\right)$$

For $s < 0$, $\lim_{R\to\infty}\left(-\dfrac{R^2}{s}e^{-sR}\right) = \infty$, and the improper integral diverges. For $s > 0$, it follows from repeated use of L'Hôpital's rule that

$$\lim_{R\to\infty} \left(-\frac{R^2}{s}e^{-sR}\right) = \lim_{R\to\infty} \frac{-R^2}{se^{sR}} = \lim_{R\to\infty} \frac{-2R}{s^2e^{sR}} = \lim_{R\to\infty} \frac{-2}{s^3e^{sR}} = 0$$

$$\lim_{R\to\infty} \left(-\frac{2R}{s}e^{-sR}\right) = \lim_{R\to\infty} \frac{-2R}{se^{sR}} = \lim_{R\to\infty} \frac{-2}{s^2e^{sR}} = 0$$

Also, $\lim_{R\to\infty}\left(-\frac{2}{s^3}e^{-sR}\right)=0$ directly; hence the integral converges, and $F(s)=2/s^3$. For the special case $s=0$, we have $\int_0^\infty e^{-sx}x^2\,dx=\int_0^\infty e^{-s(0)}x^2\,dx=\lim_{R\to\infty}\int_0^R x^2\,dx=\lim_{R\to\infty}\frac{R^3}{3}=\infty$. Finally, combining all cases, we obtain $\mathscr{L}\{x^2\}=2/s^3$, for $s>0$.

12.5 Find the Laplace transform of $f(t)=te^{at}$, where a denotes a constant.

$$\mathscr{L}\{te^{at}\}=\int_0^\infty te^{at}e^{-st}\,dt=\int_0^\infty te^{(a-s)t}\,dt=\lim_{T\to\infty}\left[\left(\frac{te^{(a-s)t}}{a-s}\right)_0^T-\int_0^T\frac{e^{(a-s)t}}{a-s}\,dt\right]$$

$$=\lim_{T\to\infty}\left[\frac{te^{(a-s)t}}{a-s}-\frac{e^{(a-s)t}}{(a-s)^2}\right]_0^T=\lim_{T\to\infty}\left\{\frac{Te^{(a-s)T}}{a-s}-\frac{0}{a-s}-\left[\frac{e^{(a-s)T}}{(a-s)^2}-\frac{1}{(a-s)^2}\right]\right\}$$

$$=\frac{1}{(a-s)^2}+\lim_{T\to\infty}\frac{Te^{(a-s)T}}{a-s}-\lim_{T\to\infty}\frac{e^{(a-s)T}}{a-s}=\frac{1}{(a-s)^2}\qquad\text{for } s>a$$

12.6 Find $\mathscr{L}\{\sin at\}$, where a denotes a constant.

Using integration by parts twice and the formulas $\int e^{\alpha t}\sin\beta t\,dt=\frac{e^{\alpha t}(\alpha\sin\beta t-\beta\cos\beta t)}{\alpha^2+\beta^2}$ and $\int e^{\alpha t}\cos\beta t\,dt=\frac{e^{\alpha t}(\alpha\cos\beta t+\beta\sin\beta t)}{\alpha^2+\beta^2}$, we obtain

$$\mathscr{L}\{\sin at\}=\int_0^\infty e^{-st}\sin at\,dt=\lim_{P\to\infty}\int_0^P e^{-st}\sin at\,dt=\lim_{P\to\infty}\left.\frac{e^{-st}(-s\sin at-a\cos at)}{s^2+a^2}\right|_0^P$$

$$=\lim_{P\to\infty}\left[\frac{a}{s^2+a^2}-\frac{e^{-sP}(s\sin aP+a\cos aP)}{s^2+a^2}\right]=\frac{a}{s^2+a^2}\qquad\text{for } s>0$$

12.7 Find $\mathscr{L}\{\cos at\}$, where a denotes a constant.

Using integration by parts twice along with the formulas given in the preceding problem, we obtain

$$\mathscr{L}\{\cos at\}=\int_0^\infty e^{-st}\cos at\,dt=\lim_{P\to\infty}\int_0^P e^{-st}\cos at\,dt=\lim_{P\to\infty}\left.\frac{e^{-st}(-s\cos at+a\sin at)}{s^2+a^2}\right|_0^P$$

$$=\lim_{P\to\infty}\left[\frac{s}{s^2+a^2}-\frac{e^{-sP}(s\cos aP-a\sin aP)}{s^2+a^2}\right]=\frac{s}{s^2+a^2}\qquad\text{for } s>0$$

12.8 Solve the previous two problems by utilizing complex numbers.

Assuming that the result of Problem 12.2 holds for complex numbers (which can be proved), we have $\mathscr{L}\{e^{iat}\}=\frac{1}{s-ia}=\frac{s+ia}{s^2+a^2}$. Using Euler's formula, we also have $e^{iat}=\cos at+i\sin at$, so

$$\mathscr{L}\{e^{iat}\}=\int_0^\infty e^{-st}(\cos at+i\sin at)$$

$$=\int_0^\infty e^{-st}\cos at\,dt+i\int_0^\infty e^{-st}\sin at\,dt=\mathscr{L}\{\cos at\}+i\mathscr{L}\{\sin at\}$$

Equating our two expressions for $\mathscr{L}\{e^{iat}\}$ and then equating real and imaginary parts, we conclude that $\mathscr{L}\{\cos at\}=\frac{s}{s^2+a^2}$ and $\mathscr{L}\{\sin at\}=\frac{a}{s^2+a^2}$.

12.9 Find the Laplace transform of $f(x)=e^{3x}$.

Using the result of Problem 12.2 with $a=3$, we have $\mathscr{L}\{e^{3x}\}=\frac{1}{s-3}$.

12.10 Find the Laplace transform $f(t)=e^{-4t}$.

The Laplace transform of $f(t)$ is the same as that of $f(x)$, so the result of Problem 12.2 (with $a=-4$) yielc $\mathscr{L}\{e^{-4t}\}=\frac{1}{s-(-4)}=\frac{1}{s+4}$.

1 Find the Laplace transform of $\sin \pi t$.

▌ Using the result of Problem 12.6 with $a = \pi$, we have $\mathscr{L}\{\sin \pi t\} = \dfrac{\pi}{s^2 + \pi^2}$.

2 Find the Laplace transform of $\cos 2x$.

▌ The Laplace transform of $f(x)$ is the same as that of $f(t)$, so the result of Problem 12.7 (with $a = 2$) gives $\mathscr{L}\{\cos 2t\} = \dfrac{s}{s^2 + 2^2} = \dfrac{s}{s^2 + 4}$.

3 Find the Laplace transform of $f(x) = \sin(-3x)$.

▌ The Laplace transform of $f(x)$ is the same as that of $f(t)$, so the result of Problem 12.6 (with $a = -3$) yields $\mathscr{L}\{\sin(-3x)\} = \dfrac{-3}{s^2 + (-3)^2} = -\dfrac{3}{s^2 + 9}$.

4 Find the Laplace transform of $f(t) = \cos(-5t)$.

▌ Using the result of Problem 12.7 with $a = -5$, we obtain $\mathscr{L}\{\cos(-5t)\} = \dfrac{s}{s^2 + (-5)^2} = \dfrac{s}{s^2 + 25}$.

5 Find the Laplace transform of $f(t) = te^{-4t}$.

▌ Using the result of Problem 12.5 with $a = -4$, we have $\mathscr{L}\{te^{-4t}\} = \dfrac{1}{(-4 - s)^2} = \dfrac{1}{(s + 4)^2}$.

6 Find the Laplace transform of $f(x) = xe^x$.

▌ Using the result of Problem 12.5 with $a = 1$, we have $\mathscr{L}\{xe^x\} = \dfrac{1}{(1 - s)^2} = \dfrac{1}{(s - 1)^2}$.

7 Find $\mathscr{L}\{f(x)\}$ if $f(x) = \begin{cases} -1 & x \le 4 \\ 1 & x > 4 \end{cases}$.

▌
$$\mathscr{L}\{f(x)\} = \int_0^\infty e^{-sx} f(x)\,dx = \int_0^4 e^{-sx}(-1)\,dx + \int_4^\infty e^{-sx}(1)\,dx$$
$$= \frac{e^{-sx}}{s}\Big|_{x=0}^{x=4} + \lim_{R\to\infty} \int_4^R e^{-sx}\,dx$$
$$= \frac{e^{-4s}}{s} - \frac{1}{s} + \lim_{R\to\infty}\left(\frac{-1}{s}e^{-Rs} + \frac{1}{s}e^{-4s}\right) = \frac{2e^{-4s}}{s} - \frac{1}{s} \quad \text{for } s > 0$$

18 Find $\mathscr{L}\{f(t)\}$ if $f(t) = \begin{cases} 5 & 0 < t < 3 \\ 0 & t > 3 \end{cases}$.

▌
$$\mathscr{L}\{f(t)\} = \int_0^\infty e^{-st} f(t)\,dt = \int_0^3 e^{-st}(5)\,dt + \int_3^\infty e^{-st}(0)\,dt$$
$$= 5\int_0^3 e^{-st}\,dt = 5\,\frac{e^{-st}}{-s}\Big|_0^3 = \frac{5(1 - e^{-3s})}{s}$$

19 Find the Laplace transform of $f(x) = \begin{cases} x & 0 \le x \le 2 \\ 2 & x > 2 \end{cases}$.

▌
$$\mathscr{L}\{f(x)\} = \int_0^\infty e^{-sx} f(x)\,dx = \int_0^2 e^{-sx} x\,dx + \int_2^\infty e^{-sx}(2)\,dx$$
$$= \frac{e^{-sx}}{s^2}(-sx - 1)\Big|_0^2 + \lim_{M\to\infty} 2\int_2^M e^{-sx}\,dx$$
$$= \frac{e^{-2s}}{s^2}(-2s - 1) + \frac{1}{s^2} + \lim_{M\to\infty}\left(-\frac{2e^{-Ms}}{s} + \frac{2e^{-2s}}{s.}\right) = \frac{1 - e^{-2s}}{s^2}$$

12.20 Find the Laplace transform of $f(x) = \begin{cases} 1 & 0 \le x \le 1 \\ e^x & 1 < x \le 4 \\ 0 & x > 4 \end{cases}$

▮
$$\mathscr{L}\{f(x)\} = \int_0^\infty e^{-sx} f(x)\,dx = \int_0^1 e^{-sx}(1)\,dx + \int_1^4 e^{-sx}e^x\,dx + \int_4^\infty e^{-sx}(0)\,dx$$
$$= \int_0^1 e^{-sx}\,dx + \int_1^4 e^{-(s-1)x}\,dx = -\frac{e^{-sx}}{s}\bigg|_0^1 - \frac{e^{-(s-1)x}}{s-1}\bigg|_1^4$$
$$= \frac{1-e^{-s}}{s} + \frac{e^{-(s-1)} - e^{-4(s-1)}}{s-1}$$

12.21 Find the Laplace transform of $f(x) = \begin{cases} 0 & 0 < x < 2 \\ 4 & x > 2 \end{cases}$.

▮
$$\mathscr{L}\{f(x)\} = \int_0^\infty e^{-sx} f(x)\,dx = \int_0^2 e^{-sx}(0)\,dx + \int_2^\infty e^{-sx}(4)\,dx$$
$$= \lim_{P\to\infty} \int_2^P 4e^{-sx}\,dx = \lim_{P\to\infty}\left(-\frac{4e^{-sP}}{s} + \frac{4e^{-s2}}{s}\right) = 4\frac{e^{-2s}}{s}$$

12.22 Find the Laplace transform for the (Heaviside) *unit step function* about the point c, defined by $u(x-c) = \begin{cases} 0 & x < c \\ 1 & x \ge c \end{cases}$.

▮
$$\mathscr{L}\{u(x-c)\} = \int_0^\infty e^{-sx}u(x-c)\,dx = \int_0^c e^{-sx}(0)\,dx + \int_c^\infty e^{-sx}(1)\,dx = \int_c^\infty e^{-sx}\,dx$$
$$= \lim_{R\to\infty} \int_c^R e^{-sx}\,dx = \lim_{R\to\infty} \frac{e^{-sR} - e^{-sc}}{-s} = \frac{1}{s}e^{-sc} \qquad \text{for } s > 0$$

12.23 Find the Laplace transform of $f(x) = \sin^2 ax$, where a denotes a constant.

▮
$$\mathscr{L}\{\sin^2 ax\} = \int_0^\infty e^{-sx}\sin^2 ax\,dx = \lim_{L\to\infty}\int_0^L e^{-sx}\sin^2 ax\,dx = \lim_{L\to\infty}\left\{e^{-sx}\left[\frac{-1}{2s} - \frac{2a\sin 2ax - s\cos 2ax}{2(s^2+4a^2)}\right]\right\}_0^L$$
$$= \lim_{L\to\infty}\left[\frac{-1}{2s}e^{-sL} + \frac{1}{2s} - \frac{2a(\sin 2aL)e^{-sL}}{2(s^2+4a^2)} + \frac{s(\cos 2aL)e^{-sL}}{2(s^2+4a^2)} - \frac{s}{2(s^2+4a^2)}\right]$$
$$= \frac{1}{2s} - \frac{s}{2(s^2+4a^2)} = \frac{2a^2}{s(s^2+4a^2)} \qquad \text{for } s > 0$$

12.24 Find the Laplace transform of $f(x) = \sin(ax+b)$, where both a and b denote constants.

▮
$$\mathscr{L}\{\sin(ax+b)\} = \lim_{P\to\infty}\int_0^P e^{-sx}\sin(ax+b)\,dx = \lim_{P\to\infty}\left[\frac{-se^{-sx}\sin(ax+b) - ae^{-sx}\cos(ax+b)}{s^2+a^2}\right]_0^P$$
$$= \lim_{P\to\infty}\left[\frac{-se^{-sP}\sin(aP+b) - ae^{-sP}\cos(aP+b)}{s^2+a^2} + \frac{s\sin b + a\cos b}{s^2+a^2}\right]$$
$$= \frac{s\sin b + a\cos b}{s^2+a^2} \qquad \text{for } s > 0$$

12.25 Find the Laplace transform of $\cos(ax+b)$, where both a and b denote constants.

▮
$$\mathscr{L}\{\cos(ax+b)\} = \lim_{P\to\infty}\int_0^P e^{-sx}\cos(ax+b)\,dx = \lim_{P\to\infty}\left[\frac{-se^{-sx}\cos(ax+b) + ae^{-sx}\sin(ax+b)}{s^2+a^2}\right]_0^P$$
$$= \lim_{P\to\infty}\left[\frac{-se^{-sP}\cos(aP+b) + ae^{-sP}\sin(aP+b)}{s^2+a^2} + \frac{s\cos b - a\sin b}{s^2+a^2}\right]$$
$$= \frac{s\cos b - a\sin b}{s^2+a^2} \qquad \text{for } s > 0$$

12.26 Find $\mathscr{L}\{f_\epsilon(t)\}$, where $f_\epsilon(t) = \begin{cases} 1/\epsilon & 0 \le t \le \epsilon \\ 0 & t > \epsilon \end{cases}$.

$$\mathscr{L}\{f_\epsilon(t)\} = \int_0^\infty e^{-st} f_\epsilon(t)\,dt = \int_0^\epsilon e^{-st}\frac{1}{\epsilon}\,dt + \int_\epsilon^\infty e^{-st}(0)\,dt$$

$$= \frac{1}{\epsilon}\int_0^\epsilon e^{-st}\,dt = \frac{1-e^{-s\epsilon}}{\epsilon s}$$

12.27 Show that $\lim_{\epsilon\to 0} \mathscr{L}\{f_\epsilon(t)\} = 1$ in Problem 12.26. Is this limit the same as $\mathscr{L}\{\lim_{\epsilon\to 0} f_\epsilon(t)\}$? Explain.

The required result follows at once, since

$$\lim_{\epsilon\to 0}\frac{1-e^{-s\epsilon}}{s\epsilon} = \lim_{\epsilon\to\infty}\frac{1-(1-s\epsilon+s^2\epsilon^2/2!-\cdots)}{s\epsilon} = \lim_{\epsilon\to 0}\left(1-\frac{s\epsilon}{2!}+\cdots\right) = 1$$

It also follows by use of L'Hôspital's rule.

Mathematically speaking, $\lim_{\epsilon\to 0} f_\epsilon(t)$ does not exist, so that $\mathscr{L}\{\lim_{\epsilon\to 0} f_\epsilon(t)\}$ is not defined. Nevertheless it proves useful to consider $\delta(t) = \lim_{\epsilon\to 0} f_\epsilon(t)$ to be such that $\mathscr{L}\{\delta(t)\} = 1$. We call $\delta(t)$ the *Dirac delta function* or *impulse function.*

TRANSFORMS INVOLVING GAMMA FUNCTIONS

12.28 Define the *gamma function* $\Gamma(p)$, and then show that $\Gamma(p+1) = p\Gamma(p)$ for $p > 0$.

The *gamma function* is defined, for any positive real number p, as $\Gamma(p) = \int_0^\infty x^{p-1}e^{-x}\,dx$.

Using integration by parts, we have

$$\Gamma(p+1) = \int_0^\infty x^{(p+1)-1}e^{-x}\,dx = \lim_{r\to\infty}\int_0^r x^p e^{-x}\,dx = \lim_{r\to\infty}\left(-x^p e^{-x}\Big|_0^r + \int_0^r p x^{p-1}e^{-x}\,dx\right)$$

$$= \lim_{r\to\infty}(-r^p e^{-r}+0) + p\int_0^\infty x^{p-1}e^{-x}\,dx = p\Gamma(p)$$

The result $\lim_{r\to\infty} r^p e^{-r} = 0$ is easily obtained by first writing $r^p e^{-r}$ as r^p/e^r and then using L'Hôspital's rule.

12.29 Prove that $\Gamma(1) = 1$.

$$\Gamma(1) = \int_0^\infty x^{1-1}e^{-x}\,dx = \lim_{r\to\infty}\int_0^r e^{-x}\,dx = \lim_{r\to\infty}\left(-e^{-x}\right)_0^r = \lim_{r\to\infty}(-e^{-r}+1) = 1$$

12.30 Prove that if n is a positive integer, then $\Gamma(n+1) = n!$.

The proof is by induction. First we consider $n = 1$. Using Problem 12.28 with $p = 1$ and then Problem 12.29, we have $\Gamma(1+1) = 1\Gamma(1) = 1(1) = 1 = 1!$.

Next we assume that $\Gamma(n+1) = n!$ holds for $n = k$ and then try to prove its validity for $n = k+1$. From Problem 12.28 with $p = k+1$ and using the induction hypothesis, we have

$$\Gamma[(k+1)+1] = (k+1)\Gamma(k+1) = (k+1)(k!) = (k+1)!$$

Thus, $\Gamma(n+1) = n!$ is true by induction.

Note that we can now use this equality to define $0!$; that is, $0! = \Gamma(0+1) = \Gamma(1) = 1$.

12.31 Prove that $\Gamma(p+k+1) = (p+k)(p+k-1)\cdots(p+2)(p+1)\Gamma(p+1)$.

Using Problem 12.28 repeatedly, where p is replaced first by $p+k$, then by $p+k-1$, and so on, we obtain

$$\Gamma(p+k+1) = \Gamma[(p+k)+1] = (p+k)\Gamma(p+k)$$

$$= (p+k)\Gamma[(p+k-1)+1] = (p+k)(p+k-1)\Gamma(p+k-1) = \cdots$$

$$= (p+k)(p+k-1)\cdots(p+2)(p+1)\Gamma(p+1)$$

12.32 Evaluate $\Gamma(6)/2\Gamma(3)$.

$$\frac{\Gamma(6)}{2\Gamma(3)} = \frac{5!}{2(2!)} = \frac{(5)(4)(3)(2)}{(2)(2)} = 30$$

12.33 Evaluate $\dfrac{\Gamma(5/2)}{\Gamma(1/2)}$.

▮
$$\frac{\Gamma(5/2)}{\Gamma(1/2)} = \frac{(3/2)\Gamma(3/2)}{\Gamma(1/2)} = \frac{(3/2)(1/2)\Gamma(1/2)}{\Gamma(1/2)} = \frac{3}{4}$$

12.34 Evaluate $\dfrac{\Gamma(3)\Gamma(2.5)}{\Gamma(5.5)}$.

▮
$$\frac{\Gamma(3)\Gamma(2.5)}{\Gamma(5.5)} = \frac{2!(1.5)(0.5)\Gamma(0.5)}{(4.5)(3.5)(2.5)(1.5)(0.5)\Gamma(0.5)} = \frac{16}{315}$$

12.35 Evaluate $\dfrac{6\Gamma(8/3)}{5\Gamma(2/3)}$.

▮
$$\frac{6\Gamma(8/3)}{5\Gamma(2/3)} = \frac{6(5/3)(2/3)\Gamma(2/3)}{5\Gamma(2/3)} = \frac{4}{3}$$

12.36 Evaluate $\int_0^\infty x^3 e^{-x}\,dx$.

▮
$$\int_0^\infty x^3 e^{-x}\,dx = \Gamma(4) = 3! = 6$$

12.37 Evaluate $\int_0^\infty x^6 e^{-2x}\,dx$.

▮ Let $2x = y$. Then the integral becomes $\displaystyle\int_0^\infty \left(\frac{y}{2}\right)^6 e^{-y}\frac{dy}{2} = \frac{1}{2^7}\int_0^\infty y^6 e^{-y}\,dy = \frac{\Gamma(7)}{2^7} = \frac{6!}{2^7} = \frac{45}{8}$.

12.38 Express $\int_0^\infty e^{-x^2}\,dx$ as a gamma function.

▮ Let $z = x^2$; then $x = z^{1/2}$ and $dx = \frac{1}{2}z^{-1/2}\,dz$. Substituting these values into the integral and noting that as x goes from 0 to ∞ so does z, we have

$$\int_0^\infty e^{-x^2}\,dx = \int_0^\infty e^{-z}\left(\frac{1}{2}z^{-1/2}\right)dz = \frac{1}{2}\int_0^\infty z^{(1/2)-1}e^{-z}\,dz = \frac{1}{2}\Gamma\left(\frac{1}{2}\right)$$

12.39 Prove that $\Gamma(\frac{1}{2}) = \sqrt{\pi}$.

▮ We have $\Gamma(\frac{1}{2}) = \int_0^\infty x^{-1/2}e^{-x}\,dx = 2\int_0^\infty e^{-u^2}\,du$, where we have substituted $x = u^2$. It follows that

$$[\Gamma(\tfrac{1}{2})]^2 = \left\{2\int_0^\infty e^{-u^2}\,du\right\}\left\{2\int_0^\infty e^{-v^2}\,dv\right\} = 4\int_0^\infty\int_0^\infty e^{-(u^2+v^2)}\,du\,dv$$

If we change now to polar coordinates (ρ, ϕ), substituting $u = \rho\cos\phi$ and $v = \rho\sin\phi$, the last integral becomes $4\int_{\phi=0}^{\pi/2}\int_{\rho=0}^{\infty} e^{-\rho^2}\rho\,d\rho\,d\phi = 4\int_{\phi=0}^{\pi/2} -\frac{1}{2}e^{-\rho^2}\Big|_{\rho=0}^{\infty}\,d\phi = \pi$, and so $\Gamma(\frac{1}{2}) = \sqrt{\pi}$.

12.40 Evaluate $\int_0^\infty \sqrt{y}e^{-y^3}\,dy$.

▮ If we let $y^3 = x$, the integral becomes $\int_0^\infty \sqrt{x^{1/3}}e^{-x}(\frac{1}{3}x^{-2/3}\,dx) = \frac{1}{3}\int_0^\infty x^{-1/2}e^{-x}\,dx = \frac{1}{3}\Gamma(\frac{1}{2}) = \sqrt{\pi}/3$.

12.41 Evaluate $\int_0^\infty 3^{-4z^2}\,dz$.

▮ We write the integral as $\int_0^\infty 3^{-4z^2}\,dz = \int_0^\infty (e^{\ln 3})^{-4z^2}\,dz = \int_0^\infty e^{-(4\ln 3)z^2}\,dz$. Now we let $(4\ln 3)z^2 = x$, and the integral becomes $\displaystyle\int_0^\infty e^{-x}d\left(\frac{x^{1/2}}{\sqrt{4\ln 3}}\right) = \frac{1}{2\sqrt{4\ln 3}}\int_0^\infty x^{-1/2}e^{-x}\,dx = \frac{\Gamma(1/2)}{2\sqrt{4\ln 3}} = \frac{\sqrt{\pi}}{4\sqrt{\ln 3}}$.

12.42 Evaluate $\displaystyle\int_0^1 \frac{dx}{\sqrt{-\ln x}}$.

Let $-\ln x = u$. Then $x = e^{-u}$. When $x = 1$, $u = 0$; and when $x = 0$, $u = \infty$. Thus, the given integral becomes $\int_0^\infty \frac{e^{-u}}{\sqrt{u}}\,du = \int_0^\infty u^{-1/2}e^{-u}\,du = \Gamma(\tfrac{1}{2}) = \sqrt{\pi}$.

12.43 Using the relationship $\Gamma(p+1) = p\Gamma(p)$ of Problem 12.28 as the definition of $\Gamma(p)$ for nonpositive p, find (a) $\Gamma(-\tfrac{1}{2})$; (b) $\Gamma(-\tfrac{3}{2})$; (c) $\Gamma(-\tfrac{5}{2})$; (d) $\Gamma(0)$; (e) $\Gamma(-1)$; (f) $\Gamma(-2)$.

(a) For $p = -\tfrac{1}{2}$, we have $\Gamma(\tfrac{1}{2}) = -\tfrac{1}{2}\Gamma(-\tfrac{1}{2})$. Then $\Gamma(-\tfrac{1}{2}) = -2\Gamma(\tfrac{1}{2}) = -2\sqrt{\pi}$.
(b) For $p = -\tfrac{3}{2}$, we have $\Gamma(-\tfrac{1}{2}) = -\tfrac{3}{2}\Gamma(-\tfrac{3}{2})$. Then $\Gamma(-\tfrac{3}{2}) = -\tfrac{2}{3}\Gamma(-\tfrac{1}{2}) = \tfrac{4}{3}\sqrt{\pi}$.
(c) For $p = -\tfrac{5}{2}$, we have $\Gamma(-\tfrac{3}{2}) = -\tfrac{5}{2}\Gamma(-\tfrac{5}{2})$. Then $\Gamma(-\tfrac{5}{2}) = -\tfrac{2}{5}\Gamma(-\tfrac{3}{2}) = -\tfrac{8}{15}\sqrt{\pi}$.
(d) For $p = 0$, $\Gamma(1) = 0\Gamma(0)$. It follows that $\Gamma(0)$ must be infinite, since $\Gamma(1) = 1$.
(e) For $p = -1$, $\Gamma(0) = -1\Gamma(-1)$, and it follows that $\Gamma(-1)$ must be infinite.
(f) For $p = -2$, $\Gamma(-1) = -2\Gamma(-2)$, and it follows that $\Gamma(-2)$ must be infinite.

In general, if p is a positive integer or zero, $\Gamma(-p)$ is infinite and

$$\Gamma(-p-\tfrac{1}{2}) = (-1)^{p+1}\left(\frac{2}{1}\right)\left(\frac{2}{3}\right)\left(\frac{2}{5}\right)\cdots\left(\frac{2}{2p+1}\right)\sqrt{\pi}$$

12.44 Prove that $\mathscr{L}\{t^n\} = \frac{\Gamma(n+1)}{s^{n+1}}$ for $n > -1$, where $s > 0$.

We have $\mathscr{L}\{t^n\} = \int_0^\infty e^{-st}t^n\,dt$. Letting $st = u$ and assuming $s > 0$, we obtain

$$\mathscr{L}\{t^n\} = \int_0^\infty e^{-u}\left(\frac{u}{s}\right)^n d\left(\frac{u}{s}\right) = \frac{1}{s^{n+1}}\int_0^\infty u^n e^{-u}\,du = \frac{\Gamma(n+1)}{s^{n+1}}$$

12.45 Prove that $\mathscr{L}\{t^{-1/2}\} = \sqrt{\pi/s}$, where $s > 0$.

Let $n = -1/2$ in Problem 12.44. Then $\mathscr{L}\{t^{-1/2}\} = \frac{\Gamma(1/2)}{s^{1/2}} = \frac{\sqrt{\pi}}{s^{1/2}} = \sqrt{\frac{\pi}{s}}$.

12.46 Find the Laplace transform of $f(x) = \sqrt{x}$.

Using the result of Problem 12.44 with x replacing t and $n = \tfrac{1}{2}$, we have $\mathscr{L}\{\sqrt{x}\} = \frac{\Gamma(3/2)}{s^{3/2}} = \frac{(1/2)\Gamma(1/2)}{s^{3/2}} = \frac{1}{2}\sqrt{\pi}s^{-3/2}$.

12.47 Find the Laplace transform of $f(x) = x^{n-1/2}$ $(n = 1, 2, \ldots)$.

Using the result of Problem 12.44 with x replacing t and $n - 1/2$ replacing n, we have $\mathscr{L}\{x^{n-1/2}\} = \Gamma(n+\tfrac{1}{2})/s^{n+1/2}$. Then repeated use of the formula of Problem 12.28 yields

$$\mathscr{L}\{x^{n-1/2}\} = \frac{\Gamma(n+1/2)}{s^{n+1/2}} = \frac{(n-1/2)(n-3/2)\cdots(5/2)(3/2)(1/2)\Gamma(1/2)}{s^{n+1/2}} = \frac{(2n-1)(2n-3)\cdots(5)(3)(1)\sqrt{\pi}}{2^n s^{n+1/2}}$$

12.48 Find the Laplace transform of $f(x) = x^n$ for n a positive integer.

The Laplace transform of $f(x)$ is identical to the Laplace transform of $f(t)$, so it follows from Problem 12.44 that $\mathscr{L}\{x^n\} = \frac{\Gamma(n+1)}{s^{n+1}}$, which, as a result of Problem 12.30, may be written as $\mathscr{L}\{x^n\} = \frac{n!}{s^{n+1}}$.

12.49 Find the Laplace transform of x^4.

It follows from the previous problem that $\mathscr{L}\{x^4\} = \frac{4!}{s^{4+1}} = \frac{24}{s^5}$.

12.50 Find the Laplace transform of x^{14}.

It follows from Problem 12.48 with $n = 14$ that $\mathscr{L}\{x^{14}\} = \frac{14!}{s^{15}}$.

LINEARITY

12.51 Prove the linearity property of Laplace transforms: If both $f(x)$ and $g(x)$ have Laplace transforms, then for any two constants c_1 and c_2, $\mathscr{L}\{c_1 f(x) \pm c_2 g(x)\} = c_1\mathscr{L}\{f(x)\} \pm c_2\mathscr{L}\{g(x)\}$.

▮
$$\mathscr{L}\{c_1 f(x) \pm c_2 g(x)\} = \int_0^\infty e^{-sx}[c_1 f(x) \pm c_2 g(x)]\,dx = c_1\int_0^\infty e^{-sx} f(x)\,dx \pm c_2\int_0^\infty e^{-sx} g(x)\,dx$$
$$= c_1\mathscr{L}\{f(x)\} \pm c_2\mathscr{L}\{g(x)\}$$

12.52 Find the Laplace transform of $\sinh ax$, where a denotes a constant.

▮ Using the linearity property and the result of Problem 12.2, we have

$$\mathscr{L}\{\sinh ax\} = \mathscr{L}\left\{\frac{e^{ax} - e^{-ax}}{2}\right\} = \frac{1}{2}\mathscr{L}\{e^{ax}\} - \frac{1}{2}\mathscr{L}\{e^{-ax}\} = \frac{1}{2}\frac{1}{s-a} - \frac{1}{2}\frac{1}{s+a}$$
$$= \frac{1}{2}\frac{(s+a)-(s-a)}{(s-a)(s+a)} = \frac{a}{s^2 - a^2}$$

12.53 Find the Laplace transform of $\cosh ax$, where a denotes a constant.

▮ Using the linearity property and the result of Problem 12.2, we have

$$\mathscr{L}\{\cosh ax\} = \mathscr{L}\left\{\frac{e^{ax} + e^{-ax}}{2}\right\} = \frac{1}{2}\mathscr{L}\{e^{ax}\} + \frac{1}{2}\mathscr{L}\{e^{-ax}\} = \frac{1}{2}\frac{1}{s-a} + \frac{1}{2}\frac{1}{s+a} = \frac{s}{s^2 - a^2}$$

12.54 Find the Laplace transform of $\cos^2 ax$, where a denotes a constant.

▮ Using the linearity property and the results of Problems 12.1 and 12.23, we have

$$\mathscr{L}\{\cos^2 ax\} = \mathscr{L}\{1 - \sin^2 ax\} = \mathscr{L}\{1\} - \mathscr{L}\{\sin^2 ax\} = \frac{1}{s} - \frac{2a^2}{s(s^2 + 4a^2)} = \frac{s^2 - 2a^2}{s(s^2 + 4a^2)}$$

12.55 Find $\mathscr{L}\{3 + 2x^2\}$.

▮
$$\mathscr{L}\{3 + 2x^2\} = 3\mathscr{L}\{1\} + 2\mathscr{L}\{x^2\} = 3\frac{1}{s} + 2\frac{2}{s^3} = \frac{3}{s} + \frac{4}{s^3}$$

12.56 Find the Laplace transform of $f(x) = x + x^2$.

▮
$$\mathscr{L}\{x + x^2\} = \mathscr{L}\{x\} + \mathscr{L}\{x^2\} = \frac{1}{s^2} + \frac{2}{s^3}$$

12.57 Find $\mathscr{L}\{20x + 4x^2\}$.

▮
$$\mathscr{L}\{20x + 4x^2\} = 20\mathscr{L}\{x\} + 4\mathscr{L}\{x^2\} = 20\frac{1}{s^2} + 4\frac{2!}{s^3} = \frac{20}{s^2} + \frac{8}{s^3}$$

12.58 Find $\mathscr{L}\{-15x^2 + 3x\}$.

▮
$$\mathscr{L}\{-15x^2 + 3x\} = -15\mathscr{L}\{x^2\} + 3\mathscr{L}\{x\} = -15\frac{2!}{s^3} + 3\frac{1}{s^2} = \frac{3}{s^2} - \frac{30}{s^3}$$

12.59 Find the Laplace transform of $f(x) = 15x^4 - x^2$.

▮
$$\mathscr{L}\{15x^4 - x^2\} = 15\mathscr{L}\{x^4\} - \mathscr{L}\{x^2\} = 15\frac{4!}{s^5} - \frac{2!}{s^3} = \frac{360}{s^5} - \frac{2}{s^3}$$

12.60 Find the Laplace transform of $f(x) = 2x^2 - 3x + 4$.

▮
$$\mathscr{L}\{2x^2 - 3x + 4\} = 2\mathscr{L}\{x^2\} - 3\mathscr{L}\{x\} + 4\mathscr{L}\{1\} = 2\frac{2!}{s^3} - 3\frac{1}{s^2} + 4\frac{1}{s} = \frac{4}{s^3} - \frac{3}{s^2} + \frac{4}{s}$$

12.61 Find $\mathscr{L}\{-7x + 4x^2 + 1\}$.

▮
$$\mathscr{L}\{-7x + 4x^2 + 1\} = -7\mathscr{L}\{x\} + 4\mathscr{L}\{x^2\} + \mathscr{L}\{1\} = -7\frac{1}{s^2} + 4\frac{2!}{s^3} + \frac{1}{s} = \frac{8}{s^3} - \frac{7}{s^2} + \frac{1}{s}$$

12.62 Find the Laplace transform of $f(x) = 79x^{14} - 8x^2 + 32$.

$$\mathscr{L}\{79x^{14} - 8x^2 + 32\} = 79\mathscr{L}\{x^{14}\} - 8\mathscr{L}\{x^2\} + 32\mathscr{L}\{1\} = 79\frac{14!}{s^{15}} - 8\frac{2!}{s^3} + 32\frac{1}{s}$$

$$= \frac{79(14!)}{s^{15}} - \frac{16}{s^3} + \frac{32}{s}$$

12.63 Find $\mathscr{L}\{9x^4 - 16 + 6x^2\}$.

$$\mathscr{L}\{9x^4 - 16 + 6x^2\} = 9\mathscr{L}\{x^4\} - 16\mathscr{L}\{1\} + 6\mathscr{L}\{x^2\} = 9\frac{4!}{s^5} - 16\frac{1}{s} + 6\frac{2!}{s^3} = \frac{216}{s^5} - \frac{16}{s} + \frac{12}{s^3}$$

12.64 Find $\mathscr{L}\{x^8 + x^3 - 26 + 40x^2\}$.

$$\mathscr{L}\{x^8 + x^3 - 26 + 40x^2\} = \mathscr{L}\{x^8\} + \mathscr{L}\{x^3\} - 26\mathscr{L}\{1\} + 40\mathscr{L}\{x^2\} = \frac{8!}{s^9} + \frac{3!}{s^4} - 26\frac{1}{s} + 40\frac{2!}{s^3}$$

$$= \frac{8!}{s^9} + \frac{6}{s^4} - \frac{26}{s} + \frac{80}{s^3}$$

12.65 Find $\mathscr{L}\{-14x^5 + 6x^4 - 100x\}$.

$$\mathscr{L}\{-14x^5 + 6x^4 - 100x\} = -14\mathscr{L}\{x^5\} + 6\mathscr{L}\{x^4\} - 100\mathscr{L}\{x\} = -14\frac{5!}{s^6} + 6\frac{4!}{s^5} - 100\frac{1}{s^2}$$

$$= -\frac{1680}{s^6} + \frac{144}{s^5} - \frac{100}{s^2}$$

12.66 Find $\mathscr{L}\{19x^3 - 40\sqrt{x}\}$.

$$\mathscr{L}\{19x^3 - 40\sqrt{x}\} = 19\mathscr{L}\{x^3\} - 40\mathscr{L}\{\sqrt{x}\} = 19\frac{3!}{s^4} - 40\frac{1}{2}\sqrt{\pi}s^{-3/2} = \frac{114}{s^4} - \frac{20\sqrt{\pi}}{s^{3/2}}$$

12.67 Find $\mathscr{L}\{17\sqrt{x} - 10/\sqrt{x} + 25x^2\}$.

$$\mathscr{L}\left\{17\sqrt{x} - \frac{10}{\sqrt{x}} + 25x^2\right\} = 17\mathscr{L}\{\sqrt{x}\} - 10\mathscr{L}\left\{\frac{1}{\sqrt{x}}\right\} + 25\mathscr{L}\{x^2\} = 17\frac{1}{2}\sqrt{\pi}s^{-3/2} - 10\sqrt{\pi}s^{-1/2} + 25\frac{2!}{s^3}$$

$$= \frac{17}{2}\sqrt{\pi}s^{-3/2} - \frac{10\sqrt{\pi}}{\sqrt{s}} + \frac{50}{s^3}$$

12.68 Find $\mathscr{L}\{14x^{3/2} + 13x - 10x^{1/2}\}$.

$$\mathscr{L}\{14x^{3/2} + 13x - 10x^{1/2}\} = 14\mathscr{L}\{x^{3/2}\} + 13\mathscr{L}\{x\} - 10\mathscr{L}\{x^{1/2}\} = 14\frac{3\sqrt{\pi}}{4s^{5/2}} + 13\frac{1}{s^2} - 10\frac{\sqrt{\pi}}{2s^{3/2}}$$

$$= \frac{21\sqrt{\pi}}{2s^{5/2}} + \frac{13}{s^2} - \frac{5\sqrt{\pi}}{s^{3/2}}$$

12.69 Find the Laplace transform of $2\sin x + 3\cos 2x$.

$$\mathscr{L}\{2\sin x + 3\cos 2x\} = 2\mathscr{L}\{\sin x\} + 3\mathscr{L}\{\cos 2x\} = 2\frac{1}{s^2+1} + 3\frac{s}{s^2+4} = \frac{2}{s^2+1} + \frac{3s}{s^2+4}$$

12.70 Find the Laplace transform of $65\sin 7x - 8\cos(-3x)$.

$$\mathscr{L}\{65\sin 7x - 8\cos(-3x)\} = 65\mathscr{L}\{\sin 7x\} - 8\mathscr{L}\{\cos(-3x)\} = 65\frac{7}{s^2+49} - 8\frac{s}{s^2+9} = \frac{455}{s^2+49} - \frac{8s}{s^2+9}$$

12.71 Find the Laplace transform of $6\cos 4x - 3\sin(-5x)$.

$$\mathscr{L}\{6\cos 4x - 3\sin(-5x)\} = 6\mathscr{L}\{\cos 4x\} - 3\mathscr{L}\{\sin(-5x)\} = 6\frac{s}{s^2+16} - 3\frac{-5}{s^2+25} = \frac{6s}{s^2+16} + \frac{15}{s^2+25}$$

12.72 Find $\mathscr{L}\{5\sin x + 10\cos x\}$.

$$\mathscr{L}\{5\sin x + 10\cos x\} = 5\mathscr{L}\{\sin x\} + 10\mathscr{L}\{\cos x\} = 5\frac{1}{s^2+1} + 10\frac{s}{s^2+1} = \frac{5+10s}{s^2+1}$$

12.73 Find the Laplace transform of $10\cos 10x - \sin(-10x)$.

$$\mathscr{L}\{10\cos 10x - \sin(-10x)\} = 10\mathscr{L}\{\cos 10x\} - \mathscr{L}\{\sin(-10x)\} = 10\frac{s}{s^2+100} - \frac{-10}{s^2+100} = \frac{10s+10}{s^2+100}$$

12.74 Find $\mathscr{L}\{f(x)\}$ if $f(x) = \sin 3x + x^3 - 25x$.

$$\mathscr{L}\{f(x)\} = \mathscr{L}\{\sin 3x\} + \mathscr{L}\{x^3\} - 25\mathscr{L}\{x\} = \frac{3}{s^2+9} + \frac{3!}{s^4} - \frac{25}{s}$$

12.75 Find $\mathscr{L}\{9\sin 4x + 20\cos(-5x) + 10e^{10x}\}$.

$$\mathscr{L}\{9\sin 4x + 20\cos(-5x) + 10e^{10x}\} = 9\mathscr{L}\{\sin 4x\} + 20\mathscr{L}\{\cos(-5x)\} + 10\mathscr{L}\{e^{10x}\}$$
$$= 9\frac{4}{s^2+16} + 20\frac{s}{s^2+25} + 10\frac{1}{s-10} = \frac{36}{s^2+16} + \frac{20s}{s^2+25} + \frac{10}{s-10}$$

12.76 Find $\mathscr{L}\{103e^{-6x} - 18\cos 5x - 9x^{14} + 2\}$.

$$\mathscr{L}\{103e^{-6x} - 18\cos 5x - 9x^{14} + 2\} = 103\mathscr{L}\{e^{-6x}\} - 18\mathscr{L}\{\cos 5x\} - 9\mathscr{L}\{x^{14}\} + 2\mathscr{L}\{1\}$$
$$= 103\frac{1}{s+6} - 18\frac{s}{s^2+25} - 9\frac{14!}{s^{15}} + 2\frac{1}{s} = \frac{103}{s+6} - \frac{18s}{s^2+25} - \frac{9(14!)}{s^{15}} + \frac{2}{s}$$

12.77 Find $\mathscr{L}\{2x^{9/2} + 16e^x\}$.

Using the results of Problems 12.2 and 12.47, we have

$$\mathscr{L}\{2x^{9/2} + 16e^x\} = 2\mathscr{L}\{x^{9/2}\} + 16\mathscr{L}\{e^x\} = 2\frac{(9)(7)(5)(3)(1)}{2^5 s^{11/2}}\sqrt{\pi} + 16\frac{1}{s-1} = \frac{945\sqrt{\pi}}{32s^{11/2}} + \frac{16}{s-1}$$

12.78 Find $\mathscr{L}\{f(t)\}$ where $f(t) = 4e^{5t} + 6t^3 - 3\sin 4t + 2\cos 2t$.

$$\mathscr{L}\{f(t)\} = 4\mathscr{L}\{e^{5t}\} + 6\mathscr{L}\{t^3\} - 3\mathscr{L}\{\sin 4t\} + 2\mathscr{L}\{\cos 2t\} = 4\frac{1}{s-5} + 6\frac{3!}{s^4} - 3\frac{4}{s^2+16} + 2\frac{s}{s^2+4}$$
$$= \frac{4}{s-5} + \frac{36}{s^4} - \frac{12}{s^2+16} + \frac{2s}{s^2+4} \qquad \text{where } s > 5.$$

12.79 Find the Laplace transform of $f(t) = \begin{cases} 1 & \text{if } 0 \le t < 1 \\ 2 & \text{if } 1 \le t < 3 \\ 4 & \text{if } 3 \le t < 4 \\ -2 & \text{if } 4 \le t \end{cases}$.

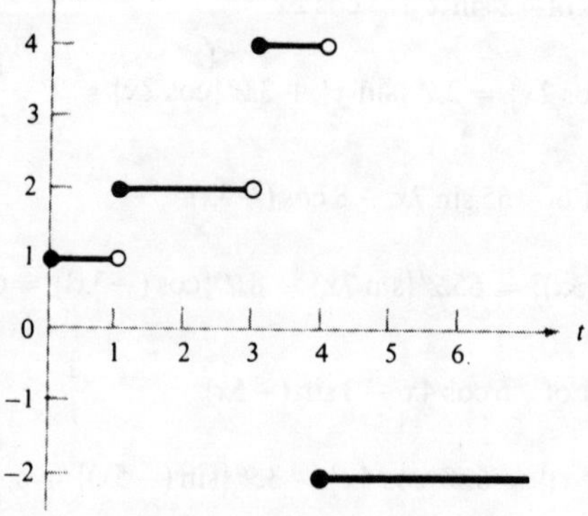

Fig. 12.1

▮ The graph of this function is shown in Fig. 12.1. Using the unit step function (see Problem 12.22), we can write $f(t) = 1 + u(t-1) + 2u(t-3) - 6u(t-4)$. It then follows from linearity that

$$\mathscr{L}\{f(t)\} = \mathscr{L}\{1\} + \mathscr{L}\{u(t-1)\} + 2\mathscr{L}\{u(t-3)\} - 6\mathscr{L}\{u(t-4)\}$$
$$= \frac{1}{2} + \frac{e^{-s}}{s} + 2\frac{e^{-3s}}{s} - 6\frac{e^{-4s}}{s} \qquad \text{for} \quad s > 0$$

12.80 Find the Laplace transform of $g(t) = \begin{cases} 0 & \text{if} \quad 0 \le t < a \\ 1 & \text{if} \quad a \le t < b \\ 0 & \text{if} \quad b \le t \end{cases}$ for $0 < a < b$.

▮ This function is known as a *square pulse*; its graph is given in Fig. 12.2. Since $g(t) = u(t-a) - u(t-b)$, it follows from linearity and Problem 12.22 that

$$\mathscr{L}\{g(t)\} = \mathscr{L}\{u(t-a)\} - \mathscr{L}\{u(t-b)\} = \frac{e^{-as}}{s} - \frac{e^{-bs}}{s} = \frac{e^{-as} - e^{-bs}}{s} \qquad s > 0$$

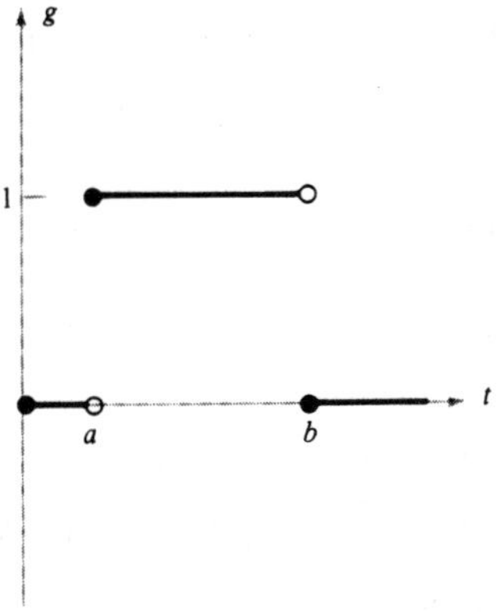

Fig. 12.2

12.81 Find $\mathscr{L}\{\frac{1}{2}(\sinh at - \sin at)\}$, where a denotes a constant.

▮
$$\mathscr{L}\{\tfrac{1}{2}(\sinh at - \sin at)\} = \tfrac{1}{2}\mathscr{L}\{\sinh at\} - \tfrac{1}{2}\mathscr{L}\{\sin at\} = \frac{1}{2}\frac{a}{s^2 - a^2} - \frac{1}{2}\frac{a}{s^2 + a^2} = \frac{a^3}{s^4 - a^4}$$

12.82 Find $\mathscr{L}\{\frac{1}{2}(\cosh at - \cos at)\}$, where a denotes a constant.

▮
$$\mathscr{L}\{\tfrac{1}{2}(\cosh at - \cos at)\} = \tfrac{1}{2}\mathscr{L}\{\cosh at\} - \tfrac{1}{2}\mathscr{L}\{\cos at\} = \frac{1}{2}\frac{s}{s^2 - a^2} - \frac{1}{2}\frac{s}{s^2 + a^2} = \frac{a^2 s}{s^4 - a^4}$$

12.83 Find $\mathscr{L}\{\frac{1}{2}(\sinh at + \sin at)\}$, where a denotes a constant.

▮
$$\mathscr{L}\{\tfrac{1}{2}(\sinh at + \sin at)\} = \tfrac{1}{2}\mathscr{L}\{\sinh at\} + \tfrac{1}{2}\mathscr{L}\{\sin at\} = \frac{1}{2}\frac{a}{s^2 - a^2} + \frac{1}{2}\frac{a}{s^2 + a^2} = \frac{as^2}{s^4 - a^4}$$

12.84 Find $\mathscr{L}\{\frac{1}{2}(\cosh at + \cos at)\}$, where a denotes a constant.

▮
$$\mathscr{L}\{\tfrac{1}{2}(\cosh at + \cos at)\} = \tfrac{1}{2}\mathscr{L}\{\cosh at\} + \tfrac{1}{2}\mathscr{L}\{\cos at\} = \frac{1}{2}\frac{s}{s^2 - a^2} + \frac{1}{2}\frac{s}{s^2 + a^2} = \frac{s^3}{s^4 - a^4}$$

12.85 Use the linearity property to find $\mathscr{L}\{\sin^2 ax\}$, where a denotes a contant.

▮
$$\mathscr{L}\{\sin^2 ax\} = \mathscr{L}\{\tfrac{1}{2}(\cos 2ax - 1)\} = \tfrac{1}{2}\mathscr{L}\{\cos 2ax\} - \tfrac{1}{2}\mathscr{L}\{1\} = \frac{1}{2}\frac{s}{s^2 + (2a)^2} - \frac{1}{2}\frac{1}{s} = \frac{2a^2}{s(s^2 + 4a^2)}$$

(Compare this with Problem 12.23.)

12.86 Find $\mathscr{L}\left\{\dfrac{ae^{ax} - be^{bx}}{a - b}\right\}$, where a and b are constants.

$$\mathscr{L}\left\{\frac{ae^{ax}-be^{bx}}{a-b}\right\}=\frac{a}{a-b}\mathscr{L}\{e^{ax}\}-\frac{b}{a-b}\mathscr{L}\{e^{bx}\}=\frac{a}{a-b}\frac{1}{s-a}-\frac{b}{a-b}\frac{1}{s-b}=\frac{s}{(s-a)(s-b)}$$

12.87 Find $\mathscr{L}\left\{\frac{ae^{-x/b}-be^{-x/a}}{ab(a-b)}\right\}$, where a and b denote constants.

$$\mathscr{L}\left\{\frac{ae^{-x/b}-be^{-x/a}}{ab(a-b)}\right\}=\frac{1}{b(a-b)}\mathscr{L}\{e^{-x/b}\}-\frac{1}{a(a-b)}\mathscr{L}\{e^{-x/a}\}$$

$$=\frac{1}{b(a-b)}\frac{1}{s-(-1/b)}-\frac{1}{a(a-b)}\frac{1}{s-(-1/a)}=\frac{s}{(1+as)(1+bs)}$$

12.88 Find $\mathscr{L}\left\{\frac{1}{a^2}(e^{at}-1-at)\right\}$, where a denotes a constant.

$$\mathscr{L}\left\{\frac{1}{a^2}(e^{at}-1-at)\right\}=\frac{1}{a^2}\mathscr{L}\{e^{at}\}-\frac{1}{a^2}\mathscr{L}\{1\}-\frac{1}{a}\mathscr{L}\{t\}=\frac{1}{a^2}\frac{1}{s-a}-\frac{1}{a^2}\frac{1}{s}-\frac{1}{a}\frac{1}{s^2}=\frac{1}{s^2(s-a)}$$

FUNCTIONS MULTIPLIED BY A POWER OF THE INDEPENDENT VARIABLE

12.89 Prove that if $\mathscr{L}\{f(t)\}=F(s)$, then $\mathscr{L}\{t^nf(t)\}=(-1)^n\frac{d^n}{ds^n}F(s)=(-1)^nF^{(n)}(s)$, where $n=1,2,3,\ldots$.

We have $F(s)=\int_0^\infty e^{-st}f(t)\,dt$. Then by Leibnitz's rule for differentiating under the integral sign,

$$\frac{dF}{ds}=F'(s)=\frac{d}{ds}\int_0^\infty e^{-st}f(t)\,dt=\int_0^\infty\frac{\partial}{\partial s}e^{-st}f(t)\,dt=\int_0^\infty -te^{-st}f(t)\,dt=-\int_0^\infty e^{-st}\{tf(t)\}\,dt=-\mathscr{L}\{tf(t)\}$$

Thus $\mathscr{L}\{tf(t)\}=-\frac{dF}{ds}=-F'(s)$, which proves the theorem for $n=1$.

To establish the theorem in general, we use *mathematical induction.* We assume the theorem is true for $n=k$; that is, we assume $\int_0^\infty e^{-st}\{t^kf(t)\}\,dt=(-1)^kF^{(k)}(s)$. Then $\frac{d}{ds}\int_0^\infty e^{-st}\{t^kf(t)\}\,dt=(-1)^kF^{(k+1)}(s)$ or, by Leibnitz's rule,

$$-\int_0^\infty e^{-st}\{t^{k+1}f(t)\}\,dt=(-1)^kF^{(k+1)}(s)$$

That is, $\int_0^\infty e^{-st}\{t^{k+1}f(t)\}\,dt=(-1)^{k+1}F^{(k+1)}(s)$. Thus the theorem is true by induction.

12.90 Find $\mathscr{L}\{te^{2t}\}$.

Since $\mathscr{L}\{e^{2t}\}=\frac{1}{s-2}$, $\mathscr{L}\{te^{2t}\}=-\frac{d}{ds}\left(\frac{1}{s-2}\right)=\frac{1}{(s-2)^2}$.

12.91 Find $\mathscr{L}\{t^2e^{2t}\}$.

Since $\mathscr{L}\{e^{2t}\}=\frac{1}{s-2}$, $\mathscr{L}\{t^2e^{2t}\}=\frac{d^2}{ds^2}\left(\frac{1}{s-2}\right)=\frac{2}{(s-2)^3}$.

12.92 Find $\mathscr{L}\{t^3e^{2t}\}$.

Since $\mathscr{L}\{e^{2t}\}=\frac{1}{s-2}$, $\mathscr{L}\{t^3e^{2t}\}=-\frac{d^3}{ds^3}\left(\frac{1}{s-2}\right)=\frac{6}{(s-2)^4}$.

12.93 Find $\mathscr{L}\{xe^{4x}\}$.

Since $\mathscr{L}\{e^{4x}\}=\frac{1}{s-4}$, $\mathscr{L}\{xe^{4x}\}=-\frac{d}{ds}\left(\frac{1}{s-4}\right)=\frac{1}{(s-4)^2}$.

12.94 Find $\mathscr{L}\{x^3e^{4x}\}$.

Since $\mathscr{L}\{e^{4x}\}=\frac{1}{s-4}$, $\mathscr{L}\{x^3e^{4x}\}=-\frac{d^3}{ds^3}\left(\frac{1}{s-4}\right)=\frac{6}{(s-4)^4}$.

12.95 Find $\mathscr{L}\{x^6e^{4x}\}$.

▌ Since $\mathscr{L}\{e^{4x}\} = \dfrac{1}{s-4}$, $\mathscr{L}\{x^6e^{4x}\} = -\dfrac{d^6}{ds^6}\left(\dfrac{1}{s-4}\right) = \dfrac{720}{(s-4)^7}$.

12.96 Find $\mathscr{L}\{x^5e^{-3x}\}$.

▌ Since $\mathscr{L}\{e^{-3x}\} = \dfrac{1}{s+3}$, $\mathscr{L}\{x^5e^{-3x}\} = -\dfrac{d^5}{ds^5}\left(\dfrac{1}{s+3}\right) = \dfrac{120}{(s+3)^6}$.

12.97 Find $\mathscr{L}\{x\cos ax\}$, where a is a constant.

▌ Taking $f(x) = \cos ax$, we have $F(s) = \mathscr{L}\{f(x)\} = \dfrac{s}{s^2+a^2}$. Then

$$\mathscr{L}\{x\cos ax\} = -\frac{d}{ds}\left(\frac{s}{s^2+a^2}\right) = \frac{s^2-a^2}{(s^2+a^2)^2}.$$

12.98 Find $\mathscr{L}\{x^2\cos ax\}$, where a is a constant.

▌ Since $\mathscr{L}\{\cos ax\} = \dfrac{s}{s^2+a^2}$, $\mathscr{L}\{x^2\cos ax\} = \dfrac{d^2}{ds^2}\left(\dfrac{s}{s^2+a^2}\right) = \dfrac{2s^3-6sa^2}{(s^2+a^2)^3}$.

12.99 Find $\mathscr{L}\{t\sin at\}$, where a denotes a constant.

▌ Since $\mathscr{L}\{\sin at\} = \dfrac{a}{s^2+a^2}$, $\mathscr{L}\{t\sin at\} = -\dfrac{d}{ds}\left(\dfrac{a}{s^2+a^2}\right) = \dfrac{2as}{(s^2+a^2)^2}$.

12.100 Find $\mathscr{L}\{t^2\sin at\}$, where a denotes a constant.

▌ Since $\mathscr{L}\{\sin at\} = \dfrac{a}{s^2+a^2}$, $\mathscr{L}\{t^2\sin at\} = \dfrac{d^2}{ds^2}\left(\dfrac{a}{s^2+a^2}\right) = \dfrac{6as^2-2a^3}{(s^2+a^2)^3}$.

12.101 Find $\mathscr{L}\{x^{7/2}\}$.

▌ Define $f(x) \equiv \sqrt{x}$. Then $x^{7/2} = x^3\sqrt{x} = x^3f(x)$ and, from Problem 12.46, $\mathscr{L}\{f(x)\} = \mathscr{L}\{\sqrt{x}\} = \frac{1}{2}\sqrt{\pi}s^{-3/2}$. Therefore, $\mathscr{L}\{x^3\sqrt{x}\} = (-1)^3\dfrac{d^3}{ds^3}(\frac{1}{2}\sqrt{\pi}s^{-3/2}) = \frac{105}{16}\sqrt{\pi}s^{-9/2}$. (Compare this with the result of Problem 12.47 for $n = 4$.)

12.102 Find $\mathscr{L}\{x^4/\sqrt{x}\}$.

▌ From Problem 12.45, we have $\mathscr{L}\{1/\sqrt{x}\} = \sqrt{\pi}s^{-1/2}$.

Therefore, $\mathscr{L}\left\{\dfrac{x^4}{\sqrt{x}}\right\} = (-1)^4\dfrac{d^4}{ds^4}(\sqrt{\pi}s^{-1/2}) = \left(\dfrac{-1}{2}\right)\left(\dfrac{-3}{2}\right)\left(\dfrac{-5}{2}\right)\left(\dfrac{-7}{2}\right)\sqrt{\pi}s^{-9/2} = \dfrac{105}{16}\sqrt{\pi}s^{-9/2}$.

12.103 Find $\mathscr{L}\{x\cosh 3x\}$.

▌ We know that $\mathscr{L}\{\cosh 3x\} = \dfrac{s}{s^2-9}$. Therefore, $\mathscr{L}\{x\cosh 3x\} = -\dfrac{d}{ds}\left(\dfrac{s}{s^2-9}\right) = \dfrac{s^2+9}{(s^2-9)^2}$.

12.104 Find $\mathscr{L}\{t\sinh 5t\}$.

▌ Since $\mathscr{L}\{\sinh 5t\} = \dfrac{5}{s^2-25}$, $\mathscr{L}\{t\sinh 5t\} = -\dfrac{d}{ds}\left(\dfrac{5}{s^2-25}\right) = \dfrac{10s}{(s^2-25)^2}$.

12.105 Find $\mathscr{L}\{t^2\sinh 4t\}$.

▌ $\mathscr{L}\{t^2\sinh 4t\} = \dfrac{d^2}{ds^2}\left(\dfrac{4}{s^2-16}\right) = \dfrac{24s^2+128}{(s^2-16)^3}$.

TRANSLATIONS

12.106 Prove the *first translation* or *shifting* property: If $\mathscr{L}\{f(t)\} = F(s)$, then $\mathscr{L}\{e^{at}f(t)\} = F(s-a)$.

▌ We have $\mathscr{L}\{f(t)\} = \int_0^\infty e^{-st}f(t)\,dt = F(s)$. Then

$$\mathscr{L}\{e^{at}f(t)\} = \int_0^\infty e^{-st}\{e^{at}f(t)\}\,dt = \int_0^\infty e^{-(s-a)t}f(t)\,dt = F(s-a)$$

12.107 Find $\mathscr{L}\{t^2e^{3t}\}$.

▌ We have $\mathscr{L}\{t^2\} = \dfrac{2!}{s^3} = \dfrac{2}{s^3}$. Then $\mathscr{L}\{t^2e^{3t}\} = \dfrac{2}{(s-3)^3}$.

12.108 Find $\mathscr{L}\{e^{-2t}\sin 4t\}$.

▌ We have $\mathscr{L}\{\sin 4t\} = \dfrac{4}{s^2+16}$. Then $\mathscr{L}\{e^{-2t}\sin 4t\} = \dfrac{4}{(s+2)^2+16} = \dfrac{4}{s^2+4s+20}$.

12.109 Find $\mathscr{L}\{e^{4t}\cosh 5t\}$.

▌ Since $\mathscr{L}\{\cosh 5t\} = \dfrac{s}{s^2-25}$, $\mathscr{L}\{e^{4t}\cosh 5t\} = \dfrac{s-4}{(s-4)^2-25} = \dfrac{s-4}{s^2-8s-9}$.

Alternative Method:

$$\mathscr{L}\{e^{4t}\cosh 5t\} = \mathscr{L}\left\{e^{4t}\left(\frac{e^{5t}+e^{-5t}}{2}\right)\right\} = \frac{1}{2}\mathscr{L}\{e^{9t}+e^{-t}\} = \frac{1}{2}\left\{\frac{1}{s-9}+\frac{1}{s+1}\right\} = \frac{s-4}{s^2-8s-9}$$

12.110 Find $\mathscr{L}\{e^{2t}\cos 7t\}$.

▌ Since $\mathscr{L}\{\cos 7t\} = \dfrac{s}{s^2+49} = F(s)$, $\mathscr{L}\{e^{2t}\cos 7t\} = F(s-2) = \dfrac{s-2}{(s-2)^2+49}$.

12.111 Find $\mathscr{L}\{e^{3t}\cos 7t\}$.

▌ With $F(s)$ as defined in Problem 12.110, we have $\mathscr{L}\{e^{3t}\cos 7t\} = F(s-3) = \dfrac{s-3}{(s-3)^2+49}$.

12.112 Find $\mathscr{L}\{e^{-3t}\cos 7t\}$.

▌ With $F(s)$ as defined in Problem 12.110, we have $\mathscr{L}\{e^{-3t}\cos 7t\} = F[s-(-3)] = F(s+3) = \dfrac{s+3}{(s+3)^2+49}$.

12.113 Find $\mathscr{L}\{e^{-5t}\cos 7t\}$

▌ With $F(s)$ as defined in Problem 12.110, we have $\mathscr{L}\{e^{-5t}\cos 7t\} = F[s-(-5)] = F(s+5) = \dfrac{s+5}{(s+5)^2+49}$.

12.114 Find $\mathscr{L}\{e^{-5t}\cos 6t\}$.

▌ Setting $F(s) = \mathscr{L}\{\cos 6t\} = \dfrac{s}{s^2+36}$, we have $\mathscr{L}\{e^{-5t}\cos 6t\} = F[s-(-5)] = F(s+5) = \dfrac{s+5}{(s+5)^2+36}$.

12.115 Find $\mathscr{L}\{e^{-5t}\cos 5t\}$.

▌ Setting $F(s) = \mathscr{L}\{\cos 5t\} = \dfrac{s}{s^2+25}$, we have $\mathscr{L}\{e^{-5t}\cos 5t\} = F[s-(-5)] = F(s+5) = \dfrac{s+5}{(s+5)^2+25}$.

12.116 Find $\mathscr{L}\{e^{t}\cos 5t\}$.

▌ With $F(s)$ as defined in Problem 12.115, we have $\mathscr{L}\{e^{t}\cos 5t\} = F(s-1) = \dfrac{s-1}{(s-1)^2+25}$.

12.117 Find $\mathscr{L}\{e^t \sin 5t\}$.

Setting $F(s) = \mathscr{L}\{\sin 5t\} = \dfrac{5}{s^2+25}$, we have $\mathscr{L}\{e^t \sin 5t\} = F(s-1) = \dfrac{5}{(s-1)^2+25}$.

12.118 Find $\mathscr{L}\{e^{-5t} \sin 5t\}$.

Using $F(s)$ as defined in the previous problem, we have $\mathscr{L}\{e^{-5t} \sin 5t\} = F[s-(-5)] = F(s+5) = \dfrac{5}{(s+5)^2+25}$.

12.119 Find $\mathscr{L}\{e^{-5t} \sin 6t\}$.

Setting $F(s) = \mathscr{L}\{\sin 6t\} = \dfrac{6}{s^2+36}$, we have $\mathscr{L}\{e^{-5t} \sin 6t\} = F[s-(-5)] = \dfrac{6}{(s+5)^2+36}$.

12.120 Find $\mathscr{L}\{e^{-2t}(3\cos 6t - 5\sin 6t)\}$.

$$\mathscr{L}\{3\cos 6t - 5\sin 6t\} = 3\mathscr{L}\{\cos 6t\} - 5\mathscr{L}\{\sin 6t\} = 3\frac{s}{s^2+36} - 5\frac{6}{s^2+36} = \frac{3s-30}{s^2+36}$$

so that $\mathscr{L}\{e^{-2t}(3\cos 6t - 5\sin 6t\} = \dfrac{3(s+2)-30}{(s+2)^2+36} = \dfrac{3s-24}{s^2+4s+40}$.

12.121 Find $\mathscr{L}\{e^{-2x} \sin 5x\}$.

Setting $F(s) = \mathscr{L}\{\sin 5x\} = \dfrac{5}{s^2+25}$, we have $\mathscr{L}\{e^{-2x} \sin 5x\} = F(s+2) = \dfrac{5}{(s+2)^2+25}$.

12.122 Find $\mathscr{L}\{e^{-x} x \cos 2x\}$.

Let $f(x) = x\cos 2x$. From Problem 12.97 with $a = 2$, we obtain $F(s) = \dfrac{s^2-4}{(s^2+4)^2}$. Then $\mathscr{L}\{e^{-x} x \cos 2x\} = F(s+1) = \dfrac{(s+1)^2-4}{[(s+1)^2+4]^2}$.

12.123 Find $\mathscr{L}\{xe^{4x}\}$.

Setting $F(s) = \mathscr{L}\{x\} = \dfrac{1}{s^2}$, we have $\mathscr{L}\{e^{4x}x\} = F(s-4) = \dfrac{1}{(s-4)^2}$. (Compare this with Problem 12.93.)

12.124 Find $\mathscr{L}\{x^3 e^{4x}\}$.

Setting $f(x) = x^3$, we have $F(s) = \mathscr{L}\{x^3\} = 3!/s^4$. Then $\mathscr{L}\{x^3 e^{4x}\} = F(s-4) = \dfrac{3!}{(s-4)^4}$. (Compare with Problem 12.94.)

12.125 Find $\mathscr{L}\{x^6 e^{4x}\}$.

Setting $f(x) = x^6$, we have $F(s) = \mathscr{L}\{x^6\} = 6!/s^7$. Then $\mathscr{L}\{x^6 e^{4x}\} = F(s-4) = \dfrac{6!}{(s-4)^7}$. (Compare with Problem 12.95.)

12.126 Find $\mathscr{L}\{e^{3x}\sqrt{x}\}$.

From Problem 12.46 we have $\mathscr{L}\{\sqrt{x}\} = \frac{1}{2}\sqrt{\pi}s^{-3/2}$, so $\mathscr{L}\{e^{3x}\sqrt{x}\} = \frac{1}{2}\sqrt{\pi}(s-3)^{-3/2}$.

12.127 Find $\mathscr{L}\{e^{-4x}\sqrt{x}\}$.

Since $\mathscr{L}\{\sqrt{x}\} = \frac{1}{2}\sqrt{\pi}s^{-3/2}$, we have $\mathscr{L}\{e^{-4x}\sqrt{x}\} = \frac{1}{2}\sqrt{\pi}(s+4)^{-3/2}$.

12.128 Find $\mathscr{L}\{e^{2t}/\sqrt{t}\}$.

From Problem 12.45 we have $\mathscr{L}\{1/\sqrt{t}\} = \sqrt{\pi/s}$. Then $\mathscr{L}\{e^{2t}/\sqrt{t}\} = \sqrt{\pi/(s-2)}$.

12.129 Find $\mathscr{L}\{e^{-2t}t^{7.5}\}$.

▌ It follows from Problem 12.44 with $n = 7.5$ that $\mathscr{L}\{t^{7.5}\} = \dfrac{\Gamma(8.5)}{s^{8.5}}$. Then

$$\mathscr{L}\{e^{-2t}t^{7.5}\} = \frac{\Gamma(8.5)}{(s+2)^{8.5}} = \frac{(7.5)(6.5)(5.5)(4.5)(3.5)(2.5)(1.5)(0.5)\Gamma(0.5)}{(s+2)^{8.5}} = \frac{(15)(13)(11)(9)(7)(5)(3)(1)\sqrt{\pi}}{2^8(s+2)^{8.5}}$$

12.130 Find $\mathscr{L}\{te^{2t}\sin t\}$.

▌ If we set $f(t) = t\sin t$, then it follows from Problem 12.99 with $a = 1$ that $F(s) = \mathscr{L}\{t\sin t\} = \dfrac{2s}{(s^2+1)^2}$.

Therefore, $\mathscr{L}\{te^{2t}\sin t\} = F(s-2) = \dfrac{2(s-2)}{[(s-2)^2+1]^2}$.

12.131 Find $\mathscr{L}\{t^2e^{-t}\sin 3t\}$.

▌ If we set $f(t) = t^2\sin 3t$, then it follows from Problem 12.100 with $a = 3$ that $F(s) = \mathscr{L}\{t^2\sin 3t\} = \dfrac{18s^2-54}{(s^2+9)^3}$. Therefore, $\mathscr{L}\{t^2e^{-t}\sin 3t\} = F(s+1) = \dfrac{18(s+1)^2-54}{[(s+1)^2+9]^3}$.

12.132 Find $\mathscr{L}\{\sin ax\sinh ax\}$, where a denotes a constant.

▌ Setting $f(x) = \sin ax$, we have $F(s) = \mathscr{L}\{\sin ax\} = \dfrac{a}{s^2+a^2}$. Then, using the principle of linearity, we obtain

$$\mathscr{L}\{\sin ax\sinh ax\} = \mathscr{L}\left\{(\sin ax)\frac{e^{ax}-e^{-ax}}{2}\right\} = \tfrac{1}{2}\mathscr{L}\{e^{ax}\sin ax\} - \tfrac{1}{2}\mathscr{L}\{e^{-ax}\sin ax\} = \tfrac{1}{2}F(s-a) - \tfrac{1}{2}F(s+a)$$

$$= \frac{1}{2}\frac{a}{(s-a)^2+a^2} - \frac{1}{2}\frac{a}{(s+a)^2+a^2} = \frac{a}{2}\cdot\frac{(s^2+2as+2a^2)-(s^2-2as+2a^2)}{(s^2-2as+2a^2)(s^2+2as+2a^2)} = \frac{2a^2s}{s^4+4a^4}$$

12.133 Find $\mathscr{L}\{\sin ax\cosh ax\}$, where a denotes a constant.

▌ With $F(s)$ as defined in the previous problem, we have

$$\mathscr{L}\{\sin ax\cosh ax\} = \mathscr{L}\left\{(\sin ax)\frac{e^{ax}+e^{-ax}}{2}\right\} = \tfrac{1}{2}\mathscr{L}\{e^{ax}\sin ax\} + \tfrac{1}{2}\mathscr{L}\{e^{-ax}\sin ax\} = \tfrac{1}{2}F(s-a) + \tfrac{1}{2}F(s+a)$$

$$= \frac{1}{2}\frac{a}{(s-a)^2+a^2} + \frac{1}{2}\frac{a}{(s+a)^2+a^2} = \frac{a(s^2+2a^2)}{s^4+4a^4}$$

12.134 Find $\mathscr{L}\{\cos ax\cosh ax\}$, where a denotes a constant.

▌ Setting $f(x) = \cos ax$, we have $F(s) = \mathscr{L}\{\cos ax\} = \dfrac{s}{s^2+a^2}$. Then

$$\mathscr{L}\{\cos ax\cosh ax\} = \mathscr{L}\left\{(\cos ax)\frac{e^{ax}+e^{-ax}}{2}\right\} = \tfrac{1}{2}\mathscr{L}\{e^{ax}\cos ax\} + \tfrac{1}{2}\mathscr{L}\{e^{-ax}\cos ax\} = \tfrac{1}{2}F(s-a) + \tfrac{1}{2}F(s+a)$$

$$= \frac{1}{2}\frac{s+a}{(s+a)^2+a^2} + \frac{1}{2}\frac{s-a}{(s-a)^2+a^2} = \frac{1}{2}\frac{(s+a)(s^2-2as+2a^2)+(s-a)(s^2+2as+2a^2)}{(s^2+2as+2a^2)(s^2-2as+2a^2)} = \frac{s^3}{s^4+4a^4}$$

12.135 Prove the *second translation* or *shifting* property: If $\mathscr{L}\{f(t)\} = F(s)$ and $g(t) = \begin{cases} f(t-a) & t > a \\ 0 & t < a \end{cases}$ then $\mathscr{L}\{g(t)\} = e^{-as}F(s)$.

▌

$$\mathscr{L}\{g(t)\} = \int_0^\infty e^{-st}g(t)\,dt = \int_0^a e^{-st}g(t)\,dt + \int_a^\infty e^{-st}g(t)\,dt$$

$$= \int_0^a e^{-st}(0)\,dt + \int_a^\infty e^{-st}f(t-a)\,dt = \int_0^\infty e^{-s(u+a)}f(u)\,du$$

$$= e^{-as}\int_0^\infty e^{-su}f(u)\,du = e^{-as}F(s)$$

where we have used the substitution $t = u + a$.

Observe that $g(t)$ may be written compactly as $g(t) = f(t-a)u(t-a)$, where $u(t-a)$ denotes the unit step function (see Problem 12.22). Thus the second shifting property may be stated as follows: If $\mathscr{L}\{f(t)\} = F(s)$, then $\mathscr{L}\{f(t-a)u(t-a)\} = e^{-as}f(s)$.

12.136 Find the Laplace transform of $g(t) = \begin{cases} (t-2)^3 & t > 2 \\ 0 & t < 2 \end{cases}$.

▮ Since $\mathscr{L}\{t^3\} = 3!/s^4$, it follows from the previous problem (with $a = 2$) that $\mathscr{L}\{g(t)\} = 6e^{-2s}/s^4$.

12.137 Find $\mathscr{L}\{f(t)\}$ if $f(t) = \begin{cases} \cos(t - 2\pi/3) & t > 2\pi/3 \\ 0 & t < 2\pi/3 \end{cases}$.

▮ Since $\mathscr{L}\{\cos t\} = \dfrac{s}{s^2+1}$, it follows from Problem 12.135 (with $a = 2\pi/3$) that $\mathscr{L}\{f(t)\} = \dfrac{se^{-2\pi s/3}}{s^2+1}$.

12.138 Find $\mathscr{L}\{f(t)\}$ if $f(t) = \begin{cases} 0 & t < 3 \\ e^{t-3} & t > 3 \end{cases}$.

▮ Since $\mathscr{L}\{e^t\} = \dfrac{1}{s-1}$, it follows from Problem 12.135 (with $a = 3$) that $\mathscr{L}\{f(t)\} = \dfrac{1}{s-1}e^{-3s}$.

12.139 Find $\mathscr{L}\{f(t)\}$ if $f(t) = \begin{cases} 0 & t < 3 \\ e^{t} & t > 3 \end{cases}$.

▮ If we write $e^t = e^3e^{t-3}$, then $f(t) = e^3e^{t-3}u(t-3)$, and

$$\mathscr{L}\{f(t)\} = \mathscr{L}\{e^3e^{t-3}u(t-3)\} = e^3\mathscr{L}\{e^{t-3}u(t-3)\} = e^3\frac{1}{s-1}e^{-3s} = \frac{1}{s-1}e^{-3(s-1)}$$

12.140 Find the Laplace transform of $f(t) = \begin{cases} e^{4t} & t > 3 \\ 0 & t < 3 \end{cases}$.

▮ If we write $e^{4t} = e^{12}e^{4(t-3)}$, then $f(t) = e^{12}e^{4(t-3)}u(t-3)$. Since $\mathscr{L}\{e^{4t}\} = \dfrac{1}{s-4}$, it follows that

$$\mathscr{L}\{f(t)\} = \mathscr{L}\{e^{12}e^{4(t-3)}u(t-3)\} = e^{12}\mathscr{L}\{e^{4(t-3)}u(t-3)\} = e^{12}\frac{1}{s-4}e^{-3s} = \frac{e^{-3(s-4)}}{s-4}$$

12.141 Discuss the graphical relationship between an arbitrary function $f(t)$ defined for all nonnegative x and the function $u(x-c)f(x-c)$, where c is a positive constant.

▮ With $f(x)$ defined for $x \geq 0$, the function $u(x-c)f(x-c) = \begin{cases} 0 & x < c \\ f(x-c) & x \geq c \end{cases}$ represents a shift, or translation, of $f(x)$ by c units in the positive x direction. For example, if $f(x)$ is given graphically by Fig. 12.3, then $u(x-c)f(x-c)$ is given graphically by Fig. 12.4.

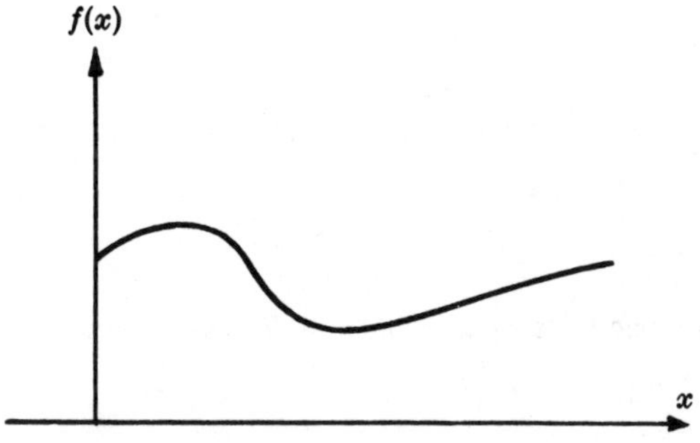

Fig. 12.3

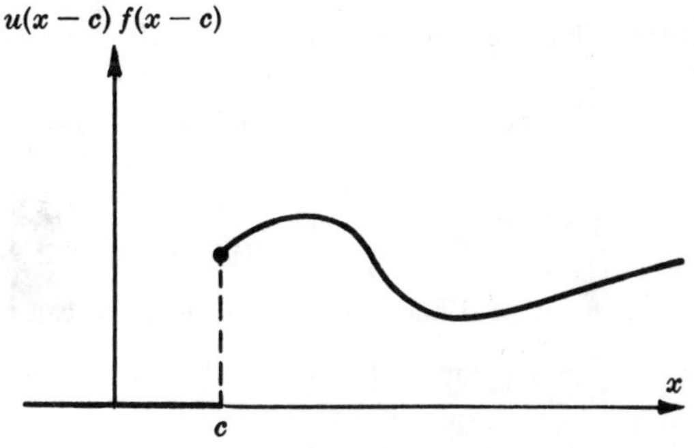

Fig. 12.4

12.142 Graph the function $f(x) = u(x - \pi)\cos 2(x - \pi)$.

▌ $f(x)$ is sketched in Fig. 12.5.

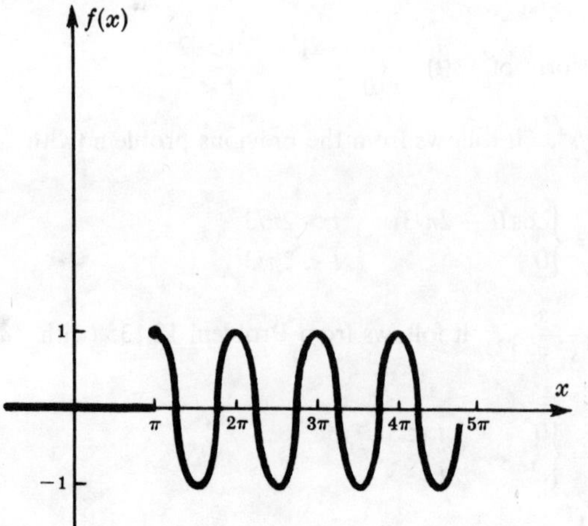

Fig. 12.5

12.143 Graph the function $f(x) = \frac{1}{2}(x - 1)^2 u(x - 1)$.

▌ $f(x)$ is sketched in Fig. 12.6.

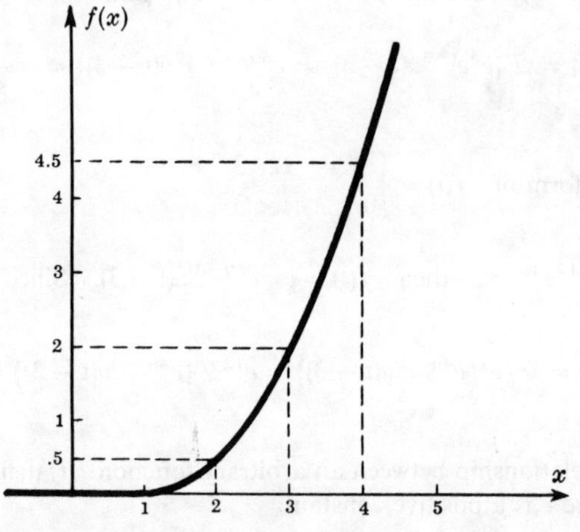

Fig. 12.6

12.144 Find the Laplace transform of $u(x - \pi)\cos 2(x - \pi)$.

▌ Since $\mathscr{L}\{\cos 2x\} = \dfrac{s}{s^2 + 4}$, it follows that $\mathscr{L}\{u(x - \pi)\cos 2(x - \pi)\} = \dfrac{s}{s^2 + 4}e^{-\pi s}$.

12.145 Find the Laplace transform of $\frac{1}{2}(x - 1)^2 u(x - 1)$.

▌ Since $\mathscr{L}\{\frac{1}{2}x^2\} = \frac{1}{2}\mathscr{L}\{x^2\} = \dfrac{1}{2}\left(\dfrac{2!}{s^3}\right) = \dfrac{1}{s^3}$, it follows that $\mathscr{L}\{\frac{1}{2}(x - 1)^2 u(x - 1)\} = \dfrac{1}{s^3}e^{-s}$.

Find $\mathscr{L}\{g(x)\}$ if $g(x) = \begin{cases} 0 & x < 4 \\ (x - 4)^2 & x \geq 4 \end{cases}$.

▌ If we define $f(x) = x^2$, then $g(x)$ can be given compactly as $g(x) = u(x - 4)f(x - 4) = u(x - 4)(x - 4)^2$. Noting that $\mathscr{L}\{f(x)\} = 2/s^3$, we conclude that $\mathscr{L}\{g(x)\} = e^{-4s}\dfrac{2}{s^3}$.

12.147 Find $\mathscr{L}\{g(x)\}$ if $g(x) = \begin{cases} 0 & x < 4 \\ x^2 & x \geq 4 \end{cases}$.

We first determine a function $f(x)$ such that $f(x-4)=x^2$. Once this has been done, $g(x)$ can be written as $g(x)=u(x-4)f(x-4)$. Now, $f(x-4)=x^2$ only if $f(x)=f(x-4+4)=(x+4)^2=x^2+8x+16$. Then

$$\mathscr{L}\{f(x)\}=\mathscr{L}\{x^2\}+8\mathscr{L}\{x\}+16\mathscr{L}\{1\}=\frac{2}{s^3}+\frac{8}{s^2}+\frac{16}{s}$$

and it follows that $\mathscr{L}\{g(x)\}=\mathscr{L}\{u(x-4)f(x-4)\}=e^{-4s}\left(\frac{2}{s^3}+\frac{8}{s^2}+\frac{16}{s}\right)$.

TRANSFORMS OF PERIODIC FUNCTIONS

12.148 Prove that if $f(t)$ has period $T>0$, then $\mathscr{L}\{f(t)\}=\dfrac{\int_0^T e^{-st}f(t)\,dt}{1-e^{-sT}}$.

We have $\mathscr{L}\{f(t)\}=\int_0^\infty e^{-st}f(t)\,dt$, which we write as

$$\mathscr{L}\{f(t)\}=\int_0^T e^{-st}f(t)\,dt+\int_T^{2T} e^{-st}f(t)\,dt+\int_{2T}^{3T} e^{-st}f(t)\,dt+\cdots$$

In the first integral let $t=u$; in the second integral let $t=u+T$; in the third integral let $t=u+2T$; and so on. Then

$$\begin{aligned}\mathscr{L}\{f(t)\}&=\int_0^T e^{-su}f(u)\,du+\int_0^T e^{-s(u+T)}f(u+T)\,du+\int_0^T e^{-s(u+2T)}f(u+2T)\,du+\cdots\\&=\int_0^T e^{-su}f(u)\,du+e^{-sT}\int_0^T e^{-su}f(u)\,du+e^{-2sT}\int_0^T e^{-su}f(u)\,du+\cdots\\&=(1+e^{-sT}+e^{-2sT}+\cdots)\int_0^T e^{-su}f(u)\,du=\frac{\int_0^T e^{-su}f(u)\,du}{1-e^{-sT}}\end{aligned}$$

where we have used the periodicity of $f(t)$ to write $f(u+T)=f(u)$, $f(u+2T)=f(u),\ldots,$ along with the fact that $1+r+r^2+r^3+\cdots=\dfrac{1}{1-r}$, for $|r|<1$.

12.149 Graph the function $f(t)=\begin{cases}\sin t & 0<t<\pi\\ 0 & \pi<t<2\pi\end{cases}$ extended periodically with period 2π, and find $\mathscr{L}\{f(t)\}$.

The graph appears in Fig. 12.7. By Problem 12.148, since $T=2\pi$, we have

$$\begin{aligned}\mathscr{L}\{f(t)\}&=\frac{1}{1-e^{-2\pi s}}\int_0^{2\pi} e^{-st}f(t)\,dt=\frac{1}{1-e^{-2\pi s}}\int_0^{\pi} e^{-st}\sin t\,dt=\frac{1}{1-e^{-2\pi s}}\left[\frac{e^{-st}(-s\sin t-\cos t)}{s^2+1}\right]_0^{\pi}\\&=\frac{1}{1-e^{-2\pi s}}\,\frac{1+e^{-\pi s}}{s^2+1}=\frac{1}{(1-e^{-\pi s})(s^2+1)}\end{aligned}$$

The graph of the function $f(t)$ is often called a *half-wave-rectified sine curve.*

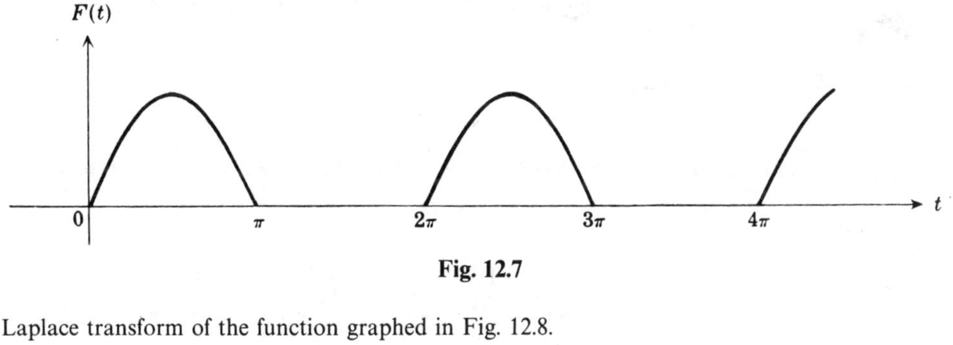

Fig. 12.7

12.150 Find the Laplace transform of the function graphed in Fig. 12.8.

Note that $f(x)$ is periodic with period $T=2\pi$, and in the interval $0\le x<2\pi$ it can be defined analytically by $f(x)=\begin{cases}x & 0\le x\le\pi\\ 2\pi-x & \pi\le x<2\pi\end{cases}$. Thus, using the formula of Problem 12.148 with x replacing t, we obtain $\mathscr{L}\{f(x)\} = \dfrac{\int_0^{2\pi} e^{-sx}f(x)\,dx}{1-e^{-2\pi s}}$. Since

$$\int_0^{2\pi} e^{-sx}f(x)\,dx = \int_0^{\pi} e^{-sx}x\,dx + \int_\pi^{2\pi} e^{-sx}(2\pi - x)\,dx = \frac{1}{s^2}(e^{-2\pi s} - 2e^{-\pi s} + 1) = \frac{1}{s^2}(e^{-\pi s} - 1)^2$$

it follows that

$$\mathscr{L}\{f(x)\} = \frac{(1/s^2)(e^{-\pi s}-1)^2}{1-e^{-2\pi s}} = \frac{(1/s^2)(e^{-\pi s}-1)^2}{(1-e^{-\pi s})(1+e^{-\pi s})} = \frac{1}{s^2}\left(\frac{1-e^{-\pi s}}{1+e^{-\pi s}}\right) = \frac{1}{s^2}\tanh\frac{\pi s}{2}$$

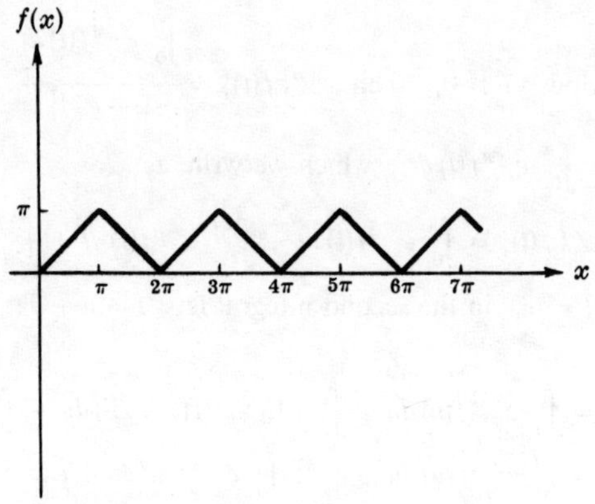

Fig. 12.8

12.151 Find $\mathscr{L}\{f(x)\}$ for the square wave shown in Fig. 12.9.

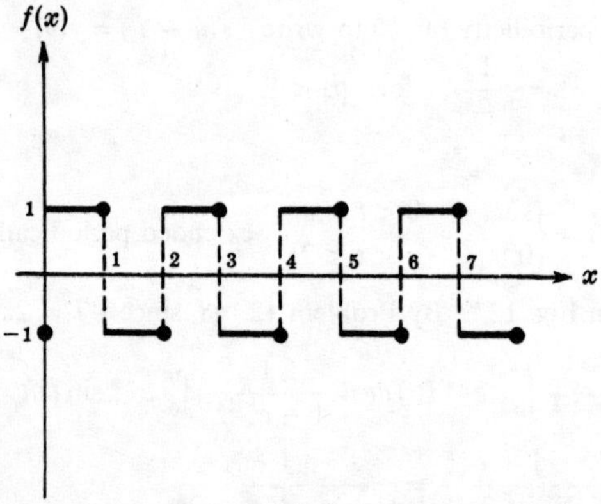

Fig. 12.9

▌ Note that $f(x)$ is periodic with period $T = 2$, and in the interval $0 < x \le 2$ it can be defined analytically by $f(x) = \begin{cases} 1 & 0 < x \le 1 \\ -1 & 1 < x \le 2 \end{cases}$. Thus, from the formula of Problem 12.148 with x replacing t, we have $\mathscr{L}\{f(x)\} = \dfrac{\int_0^2 e^{-sx}f(x)\,dx}{1-e^{-2s}}$. Since

$$\int_0^2 e^{-sx}f(x)\,dx = \int_0^1 e^{-sx}(1)\,dx + \int_1^2 e^{-sx}(-1)\,dx = \frac{1}{s}(e^{-2s} - 2e^{-s} + 1) = \frac{1}{s}(e^{-s} - 1)^2$$

it follows that

$$F(s) = \frac{(e^{-s}-1)^2}{s(1-e^{-2s})} = \frac{(1-e^{-s})^2}{s(1-e^{-s})(1+e^{-s})} = \frac{1-e^{-s}}{s(1+e^{-s})}$$
$$= \frac{e^{s/2}}{e^{s/2}}\,\frac{1-e^{-s}}{s(1+e^{-s})} = \frac{e^{s/2}-e^{-s/2}}{s(e^{s/2}+e^{-s/2})} = \frac{1}{s}\tanh\frac{s}{2}$$

12.152 Find the Laplace transform for the function shown in Fig. 12.10.

▮ We note that $f(x)$ is periodic with period $T = 2$, and that in the interval (0, 2) it is defined analytically as $f(x) = \begin{cases} 1 & 0 < x < 1 \\ 0 & 1 \le x < 2 \end{cases}$. It follows from Problem 12.148 that $\mathscr{L}\{f(x)\} = \dfrac{\int_0^2 e^{-sx} f(x)\,dx}{1 - e^{-2s}}$. But

$$\int_0^2 e^{-sx} f(x)\,dx = \int_0^1 e^{-sx}(1)\,dx + \int_1^2 e^{-sx}(0)\,dx = -\frac{1}{s}e^{-sx}\Big|_0^1 = \frac{1}{s}(1 - e^{-s})$$

Therefore, $\mathscr{L}\{f(x)\} = \dfrac{(1/s)(1 - e^{-s})}{1 - e^{-2s}} = \dfrac{(1/s)(1 - e^{-s})}{(1 - e^{-s})(1 + e^{-s})} = \dfrac{1}{s(1 + e^{-s})}$.

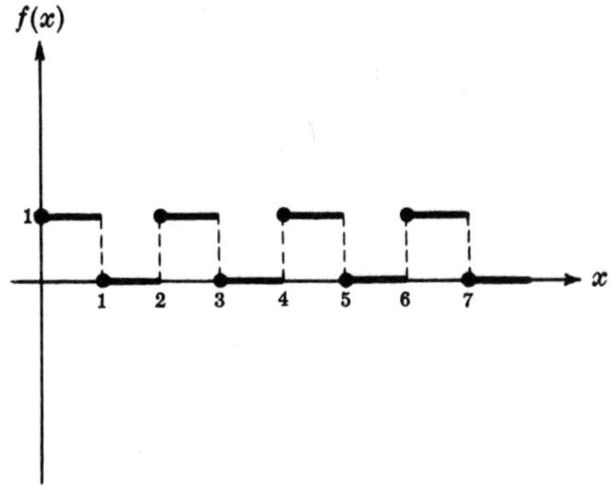

Fig. 12.10

12.153 Find the Laplace transform for the function shown in Fig. 12.11.

▮ This function is periodic with period $T = 1$, and it is defined as $f(t) = t$ on the interval (0, 1). Using the formula of Problem 12.148, we have

$$\mathscr{L}\{f(t)\} = \frac{\int_0^1 e^{-st} t\,dt}{1 - e^{-s}} = \frac{\left(-\frac{t}{s}e^{-st} - \frac{1}{s^2}e^{-st}\right)_0^1}{1 - e^{-s}} = \frac{-\frac{1}{s}e^{-s} - \frac{1}{s^2}e^{-s} + \frac{1}{s^2}}{1 - e^{-s}} = \frac{1 - e^{-s}(s + 1)}{s^2(1 - e^{-s})}$$

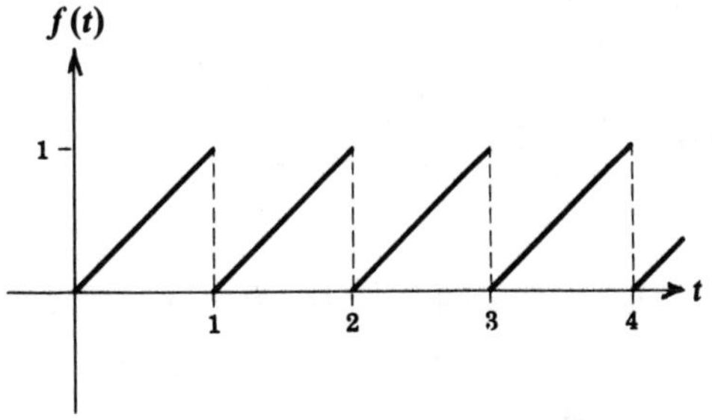

Fig. 12.11

12.154 Find the Laplace transform for the function shown in Fig. 12.12.

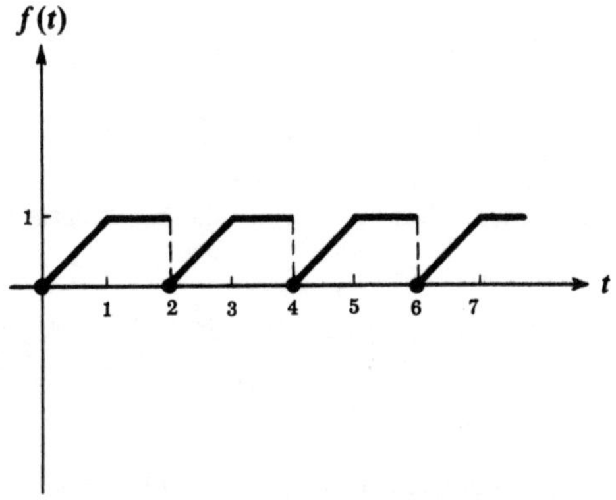

Fig. 12.12

▌ This function is periodic with period $T = 2$. In the interval (0, 2) it is defined analytically as $f(t) = \begin{cases} t & 0 < t < 1 \\ 1 & 1 \le t < 2 \end{cases}$. Using the formula of Problem 12.148, we have $\mathscr{L}\{f(t)\} = \dfrac{\int_0^2 e^{-st} f(t)\,dt}{1 - e^{-2s}}$. But

$$\int_0^2 e^{-st} f(t)\,dt = \int_0^1 e^{-st} t\,dt + \int_1^2 e^{-st}(1)\,dt = \left(-\frac{t}{s}e^{-st} - \frac{1}{s^2}e^{-st}\right)\Bigg|_0^1 + \frac{e^{-st}}{-s}\Bigg|_1^2 = \frac{1 - e^{-s} - se^{-2s}}{s^2}$$

Therefore, $\mathscr{L}\{f(t)\} = \dfrac{1 - e^{-s} - se^{-2s}}{s^2(1 - e^{-2s})}$.

12.155 Find the Laplace transform for the function shown in Fig. 12.13.

▌ This function is periodic with period $T = 2$, and it is defined on the interval (0, 2) as $f(t) = 1 - t$. Using the formula of Problem 12,148, we thus have

$$\mathscr{L}\{f(t)\} = \frac{\int_0^2 e^{-st}(1 - t)\,dt}{1 - e^{-2s}} = \frac{\left(\dfrac{1-t}{-s}e^{-st} + \dfrac{1}{s^2}e^{-st}\right)\Bigg|_0^2}{1 - e^{-2s}} = \frac{e^{-2s}(s + 1) + s - 1}{s^2(1 - e^{-2s})}$$

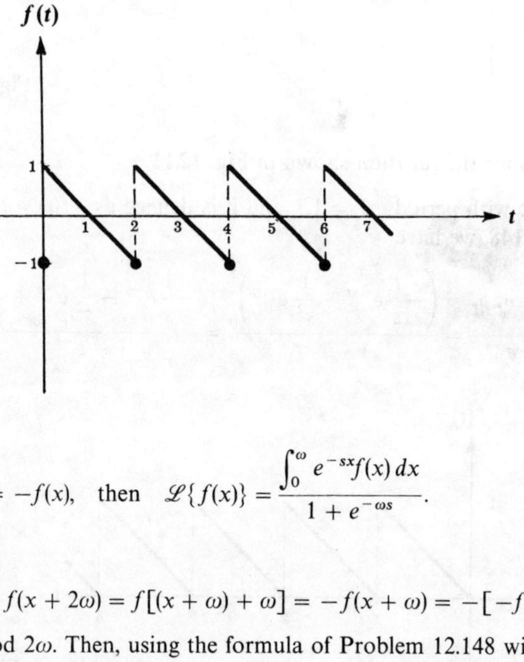

Fig. 12.13

12.156 Prove that if $f(x + \omega) = -f(x)$, then $\mathscr{L}\{f(x)\} = \dfrac{\int_0^\omega e^{-sx} f(x)\,dx}{1 + e^{-\omega s}}$.

▌ Since

$$f(x + 2\omega) = f[(x + \omega) + \omega] = -f(x + \omega) = -[-f(x)] = f(x)$$

$f(x)$ is periodic with period 2ω. Then, using the formula of Problem 12.148 with x replacing t and T replaced by 2ω, we have

$$\mathscr{L}\{f(x)\} = \frac{\int_0^{2\omega} e^{-sx} f(x)\,dx}{1 - e^{-2\omega s}} = \frac{\int_0^\omega e^{-sx} f(x)\,dx + \int_\omega^{2\omega} e^{-sx} f(x)\,dx}{1 - e^{-2\omega s}}$$

Substituting $y = x - \omega$ into the second integral, we find that

$$\int_\omega^{2\omega} e^{-sx} f(x)\,dx = \int_0^\omega e^{-s(y+\omega)} f(y + \omega)\,dy = e^{-\omega s}\int_0^\omega e^{-sy}[-f(y)]\,dy = -e^{-\omega s}\int_0^\omega e^{-sy} f(y)\,dy$$

If we change the dummy variable of integration back to x, this last integral becomes $-e^{-\omega s}\int_0^\omega e^{-sx} f(x)\,dx$. Then

$$\mathscr{L}\{f(x)\} = \frac{(1 - e^{-\omega s})\int_0^\omega e^{-sx} f(x)\,dx}{1 - e^{-2\omega s}} = \frac{(1 - e^{-\omega s})\int_0^\omega e^{-sx} f(x)\,dx}{(1 - e^{-\omega s})(1 + e^{-\omega s})} = \frac{\int_0^\omega e^{-sx} f(x)\,dx}{1 + e^{-\omega s}}$$

12.157 Solve Problem 12.151 using the formula derived in the previous problem.

▌ The square wave $f(x)$ satisfies the equation $f(x+1) = -f(x)$, and on the interval (0, 1) it is defined by $f(x) = 1$. With $\omega = 1$, the formula of the previous problem becomes

$$\mathscr{L}\{f(x)\} = \frac{\int_0^1 e^{-sx} f(x)\,dx}{1+e^{-s}} = \frac{\int_0^1 e^{-sx}(1)\,dx}{1+e^{-s}} = \frac{(1/s)(1-e^{-s})}{1+e^{-s}}$$

This is the same result as is obtained in Problem 12.151. It may also be simplified to $\frac{1}{s}\tanh\frac{s}{2}$.

12.158 Use the formula of Problem 12.156 to obtain the Laplace transform of $f(x) = \sin x$.

▌ The function $f(x) = \sin x$ satisfies the equation $f(x+\pi) = -f(x)$, so the formula of Problem 12.156 (with $\omega = \pi$) becomes

$$\mathscr{L}\{f(x)\} = \frac{\int_0^\pi e^{-sx}\sin x\,dx}{1+e^{-\pi s}} = \frac{\left[\dfrac{e^{-sx}}{s^2+1}(-s\sin x - \cos x)\right]_0^\pi}{1+e^{-\pi s}} = \frac{\dfrac{1}{s^2+1}(e^{-\pi s}+1)}{1+e^{-\pi s}} = \frac{1}{s^2+1}$$

CHAPTER 13

Inverse Laplace Transforms and Their Use in Solving Differential Equations

INVERSE LAPLACE TRANSFORMS BY INSPECTION

13.1 Develop a table of inverse Laplace transforms.

Since $\mathscr{L}^{-1}\{F(s)\} = f(x)$ if and only if $\mathscr{L}\{f(x)\} = F(s)$, every formula generated in Chapter 12 for a Laplace transform automatically provides us with a formula for an inverse Laplace transform. We have, for example, from Problem 12.1 that $\mathscr{L}\{1\} = 1/s$, so it follows that $\mathscr{L}^{-1}\{1/s\} = 1$. We have from Problem 12.2 that $\mathscr{L}\{e^{ax}\} = \dfrac{1}{s-a}$ for any constant a, so it follows that $\mathscr{L}^{-1}\left\{\dfrac{1}{s-a}\right\} = e^{ax}$. We have from Problem 12.23 that $\mathscr{L}\{\sin^2 ax\} = \dfrac{2a^2}{s(s^2+4a^2)}$ for any constant a, so it follows that $\mathscr{L}^{-1}\left\{\dfrac{2a^2}{s(s^2+4a^2)}\right\} = \sin^2 ax$. Continuing in this manner, we generate Table 13.1, where all the inverse Laplace transforms are given as functions of x. To obtain an inverse Laplace transform as a function of t instead, we simply replace x with t.

13.2 Find $\mathscr{L}^{-1}\{2/s^3\}$ as a function of x.

It follows from Table 13.1, entry 3, with $n = 3$ that $\mathscr{L}^{-1}\{2/s^3\} = x^2$.

13.3 Find $\mathscr{L}^{-1}\left\{\dfrac{2}{s^2+4}\right\}$ as a function of x.

It follows from Table 13.1, entry 8, with $a = 2$ that $\mathscr{L}^{-1}\left\{\dfrac{2}{s^2+4}\right\} = \sin 2x$.

13.4 Find $\mathscr{L}^{-1}\left\{\dfrac{s}{s^2+25}\right\}$ as a function of x.

It follows from Table 13.1, entry 9, with $a = 5$ that $\mathscr{L}^{-1}\left\{\dfrac{s}{s^2+25}\right\} = \cos 5x$.

13.5 Find $\mathscr{L}^{-1}\left\{\dfrac{s}{s^2-25}\right\}$ as a function of x.

It follows from Table 13.1, entry 11, with $a = 5$ that $\mathscr{L}^{-1}\left\{\dfrac{s}{s^2-25}\right\} = \cosh 5x$.

13.6 Find $\mathscr{L}^{-1}\left\{\dfrac{2s}{(s^2+1)^2}\right\}$ as a function of x.

It follows from Table 13.1, entry 12, with $a = 1$ that $\mathscr{L}^{-1}\left\{\dfrac{2s}{(s^2+1)^2}\right\} = x \sin x$.

13.7 Find $\mathscr{L}^{-1}\left\{\dfrac{s}{s^2+3}\right\}$ as a function of x.

It follows from Table 13.1, entry 9, with $a = \sqrt{3}$ that $\mathscr{L}^{-1}\left\{\dfrac{s}{s^2+3}\right\} = \cos \sqrt{3}x$.

13.8 Find $\mathscr{L}^{-1}\left\{\dfrac{1}{s+5}\right\}$ as a function of t.

It follows from Table 13.1, entry 7, with $a = -5$ that $\mathscr{L}^{-1}\left\{\dfrac{1}{s+5}\right\} = e^{-5t}$.

Table 13-1

	$F(s)$	$f(x) = \mathscr{L}^{-1}\{F(s)\}$
1.	$\frac{1}{s} \quad (s > 0)$	1
2.	$\frac{1}{s^2} \quad (s > 0)$	x
3.	$\frac{(n-1)!}{s^n} \quad (s > 0)$	$x^{n-1} \; (n = 1, 2, \ldots)$
4.	$\frac{1}{2}\sqrt{\pi}\, s^{-3/2} \quad (s > 0)$	$\sqrt{x}$
5.	$\sqrt{\pi}\, s^{-1/2} \quad (s > 0)$	$1/\sqrt{x}$
6.	$\frac{(1)(3)(5)\cdots(2n-1)\sqrt{\pi}}{2^n} s^{-n-1/2} \quad (s > 0)$	$x^{n-1/2} \; (n = 1, 2, \ldots)$
7.	$\frac{1}{s-a} \quad (s > a)$	e^{ax}
8.	$\frac{a}{s^2+a^2} \quad (s > 0)$	$\sin ax$
9.	$\frac{s}{s^2+a^2} \quad (s > 0)$	$\cos ax$
10.	$\frac{a}{s^2-a^2} \quad (s > \lvert a \rvert)$	$\sinh ax$
11.	$\frac{s}{s^2-a^2} \quad (s > \lvert a \rvert)$	$\cosh ax$
12.	$\frac{2as}{(s^2+a^2)^2} \quad (s > 0)$	$x \sin ax$
13.	$\frac{s^2-a^2}{(s^2+a^2)^2} \quad (s > 0)$	$x \cos ax$
14.	$\frac{(n-1)!}{(s-a)^n} \quad (s > a)$	$x^{n-1}e^{ax} \; (n = 1, 2, \ldots)$
15.	$\frac{a}{(s-b)^2+a^2} \quad (s > b)$	$e^{bx} \sin ax$
16.	$\frac{s-b}{(s-b)^2+a^2} \quad (s > b)$	$e^{bx} \cos ax$
17.	$\frac{2a^3}{(s^2+a^2)^2} \quad (s > 0)$	$\sin ax - ax \cos ax$

Table 13-1 *(continued)*

	$F(s)$	$f(x) = \mathcal{L}^{-1}\{F(s)\}$
18.	$\dfrac{1}{1+as}$	$\dfrac{1}{a}e^{-x/a}$
19.	$\dfrac{1}{s(s-a)}$	$\dfrac{1}{a}(e^{ax}-1)$
20.	$\dfrac{1}{s(1+as)}$	$1-e^{-x/a}$
21.	$\dfrac{1}{(1+as)^2}$	$\dfrac{1}{a^2}xe^{-x/a}$
22.	$\dfrac{1}{(s-a)(s-b)}$	$\dfrac{e^{ax}-e^{bx}}{a-b}$
23.	$\dfrac{1}{(1+as)(1+bs)}$	$\dfrac{e^{-x/a}-e^{-x/b}}{a-b}$
24.	$\dfrac{s}{(s-a)^2}$	$(1+ax)e^{ax}$
25.	$\dfrac{s}{(1+as)^2}$	$\dfrac{1}{a^3}(a-x)e^{-x/a}$
26.	$\dfrac{s}{(s-a)(s-b)}$	$\dfrac{ae^{ax}-be^{bx}}{a-b}$
27.	$\dfrac{s}{(1+as)(1+bs)}$	$\dfrac{ae^{-x/b}-be^{-x/a}}{ab(a-b)}$
28.	$\dfrac{1}{s^2(s-a)}$	$\dfrac{1}{a^2}(e^{ax}-1-ax)$
29.	$\dfrac{2a^2}{s(s^2+4a^2)}$	$\sin^2 ax$
30.	$\dfrac{2a^2}{s(s^2-4a^2)}$	$\sinh^2 ax$
31.	$\dfrac{a^3}{s^4+a^4}$	$\dfrac{1}{\sqrt{2}}\left(\cosh\dfrac{ax}{\sqrt{2}}\sin\dfrac{ax}{\sqrt{2}} - \sinh\dfrac{ax}{\sqrt{2}}\cos\dfrac{ax}{\sqrt{2}}\right)$

Table 13-1 *(continued)*

	$F(s)$	$f(x) = \mathscr{L}^{-1}\{F(s)\}$
32.	$\dfrac{a^2 s}{s^4 + a^4}$	$\sin \dfrac{ax}{\sqrt{2}} \sinh \dfrac{ax}{\sqrt{2}}$
33.	$\dfrac{as^2}{s^4 + a^4}$	$\dfrac{1}{\sqrt{2}}\left(\cos \dfrac{ax}{\sqrt{2}} \sinh \dfrac{ax}{\sqrt{2}} + \sin \dfrac{ax}{\sqrt{2}} \cosh \dfrac{ax}{\sqrt{2}}\right)$
34.	$\dfrac{s^3}{s^4 + a^4}$	$\cos \dfrac{ax}{\sqrt{2}} \cosh \dfrac{ax}{\sqrt{2}}$
35.	$\dfrac{a^3}{s^4 - a^4}$	$\dfrac{1}{2}(\sinh ax - \sin ax)$
36.	$\dfrac{a^2 s}{s^4 - a^4}$	$\dfrac{1}{2}(\cosh ax - \cos ax)$
37.	$\dfrac{as^2}{s^4 - a^4}$	$\dfrac{1}{2}(\sinh ax + \sin ax)$
38.	$\dfrac{s^3}{s^4 - a^4}$	$\dfrac{1}{2}(\cosh ax + \cos ax)$
39.	$\dfrac{2a^2 s}{s^4 + 4a^4}$	$\sin ax \sinh ax$
40.	$\dfrac{a(s^2 - 2a^2)}{s^4 + 4a^4}$	$\cos ax \sinh ax$
41.	$\dfrac{a(s^2 + 2a^2)}{s^4 + 4a^4}$	$\sin ax \cosh ax$
42.	$\dfrac{s^3}{s^4 + 4a^4}$	$\cos ax \cosh ax$
43.	$\dfrac{as^2}{(s^2 + a^2)^2}$	$\dfrac{1}{2}(\sin ax + ax \cos ax)$
44.	$\dfrac{s^3}{(s^2 + a^2)^2}$	$\cos ax - \dfrac{ax}{2} \sin ax$
45.	$\dfrac{a^3}{(s^2 - a^2)^2}$	$\dfrac{1}{2}(ax \cosh ax - \sinh ax)$
46.	$\dfrac{as}{(s^2 - a^2)^2}$	$\dfrac{x}{2} \sinh ax$
47.	$\dfrac{as^2}{(s^2 - a^2)^2}$	$\dfrac{1}{2}(\sinh ax + ax \cosh ax)$
48.	$\dfrac{s^3}{(s^2 - a^2)^2}$	$\cosh ax + \dfrac{ax}{2} \sinh ax$

13.9 Find $\mathscr{L}^{-1}\left\{\dfrac{1}{s(1+2s)}\right\}$ as a function of t.

▮ It follows from Table 13.1, entry 20, with $a = 2$ that $\mathscr{L}^{-1}\left\{\dfrac{1}{s(1+2s)}\right\} = 1 - e^{-t/2}$.

13.10 Find $\mathscr{L}^{-1}\left\{\dfrac{6}{(s-9)^4}\right\}$ as a function of t.

▮ It follows from Table 13.1, entry 14, with $a = 9$ and $n = 4$ that $\mathscr{L}^{-1}\left\{\dfrac{6}{(s-9)^4}\right\} = t^3e^{9t}$.

13.11 Find $\mathscr{L}^{-1}\left\{\dfrac{2s}{s^4+4}\right\}$ as a function of t.

▮ It follows from Table 13.1, entry 36, with $a = \sqrt{2}$ that $\mathscr{L}^{-1}\left\{\dfrac{2s}{s^4-4}\right\} = \dfrac{1}{2}(\cosh\sqrt{2}t - \cos\sqrt{2}t)$.

13.12 Find $\mathscr{L}^{-1}\left\{\dfrac{2s}{(s^2-4)^2}\right\}$ as a function of t.

▮ It follows from Table 13.1, entry 46, with $a = 2$ that $\mathscr{L}^{-1}\left\{\dfrac{2s}{(s^2-4)^2}\right\} = \dfrac{t}{2}\sinh 2t$.

LINEARITY

13.13 Find $\mathscr{L}^{-1}\left\{\dfrac{4}{s-2}\right\}$ as a function of t.

▮ It follows from Table 13.1, entry 7, with $a = 2$ that $\mathscr{L}^{-1}\left\{\dfrac{4}{s-2}\right\} = 4\mathscr{L}^{-1}\left\{\dfrac{1}{s-2}\right\} = 4e^{2t}$.

13.14 Find $\mathscr{L}^{-1}\left\{\dfrac{1}{s^2+9}\right\}$ as a function of t.

▮ It follows from Table 13.1, entry 8, with $a = 3$ that

$$\mathscr{L}^{-1}\left\{\frac{1}{s^2+9}\right\} = \frac{3}{3}\mathscr{L}^{-1}\left\{\frac{1}{s^2+9}\right\} = \frac{1}{3}\mathscr{L}^{-1}\left\{\frac{3}{s^2+9}\right\} = \frac{1}{3}\sin 3t$$

13.15 Find $\mathscr{L}^{-1}\left\{\dfrac{1}{s^2-3}\right\}$ as a function of t.

▮ It follows from Table 13.1, entry 10, with $a = \sqrt{3}$ that

$$\mathscr{L}^{-1}\left\{\frac{1}{s^2-3}\right\} = \frac{\sqrt{3}}{\sqrt{3}}\mathscr{L}^{-1}\left\{\frac{1}{s^2-3}\right\} = \frac{1}{\sqrt{3}}\mathscr{L}^{-1}\left\{\frac{\sqrt{3}}{s^2-(\sqrt{3})^2}\right\} = \frac{1}{\sqrt{3}}\sinh\sqrt{3}t$$

13.16 Find $\mathscr{L}^{-1}\{1/s^3\}$ as a function of t.

▮ It follows from Table 13.1, entry 3, with $n = 2$ that $\mathscr{L}^{-1}\left\{\dfrac{1}{s^3}\right\} = \dfrac{2}{2}\mathscr{L}^{-1}\left\{\dfrac{1}{s^3}\right\} = \dfrac{1}{2}\mathscr{L}^{-1}\left\{\dfrac{2}{s^3}\right\} = \dfrac{1}{2}t^2$.

13.17 Find $\mathscr{L}^{-1}\{1/\sqrt{s}\}$ as a function of x.

▮ It follows from Table 13.1, entry 5, that $\mathscr{L}^{-1}\left\{\dfrac{1}{\sqrt{s}}\right\} = \dfrac{\sqrt{\pi}}{\sqrt{\pi}}\mathscr{L}^{-1}\left\{\dfrac{1}{\sqrt{s}}\right\} = \dfrac{1}{\sqrt{\pi}}\mathscr{L}^{-1}\left\{\dfrac{\sqrt{\pi}}{\sqrt{s}}\right\} = \dfrac{1}{\sqrt{\pi}}\dfrac{1}{\sqrt{x}}$.

13.18 Find $\mathscr{L}^{-1}\left\{\dfrac{5s+4}{s^3}\right\}$ as a function of x.

▮ We use Table 13.1, entries 2 and 3, to write

$$\mathscr{L}^{-1}\left\{\frac{5s+4}{s^3}\right\} = \mathscr{L}^{-1}\left\{\frac{5s}{s^3}+\frac{4}{s^3}\right\} = 5\mathscr{L}^{-1}\left\{\frac{1}{s^2}\right\} + 2\mathscr{L}^{-1}\left\{\frac{2}{s^3}\right\} = 5x + 2x^2.$$

13.19 Find $\mathscr{L}^{-1}\left\{\dfrac{s^3+s}{s^6}\right\}$ as a function of x.

▌ We use Table 13.1, entry 3, first with $n=2$ then with $n=4$ to write

$$\mathscr{L}^{-1}\left\{\frac{s^3+s}{s^6}\right\}=\mathscr{L}^{-1}\left\{\frac{1}{s^3}+\frac{1}{s^5}\right\}=\frac{1}{2!}\mathscr{L}^{-1}\left\{\frac{2!}{s^3}\right\}+\frac{1}{4!}\mathscr{L}^{-1}\left\{\frac{4!}{s^5}\right\}=\frac{1}{2!}x^2+\frac{1}{4!}x^4=\frac{1}{2}x^2+\frac{1}{24}x^4$$

13.20 Find $\mathscr{L}^{-1}\left\{\dfrac{24-30\sqrt{s}}{s^4}\right\}$ as a function of t.

▌ We use Table 13.1, entry 3 with $n=4$ and entry 6 with $n=3$, to write

$$\mathscr{L}^{-1}\left\{\frac{24-30\sqrt{s}}{s^4}\right\}=\mathscr{L}^{-1}\left\{\frac{4!}{s^4}-30s^{-7/2}\right\}=4\mathscr{L}^{-1}\left\{\frac{3!}{s^4}\right\}-\frac{16}{\sqrt{\pi}}\mathscr{L}^{-1}\left\{\frac{15}{8}\sqrt{\pi}s^{-7/2}\right\}=4t^3-\frac{16}{\sqrt{\pi}}t^{5/2}$$

13.21 Find $\mathscr{L}^{-1}\left\{\dfrac{6}{2s-3}\right\}$ as a function of t.

▌ From Table 13.1, entry 7, with $a=\frac{3}{2}$, we have

$$\mathscr{L}^{-1}\left\{\frac{6}{2s-3}\right\}=\mathscr{L}^{-1}\left\{\frac{6}{2}\frac{1}{s-3/2}\right\}=3\mathscr{L}^{-1}\left\{\frac{1}{s-3/2}\right\}=3e^{3t/2}.$$

13.22 Find $\mathscr{L}^{-1}\left\{\dfrac{4}{3s+5}\right\}$ as a function of x.

▌ It follows from Table 13.1, entry 7, with $a=-\frac{5}{3}$ that

$$\mathscr{L}^{-1}\left\{\frac{4}{3s+5}\right\}=\mathscr{L}^{-1}\left\{\frac{4}{3}\frac{1}{s+5/3}\right\}=\frac{4}{3}\mathscr{L}^{-1}\left\{\frac{1}{s-(-5/3)}\right\}=\frac{4}{3}e^{-5x/3}.$$

13.23 Find $\mathscr{L}^{-1}\left\{\dfrac{2s-18}{s^2+9}\right\}$ as a function of x.

▌ From Table 13.1, entries 8 and 9, both with $a=3$, we have

$$\mathscr{L}^{-1}\left\{\frac{2s-18}{s^2+9}\right\}=2\mathscr{L}^{-1}\left\{\frac{s}{s^2+9}\right\}-6\mathscr{L}^{-1}\left\{\frac{3}{s^2+9}\right\}=2\cos 3x-6\sin 3x.$$

13.24 Find $\mathscr{L}^{-1}\left\{\dfrac{2s+18}{s^2+25}\right\}$ as a function of x.

▌ From Table 13.1, entries 8 and 9, both with $a=5$, we have

$$\mathscr{L}^{-1}\left\{\frac{2s+18}{s^2+25}\right\}=2\mathscr{L}^{-1}\left\{\frac{s}{s^2+25}\right\}+\frac{18}{5}\mathscr{L}^{-1}\left\{\frac{5}{s^2+25}\right\}=2\cos 5t+\frac{18}{5}\sin 5t.$$

13.25 Find $\mathscr{L}^{-1}\left\{\dfrac{s+3}{s^2+5}\right\}$ as a function of x.

▌ From Table 13.1, entries 8 and 9, both with $a=\sqrt{5}$, we have

$$\mathscr{L}^{-1}\left\{\frac{s+3}{s^2+5}\right\}=\mathscr{L}^{-1}\left\{\frac{s}{s^2+5}\right\}+\frac{3}{\sqrt{5}}\mathscr{L}^{-1}\left\{\frac{\sqrt{5}}{s^2+5}\right\}=\cos\sqrt{5}x+\frac{3}{\sqrt{5}}\sin\sqrt{5}x.$$

13.26 Find $\mathscr{L}^{-1}\left\{\dfrac{8-6s}{16s^2+9}\right\}$ as a function of t.

▌ $$\mathscr{L}^{-1}\left\{\frac{8-6s}{16s^2+9}\right\}=\mathscr{L}^{-1}\left\{\frac{1}{16}\frac{8-6s}{s^2+9/16}\right\}=\frac{2}{3}\mathscr{L}^{-1}\left\{\frac{3/4}{s^2+(3/4)^2}\right\}-\frac{3}{8}\mathscr{L}^{-1}\left\{\frac{s}{s^2+(3/4)^2}\right\}=\frac{2}{3}\sin\frac{3t}{4}-\frac{3}{8}\cos\frac{3t}{4}$$

13.27 Find $\mathscr{L}^{-1}\left\{\dfrac{2s+3}{4s^2+20}\right\}$ as a function of t.

$$\mathscr{L}^{-1}\left\{\frac{2s+3}{4s^2+20}\right\}=\mathscr{L}^{-1}\left\{\frac{1}{4}\frac{2s+3}{s^2+5}\right\}=\frac{1}{2}\mathscr{L}^{-1}\left\{\frac{s}{s^2+(\sqrt{5})^2}\right\}+\frac{3}{4\sqrt{5}}\mathscr{L}^{-1}\left\{\frac{\sqrt{5}}{s^2+(\sqrt{5})^2}\right\}$$
$$=\frac{1}{2}\cos\sqrt{5}t+\frac{3}{4\sqrt{5}}\sin\sqrt{5}t.$$

13.28 Find $\mathscr{L}^{-1}\left\{\dfrac{1-3s}{2s^2-7}\right\}$ as a function of t.

$$\mathscr{L}^{-1}\left\{\frac{1-3s}{2s^2-7}\right\}=\mathscr{L}^{-1}\left\{\frac{1}{2}\frac{1-3s}{s^2-7/2}\right\}=\frac{1}{2\sqrt{7/2}}\mathscr{L}^{-1}\left\{\frac{\sqrt{7/2}}{s^2-(\sqrt{7/2})^2}\right\}-\frac{3}{2}\mathscr{L}^{-1}\left\{\frac{s}{s^2-(\sqrt{7/2})^2}\right\}$$
$$=\frac{1}{\sqrt{14}}\sinh\sqrt{\frac{7}{2}}\,t-\frac{3}{2}\cosh\sqrt{\frac{7}{2}}\,t$$

13.29 Find $\mathscr{L}^{-1}\left\{\dfrac{2s+7}{3s^2+5}\right\}$ as a function of t.

$$\mathscr{L}^{-1}\left\{\frac{2s+7}{3s^2+5}\right\}=\mathscr{L}^{-1}\left\{\frac{1}{3}\frac{2s+7}{s^2+5/3}\right\}=\frac{2}{3}\mathscr{L}^{-1}\left\{\frac{s}{s^2+(\sqrt{5/3})^2}\right\}+\frac{7}{3\sqrt{5/3}}\mathscr{L}^{-1}\left\{\frac{\sqrt{5/3}}{s^2+(\sqrt{5/3})^2}\right\}$$
$$=\frac{2}{3}\cos\sqrt{\frac{5}{3}}\,t+\frac{7}{\sqrt{15}}\sin\sqrt{\frac{5}{3}}\,t$$

13.30 Find $\mathscr{L}^{-1}\left\{\dfrac{3s+2}{(s-1)^5}\right\}$ as a function of t.

$$\mathscr{L}^{-1}\left\{\frac{3s+2}{(s-1)^5}\right\}=\mathscr{L}^{-1}\left\{\frac{3(s-1)+3+2}{(s-1)^5}\right\}=\mathscr{L}^{-1}\left\{\frac{3(s-1)}{(s-1)^5}+\frac{5}{(s-1)^5}\right\}$$
$$=\frac{1}{2!}\mathscr{L}^{-1}\left\{\frac{3!}{(s-1)^4}\right\}+\frac{5}{4!}\mathscr{L}^{-1}\left\{\frac{4!}{(s-1)^5}\right\}=\frac{1}{2}t^3e^t+\frac{5}{24}t^4e^t$$

COMPLETING THE SQUARE AND TRANSLATIONS

13.31 Develop a method for *completing the square* for a quadratic polynomial.

Every real quadratic polynomial in s can be put into the form $a(s+k)^2+h^2$. To do so, we write

$$as^2+bs+c=a\left(s^2+\frac{b}{a}s\right)+c=a\left[s^2+\frac{b}{a}s+\left(\frac{b}{2a}\right)^2\right]+\left(c-\frac{b^2}{4a}\right)=a\left(s+\frac{b}{2a}\right)^2+\left(c-\frac{b^2}{4a}\right)\equiv a(s+k)^2+h^2$$

where $k=b/2a$ and $h=\sqrt{c-b^2/4a}$.

13.32 Find $\mathscr{L}^{-1}\left\{\dfrac{1}{s^2-2s+9}\right\}$ as a function of x.

No function of this form appears in Table 13.1. But, by completing the square, we obtain $s^2-2s+9=(s^2-2s+1)+(9-1)=(s-1)^2+(\sqrt{8})^2$. Hence,

$$\frac{1}{s^2-2s+9}=\frac{1}{(s-1)^2+(\sqrt{8})^2}=\frac{1}{\sqrt{8}}\frac{\sqrt{8}}{(s-1)^2+(\sqrt{8})^2}$$

Then, using linearity and Table 13.1, entry 15, with $a=\sqrt{8}$ and $b=1$, we find that

$$\mathscr{L}^{-1}\left\{\frac{1}{s^2-2s+9}\right\}=\frac{1}{\sqrt{8}}\mathscr{L}^{-1}\left\{\frac{\sqrt{8}}{(s-1)^2+(\sqrt{8})^2}\right\}=\frac{1}{\sqrt{8}}e^x\sin\sqrt{8}x$$

13.33 Find $\mathscr{L}^{-1}\left\{\dfrac{3}{s^2+4s+6}\right\}$ as a function of x.

No function of this form appears in Table 13.1. However, by completing the square of the denominator, we obtain $s^2+4s+6=(s^2+4s+4)+(6-4)=(s+2)^2+2$. Then, from Table 13.1, entry 15, with $a=\sqrt{2}$

and $b = -2$, we have

$$\mathscr{L}^{-1}\left\{\frac{3}{s^2+4s+6}\right\} = \frac{3}{\sqrt{2}}\mathscr{L}^{-1}\left\{\frac{\sqrt{2}}{(s+2)^2+2}\right\} = \frac{3}{\sqrt{2}}e^{-2x}\sin\sqrt{2}x$$

13.34 Find $\mathscr{L}^{-1}\left\{\dfrac{s-3}{s^2-6s+25}\right\}$ as a function of x.

Completing the square of the denominator, we obtain $s^2-6s+25 = (s^2-6s+9)+(25-9) = (s-3)^2+16$. Then, from Table 13.1, entry 16, with $a = 4$ and $b = 3$, we have $\mathscr{L}^{-1}\left\{\dfrac{s-3}{s^2-6s+25}\right\} = \mathscr{L}^{-1}\left\{\dfrac{s-3}{(s-3)^2+16}\right\} = e^{3x}\cos 4x.$

13.35 Find $\mathscr{L}^{-1}\left\{\dfrac{s+1}{2s^2+4s+7}\right\}$ as a function of x.

Completing the square of the denominator yields $2s^2+4s+7 = 2(s^2+2s)+7 = 2(s^2+2s+1)+(5-2) = 2(s+1)^2+5$. Then, from Table 13.1, entry 16, with $a = \sqrt{5/2}$ and $b = -1$, we get

$$\mathscr{L}^{-1}\left\{\frac{s+1}{2s^2+4s+7}\right\} = \mathscr{L}^{-1}\left\{\frac{s+1}{2(s+1)^2+5}\right\} = \mathscr{L}^{-1}\left\{\frac{1}{2}\frac{s+1}{(s+1)^2+5/2}\right\}$$

$$= \frac{1}{2}\mathscr{L}^{-1}\left\{\frac{s+1}{(s+1)^2+(\sqrt{5/2})^2}\right\} = \frac{1}{2}e^{-x}\cos\sqrt{\frac{5}{2}}\,t$$

13.36 Find $\mathscr{L}^{-1}\left\{\dfrac{1}{4s^2-8s}\right\}$ as a function of x.

Completing the square of the denominator, we obtain $4s^2-8s = 4(s^2-2s) = 4(s^2-2s+1)-4 = 4(s-1)^2-4$. Then

$$\mathscr{L}^{-1}\left\{\frac{1}{4s^2-8s}\right\} = \mathscr{L}^{-1}\left\{\frac{1}{4(s-1)^2-4}\right\} = \mathscr{L}^{-1}\left\{\frac{1}{4}\frac{1}{(s-1)^2-1}\right\} = \frac{1}{4}\mathscr{L}^{-1}\left\{\frac{1}{(s-1)^2-1}\right\} = \frac{1}{4}e^x\sinh x$$

This last result follows from Table 13.1, entry 10, with $a = 1$, coupled with the first translation property (see Problem 12.106): Since $\mathscr{L}\{\sinh x\} = \dfrac{1}{s^2-1}$, we have $\mathscr{L}\{e^x\sinh x\} = \dfrac{1}{(s-1)^2-1}$, so that $\mathscr{L}^{-1}\left\{\dfrac{1}{(s-1)^2-1}\right\} = e^x\sinh x.$

13.37 Find $\mathscr{L}^{-1}\left\{\dfrac{1}{\sqrt{s-3}}\right\}$ as a function of x.

It follows from entry 5 of Table 13.1 that $\mathscr{L}\left\{\dfrac{1}{\sqrt{\pi}\sqrt{x}}\right\} = \dfrac{1}{\sqrt{\pi}}\mathscr{L}\left\{\dfrac{1}{\sqrt{x}}\right\} = \dfrac{1}{\sqrt{\pi}}\sqrt{\pi}s^{-1/2} = \dfrac{1}{\sqrt{s}}$. It follows from the first translation property that $\mathscr{L}\left\{e^{3x}\dfrac{1}{\sqrt{\pi}\sqrt{x}}\right\} = \dfrac{1}{\sqrt{s-3}}$, so we conclude that $\mathscr{L}^{-1}\left\{\dfrac{1}{\sqrt{s-3}}\right\} = e^{3x}\dfrac{1}{\sqrt{\pi x}}.$

13.38 Find $\mathscr{L}^{-1}\left\{\dfrac{1}{\sqrt{2s+3}}\right\}$ as a function of t.

$$\mathscr{L}^{-1}\left\{\frac{1}{\sqrt{2s+3}}\right\} = \frac{1}{\sqrt{2}}\mathscr{L}^{-1}\left\{\frac{1}{(s+3/2)^{1/2}}\right\} = \frac{1}{\sqrt{2}}e^{-3t/2}\frac{t^{-1/2}}{\sqrt{\pi}} = \frac{1}{\sqrt{2\pi}}t^{-1/2}e^{-3t/2}$$

13.39 Find $\mathscr{L}^{-1}\left\{\dfrac{s+4}{s^2+4s+8}\right\}$ as a function of x.

▌ No function of this form appears in Table 13.1. But completing the square in the denominator yields $s^2 + 4s + 8 = (s^2 + 4s + 4) + (8 - 4) = (s + 2)^2 + (2)^2$. Hence, $\dfrac{s+4}{s^2+4s+8} = \dfrac{s+4}{(s+2)^2+(2)^2}$.

Table 13.1 does not contain this expression either. However, if we rewrite the numerator as $s + 4 = (s + 2) + 2$ and then decompose the fraction, we get $\dfrac{s+4}{s^2+4s+8} = \dfrac{s+2}{(s+2)^2+(2)^2} + \dfrac{2}{(s+2)^2+(2)^2}$. Then, from entries 15 and 16 of Table 13.1,

$$\mathcal{L}^{-1}\left\{\frac{s+4}{s^2+4s+8}\right\} = \mathcal{L}^{-1}\left\{\frac{s+2}{(s+2)^2+(2)^2}\right\} + \mathcal{L}^{-1}\left\{\frac{2}{(s+2)^2+(2)^2}\right\} = e^{-2x}\cos 2x + e^{-2x}\sin 2x$$

13.40 Find $\mathcal{L}^{-1}\left\{\dfrac{s+2}{s^2-3s+4}\right\}$ as a function of x.

▌ No function of this form appears in Table 13.1. But completing the square in the denominator yields $s^2 - 3s + 4 = \left(s^2 - 3s + \frac{9}{4}\right) + \left(4 - \frac{9}{4}\right) = \left(s - \frac{3}{2}\right)^2 + \left(\frac{\sqrt{7}}{2}\right)^2$, so that $\dfrac{s+2}{s^2-3s+4} = \dfrac{s+2}{(s-3/2)^2+(\sqrt{7}/2)^2}$.

We now rewrite the numerator as $s + 2 = \left(s - \frac{3}{2}\right) + \frac{7}{2} = \left(s - \frac{3}{2}\right) + \sqrt{7}\frac{\sqrt{7}}{2}$, so that

$$\frac{s+2}{s^2-3s+4} = \frac{s-3/2}{(s-3/2)^2+(\sqrt{7}/2)^2} + \sqrt{7}\,\frac{\sqrt{7}/2}{(s-3/2)^2+(\sqrt{7}/2)^2}$$

and

$$\mathcal{L}^{-1}\left\{\frac{s+2}{s^2-3s+4}\right\} = \mathcal{L}^{-1}\left\{\frac{s-3/2}{(s-3/2)^2+(\sqrt{7}/2)^2}\right\} + \sqrt{7}\mathcal{L}^{-1}\left\{\frac{\sqrt{7}/2}{(s-3/2)^2+(\sqrt{7}/2)^2}\right\}$$

$$= e^{(3/2)x}\cos\frac{\sqrt{7}}{2}x + \sqrt{7}e^{(3/2)x}\sin\frac{\sqrt{7}}{2}x$$

13.41 Find $\mathcal{L}^{-1}\left\{\dfrac{6s-4}{s^2-4s+20}\right\}$ as a function of t.

▌

$$\mathcal{L}^{-1}\left\{\frac{6s-4}{s^2-4s+20}\right\} = \mathcal{L}^{-1}\left\{\frac{6(s-2)+8}{(s-2)^2+16}\right\} = 6\,\mathcal{L}^{-1}\left\{\frac{s-2}{(s-2)^2+16}\right\} + 2\,\mathcal{L}^{-1}\left\{\frac{4}{(s-2)^2+16}\right\}$$

$$= 6\,e^{2t}\cos 4t + 2e^{2t}\sin 4t = 2e^{2t}(3\cos 4t + \sin 4t)$$

13.42 Find $\mathcal{L}^{-1}\left\{\dfrac{4s+12}{s^2+8s+16}\right\}$ as a function of t.

▌

$$\mathcal{L}^{-1}\left\{\frac{4s+12}{s^2+8s+16}\right\} = \mathcal{L}^{-1}\left\{\frac{4(s+4)-4}{(s+4)^2}\right\} = 4\mathcal{L}^{-1}\left\{\frac{1}{s+4}\right\} - 4\mathcal{L}^{-1}\left\{\frac{1}{(s+4)^2}\right\}$$

$$= 4e^{-4t} - 4te^{-4t} = 4e^{-4t}(1 - t)$$

13.43 Find $\mathcal{L}^{-1}\left\{\dfrac{3s+7}{s^2-2s-3}\right\}$ as a function of t.

▌

$$\mathcal{L}^{-1}\left\{\frac{3s+7}{s^2-2s-3}\right\} = \mathcal{L}^{-1}\left\{\frac{3(s-1)+10}{(s-1)^2-4}\right\} = 3\,\mathcal{L}^{-1}\left\{\frac{s-1}{(s-1)^2-4}\right\} + 5\,\mathcal{L}^{-1}\left\{\frac{2}{(s-1)^2-4}\right\}$$

$$= 3e^t\cosh 2t + 5e^t\sinh 2t = e^t(3\cosh 2t + 5\sinh 2t) = 4e^{3t} - e^{-t}$$

13.44 Find $\mathcal{L}^{-1}\left\{\dfrac{s+1}{s^2+s+1}\right\}$ as a function of t.

▌

$$\mathcal{L}^{-1}\left\{\frac{s+1}{s^2+s+1}\right\} = \mathcal{L}^{-1}\left\{\frac{s+1}{(s+\frac{1}{2})^2+\frac{3}{4}}\right\} = \mathcal{L}^{-1}\left\{\frac{s+\frac{1}{2}+\frac{1}{2}}{(s+\frac{1}{2})^2+\frac{3}{4}}\right\}$$

$$= \mathcal{L}^{-1}\left\{\frac{s+\frac{1}{2}}{(s+\frac{1}{2})^2+\frac{3}{4}}\right\} + \frac{1}{\sqrt{3}}\mathcal{L}^{-1}\left\{\frac{\sqrt{3}/2}{(s+\frac{1}{2})^2+\frac{3}{4}}\right\} = e^{-t/2}\cos\frac{\sqrt{3}t}{2} + \frac{1}{\sqrt{3}}e^{-t/2}\sin\frac{\sqrt{3}t}{2}$$

$$= \frac{e^{-t/2}}{\sqrt{3}}\left(\sqrt{3}\cos\frac{\sqrt{3}t}{2} + \sin\frac{\sqrt{3}t}{2}\right)$$

13.45 Find $\mathscr{L}^{-1}\left\{\dfrac{s}{s^2-2s+3}\right\}$ as a function of t.

▌ Completing the square of the denominator and rewriting yield

$$\frac{s}{s^2-2s+3}=\frac{s}{(s-1)^2+2}=\frac{s-1}{(s-1)^2+(\sqrt{2})^2}+\frac{1}{\sqrt{2}}\frac{\sqrt{2}}{(s-1)^2+(\sqrt{2})^2}.\quad \text{Then}$$

$$\mathscr{L}^{-1}\left\{\frac{s}{s^2-2s+3}\right\}=\mathscr{L}^{-1}\left\{\frac{s-1}{(s-1)^2+(\sqrt{2})^2}\right\}+\frac{1}{\sqrt{2}}\mathscr{L}^{-1}\left\{\frac{\sqrt{2}}{(s-1)^2+(\sqrt{2})^2}\right\}$$

$$=e^t\cos\sqrt{2}t+\frac{1}{\sqrt{2}}e^t\sin\sqrt{2}t$$

13.46 Find $\mathscr{L}^{-1}\left\{\dfrac{7s+4}{4s^2+4s+9}\right\}$ as a function of t.

▌

$$\frac{7s+4}{4s^2+4s+9}=\frac{7s/4+1}{s^2+s+9/4}=\frac{7s/4+1}{(s+1/2)^2+2}=\frac{7s/4+1}{(s+1/2)^2+(\sqrt{2})^2}=\frac{(7/4)(s+1/2)+1-7/8}{(s+1/2)^2+(\sqrt{2})^2}$$

$$=\mathscr{L}\left\{\frac{7}{4}e^{-t/2}\cos\sqrt{2}t\right\}+\frac{1}{8}\frac{1}{(s+1/2)^2+(\sqrt{2})^2}=\mathscr{L}\left\{\frac{7}{4}e^{-t/2}\cos\sqrt{2}t\right\}+\frac{1}{8\sqrt{2}}\frac{\sqrt{2}}{(s+1/2)^2+(\sqrt{2})^2}$$

$$=\mathscr{L}\left\{\frac{7}{4}e^{-t/2}\cos\sqrt{2}t+\frac{1}{8\sqrt{2}}e^{-t/2}\sin\sqrt{2}t\right\}$$

Therefore, $\mathscr{L}^{-1}\left\{\dfrac{7s+4}{4s^2+4s+9}\right\}=e^{-t/2}\left(\dfrac{7}{4}\cos\sqrt{2}t+\dfrac{1}{8\sqrt{2}}\sin\sqrt{2}t\right).$

3.47 Find $\mathscr{L}^{-1}\left\{\dfrac{s+3}{4s^2+4s+9}\right\}$ as a function of t.

▌

$$\frac{s+3}{4s^2+4s+9}=\frac{\frac{1}{4}s+\frac{3}{4}}{s^2+s+\frac{9}{4}}=\frac{\frac{1}{4}s+\frac{3}{4}}{(s+\frac{1}{2})^2+(\frac{9}{4}-\frac{1}{4})}=\frac{\frac{1}{4}s+\frac{3}{4}}{(s+\frac{1}{2})^2+2}$$

$$=\frac{(s+\frac{1}{2})/4+(3-\frac{1}{2})/4}{(s+\frac{1}{2})^2+\sqrt{2}^2}=\frac{1}{4}\frac{s+\frac{1}{2}}{(s+\frac{1}{2})^2+\sqrt{2}^2}+\frac{1}{4}\frac{3-\frac{1}{2}}{\sqrt{2}}\frac{\sqrt{2}}{(s+\frac{1}{2})^2+(\sqrt{2})^2}$$

$$=\frac{1}{4}\mathscr{L}\{e^{-t/2}\cos\sqrt{2}t\}+\frac{1}{4}\frac{5}{2\sqrt{2}}\mathscr{L}\{e^{-t/2}\sin\sqrt{2}t\}$$

and therefore $\mathscr{L}^{-1}\left\{\dfrac{s+3}{4s^2+4s+9}\right\}=\frac{1}{4}e^{-t/2}\cos\sqrt{2}t+\dfrac{5}{8\sqrt{2}}e^{-t/2}\sin\sqrt{2}t.$

.48 Find $\mathscr{L}^{-1}\left\{\dfrac{s+3}{4s^2+4s+1}\right\}$ as a function of t.

▌

$$\frac{s+3}{4s^2+4s+1}=\frac{\frac{1}{4}s+\frac{3}{4}}{s^2+s+\frac{1}{4}}=\frac{\frac{1}{4}s+\frac{3}{4}}{(s+\frac{1}{2})^2}=\frac{(s+\frac{1}{2})/4+(3-\frac{1}{2})/4}{(s+\frac{1}{2})^2}$$

$$=\frac{(s+\frac{1}{2})/4+\frac{5}{8}}{(s+\frac{1}{2})^2}=\frac{1}{4}\frac{s+\frac{1}{2}}{(s+\frac{1}{2})^2}+\frac{5}{8}\frac{1}{(s+\frac{1}{2})^2}=\frac{1}{4}\frac{1}{s+\frac{1}{2}}+\frac{5}{8}\frac{1}{(s+\frac{1}{2})^2}$$

$$=\tfrac{1}{4}\mathscr{L}\{e^{-t/2}\}+\tfrac{5}{8}\mathscr{L}\{te^{-t/2}\}$$

and therefore $\mathscr{L}^{-1}\left\{\dfrac{s+3}{4s^2+4s+1}\right\}=\frac{1}{4}e^{-t/2}+\frac{5}{8}te^{-t/2}.$

49 Find $\mathscr{L}^{-1}\left\{\dfrac{2}{s^3}e^{-4s}\right\}$ as a function of x.

▮ Since $\mathscr{L}^{-1}\left\{\frac{2}{s^3}\right\} = x^2$, it follows from the second translation property (Problem 12.135) that $\mathscr{L}^{-1}\left\{\frac{2}{s^3}e^{-4s}\right\} = (x-4)^2 u(x-4)$.

13.50 Find $\mathscr{L}^{-1}\left\{\frac{s}{s^2+4}e^{-\pi s}\right\}$ as a function of x.

▮ Since $\mathscr{L}^{-1}\left\{\frac{s}{s^2+4}\right\} = \cos 2x$, it follows from the second translation property that

$$\mathscr{L}^{-1}\left\{\frac{s}{s^2+4}e^{-\pi s}\right\} = \begin{cases}\cos 2(x-\pi) & x > \pi \\ 0 & x < \pi\end{cases} = u(x-\pi)\cos 2(x-\pi) = u(x-\pi)\cos 2x.$$ (See Problem 12.144.)

13.51 Find $\mathscr{L}^{-1}\left\{\frac{e^{-\pi s/3}}{s^2+1}\right\}$ as a function of t.

▮ Since $\mathscr{L}^{-1}\left\{\frac{1}{s^2+1}\right\} = \sin t$, we have $\mathscr{L}^{-1}\left\{\frac{e^{-\pi s/3}}{s^2+1}\right\} = \begin{cases}\sin(t-\pi/3) & t > \pi/3 \\ 0 & t < \pi/3\end{cases}$.

13.52 Find $\mathscr{L}^{-1}\left\{\frac{e^{-5s}}{(s-2)^4}\right\}$ as a function of t.

▮ Since $\mathscr{L}^{-1}\left\{\frac{1}{(s-2)^4}\right\} = e^{2t}\mathscr{L}^{-1}\left\{\frac{1}{s^4}\right\} = \frac{t^3e^{2t}}{3!} = \frac{1}{6}t^3e^{2t}$, we have

$$\mathscr{L}^{-1}\left\{\frac{e^{-5s}}{(s-2)^4}\right\} = \begin{cases}\frac{1}{6}(t-5)^3e^{2(t-5)} & t > 5 \\ 0 & t < 5\end{cases} = \tfrac{1}{6}(t-5)^3e^{2(t-5)}u(t-5)$$

13.53 Find $\mathscr{L}^{-1}\left\{\frac{se^{-4\pi s/5}}{s^2+25}\right\}$ as a function of t.

▮ Since $\mathscr{L}^{-1}\left\{\frac{s}{s^2+25}\right\} = \cos 5t$,

$$\mathscr{L}^{-1}\left\{\frac{se^{-4\pi s/5}}{s^2+25}\right\} = \begin{cases}\cos 5(t-4\pi/5) & t > 4\pi/5 \\ 0 & t < 4\pi/5\end{cases} = \begin{cases}\cos 5t & t > 4\pi/5 \\ 0 & t < 4\pi/5\end{cases}$$
$$= \cos 5t\, u(t-4\pi/5)$$

13.54 Find $\mathscr{L}^{-1}\left\{\frac{(s+1)e^{-\pi s}}{s^2+s+1}\right\}$ as a function of t.

▮ Using the result of Problem 13.44, we conclude that

$$\mathscr{L}^{-1}\left\{\frac{(s+1)e^{-\pi s}}{s^2+s+1}\right\} = \begin{cases}\frac{e^{(t-\pi)/2}}{\sqrt{3}}\left[\sqrt{3}\cos\frac{\sqrt{3}}{2}(t-\pi) + \sin\frac{\sqrt{3}}{2}(t-\pi)\right] & t > \pi \\ 0 & t < \pi\end{cases}$$
$$= \frac{e^{(t-\pi)/2}}{\sqrt{3}}\left[\sqrt{3}\cos\frac{\sqrt{3}}{2}(t-\pi) + \sin\frac{\sqrt{3}}{2}(t-\pi)\right]u(t-\pi)$$

13.55 Find $\mathscr{L}^{-1}\left\{\frac{e^{4-3s}}{(s+4)^{5/2}}\right\}$ as a function of t.

▮ We have $\mathscr{L}^{-1}\left\{\frac{1}{(s+4)^{5/2}}\right\} = e^{-4t}\mathscr{L}^{-1}\left\{\frac{1}{s^{5/2}}\right\} = \frac{4t^{3/2}e^{-4t}}{3\sqrt{\pi}}$. Thus,

$$\mathscr{L}^{-1}\left\{\frac{e^{4-3s}}{(s+4)^{5/2}}\right\} = e^4\mathscr{L}^{-1}\left\{\frac{e^{-3s}}{(s+4)^{5/2}}\right\} = \begin{cases}\frac{4e^4(t-3)^{3/2}e^{-4(t-3)}}{3\sqrt{\pi}} & t > 3 \\ 0 & t < 3\end{cases}$$
$$= \frac{4(t-3)^{3/2}e^{-4(t-4)}}{3\sqrt{\pi}}u(t-3)$$

PARTIAL-FRACTION DECOMPOSITIONS

13.56 Develop the method of partial-fraction decomposition.

▌ Every function of the form $v(s)/w(s)$, where $v(s)$ and $w(s)$ are polynomials in s, can be reduced to the sum of other fractions such that the denominator of each new fraction is either a first-degree or a quadratic polynomial raised to some power. The method requires only that (1) the degree of $v(s)$ be less than the degree of $w(s)$ (if this is not the case, first perform long division, and consider the remainder term) and (2) $w(s)$ be factored into the product of distinct linear and quadratic polynomials raised to various powers.

The method is carried out as follows: To each factor of $w(s)$ of the form $(s-a)^m$, assign a sum of m fractions, of the form

$$\frac{A_1}{s-a}+\frac{A_2}{(s-a)^2}+\cdots+\frac{A_m}{(s-a)^m}$$

To each factor of $w(s)$ of the form $(s^2+bs+c)^p$, assign a sum of p fractions, of the form

$$\frac{B_1s+C_1}{s^2+bs+c}+\frac{B_2s+C_2}{(s^2+bs+c)^2}+\cdots+\frac{B_ps+C_p}{(s^2+bs+c)^p}$$

where A_i, B_j, and C_k $(i=1,2,\ldots,m;\, j,k=1,2,\ldots,p)$ are constants which still must be determined.

Set the original fraction $v(s)/w(s)$ equal to the sum of the new fractions just constructed. Clear the resulting equation of fractions, and then equate coefficients of like powers of s, thereby obtaining a set of simultaneous linear equations in the unknown constants A_i, B_j, and C_k. Finally, solve these equations for A_i, B_j, and C_k.

13.57 Use partial fractions to decompose $\dfrac{1}{(s+1)(s^2+1)}$.

▌ To the linear factor $s+1$, we assign the fraction $A/(s+1)$; to the quadratic factor s^2+1, we assign the fraction $(Bs+C)/(s^2+1)$. We then set

$$\frac{1}{(s+1)(s^2+1)}\equiv\frac{A}{s+1}+\frac{Bs+C}{s^2+1} \tag{1}$$

Clearing fractions, we obtain

$$1\equiv A(s^2+1)+(Bs+C)(s+1) \tag{2}$$

or

$$s^2(0)+s(0)+1\equiv s^2(A+B)+s(B+C)+(A+C)$$

Equating coefficients of like powers of s, we conclude that $A+B=0$, $B+C=0$, and $A+C=1$. The solution of this set of equations is $A=\frac{1}{2}$, $B=-\frac{1}{2}$, and $C=\frac{1}{2}$. Substituting these values into (1), we obtain the partial-fractions decomposition, $\dfrac{1}{(s+1)(s^2+1)}\equiv\dfrac{1/2}{s+1}+\dfrac{-s/2+1/2}{s^2+1}$.

The following is an alternative procedure for finding the constants A, B, and C in (1): Since (2) must hold for all s, it must in particular hold for $s=-1$. Substituting this value into (2), we immediately find $A=\frac{1}{2}$. Equation (2) must also hold for $s=0$. Substituting this value along with $A=\frac{1}{2}$ into (2), we obtain $C=\frac{1}{2}$. Finally, substituting any other value of s into (2), we find that $B=-\frac{1}{2}$.

13.58 Use partial fractions to decompose $\dfrac{1}{(s^2+1)(s^2+4s+8)}$.

▌ To the quadratic factors s^2+1 and s^2+4s+8, we assign the fractions $(As+B)/(s^2+1)$ and $(Cs+D)/(s^2+4s+8)$. We then set $\dfrac{1}{(s^2+1)(s^2+4s+8)}\equiv\dfrac{As+B}{s^2+1}+\dfrac{Cs+D}{s^2+4s+8}$ and clear fractions to obtain $1\equiv(As+B)(s^2+4s+8)+(Cs+D)(s^2+1)$, or

$$s^3(0)+s^2(0)+s(0)+1\equiv s^3(A+C)+s^2(4A+B+D)+s(8A+4B+C)+(8B+D)$$

Equating coefficients of like powers of s, we obtain

$$A+C=0 \qquad 4A+B+D=0 \qquad 8A+4B+C=0 \qquad 8B+D=1$$

The solution of this set of equations is $A=-\frac{4}{65}$, $B=\frac{7}{65}$, $C=\frac{4}{65}$, $D=\frac{9}{65}$. Therefore, $\dfrac{1}{(s^2+1)(s^2+4s+8)}\equiv\dfrac{-4s/65+7/65}{s^2+1}+\dfrac{4s/65+9/65}{s^2+4s+8}$.

13.59 Use partial fractions to decompose $\dfrac{s+3}{(s-2)(s+1)}$.

▌ To the linear factors $s-2$ and $s+1$, we assign the fractions $A/(s-2)$ and $B/(s+1)$, respectively. We then set $\dfrac{s+3}{(s-2)(s+1)} \equiv \dfrac{A}{s-2} + \dfrac{B}{s+1}$ and, upon clearing fractions, obtain $s+3 \equiv A(s+1) + B(s-2)$.

To find A and B, we use the alternative procedure suggested in Problem 13.57. Substituting $s=-1$ and then $s=2$ into the last equation, we immediately obtain $A = 5/3$ and $B = -2/3$. Thus, $\dfrac{s+3}{(s-2)(s+1)} \equiv \dfrac{5/3}{s-2} - \dfrac{2/3}{s+1}$.

13.60 Use partial fractions to decompose $\dfrac{s-5}{s^2+6s+5}$.

▌ This fraction has the partial-fraction expansion $\dfrac{s-5}{s^2+6s+5} = \dfrac{s-5}{(s+5)(s+1)} = \dfrac{d_1}{s+5} + \dfrac{d_2}{s+1}$ for some constants d_1 and d_2. Multiplying by $(s+5)(s+1)$, we obtain $s-5 = d_1(s+1) + d_2(s+5)$. Setting $s=-5$ shows that $d_1 = \frac{5}{2}$, and setting $s=-1$ shows that $d_2 = -\frac{3}{2}$. Therefore, $\dfrac{s-5}{s^2+6s+5} = \dfrac{5/2}{s+5} + \dfrac{-3/2}{s+1}$.

13.61 Use partial fractions to decompose $\dfrac{3s+7}{s^2-2s-3}$.

▌ We let $\dfrac{3s+7}{s^2-2s-3} = \dfrac{3s+7}{(s-3)(s+1)} = \dfrac{A}{s-3} + \dfrac{B}{s+1}$. Multiplying by $(s-3)(s+1)$, we obtain $3s+7 = A(s+1) + B(s-3)$. Setting $s=-1$ yields $B=-1$; setting $s=3$ yields $A=4$. Therefore, $\dfrac{3s+7}{(s+3)(s+1)} = \dfrac{4}{s-3} - \dfrac{1}{s+1}$.

13.62 Use partial fractions to decompose $\dfrac{2s^2-4}{(s+1)(s-2)(s-3)}$.

▌ We let $\dfrac{2s^2-4}{(s+1)(s-2)(s-3)} = \dfrac{A}{s+1} + \dfrac{B}{s-2} + \dfrac{C}{s-3}$. Clearing fractions, we obtain

$$2s^2 - 4 = A(s-2)(s-3) + B(s+1)(s-3) + C(s+1)(s-2)$$

Then setting $s=2$ yields $B = -\frac{4}{3}$; setting $s=3$ yields $C = \frac{7}{2}$; and setting $s=-1$ yields $A = -\frac{1}{6}$. Therefore, $\dfrac{2s^2-4}{(s+1)(s-2)(s-3)} = \dfrac{-1/6}{s+1} + \dfrac{-4/3}{s-2} + \dfrac{7/2}{s-3}$.

13.63 Use partial fractions to decompose $\dfrac{8}{s^3(s^2-s-2)}$.

▌ Note that s^2-s-2 factors into $(s-2)(s+1)$. To the factor $s^3 = (s-0)^3$, which is a linear polynomial raised to the third power, we assign the sum $A_1/s + A_2/s^2 + A_3/s^3$. To the linear factors $(s-2)$ and $(s+1)$, we assign the fractions $B/(s-2)$ and $C/(s+1)$. Then $\dfrac{8}{s^3(s^2-s-2)} \equiv \dfrac{A_1}{s} + \dfrac{A_2}{s^2} + \dfrac{A_3}{s^3} + \dfrac{B}{s-2} + \dfrac{C}{s+1}$, or, after fractions are cleared,

$$8 \equiv A_1s^2(s-2)(s+1) + A_2s(s-2)(s+1) + A_3(s-2)(s+1) + Bs^3(s+1) + Cs^3(s-2)$$

Letting $s=-1$, 2, and 0 consecutively, we obtain $C = \frac{8}{3}$, $B = \frac{1}{3}$, and $A_3 = -4$. Then choosing $s=1$ and $s=-2$ and simplifying, we obtain the equations $A_1 + A_2 = -1$ and $2A_1 - A_2 = -8$, which have the solution $A_1 = -3$ and $A_2 = 2$. Note that any other two values for s (not -1, 2, or 0) will also do; the resulting equations may be different, but the solution will be identical. Finally, we have $\dfrac{2}{s^3(s^2-s-2)} \equiv -\dfrac{3}{s} + \dfrac{2}{s^2} - \dfrac{4}{s^3} + \dfrac{1/3}{s-2} + \dfrac{8/3}{s+1}$.

13.64 Use partial fractions to decompose $\dfrac{5s^2-15s-11}{(s+1)(s-2)^3}$.

▌ We write $\dfrac{5s^2-15s-11}{(s+1)(s-2)^3}=\dfrac{A}{s+1}+\dfrac{B}{(s-2)^3}+\dfrac{C}{(s-2)^2}+\dfrac{D}{s-2}$. Clearing fractions, we obtain

$$5s^2-15s-11=A(s-2)^3+B(s+1)+C(s+1)(s-2)+D(s+1)(s-2)^2$$

Setting $s=-1$ and $s=2$ in turn yields $A=-\frac{1}{3}$ and $B=-7$.

This procedure does not determine C and D. However, since we know A and B, we have

$$\frac{5s^2-15s-11}{(s+1)(s-2)^3}=\frac{-1/3}{s+1}+\frac{-7}{(s-2)^3}+\frac{C}{(s-2)^2}+\frac{D}{s-2}$$

Now, to find C and D we can substitute two values for s, say $s=0$ and $s=1$, from which we have

$$\frac{11}{8}=-\frac{1}{3}+\frac{7}{8}+\frac{C}{4}-\frac{D}{2}\qquad\text{and}\qquad\frac{21}{2}=-\frac{1}{6}+7+C-D$$

Then $C=4$ and $D=\frac{1}{3}$. Thus $\dfrac{5s^2-15s-11}{(s+1)(s-2)^3}=\dfrac{-1/3}{s+1}+\dfrac{-7}{(s-2)^3}+\dfrac{4}{(s-2)^2}+\dfrac{1/3}{s-2}$.

13.65 Use partial fractions to decompose $\dfrac{3s+1}{(s-1)(s^2+1)}$.

▌ We write $\dfrac{3s+1}{(s-1)(s^2+1)}=\dfrac{A}{s-1}+\dfrac{Bs+C}{s^2+1}$. Clearing fractions, we obtain $3s+1=A(s^2+1)+(Bs+C)(s-1)$. Setting $s=1$ yields $A=2$, so that $\dfrac{3s+1}{(s-1)(s^2+1)}=\dfrac{2}{s-1}+\dfrac{Bs+C}{s^2+1}$.

To determine B and C, we let $s=0$ and 2; then $-1=-2+C$ and $\dfrac{7}{5}=2+\dfrac{2B+C}{5}$, from which $C=1$ and $B=-2$. Thus we have $\dfrac{3s+1}{(s-1)(s^2+1)}=\dfrac{2}{s-1}+\dfrac{-2s+1}{s^2+1}$.

13.66 Use partial fractions to decompose $\dfrac{s^2+2s+3}{(s^2+2s+2)(s^2+2s+5)}$.

▌ We let $\dfrac{s^2+2s+3}{(s^2+2s+2)(s^2+2s+5)}=\dfrac{As+B}{s^2+2s+2}+\dfrac{Cs+D}{s^2+2s+5}$, from which

$$\begin{aligned}s^2+2s+3&=(As+B)(s^2+2s+5)+(Cs+D)(s^2+2s+2)\\&=(A+C)s^3+(2A+B+2C+D)s^2+(5A+2B+2C+2D)s+5B+2D\end{aligned}$$

Then $A+C=0$, $2A+B+2C+D=1$, $5A+2B+2C+2D=2$, and $5B+2D=3$. Thus $A=0$, $B=\frac{1}{3}$, $C=0$, and $D=\frac{2}{3}$, and $\dfrac{s^2+2s+3}{(s^2+2s+2)(s^2+2s+5)}=\dfrac{1/3}{s^2+2s+2}+\dfrac{2/3}{s^2+2s+5}$.

13.67 Use partial fraction to decompose $\dfrac{s+3}{4s^2+4s-3}$.

▌ We let $\dfrac{s+3}{4s^2+4s-3}=\dfrac{s+3}{(2s+3)(2s-1)}=\dfrac{A}{2s+3}+\dfrac{B}{2s-1}$. Clearing fractions then yields $s+3=A(2s-1)+B(2s+3)$. Setting $s=\frac{1}{2}$, we find $B=\frac{7}{8}$; setting $s=-\frac{3}{2}$, we find $A=-\frac{3}{8}$.

Therefore, $\dfrac{s+3}{4s^2+4s-3}=\dfrac{-3/8}{2s+3}+\dfrac{7/8}{2s-1}$.

13.68 Find $\mathscr{L}^{-1}\left\{\dfrac{1}{(s+1)(s^2+1)}\right\}$ as a function of x.

▌ Using the result of Problem 13.57 and noting that $\dfrac{-\frac{1}{2}s+\frac{1}{2}}{s^2+1}=-\dfrac{1}{2}\left(\dfrac{s}{s^2+1}\right)+\dfrac{1}{2}\left(\dfrac{1}{s^2+1}\right)$, we find that

$$\mathscr{L}^{-1}\left\{\frac{1}{(s+1)(s^2+1)}\right\}=\frac{1}{2}\mathscr{L}^{-1}\left\{\frac{1}{s+1}\right\}-\frac{1}{2}\mathscr{L}^{-1}\left\{\frac{s}{s^2+1}\right\}+\frac{1}{2}\mathscr{L}^{-1}\left\{\frac{1}{s^2+1}\right\}=\frac{1}{2}e^{-x}-\frac{1}{2}\cos x+\frac{1}{2}\sin x$$

13.69 Find $\mathscr{L}^{-1}\left\{\dfrac{s+3}{(s-2)(s+1)}\right\}$ as a function of x.

▮ No function of this form appears in Table 13.1. However, by the result of Problem 13.59,
$$\mathscr{L}^{-1}\left\{\frac{s+3}{(s-2)(s+1)}\right\}=\frac{5}{3}\mathscr{L}^{-1}\left\{\frac{1}{s-2}\right\}-\frac{2}{3}\mathscr{L}^{-1}\left\{\frac{1}{s+1}\right\}=\frac{5}{3}e^{2x}-\frac{2}{3}e^{-x}.$$

13.70 Find $\mathscr{L}^{-1}\left\{\dfrac{8}{s^3(s^2-s-2)}\right\}$ as a function of x.

▮ No function of this form appears in Table 13.1. However, by the result of Problem 13.63,
$$\mathscr{L}^{-1}\left\{\frac{8}{s^3(s^2-s-2)}\right\}=-3\mathscr{L}^{-1}\left\{\frac{1}{s}\right\}+2\mathscr{L}^{-1}\left\{\frac{1}{s^2}\right\}-2\mathscr{L}^{-1}\left\{\frac{2}{s^3}\right\}+\frac{1}{3}\mathscr{L}^{-1}\left\{\frac{1}{s-2}\right\}+\frac{8}{3}\mathscr{L}^{-1}\left\{\frac{1}{s+1}\right\}$$
$$=-3+2x-2x^2+\frac{1}{3}e^{2x}+\frac{8}{3}e^{-x}$$

13.71 Find $\mathscr{L}^{-1}\left\{\dfrac{3s+7}{s^2-2s-3}\right\}$ as a function of t.

▮ Using the result of Problem 13.61 and Table 13.1, entry 7, first with $a=3$ and then with $a=-1$, we have
$$\mathscr{L}^{-1}\left\{\frac{3s+7}{(s-3)(s+1)}\right\}=4\mathscr{L}^{-1}\left\{\frac{1}{s-3}\right\}-\mathscr{L}^{-1}\left\{\frac{1}{s+1}\right\}=4e^{3t}-e^{-t}.$$

13.72 Find $\mathscr{L}^{-1}\left\{\dfrac{s-5}{s^2+6s+5}\right\}$ as a function of t.

▮ Using the result of Problem 13.60 and Table 13.1, entry 7, we have
$$\mathscr{L}^{-1}\left\{\frac{s-5}{s^2+6s+5}\right\}=\frac{5}{2}\mathscr{L}^{-1}\left\{\frac{1}{s+5}\right\}-\frac{3}{2}\mathscr{L}^{-1}\left\{\frac{1}{s+1}\right\}=\frac{5}{2}e^{-5t}-\frac{3}{2}e^{-t}.$$

13.73 Find $\mathscr{L}^{-1}\left\{\dfrac{2s^2-4}{(s+1)(s-2)(s-3)}\right\}$ as a function of t.

▮ Using the result of Problem 13.62 and Table 13.1, we determine
$$\mathscr{L}^{-1}\left\{\frac{2s^2-4}{(s+1)(s-2)(s-3)}\right\}=\mathscr{L}^{-1}\left\{\frac{-1/6}{s+1}+\frac{-4/3}{s-2}+\frac{7/2}{s-3}\right\}=-\frac{1}{6}e^{-t}-\frac{4}{3}e^{2t}+\frac{7}{2}e^{3t}.$$

13.74 Find $\mathscr{L}^{-1}\left\{\dfrac{1}{(s^2+1)(s^2+4s+8)}\right\}$ as a function of x.

▮ From Problem 13.58 we have
$$\mathscr{L}^{-1}\left\{\frac{1}{(s^2+1)(s^2+4s+8)}\right\}=\mathscr{L}^{-1}\left\{\frac{-\frac{4}{65}s+\frac{7}{65}}{s^2+1}\right\}+\mathscr{L}^{-1}\left\{\frac{\frac{4}{65}s+\frac{9}{65}}{s^2+4s+8}\right\}$$

The first term on the right can be evaluated easily if we note that $\dfrac{-\frac{4}{65}s+\frac{7}{65}}{s^2+1}=-\dfrac{4}{65}\dfrac{s}{s^2+1}+\dfrac{7}{65}\dfrac{1}{s^2+1}$. To evaluate the second term, we must first complete the square in the denominator, writing $s^2+4s+8=(s+2)^2+(2)^2$; then we note that $\dfrac{\frac{4}{65}s+\frac{9}{65}}{s^2+4s+8}=\dfrac{4}{65}\dfrac{s+2}{(s+2)^2+(2)^2}+\dfrac{1}{130}\dfrac{2}{(s+2)^2+(2)^2}$. Therefore,
$$\mathscr{L}^{-1}\left\{\frac{1}{(s^2+1)(s^2+4s+8)}\right\}=-\frac{4}{65}\mathscr{L}^{-1}\left\{\frac{s}{s^2+1}\right\}+\frac{7}{65}\mathscr{L}^{-1}\left\{\frac{1}{s^2+1}\right\}+\frac{4}{65}\mathscr{L}^{-1}\left\{\frac{s+2}{(s+2)^2+(2)^2}\right\}$$
$$+\frac{1}{130}\mathscr{L}^{-1}\left\{\frac{2}{(s+2)^2+(2)^2}\right\}$$
$$=-\frac{4}{65}\cos x+\frac{7}{65}\sin x+\frac{4}{65}e^{-2x}\cos 2x+\frac{1}{130}e^{-2x}\sin 2x$$

13.75 Find $\mathscr{L}^{-1}\left\{\dfrac{5s^2-15s-11}{(s+1)(s-2)^3}\right\}$ as a function of t.

▌ It follows from Problem 13.64 and Table 13.1, entries 7 and 14, that

$$\mathscr{L}^{-1}\left\{\frac{5s^2-15s-11}{(s+1)(s-2)^3}\right\}=\mathscr{L}^{-1}\left\{\frac{-1/3}{s+1}+\frac{-7}{(s-2)^3}+\frac{4}{(s-2)^3}+\frac{1/3}{s-2}\right\}$$
$$=-\frac{1}{3}e^{-t}-\frac{7}{2}t^2e^{2t}+4te^{2t}+\frac{1}{3}e^{2t}$$

13.76 Find $\mathscr{L}^{-1}\left\{\dfrac{3s+1}{(s-1)(s^2+1)}\right\}$ as a function of t.

▌ It follows from Problem 13.65 and Table 13.1, entries 7 to 9, that

$$\mathscr{L}^{-1}\left\{\frac{3s+1}{(s-1)(s^2+1)}\right\}=\mathscr{L}^{-1}\left\{\frac{2}{s-1}+\frac{-2s+1}{s^2+1}\right\}=2\,\mathscr{L}^{-1}\left\{\frac{1}{s-1}\right\}-2\,\mathscr{L}^{-1}\left\{\frac{s}{s^2+1}\right\}+\mathscr{L}^{-1}\left\{\frac{1}{s^2+1}\right\}$$
$$=2e^t-2\cos t+\sin t$$

13.77 Find $\mathscr{L}^{-1}\left\{\dfrac{s^2+2s+3}{(s^2+2s+2)(s^2+2s+5)}\right\}$ as a function of t.

▌ Using the result of Problem 13.66 and Table 13.1, entry 15, we have

$$\mathscr{L}^{-1}\left\{\frac{s^2+2s+3}{(s^2+2s+2)(s^2+2s+5)}\right\}=\mathscr{L}^{-1}\left\{\frac{1/3}{s^2+2s+2}+\frac{2/3}{s^2+2s+5}\right\}$$
$$=\frac{1}{3}\mathscr{L}^{-1}\left\{\frac{1}{(s+1)^2+1}\right\}+\frac{2}{3}\mathscr{L}^{-1}\left\{\frac{1}{(s+1)^2+4}\right\}$$
$$=\tfrac{1}{3}e^{-t}\sin t+\tfrac{2}{3}(\tfrac{1}{2})e^{-t}\sin 2t=\tfrac{1}{3}e^{-t}(\sin t+\sin 2t)$$

13.78 Find $\mathscr{L}^{-1}\left\{\dfrac{1}{s(s^2+4)}\right\}$ as a function of x.

▌ Using the method of partial fractions, we obtain $\dfrac{1}{s(s^2+4)}\equiv\dfrac{1/4}{s}+\dfrac{(-1/4)s}{s^2+4}$. Thus,
$$\mathscr{L}^{-1}\left\{\frac{1}{s(s^2+4)}\right\}=\frac{1}{4}\mathscr{L}^{-1}\left\{\frac{1}{s}\right\}-\frac{1}{4}\mathscr{L}^{-1}\left\{\frac{s}{s^2+4}\right\}=\frac{1}{4}-\frac{1}{4}\cos 2x.$$

13.79 Find $\mathscr{L}^{-1}\left\{\dfrac{s+3}{4s^2+4s-3}\right\}$ as a function of t.

▌ Using the result of Problem 13.67, we have

$$\mathscr{L}^{-1}\left\{\frac{s+3}{4s^2+4s-3}\right\}=\mathscr{L}^{-1}\left\{\frac{-3/8}{2s+3}+\frac{7/8}{2s-1}\right\}=\mathscr{L}^{-1}\left\{\frac{1}{2}\frac{-3/8}{s+3/2}+\frac{1}{2}\frac{7/8}{s-1/2}\right\}$$
$$=-\frac{3}{16}\mathscr{L}^{-1}\left\{\frac{1}{s+3/2}\right\}+\frac{7}{16}\mathscr{L}^{-1}\left\{\frac{1}{s-1/2}\right\}=-\frac{3}{16}e^{-3t/2}+\frac{7}{16}e^{t/2}.$$

13.80 Find $\mathscr{L}^{-1}\left\{\dfrac{s^2+2s-4}{s^4+2s^3+s^2}\right\}$ as a function of t.

▌ Using a partial-fraction decomposition, we obtain

$$\frac{s^2+2s-4}{s^4+2s^3+s^2}=\frac{s^2+2s-4}{s^2(s+1)^2}=\frac{A}{s}+\frac{B}{s^2}+\frac{C}{s+1}+\frac{D}{(s+1)^2}$$
$$=\frac{A(s^3+2s^2+s)+B(s^2+2s+1)+C(s^3+s^2)+Ds^2}{s^2(s+1)^2}$$
$$=\frac{s^3(A+C)+s^2(2A+B+C+D)+s(A+2B)+1(B)}{s^2(s+1)^2}$$

Clearing fractions, we obtain $s^2 + 2s - 4 = s^3(A + C) + s^2(2A + B + C + D) + s(A + 2B) + 1(B)$. Equating coefficients of like powers of s then yields

$$A + C = 0 \qquad 2A + B + C + D = 1 \qquad A + 2B = 2 \qquad B = -4$$

The solution to this set of equations is $A = 10$, $B = -4$, $C = -10$, and $D = -5$, so that $\dfrac{s^2 + 2s - 4}{s^4 + 2s^3 + s^2} = 10\dfrac{1}{s} - 4\dfrac{1}{s^2} - 10\dfrac{1}{s+1} - 5\dfrac{1}{(s+1)^2}$. Taking the inverse Laplace transform of both sides then shows that $\mathscr{L}^{-1}\left\{\dfrac{s^2 + 2s - 4}{s^4 + 2s^3 + s^2}\right\} = 10 - 4t - 10e^{-t} - 5te^{-t}$.

13.81 Find $\mathscr{L}^{-1}\left\{\dfrac{12}{(s + 20)(s^2 + 4)}\right\}$ as a function of t.

▌ Using partial-fraction decomposition, we write $\dfrac{12}{(s + 20)(s^2 + 4)} = \dfrac{A}{s + 20} + \dfrac{Bs + C}{s^2 + 4}$. Clearing fractions, we have $12 = A(s^2 + 4) + (Bs + C)(s + 20)$. Setting $s = -20$ in this equation yields $A = \frac{3}{101}$; setting $s = 0$, yields $C = \frac{60}{101}$, and setting $s = 1$ yields $B = -\frac{3}{101}$. Thus, $\dfrac{12}{(s + 20)(s^2 + 4)} = \dfrac{3/101}{s + 20} + \dfrac{(-3/101)s + 60/101}{s^2 + 4}$. From the property of linearity, it follows that

$$\mathscr{L}^{-1}\left\{\frac{12}{(s + 2)(s^2 + 4)}\right\} = \frac{3}{101}\mathscr{L}^{-1}\left\{\frac{1}{s + 20}\right\} - \frac{3}{101}\mathscr{L}^{-1}\left\{\frac{s}{s^2 + 4}\right\} + \frac{30}{101}\mathscr{L}^{-1}\left\{\frac{2}{s^2 + 4}\right\}$$

$$= \frac{3}{101}e^{-20t} - \frac{3}{101}\cos 2t + \frac{30}{101}\sin 2t$$

13.82 Find $\mathscr{L}^{-1}\left\{\dfrac{s + 1}{s^3 + s}\right\}$ as a function of t.

▌ The denominator may be factored into $s^3 + s = s(s^2 + 1)$. By the method of partial fractions, we then obtain $\dfrac{s + 1}{s^3 + s} = \dfrac{1}{s} + \dfrac{-s + 1}{s^2 + 1}$. Therefore, using the linearity property, we have

$$\mathscr{L}^{-1}\left\{\frac{s + 1}{s^3 + s}\right\} = \mathscr{L}^{-1}\left\{\frac{1}{s}\right\} - \mathscr{L}^{-1}\left\{\frac{s}{s^2 + 1}\right\} + \mathscr{L}^{-1}\left\{\frac{1}{s^2 + 1}\right\} = 1 - \cos t + \sin t.$$

13.83 Find $\mathscr{L}^{-1}\left\{\dfrac{1}{(s - 1)(s + 2)(s^2 + 1)}\right\}$ as a function of t.

▌ Using the partial-fraction expansion $\dfrac{1}{(s - 1)(s + 2)(s^2 + 1)} = \dfrac{d_1}{s - 1} + \dfrac{d_2}{s + 2} + \dfrac{d_3 s + d_4}{s^2 + 1}$, we find $d_1 = \frac{1}{6}$, $d_2 = -\frac{1}{15}$, $d_3 = -\frac{1}{10}$, and $d_4 = -\frac{3}{10}$. Therefore,

$$\mathscr{L}^{-1}\left\{\frac{1}{(s - 1)(s + 2)(s^2 + 1)}\right\} = \frac{1}{6}\mathscr{L}^{-1}\left\{\frac{1}{s - 1}\right\} - \frac{1}{15}\mathscr{L}^{-1}\left\{\frac{1}{s + 2}\right\} - \frac{1}{10}\mathscr{L}^{-1}\left\{\frac{s}{s^2 + 1}\right\} - \frac{3}{10}\mathscr{L}^{-1}\left\{\frac{1}{s^2 + 1}\right\}$$

$$= \frac{1}{6}e^t - \frac{1}{15}e^{-2t} - \frac{1}{10}\cos t - \frac{3}{10}\sin t$$

CONVOLUTIONS

13.84 Find the convolution $f(x) * g(x)$ if $f(x) = e^{3x}$ and $g(x) = e^{2x}$.

▌ If $f(x) = e^{3x}$ and $g(x) = e^{2x}$, then $f(v) = e^{3v}$, $g(x - v) = e^{2(x-v)}$, and

$$f(x) * g(x) = \int_0^x f(v)g(x - v)\,dv = \int_0^x e^{3v}e^{2(x-v)}\,dv = \int_0^x e^{3v}e^{2x}e^{-2v}\,dv$$

$$= e^{2x}\int_0^x e^v\,dv = e^{2x}\left(e^v\right)\Bigg|_{v=0}^{v=x} = e^{2x}(e^x - 1) = e^{3x} - e^{2x}$$

13.85 Find $f(x) * g(x)$ if $f(x) = x$ and $g(x) = x^2$.

▌ Here $f(v) = v$ and $g(x - v) = (x - v)^2 = x^2 - 2xv + v^2$. Thus,

$$f(x) * g(x) = \int_0^x v(x^2 - 2xv + v^2)\,dv = x^2 \int_0^x v\,dv - 2x \int_0^x v^2\,dv + \int_0^x v^3\,dv$$
$$= x^2 \frac{x^2}{2} - 2x\frac{x^3}{3} + \frac{x^4}{4} = \frac{1}{12}x^4$$

13.86 Prove $f(x) * g(x) = g(x) * f(x)$.

▌ Making the substitution $\tau = x - v$, we have

$$f(x) * g(x) = \int_0^x f(v)g(x - v)\,dv = \int_x^0 f(x - \tau)g(\tau)(-d\tau) = -\int_x^0 g(\tau)f(x - \tau)\,d\tau$$
$$= \int_0^x g(\tau)f(x - \tau)\,d\tau = g(x) * f(x)$$

13.87 Determine $g(x) * f(x)$ for the functions defined in Problem 13.84, and then use the result to verify that convolutions are commutative.

▌ With $f(x - v) = e^{3(x-v)}$ and $g(v) = e^{2v}$, we have

$$g(x) * f(x) = \int_0^x g(v)f(x - v)\,dv = \int_0^x e^{2v}e^{3(x-v)}\,dv = e^{3x}\int_0^x e^{-v}\,dv = e^{3x}\left(-e^{-v}\right)\Big|_{v=0}^{v=x}$$
$$= e^{3x}(-e^{-x} + 1) = e^{3x} - e^{2x}$$

which, from Problem 13.84, equals $f(x) * g(x)$.

13.88 Prove that $f(x) * [g(x) + h(x)] = f(x) * g(x) + f(x) * h(x)$.

▌
$$f(x) * [g(x) + h(x)] = \int_0^x f(v)[g(x - v) + h(x - v)]\,dv = \int_0^x [f(v)g(x - v) + f(v)h(x - v)]\,dv$$
$$= \int_0^x f(v)g(x - v)\,dv + \int_0^x f(v)h(x - v)\,dv = f(x) * g(x) + f(x) * h(x)$$

13.89 Find $\mathscr{L}^{-1}\left\{\frac{1}{s(s^2 + 4)}\right\}$ as a function of x by convolution.

▌ We first note that $\frac{1}{s(s^2 + 4)} = \frac{1}{s}\frac{1}{s^2 + 4}$. Then, defining $F(s) = 1/s$ and $G(s) = 1/(s^2 + 4)$, we have, from Table 13.1, $f(x) = 1$ and $g(x) = \frac{1}{2}\sin 2x$. It now follows from Problem 13.86 that

$$\mathscr{L}^{-1}\left\{\frac{1}{s(s^2 + 4)}\right\} = \mathscr{L}^{-1}\{F(s)G(s)\} = g(x) * f(x) = \int_0^x g(v)f(x - v)\,dv$$
$$= \int_0^x (\tfrac{1}{2}\sin 2v)(1)\,dv = \tfrac{1}{4}(1 - \cos 2x)$$

Observe that in this problem it is easier to evaluate $g(x) * f(x)$ than $f(x) * g(x)$. See also Problem 13.78.

13.90 Find $\mathscr{L}^{-1}\left\{\frac{1}{(s - 1)^2}\right\}$ as a function of x by convolution.

▌ If we define $F(s) = G(s) = 1/(s - 1)$, then $f(x) = g(x) = e^x$ and

$$\mathscr{L}^{-1}\left\{\frac{1}{(s - 1)^2}\right\} = \mathscr{L}^{-1}\{F(s)G(s)\} = f(x) * g(x) = \int_0^x f(v)g(x - v)\,dv = \int_0^x e^v e^{x-v}\,dv = e^x\int_0^x (1)\,dv = xe^x$$

13.91 Find $\mathscr{L}^{-1}\{1/s^2\}$ as a function of x by convolution.

▌ Defining $F(s) = G(s) = 1/s$, we have from Table 13.1 that $f(x) = g(x) = 1$. It now follows that
$$\mathscr{L}^{-1}\left\{\frac{1}{s^2}\right\} = f(x) * g(x) = \int_0^x f(v)g(x - v)\,dv = \int_0^x (1)(1)\,dv = x.$$

13.92 Find $\mathscr{L}^{-1}\left\{\frac{1}{(s - 1)(s - 2)}\right\}$ as a function of x by convolution.

▮ Defining $F(s) = 1/(s-1)$ and $G(s) = 1/(s-2)$, we have from Table 13.1, entry 7, that $f(x) = e^x$ and $g(x) = e^{2x}$. Then

$$\mathscr{L}^{-1}\left\{\frac{1}{(s-1)(s-2)}\right\} = f(x) * g(x) = \int_0^x f(v)g(x-v)\,dv = \int_0^x e^v e^{2(x-v)}\,dv = e^{2x}\int_0^x e^{-v}\,dv$$
$$= e^{2x}(1-e^{-x}) = e^{2x} - e^x$$

13.93 Find $\mathscr{L}^{-1}\left\{\frac{2}{s(s+1)}\right\}$ as a function of x by convolution.

▮ Defining $F(s) = 2/s$ and $G(s) = 1/(s+1)$, we have $f(x) = \mathscr{L}^{-1}\{2/s\} = 2\mathscr{L}^{-1}\{1/s\} = 2$, and $g(x) = \mathscr{L}^{-1}\{1/(s+1) = e^{-x}$. Then

$$\mathscr{L}^{-1}\left\{\frac{2}{s(s+1)}\right\} = f(x) * g(x) = \int_0^x f(v)g(x-v)\,dv = \int_0^x 2e^{-(x-v)}\,dv = 2e^{-x}\int_0^x e^v\,dv$$
$$= 2e^{-x}(e^x - 1) = 2 - 2e^{-x}$$

13.94 Find $\mathscr{L}^{-1}\left\{\frac{s}{(s^2+a^2)^2}\right\}$ as a function of t by convolution.

▮ We can write $\frac{s}{(s^2+a^2)^2} = \frac{s}{s^2+a^2}\,\frac{1}{s^2+a^2}$. Then since $\mathscr{L}^{-1}\left\{\frac{s}{s^2+a^2}\right\} = \cos at$ and $\mathscr{L}^{-1}\left\{\frac{s}{s^2+a^2}\right\} = \frac{\sin at}{a}$, we have

$$\mathscr{L}^{-1}\left\{\frac{s}{(s^2+a^2)^2}\right\} = \int_0^t (\cos av)\frac{\sin a(t-v)}{a}\,dv = \frac{1}{a}\int_0^t (\cos av)(\sin at\cos av - \cos at\sin av)\,dv$$
$$= \frac{1}{a}\sin at\int_0^t \cos^2 av\,dv - \frac{1}{a}\cos at\int_0^t \sin av\cos av\,dv$$
$$= \frac{1}{a}\sin at\int_0^t\left(\frac{1+\cos 2av}{2}\right)dv - \frac{1}{a}\cos at\int_0^t \frac{\sin 2av}{2}\,dv$$
$$= \frac{1}{a}(\sin at)\left(\frac{t}{2} + \frac{\sin 2at}{4a}\right) - \frac{1}{a}(\cos at)\frac{1-\cos 2at}{4a}$$
$$= \frac{1}{a}(\sin at)\left(\frac{t}{2} + \frac{\sin at\cos at}{2a}\right) - \frac{1}{a}(\cos at)\frac{\sin^2 at}{2a} = \frac{t\sin at}{2a}$$

13.95 Find $\mathscr{L}^{-1}\left\{\frac{1}{s^2(s+1)^2}\right\}$ as a function of t by convolution.

▮ We have $\mathscr{L}^{-1}\left\{\frac{1}{s^2}\right\} = t$ and $\mathscr{L}^{-1}\left\{\frac{1}{(s+1)^2}\right\} = te^{-t}$. Then

$$\mathscr{L}^{-1}\left\{\frac{1}{s^2(s+1)^2}\right\} = \int_0^t (ve^{-v})(t-v)\,dv = \int_0^t (vt - v^2)e^{-v}\,dv$$
$$= [(vt - v^2)(-e^{-v}) - (t-2v)(e^{-v}) + (-2)(-e^{-v})]_0^t$$
$$= te^{-t} + 2e^{-t} + t - 2$$

13.96 Find $\mathscr{L}^{-1}\left\{\frac{1}{s^2(s^2+3)}\right\}$ as a function of t using convolution.

▮ Since $\mathscr{L}^{-1}\{1/s^2\} = t$ and $\mathscr{L}^{-1}\left\{\frac{1}{s^2+3}\right\} = \frac{1}{\sqrt{3}}\mathscr{L}^{-1}\left\{\frac{\sqrt{3}}{s^2+3}\right\} = \frac{1}{\sqrt{3}}\sin\sqrt{3}t$, it follows that $f(t) = t$, $g(t) = \frac{1}{\sqrt{3}}\sin\sqrt{3}t$, and the required inverse Laplace transform is $f(t) * g(t)$. In this case, it is easier to evaluate $g(t) * f(t)$ (see Problem 13.86), so we have

$$\mathscr{L}^{-1}\left\{\frac{1}{s^2(s^2+3)}\right\} = \int_0^t (t-v)\frac{1}{\sqrt{3}}\sin\sqrt{3}v\,dv = \left[(t-v)\left(-\frac{1}{3}\cos\sqrt{3}v\right)\right]_{v=0}^t - \int_0^t (-1)\left(-\frac{1}{3}\cos\sqrt{3}v\right)dv$$
$$= \frac{t}{3} - \left(\frac{1}{3\sqrt{3}}\sin\sqrt{3}v\right)_{v=0}^t = \frac{1}{3}t - \frac{1}{3\sqrt{3}}\sin\sqrt{3}t$$

SOLUTIONS USING LAPLACE TRANSFORMS

13.97 Solve $y' - 5y = 0$; $y(0) = 2$.

▌ Taking the Laplace transforms of both sides of this differential equation, we get $\mathscr{L}\{y'\} - 5\mathscr{L}\{y\} = \mathscr{L}\{0\}$. With $y(0) = 2$ and $\mathscr{L}(y) = Y(s)$, we have $[sY(s) - 2] - 5Y(s) = 0$, from which $Y(s) = \dfrac{2}{s-5}$.

Then, taking the inverse Laplace transform of $Y(s)$, we obtain

$$y(x) = \mathscr{L}^{-1}\{Y(s)\} = \mathscr{L}^{-1}\left\{\frac{2}{s-5}\right\} = 2\mathscr{L}^{-1}\left\{\frac{1}{s-5}\right\} = 2e^{5x}.$$

13.98 Solve $y' - 5y = e^{5x}$; $y(0) = 0$.

▌ Taking the Laplace transforms of both sides of this differential equation, we find that $\mathscr{L}\{y'\} - 5\mathscr{L}\{y\} = \mathscr{L}\{e^{5x}\}$, so that $[sY(s) - 0] - 5Y(s) = \dfrac{1}{s-5}$, from which $Y(s) = \dfrac{1}{(s-5)^2}$.

Then, taking the inverse Laplace transform of $Y(s)$, we obtain $y(x) = \mathscr{L}^{-1}\{Y(s)\} = \mathscr{L}^{-1}\left\{\dfrac{1}{(s-5)^2}\right\} = xe^{5x}$, where we have used Table 13.1, entry 14.

13.99 Solve $y' - 5y = e^{5x}$; $y(0) = 2$.

▌ Taking the Laplace transforms of both sides of this differential equation yields $[sY(s) - 2] - 5Y(s) = \dfrac{1}{s-5}$, or $Y(s) = \dfrac{2}{s-5} + \dfrac{1}{(s-5)^2}$. Taking inverse Laplace transforms, we then obtain

$$y = \mathscr{L}^{-1}\left\{\frac{2}{s-5} + \frac{1}{(s-5)^2}\right\} = 2\mathscr{L}^{-1}\left\{\frac{1}{s-5}\right\} + \mathscr{L}^{-1}\left\{\frac{1}{(s-5)^2}\right\} = 2e^{5x} + xe^{5x}$$

13.100 Solve $y' - 5y = 0$; $y(\pi) = 2$.

▌ Taking the Laplace transforms of both sides of the differential equation, we obtain $\mathscr{L}\{y'\} - 5\mathscr{L}\{y\} = \mathscr{L}\{0\}$. Then, with $c_0 = y(0)$ kept arbitrary, we have $[sY(s) - c_0] - 5Y(s) = 0$ or $Y(s) = \dfrac{c_0}{s-5}$. Taking inverse Laplace transforms, we find that $y(x) = \mathscr{L}^{-1}\{Y(s)\} = c_0\mathscr{L}^{-1}\left\{\dfrac{1}{s-5}\right\} = c_0 e^{5x}$.

Now we use the initial condition to solve for c_0. The result is $c_0 = 2e^{-5\pi}$, so $y(x) = 2e^{5(x-\pi)}$.

13.101 Solve $y' + y = xe^{-x}$.

▌ Since no initial condition is given, we set $y(0) = c$, where c denotes an arbitrary constant. Taking the Laplace transforms of both sides of this differential equation (using Table 13.1, entry 14), we obtain $[sY(s) - c] + Y(s) = \dfrac{1}{(s+1)^2}$, from which $Y(s) = \dfrac{c}{s+1} + \dfrac{1}{(s+1)^3}$. Taking inverse Laplace transforms, we then have

$$y = \mathscr{L}^{-1}\left\{\frac{c}{s+1} + \frac{1}{(s+1)^3}\right\} = c\mathscr{L}^{-1}\left\{\frac{1}{s+1}\right\} + \frac{1}{2}\mathscr{L}^{-1}\left\{\frac{2}{(s+1)^3}\right\} = ce^{-x} + \frac{1}{2}x^2e^{-x}$$

The constant c can be determined only if an initial condition is prescribed.

13.102 Solve $y' + y = \sin x$; $y(0) = 0$.

▌ Taking the Laplace transforms of both sides of this differential equation, we obtain $\mathscr{L}\{y'\} + \mathscr{L}\{y\} = \mathscr{L}\{\sin x\}$. This yields $[sY(s) - 0] + Y(s) = \dfrac{1}{s^2+1}$, from which $Y(s) = \dfrac{1}{(s+1)(s^2+1)}$.

Taking the inverse Laplace transforms of both sides and using the result of Problem 13.68, we then obtain

$$y(x) = \mathscr{L}^{-1}\{Y(s)\} = \mathscr{L}^{-1}\left\{\frac{1}{(s+1)(s^2+1)}\right\} = \frac{1}{2}e^{-x} - \frac{1}{2}\cos x + \frac{1}{2}\sin x.$$

13.103 Solve $y' + y = \sin x$; $y(0) = 1$.

▌ Taking the Laplace transforms of both sides of the differential equation, we obtain $\mathscr{L}\{y'\} + \mathscr{L}\{y\} = \mathscr{L}\{\sin x\}$, which yields $[sY(s) - 1] + Y(s) = \dfrac{1}{s^2+1}$. Solving for $Y(s)$, we find $Y(s) = \dfrac{1}{(s+1)(s^2+1)} + \dfrac{1}{s+1}$. Taking the inverse Laplace transforms and using the result of Problem 13.68, we then obtain

$$y(x) = \mathscr{L}^{-1}\{Y(s)\} = \mathscr{L}^{-1}\left\{\frac{1}{(s+1)(s^2+1)}\right\} + \mathscr{L}^{-1}\left\{\frac{1}{s+1}\right\}$$

$$= \left(\frac{1}{2}e^{-x} - \frac{1}{2}\cos x + \frac{1}{2}\sin x\right) + e^{-x} = \frac{3}{2}e^{-x} - \frac{1}{2}\cos x + \frac{1}{2}\sin x$$

13.104 Solve $dN/dt = kN$; $N(0) = 250$, k constant.

▌ Taking the Laplace transforms of both sides of this differential equation and denoting $\mathscr{L}\{N(t)\} = n(s)$, we get $\mathscr{L}\left\{\dfrac{dN}{dt}\right\} = \mathscr{L}\{kN\} = k\mathscr{L}\{N\}$, from which $sn(s) - 250 = kn(s)$, so that $n(s) = \dfrac{250}{s-k}$.

Then, taking inverse Laplace transforms yields $N(t) = \mathscr{L}^{-1}\{n(s)\} = \mathscr{L}^{-1}\left\{\dfrac{250}{s-k}\right\} = 250\,\mathscr{L}^{-1}\left\{\dfrac{1}{s-k}\right\} = 250e^{kt}$

(Compare with Problem 6.1.)

13.105 Solve $dP/dt = 0.05P$; $P(0) = 2000$.

▌ Taking the Laplace transforms of both sides of this differential equation and denoting $\mathscr{L}\{P(t)\} = p(s)$, we obtain $\mathscr{L}\left\{\dfrac{dP}{dt}\right\} = \mathscr{L}\{0.05P\} = 0.05\mathscr{L}\{P\}$, from which $sp(s) - 2000 = 0.05p(s)$, so that $p(s) = \dfrac{2000}{s-0.05}$.

Then, taking inverse Laplace transforms, we obtain $P(t) = \mathscr{L}^{-1}\left\{\dfrac{2000}{s-0.05}\right\} = 2000\mathscr{L}^{-1}\left\{\dfrac{1}{s-0.05}\right\} = 2000e^{0.05t}$

(Compare with Problem 6.41.)

13.106 Solve $dQ/dt + 0.2Q = 1$; $Q(0) = 0$.

▌ Taking the Laplace transforms of both sides of this differential equation and denoting $\mathscr{L}\{Q\} = q(s)$, we have $\mathscr{L}\left\{\dfrac{dQ}{dt}\right\} + 0.2\mathscr{L}\{Q\} = \mathscr{L}\{1\}$, so that $[sq(s) - 0] + 0.2q(s) = \dfrac{1}{s}$. This yields $q(s) = \dfrac{1}{s(s+0.2)}$.

Then, taking inverse Laplace transforms and using partial fractions to decompose the fraction on the right, we obtain

$$Q = \mathscr{L}^{-1}\{q(s)\} = \mathscr{L}^{-1}\left\{\frac{1}{s(s+0.2)}\right\} = \mathscr{L}^{-1}\left\{\frac{5}{s} + \frac{-5}{s+0.2}\right\} = 5\mathscr{L}^{-1}\left\{\frac{1}{s}\right\} - 5\mathscr{L}^{-1}\left\{\frac{1}{s+0.2}\right\} = 5 - 5e^{-0.2t}$$

(Compare with Problem 6.80.)

13.107 Solve $dI/dt + 50I = 5$; $I(0) = 0$.

▌ Taking the Laplace transforms of both sides of this differential equation and denoting $\mathscr{L}\{I\} = i(s)$, we obtain $\mathscr{L}\left\{\dfrac{dI}{dt}\right\} + 50\mathscr{L}\{I\} = 5\mathscr{L}\{I\}$, so that $[si(s) - 0] + 50i(s) = 5\left(\dfrac{1}{s}\right)$, or $i(s) = \dfrac{5}{s(s+50)}$.

Then, taking inverse Laplace transforms and using partial fractions to decompose the fraction on the right, we have

$$I = \mathscr{L}^{-1}\{i(s)\} = \mathscr{L}^{-1}\left\{\frac{5}{s(s+50)}\right\} = \mathscr{L}^{-1}\left\{\frac{0.1}{s} + \frac{-0.1}{s+50}\right\} = 0.1\mathscr{L}^{-1}\left\{\frac{1}{s}\right\} - 0.1\mathscr{L}^{-1}\left\{\frac{1}{s+50}\right\} = 0.1 - 0.1e^{-50t}$$

(Compare with Problem 6.104.)

13.108 Solve $dT/dt + kT = 100k$; $T(0) = 50$, k constant.

▌ Taking the Laplace transforms of both sides of this differential equation and denoting $\mathscr{L}\{T\} = t(s)$, we have $\mathscr{L}\left\{\dfrac{dT}{dt}\right\} + k\mathscr{L}\{T\} = 100k\mathscr{L}\{1\}$, so that

$$[st(s) - 50] + kt(s) = 100k\left(\frac{1}{s}\right) \qquad \text{or} \qquad t(s) = \frac{50}{s+k} + \frac{100k}{s(s+k)}$$

Then, taking inverse Laplace transforms, and using partial fractions, we obtain

$$T = \mathscr{L}^{-1}\{t(s)\} = \mathscr{L}^{-1}\left\{\frac{50}{s+k} + \frac{100k}{s(s+k)}\right\} = \mathscr{L}^{-1}\left\{\frac{50}{s+k} + \frac{100}{s} + \frac{-100}{s+k}\right\}$$

$$= \mathscr{L}^{-1}\left\{\frac{-50}{s+k} + \frac{100}{s}\right\} = -50\mathscr{L}^{-1}\left\{\frac{1}{s+k}\right\} + 100\mathscr{L}^{-1}\left\{\frac{1}{s}\right\} = -50e^{-kt} + 100$$

(Compare with Problem 6.55.)

13.109 Solve $dQ/dt + 0.04Q = 3.2e^{-0.04t}$; $Q(0) = 0$.

▮ Taking the Laplace transforms of both sides of this differential equation and denoting $\mathscr{L}(Q) = q(s)$, we have $\mathscr{L}\left\{\frac{dQ}{dt}\right\} + 0.04\mathscr{L}\{Q\} = 3.2\mathscr{L}\{e^{-0.04t}\}$, so that

$$[sq(s) - 0] + 0.04q(s) = 3.2\frac{1}{s+0.04} \qquad \text{or} \qquad q(s) = 3.2\frac{1}{(s+0.04)^2}$$

Taking inverse Laplace transforms (see Table 13.1, entry 14), we then obtain $Q = 3.2\mathscr{L}^{-1}\left\{\frac{1}{(s+0.04)^2}\right\} = 3.2te^{-0.04t}$. (Compare with Problem 6.89.)

13.110 Solve $dI/dt + 20I = 6\sin 2t$; $I(0) = 6$.

▮ Taking the Laplace transforms of both sides of this differential equation and denoting $\mathscr{L}\{I\} = i(s)$, we get $\mathscr{L}\left\{\frac{dI}{dt}\right\} + 20\mathscr{L}\{I\} = 6\mathscr{L}\{\sin 2t\}$, so that

$$[si(s) - 6] + 20i(s) = 6\frac{2}{s^2+4} \qquad \text{or} \qquad i(s) = \frac{6}{s+20} + \frac{12}{(s+20)(s^2+4)}$$

Then, taking inverse Laplace transforms and using the result of Problem 13.81, we obtain

$$I = \mathscr{L}^{-1}\{i(s)\} = \mathscr{L}^{-1}\left\{\frac{6}{s+20}\right\} + \mathscr{L}^{-1}\left\{\frac{12}{(s+20)(s^2+4)}\right\} = 6e^{-20t} + \left(\tfrac{3}{101}e^{-20t} - \tfrac{3}{101}\cos 2t + \tfrac{30}{101}\sin 2t\right)$$

$$= \tfrac{609}{101}e^{-20t} - \tfrac{3}{101}\cos 2t + \tfrac{30}{101}\sin 2t$$

(Compare with Problem 6.105.)

13.111 Solve $y'' + 4y = 0$; $y(0) = 2$, $y'(0) = 2$.

▮ Taking Laplace transforms yields $\mathscr{L}\{y''\} + 4\mathscr{L}\{y\} = \mathscr{L}\{0\}$, so that

$$[s^2Y(s) - 2s - 2] + 4Y(s) = 0 \qquad \text{or} \qquad Y(s) = \frac{2s+2}{s^2+4} = \frac{2s}{s^2+4} + \frac{2}{s^2+4}$$

Then, taking inverse Laplace transforms yields

$$y(x) = \mathscr{L}^{-1}\{Y(s)\} = 2\mathscr{L}^{-1}\left\{\frac{s}{s^2+4}\right\} + \mathscr{L}^{-1}\left\{\frac{2}{s^2+4}\right\} = 2\cos 2x + \sin 2x$$

13.112 Solve $y'' + 9y = 0$; $y(0) = 3$; $y'(0) = -5$.

▮ Taking the Laplace transforms of both sides of this differential equation, we obtain $\mathscr{L}\{y''\} + 9\mathscr{L}\{y\} = \mathscr{L}\{0\}$, so that $[s^2Y(s) - s(3) - (-5)] + 9Y(s) = 0$, or $Y(s) = \frac{3s-5}{s^2+9}$. Taking inverse Laplace transforms then gives

$$y = \mathscr{L}^{-1}\left\{\frac{3s-5}{s^2+9}\right\} = \mathscr{L}^{-1}\left\{3\frac{s}{s^2+9} - \frac{5}{3}\frac{3}{s^2+9}\right\} = 3\mathscr{L}^{-1}\left\{\frac{s}{s^2+3^2}\right\} - \frac{5}{3}\mathscr{L}^{-1}\left\{\frac{3}{s^2+3^2}\right\} = 3\cos 3x - \frac{5}{3}\sin 3x$$

13.113 Solve $y'' - 3y' + 4y = 0$; $y(0) = 1$, $y'(0) = 5$.

▮ Taking Laplace transforms, we obtain $\mathscr{L}\{y''\} - 3\mathscr{L}\{y'\} + 4\mathscr{L}\{y\} = \mathscr{L}\{0\}$, from which

$$[s^2Y(s) - s - 5] - 3[sY(s) - 1] + 4Y(s) = 0 \qquad \text{or} \qquad Y(s) = \frac{s+2}{s^2-3s+4}$$

Then, taking inverse Laplace transforms and using the result of Problem 13.40, we obtain
$y(x) = e^{(3/2)x}\cos\frac{\sqrt{7}}{2}x + \sqrt{7}\,e^{(3/2)x}\sin\frac{\sqrt{7}}{2}x.$

13.114 Solve $y'' - y' - 2y = 4x^2;\quad y(0) = 1,\quad y'(0) = 4.$

▌ Taking Laplace transforms, we have $\mathscr{L}\{y''\} - \mathscr{L}\{y'\} - 2\mathscr{L}\{y\} = 4\mathscr{L}\{x^2\}$, so that

$$[s^2Y(s) - s - 4] - [sY(s) - 1] - 2Y(s) = \frac{8}{s^3} \quad\text{or}\quad Y(s) = \frac{s+3}{s^2 - s - 2} + \frac{8}{s^3(s^2 - s - 2)}$$

Then, taking the inverse Laplace transform and using the results of Problems 13.69 and 13.70, we obtain

$$y(x) = (\tfrac{5}{3}e^{2x} - \tfrac{2}{3}e^{-x}) + (-3 + 2x - 2x^2 + \tfrac{1}{3}e^{2x} + \tfrac{8}{3}e^{-x}) = 2e^{2x} + 2e^{-x} - 2x^2 + 2x - 3$$

13.115 Solve $y'' + 4y' + 8y = \sin x;\quad y(0) = 1,\quad y'(0) = 0.$

▌ Taking Laplace transforms, we obtain $\mathscr{L}\{y''\} + 4\mathscr{L}\{y'\} + 8\mathscr{L}\{y\} = \mathscr{L}\{\sin x\}$, so that

$$[s^2Y(s) - s - 0] + 4[sY(s) - 1] + 8Y(s) = \frac{1}{s^2+1} \quad\text{or}\quad Y(s) = \frac{s+4}{s^2 + 4s + 8} + \frac{1}{(s^2+1)(s^2 + 4s + 8)}$$

Then, taking the inverse Laplace transform and using the results of Problems 13.39 and 13.74, we obtain

$$\begin{aligned} y(x) &= (e^{-2x}\cos 2x + e^{-2x}\sin 2x) + (-\tfrac{4}{65}\cos x + \tfrac{7}{65}\sin x + \tfrac{4}{65}e^{-2x}\cos 2x + \tfrac{1}{130}e^{-2x}\sin 2x) \\ &= e^{-2x}(\tfrac{69}{65}\cos 2x + \tfrac{131}{130}\sin 2x) + \tfrac{7}{65}\sin x - \tfrac{4}{65}\cos x \end{aligned}$$

13.116 Solve $\ddot{y} + \dot{y} - 2y = \sin t;\quad y(0) = 0,\quad \dot{y}(0) = 0.$

▌ Taking the Laplace transforms of both sides of this differential equation, we obtain
$\mathscr{L}\{\ddot{y}\} + \mathscr{L}\{\dot{y}\} - 2\mathscr{L}\{y\} = \mathscr{L}\{\sin t\}$, from which we write $[s^2Y(s) - s(0) - (0)] + [sY(s) - 0] - 2Y(s) = \dfrac{1}{s^2+1}$
Solving for $Y(s)$ then yields $Y(s) = \dfrac{1}{(s-1)(s+2)(s^2+1)}$. Taking the inverse Laplace transform of both sides of this equation gives $y(t) = \frac{1}{6}e^t - \frac{1}{15}e^{-2t} - \frac{1}{10}\cos t - \frac{1}{10}\sin t$ (see Problem 13.83).

13.117 Solve $\ddot{y} - y = t;\quad y(0) = -1,\quad \dot{y}(0) = 1.$

▌ Taking the Laplace transforms of both sides of this differential equation, we obtain $\mathscr{L}\{\ddot{y}\} - \mathscr{L}\{y\} = \mathscr{L}\{t\}$, from which we write $[s^2Y(s) - s(-1) - 1] - Y(s) = 1/s^2$. Solving for $Y(s)$ gives $(s^2 - 1)Y(s) = -s + 1 + 1/s^2$, from which we eventually find that $Y(s) = \dfrac{s^2 - s^3 + 1}{s^2(s-1)(s+1)}$.

The partial-fraction decomposition $\dfrac{s^2 - s^3 + 1}{s^2(s-1)(s+1)} = \dfrac{d_1}{s} + \dfrac{d_2}{s^2} + \dfrac{d_3}{s-1} + \dfrac{d_4}{s+1}$ yields $d_1 = 0$, $d_2 = -1$, $d_3 = \frac{1}{2}$, and $d_4 = -\frac{3}{2}$, and it follows that

$$y = \mathscr{L}^{-1}\left\{\frac{s^2 - s^3 + 1}{s^2(s-1)(s+1)}\right\} = -1\mathscr{L}^{-1}\left\{\frac{1}{s^2}\right\} + \frac{1}{2}\mathscr{L}^{-1}\left\{\frac{1}{s-1}\right\} - \frac{3}{2}\mathscr{L}^{-1}\left\{\frac{1}{s+1}\right\} = -t + \frac{1}{2}e^t - \frac{3}{2}e^{-t}$$

13.118 Solve $\ddot{x} + 4\dot{x} + 4x = 0;\quad x(0) = 10,\quad \dot{x}(0) = 0.$

▌ Taking the Laplace transforms of both sides of this differential equation, we obtain
$\mathscr{L}\{\ddot{x}\} + 4\mathscr{L}\{\dot{x}\} + 4\mathscr{L}\{x\} = \mathscr{L}\{0\}$, from which

$$[s^2X(s) - s(10) - 0] + 4[sX(s) - 10] + 4X(s) = 0 \quad\text{or}\quad X(s) = \frac{10s + 40}{s^2 + 4s + 4}$$

Then
$$\begin{aligned} x &= \mathscr{L}^{-1}\left\{\frac{10s+40}{(s+2)^2}\right\} = \mathscr{L}^{-1}\left\{\frac{10(s+2)+20}{(s+2)^2}\right\} = 10\mathscr{L}^{-1}\left\{\frac{1}{s+2}\right\} + 20\mathscr{L}^{-1}\left\{\frac{1}{(s+2)^2}\right\} \\ &= 10e^{-2t} + 20te^{-2t} = 10e^{-2t}(1 + 2t) \end{aligned}$$

(Compare with Problem 11.55.)

13.119 Solve $\ddot{x} + 4\dot{x} + 4x = 0;\quad x(0) = 2,\quad \dot{x}(0) = -2.$

▮ Taking the Laplace transforms of both sides of this differential equation, we obtain $\mathscr{L}\{\ddot{x}\} + 4\mathscr{L}\{\dot{x}\} + 4\mathscr{L}\{x\} = \mathscr{L}\{0\}$, from which

$$[s^2X(s) - s(2) - (-2)] + 4[sX(s) - 2] + 4X(s) = 0 \qquad \text{or} \qquad X(s) = \frac{2s+6}{s^2+4s+4}$$

Then

$$x = \mathscr{L}^{-1}\left\{\frac{2s+6}{(s+2)^2}\right\} = \mathscr{L}^{-1}\left\{\frac{2(s+2)+2}{(s+2)^2}\right\}$$

$$= 2\mathscr{L}^{-1}\left\{\frac{1}{s+2}\right\} + 2\mathscr{L}^{-1}\left\{\frac{1}{(s+2)^2}\right\} = 2e^{-t} + 2te^{-t}$$

(Compare with Problem 11.21.)

13.120 Solve $\dfrac{d^2I}{dt^2} + 20\dfrac{dI}{dt} + 200I = 0;\quad I(0) = 0,\quad I'(0) = 24.$

▮ Taking the Laplace transforms of both sides of the equation, $[s^2i(s) - s(0) - 24] + 20[si(s) - 0] + 200i(s) = 0$, which yields $i(s) = \dfrac{24}{s^2+20s+200}$. Completing the square of the denominator and taking inverse Laplace transforms then yield

$$I = \mathscr{L}^{-1}\{i(s)\} = \mathscr{L}^{-1}\left\{\frac{(12/5)(10)}{(s^2+20s+100)+(200-100)}\right\} = \frac{12}{5}\mathscr{L}^{-1}\left\{\frac{10}{(s+10)^2+10^2}\right\} = \frac{12}{5}e^{-10t}\sin 10t$$

(Compare with Problem 11.112.)

3.121 Solve $Y'' + Y = t;\quad Y(0) = 1,\quad Y'(0) = -2.$

▮ Taking the Laplace transforms of both sides of the differential equation, we have $\mathscr{L}\{Y''\} + \mathscr{L}\{Y\} = \mathscr{L}\{t\}$, from which $s^2y(s) - s + 2 + y(s) = 1/s^2$. Then partial-fraction decomposition gives

$$y(s) = \frac{1}{s^2(s^2+1)} + \frac{s-2}{s^2+1} = \frac{1}{s^2} + \frac{s}{s^2+1} - \frac{3}{s^2+1}$$

and $Y = \mathscr{L}^{-1}\left\{\dfrac{1}{s^2} + \dfrac{s}{s^2+1} - \dfrac{3}{s^2+1}\right\} = t + \cos t - 3\sin t.$

3.122 Solve $Y'' - 3Y' + 2Y = 4e^{2t};\quad Y(0) = -3,\quad Y'(0) = 5.$

▮ We have $\mathscr{L}\{Y''\} - 3\mathscr{L}\{Y'\} + 2\mathscr{L}\{Y\} = 4\mathscr{L}\{e^{2t}\}$, from which $[s^2y(s) + 3s - 5] - 3[sy(s) + 3] + 2y(s) = \dfrac{4}{s-2}$. Then partial-fraction decomposition yields

$$y(s) = \frac{4}{(s^2-3s+2)(s-2)} + \frac{14-3s}{s^2-3s+2} = \frac{-7}{s-1} + \frac{4}{s-2} + \frac{4}{(s-2)^2}$$

and $Y = \mathscr{L}^{-1}\left\{\dfrac{-7}{s-1} + \dfrac{4}{s-2} + \dfrac{4}{(s-2)^2}\right\} = -7e^t + 4e^{2t} + 4te^{2t}.$

.123 Solve $Y'' + 2Y' + 5Y = e^{-t}\sin t;\quad Y(0) = 0,\quad Y'(0) = 1.$

▮ We have $\mathscr{L}\{Y''\} + 2\mathscr{L}\{Y'\} + 5\mathscr{L}\{Y\} = \mathscr{L}\{e^{-t}\sin t\}$, so that $[s^2y(s) - s(0) - 1] + 2[sy(s) - 0] + 5y(s) = \dfrac{1}{(s+1)^2+1} = \dfrac{1}{s^2+2s+2}$. Then

$$y(s) = \frac{1}{s^2+2s+5} + \frac{1}{(s^2+2s+2)(s^2+2s+5)} = \frac{s^2+2s+3}{(s^2+2s+2)(s^2+2s+5)}$$

and $Y = \mathscr{L}^{-1}\{y(s)\} = \frac{1}{3}e^{-t}(\sin t + \sin 2t)$ (see Problem 13.77).

13.124 Solve $\dfrac{d^2Q}{dt^2} + 8\dfrac{dQ}{dt} + 25Q = 150$; $Q(0) = 0$, $Q'(0) = 0$.

∎ Taking Laplace transforms yields $\mathscr{L}\left\{\dfrac{d^2Q}{dt^2}\right\} + 8\mathscr{L}\left\{\dfrac{dQ}{dt}\right\} + 25\mathscr{L}\{Q\} = 150\mathscr{L}\{1\}$, so that $[s^2q(s) - s(0) - 0] + 8[sq(s) - 0] + 25q(s) = 150/s$. Then

$$q(s) = \frac{150}{s(s^2 + 8s + 25)} = \frac{6}{s} - \frac{6s + 48}{s^2 + 8s + 25} = \frac{6}{s} - \frac{6(s + 4) + 24}{(s + 4)^2 + 9}$$
$$= \frac{6}{s} - \frac{6(s + 4)}{(s + 4)^2 + 9} - \frac{24}{(s + 4)^2 + 9}$$

and $Q = 6 - 6e^{-4t}\cos 3t - 8e^{-4t}\sin 3t$.

13.125 Solve $\dfrac{d^2Q}{dt^2} + 8\dfrac{dQ}{dt} + 25Q = 50\sin 3t$; $Q(0) = 0$, $Q'(0) = 0$.

∎ Taking Laplace transforms yields $(s^2 + 8s + 25)q(s) = \dfrac{150}{s^2 + 9}$, so that

$$q(s) = \frac{150}{(s^2 + 9)(s^2 + 8s + 25)} = \frac{75}{26}\frac{1}{s^2 + 9} - \frac{75}{52}\frac{s}{s^2 + 9} + \frac{75}{26}\frac{1}{(s + 4)^2 + 9} + \frac{75}{52}\frac{s + 4}{(s + 4)^2 + 9}$$

Thus
$$Q = \tfrac{25}{26}\sin 3t - \tfrac{75}{52}\cos 3t + \tfrac{25}{26}e^{-4t}\sin 3t + \tfrac{75}{52}e^{-4t}\cos 3t$$
$$= \tfrac{25}{52}(2\sin 3t - 3\cos 3t) + \tfrac{25}{52}e^{-4t}(3\cos 3t + 2\sin 3t)$$

(Compare with Problem 11.130.)

13.126 Solve $\dfrac{d^2X}{dt^2} + 2\alpha\dfrac{dX}{dt} + \omega^2 X = 0$; $X(0) = X_0$, $X'(0) = V_0$, where X_0, V_0, α, and ω are all constants and $\omega^2 > \alpha^2$.

∎ Taking Laplace transforms yields $s^2x(s) - X_0 s - V_0 + 2\alpha[sx(s) - X_0] + \omega^2 x(s) = 0$, so that

$$x(s) = \frac{sX_0 + (V_0 + 2\alpha X_0)}{s^2 + 2\alpha s + \omega^2} = \frac{(s + \alpha)X_0}{(s + \alpha)^2 + \omega^2 - \alpha^2} + \frac{V_0 + \alpha X_0}{(s + \alpha)^2 + \omega^2 - \alpha^2}$$

and
$$X = \mathscr{L}^{-1}\{x(s)\} = X_0 e^{-\alpha t}\cos\sqrt{\omega^2 - \alpha^2}\,t + \frac{V_0 + \alpha X_0}{\sqrt{\omega^2 - \alpha^2}}\,e^{-\alpha t}\sin\sqrt{\omega^2 - \alpha^2}\,t$$

13.127 Rework the previous problem for $\omega^2 = \alpha^2$.

∎ In this case, $X = \mathscr{L}^{-1}\{x(s)\} = \mathscr{L}^{-1}\left\{\dfrac{X_0}{s + \alpha} + \dfrac{V_0 + \alpha X_0}{(s + \alpha)^2}\right\} = X_0 e^{-\alpha t} + (V_0 + \alpha X_0)te^{-\alpha t}$.

13.128 Rework Problem 13.126 for $\omega^2 < \alpha^2$.

∎ In this case,

$$X = \mathscr{L}^{-1}\{x\} = \mathscr{L}^{-1}\left\{\frac{(s + \alpha)X_0}{(s + \alpha)^2 - (\alpha^2 - \omega^2)} + \frac{V_0 + \alpha X_0}{(s + \alpha)^2 - (\alpha^2 - \omega^2)}\right\}$$
$$= X_0\cosh\sqrt{\alpha^2 - \omega^2}\,t + \frac{V_0 + \alpha X_0}{\sqrt{\alpha^2 + \omega^2}}\sinh\sqrt{\alpha^2 - \omega^2}\,t.$$

(Compare with Problem 11.23 for $\alpha = a/2m$ and $\omega = \sqrt{k/m}$.)

13.129 Solve $y'' - 3y' + 2y = e^{-x}$; $y(1) = 0$, $y'(1) = 0$.

∎ Taking Laplace transforms, we have $\mathscr{L}\{y''\} - 3\mathscr{L}\{y'\} + 2\mathscr{L}\{y\} = \mathscr{L}\{e^{-x}\}$, or $[s^2Y(s) - sc_0 - c_1] - 3[sY(s) - c_0] + 2Y(s) = 1/(s + 1)$. Here c_0 and c_1 must remain arbitrary, since they represent $y(0)$ and $y'(0)$, respectively, which are unknown. Thus,

$$Y(s) = c_0\frac{s - 3}{s^2 - 3s + 2} + c_1\frac{1}{s^2 - 3s + 2} + \frac{1}{(s + 1)(s^2 - 3s + 2)}$$

Using the method of partial fractions and noting that $s^2 - 3s + 2 = (s-1)(s-2)$, we obtain

$$\begin{aligned} y(x) &= c_0\mathscr{L}^{-1}\left\{\frac{2}{s-1} + \frac{-1}{s-2}\right\} + c_1\mathscr{L}^{-1}\left\{\frac{-1}{s-1} + \frac{1}{s-2}\right\} + \mathscr{L}^{-1}\left\{\frac{1/6}{s+1} + \frac{-1/2}{s-1} + \frac{1/3}{s-2}\right\} \\ &= c_0(2e^x - e^{2x}) + c_1(-e^x + e^{2x}) + (\tfrac{1}{6}e^{-x} - \tfrac{1}{2}e^x + \tfrac{1}{3}e^{2x}) \\ &= (2c_0 - c_1 - \tfrac{1}{2})e^x + (-c_0 + c_1 + \tfrac{1}{3})e^{2x} + \tfrac{1}{6}e^{-x} \equiv d_0 e^x + d_1 e^{2x} + \tfrac{1}{6}e^{-x} \end{aligned}$$

where $d_0 = 2c_0 - c_1 - \frac{1}{2}$ and $d_1 = -c_0 + c_1 + \frac{1}{3}$.

Applying the initial conditions to the last displayed equation, we find that $d_0 = -\frac{1}{2}e^{-2}$ and $d_1 = \frac{1}{3}e^{-3}$; hence, $y(x) = -\frac{1}{2}e^{x-2} + \frac{1}{3}e^{2x-3} + \frac{1}{6}e^{-x}$.

13.130 Solve $y'' - 2y' + y = f(x)$; $y(0) = 0$, $y'(0) = 0$.

▌ In this equation $f(x)$ is unspecified. Taking Laplace transforms and designating $\mathscr{L}\{f(x)\}$ by $F(s)$, we obtain $[s^2Y(s) - (0)s - 0] - 2[sY(s) - 0] + Y(s) = F(s)$ or $Y(s) = \dfrac{F(s)}{(s-1)^2}$.

From Table 13.1, entry 14, $\mathscr{L}^{-1}\{1/(s-1)^2\} = xe^x$. Taking the inverse transform of $Y(s)$ and using convolution, we conclude that $y(x) = xe^x * f(x) = \int_0^x te^t f(x-t)\,dt$.

13.131 Solve $y'' + y = f(x)$; $y(0) = 0$, $y'(0) = 0$, if $f(x) = \begin{cases} 0 & x < 1 \\ 2 & x \ge 1 \end{cases}$.

▌ We note that $f(x) = 2u(x-1)$. Taking Laplace transforms, we obtain $[s^2Y(s) - (0)s - 0] + Y(s) = \mathscr{L}\{f(x)\} = 2\mathscr{L}\{u(x-1)\} = \dfrac{2e^{-s}}{s}$, so that $Y(s) = e^{-s}\dfrac{2}{s(s^2+1)}$. Since

$$\mathscr{L}^{-1}\left\{\frac{2}{s(s^2+1)}\right\} = 2\mathscr{L}^{-1}\left\{\frac{1}{s}\right\} - 2\mathscr{L}^{-1}\left\{\frac{s}{s^2+1}\right\} = 2 - 2\cos x$$

it follows that $y(x) = \mathscr{L}^{-1}\left\{e^{-s}\dfrac{2}{s(s^2+1)}\right\} = [2 - 2\cos(x-1)]u(x-1)$.

13.132 Solve $\dfrac{d^2x}{dt^2} + 16x = -16 + 16u(t-3)$; $x(0) = x'(0) = 0$.

▌ Taking the Laplace transforms of both sides of this differential equation, we obtain

$$[s^2X(s) - s(0) - 0] + 16X(s) = -16\frac{1}{s} + 16\frac{e^{-3s}}{s} \qquad \text{or} \qquad X(s) = \frac{16}{s(s^2+16)}(e^{-3s} - 1)$$

Applying the method of partial fractions to the fraction on the right side and taking inverse Laplace transforms, we get

$$\begin{aligned} x &= \mathscr{L}^{-1}\left\{\frac{16}{s(s^2+16)}(e^{-3s} - 1)\right\} = \mathscr{L}^{-1}\left\{\left(\frac{1}{s} - \frac{s}{s^2+16}\right)(e^{-3s} - 1)\right\} \\ &= \mathscr{L}^{-1}\left\{\frac{1}{s}e^{-3s}\right\} - \mathscr{L}^{-1}\left\{\frac{1}{s}\right\} - \mathscr{L}^{-1}\left\{\frac{s}{s^2+16}e^{-3s}\right\} + \mathscr{L}^{-1}\left\{\frac{s}{s^2+16}\right\} \\ &= u(t-3) - 1 - u(t-3)\cos 4(t-3) + \cos 4t = -1 + \cos 4t + u(t-3)[1 - \cos 4(t-3)] \end{aligned}$$

13.133 Solve $y'' - 3y' - 4y = f(t) \equiv \begin{cases} e^t & 0 \le t < 2 \\ 0 & 2 \le t \end{cases}$; $y(0) = 0$, $y'(0) = 0$.

▌ We first note that $f(t) = e^t - u(t-2)e^t = e^t - u(t-2)e^{t-2}e^2$. Then, taking the Laplace transforms of both sides of the differential equation, we obtain

$$[s^2Y(s) - s(0) - 0] - 3[sY(s) - 0] - 4Y(s) = \frac{1}{s-1} - e^2\frac{1}{s-1}e^{-2s}$$

or

$$Y(s) = \frac{1}{(s-1)(s+1)(s-4)} - e^2e^{-2s}\frac{1}{(s-1)(s+1)(s-4)}$$

where we have used the fact that $s^2 - 3s - 4 = (s+1)(s-4)$.

Now partial fractions yields $\dfrac{1}{(s-1)(s+1)(s-4)} = \dfrac{-1/6}{s-1} + \dfrac{1/10}{s+1} + \dfrac{1/15}{s-4}$, so it follows that

$$y = -\tfrac{1}{6}e^t + \tfrac{1}{10}e^{-t} + \tfrac{1}{15}e^{4t} - e^2 u(t-2)[-\tfrac{1}{6}e^{t-2} + \tfrac{1}{10}e^{-(t-2)} + \tfrac{1}{15}e^{4(t-2)}]$$
$$= -\tfrac{1}{6}e^t + \tfrac{1}{10}e^{-t} + \tfrac{1}{15}e^{4t} + u(t-2)[\tfrac{1}{6}e^t - \tfrac{1}{10}e^4 e^{-t} + \tfrac{1}{15}e^{-6}e^{4t}]$$

13.134 Solve $Y'' + 9Y = \cos 2t$; $Y(0) = 1$, $Y(\pi/2) = -1$.

▌ Since $Y'(0)$ is not known, let $Y'(0) = c$. Then $s^2 y(s) - s(1) - c + 9y(s) = \dfrac{s}{s^2+4}$. Thus,

$$y(s) = \frac{s+c}{s^2+9} + \frac{s}{(s^2+9)(s^2+4)} = \frac{s}{s^2+9} + \frac{c}{s^2+9} + \frac{s}{5(s^2+4)} - \frac{s}{5(s^2+9)}$$
$$= \frac{4}{5}\left(\frac{s}{s^2+9}\right) + \frac{c}{s^2+9} + \frac{s}{5(s^2+4)}$$

and $Y = \dfrac{4}{5}\cos 3t + \dfrac{c}{3}\sin 3t + \dfrac{1}{5}\cos 2t$.

To determine c, we note that $Y(\pi/2) = -1$ so that $-1 = -c/3 - 1/5$ or $c = 12/5$. Then $Y = \frac{4}{5}\cos 3t + \frac{4}{5}\sin 3t + \frac{1}{5}\cos 2t$.

13.135 Solve $Y'' + a^2 Y = F(t)$; $Y(0) = 1$, $Y'(0) = -2$.

▌ We have $s^2 y(s) - sY(0) - Y'(0) + a^2 y(s) = f(s)$, from which $s^2 y(s) - s + 2 + a^2 y(s) = f(s)$, and so $y(s) = \dfrac{s-2}{s^2+a^2} + \dfrac{f(s)}{s^2+a^2}$. Then, using the convolution theorem, we find that

$$Y = \mathscr{L}^{-1}\left\{\frac{s-2}{s^2+a^2}\right\} + \mathscr{L}^{-1}\left\{\frac{f(s)}{s^2+a^2}\right\} = \cos at - \frac{2\sin at}{a} + F(t) * \frac{\sin at}{a}$$
$$= \cos at - \frac{2\sin at}{a} + \frac{1}{a}\int_0^t F(v)\sin a(t-v)\,dv$$

13.136 Find the general solution of $Y'' - a^2 Y = F(t)$.

▌ Let $Y(0) = c_1$ and $Y'(0) = c_2$. Then, taking Laplace transforms, we find $s^2 y(s) - sc_1 - c_2 - a^2 y(s) = f(s)$, so that $y(s) = \dfrac{sc_1 + c_2}{s^2 - a^2} + \dfrac{f(s)}{s^2 - a^2}$. Thus,

$$Y = c_1 \cosh at + \frac{c_2}{a}\sinh at + \frac{1}{a}\int_0^t F(v)\sinh a(t-v)\,dv$$
$$= A\cosh at + B\sinh at + \frac{1}{a}\int_0^t F(v)\sinh a(t-v)\,dv$$

which is the required general solution.

13.137 Solve $y''' + y' = e^x$; $y(0) = y'(0) = y''(0) = 0$.

▌ Taking Laplace transforms, we obtain

$$[s^3 Y(s) - (0)s^2 - (0)s - 0] + [sY(s) - 0] = \frac{1}{s-1} \qquad \text{or} \qquad Y(s) = \frac{1}{(s-1)(s^3+s)}$$

Then, using the method of partial fractions, we obtain

$$y(x) = \mathscr{L}^{-1}\left\{-\frac{1}{s} + \frac{1/2}{s-1} + \frac{(1/2)s - 1/2}{s^2+1}\right\} = -1 + \frac{1}{2}e^x + \frac{1}{2}\cos x - \frac{1}{2}\sin x$$

13.138 Solve $Y''' - 3Y'' + 3Y' - Y = t^2 e^t$; $Y(0) = 1$, $Y'(0) = 0$, $Y''(0) = -2$.

▌ We have $\mathscr{L}\{Y'''\} - 3\mathscr{L}\{Y''\} + 3\mathscr{L}\{Y'\} - \mathscr{L}\{Y\} = \mathscr{L}\{t^2 e^t\}$, from which

$$[s^3 y(s) - s^2 Y(0) - sY'(0) - Y''(0)] - 3[s^2 y(s) - sY(0) - Y'(0)] + 3[sy(s) - Y(0)] - y(s) = \frac{2}{(s-1)^3}$$

Thus $(s^3 - 3s^2 + 3s - 1)y(s) - s^2 + 3s - 1 = \dfrac{2}{(s-1)^3}$, and

$$y(s) = \frac{s^2 - 3s + 1}{(s-1)^3} + \frac{2}{(s-1)^6} = \frac{s^2 - 2s + 1 - s}{(s-1)^3} + \frac{2}{(s-1)^6}$$

$$= \frac{(s-1)^2 - (s-1) - 1}{(s-1)^3} + \frac{2}{(s-1)^6} = \frac{1}{s-1} - \frac{1}{(s-1)^2} - \frac{1}{(s-1)^3} + \frac{2}{(s-1)^6}$$

so that $Y = e^t - te^t - \dfrac{t^2e^t}{2} + \dfrac{t^5e^t}{60}$.

3.139 Find the general solution of the differential equation in the previous problem.

For the general solution, the initial conditions are arbitrary. If we let $Y(0) = A$, $Y'(0) = B$, and $Y''(0) = C$, then the Laplace transform of the differential equation becomes

$$[s^3y(s) - As^2 - Bs - C] - 3[s^2y(s) - As - B] + 3[sy(s) - A] - y(s) = \frac{2}{(s-1)^3}$$

or

$$y(s) = \frac{As^2 + (B - 3A)s + 3A - 3B + C}{(s-1)^3} + \frac{2}{(s-1)^6}$$

Since A, B, and C are arbitrary, so also is the polynomial in the numerator of the first term on the right. We can thus write $y(s) = \dfrac{c_1}{(s-1)^3} + \dfrac{c_2}{(s-1)^2} + \dfrac{c_3}{s-1} + \dfrac{2}{(s-1)^6}$, and invert to find the general solution $Y = \dfrac{c_1t^2}{2}e^t + c_2te^t + c_3e^t + \dfrac{t^5e^t}{60}$, where the c_k are arbitrary constants. (The general solution is easier to find than the particular solution, since we avoid the necessity of determining the constants in the partial-fraction expansion.)

.140 The differential equation governing the deflection $Y(x)$ of a horizontal beam of length l is known to be $\dfrac{d^4Y}{dx^4} = \dfrac{W_0}{EI}$, $0 < x < l$; $Y(0) = 0$, $Y''(0) = 0$, $Y(l) = 0$ $Y''(l) = 0$. Find $Y(x)$ if W_0, E, and I denote positive constants.

Taking the Laplace transforms of both sides of the differential equation, we have $s^4y(s) - s^3Y(0) - s^2Y'(0) - sY''(0) - Y'''(0) = \dfrac{W_0}{EIs}$. Letting the unknown conditions $Y'(0) = c_1$ and $Y'''(0) = c_2$ gives $y(s) = \dfrac{c_1}{s^2} + \dfrac{c_2}{s^4} + \dfrac{W_0}{EIs^5}$, and taking inverse Laplace transforms yields

$$Y(x) = c_1x + \frac{c_2x^3}{3!} + \frac{W_0}{EI}\frac{x^4}{4!} = c_1x + \frac{c_2x^3}{6} + \frac{W_0x^4}{24EI}$$

From the last two given conditions, we find that $c_1 = W_0l^3/24EI$ and $c_2 = -W_0l/2EI$. Thus, the required solution is $Y(x) = \dfrac{W_0}{24EI}(l^3x - 2lx^3 + x^4) = \dfrac{W_0}{24EI}x(l-x)(l^2 + lx - x^2)$.

41 Solve $\dfrac{d^4Y}{dx^4} = \dfrac{W(x)}{EI}$, $0 < x < l$; $Y(0) = 0$, $Y'(0) = 0$, $Y''(l) = 0$, $Y'''(l) = 0$.

So as to apply Laplace transforms, we extend the definition of $W(x)$ as follows:

$$W(x) = \begin{cases} W_0 & 0 < x < l/2 \\ 0 & x > l/2 \end{cases} = W_0\{u(x) - u(x - l/2)\}$$

Now taking Laplace transforms yields

$$s^4y(s) - s^3Y(0) - s^2Y'(0) - sY''(0) - Y'''(0) = \frac{W_0}{EI}\frac{1 - e^{-sl/2}}{s}$$

Letting the unknown conditions $Y''(0) = c_1$ and $Y'''(0) = c_2$, we have $y(s) = \dfrac{c_1}{s^3} + \dfrac{c_2}{s^4} + \dfrac{W_0}{EIs^5}(1 - e^{-sl/2})$.

Inverting then yields

$$Y(x) = \frac{c_1x^2}{2!} + \frac{c_2x^3}{3!} + \frac{W_0}{EI}\frac{x^4}{4!} - \frac{W_0}{EI}\frac{(x - l/2)^4}{4!}u\left(x - \frac{l}{2}\right)$$

We now use the conditions $Y''(l) = 0$ and $Y'''(l) = 0$ to find $c_1 = W_0l^2/8EI$ and $c_2 = -W_0l/2EI$. Thus, the required solution is

$$Y(x) = \frac{W_0l^2}{16EI}x^2 - \frac{W_0l}{12EI}x^3 + \frac{W_0}{24EI}x^4 - \frac{W_0}{24EI}\left(x - \frac{l}{2}\right)^4 u\left(x - \frac{l}{2}\right)$$

13.142 Solve the system $\begin{cases} y' + z = x \\ z' + 4y = 0 \end{cases}$; $y(0) = 1$, $z(0) = -1$.

▮ We denote $\mathscr{L}\{y(x)\}$ and $\mathscr{L}\{z(x)\}$ by $Y(s)$ and $Z(s)$, respectively. Then, taking the Laplace transforms of both differential equations, we obtain

$$\begin{aligned} [sY(s) - 1] + Z(s) &= \frac{1}{s^2} \\ [sZ(s) + 1] + 4Y(s) &= 0 \end{aligned} \qquad \text{or} \qquad \begin{aligned} sY(s) + Z(s) &= \frac{s^2 + 1}{s^2} \\ 4Y(s) + sZ(s) &= -1 \end{aligned}$$

The solution to this last set of simultaneous linear equations is

$$Y(s) = \frac{s^2 + s + 1}{s(s^2 - 4)} \qquad Z(s) = -\frac{s^3 + 4s^2 + 4}{s^2(s^2 - 4)}$$

Using the method of partial fractions to solve each of these equations separately, we obtain

$$y(x) = \mathscr{L}^{-1}\{Y(s)\} = \mathscr{L}^{-1}\left\{-\frac{1/4}{s} + \frac{7/8}{s - 2} + \frac{3/8}{s + 2}\right\} = -\frac{1}{4} + \frac{7}{8}e^{2x} + \frac{3}{8}e^{-2x}$$

$$z(x) = \mathscr{L}^{-1}\{Z(s)\} = \mathscr{L}^{-1}\left\{\frac{1}{s^2} - \frac{7/4}{s - 2} + \frac{3/4}{s + 2}\right\} = x - \frac{7}{4}e^{2x} + \frac{3}{4}e^{-2x}$$

13.143 Solve the system $\begin{cases} w' + y = \sin x \\ y' - z = e^x \\ z' + w + y = 1 \end{cases}$; $w(0) = 0$, $y(0) = 1$, $z(0) = 1$.

▮ We denote $\mathscr{L}\{w(x)\}$, $\mathscr{L}\{y(x)\}$, and $\mathscr{L}\{z(x)\}$ by $W(s)$, $Y(s)$, and $Z(s)$, respectively. Then, taking the Laplace transforms of all three differential equations, we obtain

$$\begin{aligned} [sW(s) - 0] + Y(s) &= \frac{1}{s^2 + 1} \\ [sY(s) - 1] - Z(s) &= \frac{1}{s - 1} \\ [sZ(s) - 1] + W(s) + Y(s) &= \frac{1}{s} \end{aligned} \qquad \text{or} \qquad \begin{aligned} sW(s) + Y(s) &= \frac{1}{s^2 + 1} \\ sY(s) - Z(s) &= \frac{s}{s - 1} \\ W(s) + Y(s) + sZ(s) &= \frac{s + 1}{s} \end{aligned}$$

The solution to this last system of simultaneous linear equations is

$$W(s) = \frac{-1}{s(s - 1)} \qquad Y(s) = \frac{s^2 + s}{(s - 1)(s^2 + 1)} \qquad Z(s) = \frac{s}{s^2 + 1}$$

Using the method of partial fractions and solving each equation separately,

$$w(x) = \mathscr{L}^{-1}\{W(s)\} = \mathscr{L}^{-1}\left\{\frac{1}{s} - \frac{1}{s - 1}\right\} = 1 - e^x$$

$$y(x) = \mathscr{L}^{-1}\{Y(s)\} = \mathscr{L}^{-1}\left\{\frac{1}{s - 1} + \frac{1}{s^2 + 1}\right\} = e^x + \sin x$$

$$z(x) = \mathscr{L}^{-1}\{Z(s)\} = \mathscr{L}^{-1}\left\{\frac{s}{s^2 + 1}\right\} = \cos x$$

13.144 Solve the system $\begin{cases} y'' + z + y = 0 \\ z' + y' = 0 \end{cases}$; $y(0) = 0$, $y'(0) = 0$, $z(0) = 1$

Taking the Laplace transforms of both differential equations, we obtain

$$[s^2Y(s)-(0)s-(0)]+Z(s)+Y(s)=0 \qquad\qquad (s^2+1)Y(s)+Z(s)=0$$
$$\text{or}$$
$$[sZ(s)-1]+[sY(s)-0]=0 \qquad\qquad Y(s)+Z(s)=\frac{1}{s}$$

Solving this last system, we find that $Y(s)=-\dfrac{1}{s^3}$ and $Z(s)=\dfrac{1}{s}+\dfrac{1}{s^3}$. Then, taking inverse transforms yields $y(x)=-\frac{1}{2}x^2$ and $z(x)=1+\frac{1}{2}x^2$.

3.145 Solve $\begin{cases} z''+y'=\cos x \\ y''-z=\sin x \end{cases}$; $z(0)=-1$, $z'(0)=-1$, $y(0)=1$, $y'(0)=0$

Taking the Laplace transforms of both differential equations, we obtain

$$[s^2Z(s)+s+1]+[sY(s)-1]=\frac{s}{s^2+1} \qquad\qquad s^2Z(s)+sY(s)=-\frac{s^3}{s^2+1}$$
$$\text{or}$$
$$[s^2Y(s)-s-0]-Z(s)=\frac{1}{s^2+1} \qquad\qquad -Z(s)+s^2Y(s)=\frac{s^3+s+1}{s^2+1}$$

Solving this last system, we find that $Z(s)=-\dfrac{s+1}{s^2+1}$ and $Y(s)=\dfrac{s}{s^2+1}$. Then, taking inverse transforms yields $z(x)=-\cos x-\sin x$ and $y(x)=\cos x$.

3.146 Solve $\begin{cases} w''-y+2z=3e^{-x} \\ -2w'+2y'+z=0 \\ 2w'-2y+z'+2z''=0 \end{cases}$; $w(0)=1$, $w'(0)=1$, $y(0)=2$, $z(0)=2$, $z'(0)=-2$.

Taking the Laplace transforms of all three differential equations yields

$$[s^2W(s)-s-1]-Y(s)+2Z(s)=\frac{3}{s+1}$$
$$-2[sW(s)-1]+2[sY(s)-2]+Z(s)=0$$
$$2[sW(s)-1]-2Y(s)+[sZ(s)-2]+2[s^2Z(s)-2s+2]=0$$

or

$$s^2W(s)-Y(s)+2Z(s)=\frac{s^2+2s+4}{s+1}$$
$$-2sW(s)+2sY(s)+Z(s)=2$$
$$2sW(s)-2Y(s)+(2s^2+s)Z(s)=4s$$

The solution to this system is $W(s)=\dfrac{1}{s-1}$, $Y(s)=\dfrac{2s}{(s-1)(s+1)}$, and $Z(s)=\dfrac{2}{s+1}$, so that

$$w(x)=e^x \qquad y(x)=\mathscr{L}^{-1}\left\{\frac{1}{s-1}+\frac{1}{s+1}\right\}=e^x+e^{-x} \qquad z(x)=2e^{-x}$$

3.147 Solve $\begin{cases} dX/dt=2X-3Y \\ dY/dt=Y-2X \end{cases}$; $X(0)=8$, $Y(0)=3$.

Taking Laplace transforms, we have, with $\mathscr{L}\{X\}=x$ and $\mathscr{L}\{Y\}=y$,

$$sx-8=2x-3y \qquad\qquad (s-2)x+3y=8$$
$$\text{or}$$
$$sy-3=y-2x \qquad\qquad 2x+(s-1)y=3$$

Solving this last system, we obtain

$$x=\frac{5}{s+1}+\frac{3}{s-4} \qquad \text{and} \qquad y=\frac{5}{s+1}-\frac{2}{s-4}$$

which yield $X=5e^{-t}+3e^{4t}$ and $Y=5e^{-t}-2e^{4t}$.

3.148 Solve $\begin{cases} X''+Y'+3X=15e^{-t} \\ Y''-4X'+3Y=15\sin 2t \end{cases}$; $X(0)=35$, $X'(0)=-48$, $Y(0)=27$, $Y'(0)=-55$.

▌ Taking Laplace transforms yields, in the notation of the preceding problem,

$$s^2x - s(35) - (-48) + sy - 27 + 3x = \frac{15}{s+1}$$

$$s^2y - s(27) - (-55) - 4(sx - 35) + 3y = \frac{30}{s^2+4}$$

or

$$(s^2+3)x + sy = 35s - 21 + \frac{15}{s+1}$$

$$-4sx + (s^2+3)y = 27s - 195 + \frac{30}{s^2+4} \qquad (1)$$

Solving system (1), we then obtain

$$x = \frac{35s^3 - 48s^2 + 300s - 63}{(s^2+1)(s^2+9)} + \frac{15(s^2+3)}{(s+1)(s^2+1)(s^2+9)} - \frac{30s}{(s^2+1)(s^2+4)(s^2+9)}$$

$$= \frac{30s}{s^2+1} - \frac{45}{s^2+9} + \frac{3}{s+1} + \frac{2s}{s^2+4}$$

and

$$y = \frac{27s^3 - 55s^2 + 3s - 585}{(s^2+1)(s^2+9)} + \frac{60s}{(s+1)(s^2+1)(s^2+9)} - \frac{30(s^2+3)}{(s^2+1)(s^2+4)(s^2+9)}$$

$$= \frac{30s}{s^2+9} - \frac{60}{s^2+1} - \frac{3}{s+1} + \frac{2}{s^2+4}$$

Thus $\qquad X = \mathcal{L}^{-1}\{x\} = 30\cos t - 15\sin 3t + 3e^{-t} + 2\cos 2t$

and $\qquad Y = \mathcal{L}^{-1}\{y\} = 30\cos 3t - 60\sin t - 3e^{-t} + \sin 2t$

13.149 Solve $\begin{cases} -5I_1 - \dfrac{dI_1}{dt} + 2\dfrac{dI_2}{dt} + 10I_2 = 0 \\ \dfrac{dI_1}{dt} + 20I_1 + 15I_2 = 55 \end{cases}$; $\quad I_1(0) = I_2(0) = 0.$

▌ Taking the Laplace transforms of both equations, we find

$$-5i_1 - [si_1 - I_1(0)] + 2[si_2 - I_2(0)] + 10i_2 = 0 \qquad (s+5)i_1 - (2s+10)i_2 = 0$$

$$\{si_1 - I_1(0)\} + 20i_1 + 15i_2 = \frac{55}{s} \qquad \text{or} \qquad (s+20)i_1 + 15i_2 = \frac{55}{s}$$

From the first equation, $i_1 = 2i_2$, so that the second equation yields $i_2 = \dfrac{55}{s(2s+55)} = \dfrac{1}{s} - \dfrac{2}{2s+55}$.

Then inverting gives $I_2 = 1 - e^{-55t/2}$ and $I_1 = 2I_2 = 2 - 2e^{-55t/2}$.

CHAPTER 14
Matrix Methods

FINDING $e^{\mathbf{A}t}$

14.1 Develop a method for calculating $e^{\mathbf{A}t}$ when $\mathbf{A}$ is a square matrix having numbers as its elements.

▌ If $\mathbf{A}$ is a matrix having n rows and n columns, then

$$e^{\mathbf{A}t} = \alpha_{n-1}\mathbf{A}^{n-1}t^{n-1} + \alpha_{n-2}\mathbf{A}^{n-2}t^{n-2} + \cdots + \alpha_2\mathbf{A}^2t^2 + \alpha_1\mathbf{A}t + \alpha_0\mathbf{I} \tag{1}$$

where $\alpha_0, \alpha_1, \ldots, \alpha_{n-1}$ are functions of t which must be determined for each $\mathbf{A}$.

To determine the α's, we define $r(\lambda) \equiv \alpha_{n-1}\lambda^{n-1} + \alpha_{n-2}\lambda^{n-2} + \cdots + \alpha_2\lambda^2 + \alpha_1\lambda + \alpha_0$. Now if λ_i is an eigenvalue of $\mathbf{A}t$, then $e^{\lambda_i} = r(\lambda_i)$. Furthermore, if λ_i is an eigenvalue of multiplicity k, for $k > 1$, then the following equations are also valid:

$$e^{\lambda_i} = \frac{d}{d\lambda} r(\lambda)\bigg|_{\lambda=\lambda_i}$$

$$e^{\lambda_i} = \frac{d^2}{d\lambda^2} r(\lambda)\bigg|_{\lambda=\lambda_i}$$

$$\cdots\cdots\cdots\cdots\cdots\cdots$$

$$e^{\lambda_i} = \frac{d^{k-1}}{d\lambda^{k-1}} r(\lambda)\bigg|_{\lambda=\lambda_i}$$

When such a set of equations is found for each eigenvalue of $\mathbf{A}t$, the result is a set of n linear equations, all containing e^{λ_i} on the left side, which may be solved for $\alpha_0, \alpha_1, \ldots, \alpha_n$. These values may then be substituted into (1) to compute $e^{\mathbf{A}t}$.

14.2 Find $e^{\mathbf{A}t}$ for $\mathbf{A} = \begin{bmatrix} 0 & 1 \\ -14 & -9 \end{bmatrix}$.

▌ Here $\mathbf{A}t = \begin{bmatrix} 0 & t \\ -14t & -9t \end{bmatrix}$, with characteristic equation $\lambda^2 + 9t\lambda + 14t^2 = 0$. The eigenvalues are $-2t$ and $-7t$. Since $\mathbf{A}$ has order 2×2, it follows from Problem 14.1 that $e^{\mathbf{A}t} = \alpha_1\mathbf{A}t + \alpha_0\mathbf{I}$. Then $r(\lambda) = \alpha_1\lambda + \alpha_0$, where α_1 and α_0 satisfy the equations

$$e^{-2t} = r(-2t) = \alpha_1(-2t) + \alpha_0$$
$$e^{-7t} = r(-7t) = \alpha_1(-7t) + \alpha_0$$

Solving this set of equations, we obtain $\alpha_1 = \dfrac{e^{-2t} - e^{-7t}}{5t}$ and $\alpha_0 = \dfrac{7e^{-2t} - 2e^{-7t}}{5}$. Then

$$e^{\mathbf{A}t} = \frac{e^{-2t} - e^{-7t}}{5t}\begin{bmatrix} 0 & t \\ -14t & -9t \end{bmatrix} + \frac{7e^{-2t} - 2e^{-7t}}{5}\begin{bmatrix} 1 & 0 \\ 0 & 1 \end{bmatrix} = \frac{1}{5}\begin{bmatrix} 7e^{-2t} - 2e^{-7t} & e^{-2t} - e^{-7t} \\ -14e^{-2t} + 14e^{-7t} & -2e^{-2t} + 7e^{-7t} \end{bmatrix}$$

14.3 Find $e^{\mathbf{A}t}$ for $\mathbf{A} = \begin{bmatrix} 0 & 1 \\ -64 & -20 \end{bmatrix}$.

▌ Here $\mathbf{A}t = \begin{bmatrix} 0 & t \\ -64t & -20t \end{bmatrix}$, with characteristic equation $\lambda^2 + 20t\lambda + 64t^2 = 0$. The eigenvalues are $-4t$ and $-16t$. Since $\mathbf{A}$ has order 2×2, it follows from Problem 14.1 that $e^{\mathbf{A}t} = \alpha_1\mathbf{A}t + \alpha_0\mathbf{I}$. Then $r(\lambda) = \alpha_1\lambda + \alpha_0$, where α_1 and α_0 satisfy the equations

$$e^{-4t} = r(-4t) = \alpha_1(-4t) + \alpha_0$$
$$e^{-16t} = r(-16t) = \alpha_1(-16t) + \alpha_0$$

Solving this set of equations, we obtain $\alpha_1 = \dfrac{e^{-4t} - e^{-16t}}{12t}$ and $\alpha_0 = \dfrac{4e^{-4t} - e^{-16t}}{3}$. Then

$$e^{\mathbf{A}t} = \frac{e^{-4t} - e^{-16t}}{12t}\begin{bmatrix} 0 & t \\ -64t & -20t \end{bmatrix} + \frac{4e^{-4t} - e^{-16t}}{3}\begin{bmatrix} 1 & 0 \\ 0 & 1 \end{bmatrix} = \frac{1}{12}\begin{bmatrix} 16e^{-4t} - 4e^{-16t} & e^{-4t} - e^{-16t} \\ -64e^{-4t} + 64e^{-16t} & -4e^{-4t} + 16e^{-16t} \end{bmatrix}$$

14.4 Find $e^{\mathbf{A}(t-2)}$ for the matrix of the previous problem.

▮ In that problem, we found $e^{\mathbf{A}t}$. If we replace t with the quantity $t - 2$, we obtain

$$e^{\mathbf{A}(t-2)} = \frac{1}{12}\begin{bmatrix} 16e^{-4(t-2)} - 4e^{-16(t-2)} & e^{-4(t-2)} - e^{-16(t-2)} \\ -64e^{-4(t-2)} + 64e^{-16(t-2)} & -4e^{-4(t-2)} + 16e^{-16(t-2)} \end{bmatrix}$$

14.5 Find $e^{\mathbf{A}(t-s)}$ for the matrix $\mathbf{A}$ in Problem 14.2.

▮ In that problem, we found $e^{\mathbf{A}t}$. If we replace t with the quantity $t - s$, we obtain

$$e^{\mathbf{A}(t-s)} = \frac{1}{5}\begin{bmatrix} 7e^{-2(t-s)} - 2e^{-7(t-s)} & e^{-2(t-s)} - e^{-7(t-s)} \\ -14e^{-2(t-s)} + 14e^{-7(t-s)} & -2e^{-2(t-s)} + 7e^{-7(t-s)} \end{bmatrix}$$

14.6 Find $e^{\mathbf{A}t}$ for $\mathbf{A} = \begin{bmatrix} 1 & 2 \\ 4 & 3 \end{bmatrix}$.

▮ Here, $n = 2$, $e^{\mathbf{A}t} = \alpha_1\mathbf{A}t + \alpha_0\mathbf{I} = \begin{bmatrix} \alpha_1 t + \alpha_0 & 2\alpha_1 t \\ 4\alpha_1 t & 3\alpha_1 t + \alpha_0 \end{bmatrix}$ and $r(\lambda) = \alpha_1\lambda + \alpha_0$. The eigenvalues of $\mathbf{A}t$ are $\lambda_1 = -t$ and $\lambda_2 = 5t$, which are both of multiplicity one. Thus, α_1 and α_0 satisfy the equations

$$e^{-t} = \alpha_1(-t) + \alpha_0 \qquad e^{5t} = \alpha_1(5t) + \alpha_0$$

Solving these equations, we find that $\alpha_1 = \dfrac{1}{6t}(e^{5t} - e^{-t})$ and $\alpha_0 = \dfrac{1}{6}(e^{5t} + 5e^{-t})$. Then substitution of these values and simplification yield $e^{\mathbf{A}t} = \dfrac{1}{6}\begin{bmatrix} 2e^{5t} + 4e^{-t} & 2e^{5t} - 2e^{-t} \\ 4e^{5t} - 4e^{-t} & 4e^{5t} + 2e^{-t} \end{bmatrix}$.

14.7 Find $e^{\mathbf{A}t}$ for $\mathbf{A} = \begin{bmatrix} 0 & 1 \\ 8 & -2 \end{bmatrix}$.

▮ Since $n = 2$, it follows that $e^{\mathbf{A}t} = \alpha_1\mathbf{A}t + \alpha_0\mathbf{I} = \begin{bmatrix} \alpha_0 & \alpha_1 t \\ 8\alpha_1 t & -2\alpha_1 t + \alpha_0 \end{bmatrix}$ and $r(\lambda) = \alpha_1\lambda + \alpha_0$. The eigenvalues of $\mathbf{A}t$ are $\lambda_1 = 2t$ and $\lambda_2 = -4t$, which are both of multiplicity one. Thus, we have

$$e^{2t} = \alpha_1(2t) + \alpha_0 \qquad e^{-4t} = \alpha_1(-4t) + \alpha_0$$

Solving these equations for α_1 and α_0, we find that $\alpha_1 = \dfrac{1}{6t}(e^{2t} - e^{-4t})$ and $\alpha_0 = \dfrac{1}{3}(2e^{2t} + e^{-4t})$. Then substituting these values and simplifying yield $e^{\mathbf{A}t} = \dfrac{1}{6}\begin{bmatrix} 4e^{2t} + 2e^{-4t} & e^{2t} - e^{-4t} \\ 8e^{2t} - 8e^{-4t} & 2e^{2t} + 4e^{-4t} \end{bmatrix}$.

14.8 Find $e^{\mathbf{A}(t-1)}$ for the matrix of the previous problem.

▮ In that problem we found $e^{\mathbf{A}t}$. If we replace t with the quantity $t - 1$, we obtain

$$e^{\mathbf{A}(t-1)} = \frac{1}{6}\begin{bmatrix} 4e^{2(t-1)} + 2e^{-4(t-1)} & e^{2(t-1)} - e^{-4(t-1)} \\ 8e^{2(t-1)} - 8e^{-4(t-1)} & 2e^{2(t-1)} + 4e^{-4(t-1)} \end{bmatrix}$$

14.9 Find $e^{\mathbf{A}(t-s)}$ for the matrix of Problem 14.7.

▮ In that problem we found $e^{\mathbf{A}t}$. If we replace t with the quantity $t - s$, we obtain

$$e^{\mathbf{A}(t-s)} = \frac{1}{6}\begin{bmatrix} 4e^{2(t-s)} + 2e^{-4(t-s)} & e^{2(t-s)} - e^{-4(t-s)} \\ 8e^{2(t-s)} - 8e^{-4(t-s)} & 2e^{2(t-s)} + 4e^{-4(t-s)} \end{bmatrix}$$

14.10 Find $e^{\mathbf{A}t}$ for $\mathbf{A} = \begin{bmatrix} 0 & 1 \\ -16 & 0 \end{bmatrix}$.

Here $\mathbf{A}t = \begin{bmatrix} 0 & t \\ -16t & 0 \end{bmatrix}$, with characteristic equation $\lambda^2 + 16t^2 = 0$. The eigenvalues are $\pm i4t$. Since $\mathbf{A}$ has order 2×2, it follows from Problem 14.1 that $e^{\mathbf{A}t} = \alpha_1 \mathbf{A}t + \alpha_0 \mathbf{I}$. Then $r(\lambda) = \alpha_1\lambda + \alpha_0$, where α_1 and α_0 satisfy the equations

$$e^{i4t} = r(i4t) = \alpha_1(i4t) + \alpha_0 \qquad e^{-i4t} = r(-i4t) = \alpha_1(-i4t) + \alpha_0$$

Solving this set of equations, we obtain $\alpha_1 = \dfrac{e^{i4t} - e^{-i4t}}{i8t} = \dfrac{1}{4t}\sin 4t$ and $\alpha_0 = \dfrac{e^{i4t} + e^{-i4t}}{2} = \cos 4t$. Then

$$e^{\mathbf{A}t} = \frac{1}{4t}\sin 4t \begin{bmatrix} 0 & t \\ -16t & 0 \end{bmatrix} + \cos 4t \begin{bmatrix} 1 & 0 \\ 0 & 1 \end{bmatrix} = \begin{bmatrix} \cos 4t & \frac{1}{4}\sin 4t \\ -4t\sin 4t & \cos 4t \end{bmatrix}$$

4.11 Find $e^{\mathbf{A}t}$ for $\mathbf{A} = \begin{bmatrix} 0 & 1 \\ -96 & 0 \end{bmatrix}$.

Here $\mathbf{A}t = \begin{bmatrix} 0 & t \\ -96t & 0 \end{bmatrix}$, with characteristic equation $\lambda^2 + 96t^2 = 0$. The eigenvalues are $\pm i\sqrt{96}t$. Since $\mathbf{A}$ has order 2×2, it follows from Problem 14.1 that $e^{\mathbf{A}t} = \alpha_1 \mathbf{A}t + \alpha_0 \mathbf{I}$. Then $r(\lambda) = \alpha_1\lambda + \alpha_0$, where α_1 and α_0 satisfy the equations

$$e^{i\sqrt{96}t} = r(i\sqrt{96}t) = \alpha_1(i\sqrt{96}t) + \alpha_0 \qquad e^{-i\sqrt{96}t} = r(-i\sqrt{96}t) = \alpha_1(-i\sqrt{96}t) + \alpha_0$$

Solving this set of equations, we obtain $\alpha_1 = \dfrac{e^{i\sqrt{96}t} - e^{-i\sqrt{96}t}}{i^2\sqrt{96}t} = \dfrac{1}{\sqrt{96}t}\sin\sqrt{96}t$ and

$\alpha_0 = \dfrac{e^{i\sqrt{96}t} + e^{-i\sqrt{96}t}}{2} = \cos\sqrt{96}t$. Then

$$e^{\mathbf{A}t} = \frac{1}{\sqrt{96}t}\sin\sqrt{96}t \begin{bmatrix} 0 & t \\ -96t & 0 \end{bmatrix} + (\cos\sqrt{96}t)\begin{bmatrix} 1 & 0 \\ 0 & 1 \end{bmatrix} = \begin{bmatrix} \cos\sqrt{96}t & \frac{1}{\sqrt{96}}\sin\sqrt{96}t \\ -\sqrt{96}\sin\sqrt{96}t & \cos\sqrt{96}t \end{bmatrix}$$

4.12 Find $e^{\mathbf{A}t}$ for $\mathbf{A} = \begin{bmatrix} 0 & 1 \\ -64 & 0 \end{bmatrix}$.

Here $\mathbf{A}t = \begin{bmatrix} 0 & t \\ -64t & 0 \end{bmatrix}$, with characteristic equation $\lambda_2 + 64t^2 = 0$. The eigenvalues are $\pm i8t$. Thus, $e^{\mathbf{A}t} = \alpha_1 \mathbf{A}t + \alpha_0 \mathbf{I}$, and $r(\lambda) = \alpha_1\lambda + \alpha_0$, where α_1 and α_0 satisfy the equations

$$e^{i8t} = r(i8t) = \alpha_1(i8t) + \alpha_0 \qquad e^{-i8t} = r(-i8t) = \alpha_1(-i8t) + \alpha_0$$

Solving this set of equations, we get $\alpha_1 = \dfrac{e^{i8t} - e^{-i8t}}{i16t} = \dfrac{1}{8t}\sin 8t$ and $\alpha_0 = \dfrac{e^{i8t} + e^{-i8t}}{2} = \cos 8t$. Then

$$e^{\mathbf{A}t} = \frac{1}{8t}\sin 8t \begin{bmatrix} 0 & t \\ -64t & 0 \end{bmatrix} + \cos 8t \begin{bmatrix} 1 & 0 \\ 0 & 1 \end{bmatrix} = \begin{bmatrix} \cos 8t & \frac{1}{8}\sin 8t \\ -8\sin 8t & \cos 8t \end{bmatrix}$$

.13 Find $e\mathbf{A}^t$ for $A = \begin{bmatrix} 0 & 1 \\ -1 & 0 \end{bmatrix}$.

Here $n = 2$; hence, $e^{\mathbf{A}t} = \alpha_1 \mathbf{A}t + \alpha_0 \mathbf{I} = \begin{bmatrix} \alpha_0 & \alpha_1 t \\ -\alpha_1 t & \alpha_0 \end{bmatrix}$ and $r(\lambda) = \alpha_1\lambda + \alpha_0$. The eigenvalues of $\mathbf{A}t$ are $\lambda_1 = it$ and $\lambda_2 = -it$, which are both of multiplicity one. Thus,

$$e^{it} = \alpha_1(it) + \alpha_0 \qquad e^{-it} = \alpha_1(-it) + \alpha_0$$

Solving these equations for α_1 and α_0, we find that $\alpha_1 = \dfrac{1}{2it}(e^{it} - e^{-it}) = \dfrac{\sin t}{t}$ and $\alpha_o = \dfrac{1}{2}(e^{it} + e^{-it}) = \cos t$.

Substituting these values above, we obtain $e^{\mathbf{A}t} = \begin{bmatrix} \cos t & \sin t \\ -\sin t & \cos t \end{bmatrix}$.

14.14 Find $e^{\mathbf{A}(t-\pi)}$ for the matrix of the previous problem.

▌ In that problem, we found $e^{\mathbf{A}t}$. If we replace t with the quantity $t-\pi$, we obtain

$$e^{\mathbf{A}(t-\pi)} = \begin{bmatrix} \cos(t-\pi) & \sin(t-\pi) \\ -\sin(t-\pi) & \cos(t-\pi) \end{bmatrix}.$$

14.15 Find $e^{\mathbf{A}(t-s)}$ for the matrix of Problem 14.13.

▌ In that problem we found $e^{\mathbf{A}t}$. If we replace t with the quantity $t-s$, we obtain

$$e^{\mathbf{A}(t-s)} = \begin{bmatrix} \cos(t-s) & \sin(t-s) \\ -\sin(t-s) & \cos(t-s) \end{bmatrix}.$$

14.16 Find $e^{\mathbf{A}t}$ for $\mathbf{A} = \begin{bmatrix} 0 & 1 \\ -25 & -8 \end{bmatrix}$.

▌ Here $\mathbf{A}t = \begin{bmatrix} 0 & t \\ -25t & -8t \end{bmatrix}$, with characteristic equation $\lambda^2 + 8t\lambda + 25t^2 = 0$. The eigenvalues are $-4t \pm i3t$. Since $\mathbf{A}$ has order 2×2, it follows that $e^{\mathbf{A}t} = \alpha_1 \mathbf{A}t + \alpha_0 \mathbf{I}$. Then $r(\lambda) = \alpha_1 \lambda_1 + \alpha_0$, where α_1 and α_0 satisfy the equations

$$e^{-4t+i3t} = r(-4t+i3t) = \alpha_1(-4t+i3t) + \alpha_0 \qquad e^{-4t-i3t} = r(-4t-i3t) = \alpha_1(-4t-i3t) + \alpha_0$$

The solution to this set of equations is

$$\alpha_1 = \frac{e^{-4i+i3t} - e^{-4i-i3t}}{i6t} = \frac{e^{-4t}}{3t}\left(\frac{e^{i3t} - e^{-i3t}}{i2}\right) = \frac{e^{-4t}\sin 3t}{3t}$$

$$\alpha_0 = \frac{(-4t-i3t)e^{-4t+i3t} - (-4t+i3t)e^{-4t-i3t}}{-i6t} = e^{-4t}\frac{-4t(e^{i3t} - e^{-i3t}) - i3t(e^{i3t} + e^{-i3t})}{-i6t} = e^{-4t}\left(\frac{4}{3}\sin 3t + \cos 3t\right)$$

Then

$$e^{\mathbf{A}t} = \frac{e^{-4t}\sin 3t}{3t}\begin{bmatrix} 0 & t \\ -25t & -8t \end{bmatrix} + e^{-4t}\left(\frac{4}{3}\sin 3t + \cos 3t\right)\begin{bmatrix} 1 & 0 \\ 0 & 1 \end{bmatrix} = e^{-4t}\begin{bmatrix} \frac{4}{3}\sin 3t + \cos 3t & \frac{1}{3}\sin 3t \\ -\frac{25}{3}\sin 3t & -\frac{4}{3}\sin 3t + \cos 3t \end{bmatrix}$$

14.17 Find $e^{\mathbf{A}t}$ for $\mathbf{A} = \begin{bmatrix} 0 & 1 \\ -64 & -12.8 \end{bmatrix}$.

▌ Here $\mathbf{A}t = \begin{bmatrix} 0 & t \\ -64t & -12.8t \end{bmatrix}$, with characteristic equation $\lambda^2 + 12.8t\lambda + 64t^2 = 0$. The eigenvalues are $-6.4t \pm i4.8t$. Since $\mathbf{A}$ has order 2×2, it follows that $e^{\mathbf{A}t} + \alpha_1 \mathbf{A}t + \alpha_0 \mathbf{I}$. Then $r(\lambda) = \alpha_1 \lambda + \alpha_0$, where α_1 and α_0 satisfy the equations

$$e^{-6.4t+i4.8t} = \alpha_1(-6.4t+i4.8t) + \alpha_0 \qquad e^{-6.4t-i4.8t} = \alpha_1(-6.4t-i4.8t) + \alpha_0$$

This solution to this system is

$$\alpha_1 = \frac{e^{-6.4t+i4.8t} - e^{-6.4t-i4.8t}}{-i9.6t} = \frac{e^{-6.4t}\sin 4.8t}{4.8t}$$

$$\alpha_0 = \frac{(-6.4t-i4.8t)e^{-6.4t+i4.8t} - (-6.4t+i4.8t)e^{-6.4t-i4.8t}}{-i9.6t} = e^{-6.4t}\left(\frac{4}{3}\sin 4.8t + \cos 4.8t\right)$$

and

$$e^{\mathbf{A}t} = \frac{e^{-6.4t}\sin 4.8t}{4.8t}\begin{bmatrix} 0 & t \\ -64t & -12.8t \end{bmatrix} + e^{-6.4}\left(\frac{4}{3}\sin 4.8t + \cos 4.8t\right)\begin{bmatrix} 1 & 0 \\ 0 & 1 \end{bmatrix}$$

$$= e^{-6.4t}\begin{bmatrix} \frac{4}{3}\sin 4.8t + \cos 4.8t & \frac{5}{24}\sin 4.8t \\ -\frac{40}{3}\sin 4.8t & -\frac{4}{3}\sin 4.8t + \cos 4.8t \end{bmatrix}$$

14.18 Find $e^{\mathbf{A}t}$ for $\mathbf{A} = \begin{bmatrix} 0 & 1 \\ -64 & -16 \end{bmatrix}$.

▌ Here $\mathbf{A}t = \begin{bmatrix} 0 & t \\ -64t & -16t \end{bmatrix}$, with characteristic equation $\lambda^2 + 16t\lambda + 64t^2 = 0$. There is only one eigenvalue, $-8t$, of multiplicity two. Since $\mathbf{A}$ has order 2×2, we have $e^{\mathbf{A}t} = \alpha_1 \mathbf{A}t + \alpha_0 \mathbf{I}$. Then

$r(\lambda) = \alpha_1\lambda + \alpha_0$ and $r'(\lambda) = \alpha_1$, where α_1 and α_0 satisfy the equations

$$e^{-8t} = r(-8t) = \alpha_1(-8t) + \alpha_0 \qquad e^{-8t} = r'(-8t) = \alpha_1$$

The solution to this system is $\alpha_1 = e^{-8t}$ and $\alpha_0 = (1 + 8t)e^{-8t}$, so that

$$e^{\mathbf{A}t} = e^{-8t}\begin{bmatrix} 0 & t \\ -64t & -16t \end{bmatrix} + (1 + 8t)e^{-8t}\begin{bmatrix} 1 & 0 \\ 0 & 1 \end{bmatrix} = e^{-8t}\begin{bmatrix} 1 + 8t & t \\ -64t & 1 - 8t \end{bmatrix}$$

14.19 Find $e^{\mathbf{A}(t-3)}$ for the matrix **A** of the previous problem.

▌ In the previous problem, we found $e^{\mathbf{A}t}$. If we replace t with the quantity $t - 3$, we obtain

$$e^{\mathbf{A}(t-3)} = e^{-8(t-3)}\begin{bmatrix} 1 + 8(t-3) & t - 3 \\ -64(t-3) & 1 - 8(t-3) \end{bmatrix} = e^{-8(t-3)}\begin{bmatrix} 8t - 23 & t - 3 \\ -64t + 192 & -8t + 25 \end{bmatrix}$$

14.20 Find $e^{\mathbf{A}t}$ for $\mathbf{A} = \begin{bmatrix} 0 & 1 \\ -4 & -4 \end{bmatrix}$.

▌ Here $\mathbf{A}t = \begin{bmatrix} 0 & t \\ -4t & -4t \end{bmatrix}$. Its characteristic equation is $\lambda^2 + 4t\lambda + 4t^2 = 0$, which has $\lambda = -2t$ as an eigenvalue of multiplicity two. Since **A** has order 2×2, we have $e^{\mathbf{A}t} = \alpha_1\mathbf{A}t + \alpha_0\mathbf{I}$. Then $r(\lambda) = \alpha_1\lambda + \alpha_0$ and $r'(\lambda) = \alpha_1$, where α_1 and α_0 satisfy the equations

$$e^{-2t} = r(-2t) = \alpha_1(-2t) + \alpha_0 \qquad e^{-2t} = r'(-2t) = \alpha_1$$

The solution to this system is $\alpha_1 = e^{-2t}$ and $\alpha_0 = (1 + 2t)e^{-2t}$, so that

$$e^{\mathbf{A}t} = e^{-2t}\begin{bmatrix} 0 & t \\ -4t & -4t \end{bmatrix} + (1 + 2t)e^{-2t}\begin{bmatrix} 1 & 0 \\ 0 & 1 \end{bmatrix} = \begin{bmatrix} (1 + 2t)e^{-2t} & te^{-2t} \\ -4te^{-2t} & (1 - 2t)e^{-2t} \end{bmatrix}$$

14.21 Find $e^{\mathbf{A}(t+1)}$ for the matrix **A** given in the previous problem.

▌ In the previous problem, we found $e^{\mathbf{A}t}$. If we replace t with the quantity $t + 1$, we obtain

$$e^{\mathbf{A}(t+1)} = \begin{bmatrix} [1 + 2(t+1)]e^{-2(t+1)} & (t + 1)e^{-2(t+1)} \\ -4(t + 1)e^{-2(t+1)} & [1 - 2(t+1)]e^{-2(t+1)} \end{bmatrix} = e^{-2(t+1)}\begin{bmatrix} 2t + 3 & t + 1 \\ -4t - 4 & -2t - 1 \end{bmatrix}$$

14.22 Find $e^{\mathbf{A}t}$ for $\mathbf{A} = \begin{bmatrix} 0 & 1 \\ -9 & 6 \end{bmatrix}$.

▌ Here $\mathbf{A}t = \begin{bmatrix} 0 & t \\ -9t & 6t \end{bmatrix}$. Its characteristic equation is $\lambda^2 - 6t\lambda + 9t^2 = 0$, which has $\lambda = 3t$ as an eigenvalue of multiplicity two. Since **A** has order 2×2, we have $e^{\mathbf{A}t} = \alpha_1\mathbf{A}t + \alpha_0\mathbf{I}$. Then $r(\lambda) = \alpha_1\lambda + \alpha_0$ and $r'(\lambda) = \alpha_1$, where α_1 and α_0 satisfy the equations

$$e^{3t} = r(3t) = \alpha_1(3t) + \alpha_0 \qquad e^{3t} = r'(3t) = \alpha_1$$

The solution to this system is $\alpha_1 = e^{3t}$ and $\alpha_0 = (1 - 3t)e^{3t}$. Then

$$e^{\mathbf{A}t} = e^{3t}\begin{bmatrix} 0 & t \\ -9t & 6t \end{bmatrix} + (1 - 3t)e^{3t}\begin{bmatrix} 1 & 0 \\ 0 & 1 \end{bmatrix} = \begin{bmatrix} (1 - 3t)e^{3t} & te^{3t} \\ -9te^{3t} & (1 + 3t)e^{3t} \end{bmatrix}$$

14.23 Find $e^{-\mathbf{A}t}$ for the matrix **A** in the previous problem.

▌ In the previous problem, we found $e^{\mathbf{A}t}$. If we replace t with $-t$, we obtain

$$e^{-\mathbf{A}t} = \begin{bmatrix} (1 + 3t)e^{-3t} & -te^{-3t} \\ 9te^{-3t} & (1 - 3t)e^{-3t} \end{bmatrix}.$$

14.24 Find $e^{\mathbf{A}t}$ for $\mathbf{A} = \begin{bmatrix} 0 & 1 & 0 \\ 0 & -2 & -5 \\ 0 & 1 & 2 \end{bmatrix}$.

▌ Here $\mathbf{A}t = \begin{bmatrix} 0 & t & 0 \\ 0 & -2t & -5t \\ 0 & t & 2t \end{bmatrix}$, which has as its characteristic equation $\lambda^3 + \lambda t^2 = 0$. Its eigenvalues are 0 and $\pm it$. Since **A** has order 3×3, we have $e^{\mathbf{A}t} = \alpha_2\mathbf{A}^2t^2 + \alpha_1\mathbf{A}t + \alpha_0\mathbf{I}$. Then $r(\lambda) = \alpha_2\lambda^2 + \alpha_1\lambda + \alpha_0$, where α_2, α_1, and α_0 satisfy the equations

$$e^0 = r(0) = \alpha_2(0)^2 + \alpha_1(0) + \alpha_0$$
$$e^{it} = r(it) = \alpha_2(it)^2 + \alpha_1(it) + \alpha_0$$
$$e^{-it} = r(-it) = \alpha_2(-it)^2 + \alpha_1(-it) + \alpha_0$$

The solution to this set of equations is $\alpha_2 = \dfrac{e^{it} + e^{-it} - 2}{-2t^2} = \dfrac{1 - \cos t}{t^2}$, $\alpha_1 = \dfrac{e^{it} - e^{-it}}{2it} = \dfrac{1}{t}\sin t$, and $\alpha_0 = 1$. Then

$$e^{\mathbf{A}t} = \frac{1 - \cos t}{t^2}\begin{bmatrix} 0 & -2 & -5 \\ 0 & -1 & 0 \\ 0 & 0 & -1 \end{bmatrix} t^2 + \frac{1}{t}(\sin t)\begin{bmatrix} 0 & 1 & 0 \\ 0 & -2 & -5 \\ 0 & 1 & 2 \end{bmatrix} t + 1\begin{bmatrix} 1 & 0 & 0 \\ 0 & 1 & 0 \\ 0 & 0 & 1 \end{bmatrix}$$
$$= \begin{bmatrix} 1 & -2 + 2\cos t + \sin t & -5 + 5\cos t \\ 0 & \cos t - 2\sin t & -5\sin t \\ 0 & \sin t & \cos t + 2\sin t \end{bmatrix}$$

14.25 Find $e^{\mathbf{A}(t-s)}$ for matrix $\mathbf{A}$ of the previous problem.

▌ Replacing t with the quantity $t - s$ in the result of the previous problem yields

$$e^{\mathbf{A}(t-s)} = \begin{bmatrix} 1 & -2 + 2\cos(t-s) + \sin(t-s) & -5 + 5\cos(t-s) \\ 0 & \cos(t-s) - 2\sin(t-s) & -5\sin(t-s) \\ 0 & \sin(t-s) & \cos(t-s) + 2\sin(t-s) \end{bmatrix}$$

14.26 Find $e^{\mathbf{A}t}$ for $\mathbf{A} = \begin{bmatrix} 3 & 1 & 0 \\ 0 & 3 & 1 \\ 0 & 0 & 3 \end{bmatrix}$.

▌ Here $n = 3$, so

$$e^{\mathbf{A}t} = \alpha_2\mathbf{A}^2t^2 + \alpha_1\mathbf{A}t + \alpha_0\mathbf{I} = \alpha_2\begin{bmatrix} 9 & 6 & 1 \\ 0 & 9 & 6 \\ 0 & 0 & 9 \end{bmatrix} t^2 + \alpha_1\begin{bmatrix} 3 & 1 & 0 \\ 0 & 3 & 1 \\ 0 & 0 & 3 \end{bmatrix} t + \alpha_0\begin{bmatrix} 1 & 0 & 0 \\ 0 & 1 & 0 \\ 0 & 0 & 1 \end{bmatrix}$$
$$= \begin{bmatrix} 9\alpha_2t^2 + 3\alpha_1t + \alpha_0 & 6\alpha_2t^2 + \alpha_1t & \alpha_2t^2 \\ 0 & 9\alpha_2t^2 + 3\alpha_1t + \alpha_0 & 6\alpha_2t^2 + \alpha_1t \\ 0 & 0 & 9\alpha_2t^2 + 3\alpha_1t + \alpha_0 \end{bmatrix}$$

Then $r(\lambda) = \alpha_2\lambda^2 + \alpha_1\lambda + \alpha_0$, $r'(\lambda) = 2\alpha_2\lambda + \alpha_1$, and $r''(\lambda) = 2\alpha_2$. Since the eigenvalues of $\mathbf{A}t$ are $\lambda_1 = \lambda_2 = \lambda_3 = 3t$, an eigenvalue of multiplicity three, it follows that

$$e^{3t} = \alpha_2 9t^2 + \alpha_1 3t + \alpha_0$$
$$e^{3t} = \alpha_2 6t + \alpha_1$$
$$e^{3t} = 2\alpha_2$$

The solution to this set of equations is $\alpha_2 = \frac{1}{2}e^{3t}$, $\alpha_1 = (1 - 3t)e^{3t}$, and $\alpha_0 = (1 - 3t + \frac{9}{2}t^2)e^{3t}$. Substituting these values above and simplifying, we obtain

$$e^{\mathbf{A}t} = e^{3t}\begin{bmatrix} 1 & t & t^2/2 \\ 0 & 1 & t \\ 0 & 0 & 1 \end{bmatrix}$$

14.27 Find $e^{\mathbf{A}t}$ for $\mathbf{A} = \begin{bmatrix} 0 & 0 & 0 \\ 1 & 0 & 0 \\ 1 & 0 & 1 \end{bmatrix}$.

▌ Here $n = 3$, so $e^{\mathbf{A}t} = \alpha_2\mathbf{A}^2t^2 + \alpha_1\mathbf{A}t + \alpha_0\mathbf{I} = \begin{bmatrix} \alpha_0 & 0 & 0 \\ \alpha_1t & \alpha_0 & 0 \\ \alpha_2t^2 + \alpha_1t & 0 & \alpha_2t^2 + \alpha_1t + \alpha_0 \end{bmatrix}$

and $r(\lambda) = \alpha_2\lambda^2 + \alpha_1\lambda + \alpha_0$. The eigenvalues of $\mathbf{A}t$ are $\lambda_1 = \lambda_2 = 0$ and $\lambda_3 = t$. It then follows that $e^0 = r(0)$, $e^0 = r'(0)$, and $e^t = r(t)$. Since $r'(\lambda) = 2\alpha_2\lambda + \alpha_1$, these equations become

$$1 = \alpha_0 \qquad 1 = \alpha_1 \qquad e^t = \alpha_2 t^2 + \alpha_1 t + \alpha_0$$

from which $\alpha_2 = (e^t - t - 1)/t^2$, $\alpha_1 = 1$, and $\alpha_0 = 1$. Substituting these results above and simplifying, we obtain

$$e^{\mathbf{A}t} = \begin{bmatrix} 1 & 0 & 0 \\ t & 1 & 0 \\ e^t - 1 & 0 & e^t \end{bmatrix}$$

14.28 Establish the equations required to find $e^{\mathbf{A}t}$ if $\mathbf{A}$ is a 4×4 constant matrix and $\mathbf{A}t$ has eigenvalues $2t$ and $3t$, both of multiplicity two.

▌ Since the order of the matrix is 4×4, it follows that $e^{\mathbf{A}t} = \alpha_3\mathbf{A}^3t^3 + \alpha_2\mathbf{A}^2t^2 + \alpha_1\mathbf{A}t + \alpha_0\mathbf{I}$. Then, also,

$$r(\lambda) = \alpha_3\lambda^3 + \alpha_2\lambda^2 + \alpha_1\lambda + \alpha_0 \qquad \text{and} \qquad r'(\lambda) = 3\alpha_3\lambda^2 + 2\alpha_2\lambda + \alpha_1$$

We then use

$$\begin{aligned} e^{2t} &= r(2t) = \alpha_3(2t)^3 + \alpha_2(2t)^2 + \alpha_1(2t) + \alpha_0 \\ e^{2t} &= r'(2t) = 3\alpha_3(2t)^2 + 2\alpha_2(2t) + \alpha_1 \\ e^{3t} &= r(3t) = \alpha_3(3t)^3 + \alpha_2(3t)^2 + \alpha_1(3t) + \alpha_0 \\ e^{3t} &= r'(3t) = 3\alpha_3(3t)^2 + 2\alpha_2(3t) + \alpha_1 \end{aligned} \qquad \text{or} \qquad \begin{aligned} e^{2t} &= 8t^3\alpha_3 + 4t^2\alpha_2 + 2t\alpha_1 + \alpha_0 \\ e^{2t} &= 12t^2\alpha_3 + 4t\alpha_2 + \alpha_1 \\ e^{3t} &= 27t^3\alpha_3 + 9t^2\alpha_2 + 3t\alpha_1 + \alpha_0 \\ e^{3t} &= 27t^2\alpha_3 + 6t\alpha_2 + \alpha_1 \end{aligned}$$

to solve for the α's, substituting the results in the first equation above.

14.29 Establish the equations required to find $e^{\mathbf{A}t}$ if $\mathbf{A}$ is a 4×4 constant matrix and $\mathbf{A}t$ has as its eigenvalues $-t$ of multiplicity three and $4t$ of multiplicity one.

▌ Since $\mathbf{A}$ is of order 4×4, the formulas for $e^{\mathbf{A}t}$, $r(\lambda)$, and $r'(\lambda)$ are identical to those in the previous problem. In addition, $r''(\lambda) = 6\alpha_3\lambda + 2\alpha_2$. Now we must solve

$$\begin{aligned} e^{-t} &= r(-t) = \alpha_3(-t)^3 + \alpha_2(-t)^2 + \alpha_1(-t) + \alpha_0 \\ e^{-t} &= r'(-t) = 3\alpha_3(-t)^2 + 2\alpha_2(-t) + \alpha_1 \\ e^{-t} &= r''(-t) = 6\alpha_3(-t) + 2\alpha_2 \\ e^{4t} &= r(4t) = \alpha_3(4t)^3 + \alpha_2(4t)^2 + \alpha_1(4t) + \alpha_0 \end{aligned} \qquad \text{or} \qquad \begin{aligned} e^{-t} &= -t^3\alpha_3 + t^2\alpha_2 - t\alpha_1 + \alpha_0 \\ e^{-t} &= 3t^2\alpha_3 - 2t\alpha_2 + \alpha_1 \\ e^{-t} &= -6t\alpha_3 + 2\alpha_2 \\ e^{4t} &= 64t^3\alpha_3 + 16t^2\alpha_2 + 4t\alpha_1 + \alpha_0 \end{aligned}$$

for the α's.

14.30 Establish the equations needed to find $e^{\mathbf{A}t}$ if

$$\mathbf{A} = \begin{bmatrix} 1 & 2 & 3 & 4 & 5 & 6 \\ 0 & 1 & 2 & 3 & 4 & 5 \\ 0 & 0 & 2 & 3 & 4 & 5 \\ 0 & 0 & 0 & 2 & 3 & 4 \\ 0 & 0 & 0 & 0 & 0 & 0 \\ 0 & 0 & 0 & 0 & 0 & 1 \end{bmatrix}$$

▌ Here $n = 6$, so $e^{\mathbf{A}t} = \alpha_5\mathbf{A}^5t^5 + \alpha_4\mathbf{A}^4t^4 + \alpha_3\mathbf{A}^3t^3 + \alpha_2\mathbf{A}^2t^2 + \alpha_1\mathbf{A}t + \alpha_0\mathbf{I}$ and

$$\begin{aligned} r(\lambda) &= \alpha_5\lambda^5 + \alpha_4\lambda^4 + \alpha_3\lambda^3 + \alpha_2\lambda^2 + \alpha_1\lambda + \alpha_0 \\ r'(\lambda) &= 5\alpha_5\lambda^4 + 4\alpha_4\lambda^3 + 3\alpha_3\lambda^2 + 2\alpha_2\lambda + \alpha_1 \\ r''(\lambda) &= 20\alpha_5\lambda^3 + 12\alpha_4\lambda^2 + 6\alpha_3\lambda + 2\alpha_2 \end{aligned}$$

The eigenvalues of $\mathbf{A}t$ are $\lambda_1 = \lambda_2 = \lambda_3 = t$, $\lambda_4 = \lambda_5 = 2t$, and $\lambda_6 = 0$. It now follows that

$$\begin{aligned} e^{2t} &= r(2t) = \alpha_5(2t)^5 + \alpha_4(2t)^4 + \alpha_3(2t)^3 + \alpha_2(2t)^2 + \alpha_1(2t) + \alpha_0 \\ e^{2t} &= r'(2t) = 5\alpha_5(2t)^4 + 4\alpha_4(2t)^3 + 3\alpha_3(2t)^2 + 2\alpha_2(2t) + \alpha_1 \\ e^{2t} &= r''(2t) = 20\alpha_5(2t)^3 + 12\alpha_4(2t)^2 + 6\alpha_3(2t) + 2\alpha_2 \\ e^t &= r(t) = \alpha_5(t)^5 + \alpha_4(t)^4 + \alpha_3(t)^3 + \alpha_2(t)^2 + \alpha_1(t) + \alpha_0 \\ e^t &= r'(t) = 5\alpha_5(t)^4 + 4\alpha_4(t)^3 + 3\alpha_3(t)^2 + 2\alpha_2(t) + \alpha_1 \\ e^0 &= r(0) = \alpha_5(0)^5 + \alpha_4(0)^4 + \alpha_3(0)^3 + \alpha_2(0)^2 + \alpha_1(0) + \alpha_0 \end{aligned}$$

or, more simply,

$$\begin{aligned} e^{2t} &= 32t^5\alpha_5 + 16t^4\alpha_4 + 8t^3\alpha_3 + 4t^2\alpha_2 + 2t\alpha_1 + \alpha_0 \\ e^{2t} &= 80t^4\alpha_5 + 32t^3\alpha_4 + 12t^2\alpha_3 + 4t\alpha_2 + \alpha_1 \\ e^{2t} &= 160t^3\alpha_5 + 48t^2\alpha_4 + 12t\alpha_3 + 2\alpha_2 \\ e^{t} &= t^5\alpha_5 + t^4\alpha_4 + t^3\alpha_3 + t^2\alpha_2 + t\alpha_1 + \alpha_0 \\ e^{t} &= 5t^4\alpha_5 + 4t^3\alpha_4 + 3t^2\alpha_3 + 2t\alpha_2 + \alpha_1 \\ 1 &= \alpha_0 \end{aligned}$$

must be solved for the α's.

MATRIX DIFFERENTIAL EQUATIONS

14.31 Transform the initial-value problem $\ddot{x} + 16\dot{x} + 64x = 0$; $x(0) = \frac{1}{6}$, $\dot{x}(0) = 0$ into matrix form.

▎ Solving the differential equation for its highest derivative, we obtain $\ddot{x} = -16\dot{x} - 64x$. This equation has order 2, so we define two new variables: $x_1 = x$ and $x_2 = \dot{x}$. Differentiating each of these equations once yields

$$\begin{aligned} \dot{x}_1 &= \dot{x} = x_2 \\ \dot{x}_2 &= \ddot{x} = -16\dot{x} - 64x = -16x_2 - 64x_1 \end{aligned} \qquad \text{or} \qquad \begin{aligned} \dot{x}_1 &= 0x_1 + 1x_2 \\ \dot{x}_2 &= -64x_1 - 16x_2 \end{aligned}$$

This system has the matrix form $\dot{\mathbf{X}} = \begin{bmatrix} \dot{x}_1 \\ \dot{x}_2 \end{bmatrix} = \begin{bmatrix} 0 & 1 \\ -64 & -16 \end{bmatrix} \begin{bmatrix} x_2 \\ x_1 \end{bmatrix}$. The initial conditions may be written $\mathbf{C} = \mathbf{X}(0) = \begin{bmatrix} x_1(0) \\ x_2(0) \end{bmatrix} = \begin{bmatrix} x(0) \\ \dot{x}(0) \end{bmatrix} = \begin{bmatrix} 1/6 \\ 0 \end{bmatrix}$.

14.32 Transform the initial-value problem $\ddot{x} + 8\dot{x} + 25x = 0$; $x(0) = 0$; $\dot{x}(0) = 4$ into matrix form.

▎ Solving the differential equation for its highest derivative, we obtain $\ddot{x} = -8\dot{x} - 25x$. This equation has order 2, so we define two new variables: $x_1 = x$ and $x_2 = \dot{x}$. Differentiating each of these equation once yields

$$\begin{aligned} \dot{x}_1 &= \dot{x} = x_2 \\ \dot{x}_2 &= \ddot{x} = -8\dot{x} - 25x = -8x_2 - 25x_1 \end{aligned} \qquad \text{or} \qquad \begin{aligned} \dot{x}_1 &= 0x_1 + 1x_2 \\ \dot{x}_2 &= -25x_1 - 8x_2 \end{aligned}$$

This system has the matrix form $\dot{\mathbf{X}} = \begin{bmatrix} \dot{x}_1 \\ \dot{x}_2 \end{bmatrix} = \begin{bmatrix} 0 & 1 \\ -25 & -8 \end{bmatrix} \begin{bmatrix} x_1 \\ x_2 \end{bmatrix}$. The initial conditions may be written $\mathbf{C} = \mathbf{X}(0) = \begin{bmatrix} x_1(0) \\ x_2(0) \end{bmatrix} = \begin{bmatrix} x(0) \\ \dot{x}(0) \end{bmatrix} = \begin{bmatrix} 0 \\ 4 \end{bmatrix}$.

14.33 Transform the initial-value problem $\ddot{x} + 20\dot{x} + 64x = 0$; $x(0) = \frac{1}{6}$, $\dot{x}(0) = 0$ into matrix form.

▎ Solving the differential equation for its highest derivative, we obtain $\ddot{x} = -20\dot{x} - 64x$. Since the differential equation has order 2, we define two new variables: $x_1 = x$ and $x_2 = \dot{x}$. Differentiating each of these equations once yields

$$\begin{aligned} \dot{x}_1 &= \dot{x} = x_2 \\ \dot{x}_2 &= \ddot{x} = -20\dot{x} - 64x = -20x_2 - 64x_1 \end{aligned} \qquad \text{or} \qquad \begin{aligned} \dot{x}_1 &= 0x_1 + 1x_2 \\ \dot{x}_2 &= -64x_1 - 20x_2 \end{aligned}$$

This system has the matrix form $\dot{\mathbf{X}} = \begin{bmatrix} \dot{x}_1 \\ \dot{x}_2 \end{bmatrix} = \begin{bmatrix} 0 & 1 \\ -64 & -20 \end{bmatrix} \begin{bmatrix} x_1 \\ x_2 \end{bmatrix}$. The initial conditions may be written $\mathbf{C} = \mathbf{X}(0) = \begin{bmatrix} x_1(0) \\ x_2(0) \end{bmatrix} = \begin{bmatrix} x(0) \\ \dot{x}(0) \end{bmatrix} = \begin{bmatrix} 1/6 \\ 0 \end{bmatrix}$.

14.34 Transform the initial-value problem $\ddot{x} + 16x = 0$; $x(0) = -\frac{1}{2}$, $\dot{x}(0) = 0$ into matrix form.

▎ Solving the differential equation for its highest derivative, we obtain $\ddot{x} = -16x$. Since the differential equation has order 2, we define two new variables: $x_1 = x$ and $x_2 = \dot{x}$. Differentiating each of these equations once yields

$$\begin{aligned} \dot{x}_1 &= \dot{x} = x_2 \\ \dot{x}_2 &= \ddot{x} = -16x = -16x_1 \end{aligned} \qquad \text{or} \qquad \begin{aligned} \dot{x}_1 &= 0x_1 + 1x_2 \\ \dot{x}_2 &= -16x_1 + 0x_2 \end{aligned}$$

This system has the matrix form $\dot{\mathbf{X}} = \begin{bmatrix} \dot{x}_1 \\ \dot{x}_2 \end{bmatrix} = \begin{bmatrix} 0 & 1 \\ -16 & 0 \end{bmatrix} \begin{bmatrix} x_1 \\ x_2 \end{bmatrix}$. The initial conditions may be written as $\mathbf{C} = \mathbf{X}(0) = \begin{bmatrix} x_1(0) \\ x_2(0) \end{bmatrix} = \begin{bmatrix} x(0) \\ \dot{x}(0) \end{bmatrix} = \begin{bmatrix} -1/2 \\ 0 \end{bmatrix}$.

4.35 Transform the initial-value problem $\ddot{x} + 96x = 0$; $x(0) = \frac{1}{6}$, $\dot{x}(0) = 0$ into matrix form.

▮ Solving the differential equation for its highest derivative, we obtain $\ddot{x} = -96x$. Since the differential equation has order 2, we define two new variables: $x_1 = x$ and $x_2 = \dot{x}$. Differentiating each of these equations once yields

$$\begin{aligned} \dot{x}_1 &= \dot{x} = x_2 \\ \dot{x}_2 &= \ddot{x} = -96x = -96x_1 \end{aligned} \qquad \text{or} \qquad \begin{aligned} \dot{x}_1 &= 0x_1 + 1x_2 \\ \dot{x}_2 &= -96x_1 + 0x_2 \end{aligned}$$

This system has the matrix form $\dot{\mathbf{X}} = \begin{bmatrix} \dot{x}_1 \\ \dot{x}_2 \end{bmatrix} = \begin{bmatrix} 0 & 1 \\ -96 & 0 \end{bmatrix} \begin{bmatrix} x_1 \\ x_2 \end{bmatrix}$. The initial conditions may be written as $\mathbf{C} = \mathbf{X}(0) = \begin{bmatrix} x_1(0) \\ x_2(0) \end{bmatrix} = \begin{bmatrix} x(0) \\ \dot{x}(0) \end{bmatrix} = \begin{bmatrix} 1/6 \\ 0 \end{bmatrix}$.

.36 Transform the initial-value problem $\ddot{x} + 64x = 0$; $x(0) = \frac{1}{4}$, $x(0) = 2$ into matrix form.

▮ Solving the differential equation for its highest derivative, we obtain $\ddot{x} = -64x$. Since the differential equation has order 2, we define two new variables: $x_1 = x$ and $x_2 = \dot{x}$. Differentiating each of these equations once yields

$$\begin{aligned} \dot{x}_1 &= \dot{x} = x_2 \\ \dot{x}_2 &= \ddot{x} = -64x = -64x_1 \end{aligned} \qquad \text{or} \qquad \begin{aligned} \dot{x}_1 &= 0x_1 + 1x_2 \\ \dot{x}_2 &= -64x_1 + 0x_2 \end{aligned}$$

This system has the matrix form $\dot{\mathbf{X}} = \begin{bmatrix} \dot{x}_1 \\ \dot{x}_2 \end{bmatrix} = \begin{bmatrix} 0 & 1 \\ -64 & 0 \end{bmatrix} \begin{bmatrix} x_1 \\ x_2 \end{bmatrix}$. The initial conditions may be written as $\mathbf{C} = \mathbf{X}(0) = \begin{bmatrix} x_1(0) \\ x_2(0) \end{bmatrix} = \begin{bmatrix} x(0) \\ \dot{x}(0) \end{bmatrix} = \begin{bmatrix} 1/4 \\ 2 \end{bmatrix}$.

37 Transform the initial-value problem $\ddot{x} + x = 3$; $x(\pi) = 1$, $\dot{x}(\pi) = 2$ into matrix form.

▮ Solving the differential equation for its highest derivative, we obtain $\ddot{x} = -x + 3$. This equation has order 2, so we introduce two new variables: $x_1 = x$ and $x_2 = \dot{x}$. Differentiating each of these equations once yields

$$\begin{aligned} \dot{x}_1 &= \dot{x} = x_2 \\ \dot{x}_2 &= \ddot{x} = -x + 3 = -x_1 + 3 \end{aligned} \qquad \text{or} \qquad \begin{aligned} \dot{x}_1 &= 0x_1 + 1x_2 \\ \dot{x}_2 &= -1x_1 + 0x_2 + 3 \end{aligned}$$

This system has the matrix form $\dot{\mathbf{X}} = \begin{bmatrix} \dot{x}_1 \\ \dot{x}_2 \end{bmatrix} = \begin{bmatrix} 0 & 1 \\ -1 & 0 \end{bmatrix} \begin{bmatrix} x_1 \\ x_2 \end{bmatrix} + \begin{bmatrix} 0 \\ 3 \end{bmatrix}$. The initial conditions may be written as $\mathbf{C} = \mathbf{X}(\pi) = \begin{bmatrix} x_1(\pi) \\ x_2(\pi) \end{bmatrix} = \begin{bmatrix} x(\pi) \\ \dot{x}(\pi) \end{bmatrix} = \begin{bmatrix} 1 \\ 2 \end{bmatrix}$.

38 Transform the initial-value problem $\ddot{x} + 4\dot{x} + 4x = 0$; $x(0) = 2$, $\dot{x}(0) = -2$ into matrix form.

▮ Solving the differential equation for its highest derivative, we obtain $\ddot{x} = -4\dot{x} - 4x$. This equation has order 2, so we introduce two new variables: $x_1 = x$ and $x_2 = \dot{x}$. Differentiating each of these equations once yields

$$\begin{aligned} \dot{x}_1 &= \dot{x} = x_2 \\ \dot{x}_2 &= \ddot{x} = -4\dot{x} - 4x = -4x_2 - 4x_1 \end{aligned} \qquad \text{or} \qquad \begin{aligned} \dot{x}_1 &= 0x_1 + 1x_2 \\ \dot{x}_2 &= -4x_1 - 4x_2 \end{aligned}$$

This system has the matrix form $\dot{\mathbf{X}} = \begin{bmatrix} \dot{x}_1 \\ \dot{x}_2 \end{bmatrix} = \begin{bmatrix} 0 & 1 \\ -4 & -4 \end{bmatrix} \begin{bmatrix} x_1 \\ x_2 \end{bmatrix}$. The initial conditions may be written as $\mathbf{C} = \mathbf{X}(0) = \begin{bmatrix} x_1(0) \\ x_2(0) \end{bmatrix} = \begin{bmatrix} x(0) \\ \dot{x}(0) \end{bmatrix} = \begin{bmatrix} 2 \\ -2 \end{bmatrix}$.

39 Transform the initial-value problem $\ddot{x} + 12.8\dot{x} + 64x = 0$; $x(0) = \frac{1}{6}$, $\dot{x}(0) = 0$ into matrix form.

▮ Solving the differential equation for its highest derivative, we obtain $\ddot{x} = -12.8\dot{x} - 64x$. This equation has order 2, so we define two new variables: $x_1 = x$ and $x_2 = \dot{x}$. Differentiating each of these equations once yield

$$\begin{aligned}\dot{x}_1 &= \dot{x} = x_2\\ \dot{x}_2 &= \ddot{x} = -12.8\dot{x} - 64x = -12.8x_2 - 64x_1\end{aligned} \quad \text{or} \quad \begin{aligned}\dot{x}_1 &= 0x_1 + 1x_2\\ \dot{x}_2 &= -64x_1 - 12.8x_2\end{aligned}$$

This system has the matrix form $\dot{\mathbf{X}} = \begin{bmatrix}\dot{x}_1\\ \dot{x}_2\end{bmatrix} = \begin{bmatrix}0 & 1\\ -64 & -12.8\end{bmatrix}\begin{bmatrix}x_1\\ x_2\end{bmatrix}$. The initial conditions may be written as $\mathbf{C} = \mathbf{X}(0) = \begin{bmatrix}x_1(0)\\ x_2(0)\end{bmatrix} = \begin{bmatrix}x(0)\\ \dot{x}(0)\end{bmatrix} = \begin{bmatrix}1/6\\ 0\end{bmatrix}$.

14.40 Transform the initial-value problem $\ddot{x} + 9\dot{x} + 14x = 0$; $x(0) = 0$, $\dot{x}(0) = -1$ into the matrix form.

▮ Solving the differential equation for its highest derivative, we obtain $\ddot{x} = -9\dot{x} - 14x$. Since the differential equation is of order 2, we define two new variables: $x_1 = x$ and $x_2 = \dot{x}$. Then differentiation yields

$$\begin{aligned}\dot{x}_1 &= \dot{x} = x_2\\ \dot{x}_2 &= \ddot{x} = -9\dot{x} - 14x = -9x_2 - 14x_1\end{aligned} \quad \text{or} \quad \begin{aligned}\dot{x}_1 &= 0x_1 + 1x_2\\ \dot{x}_2 &= -14x_1 - 9x_2\end{aligned}$$

This system has the matrix form $\dot{\mathbf{X}} = \begin{bmatrix}\dot{x}_1\\ \dot{x}_2\end{bmatrix} = \begin{bmatrix}0 & 1\\ -14 & -9\end{bmatrix}\begin{bmatrix}x_1\\ x_2\end{bmatrix}$. The initial conditions may be written as $\mathbf{C} = \mathbf{X}(0) = \begin{bmatrix}x_1(0)\\ x_2(0)\end{bmatrix} = \begin{bmatrix}x(0)\\ \dot{x}(0)\end{bmatrix} = \begin{bmatrix}0\\ -1\end{bmatrix}$.

14.41 Transform the initial-value problem $\ddot{x} + 9\dot{x} + 14x = \frac{1}{2}\sin t$; $x(0) = 0$, $\dot{x}(0) = -1$ into matrix form.

▮ This problem is similar to the previous problem, except now the differential equation is nonhomogeneous. Solving for the highest derivative, we obtain $\ddot{x} = -9\dot{x} - 14x + \frac{1}{2}\sin t$. We define x_1 and x_2 as in the previous problem, and then differentiate to obtain

$$\begin{aligned}\dot{x}_1 &= \dot{x} = x_2\\ \dot{x}_2 &= \ddot{x} = -9x_2 - 14x_1 + \tfrac{1}{2}\sin t\end{aligned} \quad \text{or} \quad \begin{aligned}\dot{x}_1 &= 0x_1 + 1x_2\\ \dot{x}_2 &= -14x_1 - 9x_2 + \tfrac{1}{2}\sin t\end{aligned}$$

This system has the matrix form $\dot{\mathbf{X}} = \begin{bmatrix}\dot{x}_1\\ \dot{x}_2\end{bmatrix} = \begin{bmatrix}0 & 1\\ -14 & -9\end{bmatrix}\begin{bmatrix}x_1\\ x_2\end{bmatrix} + \begin{bmatrix}0\\ \frac{1}{2}\sin t\end{bmatrix}$. The initial conditions take the same form as in the previous problem.

14.42 Transform the initial-value problem $\ddot{x} + 2\dot{x} - 8x = e^t$; $\dot{x}(0) = 1$, $\dot{x}(0) = -4$ into matrix form.

▮ Solving the differential equation for its highest derivative, we obtain $\ddot{x} = -2\dot{x} - 8x + e^t$. This equation ha order 2, so we introduce two new variables: $x_1 = x$ and $x_2 = \dot{x}$. Differentiating each of these equations once yields

$$\begin{aligned}\dot{x}_1 &= \dot{x} = x_2\\ \dot{x}_2 &= \ddot{x} = -2\dot{x} + 8x + e^t = -2x_2 + 8x_1 + e^t\end{aligned} \quad \text{or} \quad \begin{aligned}\dot{x}_1 &= 0x_1 + 1x_2 + 0\\ \dot{x}_2 &= 8x_1 - 2x_2 + e^t\end{aligned}$$

This system is equivalent to the matrix equation $\dot{\mathbf{X}}(t) = \mathbf{A}(t)\mathbf{X}(t) + \mathbf{F}(t)$, where $\mathbf{X}(t) \equiv \begin{bmatrix}x_1(t)\\ x_2(t)\end{bmatrix}$, $\mathbf{A}(t) \equiv \begin{bmatrix}0 & 1\\ 8 & -2\end{bmatrix}$, and $\mathbf{F}(t) \equiv \begin{bmatrix}0\\ e^t\end{bmatrix}$. Furthermore, if we define $\mathbf{C} \equiv \begin{bmatrix}1\\ -4\end{bmatrix}$, then the initial conditions are given by $\mathbf{X}(t_0) = \mathbf{C}$, where $t_0 = 0$.

14.43 Transform the differential equation $\ddot{x} - 6\dot{x} + 9x = t$ into matrix form.

▮ Solving for the highest-order derivative, we find that $\ddot{x} = 6\dot{x} - 9x + t$. Since this equation has order 2, so introduce two new variables: $x_1 = x$ and $x_2 = \dot{x}$. Differentiation then yields

$$\begin{aligned}\dot{x}_1 &= \dot{x} = x_2\\ \dot{x}_2 &= \ddot{x} = 6\dot{x} - 9x + t = 6x_2 - 9x_1 + t\end{aligned} \qquad \begin{aligned}\dot{x}_1 &= 0x_1 + 1x_2 + 0\\ \dot{x}_2 &= -9x_1 + 6x_2 + t\end{aligned}$$

These equations are equivalent to the matrix equation $\dot{\mathbf{X}}(t) = \mathbf{A}(t)\mathbf{X}(t) + \mathbf{F}(t)$, where $\mathbf{X}(t) \equiv \begin{bmatrix}x_1(t)\\ x_2(t)\end{bmatrix}$, $\mathbf{A}(t) \equiv \begin{bmatrix}0 & 1\\ -9 & 6\end{bmatrix}$, and $\mathbf{F}(t) \equiv \begin{bmatrix}0\\ t\end{bmatrix}$.

14.44 Put the initial-value problem $\frac{d^3x}{dt^3} + 2\frac{d^2x}{dt^2} - 3\frac{dx}{dt} + 4x = t^2 + 5$; $x(2) = 10$, $\dot{x}(2) = 11$, $\ddot{x}(2) = 12$ into matrix form.

■ Solving this differential equation for its highest derivative, we obtain $d^3x/dt^3 = -2\ddot{x} + 3\dot{x} - 4x + t^2 + 5$. This equation has order 3, so we introduce three new variables: $x_1 = x$, $x_2 = \dot{x}$, and $x_3 = \ddot{x}$. Differentiating each of these equations once yields

$$\begin{aligned} \dot{x}_1 &= \dot{x} \\ \dot{x}_2 &= \ddot{x} \\ \dot{x}_3 &= \dddot{x} = -2\ddot{x} + 3\dot{x} - 4x + t^2 + 5 \end{aligned} \qquad \text{or} \qquad \begin{aligned} \dot{x}_1 &= 0x_1 + 1x_2 + 0x_3 \\ \dot{x}_2 &= 0x_1 + 0x_2 + 1x_3 \\ \dot{x}_3 &= -4x_1 + 3x_2 - 2x_3 + (t^2 + 5) \end{aligned}$$

This system has the matrix form $\dot{\mathbf{X}} = \begin{bmatrix} \dot{x}_1 \\ \dot{x}_2 \\ \dot{x}_3 \end{bmatrix} = \begin{bmatrix} 0 & 1 & 0 \\ 0 & 0 & 1 \\ -4 & 3 & -2 \end{bmatrix} \begin{bmatrix} x_1 \\ x_2 \\ x_3 \end{bmatrix} + \begin{bmatrix} 0 \\ 0 \\ t^2 + 5 \end{bmatrix}$.

The initial conditions may be written as $\mathbf{C} = \mathbf{X}(2) = \begin{bmatrix} x_1(2) \\ x_2(2) \\ x_3(2) \end{bmatrix} = \begin{bmatrix} x(2) \\ \dot{x}(2) \\ \ddot{x}(2) \end{bmatrix} = \begin{bmatrix} 10 \\ 11 \\ 12 \end{bmatrix}$.

14.45 Put the initial-value problem $2\frac{d^3x}{dt^3} - 4\frac{d^2x}{dt^2} + x = 8e^{-3t}$; $x(1) = 2$, $\dot{x}(1) = -2$, $\ddot{x}(-1) = 0$ into matrix form.

■ Solving the differential equation for its highest derivative, we obtain $\dddot{x} = 2\ddot{x} - \frac{1}{2}x + 4e^{-3t}$. This equation has order 3, so we introduce three new variables: $x_1 = x$, $x_2 = \dot{x}$, and $x_3 = \ddot{x}$. Differentiating each of these equations once yields

$$\begin{aligned} \dot{x}_1 &= \dot{x} \\ \dot{x}_2 &= \ddot{x} \\ \dot{x}_3 &= \dddot{x} = 2\ddot{x} - \tfrac{1}{2}x + 4e^{-3t} \end{aligned} \qquad \text{or} \qquad \begin{aligned} \dot{x}_1 &= 0x_1 + 1x_2 + 0x_3 \\ \dot{x}_2 &= 0x_1 + 0x_2 + 1x_3 \\ \dot{x}_3 &= -\tfrac{1}{2}x_1 + 0x_2 + 2x_3 + 4e^{-3t} \end{aligned}$$

This system has the matrix form $\dot{\mathbf{X}} = \begin{bmatrix} \dot{x}_1 \\ \dot{x}_2 \\ \dot{x}_3 \end{bmatrix} = \begin{bmatrix} 0 & 1 & 0 \\ 0 & 0 & 1 \\ -1/2 & 0 & 2 \end{bmatrix} \begin{bmatrix} x_1 \\ x_2 \\ x_3 \end{bmatrix} + \begin{bmatrix} 0 \\ 0 \\ 4e^{-3t} \end{bmatrix}$.

The initial conditions may be written as $\mathbf{C} = \mathbf{X}(1) = \begin{bmatrix} x_1(1) \\ x_2(1) \\ x_3(1) \end{bmatrix} = \begin{bmatrix} x(1) \\ \dot{x}(1) \\ \ddot{x}(1) \end{bmatrix} = \begin{bmatrix} 2 \\ -2 \\ 0 \end{bmatrix}$.

14.46 Transform the initial-value problem $e^{-t}\frac{d^4x}{dt^4} - \frac{d^2x}{dt^2} + e^t t^2 \frac{dx}{dt} = 5e^{-t}$; $x(1) = 2$ $\dot{x}(1) = 3$, $\ddot{x}(1) = 4$, $\dddot{x}(1) = 5$ into matrix form.

■ Solving the differential equation for its highest derivative, we obtain $\frac{d^4x}{dt^4} = e^t\frac{d^2x}{dt^2} - t^2e^{2t}\frac{dx}{dt} + 5$. This equation has order 4, so we introduce four new variables: $x_1 = x$, $x_2 = \dot{x}$, $x_3 = \ddot{x}$, and $x_4 = \dddot{x}$. Differentiating each of these equations once yields

$$\begin{aligned} \dot{x}_1 &= \dot{x} \\ \dot{x}_2 &= \ddot{x} \\ \dot{x}_3 &= \dddot{x} \\ \dot{x}_4 &= e^t\ddot{x} + e^{2t}t^2\dot{x} + 5 \end{aligned} \qquad \text{or} \qquad \begin{aligned} \dot{x}_1 &= 0x_1 + 1x_2 + 0x_3 + 0x_4 + 0 \\ \dot{x}_2 &= 0x_1 + 0x_2 + 1x_3 + 0x_4 + 0 \\ \dot{x}_3 &= 0x_1 + 0x_2 + 0x_3 + 1x_4 + 0 \\ \dot{x}_4 &= 0x_1 - t^2e^{2t}x_2 + e^tx_3 + 0x_4 + 5 \end{aligned}$$

These equations are equivalent to the matrix equation $\dot{\mathbf{X}}(t) = \mathbf{A}(t)\mathbf{X}(t) + \mathbf{F}(t)$, where

$$\mathbf{X}(t) \equiv \begin{bmatrix} x_1(t) \\ x_2(t) \\ x_3(t) \\ x_4(t) \end{bmatrix} \qquad \mathbf{A}(t) \equiv \begin{bmatrix} 0 & 1 & 0 & 0 \\ 0 & 0 & 1 & 0 \\ 0 & 0 & 0 & 1 \\ 0 & -t^2e^{2t} & e^t & 0 \end{bmatrix} \qquad \mathbf{F}(t) \equiv \begin{bmatrix} 0 \\ 0 \\ 0 \\ 5 \end{bmatrix}$$

Furthermore, if we define $\mathbf{C} \equiv [2, 3, 4, 5]^T$, then the initial conditions are given by $\mathbf{X}(t_0) = \mathbf{C}$, where $t_0 = 1$.

14.47 Put the following system into matrix form: $\dddot{x} = t\ddot{x} + x - \dot{y} + t + 1$; $\ddot{y} = (\sin t)\dot{x} + x - y + t^2$; $x(1) = 2$, $\dot{x}(1) = 3$, $\ddot{x}(1) = 4$, $y(1) = 5$, $\dot{y}(1) = 6$.

Since this system contains a third-order differential equation in x *and* a second-order differential equation in y, we will need three new x-variables and two new y-variables. We therefore define $x_1(t) = x$; $x_2(t) = \dot{x}$, $x_3(t) = \ddot{x}$, $y_1(t) = y$, and $y_2(t) = \dot{y}$. Then differentiation yields

$$\begin{aligned}
\dot{x}_1 &= x_2 \\
\dot{x}_2 &= x_3 \\
\dot{x}_3 &= \dddot{x} = t\ddot{x} + x - \dot{y} + t + 1 = tx_3 + x_1 - y_2 + t + 1 \\
\dot{y}_1 &= y_2 \\
\dot{y}_2 &= \ddot{y} = (\sin t)\dot{x} + x - y + t^2 = (\sin t)x_2 + x_1 - y_1 + t^2 \\
\dot{x}_1 &= 0x_1 + 1x_2 + 0x_3 + 0y_1 + 0y_2 + 0 \\
\dot{x}_2 &= 0x_1 + 0x_2 + 1x_3 + 0y_1 + 0y_2 + 0 \\
\dot{x}_3 &= 1x_1 + 0x_2 + tx_3 + 0y_1 - 1y_2 + (t + 1) \\
\dot{y}_1 &= 0x_1 + 0x_2 + 0x_3 + 0y_1 + 1y_2 + 0 \\
\dot{y}_2 &= 1x_1 + (\sin t)x_2 + 0x_3 - 1y_1 + 0y_2 + t^2
\end{aligned}$$

These equations are equivalent to the matrix equation $\dot{\mathbf{X}}(t) = \mathbf{A}(t)\mathbf{X}(t) + \mathbf{F}(t)$, where

$$\mathbf{X}(t) \equiv \begin{bmatrix} x_1(t) \\ x_2(t) \\ x_3(t) \\ y_1(t) \\ y_2(t) \end{bmatrix} \qquad \mathbf{A}(t) \equiv \begin{bmatrix} 0 & 1 & 0 & 0 & 0 \\ 0 & 0 & 1 & 0 & 0 \\ 1 & 0 & t & 0 & -1 \\ 0 & 0 & 0 & 0 & 1 \\ 1 & \sin t & 0 & -1 & 0 \end{bmatrix} \qquad \mathbf{F}(t) = \begin{bmatrix} 0 \\ 0 \\ t+1 \\ 0 \\ t^2 \end{bmatrix}$$

Furthermore, if we define $\mathbf{C} \equiv [2, 3, 4, 5, 6]^T$ and $t_0 = 1$, then the initial conditions are given by $\mathbf{X}(t_0) = \mathbf{C}$

14.48 Put the following system into matrix form: $\ddot{x} = -2\dot{x} - 5y + 3$; $\dot{y} = \dot{x} + 2y$; $x(0) = 0$, $\dot{x}(0) = 0$, $y(0) = 1$.

Since the system contains a second-order differential equation in x and a first-order differential equation in y, we define the three new variables: $x_1(t) = x$, $x_2(t) = \dot{x}$, and $y_1(t) = y$. Then differentiation yields

$$\begin{aligned}
\dot{x}_1 &= x_2 & \qquad && \dot{x}_1 &= 0x_1 + 1x_2 + 0y_1 + 0 \\
\dot{x}_2 &= \ddot{x} = -2\dot{x} - 5y + 3 = -2x_2 - 5y_1 + 3 & \text{or} && \dot{x}_2 &= 0x_1 - 2x_2 - 5y_1 + 3 \\
\dot{y}_1 &= \dot{y} = \dot{x} + 2y = x_2 + 2y_1 & && \dot{y}_1 &= 0x_1 + 1x_2 + 2y_1 + 0
\end{aligned}$$

These equations are equivalent to the matrix equation $\dot{\mathbf{X}}(t) = \mathbf{A}(t)\mathbf{X}(t) + \mathbf{F}(t)$, where

$$\mathbf{X}(t) \equiv \begin{bmatrix} x_1(t) \\ x_2(t) \\ y_1(t) \end{bmatrix} \qquad \mathbf{A}(t) = \begin{bmatrix} 0 & 1 & 0 \\ 0 & -2 & -5 \\ 0 & 1 & 2 \end{bmatrix} \qquad \mathbf{F}(t) = \begin{bmatrix} 0 \\ 3 \\ 0 \end{bmatrix}$$

If we also define $t_0 = 0$ and $\mathbf{C} \equiv \begin{bmatrix} 0 \\ 0 \\ 1 \end{bmatrix}$, then the initial conditions are given by $\mathbf{X}(t_0) = \mathbf{C}$.

14.49 Put the following system into matrix form: $\ddot{x} = \dot{x} + \dot{y} - z + t$; $\ddot{y} = tx + \dot{y} - 2y + t^2 + 1$; $\dot{z} = x - y + \dot{y} + z$; $x(1) = 1$, $\dot{x}(1) = 15$, $y(1) = 0$, $\dot{y}(1) = -7$, $z(1) = 4$.

Since this system contains second-order differential equations in x and y and a first-order differential equation z, we define five new variables—two for x, two for y, and one for z: $x_1 = x$, $x_2 = \dot{x}$, $y_1 = y$, $y_2 = \dot{y}$, and $z_1 = z$. Differentiating each of these variables once and using the original set of differential equations, we obtai

$$\begin{aligned}
\dot{x}_1 &= \dot{x} = x_2 \\
\dot{x}_2 &= \ddot{x} = \dot{x} + \dot{y} - z + t = x_2 + y_2 - z_1 + t \\
\dot{y}_1 &= \dot{y} = y_2 \\
\dot{y}_2 &= \ddot{y} = tx + \dot{y} - 2y + t^2 + 1 = tx_1 + y_2 - 2y_1 + t^2 + 1 \\
\dot{z}_1 &= \dot{z} = x - y + \dot{y} + z = x_1 - y_1 + y_2 + z_1
\end{aligned}$$

or

$$\begin{aligned}\dot{x}_1 &= 0x_1 + 1x_2 + 0y_1 + 0y_2 + 0z_1 + 0\\ \dot{x}_2 &= 0x_1 + 1x_2 + 0y_1 + 1y_2 - 1z_1 + t\\ \dot{y}_1 &= 0x_1 + 0x_2 + 0y_1 + 1y_2 + 0z_1 + 0\\ \dot{y}_2 &= tx_1 + 0x_2 - 2y_1 + 1y_2 + 0z_1 + (t^2+1)\\ \dot{z}_1 &= 1x_1 + 0x_2 - 1y_1 + 1y_2 + 1z_1 + 0\end{aligned}$$

These equations are equivalent to the matrix initial-value problem $\dot{\mathbf{X}} = \mathbf{A}(t)\mathbf{X}(t) + \mathbf{F}(t)$; $\mathbf{X}(1) = \mathbf{C}$, where

$$\mathbf{X}(t) \equiv \begin{bmatrix} x_1(t)\\ x_2(t)\\ y_1(t)\\ y_2(t)\\ z_1(t)\end{bmatrix} \quad \mathbf{A}(t) \equiv \begin{bmatrix} 0 & 1 & 0 & 0 & 0\\ 0 & 1 & 0 & 1 & -1\\ 0 & 0 & 0 & 1 & 0\\ t & 0 & -2 & 1 & 0\\ 1 & 0 & -1 & 1 & 1\end{bmatrix} \quad \mathbf{F}(t) \equiv \begin{bmatrix} 0\\ t\\ 0\\ t^2+1\\ 0\end{bmatrix} \quad \mathbf{C} \equiv \begin{bmatrix} 1\\ 15\\ 0\\ -7\\ 4\end{bmatrix}$$

14.50 Put the following system into matrix form: $\dot{x} = x + y$; $\dot{y} = 9x + y$.

▌ Since the system consists of two first-order differential equations, we define two new variables: $x_1(t) = x$ and $y_1(t) = y$. Thus,

$$\begin{aligned}\dot{x}_1 &= \dot{x} = x + y = x_1 + y_1\\ \dot{y}_1 &= \dot{y} = 9x + y = 9x_1 + y_1\end{aligned}$$

If we now define $\mathbf{X}(t) \equiv \begin{bmatrix} x_1(t)\\ y_1(t)\end{bmatrix}$ and $\mathbf{A}(t) \equiv \begin{bmatrix} 1 & 1\\ 9 & 1\end{bmatrix}$, then this last set of equations is equivalent to the matrix equation $\dot{\mathbf{X}}(t) = \mathbf{A}(t)\mathbf{X}(t)$.

SOLUTIONS

14.51 Solve $\ddot{x} + 9\dot{x} + 14x = 0$; $x(0) = 0$, $\dot{x}(0) = -1$.

▌ This homogeneous differential equation has the matrix form $\dot{\mathbf{X}} = \mathbf{AX}$; $\mathbf{X}(0) = \mathbf{C}$ (see Problem 14.40), where $\mathbf{X} = \begin{bmatrix} x_1\\ x_2\end{bmatrix}$, $\mathbf{A} = \begin{bmatrix} 0 & 1\\ -14 & -9\end{bmatrix}$, and $\mathbf{C} = \begin{bmatrix} 0\\ -1\end{bmatrix}$; $x_1 = x$ and $x_2 = \dot{x}$. Using the result of Problem 14.2, we may write its solution as

$$\mathbf{X} = e^{\mathbf{A}t}\mathbf{C} = \frac{1}{5}\begin{bmatrix} 7e^{-2t} - 2e^{-7t} & e^{-2t} - e^{-7t}\\ -14e^{-2t} + 14e^{-7t} & -2e^{-2t} + 7e^{-7t}\end{bmatrix}\begin{bmatrix} 0\\ -1\end{bmatrix} = \frac{1}{5}\begin{bmatrix} -e^{-2t} + e^{-7t}\\ 2e^{-2t} - 7e^{-7t}\end{bmatrix}$$

Therefore, $x(t) = x_1(t) = \frac{1}{5}(-e^{-2t} + e^{-7t})$. (Compare this with Problem 11.17.)

14.52 Solve $\ddot{x} + 20\dot{x} + 64x = 0$; $x(0) = \frac{1}{6}$, $\dot{x}(0) = 0$.

▌ This homogeneous differential equation has the matrix form $\dot{\mathbf{X}} = \mathbf{AX}$; $\mathbf{X}(0) = \mathbf{C}$ (see Problem 14.33), where $\mathbf{X} = \begin{bmatrix} x_1\\ x_2\end{bmatrix}$, $\mathbf{A} = \begin{bmatrix} 0 & 1\\ -64 & -20\end{bmatrix}$, and $\mathbf{C} = \begin{bmatrix} \frac{1}{6}\\ 0\end{bmatrix}$; $x_1 = x$ and $x_2 = \dot{x}$. Using the result of Problem 14.3, we write its solution as

$$\mathbf{X} = e^{\mathbf{A}t}\mathbf{C} = \frac{1}{12}\begin{bmatrix} 16e^{-4t} - 4e^{-16t} & e^{-4t} - e^{-16t}\\ -64e^{-4t} + 64e^{-16t} & -4e^{-4t} + 16e^{-16t}\end{bmatrix}\begin{bmatrix} \frac{1}{6}\\ 0\end{bmatrix} = \begin{bmatrix} \frac{2}{9}e^{-4t} - \frac{1}{18}e^{-16t}\\ -\frac{8}{9}e^{-4t} + \frac{8}{9}e^{-16t}\end{bmatrix}$$

Therefore, $x(t) = x_1(t) = \frac{2}{9}e^{-4t} - \frac{1}{18}e^{-16t}$. (Compare with Problem 11.26.)

14.53 Solve $\ddot{x} + 20\dot{x} + 64x = 0$; $x(0) = -1$, $\dot{x}(0) = 4$.

▌ This homogeneous differential equation has the matrix form $\dot{\mathbf{X}} = \mathbf{AX}$ (see Problem 14.33), where $\mathbf{X} = \begin{bmatrix} x_1\\ x_2\end{bmatrix}$ and $\mathbf{A} = \begin{bmatrix} 0 & 1\\ -64 & -20\end{bmatrix}$; $x_1 = x$ and $x_2 = \dot{x}$. Here the initial conditions are $\mathbf{C} = \mathbf{X}(0) = \begin{bmatrix} x_1(0)\\ x_2(0)\end{bmatrix} = \begin{bmatrix} x(0)\\ \dot{x}(0)\end{bmatrix} = \begin{bmatrix} -1\\ 4\end{bmatrix}$. Using the result of Problem 14.3, we can write the solution as

$$\mathbf{X} = e^{\mathbf{A}t}\mathbf{C} = \frac{1}{12}\begin{bmatrix} 16e^{-4t} - 4e^{-16t} & e^{-4t} - e^{-16t}\\ -64e^{-4t} + 64e^{-16t} & -4e^{-4t} + 16e^{-16t}\end{bmatrix}\begin{bmatrix} -1\\ 4\end{bmatrix} = \begin{bmatrix} -e^{-4t}\\ 4e^{-4t}\end{bmatrix}$$

Therefore, $x(t) = x_1(t) = -e^{-4t}$.

14.54 Solve $\ddot{x} + 20\dot{x} + 64x = 0$; $x(2) = 0$. $\dot{x}(2) = 4$.

▌ This differential equation has the matrix form $\dot{\mathbf{X}} = \mathbf{AX}$, with $\mathbf{X}$ and $\mathbf{A}$ as in the previous two problems. Here, however, the initial conditions take the form $\mathbf{C} = \mathbf{X}(2) = \begin{bmatrix} x_1(2) \\ x_2(2) \end{bmatrix} = \begin{bmatrix} x(2) \\ \dot{x}(2) \end{bmatrix} = \begin{bmatrix} 0 \\ 4 \end{bmatrix}$. Using the result of Problem 14.4, we write the solution as

$$\mathbf{X} = e^{\mathbf{A}(t-2)}\mathbf{C} = \frac{1}{12}\begin{bmatrix} 16e^{-4(t-2)} - 4e^{-16(t-2)} & e^{-4(t-2)} - e^{-16(t-2)} \\ -64e^{-4(t-2)} + 64e^{-16(t-2)} & -4e^{-4(t-2)} + 16^{-16(t-2)} \end{bmatrix}\begin{bmatrix} 0 \\ 4 \end{bmatrix} = \begin{bmatrix} \frac{1}{3}e^{-4(t-2)} - \frac{1}{3}e^{-16(t-2)} \\ -\frac{4}{3}e^{-4(t-2)} + \frac{16}{3}e^{-16(t-2)} \end{bmatrix}$$

Thus, $x(t) = x_1(t) = \frac{1}{3}e^{-4(t-2)} - \frac{1}{3}e^{-16(t-2)}$.

14.55 Solve $\ddot{x} + 16x = 0$; $x(0) = -\frac{1}{2}$, $\dot{x}(0) = 0$.

▌ This homogeneous differential equation has the matrix form $\dot{\mathbf{X}} = \mathbf{AX}$; $\mathbf{X}(0) = \mathbf{C}$ (see Problem 14.34), where $\mathbf{X} = \begin{bmatrix} x_1 \\ x_2 \end{bmatrix}$, $\mathbf{A} = \begin{bmatrix} 0 & 1 \\ -16 & 0 \end{bmatrix}$, and $\mathbf{C} = \begin{bmatrix} -\frac{1}{2} \\ 0 \end{bmatrix}$; $x_1 = x$ and $x_2 = \dot{x}$. Using the result of Problem 14.10, we can write its solution as $\mathbf{X} = e^{\mathbf{A}t}\mathbf{C} = \begin{bmatrix} \cos 4t & \frac{1}{4}\sin 4t \\ -4\sin 4t & \cos 4t \end{bmatrix}\begin{bmatrix} -\frac{1}{2} \\ 0 \end{bmatrix} = \begin{bmatrix} -\frac{1}{2}\cos 4t \\ 2\sin 4t \end{bmatrix}$.
Therefore, $x(t) = x_1(t) = -\frac{1}{2}\cos 4t$. (Compare with Problem 11.1.)

14.56 Solve $\ddot{x} + 96x = 0$; $x(0) = \frac{1}{6}$, $\dot{x}(0) = 0$.

▌ This differential equation has the matrix form $\dot{\mathbf{X}} = \mathbf{AX}$; $\mathbf{X}(0) = \mathbf{C}$ (see Problem 14.35), where $\mathbf{X} = \begin{bmatrix} x_1 \\ x_2 \end{bmatrix}$, $\mathbf{A} = \begin{bmatrix} 0 & 1 \\ -96 & 0 \end{bmatrix}$, and $\mathbf{C} = \begin{bmatrix} 1/6 \\ 0 \end{bmatrix}$; $x_1 = x$ and $x_2 = \dot{x}$. Using the result of Problem 14.11, we can write its solution as

$$\mathbf{X} = e^{\mathbf{A}t}\mathbf{C} = \begin{bmatrix} \cos\sqrt{96}t & (1/\sqrt{96})\sin\sqrt{96}t \\ \sqrt{96}\sin\sqrt{96}t & \cos\sqrt{96}t \end{bmatrix}\begin{bmatrix} 1/6 \\ 0 \end{bmatrix} = \begin{bmatrix} (1/6)\cos\sqrt{96}t \\ (\sqrt{96}/6)\sin\sqrt{96}t \end{bmatrix}$$

Therefore, $x(t) = x_1(t) = \frac{1}{6}\cos\sqrt{96}t$. (Compare with Problem 11.2.)

14.57 Solve $\ddot{x} + 4\dot{x} + 4x = 0$; $x(0) = 2$, $\dot{x}(0) = -2$.

▌ This differential equation has the matrix form $\dot{\mathbf{X}} = \mathbf{AX}$; $\mathbf{X}(0) = \mathbf{C}$ (see Problem 14.38), where $\mathbf{X} = \begin{bmatrix} x_1 \\ x_2 \end{bmatrix}$, $\mathbf{A} = \begin{bmatrix} 0 & 1 \\ -4 & -4 \end{bmatrix}$, and $\mathbf{C} = \begin{bmatrix} 2 \\ -2 \end{bmatrix}$; $x_1 = x$ and $x_2 = \dot{x}$. Using the result of Problem 14.20, we write its solution as

$$\mathbf{X} = e^{\mathbf{A}t}\mathbf{C} = \begin{bmatrix} (1+2t)e^{-2t} & te^{-2t} \\ -4te^{-2t} & (1-2t)e^{-2t} \end{bmatrix}\begin{bmatrix} 2 \\ -2 \end{bmatrix} = \begin{bmatrix} (2+2t)e^{-2t} \\ (-2-4t)e^{-2t} \end{bmatrix}$$

Therefore, $x(t) = x_1(t) = (2 + 2t)e^{-2t}$. (Compare with Problem 11.21.)

14.58 Solve $\ddot{x} + 4\dot{x} + 4x = 0$; $x(-1) = 2$, $\dot{x}(-1) = -2$.

▌ This differential equation has the same matrix form as the differential equation in the previous problem, except now the initial time is $t = -1$ rather than $t = 0$. That is, now $\mathbf{X}(-1) = \mathbf{C}$. Using the result of Problem 14.21, we write the solution as

$$\mathbf{X} = e^{\mathbf{A}[t-(-1)]}\mathbf{C} = e^{\mathbf{A}(t+1)}\mathbf{C} = e^{-2(t+1)}\begin{bmatrix} 2t+3 & t+1 \\ -4t-4 & -2t-1 \end{bmatrix}\begin{bmatrix} 2 \\ -2 \end{bmatrix} = e^{-2(t+1)}\begin{bmatrix} 2t+4 \\ -4t-6 \end{bmatrix}$$

Then $x(t) = x_1(t) = (2t + 4)e^{-2(t+1)}$.

14.59 Solve $\ddot{x} + 64x = 0$; $x(0) = \frac{1}{4}$, $x(0) = 2$.

▌ This differential equation has the matrix form $\dot{\mathbf{X}} = \mathbf{AX}$; $\mathbf{X}(0) = \mathbf{C}$ (see Problem 14.36), where $\mathbf{X} = \begin{bmatrix} x_1 \\ x_2 \end{bmatrix}$, $\mathbf{A} = \begin{bmatrix} 0 & 1 \\ -64 & 0 \end{bmatrix}$, and $\mathbf{C} = \begin{bmatrix} 1/4 \\ 2 \end{bmatrix}$; $x_1 = x$ and $x_2 = \dot{x}$. Using the result of Problem 14.12, we write its solution as

$$\mathbf{X} = e^{\mathbf{A}t}\mathbf{C} = \begin{bmatrix} \cos 8t & \frac{1}{8}\sin 8t \\ -8\sin 8t & \cos 8t \end{bmatrix}\begin{bmatrix} 1/4 \\ 2 \end{bmatrix} = \begin{bmatrix} \frac{1}{4}\cos 8t + \frac{1}{4}\sin 8t \\ -2\sin 8t + 2\cos 8t \end{bmatrix}$$

Therefore, $x(t) = x_1(t) = \frac{1}{4}\cos 8t + \frac{1}{4}\sin 8t$. (Compare with Problem 11.4.)

14.60 Solve $\ddot{x} + 64x = 0$.

▮ This differential equation is the same as that of the previous problem, and so it has the same matrix form $\dot{\mathbf{X}} = \mathbf{AX}$, with $\mathbf{X}$ and $\mathbf{A}$ as defined in Problem 14.59. The difference here is that there are no initial conditions. Nonetheless, we may set $x(0) = k_1$ and $\dot{x}(0) = k_2$, where k_1 and k_2 denote unknown numbers, and write $\mathbf{C} = \mathbf{X}(0) = \begin{bmatrix} x(0) \\ \dot{x}(0) \end{bmatrix} = \begin{bmatrix} x_1(0) \\ x_2(0) \end{bmatrix} = \begin{bmatrix} k_1 \\ k_2 \end{bmatrix}$. Then, using the result of Problem 14.12, we write the solution as

$$\mathbf{X} = e^{\mathbf{A}t}\mathbf{C} = \begin{bmatrix} \cos 8t & \frac{1}{8}\sin 8t \\ -8\sin 8t & \cos 8t \end{bmatrix}\begin{bmatrix} k_1 \\ k_2 \end{bmatrix} = \begin{bmatrix} k_1\cos 8t + \frac{1}{8}k_2\sin 8t \\ -8k_1\sin 8t + k_2\cos 8t \end{bmatrix}$$

Therefore, $x(t) = x_1(t) = k_1\cos 8t + k_3\sin 8t$, where $k_3 = k_2/8$.

14.61 Solve $\ddot{x} + 8\dot{x} + 25x = 0$; $x(0) = 0$, $\dot{x}(0) = 4$.

▮ This differential equation has the matrix form $\dot{\mathbf{X}} = \mathbf{AX}$; $\mathbf{X}(0) = \mathbf{C}$ (see Problem 14.32), where $\mathbf{X} = \begin{bmatrix} x_1 \\ x_2 \end{bmatrix}$, $\mathbf{A} = \begin{bmatrix} 0 & 1 \\ -25 & -8 \end{bmatrix}$, and $\mathbf{C} = \begin{bmatrix} 0 \\ 4 \end{bmatrix}$; $x_1 = x$ and $x_2 = \dot{x}$. Using the result of Problem 14.16, we can write its solution as

$$\mathbf{X} = e^{\mathbf{A}t}\mathbf{C} = e^{-4t}\begin{bmatrix} \frac{4}{3}\sin 3t + \cos 3t & \frac{1}{3}\sin 3t \\ -\frac{25}{3}\sin 3t & -\frac{4}{3}\sin 3t + \cos 3t \end{bmatrix}\begin{bmatrix} 0 \\ 4 \end{bmatrix} = e^{-4t}\begin{bmatrix} \frac{4}{3}\sin 3t \\ -\frac{16}{3}\sin 3t + 4\cos 3t \end{bmatrix}$$

Therefore, $x(t) = x_1(t) = \frac{4}{3}e^{-4t}\sin 3t$. (Compare with Problem 11.19.)

14.62 Solve $\ddot{x} + 8\dot{x} + 25x = 0$.

▮ This differential equation is the same as that of the previous problem, and so it has the matrix form $\dot{\mathbf{X}} = \mathbf{AX}$, with $\mathbf{X}$ and $\mathbf{A}$ as defined in Problem 14.61. Since no initial conditions are specified, we may set $x(0) = k_1$ and $\dot{x}(0) = k_2$, where k_1 and k_2 denote unknown numbers; then $\mathbf{C} = \mathbf{X}(0) = \begin{bmatrix} x(0) \\ \dot{x}(0) \end{bmatrix} = \begin{bmatrix} x_1(0) \\ x_2(0) \end{bmatrix} = \begin{bmatrix} k_1 \\ k_2 \end{bmatrix}$. Using the result of Problem 14.16, we may write the solution as

$$\mathbf{X} = e^{\mathbf{A}t}\mathbf{C} = e^{-4t}\begin{bmatrix} \frac{4}{3}\sin 3t + \cos 3t & \frac{1}{3}\sin 3t \\ -\frac{25}{3}\sin 3t & -\frac{4}{3}\sin 3t + \cos 3t \end{bmatrix}\begin{bmatrix} k_1 \\ k_2 \end{bmatrix} = e^{-4t}\begin{bmatrix} (\frac{4}{3}k_1 + \frac{1}{3}k_2)\sin 3t + k_1\cos 3t \\ (-\frac{25}{3}k_1 - \frac{4}{3}k_2)\sin 3t + k_2\cos 3t \end{bmatrix}$$

Therefore, $x(t) = x_1(t) = k_3e^{-4t}\sin 3t + k_1e^{-4t}\cos 3t$, where $k_3 = \frac{4}{3}k_1 + \frac{1}{3}k_2$.

14.63 Solve $\ddot{x} + 12.8\dot{x} + 64x = 0$; $x(0) = \frac{1}{6}$, $\dot{x}(0) = 0$.

▮ This differential equation has the matrix form $\dot{\mathbf{X}} = \mathbf{AX}$; $\mathbf{X}(0) = \mathbf{C}$ (see Problem 14.39), where $\mathbf{X} = \begin{bmatrix} x_1 \\ x_2 \end{bmatrix}$, $\mathbf{A} = \begin{bmatrix} 0 & 1 \\ -64 & -12.8 \end{bmatrix}$, and $\mathbf{C} = \begin{bmatrix} 1/6 \\ 0 \end{bmatrix}$; $x_1 = x$ and $x_2 = \dot{x}$. Using the result of Problem 14.17, we may write its solution as

$$\mathbf{X} = e^{\mathbf{A}t}\mathbf{C} = e^{-6.4t}\begin{bmatrix} \frac{4}{3}\sin 4.8t + \cos 4.8t & \frac{5}{24}\sin 4.8t \\ -\frac{40}{3}\sin 4.8t & -\frac{4}{3}\sin 4.8t + \cos 4.8t \end{bmatrix}\begin{bmatrix} 1/6 \\ 0 \end{bmatrix} = e^{-6.4t}\begin{bmatrix} \frac{2}{9}\sin 4.8t + \frac{1}{6}\cos 4.8t \\ -\frac{20}{9}\sin 4.8t \end{bmatrix}$$

Therefore, $x(t) = x_1(t) = e^{-6.4t}(\frac{2}{9}\sin 4.8t + \frac{1}{6}\cos 4.8t)$. (Compare with Problem 11.24.)

14.64 Solve $\ddot{x} + 16\dot{x} + 64x = 0$; $x(0) = \frac{1}{6}$, $\dot{x}(0) = 0$.

▮ This differential equation has the matrix form $\dot{\mathbf{X}} = \mathbf{AX}$; $\mathbf{X}(0) = \mathbf{C}$ (see Problem 14.31), where $\mathbf{X} = \begin{bmatrix} x_1 \\ x_2 \end{bmatrix}$, $\mathbf{A} = \begin{bmatrix} 0 & 1 \\ -64 & -16 \end{bmatrix}$, and $\mathbf{C} = \begin{bmatrix} 1/6 \\ 0 \end{bmatrix}$; $x_1 = x$ and $x_2 = \dot{x}$. Using the result of Problem 14.18, we write its solution as

$$\mathbf{X} = e^{\mathbf{A}t}\mathbf{C} = e^{-8t}\begin{bmatrix} 1+8t & t \\ -64t & 1-8t \end{bmatrix}\begin{bmatrix} 1/6 \\ 0 \end{bmatrix} = \begin{bmatrix} \frac{1}{6}(1+8t)e^{-8t} \\ \frac{1}{6}(-64t)e^{-8t} \end{bmatrix}$$

Therefore, $x(t) = x_1(t) = \frac{1}{6}(1+8t)e^{-8t}$. (Compare with Problem 11.28.)

14.65 Solve $\ddot{x} + 2\dot{x} - 8x = 0;\ \ x(1) = 2,\ \ \dot{x}(1) = 3.$

This differential equation has the matrix form $\dot{\mathbf{X}} = \mathbf{AX}$; $\mathbf{X}(1) = \mathbf{C}$ (see Problem 14.42 with e^t replaced by 0), where $\mathbf{X}(t) = \begin{bmatrix} x_1(t) \\ x_2(t) \end{bmatrix}$, $\mathbf{A} = \begin{bmatrix} 0 & 1 \\ 8 & -2 \end{bmatrix}$, and $\mathbf{C} = \begin{bmatrix} 2 \\ 3 \end{bmatrix}$; $x_1 = x$ and $x_2 = \dot{x}$. Using the result of Problem 14.8, we may write its solution as

$$\mathbf{X} = e^{\mathbf{A}(t-1)}\mathbf{C} = \frac{1}{6}\begin{bmatrix} 4e^{2(t-1)} + 2e^{-4(t-1)} & e^{2(t-1)} - e^{-4(t-1)} \\ 8e^{2(t-1)} - 8e^{-4(t-1)} & 2e^{2(t-1)} + 4e^{-4(t-1)} \end{bmatrix}\begin{bmatrix} 2 \\ 3 \end{bmatrix} = \begin{bmatrix} \frac{11}{6}e^{2(t-1)} + \frac{1}{6}e^{-4(t-1)} \\ \frac{22}{6}e^{2(t-1)} - \frac{4}{6}e^{-4(t-1)} \end{bmatrix}$$

The solution to the initial-value problem is $x(t) = x_1(t) = \frac{11}{6}e^{2(t-1)} + \frac{1}{6}e^{-4(t-1)}$.

14.66 Solve $\ddot{x} + 2\dot{x} - 8x = e^t;\ \ x(0) = 1,\ \ \dot{x}(0) = 4.$

This differential equation has the matrix form $\dot{\mathbf{X}} = \mathbf{AX} + \mathbf{F}(t)$; $\mathbf{X}(0) = \mathbf{C}$ (see Problem 14.42), where $\mathbf{X}(t) = \begin{bmatrix} x_1(t) \\ x_2(t) \end{bmatrix}$, $\mathbf{A} = \begin{bmatrix} 0 & 1 \\ 8 & -2 \end{bmatrix}$, $\mathbf{F}(t) = \begin{bmatrix} 0 \\ e^t \end{bmatrix}$; and $\mathbf{C} = \begin{bmatrix} 1 \\ -4 \end{bmatrix}$; $x_1 = x$ and $x_2 = \dot{x}$. Its solution is $\mathbf{X} = e^{\mathbf{A}t}\mathbf{C} + \int_0^t e^{\mathbf{A}(t-s)}\mathbf{F}(s)\,ds$. Using the results of Problems 14.7 and 14.9, we have

$$e^{\mathbf{A}t}\mathbf{C} = \frac{1}{6}\begin{bmatrix} 4e^{2t} + 2e^{-4t} & e^{2t} - e^{-4t} \\ 8e^{2t} - 8e^{-4t} & 2e^{2t} + 4e^{-4t} \end{bmatrix}\begin{bmatrix} 1 \\ -4 \end{bmatrix} = \begin{bmatrix} e^{-4t} \\ -4e^{-4t} \end{bmatrix}$$

and

$$e^{\mathbf{A}(t-s)}\mathbf{F}(s) = \frac{1}{6}\begin{bmatrix} 4e^{2(t-s)} + 2e^{-4(t-s)} & e^{2(t-s)} - e^{-4(t-s)} \\ 8e^{2(t-s)} - 8e^{-4(t-s)} & 2e^{2(t-s)} + 4e^{-4(t-s)} \end{bmatrix}\begin{bmatrix} 0 \\ e^s \end{bmatrix} = \frac{1}{6}\begin{bmatrix} e^{(2t-s)} - e^{(-4t+5s)} \\ 2e^{(2t-s)} + 4e^{(-4t+5s)} \end{bmatrix}$$

so

$$\int_0^t e^{\mathbf{A}(t-s)}\mathbf{F}(s)\,ds = \frac{1}{6}\begin{bmatrix} \int_0^t [e^{(2t-s)} - e^{(-4t+5s)}]\,ds \\ \int_0^t [2e^{(2t-s)} + 4e^{(-4t+5s)}]\,ds \end{bmatrix} = \frac{1}{6}\begin{bmatrix} (-e^{(2t-s)} - \frac{1}{5}e^{(-4t+5s)})\big|_{s=0}^{s=t} \\ (-2e^{(2t-s)} + \frac{4}{5}e^{(-4t+5s)})\big|_{s=0}^{s=t} \end{bmatrix}$$

$$= \frac{1}{6}\begin{bmatrix} -\frac{6}{5}e^t + e^{2t} + \frac{1}{5}e^{-4t} \\ -\frac{6}{5}e^t + 2e^{2t} - \frac{4}{5}e^{-4t} \end{bmatrix}$$

Thus,

$$\mathbf{X} = \begin{bmatrix} e^{-4t} \\ -4e^{-4t} \end{bmatrix} + \frac{1}{6}\begin{bmatrix} -\frac{6}{5}e^t + e^{2t} + \frac{1}{5}e^{-4t} \\ -\frac{6}{5}e^t + 2e^{2t} - \frac{4}{5}e^{-4t} \end{bmatrix} = \begin{bmatrix} \frac{31}{30}e^{-4t} + \frac{1}{6}e^{2t} - \frac{1}{5}e^t \\ -\frac{62}{15}e^{-4t} + \frac{1}{3}e^{2t} - \frac{1}{5}e^t \end{bmatrix}$$

and $x(t) = x_1(t) = \frac{31}{30}e^{-4t} + \frac{1}{6}e^{2t} - \frac{1}{5}e^t$.

14.67 Solve $\ddot{x} + x = 3;\ \ x(\pi) = 1,\ \ \dot{x}(\pi) = 2.$

This differential equation has the matrix form $\dot{\mathbf{X}} = \mathbf{AX} + \mathbf{F}(t)$; $\mathbf{X}(\pi) = \mathbf{C}$ (see Problem 14.37), where $\mathbf{X}(t) = \begin{bmatrix} x_1(t) \\ x_2(t) \end{bmatrix}$, $\mathbf{A} = \begin{bmatrix} 0 & 1 \\ -1 & 0 \end{bmatrix}$, $\mathbf{F}(t) = \begin{bmatrix} 0 \\ 3 \end{bmatrix}$; and $\mathbf{C} = \begin{bmatrix} 1 \\ 2 \end{bmatrix}$; $x_1 = x$ and $x_2 = \dot{x}$. Its solution is $\mathbf{X} = e^{\mathbf{A}(t-\pi)}\mathbf{C} + \int_\pi^t e^{\mathbf{A}(t-s)}\mathbf{F}(s)\,ds$. Using the results of Problems 14.14 and 14.15, we have

$$e^{\mathbf{A}(t-\pi)}\mathbf{C} = \begin{bmatrix} \cos(t-\pi) & \sin(t-\pi) \\ -\sin(t-\pi) & \cos(t-\pi) \end{bmatrix}\begin{bmatrix} 1 \\ 2 \end{bmatrix} = \begin{bmatrix} \cos(t-\pi) + 2\sin(t-\pi) \\ -\sin(t-\pi) + 2\cos(t-\pi) \end{bmatrix}$$

and

$$e^{\mathbf{A}(t-s)}\mathbf{F}(s) = \begin{bmatrix} \cos(t-s) & \sin(t-s) \\ -\sin(t-s) & \cos(t-s) \end{bmatrix}\begin{bmatrix} 0 \\ 3 \end{bmatrix} = \begin{bmatrix} 3\sin(t-s) \\ 3\cos(t-s) \end{bmatrix}$$

so

$$\int_\pi^t e^{\mathbf{A}(t-s)}\mathbf{F}(s)\,ds = \begin{bmatrix} \int_\pi^t 3\sin(t-s)\,ds \\ \int_\pi^t 3\cos(t-s)\,ds \end{bmatrix} = \begin{bmatrix} 3\cos(t-s)\big|_{s=\pi}^{s=t} \\ -3\sin(t-s)\big|_{s=\pi}^{s=t} \end{bmatrix} = \begin{bmatrix} 3 - 3\cos(t-\pi) \\ 3\sin(t-\pi) \end{bmatrix}$$

Thus,

$$\mathbf{X}(t) = \begin{bmatrix} \cos(t-\pi) + 2\sin(t-\pi) \\ -\sin(t-\pi) + 2\cos(t-\pi) \end{bmatrix} + \begin{bmatrix} 3 - 3\cos(t-\pi) \\ 3\sin(t-\pi) \end{bmatrix} = \begin{bmatrix} 3 - 2\cos(t-\pi) + 2\sin(t-\pi) \\ 2\cos(t-\pi) + 2\sin(t-\pi) \end{bmatrix}$$

and $x(t) = x_1(t) = 3 - 2\cos(t-\pi) + 2\sin(t-\pi)$. Noting that $\cos(t-\pi) = -\cos t$ and $\sin(t-\pi) = -\sin t$, we finally obtain $x(t) = 3 + 2\cos t - 2\sin t$.

14.68 Solve $\ddot{x} + 9\dot{x} + 14x = \frac{1}{2}\sin t$; $x(0) = 0$, $\dot{x}(0) = -1$.

▌ This differential equation has the matrix form $\dot{\mathbf{X}} = \mathbf{AX} + \mathbf{F}(t)$; $\mathbf{X}(0) = \mathbf{C}$ (see Problem 14.41), where $\mathbf{X}(t) = \begin{bmatrix} x_1(t) \\ x_2(t) \end{bmatrix}$, $\mathbf{A} = \begin{bmatrix} 0 & 1 \\ -14 & -9 \end{bmatrix}$, $\mathbf{F}(t) = \begin{bmatrix} 0 \\ \frac{1}{2}\sin t \end{bmatrix}$, and $\mathbf{C} = \begin{bmatrix} 0 \\ -1 \end{bmatrix}$; with $x_1(t) = x(t)$ and $x_2(t) = \dot{x}(t)$. Its solution is $\mathbf{X} = e^{\mathbf{A}t}\mathbf{C} + \int_0^t e^{\mathbf{A}(t-s)}\mathbf{F}(s)\,ds$.

We determined $e^{\mathbf{A}t}\mathbf{C}$ in Problem 14.51. Using the result of Problem 14.5, we have

$$e^{\mathbf{A}(t-s)}\mathbf{F}(s) = \frac{1}{5}\begin{bmatrix} 7e^{-2(t-s)} - 2e^{-7(t-s)} & e^{-2(t-s)} - e^{-7(t-s)} \\ -14e^{-2(t-s)} + 14e^{-7(t-s)} & -2e^{-2(t-s)} + 7e^{-7(t-s)} \end{bmatrix}\begin{bmatrix} 0 \\ \frac{1}{2}\sin s \end{bmatrix} = \frac{1}{10}\begin{bmatrix} e^{-2(t-s)}\sin s - e^{-7(t-s)}\sin s \\ -2e^{-2(t-s)}\sin s + 7e^{-7(t-s)}\sin s \end{bmatrix}$$

so

$$\int_0^t e^{\mathbf{A}(t-s)}\mathbf{F}(s)\,ds = \frac{1}{10}\begin{bmatrix} e^{-2t}\int_0^t e^{2s}\sin s\,ds - e^{-7t}\int_0^t e^{7s}\sin s\,ds \\ -2e^{-2t}\int_0^t e^{2s}\sin s\,ds + 7e^{-7t}\int_0^t e^{7s}\sin s\,ds \end{bmatrix}$$

$$= \frac{1}{10}\begin{bmatrix} e^{-2t}(\frac{2}{5}e^{2t}\sin t - \frac{1}{5}e^{2t}\cos t + \frac{1}{5}) - e^{-7t}(\frac{7}{50}e^{7t}\sin t - \frac{1}{50}e^{7t}\cos t + \frac{1}{50}) \\ -2e^{-2t}(\frac{2}{5}e^{2t}\sin t - \frac{1}{5}e^{2t}\cos t + \frac{1}{5}) + 7e^{-7t}(\frac{7}{50}e^{7t}\sin t - \frac{1}{50}e^{7t}\cos t + \frac{1}{50}) \end{bmatrix}$$

$$= \frac{1}{10}\begin{bmatrix} \frac{13}{50}\sin t - \frac{9}{50}\cos t + \frac{1}{5}e^{-2t} - \frac{1}{50}e^{-7t} \\ \frac{9}{50}\sin t + \frac{13}{50}\cos t - \frac{2}{5}e^{-2t} + \frac{7}{50}e^{-7t} \end{bmatrix}$$

Then

$$\mathbf{X}(t) = \frac{1}{5}\begin{bmatrix} -e^{-2t} + e^{-7t} \\ 2e^{-2t} - 7e^{-7t} \end{bmatrix} + \frac{1}{10}\begin{bmatrix} \frac{13}{50}\sin t - \frac{9}{50}\cos t + \frac{1}{5}e^{-2t} - \frac{1}{50}e^{-7t} \\ \frac{9}{50}\sin t + \frac{13}{50}\cos t - \frac{2}{5}e^{-2t} + \frac{7}{50}e^{-7t} \end{bmatrix}$$

$$= \begin{bmatrix} \frac{13}{500}\sin t - \frac{9}{500}\cos t - \frac{90}{500}e^{-2t} + \frac{99}{500}e^{-7t} \\ \frac{9}{500}\sin t + \frac{13}{500}\cos t + \frac{180}{500}e^{-2t} - \frac{693}{500}e^{-7t} \end{bmatrix}$$

and $x(t) = x_1(t) = \frac{13}{500}\sin t - \frac{9}{500}\cos t - \frac{90}{500}e^{-2t} + \frac{99}{500}e^{-7t}$. (Compare with Problem 11.38.)

14.69 Solve the system $\ddot{x} = -2\dot{x} - 5y + 3$; $\dot{y} = \dot{x} + 2y$; $x(0) = 0$, $\dot{x}(0) = 0$, $y(0) = 1$

▌ This system has the matrix form $\dot{\mathbf{X}} = \mathbf{AX} + \mathbf{F}(t)$; $\mathbf{X}(0) = \mathbf{C}$ (see Problem 14.48), where

$$\mathbf{X}(t) = \begin{bmatrix} x_1(t) \\ x_2(t) \\ y_1(t) \end{bmatrix}, \quad \mathbf{A} = \begin{bmatrix} 0 & 1 & 0 \\ 0 & -2 & -5 \\ 0 & 1 & 2 \end{bmatrix}, \quad \mathbf{F}(t) = \begin{bmatrix} 0 \\ 3 \\ 0 \end{bmatrix}, \quad \text{and} \quad \mathbf{C} = \begin{bmatrix} 0 \\ 0 \\ 1 \end{bmatrix}; \quad x_1 = x, \quad x_2 = \dot{x}, \quad \text{and} \quad y_1 = y.$$

Its solution is $\mathbf{X} = e^{\mathbf{A}t}\mathbf{C} + \int_0^t e^{\mathbf{A}(t-s)}\mathbf{F}(s)\,ds$. Using the results of Problems 14.24 and 14.25, we have

$$e^{\mathbf{A}t}\mathbf{C} = \begin{bmatrix} 1 & -2 + 2\cos t + \sin t & -5 + 5\cos t \\ 0 & \cos t - 2\sin t & -5\sin t \\ 0 & \sin t & \cos t + 2\sin t \end{bmatrix}\begin{bmatrix} 0 \\ 0 \\ 1 \end{bmatrix} = \begin{bmatrix} -5 + 5\cos t \\ -5\sin t \\ \cos t + 2\sin t \end{bmatrix}$$

and

$$e^{\mathbf{A}(t-s)}\mathbf{F}(s) = \begin{bmatrix} 1 & -2 + 2\cos(t-s) + \sin(t-s) & -5 + 5\cos(t-s) \\ 0 & \cos(t-s) - 2\sin(t-s) & -5\sin(t-s) \\ 0 & \sin(t-s) & \cos(t-s) + 2\sin(t-s) \end{bmatrix}\begin{bmatrix} 0 \\ 3 \\ 0 \end{bmatrix}$$

$$= \begin{bmatrix} -6 + 6\cos(t-s) + 3\sin(t-s) \\ 3\cos(t-s) - 6\sin(t-s) \\ 3\sin(t-s) \end{bmatrix}$$

so

$$\int_0^t e^{\mathbf{A}(t-s)}\mathbf{F}(s)\,ds = \begin{bmatrix} \int_0^t [-6 + 6\cos(t-s) + 3\sin(t-s)]\,ds \\ \int_0^t [3\cos(t-s) - 6\sin(t-s)]\,ds \\ \int_0^t 3\sin(t-s)\,ds \end{bmatrix}$$

$$= \begin{bmatrix} [-6s - 6\sin(t-s) + 3\cos(t-s)]_{s=0}^{s=t} \\ [-3\sin(t-s) - 6\cos(t-s)]_{s=0}^{s=t} \\ 3\cos(t-s)\big|_{s=0}^{s=t} \end{bmatrix} = \begin{bmatrix} -6t + 3 + 6\sin t - 3\cos t \\ -6 + 3\sin t + 6\cos t \\ 3 - 3\cos t \end{bmatrix}$$

Then

$$\mathbf{X}(t) = \begin{bmatrix} -5 + 5\cos t \\ -5\sin t \\ \cos t + 2\sin t \end{bmatrix} + \begin{bmatrix} -6t + 3 + 6\sin t - 3\cos t \\ -6 + 3\sin t + 6\cos t \\ 3 - 3\cos t \end{bmatrix} = \begin{bmatrix} -2 - 6t + 2\cos t + 6\sin t \\ -6 + 6\cos t - 2\sin t \\ 3 - 2\cos t + 2\sin t \end{bmatrix}$$

and, finally, we have $x(t) = x_1(t) = 2\cos t + 6\sin t - 2 - 6t$ and $y(t) = y_1(t) = -2\cos t + 2\sin t + 3$.

CHAPTER 15

Infinite-Series Solutions

ANALYTIC FUNCTIONS

15.1 Define "analytic function."

A function $f(x)$ is *analytic* at x_0 if its Taylor series about x_0, $\sum_{n=0}^{\infty} \frac{f^{(n)}(x_0)(x-x_0)^n}{n!}$, converges to $f(x)$ in some neighborhood of x_0.

15.2 What is an ordinary point for the differential equation $b_2(x)y'' + b_1(x)y' + b_0(x)y = 0$?

If we divide by $b_2(x)$, we transform the differential equation to the standard form $y'' + P(x)y' + Q(x)y = 0$, where $P(x) = b_1(x)/b_2(x)$ and $Q(x) = b_0(x)/b_2(x)$. A point x_0 is an *ordinary point* if both $P(x)$ and $Q(x)$ are analytic at x_0. If either $P(x)$ or $Q(x)$ is not analytic at x_0, then x_0 is a *singular point.*

15.3 Define "regular singular point" for the differential equation of the previous problem.

The point x_0 is a *regular singular point* for the differential equation in standard form $y'' + P(x)y' + Q(x)y = 0$ if x_0 is a singular point (see the previous problem) and the products $(x - x_0)P(x)$ and $(x - x_0)^2Q(x)$ are both analytic at x_0. Singular points which are not regular are called *irregular.*

15.4 What is an ordinary point for the differential equation $b_1(x)y' + b_0(x)y = 0$?

Dividing by $b_1(x)$, we transform the differential equation to the standard form $y' + P(x)y = 0$, where $P(x) = b_0(x)/b_1(x)$. A point x_0 is an *ordinary point* if $P(x)$ is analytic at x_0.

15.5 Find a Maclaurin-series expansion for $f(x) = e^x$.

Since every derivative of e^x is e^x, it follows that $f(0)$ and all the derivatives of f at $x = 0$ are equal to $e^0 = 1$. The Maclaurin series is then

$$e^x = f(0) + f'(0)x + \frac{f''(0)}{2!}x^2 + \frac{f'''(0)}{3!}x^3 + \cdots + \frac{f^{(n)}(0)}{n!}x^n + \cdots$$

$$= 1 + (1)x + \frac{1}{2!}x^2 + \frac{1}{3!}x^3 + \cdots + \frac{1}{n!}x^n + \cdots = \sum_{n=0}^{\infty} \frac{1}{n!}x^n \qquad (1$$

15.6 Determine the interval of convergence for the Maclaurin series obtained in the previous problem.

By the ratio test, the series $\sum_{n=0}^{\infty} a_n$ converges if $L = \lim_{n\to\infty} \left|\frac{a_{n+1}}{a_n}\right| < 1$. If $L > 1$ or $L = +\infty$, the series diverges; whereas if $L = 1$, no conclusion can be inferred.

Using the ratio test, we have $\lim_{n\to\infty} \left|\frac{x^{n+1}}{(n+1)!}\frac{n!}{x^n}\right| = |x| \lim_{n\to\infty} \frac{1}{n+1} = 0$. Since this ratio is less than unity for every value of x, the Maclaurin series converges everywhere. The interval of convergence is $(-\infty, \infty)$.

15.7 Find a Maclaurin-series expansion for $f(x) = e^{-2x}$.

We have

$$\begin{aligned} f(x) &= e^{-2x} & f(0) &= 1 \\ f'(x) &= -2e^{-2x} & f'(0) &= -2 \\ f''(x) &= 2^2e^{-2x} & f''(0) &= 2^2 \\ f'''(x) &= -2^3e^{-2x} & f'''(0) &= -2^3 \\ &\cdots\cdots\cdots & &\cdots\cdots \end{aligned}$$

so $$e^{-2x} = 1 - 2x + \frac{2^2}{2!}x^2 - \frac{2^3}{3!}x^3 + \frac{2^4}{4!}x^4 - \cdots + (-1)^n \frac{2^n}{n!}x^n + \cdots$$

15.8 Determine the interval of convergence for the Maclaurin series obtained in the previous problem.

▌ Using the ratio test, we have $\lim_{n\to+\infty}\left|\frac{2^{n+1}x^{n+1}}{(n+1)!}\frac{n!}{2^n x^n}\right| = |x| \lim_{n\to+\infty}\frac{2}{n+1} = 0$. The series converges for every value of x.

15.9 Solve Problem 15.7 using the result of Problem 15.5.

▌ We have from Problem 15.5 that $e^x = \sum_{n=0}^{\infty}\frac{1}{n!}x^n$. Replacing x with $-2x$, we obtain $e^{-2x} = \sum_{n=0}^{\infty}\frac{1}{n!}(-2x)^n = \sum_{n=0}^{\infty}\frac{(-1)^n 2^n}{n!}x^n$, which is the series obtained in Problem 15.7.

15.10 Find a Maclaurin series for e^{x^2}.

▌ Replacing x with x^2 in (*1*) of Problem 15.5, we get

$$e^{x^2} = \sum_{n=0}^{\infty}\frac{1}{n!}(x^2)^n = \sum_{n=0}^{\infty}\frac{1}{n!}x^{2n} = 1 + x^2 + \frac{1}{2!}x^4 + \frac{1}{3!}x^6 + \frac{1}{4!}x^8 + \cdots.$$

15.11 Determine whether $\ln x$ possesses a Taylor series about $x = 1$.

▌ Here $x_0 = 1$ and $f(x) = \ln x$. Thus,

$$f'(x) = \frac{1}{x}, \quad f''(x) = -\frac{1}{x^2}, \quad \ldots, \quad f^{(n)}(x) = \frac{(-1)^{n-1}(n-1)!}{x^n} \qquad (n \ge 1)$$

Therefore we have $f(1) = \ln 1 = 0$ and

$$f'(1) = 1, \quad f''(1) = -1, \quad \ldots, \quad f^{(n)}(1) = (-1)^{n-1}(n-1)! \qquad (n \ge 1)$$

Recalling that $0! = 1$ and $n! = n(n-1)!$, we find that

$$\sum_{n=0}^{\infty}\frac{f^{(n)}(1)(x-1)^n}{n!} = \frac{f(1)(x-1)^0}{0!} + \sum_{n=1}^{\infty}\frac{f^{(n)}(1)(x-1)^n}{n!} = 0 + \sum_{n=1}^{\infty}\frac{(-1)^{n-1}(n-1)!(x-1)^n}{n!}$$

$$= \sum_{n=1}^{\infty}\frac{(-1)^{n-1}}{n}(x-1)^n = (x-1) - \frac{1}{2}(x-1)^2 + \frac{1}{3}(x-1)^3 - \cdots$$

so that $\ln x$ does possess a Taylor series about $x = 1$.

15.12 Does $\ln x$ possess a Taylor series about $x = 0$?

▌ No. Neither $\ln x$ nor any of its derivatives exists at $x = 0$ therefore, $\ln x$ cannot possess a Taylor series at $x = 0$.

15.13 Use the ratio test to determine those values of x for which the Taylor series found in Problem 15.11 converges.

▌ For the series of Problem 15.11 we have

$$\lim_{n\to\infty}\left|\frac{a_{n+1}}{a_n}\right| = \lim_{n\to\infty}\left|\frac{\dfrac{(-1)^n(x-1)^{n+1}}{n+1}}{\dfrac{(-1)^{n-1}(x-1)^n}{n}}\right| = \lim_{n\to\infty}\frac{n}{n+1}|x-1| = |x-1|$$

We conclude from the ratio test that the series converges when $|x-1| < 1$ or, equivalently when $0 < x < 2$. The points $x = 0$ and $x = 2$ must be checked separately, since the ratio test is inconclusive at these points. For $x = 0$ we obtain $\sum_{n=1}^{\infty}\frac{(-1)^{n-1}(-1)^n}{n} = -\sum_{n=1}^{\infty}\frac{1}{n}$; this is the harmonic series, which is known to diverge.

For $x = 2$ we obtain $\sum_{n=1}^{\infty} \frac{(-1)^{n-1}(1)^n}{n} = -\sum_{n=1}^{\infty} \frac{(-1)^n}{n}$, which converges by the alternating-series test. Thus, the series of Problem 15.11 converges for $0 < x \le 2$.

15.14 Find a Taylor-series expansion around $x = 2$ for $f(x) = \ln x$.

We have

$$\begin{array}{ll} f(x) = \ln x & f(2) = \ln 2 \\ f'(x) = x^{-1} & f'(2) = \frac{1}{2} \\ f''(x) = -x^{-2} & f''(2) = -\frac{1}{4} \\ f'''(x) = 2x^{-3} & f'''(2) = \frac{1}{4} \\ f^{\text{iv}}(x) = -6x^{-4} & f^{\text{iv}}(2) = -\frac{3}{8} \\ \cdots\cdots & \cdots\cdots \end{array}$$

so

$$\ln x = \ln 2 + \frac{1}{2}(x-2) - \frac{1}{4}\frac{(x-2)^2}{2!} + \frac{1}{4}\frac{(x-2)^3}{3!} - \frac{3}{8}\frac{(x-2)^4}{4!} + \cdots$$

$$= \ln 2 + \frac{1}{2}(x-2) - \frac{1}{8}(x-2)^2 + \frac{1}{24}(x-2)^3 - \frac{1}{64}(x-2)^4 + \cdots$$

15.15 Determine the interval of convergence of the power series obtained in the previous problem.

Using the ratio test, we have

$$\lim_{n\to+\infty} \left| \frac{(x-2)^{n+1}}{2^{n+1}(n+1)} \frac{2^n n}{(x-2)^n} \right| = \frac{1}{2}|x-2| \lim_{n\to+\infty} \frac{n}{n+1} = \frac{1}{2}|x-2|$$

Thus the series converges for $|x-2| < 2$ or $0 < x < 4$.

For $x = 0$, the series is $\ln 2 -$ (harmonic series), which diverges. For $x = 4$, the series is $\ln 2 + 1 - \frac{1}{2} + \frac{1}{3} - \frac{1}{4} + \cdots$, which converges. Thus the series converges on the interval $0 < x \le 4$.

15.16 Find a Taylor-series expansion around $x = 0$ for $f(x) = \ln(1 + x)$.

We have

$$\begin{array}{ll} f(x) = \ln(1+x) & f(0) = 0 \\ f'(x) = \dfrac{1}{1+x} & f'(0) = 1 \\ f''(x) = -\dfrac{1}{(1+x)^2} & f''(0) = -1 \\ f'''(x) = \dfrac{1\cdot 2}{(1+x)^3} & f'''(0) = 2! \\ f^{\text{iv}}(x) = -\dfrac{1\cdot 2\cdot 3}{(1+x)^4} & f^{\text{iv}}(0) = -3! \\ \cdots\cdots & \cdots\cdots \end{array}$$

Hence

$$\ln(1+x) = x - \frac{x^2}{2!} + 2!\frac{x^3}{3!} - 3!\frac{x^4}{4!} + \cdots + (-1)^{n-1}(n-1)!\frac{x^n}{n!} + \cdots$$

$$= x - \frac{1}{2}x^2 + \frac{1}{3}x^3 - \frac{1}{4}x^4 + \cdots + (-1)^{n-1}\frac{1}{n}x^n + \cdots$$

15.17 Determine the interval of convergence for the power series obtained in the previous problem.

Using the ratio test, we have $\lim_{n\to+\infty} \left| \frac{x^{n+1}}{n+1} \frac{n}{x^n} \right| = |x| \lim_{n\to+\infty} \frac{n}{n+1} = |x|$. The series converges absolutely for $|x| < 1$ and diverges for $|x| > 1$. Individual tests are required at $x = 1$ and $x = -1$.

For $x = 1$, the series becomes $1 - \frac{1}{2} + \frac{1}{3} - \frac{1}{4} + \cdots$ and is conditionally convergent. For $x = -1$, the series becomes $-(1 + \frac{1}{2} + \frac{1}{3} + \frac{1}{4} + \cdots)$ and is divergent. Thus the given series converges on the interval $-1 < x \le 1$.

15.18 Find a Maclaurin-series expansion for $f(x) = \arctan x$.

We have

$$f(x) = \arctan x \qquad f(0) = 0$$
$$f'(x) = \frac{1}{1+x^2} = 1 - x^2 + x^4 - x^6 + \cdots \qquad f'(0) = 1$$
$$f''(x) = -2x + 4x^3 - 6x^5 + \cdots \qquad f''(0) = 0$$
$$f'''(x) = -2 + 12x^2 - 30x^4 + \cdots \qquad f'''(0) = -2!$$
$$f^{\text{iv}}(x) = 24x - 120x^3 + \cdots \qquad f^{\text{iv}}(0) = 0$$
$$f^{\text{v}}(x) = 24 - 360x^2 + \cdots \qquad f^{\text{v}}(0) = 4!$$
$$f^{\text{vi}}(x) = -720x + \cdots \qquad f^{\text{vi}}(0) = 0$$
$$f^{\text{vii}}(x) = -720 + \cdots \qquad f^{\text{vii}}(0) = -6!$$

and
$$\arctan x = x - \frac{2!}{3!}x^3 + \frac{4!}{5!}x^5 - \frac{6!}{7!}x^7 + \cdots = x - \frac{x^3}{3} + \frac{x^5}{5} - \frac{x^7}{7} + \cdots + (-1)^{n-1}\frac{x^{2n-1}}{2n-1} + \cdots$$

15.19 Determine the interval of convergence for the Maclaurin series obtained in the previous problem.

Using the ratio test and noting that $|(-1)^{n-1}| = |(-1)^n| = 1$, we have $\lim_{n\to+\infty}\left|\frac{x^{2n+1}}{2n+1}\frac{2n-1}{x^{2n-1}}\right| = x^2 \lim_{n\to+\infty}\frac{2n-1}{2n+1} = x^2$. The series is absolutely convergent on the interval $x^2 < 1$ or $-1 < x < 1$.

For $x = -1$, the series becomes $-1 + \frac{1}{3} - \frac{1}{5} + \frac{1}{7} - \cdots$, and for $x = 1$, it becomes $1 - \frac{1}{3} + \frac{1}{5} - \frac{1}{7} + \cdots$. Both series converge; thus the given series converges for $-1 \le x \le 1$ and diverges elsewhere.

15.20 Find a Taylor-series expansion around $x = 0$ for $f(x) = \sin x$.

We have

$$f(x) = \sin x \qquad f(0) = 0$$
$$f'(x) = \cos x \qquad f'(0) = 1$$
$$f''(x) = -\sin x \qquad f''(0) = 0$$
$$f'''(x) = -\cos x \qquad f'''(0) = -1$$
$$\cdots\cdots\cdots \qquad \cdots\cdots$$

The values of the derivatives at $x = 0$ form cycles of 0, 1, 0, -1; hence

$$\sin x = 0 + 1x + \frac{0}{2!}x^2 + \frac{-1}{3!}x^3 + \frac{0}{4!}x^4 + \frac{1}{5!}x^5 + \cdots = x - \frac{x^3}{3!} + \frac{x^5}{5!} - \frac{x^7}{7!} + \cdots + (-1)^{n-1}\frac{x^{2n-1}}{(2n-1)!} + \cdots$$

5.21 Determine the interval of convergence of the power series obtained in the previous problem.

Using the ratio test, we have $\lim_{n\to+\infty}\left|\frac{x^{2n+1}}{(2n+1)!}\frac{(2n-1)!}{x^{2n-1}}\right| = x^2 \lim_{n\to+\infty}\frac{1}{2n(2n+1)} = 0$. The series converges for every value of x.

5.22 Find a Maclaurin-series expansion for $f(t) = \cosh t$.

We have

$$f(t) = \cosh t \qquad f(0) = 1$$
$$f'(t) = \sinh t \qquad f'(0) = 0$$
$$f''(t) = \cosh t \qquad f''(0) = 1$$
$$f'''(t) = \sinh t \qquad f'''(0) = 0$$
$$\cdots\cdots\cdots \qquad \cdots\cdots$$

The values of the derivatives form the cycles 1, 0; 1, 0; . . . ; hence

$$\cosh t = 1 + 0t + \frac{1}{2!}t^2 + \frac{0}{3!}t^3 + \frac{1}{4!}t^4 + \frac{0}{5!}t^5 + \cdots = 1 + \frac{t^2}{2!} + \frac{t^4}{4!} + \frac{t^6}{6!} + \cdots = \sum_{n=0}^{\infty}\frac{t^{2n}}{(2n)!}$$

5.23 Determine the interval of convergence of the power series obtained in the previous problem.

Using the ratio test, we have $\lim_{n\to\infty}\left|\frac{t^{2n+2}}{(2n+2)!}\frac{(2n)!}{t^{2n}}\right| = t^2 \lim_{n\to\infty}\frac{1}{(2n+2)(2n+1)} = 0$. The series converges on $(-\infty, \infty)$.

15.24 Find a Taylor-series expansion around $x = 0$ for $f(x) = \cos x$.

▌ Differentiating the expansion obtained in Problem 15.20 yields

$$\cos x = 1 - \frac{3x^2}{3!} + \frac{5x^4}{5!} - \frac{7x^6}{7!} + \cdots + (-1)^{n-1}\frac{(2n-1)x^{2n-2}}{(2n-1)!} + \cdots = 1 - \frac{x^2}{2!} + \frac{x^4}{4!} - \frac{x^6}{6!} + \cdots$$

This series converges within the interval of convergence of the Maclaurin series for $\sin x$, which, as determined in Problem 15.21, is all x.

15.25 Find a Taylor-series expansion around $t = 0$ for $f(t) = \sin \frac{1}{2}t$.

▌ Setting $x = \frac{1}{2}t$ in the result of Problem 15.20 yields

$$\sin \tfrac{1}{2}t = (\tfrac{1}{2}t) - \frac{(\frac{1}{2}t)^3}{3!} + \frac{(\frac{1}{2}t)^5}{5!} - \frac{(\frac{1}{2}t)^7}{7!} + \cdots + (-1)^{n-1}\frac{(\frac{1}{2}t)^{2n-1}}{(2n-1)!} + \cdots = \sum_{n=1}^{\infty} \frac{(-1)^{n-1}}{2^{2n-1}(2n-1)!} t^{2n-1}$$

15.26 Find a Maclaurin-series expansion for $f(t) = \sinh t$.

▌ Differentiating the expansion obtained in Problem 15.22 yields

$$\sinh t = \frac{2t}{2!} + \frac{4t^3}{4!} + \frac{6t^5}{6!} + \cdots = \frac{t}{1!} + \frac{t^3}{3!} + \frac{t^5}{5!} + \cdots = \sum_{n=1}^{\infty} \frac{t^{2n-1}}{(2n-1)!}$$

This series converges within the interval of convergence of the original series, which, as determined in Problem 15.23, is everywhere.

15.27 Find a Taylor-series expansion around $x = 2$ for $f(x) = e^{x/2}$.

▌ We have

$$\begin{aligned} f(x) &= e^{x/2} & f(2) &= e \\ f'(x) &= \tfrac{1}{2}e^{x/2} & f'(2) &= \tfrac{1}{2}e \\ f''(x) &= \tfrac{1}{4}e^{x/2} & f''(2) &= \tfrac{1}{4}e \\ &\cdots\cdots & &\cdots\cdots \end{aligned}$$

so that

$$e^{x/2} = e\left[1 + \frac{1}{2}(x-2) + \frac{1}{4}\frac{(x-2)^x}{2!} + \cdots + \frac{1}{2^{n-1}}\frac{(x-2)^{n-1}}{(n-1)!} + \cdots\right]$$

15.28 Determine the interval of convergence of the power series obtained in the previous problem.

▌ Using the ratio test, we obtain $\lim_{n\to+\infty}\left|\frac{(x-2)^n}{2^n n!}\,\frac{2^{n-1}(n-1)!}{(x-2)^{n-1}}\right| = \frac{1}{2}|x-2| \lim_{n\to+\infty}\frac{1}{n} = 0.$ The series converges for every value of x.

15.29 Find a Taylor-series expansion around $t = 1$ for $f(t) = 2/t^2$.

▌ We have

$$\begin{aligned} f(t) &= 2t^{-2} & f(1) &= 2 \\ f'(t) &= (-2)(2t^{-3}) & f'(1) &= (-2)(2!) \\ f''(t) &= (2)(3!t^{-4}) & f''(1) &= (2)(3!) \\ f'''(t) &= (-2)(4!t^{-5}) & f'''(1) &= (-2)(4!) \\ &\cdots\cdots & &\cdots\cdots \end{aligned}$$

Hence

$$\begin{aligned} 2t^{-2} &= 2 + (-2)(2!)(t-1) + \frac{(2)(3!)}{2!}(t-1)^2 + \frac{(-2)(4!)}{3!}(t-1)^3 + \cdots \\ &= 2 - (2)(2)(t-1) + (2)(3)(t-1)^2 - (2)(4)(t-1)^3 + \cdots + (-1)^n(2)(n+1)(t-1)^n + \cdots \end{aligned}$$

15.30 Determine the interval of convergence of the power series obtained in the previous problem.

▌ Since $\lim_{n\to\infty}\left|\frac{2(n+2)(t-1)^{n+1}}{2(n+1)t^n}\right| = |t-1| \lim_{n\to\infty}\frac{n+2}{n+1} = |t-1|,$ the series converges for $|t-1| < 1$ or $0 < t < 2$.

For $t = 0$, the power series becomes $2(1 + 2 + 3 + 4 + \cdots)$, which diverges. For $t = 2$, the power series becomes $2(1 - 2 + 3 - 4 + \cdots)$, which also diverges. Thus, the interval of convergence is $0 < t < 2$.

15.31 Determine whether $Q(x) = -1/(x^2 - 1)$ and $P(x) = x/(x^2 - 1)$ are analytic at $x = 0$.

▮ We need to determine whether the Maclaurin series for these functions converge in some interval around $x = 0$. Using the geometric-series expansion $\dfrac{1}{1-y} = \sum_{n=0}^{\infty} y^n$ with $y = x^2$, we write

$$Q(x) = \frac{-1}{x^2 - 1} = \frac{1}{1 - x^2} = \sum_{n=0}^{\infty} (x^2)^n = 1 + x^2 + x^4 + x^6 + \cdots$$

and $$P(x) = \frac{x}{x^2 - 1} = -x\frac{1}{1 - x^2} = -x\sum_{n=0}^{\infty} (x^2)^n = -x - x^3 - x^5 - x^7 - \cdots$$

The geometric series converges when $|y| < 1$, so the series expansions for $P(x)$ and $Q(x)$ converge when $|x^2| < 1$ or when $-1 < x < 1$. Therefore, both functions are analytic at $x = 0$.

15.32 Determine whether $Q(x) = x/(x^2 + 4)$ is analytic at $x = 0$.

▮ We need to determine whether the Maclaurin series for $Q(x)$ converges in some interval around $x = 0$. Using the geometric-series expansion $\dfrac{1}{1-y} = \sum_{n=0}^{\infty} y^n$ with $y = -x^2/4$, we have

$$Q(x) = \frac{x}{x^2 + 4} = \frac{x}{4}\left[\frac{1}{1 - (-x^2/4)}\right] = \frac{x}{4}\sum_{n=0}^{\infty}\left(-\frac{x^2}{4}\right)^n = \frac{1}{4}x - \frac{1}{16}x^3 + \frac{1}{64}x^5 - \frac{1}{256}x^7 + \cdots$$

The geometric series converges when $|y| < 1$, so this Maclaurin series converges when $|-x^2/4| < 1$ or when $-2 < x < 2$. Therefore, $Q(x)$ is analytic at $x = 0$.

15.33 Determine whether $f(x) = x/(x + 1)$ is analytic at $x = 1$.

▮ Since $f(x)$ is the quotient of two polynomials with a denominator that is not zero at $x = 1$, $f(x)$ is analytic there. That is, it has a Taylor-series expansion around $x = 1$ that converges in some interval centered at $x = 1$. In particular, it follows from the geometric-series expansion of the previous problem with $y = -\frac{1}{2}(x - 1)$ that

$$\frac{x}{x+1} = \frac{(x-1)+1}{2+(x-1)} = \frac{(x-1)}{2}\,\frac{1}{1-[-(1/2)(x-1)]} + \frac{1}{2}\,\frac{1}{1-[-(1/2)(x-1)]}$$

$$= \frac{x-1}{2}\sum_{n=0}^{\infty}\left[-\frac{1}{2}(x-1)\right]^n + \frac{1}{2}\sum_{n=0}^{\infty}\left[-\frac{1}{2}(x-1)\right]^n = \sum_{n=0}^{\infty}\frac{(-1)^n(x-1)^{n+1}}{2^{n+1}} + \sum_{n=0}^{\infty}\frac{(-1)^n(x-1)^n}{2^{n+1}}$$

$$= \frac{1}{2} + \frac{x-1}{2^2} - \frac{(x-1)^2}{2^3} + \frac{(x-1)^3}{2^4} - \cdots = \frac{1}{2} + \sum_{n=1}^{\infty}\frac{(-1)^{n-1}}{2^{n+1}}(x-1)^n$$

This series converges when $|-\frac{1}{2}(x - 1)| < 1$, or when $-1 < x < 3$.

15.34 Determine whether $g(t) = (t - 1)/t$ is analytic at $t = 1$.

▮ We need to show that the Taylor-series expansion around $t = 1$ for $g(t)$ converges in some interval centered at $t = 1$. To simplify the algebra, we set $x = t - 1$, so that $\dfrac{t-1}{t} = \dfrac{x}{x+1}$. We now seek a Maclaurin-series expansion for $\dfrac{x}{x+1}$. Using the geometric-series expansion, we have

$\dfrac{x}{x+1} = x\dfrac{1}{1-(-x)} = x\sum_{n=0}^{\infty}(-x)^n = \sum_{n=0}^{\infty}(-1)^n x^{n+1}$, with convergence on the interval $|x| < 1$. Substituting $t - 1$ for x then yields $\dfrac{t-1}{t} = \sum_{n=0}^{\infty}(-1)^n(t-1)^{n+1}$, with convergence on the interval $|t - 1| < 1$ or $0 < t < 2$. Thus $g(t)$ is analytic at $t = 1$.

ORDINARY AND SINGULAR POINTS

15.35 Determine whether $x = 0$ is an ordinary point for the differential equation $y'' - xy' + 2y = 0$.

▮ This equation is in the standard form $y'' + P(x)y' + Q(x)y = 0$, with $P(x) = -x$ and $Q(x) = 2$. Each of these functions is its own Maclaurin series with infinite radius of convergence, so $x = 0$ is an ordinary point.

15.36 Determine whether $x = 0$ is an ordinary point for the differential equation $y'' - xy = 0$.

▌ This equation is in the standard form $y'' + P(x)y' + Q(x)y = 0$, with $P(x) = 0$ and $Q(x) = x$. Each of these functions is its own Maclaurin series with infinite radius of convergence, so $x = 0$ is an ordinary point.

15.37 Determine whether $x = 5$ is an ordinary point for the differential equation of the previous problem.

▌ Polynomials and constants are analytic everywhere. Since $P(x) = 0$ is a constant and $Q(x) = x$ is a first-degree polynomial, both functions are analytic everywhere, and in particular at $x = 5$; thus $x = 5$ and every other point are ordinary points.

15.38 Determine whether $x = 0$ is an ordinary point for the differential equation $(x^2 + 4)y'' + xy = 0$.

▌ Dividing the differential equation by $x^2 + 4$, we obtain $y'' + \dfrac{x}{x^2 + 4} y = 0$. This is in the standard form $y'' + P(x)y' + Q(x)y = 0$, with $P(x) = 0$ and $Q(x) = x/(x^2 + 4)$. Because $P(x)$ is analytic everywhere and $Q(x)$ is analytic at $x = 0$ (see Problem 15.32), $x = 0$ is an ordinary point.

15.39 Determine whether $x = 0$ is an ordinary point for the differential equation $(x^2 - 1)y'' + xy' - y = 0$.

▌ Dividing the differential equation by $x^2 - 1$, we obtain $y'' + \dfrac{x}{x^2 - 1} y' - \dfrac{1}{x^2 - 1} y = 0$. This has the standard form $y'' + P(x)y' + Q(x)y = 0$, with $P(x) = x/(x^2 - 1)$ and $Q(x) = -1/(x^2 - 1)$. Since both these functions are analytic at $x = 0$ (see Problem 15.31), that point is an ordinary point.

15.40 Determine whether $x = 1$ is an ordinary point for the differential equation in the previous problem.

▌ As before, $P(x) = x/(x^2 - 1)$ and $Q(x) = -1/(x^2 - 1)$. Since $P(1)$ and $Q(1)$ are undefined, it follows that neither function has a Taylor series about $x = 1$, so neither function is analytic and $x = 1$ is not an ordinary point. It is a singular point. Since $(x - 1)P(x) = x/(x + 1)$ is analytic at $x = 1$ (see Problem 15.33) and $(x - 1)^2Q(x) = -1$ is also analytic at $x = 1$, it follows that $x = 1$ is a regular singular point.

15.41 Determine whether $x = 1$ is an ordinary point for the differential equation $(x - 1)^2y'' - 2xy' = 0$.

▌ This differential equation has the standard form $y'' - \dfrac{2x}{(x - 1)^2} y' = 0$, with $P(x) = -2x/(x - 1)^2$ and $Q(x) = 0$. Since $P(1)$ in undefined, $P(x)$ does not have a Taylor series at $x = 1$ and is not analytic there. Therefore, $x = 1$ is not an ordinary point. Furthermore, $(x - 1)P(x) = -2x/(x - 1)$ also does not have a Taylor series around $x = 1$, so $x = 1$ is not a regular singular point; it is an irregular singular point.

15.42 Determine whether $x = 0$ is an ordinary point for the differential equation $e^{2x}y'' + y = 0$.

▌ Dividing the given equation by e^{2x}, we obtain $y'' + e^{-2x}y = 0$, which is in standard form. Since e^{-2x} has a Maclaurin-series expansion that converges in some interval centered at $x = 0$ (see Problems 15.7 and 15.8), e^{-2x} is analytic there and $x = 0$ is an ordinary point.

15.43 Determine whether $x = 0$ is an ordinary point for $y'' = (\ln x)y' = 0$.

▌ This differential equation is in standard form, with $P(x) = \ln x$ and $Q(x) = 0$. Since $P(0)$ is undefined, $P(x)$ does not possess a Maclaurin-series expansion and is not analytic at $x = 0$. Accordingly, $x = 0$ is not an ordinary point. In addition, $xP(x)$ is also undefined at $x = 0$, so $x = 0$ is not a regular singular point.

15.44 Determine whether $x = 2$ is an ordinary point for the differential equation of the previous problem.

▌ $P(x)$ does have a Taylor-series expansion around $x = 2$ that converges in an interval centered at $x = 2$ (see Problems 15.14 and 15.15); thus, $P(x)$ is analytic there. $Q(x) = 0$ is analytic everywhere, including at $x = 2$, so $x = 2$ is an ordinary point.

15.45 Determine whether $x = 0$ or $x = 1$ is an ordinary point for the differential equation

$$y'' + \frac{2}{x} y' + \frac{3}{x(x - 1)^3} y = 0.$$

▌ Neither point is an ordinary point: At $x = 0$, both $P(x)$ and $Q(x)$ are undefined; at $x = 1$, although $P(x)$ is analytic, $Q(x)$ is not. All other points are ordinary points.

The point $x = 0$ is a regular singular point, since $xP(x) = 2$ and
$x^2Q(x) = \dfrac{3x}{(x-1)^3} = -3x(1-x)^{-3} = -3x(1 + 3x + 6x^2 + \cdots)$ for $|x| < 1$ are analytic at $x = 0$.
The point $x = 1$ is an irregular singular point, however, because the product $(x-1)^2Q(x) = \dfrac{3}{x(x-1)}$ is undefined at $x = 1$ and hence is not analytic there.

15.46 Determine whether $t = 1$ is an ordinary point for $t^2\ddot{y} + 2\dot{y} + (t^2 - t)y = 0$.

▌ This differential equation has the standard form $\ddot{y} + \dfrac{2}{t^2}\dot{y} + \dfrac{t-1}{t}y = 0$. Since $2/t^2$ and $(t-1)/t$ are each a rational function (that is, a polynomial divided by a polynomial), and since neither denominator is zero at $t = 1$, both functions are analytic there. Thus, $t = 1$ is an ordinary point. (The Taylor-series expansions around $t = 1$ for these functions are derived explicitly in Problems 15.29 and 15.34.)

15.47 Determine whether $x = 0$ is an ordinary point for $(x+1)^2y'' + 2y' + xy = 0$.

▌ This differential equation has the standard form $y'' + \dfrac{2}{(x+1)^2}y' + \dfrac{x}{(x+1)^2}y = 0$. Since both $2/(x+1)^2$ and $x/(x+1)^2$ are rational functions, and since neither denominator is zero at $x = 0$, both functions are analytic there. Thus, $x = 0$ is an ordinary point.

15.48 Determine whether $x = 0$ is an ordinary point for $(x^2 + 2x)y'' + (x+1)y' - y = 0$.

▌ This differential equation has the standard form $y'' + \dfrac{x+1}{x^2+2x}y' - \dfrac{1}{x^2+2x}y = 0$. Since both $P(x) = \dfrac{x+1}{x^2+2x}$ and $Q(x) = \dfrac{-1}{x^2+2x}$ are undefined at $x = 0$, neither is analytic there and $x = 0$ is not an ordinary point.

The products $(x-0)P(x) = \dfrac{x+1}{x+2}$ and $(x-0)^2Q(x) = \dfrac{-x}{x+2}$ are both rational functions with nonzero denominators at $x = 0$. Thus, both products are analytic at $x = 0$, which implies that $x = 0$ is a regular singular point.

15.49 Determine whether $x = -1$ is an ordinary point for the differential equation of the previous problem.

▌ Both $P(x)$ and $Q(x)$ are rational functions with nonzero denominators at $x = -1$. Thus both functions are analytic there, so $x = -1$ is an ordinary point.

15.50 Determine whether $x = 0$ is an ordinary point for $y'' - xy' + 2y = 0$.

▌ Here $P(x) = -x$ and $Q(x) = 2$ are both polynomials and so are analytic everywhere. Therefore, every value of x, in particular $x = 0$, is an ordinary point.

15.51 Determine whether $x = 1$ or $x = 2$ is an ordinary point for $(x^2 - 4)y'' + y = 0$.

▌ We first put the differential equation into standard form by dividing by $x^2 - 4$. Then $P(x) \equiv 0$ and $Q(x) = 1/(x^2 - 4)$. Since both $P(x)$ and $Q(x)$ are analytic at $x = 1$, this point is an ordinary point. At $x = 2$, however, the denominator of $Q(x)$ is zero; hence $Q(x)$ is not analytic there. Thus, $x = 2$ is not an ordinary point but a singular point.

Note that $(x-2)P(x) \equiv 0$ and $(x-2)^2Q(x) = (x-2)/(x+2)$ are analytic at $x = 2$, so that $x = 2$ is a regular singular point.

15.52 Determine whether $x = 0$ is an ordinary point for $2x^2y'' + 7x(x+1)y' - 3y = 0$.

▌ Dividing by $2x^2$ yields $P(x) = 7(x+1)/2x$ and $Q(x) = -3/2x^2$. As neither function is analytic at $x = 0$ (both denominators are zero there), $x = 0$ is not an ordinary point but, rather, a singular point.

Note that $(x-0)P(x) = \frac{7}{2}(x+1)$ and $(x-0)^2Q(x) = -\frac{3}{2}$ are both analytic at $x = 0$; thus, $x = 0$ is a regular singular point.

15.53 Determine whether $x = 0$ is an ordinary point for $x^2y'' + 2y' + xy = 0$.

▌ Here $P(x) = 2/x^2$ and $Q(x) = 1/x$. Neither of these functions is analytic at $x = 0$, so $x = 0$ is not an ordinary point but a singular point. Furthermore, since $(x - 0)P(x) = 2/x$ is not analytic at $x = 0$, $x = 0$ is not a regular singular point either; it is an irregular singular point.

15.54 Determine which points are not ordinary points for the differential equation $(x^2 - 4x + 3)y'' + (x - 2)y' - y = 0$.

▌ This differential equation has the standard form $y'' + \dfrac{x-2}{x^2 - 4x + 3}y' - \dfrac{1}{x^2 - 4x + 3}y = 0$. Both $P(x) = \dfrac{x-2}{x^2 - 4x + 3}$ and $Q(x) = \dfrac{-1}{x^2 - 4x + 3}$ are rational functions with denominators that are zero only when $x = 1$ and $x = 3$. These are then the only two points that are not ordinary points for the given differential equation.

RECURSION FORMULAS

15.55 Find a recursion formula for the coefficients of the general power-series solution near $x = 0$ of $y'' - xy' + 2y = 0$.

▌ It is shown in Problem 15.35 that $x = 0$ is an ordinary point for this differential equation, so we may solve by the power-series method. We assume

$$y = a_0 + a_1x + a_2x^2 + a_3x^3 + a_4x^4 + \cdots + a_nx^n + a_{n+1}x^{n+1} + a_{n+2}x^{n+2} + \cdots \tag{1}$$

Differentiating termwise, we have

$$y' = a_1 + 2a_2x + 3a_3x^2 + 4a_4x^3 + \cdots + na_nx^{n-1} + (n+1)a_{n+1}x^n + (n+2)a_{n+2}x^{n+1} + \cdots \tag{2}$$

$$y'' = 2a_2 + 6a_3x + 12a_4x^2 + \cdots + n(n-1)a_nx^{n-2} + (n+1)(n)a_{n+1}x^{n-1} + (n+2)(n+1)a_{n+2}x^n + \cdots \tag{3}$$

Substituting (*1*), (*2*), and (*3*) into the differential equation, we find

$$\begin{aligned}&[2a_2 + 6a_3x + 12a_4x^2 + \cdots + n(n-1)a_nx^{n-2} + (n+1)(n)a_{n+1}x^{n-1} + (n+2)(n+1)a_{n+2}x^n + \cdots]\\ &\quad - x[a_1 + 2a_2x + 3a_3x^2 + 4a_4x^3 + \cdots + na_nx^{n-1} + (n+1)a_{n+1}x^n + (n+2)a_{n+2}x^{n+1} + \cdots]\\ &\quad + 2[a_0 + a_1x + a_2x^2 + a_3x^3 + a_4x^4 + \cdots + a_nx^n + a_{n+1}x^{n+1} + a_{n+2}x^{n+2} + \cdots] = 0\end{aligned}$$

Combining terms that contain like powers of x yields

$$\begin{aligned}&(2a_2 + 2a_0) + x(6a_3 + a_1) + x^2(12a_4) + x^3(20a_5 - a_3) + \cdots + x^n[(n+2)(n+1)a_{n+2} - na_n + 2a_n] + \cdots\\ &\quad = 0 + 0x + 0x^2 + 0x^3 + \cdots + 0x^n + \cdots\end{aligned}$$

This last equation holds if and only if each coefficient in the left-hand side is zero. Thus,

$$2a_2 + 2a_0 = 0 \qquad 6a_3 + a_1 = 0 \qquad 12a_4 = 0 \qquad 20a_5 - a_3 = 0 \qquad \cdots$$

and, in general, $(n+2)(n+1)a_{n+2} - (n-2)a_n = 0$, or $a_{n+2} = \dfrac{n-2}{(n+2)(n+1)}a_n$.

15.56 Find a recursion formula for the coefficients of the general power-series solution near $x = 0$ of $y'' + y = 0$.

▌ Since this equation has constant coefficients, every point is an ordinary point. Equations (*1*) and (*3*) of Problem 15.55 are appropriate here; substituting them into the differential equation, we obtain

$$\begin{aligned}&[2a_2 + 6a_3x + 12a_4x^2 + \cdots + n(n-1)a_nx^{n-2} + (n+1)na_{n+1}x^{n-1} + (n+2)(n+1)a_{n+2}x^n + \cdots]\\ &\quad + (a_0 + a_1x + a_2x^2 + a_3x^3 + a_4x^4 + \cdots + a_nx^n + a_{n+1}x^{n+1} + a_{n+2}x^{n+2} + \cdots) = 0\end{aligned}$$

or

$$\begin{aligned}&(2a_2 + a_0) + x(6a_3 + a_1) + x^2(12a_4 + a_2) + x^3(20a_5 + a_3) + \cdots + x^n[(n+2)(n+1)a_{n+2} + a_n] + \cdots\\ &\quad = 0 + 0x + 0x^2 + \cdots + 0x^n + \cdots\end{aligned}$$

Equating each coefficient to zero yields

$$2a_2 + a_0 = 0 \qquad 6a_3 + a_1 = 0 \qquad 12a_4 + a_2 = 0 \qquad 20a_5 + a_3 = 0 \qquad \cdots$$

and, in general, $(n+2)(n+1)a_{n+2} + a_n = 0$, which is equivalent to $a_{n+2} = \dfrac{-1}{(n+2)(n+1)}a_n$.

15.57 Find a recursion formula for the coefficients of the general power-series solution near $x = 0$ of $(x^2 + 4)y'' + xy = 0$.

It is shown in Problem 15.38 that $x = 0$ is an ordinary point for this differential equation. Substituting (*1*) and (*3*) of Problem 15.55 into the differential equation, we have

$$(x^2+4)[2a_2+6a_3x+12a_4x^2+\cdots+n(n-1)a_nx^{n-2}+(n+1)na_{n+1}x^{n-1}+(n+2)(n+1)a_{n+2}x^n+\cdots]$$
$$+x[a_0+a_1x+a_2x^2+a_3x^3+\cdots+a_{n-1}x^{n-1}+\cdots]=0$$

or
$$(8a_2)+x(24a_3+a_0)+x^2(2a_2+48a_4+a_1)+x^3(6a_3+80a_5+a_2)+\cdots$$
$$+x^n[n(n-1)a_n+4(n+2)(n+1)a_{n+2}+a_{n-1}]+\cdots=0+0x+0x^2+0x^3+\cdots+0x^n+\cdots$$

Equating coefficients of like powers of x yields

$$8a_2=0 \qquad 24a_3+a_0=0 \qquad 2a_2+48a_4+a_1=0 \qquad 6a_3+80a_5+a_2=0 \qquad \cdots$$

and, in general, $n(n-1)a_n+4(n+2)(n+1)a_{n+2}+a_{n-1}=0$, which is equivalent to
$a_{n+2}=\dfrac{-n(n-1)}{4(n+2)(n+1)}a_n-\dfrac{1}{4(n+2)(n+1)}a_{n-1}.$

15.58 Find a recursion formula for the coefficients of the general power-series solution near $x = 0$ of $(x^2-1)y''+xy'-y=0$.

It is shown in Problem 15.31 that $x = 0$ is an ordinary point for this differential equation. Substituting (*1*) through (*3*) of Problem 15.55 into the differential equation, we have

$$(x^2-1)[2a_2+6a_3x+12a_4x^2+\cdots+n(n-1)a_nx^{n-2}+(n+1)na_{n+1}x^{n-1}+(n+2)(n+1)a_{n+2}x^n+\cdots]$$
$$+x[a_1+2a_2x+3a_3x^2+4a_4x^3+\cdots+na_nx^{n-1}+(n+1)a_{n+1}x^n+(n+2)a_{n+2}x^{n+1}+\cdots]$$
$$-[a_0+a_1x+a_2x^2+a_3x^3+a_4x^4+\cdots+a_nx^n+a_{n+1}x^{n+1}+a_{n+2}x^{n+2}+\cdots]=0$$

or, after combining terms and simplifying,

$$(-2a_2-a_0)+x(-6a_3)+x^2(3a_2-12a_4)+x^3(8a_3-20a_5)+\cdots+x^n[(n+1)(n-1)a_n-(n+2)(n+1)a_{n+2}]+\cdots$$
$$=0+0x+0x^2+0x^3+\cdots+0x^n+\cdots$$

Equating coefficients of like powers of x yields

$$-2a_2-a_0=0 \qquad -6a_3=0 \qquad 3a_2-12a_4=0 \qquad 8a_3-20a_5=0 \qquad \cdots$$

and, in general, $(n+1)(n-1)a_n-(n+2)(n+1)a_{n+2}=0$, which is equivalent to $a_{n+2}=\dfrac{n-1}{n+2}a_n$.

15.59 Find a recursion formula for the coefficients of the general power-series solution near $x = 0$ of $y''-xy=0$.

It is shown in Problem 15.36 that $x = 0$ is an ordinary point for this differential equation. Substituting (*1*) and (*3*) of Problem 15.55 into this equation, we have

$$[2a_2+6a_3x+12a_4x^2+\cdots+n(n-1)a_nx^{n-2}+(n+1)na_{n+1}x^{n-1}+(n+2)(n+1)a_{n+2}x^n+\cdots]$$
$$-x[a_0+a_1x+a_2x^2+a_3x^3+a_4x^4+\cdots+a_{n-1}x^{n-1}+a_nx^n+a_{n+1}x^{n+1}+a_{n+2}x^{n+2}+\cdots]=0$$

or
$$(2a_2)+x(6a_3-a_0)+x^2(12a_4-a_1)+\cdots+x^n[(n+2)(n+1)a_{n+2}-a_{n-1}]+\cdots$$
$$=0+0x+0x^2+0x^3+\cdots+0x^n+\cdots$$

Equating coefficients of like powers of x yields

$$2a_2=0 \qquad 6a_3-a_0=0 \qquad 12a_4-a_1=0 \qquad \cdots$$

or, in general, $(n+2)(n+1)a_{n+2}-a_{n-1}=0$, which is equivalent to $a_{n+2}=\dfrac{1}{(n+2)(n+1)}a_{n-1}.$

15.60 Rework Problem 15.55 in summation notation.

Here we assume $y=\sum_{n=0}^{\infty}a_nx^n$. Differentiating termwise, we then obtain $y'=\sum_{n=0}^{\infty}na_nx^{n-1}$ and $y''=\sum_{n=0}^{\infty}n(n-1)a_nx^{n-2}$. Substituting these expressions into the differential equation $y''-xy'+2y=0$ yields

$$\sum_{n=0}^{\infty}n(n-1)a_nx^{n-2}-x\sum_{n=0}^{\infty}na_nx^{n-1}+2\sum_{n=0}^{\infty}a_nx^n=0$$

and
$$\sum_{n=0}^{\infty}n(n-1)a_nx^{n-2}-\sum_{n=0}^{\infty}na_nx^n+2\sum_{n=0}^{\infty}a_nx^n=0 \qquad (1)$$

But because the first two terms of the first sum are zero,

$$\sum_{n=0}^{\infty} n(n-1)a_n x^{n-2} = \sum_{n=2}^{\infty} n(n-1)a_n x^{n-2} = \sum_{k=0}^{\infty} (k+2)(k+1)a_{k+2}x^k = \sum_{n=0}^{\infty} (n+2)(n+1)a_{n+2}x^n \qquad (2)$$

where we have set $k = n - 2$ to obtain the third term, and replaced k with n to obtain the last term. Substituting this result into (*1*) and combining like powers of x, we have

$$\sum_{n=0}^{\infty} [(n+2)(n+1)a_{n+2} - na_n + 2a_n]x^n = 0 = \sum_{n=0}^{\infty} 0x^n$$

Equating the coefficients of x^n yields $(n+2)(n+1)a_{n+2} - na_n + 2a_n = 0$, from which we conclude that $a_{n+2} = \dfrac{n-2}{(n+2)(n+1)}a_n$.

15.61 Rework Problem 15.56 in summation notation.

▮ Substituting the expressions for y and y'' of Problem 15.60 into the differential equation $y'' + y = 0$, we get

$$\sum_{n=0}^{\infty} n(n-1)a_n x^{n-2} + \sum_{n=0}^{\infty} a_n x^n = 0 \qquad (1)$$

Since the powers of x in these two series are not identical, we must relabel the summation index in one of them. Using (*2*) of the previous problem, we can rewrite (*1*) as $\sum_{n=0}^{\infty} (n+2)(n+1)a_{n+2}x^n + \sum_{n=0}^{\infty} a_n x^n = 0$. The summations may now be combined to yield

$$\sum_{n=0}^{\infty} [(n+2)(n+1)a_{n+2} + a_n]x^n = 0 = \sum_{n=0}^{\infty} 0x^n$$

Equating coefficients of x^n, we get $(n+2)(n+1)a_{n+2} + a_n = 0$, or $a_{n+2} = \dfrac{-1}{(n+2)(n+1)}a_n$.

15.62 Rework Problem 15.58 in summation notation.

▮ Substituting the expressions for y, y', and y'' of Problem 15.60 into the differential equation $(x^2 - 1)y'' + xy' - y = 0$, we get

$$(x^2 - 1)\sum_{n=0}^{\infty} n(n-1)a_n x^{n-2} + x\sum_{n=0}^{\infty} na_n x^{n-1} - \sum_{n=0}^{\infty} a_n x^n = 0$$

or

$$\sum_{n=0}^{\infty} n(n-1)a_n x^n - \sum_{n=0}^{\infty} n(n-1)a_n x^{n-2} + \sum_{n=0}^{\infty} na_n x^n - \sum_{n=0}^{\infty} a_n x^n = 0 \qquad (1)$$

To combine the four summations, we must relabel the dummy index on the second summation so that it too contains x^n. This is done with (*2*) of Problem 15.50. Then (*1*) becomes

$$\sum_{n=0}^{\infty} n(n-1)a_n x^n - \sum_{n=0}^{\infty} (n+2)(n+1)a_{n+2}x^n + \sum_{n=0}^{\infty} na_n x^n - \sum_{n=0}^{\infty} a_n x^n = 0$$

which may be combined into

$$\sum_{n=0}^{\infty} [n(n-1)a_n + na_n - a_n - (n+2)(n+1)a_{n+2}]x^n = 0 = \sum_{n=0}^{\infty} 0x^n$$

Equating the coefficients of x^n and then simplifying the result, we obtain $(n-1)(n+1)a_n - (n+2)(n+1)a_{n+2} = 0$ or $a_{n+2} = \dfrac{n-1}{n+2}a_n$.

15.63 Rework Problem 15.57 in summation notation.

▮ Substituting the expressions for y and y'' of Problem 15.60 into the differential equation $(x^2 + 4)y'' + xy = 0$ yields $(x^2 + 4)\sum_{n=0}^{\infty} n(n-1)a_n x^{n-2} + x\sum_{n=0}^{\infty} a_n x^n = 0$ or

$$\sum_{n=0}^{\infty} n(n-1)a_n x^n + 4\sum_{n=0}^{\infty} n(n-1)a_n x^{n-2} + \sum_{n=0}^{\infty} a_n x^{n+1} = 0 \qquad (1$$

Each summation contains a different power of x, so they cannot be combined in their current forms. We shall relabel the dummy indices in the last two summations, so that each will contain x^n.

It follows from (2) of Problem 15.60 that $\sum_{n=0}^{\infty} n(n-1)a_n x^{n-2} = \sum_{n=0}^{\infty} (n+2)(n+1)a_{n+2}x^n$. In addition, by setting $k = n + 1$ in the last summation of (1), then replacing k with n, and finally incorporating the coefficient $a_{-1} = 0$, we obtain

$$\sum_{n=0}^{\infty} a_n x^{n+1} = \sum_{k=1}^{\infty} a_{k-1}x^k = \sum_{n=1}^{\infty} a_{n-1}x^n = \sum_{n=0}^{\infty} a_{n-1}x^n \tag{2}$$

Then (1) may be rewritten as

$$\sum_{n=0}^{\infty} n(n-1)a_n x^n + 4\sum_{n=0}^{\infty} (n+2)(n+1)a_{n+2}x^n + \sum_{n=0}^{\infty} a_{n-1}x^n = 0$$

which may be combined into

$$\sum_{n=0}^{\infty} [n(n-1)a_n + 4(n+2)(n+1)a_{n+2} + a_{n-1}]x^n = 0 = \sum_{n=0}^{\infty} 0x^n$$

Equating the coefficients of x^n yields $n(n-1)a_n + 4(n+2)(n+1)a_{n+2} + a_{n-1} = 0$, or $a_{n+2} = \dfrac{-n(n-1)}{4(n+2)(n+1)}a_n - \dfrac{1}{4(n+2)(n+1)}a_{n-1}$.

15.64 Rework Problem 15.59 in summation notation.

▮ Substituting the expressions for y and y'' of Problem 15.60 into the differential equation $y'' - xy = 0$, we get $\sum_{n=0}^{\infty} n(n-1)a_n x^{n-2} - x\sum_{n=0}^{\infty} a_n x^n = 0$, or

$$\sum_{n=0}^{\infty} n(n-1)a_n x^{n-2} - \sum_{n=0}^{\infty} a_n x^{n+1} = 0$$

To rewrite each of these summations in terms of x^n, we use (2) of Problem 15.60 for the first summation, and (2) of the previous problem for the second, obtaining $\sum_{n=0}^{\infty} (n+2)(n+1)a_{n+2}x^n - \sum_{n=0}^{\infty} a_{n-1}x^n = 0$ or

$$\sum_{n=0}^{\infty} [(n+2)(n+1)a_{n+2} - a_{n-1}]x^n = \sum_{n=0}^{\infty} 0x^n$$

Equating the coefficients of x^n yields $(n+2)(n+1)a_{n+2} - a_{n-1} = 0$, or $a_{n+2} = \dfrac{1}{(n+2)(n+1)}a_{n-1}$.

15.65 Find a recursion formula for the coefficients of the general power-series solution near $x = 0$ of $(x+1)^2y'' + 2y' + xy = 0$.

▮ It is shown in Problem 15.47 that $x = 0$ is an ordinary point for this equation. We assume that $y = \sum_{n=0}^{\infty} a_n x^n$, from which we obtain $y' = \sum_{n=0}^{\infty} na_n x^{n-1}$ and $y'' = \sum_{n=0}^{\infty} n(n-1)a_n x^{n-2}$. Substituting these quantities into the differential equation and writing $(x+1)^2 = x^2 + 2x + 1$ yield

$$(x^2 + 2x + 1)\sum_{n=0}^{\infty} n(n-1)a_n x^{n-2} + 2\sum_{n=0}^{\infty} na_n x^{n-1} + x\sum_{n=0}^{\infty} a_n x^n = 0$$

or
$$\sum_{n=0}^{\infty} n(n-1)a_n x^n + \sum_{n=0}^{\infty} 2n(n-1)a_n x^{n-1} + \sum_{n=0}^{\infty} n(n-1)a_n x^{n-2} + \sum_{n=0}^{\infty} 2na_n x^{n-1} + \sum_{n=0}^{\infty} a_n x^{n+1} = 0 \tag{1}$$

The second and fourth summations may be combined into $\sum_{n=0}^{\infty} 2n^2 a_n x^{n-1} = \sum_{n=1}^{\infty} 2n^2 a_n x^{n-1}$. If we set $k = n - 1$ and then replace k with n, this becomes first $\sum_{k=0}^{\infty} 2(k+1)^2 a_{k+1}x^k$ and then $\sum_{n=0}^{\infty} 2(n+1)^2 a_{n+1}x^n$.

Using (2) of Problems 15.60 and 15.63, we rewrite the third summation in (1) as $\sum_{n=0}^{\infty} (n+2)(n+1)a_{n+2}x^n$ and the last summation in (1) as $\sum_{n=0}^{\infty} a_{n-1}x^n$. Substituting these results into (1) and simplifying, we get

$$\sum_{n=0}^{\infty} [n(n-1)a_n + 2(n+1)^2 a_{n+1} + (n+2)(n+1)a_{n+2} + a_{n-1}]x^n = 0$$

Equating the coefficients of like powers of x, we conclude that
$n(n-1)a_n + 2(n+1)^2 a_{n+1} + (n+2)(n+1)a_{n+2} + a_{n-1} = 0$, which yields

$$a_{n+2} = -\frac{2(n+1)^2 a_{n+1} + n(n-1)a_n + a_{n-1}}{(n+2)(n+1)}.$$

15.66 Find a recursion formula for the coefficients of the general power-series solution near $t=0$ of $\ddot{y} - y = 0$.

▮ Since this equation has constant coefficients, every point is an ordinary point. We assume a solution of the form $y(t) = \sum_{n=0}^{\infty} a_n t^n$, from which we obtain $\dot{y} = \sum_{n=0}^{\infty} n a_n t^{n-1}$ and $\ddot{y} = \sum_{n=0}^{\infty} n(n-1)a_n t^{n-2}$. Substitution in the differential equation then yields

$$\sum_{n=0}^{\infty} n(n-1)a_n t^{n-2} - \sum_{n=0}^{\infty} a_n t^n = 0$$

Since the powers of t in these two series are not the same, we must relabel the summation index so as to combine the two series into one. Replacing n by $n-2$ in the second series, we get $\sum_{n=0}^{\infty} n(n-1)a_n t^{n-2} - \sum_{n=2}^{\infty} a_{n-2} t^{n-2} = 0$. The second summation now begins with $n=2$. However, since the first two terms in the first series are zero, we can add the two series term by term to obtain $\sum_{n=2}^{\infty} [n(n-1)a_n - a_{n-2}]t^{n-2} = 0$. Since this power series is the zero function, each of the coefficients must be zero, and it follows that $n(n-1)a_n - a_{n-2} = 0$ for $n = 2, 3, 4, \ldots$. Hence $a_n = \frac{a_{n-2}}{n(n-1)}$ for $n = 2, 3, 4, \ldots$.

15.67 Find a recursion formula for the coefficients of the general power-series solution near $t=0$ of $\ddot{y} + ty = 0$.

▮ This equation is similar to the one considered in Problem 15.36, and for the reason given there $t=0$ is an ordinary point. We assume a general solution of the form $y(t) = \sum_{k=0}^{\infty} b_k t^k$. Differentiating twice and substituting into the equation give

$$0 = \ddot{y} + ty = \sum_{k=0}^{\infty} k(k-1)b_k t^{k-2} + t\sum_{k=0}^{\infty} b_k t^k = \sum_{k=2}^{\infty} k(k-1)b_k t^{k-2} + \sum_{k=0}^{\infty} b_k t^{k+1}$$

$$= \sum_{k=0}^{\infty} (k+2)(k+1)b_{k+2}t^k + \sum_{k=1}^{\infty} b_{k-1}t^k = 2b_2 t^0 + \sum_{k=1}^{\infty} [(k+2)(k+1)b_{k+2} + b_{k-1}]t^k$$

It follows that $2b_2 = 0$, and $(k+2)(k+1)b_{k+2} + b_{k-1} = 0$ for $k \geq 1$. This last equation is also valid for $k=0$ because $b_{-1} = 0$. Thus, $b_{k+2} = \frac{-1}{(k+2)(k+1)} b_{k-1}$.

15.68 Find a recursion formula for the coefficients of the general power-series solution near $t=0$ of $\ddot{y} - 2t\dot{y} - 2y = 0$.

▮ This differential equation is in the standard form $\ddot{y} + P(t)\dot{y} + Q(t)y = 0$, with $P(t) = -2t$ and $Q(t) = -2$. Since both $P(t)$ and $Q(t)$ are analytic everywhere, every point and, in particular, $t=0$, is an ordinary point. We assume a general solution $y = \sum_{n=0}^{\infty} a_n t^n$. Substituting it and its derivatives into the left side of the differential equation and simplifying, we get

$$\sum_{n=0}^{\infty} n(n-1)a_n t^{n-2} - 2t\sum_{n=0}^{\infty} n a_n t^{n-1} - 2\sum_{n=0}^{\infty} a_n t^n$$

$$= \sum_{n=2}^{\infty} n(n-1)a_n t^{n-2} - \sum_{n=0}^{\infty} 2n a_n t^n - \sum_{n=0}^{\infty} 2a_n t^n$$

$$= \sum_{n=2}^{\infty} n(n-1)a_n t^{n-2} - \sum_{n=2}^{\infty} 2(n-2)a_{n-2}t^{n-2} - \sum_{n=2}^{\infty} 2a_{n-2}t^{n-2}$$

$$= \sum_{n=2}^{\infty} \{n(n-1)a_n - [2(n-2)+2]a_{n-2}\}t^{n-2} = \sum_{n=2}^{\infty} [n(n-1)a_n - 2(n-1)a_{n-2}]t^{n-2}$$

Note that, following the second equal sign, the summation index of two of the series was changed by replacing n with $n-2$. This was done to obtain t^{n-2} in all three series so that the addition could be performed term by term. Since the coefficient of t^{n-2} must be zero for all $n \geq 2$, we have $n(n-1)a_n - 2(n-1)a_{n-2} = 0$ for all $n \geq 2$. Canceling $n-1$ from each term yields the recursion formula $a_n = \frac{2}{n} a_{n-2}$ for $n = 2, 3, \ldots$.

15.69 Find another form for the recursion formula obtained in the previous problem.

▌ If we set $k = n - 2$, then the recursion formula in the previous problem becomes $a_{k+2} = \frac{2}{k+2} a_k$ for $k = 0, 1, 2, \ldots$. Now if we replace the dummy index k with n, we generate $a_{n+2} = \frac{2}{n+2} a_n$ for $n = 0, 1, 2, \ldots$ as a second form of the recursion formula in the index n.

15.70 Find a recursion formula for the coefficients of the general power-series solution near $x = 0$ of $y'' + x^2y' + 2xy = 0$.

▌ This equation is in the standard form $y'' + P(x)y' + Q(x)y = 0$, with $P(x) = x^2$ and $Q(x) = 2x$. Since both $P(x)$ and $Q(x)$ are polynomials, they are analytic everywhere; thus, every point including $x = 0$ is an ordinary point. We assume a solution of the form $y = \sum_{n=0}^{\infty} a_n x^n$ and substitute it and its derivatives into the given equation, getting

$$[(2)(1a_2) + (3)(2a_3x) + (4)(3a_4x^2) + \cdots] + x^2(a_1 + 2a_2x + \cdots) + 2x(a_0 + a_1x + a_2x^2 + \cdots) = 0 \qquad (1)$$

The terms common to the three series in (1) are those which contain the second and higher powers of x. Hence it is convenient to write (1) in the form

$$\left[2a_2 + 6a_3x + \sum_{m=0}^{\infty}(m+4)(m+3)a_{m+4}x^{m-2}\right] + \sum_{m=0}^{\infty}(m+1)a_{m+1}x^{m+2} + \left(2a_0x + \sum_{m=0}^{\infty} 2a_{m+1}x^{m+2}\right) = 0$$

and then combine the three sums into one, getting

$$2a_2 + (6a_3 + 2a_0)x + \sum_{m=0}^{\infty}[(m+4)(m+3)a_{m+4} + (m+3)a_{m+1}]x^{m+2} = 0$$

Setting $n = m + 2$, we may rewrite this last equation as

$$2a_2 + (6a_3 + 2a_0)x + \sum_{n=2}^{\infty}[(n+2)(n+1)a_{n+2} + (n+1)a_{n-1}]x^n = 0$$

or, more simply, as $\sum_{n=0}^{\infty}[(n+2)(n+1)a_{n+2} + (n+1)a_{n-1}]x^n = 0$, since $a_{-1} = 0$. Equating coefficients of like powers of x, we have $(n+2)(n+1)a_{n+2} + (n+1)a_{n-1} = 0$, or $a_{n+2} = \frac{-1}{n+2} a_{n-1}$ for $n = 0, 1, 2, \ldots$.

15.71 Find a second form of the recursion formula obtained in the previous problem.

▌ If we set $k = n - 1$, the recursion formula in the previous problem becomes $a_{k+3} = \frac{-1}{k+3} a_k$ for $k = -1, 0, 1, 2, \ldots$. By replacing the dummy index k by n, we generate $a_{n+3} = \frac{-1}{n+3} a_n$ for $n = -1, 0, 1, 2, \ldots$ as a second form of the recursion formula in the index n.

15.72 Find the recursion formula for the coefficients of the general power-series solution near $x = 0$ of $y'' - x^2y' - y = 0$.

▌ This differential equation is in standard form, with $P(x) = -x^2$ and $Q(x) = -1$. Since both these functions are their own Maclaurin-series expansions, every point, including $x = 0$, is an ordinary point. We assume that $y = A_0 + A_1x + A_2x^2 + A_3x^3 + \cdots + A_nx^n + \cdots$. Then

$$y'' - x^2y' - y = 0 = (2A_2 - A_0) + (6A_3 - A_1)x + (12A_4 - A_1 - A_2)x^2 + (20A_5 - 2A_2 - A_3)x^3 + \cdots$$
$$+ [(n+2)(n+1)A_{n+2} - (n-1)A_{n-1} - A_n]x^n + \cdots$$

Equating to zero the coefficient of each power of x yields $2A_2 - A_0 = 0$ or $A_2 = \frac{1}{2}A_0$; $6A_3 - A_1 = 0$ or $A_3 = \frac{1}{6}A_1$; $12A_4 - A_1 - A_2 = 0$ or $A_4 = \frac{1}{24}A_0 + \frac{1}{12}A_1$; and so on. Then $(n+2)(n+1)A_{n+2} - (n-1)A_{n-1} - A_n = 0$ and $A_{n+2} = \dfrac{(n-1)A_{n-1} + A_n}{(n+1)(n+2)}$.

15.73 Find a recursion formula for the coefficients of the general power-series solution near $x = 0$ of $(1 + x^2)y'' + xy' - y = 0$.

▌ We divide the differential equation by $1 + x^2$ to put it in standard form, with $P(x) = x/(1 + x^2)$ and $Q(x) = -1/(1 + x^2)$. These functions are similar to the functions considered in Problem 15.31, and for analogous reasons they are analytic at $x = 0$. Thus, $x = 0$ is an ordinary point. We assume that $y = A_0 + A_1x + A_2x^2 + A_3x^3 + A_4x^4 + \cdots + A_nx^n + \cdots$. Then substituting y and its derivatives into the given differential equation yields

$$(1 + x^2)[2A_2 + 6A_3x + 12A_4x^2 + \cdots + n(n-1)A_nx^{n-2} + \cdots] + x(A_1 + 2A_2x + 3A_3x^2 + 4A_4x^3 + \cdots + nA_nx^{n-1} + \cdots) - (A_0 + A_1x + A_2x^2 + A_3x^3 + A_4x^4 + \cdots + A_nx^n + \cdots) = 0$$

or
$$(2A_2 - A_0) + 6A_3x + (12A_4 + 3A_2)x^2 + \cdots + [(n+2)(n+1)A_{n+2} + (n^2 - 1)A_n]x^n + \cdots = 0$$

Equating to zero the coefficient of each power of x yields $2A_2 - A_0 = 0$ or $A_2 = \frac{1}{2}A_0$; $6A_3 = 0$ or $A_3 = 0$; $12A_4 + 3A_2 = 0$ or $A_4 = -\frac{1}{8}A_0$; and so on. Thus $(n+2)(n+1)A_{n+2} + (n^2 - 1)A_n = 0$ and $A_{n+2} = -\dfrac{n-1}{n+2}A_n$.

15.74 Find a recursion formula for the coefficients of the general power-series solution near $t = 0$ of $\dfrac{d^2y}{dt^2} + (t - 1)\dfrac{dy}{dt} + (2t - 3)y = 0$.

▌ Since $P(t) = t - 1$ and $Q(t) = 2t - 3$ are polynomials, both are analytic everywhere; so every point, and in particular $t = 0$, is an ordinary point. We assume $y = a_0 + a_1t + a_2t^2 + a_3t^3 + \cdots + a_nt^n + a_{n+1}t^{n+1} + a_{n+2}t^{n+2} + \cdots$. Substituting it and its first two derivatives into the given differential equation yields

$$[2a_2 + 6a_3t + 12a_4t^2 + \cdots + n(n-1)a_nt^{n-2} + (n+1)na_{n+1}t^{n-1} + (n+2)(n+1)a_{n+2}t^n + \cdots] + (t-1)[a_1 + 2a_2t + 3a_3t^2 + 4a_4t^3 + \cdots + na_nt^{n-1} + (n+1)a_{n+1}t^n + (n+2)a_{n+2}t^{n+1} + \cdots] + (2t - 3)[a_0 + a_1t + a_2t^2 + a_3t^3 + a_4t^4 + \cdots + a_nt^n + a_{n+1}t^{n+1} + a_{n+2}t^{n+2} + \cdots] = 0$$

or
$$(2a_2 - a_1 - 3a_0) + t(6a_3 + a_1 - 2a_2 + 2a_0 - 3a_1) + t^2(12a_4 + 2a_2 - 3a_3 + 2a_1 - 3a_2) + \cdots + t^n[(n+2)(n+1)a_{n+2} + na_n - (n+1)a_{n+1} + 2a_{n-1} - 3a_n] + \cdots = 0 + 0t + 0t^2 + \cdots + 0t^n + \cdots$$

Equating each coefficient to zero, we obtain

$$2a_2 - a_1 - 3a_0 = 0 \qquad 6a_3 - 2a_2 - 2a_1 + 2a_0 = 0 \qquad 12a_4 - 3a_3 - a_2 + 2a_1 = 0 \qquad \cdots$$

In general, then, $(n+2)(n+1)a_{n+2} - (n+1)a_{n+1} + (n-3)a_n + 2a_{n-1} = 0$, which is equivalent to $a_{n+2} = \dfrac{1}{n+2}a_{n+1} - \dfrac{n-3}{(n+2)(n+1)}a_n - \dfrac{2}{(n+2)(n+1)}a_{n-1}$.

15.75 Find a recursion formula for the coefficients of the general power-series solution near $t = 0$ of $\ddot{y} + 2t^2y = 0$.

▌ This equation is in standard form with $P(t) = 0$ and $Q(t) = 2t^2$. Since both functions are their own Maclaurin series, both are analytic and $t = 0$ is an ordinary point. We assume $y = a_0 + a_1t + a_2t^2 + a_3t^3 + \cdots + a_nt^n + a_{n+1}t^{n+1} + a_{n+2}t^{n+2} + \cdots$. Substituting it and its first two derivatives into the given differential equation yields

$$[2a_2 + 6a_3t + 12a_4t^2 + \cdots + n(n-1)a_nt^{n-2} + (n+1)na_{n+1}t^{n-1} + (n+2)(n+1)a_{n+2}t^n + \cdots] + 2t^2(a_0 + a_1t + a_2t^2 + \cdots + a_nt^n + a_{n+1}t^{n+1} + a_{n+2}t^{n+2} + \cdots) = 0$$

The two terms directly preceding a_nt^n in the power-series expansion for y are $a_{n-1}t^{n-1}$ and $a_{n-2}t^{n-2}$. Including them in the last summation and then combining coefficients of like powers of t, we obtain

$$(2a_2) + (6a_3)t + (12a_4 + 2a_0)t^2 + \cdots + [(n+2)(n+1)a_{n+2} + 2a_{n-2}]t^n + \cdots = 0$$

Therefore, $2a_2 = 0$, $6a_3 = 0$, $12a_4 + 2a_0 = 0$, and, in general, $(n+2)(n+1)a_{n+2} + 2a_{n-2} = 0$ or $a_{n+2} = \dfrac{-2}{(n+2)(n+1)} a_{n-2}$.

15.76 Find a recursion formula for the coefficients of the general power-series solution near $v = 0$ of $\dfrac{d^2y}{dv^2} + (v+1)\dfrac{dy}{dv} + y = 0$.

▍ Here $v = 0$ is an ordinary point, so we assume $y = A_0 + A_1 v + A_2 v^2 + A_3 v^3 + A_4 v^4 + \cdots + A_n v^n + \cdots$. Then

$$\frac{d^2y}{dv^2} + (v+1)\frac{dy}{dv} + y = (2A_2 + A_1 + A_0) + (6A_3 + 2A_1 + 2A_2)v + (12A_4 + 3A_2 + 3A_3)v^2 + \cdots$$
$$+ [(n+2)(n+1)A_{n+2} + (n+1)A_n + (n+1)A_{n+1}]v^n + \cdots = 0$$

Equating the coefficients of powers of v to zero, we obtain

$$A_2 = -\tfrac{1}{2}(A_0 + A_1) \qquad A_3 = -\tfrac{1}{3}(A_1 + A_2) = \tfrac{1}{6}(A_0 - A_1) \qquad A_4 = -\tfrac{1}{4}(A_2 + A_3) = \tfrac{1}{12}(A_0 + 2A_1) \qquad \cdots$$

or, in general, $(n+2)(n+1)A_{n+2} + (n+1)A_n + (n+1)A_{n+1} = 0$, and $A_{n+2} = -\dfrac{1}{n+2}(A_n + A_{n+1})$.

15.77 Find a recursion formula for the coefficients of the general power-series solution near $t = 0$ of $\dfrac{d^2y}{dt^2} + (t+1)\dfrac{dy}{dt} + 2y = 0$.

▍ This differential equation is in standard form with $P(t) = t + 1$ and $Q(t) = 2$. Since both functions are analytic everywhere, every point including $t = 0$ is an ordinary point. We assume $y = a_0 + a_1 t + a_2 t^2 + a_3 t^3 + \cdots + a_n t^n + a_{n+1}t^{n+1} + a_{n+2}t^{n+2} + \cdots$. Substituting this expression and its first two derivatives into the differential equation, we obtain

$$[2a_2 + 6a_3 t + \cdots + n(n-1)t^{n-2} + (n+1)(n)a_{n+1}t^{n-1} + (n+2)(n+1)a_{n+2}t^n + \cdots]$$
$$+ (t+1)[a_1 + 2a_2 t + 3a_3 t^2 + \cdots + na_n t^{n-1} + (n+1)a_{n+1}t^n + (n+2)a_{n+2}t^{n+1} + \cdots]$$
$$+ 2[a_0 + a_1 t + a_2 t^2 + a_3 t^3 + \cdots + a_n t^n + \cdots] = 0$$

Combining terms that contain like powers of t yields

$$(2a_2 + a_1 + 2a_0) + t(6a_3 + a_1 + 2a_2 + 2a_1) + \cdots + t^n[(n+2)(n+1)a_{n+2} + na_n + (n+1)a_{n+1} + 2a_n] + \cdots = 0$$

Setting the coefficients of powers of t equal to zero, we obtain $2a_2 + a_1 + 2a_0 = 0$, $6a_3 + 2a_2 + 3a_1 = 0$, and so on, and in general, $(n+2)(n+1)a_{n+2} + (n+1)a_{n+1} + (n+2)a_n = 0$. It follows that $a_{n+2} = \dfrac{-1}{n+2}a_{n+1} - \dfrac{1}{n+1}a_n$.

15.78 Find a recursion formula for the coefficients of the general power-series solution near $t = 0$ of $\dfrac{d^2y}{dt^2} - (t+1)\dfrac{dy}{dt} + 2y = 0$.

▍ With the exception of a single sign, this problem is identical to the previous one. The expression for y in that problem is valid here; when we substitute it into this differential equation and combine terms containing like powers of t, we get

$$(2a_2 - a_1 + 2a_0) + t(6a_3 - a_1 - 2a_2 + 2a_1) + \cdots + t^n[(n+2)(n+1)a_{n+2} - na_n - (n+1)a_{n+1} + 2a_n] + \cdots = 0$$

Setting the coefficients of powers of t equal to zero, we obtain $2a_2 - a_1 + 2a_0 = 0$, $6a_3 - 2a_2 + a_1 = 0$ and so on, and in general, $(n+2)(n+1)a_{n+2} - (n+1)a_{n+1} - (n-2)a_n = 0$. It now follows that $a_{n+2} = \dfrac{1}{n+2}a_{n+1} + \dfrac{n-2}{(n+2)(n+1)}a_n$.

15.79 Find a recursion formula for the coefficients of the general power-series solution near $t = 1$ of $t^2y'' + 2y' + (t-1)y = 0$.

▮ For reasons identical to those given in Problem 15.46, $t = 1$ is an ordinary point for this differential equation. We assume $y(t) = \sum_{n=0}^{\infty} a_n(t-1)^n$, and since $t^2 = (t-1)^2 + 2t - 1 = (t-1)^2 + 2(t-1) + 1$, the differential equation can be written in the form $[(t-1)^2 + 2(t-1) + 1]y'' + 2y' + (t-1)y = 0$. Computing y' and y'' and substituting into this equation yield

$$\sum_{n=2}^{\infty} n(n-1)a_n(t-1)^n + 2\sum_{n=2}^{\infty} n(n-1)a_n(t-1)^{n-1} + \sum_{n=2}^{\infty} n(n-1)a_n(t-1)^{n-2}$$
$$+ 2\sum_{n=1}^{\infty} a_n n(t-1)^{n-1} + \sum_{n=0}^{\infty} a_n(t-1)^{n+1} \equiv 0$$

Changing notation so that all terms inside the summation contain the factor $(t-1)^{n-2}$, we have

$$\sum_{n=4}^{\infty} (n-2)(n-3)a_{n-2}(t-1)^{n-2} + 2\sum_{n=3}^{\infty} (n-1)(n-2)a_{n-1}(t-1)^{n-2}$$
$$+ \sum_{n=2}^{\infty} n(n-1)a_n(t-1)^{n-2} + 2\sum_{n=2}^{\infty} (n-1)a_{n-1}(t-1)^{n-2}$$
$$+ \sum_{n=3}^{\infty} a_{n-3}(t-1)^{n-2} \equiv 0$$

Simplifying and beginning the summation at $n = 4$ so that the first two terms of the resulting power series are outside the summation sign, we obtain

$$(2a_2 + 2a_1) + (8a_2 + 6a_3 + a_0)(t-1) + \sum_{n=4}^{\infty} \{(n-2)(n-3)a_{n-2} + 2(n-1)^2 a_{n-1} + n(n-1)a_n + a_{n-3}\}(t-1)^{n-2} \equiv 0$$

Setting the coefficients of the powers of $(t-1)$ equal to zero yields, from the first two terms, $2a_2 + 2a_1 = 0$ and $8a_2 + 6a_3 + a_0 = 0$, and we obtain the recursion formula $a_n = -\dfrac{2(n-1)^2 a_{n-1} + (n-2)(n-3)a_{n-2} + a_{n-3}}{n(n-1)}$ for $n \geq 4$ from the coefficients inside the summation sign.

15.80 Find a second form for the recursion formula obtained in the previous problem.

▮ If we set $k = n - 2$ so that $n = k + 2$, then the recursion formula becomes $a_{k+2} = -\dfrac{2(k+1)^2 a_{k+1} + k(k-1)a_k + a_{k-1}}{(k+2)(k+1)}$ for $k \geq 2$. Then, replacing the dummy index k by n, we obtain $a_{n+2} = -\dfrac{2(n+1)^2 a_{n+1} + n(n-1)a_n + a_{n-1}}{(n+2)(n+1)}$ for $n \geq 2$ as a second form of the recursion formula. If we note that $a_{-1} = 0$, we see that this recursion formula is also valid for $n = 0$ and $n = 1$.

15.81 Find a recursion formula for the coefficients of the general power-series solution near $t = 2$ of $\ddot{y} + ty = 0$.

▮ This differential equation is in standard form, and its coefficients are analytic everywhere; hence every point, including $t = 2$, is an ordinary point. We assume a solution of the form $y = \sum_{n=0}^{\infty} a_n(t-2)^n$. Since $t = (t-2) + 2$, the differential equation has the form $\ddot{y} + [(t-2) + 2]y = 0$, which is an equation in terms of $(t-2)$. Substituting y and its second derivative into this equation, we obtain

$$\sum_{n=0}^{\infty} n(n-1)a_n(t-2)^{n-2} + (t-2)\sum_{n=0}^{\infty} a_n(t-2)^n + 2\sum_{n=0}^{\infty} a_n(t-2)^n = 0 \tag{1}$$

By reasoning similar to that of (2) of Problem 15.60 and (2) of Problem 15.63 (let $x = t - 2$), we have $\sum_{n=0}^{\infty} n(n-1)a_n(t-2)^{n-2} = \sum_{n=0}^{\infty} (n+2)(n+1)a_{n+2}(t-2)^n$ and $(t-2)\sum_{n=0}^{\infty} a_n(t-2)^n = \sum_{n=0}^{\infty} a_n(t-2)^{n+1} = \sum_{n=0}^{\infty} a_{n-1}(t-2)^n$. Equation (1) now becomes

$$\sum_{n=0}^{\infty} [(n+2)(n+1)a_{n+2} + a_{n-1} + 2a_n](t-2)^n = 0 = \sum_{n=0}^{\infty} 0(t-2)^n$$

Equating coefficients of $(t-2)^n$ yields $(n+2)(n+1)a_{n+2} + a_{n-1} + 2a_n = 0$, or $a_{n+2} = -\dfrac{2a_n + a_{n-1}}{(n+2)(n+1)}$.

5.82 Find a recursion formula for the coefficients of the general power-series solution near $x = 0$ of *Legendre's equation*, $(1 - x^2)y'' - 2xy' + p(p+1)y = 0$, where p denotes an arbitrary constant.

▌ It is clear that the coefficient functions $P(x) = \dfrac{-2x}{1 - x^2}$ and $Q(x) = \dfrac{p(p+1)}{1 - x^2}$ are analytic at the origin. The origin is therefore an ordinary point, and we expect a solution of the form $y = \sum a_n x^n$. Since $y' = \sum (n+1)a_{n+1}x^n$, we get the following expansions for the individual terms on the left side of the given equation:

$$y'' = \sum (n+1)(n+2)a_{n+2}x^n \qquad -x^2y'' = \sum -(n-1)na_nx^n$$
$$-2xy' = \sum -2na_nx^n \qquad p(p+1)y = \sum p(p+1)a_nx^n$$

The sum of these four series is required to be zero, so the coefficient of x^n must be zero for every n: $(n+1)(n+2)a_{n+2} - (n-1)na_n - 2na_n + p(p+1)a_n = 0$. Noting that the coefficients of a_n can be simplified to $-(n-1)n - 2n + p(p+1) = (p+n)(p-n+1)$, we have $a_{n+2} = -\dfrac{(p-n)(p+n+1)}{(n+1)(n+2)}a_n$.

5.83 Find a recursion formula for the coefficients of the general power-series solution near $x = 0$ of $y' = y$.

▌ We assume that this equation has a power-series solution of the form $y = a_0 + a_1x + a_2x^2 + \cdots + a_nx^n + \cdots$. A power series can be differentiated term by term in its interval of convergence, so $y' = a_1 + 2a_2x + 3a_3x^2 + \cdots + (n+1)a_{n+1}x^n + \cdots$. Since $y' = y$, we equate the coefficients of like powers of x to get $a_1 = a_0$, $2a_2 = a_1$, $3a_3 = a_2, \ldots,$ and, in general, $(n+1)a_{n+1} = a_n$, or $a_{n+1} = \dfrac{1}{n+1}a_n$.

5.84 Find a recursion formula for the coefficients of the general power-series solution near $x = 0$ of $(1+x)y' = py$, where p denotes a constant.

▌ This first-order differential equation has the standard form $y' - \dfrac{p}{1+x}y = 0$. Since $P(x) = -\dfrac{p}{1+x}$ is a rational function with a nonzero denominator at $x = 0$, it is analytic there and $x = 0$ is an ordinary point. Consequently, the general solution has the form $y = a_0 + a_1x + a_2x^2 + \cdots + a_nx^n + \cdots$. It follows that

$$y' = a_1 + 2a_2x + 3a_3x^2 + \cdots + (n+1)a_{n+1}x^n + \cdots$$
$$xy' = a_1x + 2a_2x^2 + \cdots + na_nx^n + \cdots$$

and
$$py = pa_0 + pa_1x + pa_2x^2 + \cdots + pa_nx^n + \cdots$$

Since $(1+x)y' = py$, the sum of the first two series must equal the third. Then $a_1 = pa_0$, $2a_2 + a_1 = pa_1$, $3a_3 + 2a_2 = pa_2, \ldots,$ and, in general, $(n+1)a_{n+1} + na_n = pa_n$, or $a_{n+1} = \dfrac{p-n}{n+1}a_n$.

OLUTIONS TO HOMOGENEOUS DIFFERENTIAL EQUATIONS ABOUT AN ORDINARY POINT

85 Find the general solution near $x = 0$ of $y'' - xy' + 2y = 0$.

▌ Using the result of Problem 15.55, we have $y = a_0 + a_1x + a_2x^2 + a_3x^3 + \cdots + a_nx^n + \cdots$, with $a_{n+2} = \dfrac{n-2}{(n+2)(n+1)}a_n$. By substituting successive values of n into the recursion formula, we obtain

$$\begin{aligned} a_2 &= -a_0 & a_3 &= -\tfrac{1}{6}a_1 \\ a_4 &= 0 & a_5 &= \tfrac{1}{20}a_3 = \tfrac{1}{20}(-\tfrac{1}{6}a_1) = -\tfrac{1}{120}a_1 \\ a_6 &= \tfrac{2}{30}a_4 = \tfrac{1}{15}(0) = 0 & a_7 &= \tfrac{3}{42}a_5 = \tfrac{1}{14}(-\tfrac{1}{120})a_1 = -\tfrac{1}{1680}a_1 \\ &\cdots\cdots & &\cdots\cdots \end{aligned} \tag{1}$$

Note that since $a_4 = 0$, it follows from the recursion formula that all the even coefficients beyond a_4 are also zero. Substituting (*1*) into the power series yields

$$\begin{aligned} y &= a_0 + a_1x - a_0x^2 - \tfrac{1}{6}a_1x^3 + 0x^4 - \tfrac{1}{120}a_1x^5 + 0x^6 - \tfrac{1}{1680}a_1x^7 - \cdots \\ &= a_0(1 - x^2) + a_1(x - \tfrac{1}{6}x^3 - \tfrac{1}{120}x^5 - \tfrac{1}{1680}x^7 - \cdots) \end{aligned} \tag{2}$$

If we define $y_1(x) \equiv 1 - x^2$ and $y_2(x) \equiv x - \tfrac{1}{6}x^3 - \tfrac{1}{120}x^5 - \tfrac{1}{1680}x^7 - \cdots$, then the general solution (*2*) can be rewritten as $y = a_0y_1(x) + a_1y_2(x)$.

15.86 Find the general solution near $x = 0$ of $y'' + y = 0$.

▌ Using the result of Problem 15.56, we have $y = a_0 + a_1x + a_2x^2 + a_3x^3 + \cdots + a_nx^n + \cdots$, with $a_{n+2} = \dfrac{-1}{(n+2)(n+1)}a_n$. Substituting successive values of n into the recursion formula, we obtain

$$a_2 = -\frac{1}{2}a_0 = -\frac{1}{2!}a_0$$

$$a_3 = -\frac{1}{6}a_1 = -\frac{1}{3!}a_1$$

$$a_4 = -\frac{1}{(4)(3)}a_2 = -\frac{1}{(4)(3)}\left(-\frac{1}{2!}a_0\right) = \frac{1}{4!}a_0$$

$$a_5 = -\frac{1}{(5)(4)}a_3 = -\frac{1}{(5)(4)}\left(-\frac{1}{3!}a_1\right) = \frac{1}{5!}a_1$$

$$a_6 = -\frac{1}{(6)(5)}a_4 = -\frac{1}{(6)(5)}\frac{1}{4!}a_0 = -\frac{1}{6!}a_0$$

$$a_7 = -\frac{1}{(7)(6)}a_5 = -\frac{1}{(7)(6)}\frac{1}{5!}a_1 = -\frac{1}{7!}a_1$$

$$\cdots\cdots\cdots\cdots\cdots\cdots\cdots\cdots$$

Substituting (1) into the power series yields

$$y = a_0 + a_1x - \frac{1}{2!}a_0x^2 - \frac{1}{3!}a_1x^3 + \frac{1}{4!}a_0x^4 + \frac{1}{5!}a_1x^5 - \frac{1}{6!}a_0x^6 - \frac{1}{7!}a_1x^7 + \cdots$$

$$= a_0\left(1 - \frac{1}{2!}x^2 + \frac{1}{4!}x^4 - \frac{1}{6!}x^6 + \cdots\right) + a_1\left(x - \frac{1}{3!}x^3 + \frac{1}{5!}x^5 - \frac{1}{7!}x^7 + \cdots\right)$$

Using the results of Problems 15.20 and 15.24, we may rewrite the solution as $y = a_0 \cos x + a_1 \sin x$. This solution is obtained more simply by the methods of Chapter 8, because the coefficients of the differential equation are all constants.

15.87 Find the general solution near $x = 0$ of $(x^2 + 4)y'' + xy = 0$.

▌ Using the result of Problem 15.57, we have $y = a_0 + a_1x + a_2x^2 + a_3x^3 + \cdots + a_nx^n + \cdots$, with $a_{n+2} = \dfrac{-n(n-1)}{4(n+2)(n+1)}a_n - \dfrac{1}{4(n+2)(n+1)}a_{n-1}$. Substituting successive values of n into the recursion formula yields, first, $a_2 = -\frac{1}{8}a_{-1}$. Because a_{-1} denotes the coefficient of x^{-1}, which is presumed to be zero, $a_2 = 0$. Continuing, we obtain

$$a_3 = -\tfrac{1}{24}a_0$$

$$a_4 = -\tfrac{1}{24}a_2 - \tfrac{1}{48}a_1 = -\tfrac{1}{24}(0) - \tfrac{1}{48}a_1 = -\tfrac{1}{48}a_1$$

$$a_5 = -\tfrac{3}{40}a_3 - \tfrac{1}{80}a_2 = -\tfrac{3}{40}(-\tfrac{1}{24}a_0) - \tfrac{1}{80}(0) = \tfrac{1}{320}a_0$$

$$a_6 = -\tfrac{1}{10}a_4 - \tfrac{1}{120}a_3 = -\tfrac{1}{10}(-\tfrac{1}{48}a_1) - \tfrac{1}{120}(-\tfrac{1}{24}a_0) = \tfrac{1}{480}a_1 + \tfrac{1}{2880}a_0$$

$$\cdots\cdots\cdots\cdots\cdots\cdots\cdots\cdots$$

Thus the general solution becomes

$$y = a_0 + a_1x + (0)x^2 + (-\tfrac{1}{24}a_0)x^3 + (-\tfrac{1}{48}a_1)x^4 + (\tfrac{1}{320}a_0)x^5 + (\tfrac{1}{480}a_1 + \tfrac{1}{2880}a_0)x^6 + \cdots$$

$$= a_0(1 - \tfrac{1}{24}x^3 + \tfrac{1}{320}x^5 + \tfrac{1}{2880}x^6 + \cdots) + a_1(x - \tfrac{1}{48}x^4 + \tfrac{1}{480}x^6 + \cdots)$$

15.88 Find the general solution near $x = 0$ of $(x^2 - 1)y'' + xy' - y = 0$.

▌ Using the result of Problem 15.58, we have $y = a_0 + a_1x + a_2x^2 + a_3x^3 + \cdots + a_nx^n + \cdots$, with $a_{n+2} = \dfrac{n-1}{n+2}a_n$. Substituting successive values of n into the recursion formula yields

$$a_2 = -\tfrac{1}{2}a_0 \qquad a_3 = 0$$

$$a_4 = \tfrac{1}{4}a_2 = \tfrac{1}{4}(-\tfrac{1}{2}a_0) = -\tfrac{1}{8}a_0 \qquad a_5 = \tfrac{2}{5}a_3 = \tfrac{2}{5}(0) = 0$$

$$a_6 = \tfrac{3}{6}a_4 = \tfrac{1}{2}(-\tfrac{1}{8}a_0) = -\tfrac{1}{16}a_0 \qquad a_7 = \tfrac{4}{7}a_5 = \tfrac{4}{7}(0) = 0$$

$$\cdots\cdots\cdots\cdots \qquad \cdots\cdots\cdots\cdots$$

Note that because $a_3 = 0$, it follows from the recursion formula that all odd coefficients beyond a_3 are also zero. The general solution then becomes

$$y = a_0 + a_1x + (-\tfrac{1}{2}a_0)x^2 + (0)x^3 + (-\tfrac{1}{8}a_0)x^4 + (0)x^5 + (-\tfrac{1}{16}a_0)x^6 + \cdots = a_0(1 - \tfrac{1}{2}x^2 - \tfrac{1}{8}x^4 - \tfrac{1}{16}x^6 - \cdots) + a_1x.$$

15.89 Find the general solution near $x = 0$ of $y'' - xy = 0$.

▮ Using the result of Problem 15.59, we have $y = a_0 + a_1x + a_2x^2 + a_3x^3 + \cdots + a_nx^n + \cdots$, with $a_{n+2} = \dfrac{1}{(n+2)(n+1)}a_{n-1}$. Substituting $n = 0$ into the recursion formula, we obtain $a_2 = \frac{1}{2}a_{-1}$. Since a_{-1} denotes the coefficient of x^{-1} which is presumed to be zero, it follows that $a_2 = 0$. Substituting successive values of n yields

$$\begin{aligned} a_3 &= \tfrac{1}{6}a_0 & a_4 &= \tfrac{1}{12}a_1 \\ a_5 &= \tfrac{1}{20}a_2 = \tfrac{1}{20}(0) = 0 & a_6 &= \tfrac{1}{30}a_3 = \tfrac{1}{30}(\tfrac{1}{6}a_0) = \tfrac{1}{180}a_0 \\ a_7 &= \tfrac{1}{42}a_4 = \tfrac{1}{42}(\tfrac{1}{12}a_1) = \tfrac{1}{504}a_1 & a_8 &= \tfrac{1}{56}a_5 = \tfrac{1}{56}(0) = 0 \\ &\cdots\cdots & &\cdots\cdots \end{aligned}$$

Note that since $a_2 = 0$, it follows from the recursion formula that every third coefficient beyond a_2 (that is, $a_5, a_8, a_{11}, \ldots$) is also zero. The general solution thus becomes

$$\begin{aligned} y &= a_0 + a_1x + 0x^2 + \tfrac{1}{6}a_0x^3 + \tfrac{1}{12}a_1x^4 + 0x^5 + \tfrac{1}{180}a_0x^6 + \tfrac{1}{504}a_1x^7 + 0x^8 + \cdots \\ &= a_0(1 + \tfrac{1}{6}x^3 + \tfrac{1}{180}x^6 + \cdots) + a_1(x + \tfrac{1}{12}x^4 + \tfrac{1}{504}x^7 + \cdots) \end{aligned}$$

15.90 Find the general solution near $t = 0$ of $\ddot{y} + ty = 0$.

▮ Using the result of Problem 15.67, we have $y = b_0 + b_1t + b_2t^2 + b_3t^3 + \cdots + b_nt^n + \cdots$, with $b_{k+2} = \dfrac{-1}{(k+2)(k+1)}b_{k-1}$. Substituting successive values of k into the recursion formula, we find that

$$\begin{aligned} b_2 &= -\tfrac{1}{2}b_{-1} = \tfrac{1}{2}(0) = 0 & b_3 &= -\tfrac{1}{6}b_0 \\ b_4 &= -\tfrac{1}{12}b_1 & b_5 &= -\tfrac{1}{20}b_2 = -\tfrac{1}{20}(0) = 0 \\ b_6 &= -\tfrac{1}{30}b_3 = -\tfrac{1}{30}(-\tfrac{1}{6}b_0) = \tfrac{1}{180}b_0 & b_7 &= -\tfrac{1}{42}b_4 = -\tfrac{1}{42}(-\tfrac{1}{12}b_1) = \tfrac{1}{504}b_1 \\ &\cdots\cdots & &\cdots\cdots \end{aligned}$$

Each coefficient is determined by the one that is three coefficients before it. Thus $b_3, b_6, \ldots$ are multiples of b_0; and $b_4, b_7, \ldots$ are multiples of b_1; and $b_5, b_8, \ldots$ are multiples of b_2, which is zero. There is no restriction on b_0 and b_1, which are therefore the two expected arbitrary constants, and the general solution is

$$y = b_0\left(1 - \frac{1}{3\cdot 2}t^3 + \frac{1}{6\cdot 5\cdot 3\cdot 2}t^6 - \cdots\right) + b_1\left(t - \frac{1}{4\cdot 3}t^4 + \frac{1}{7\cdot 6\cdot 4\cdot 3}t^7 - \cdots\right)$$

15.91 Find the general solution near $t = 0$ of $\dfrac{d^2y}{dt^2} + (t+1)\dfrac{dy}{dt} + 2y = 0$.

▮ Using the result of Problem 15.77, we have $y = a_0 + a_1t + a_2t^2 + a_3t^3 + \cdots + a_nt^n + \cdots$, with $a_{n+2} = \dfrac{-1}{n+2}a_{n+1} - \dfrac{1}{n+1}a_{n+1}$. Evaluating the recursion formula for successive values of n, we find

$$\begin{aligned} a_2 &= -\tfrac{1}{2}a_1 - a_0 \\ a_3 &= -\tfrac{1}{3}a_2 - \tfrac{1}{2}a_1 = -\tfrac{1}{3}(-\tfrac{1}{2}a_1 - a_0) - \tfrac{1}{2}a_1 = -\tfrac{1}{3}a_1 + \tfrac{1}{3}a_0 \\ a_4 &= -\tfrac{1}{4}a_3 - \tfrac{1}{3}a_2 = -\tfrac{1}{4}(-\tfrac{1}{3}a_1 + \tfrac{1}{3}a_0) - \tfrac{1}{3}(-\tfrac{1}{2}a_1 - a_0) = \tfrac{1}{4}a_1 + \tfrac{1}{4}a_0 \\ &\cdots\cdots \end{aligned}$$

Substituting these values yields the general solution

$$\begin{aligned} y &= a_0 + a_1t + (-\tfrac{1}{2}a_1 - a_0)t^2 + (-\tfrac{1}{3}a_1 + \tfrac{1}{3}a_0)t^3 + (\tfrac{1}{4}a_1 + \tfrac{1}{4}a_0)t^4 + \cdots \\ &= a_0(1 - t^2 + \tfrac{1}{3}t^3 + \tfrac{1}{4}t^4 + \cdots) + a_1(t - \tfrac{1}{2}t^2 - \tfrac{1}{3}t^3 + \tfrac{1}{4}t^4 + \cdots) \end{aligned}$$

15.92 Find the general solution near $t = 0$ of $\dfrac{d^2y}{dt^2} - (t+1)\dfrac{dy}{dt} + 2y = 0$.

Using the result of Problem 15.78, we have $y = a_0 + a_1t + a_2t^2 + a_3t^3 + \cdots + a_nt^n + \cdots$, with $a_{n+2} = \dfrac{1}{n+2}a_{n+1} + \dfrac{n-2}{(n+2)(n+1)}a_n$. Evaluating the recursion formula for successive values of n, we find

$$a_2 = \tfrac{1}{2}a_1 - a_0$$
$$a_3 = \tfrac{1}{3}a_2 - \tfrac{1}{6}a_1 = \tfrac{1}{3}(\tfrac{1}{2}a_1 - a_0) - \tfrac{1}{6}a_1 = -\tfrac{1}{3}a_0$$
$$a_4 = \tfrac{1}{4}a_3 + 0a_2 = \tfrac{1}{4}(-\tfrac{1}{3}a_0) = -\tfrac{1}{12}a_0$$
$$a_5 = \tfrac{1}{5}a_4 + \tfrac{1}{20}a_3 = \tfrac{1}{5}(-\tfrac{1}{12}a_0) + \tfrac{1}{20}(-\tfrac{1}{3}a_0) = -\tfrac{1}{30}a_0$$
$$\cdots\cdots\cdots\cdots\cdots\cdots\cdots\cdots$$

Substituting these values into the power series for y, we obtain

$$y = a_0 + a_1t + (\tfrac{1}{2}a_1 - a_0)t^2 + (-\tfrac{1}{3}a_0)t^3 + (-\tfrac{1}{12}a_0)t^4 + (-\tfrac{1}{30}a_0)t^5 + \cdots$$
$$= a_0(1 - t^2 - \tfrac{1}{3}t^3 - \tfrac{1}{12}t^4 - \tfrac{1}{30}t^5 - \cdots) + a_1(t + \tfrac{1}{2}t^2)$$

15.93 Find the general solution near $t = 0$ of $\ddot{y} - y = 0$.

Using the results of Problem 15.66, we have $y = a_0 + a_1t + a_2t^2 + a_3t^3 + \cdots + a_nt^n + \cdots$, with $a_n = \dfrac{a_{n-2}}{n(n-1)}$ for $n = 2, 3, 4, \ldots$. Evaluating the recursion formula for successive values of n, we obtain

$$a_2 = \frac{1}{2(1)}a_0 = \frac{1}{2!}a_0 \qquad a_3 = \frac{1}{3(2)}a_1 = \frac{1}{3!}a_1$$
$$a_4 = \frac{1}{4(3)}a_2 = \frac{1}{4(3)}\frac{1}{2!}a_0 = \frac{1}{4!}a_0 \qquad a_5 = \frac{1}{5(4)}a_3 = \frac{1}{5(4)}\frac{1}{3!}a_1 = \frac{1}{5!}a_1$$
$$a_6 = \frac{1}{6(5)}a_4 = \frac{1}{6(5)}\frac{1}{4!}a_0 = \frac{1}{6!}a_0 \qquad a_7 = \frac{1}{7(6)}a_5 = \frac{1}{7(6)}\frac{1}{5!}a_1 = \frac{1}{7!}a_1$$
$$\cdots\cdots\cdots\cdots \qquad \cdots\cdots\cdots\cdots$$

Substituting these values yields the general solution

$$y = a_0 + a_1t + \frac{1}{2!}a_0t^2 + \frac{1}{3!}a_1t^3 + \frac{1}{4!}a_0t^4 + \frac{1}{5!}a_1t^5 + \frac{1}{6!}a_0t^6 + \frac{1}{7!}a_1t^7 + \cdots$$
$$= a_0\left(1 + \frac{1}{2!}t^2 + \frac{1}{4!}t^4 + \frac{1}{6!}t^6 + \cdots\right) + a_1\left(t + \frac{1}{3!}t^3 + \frac{1}{5!}t^5 + \frac{1}{7!}t^7 + \cdots\right)$$

15.94 Reconcile the answer obtained in the previous problem with the one that would have been obtained had the differential equation been solved by the techniques of Chapter 8.

Since the differential equation is linear and homogeneous with constant coefficients, its characteristic equatio[n] is $\lambda^2 - 1 = 0$, which has as its roots $\lambda = \pm 1$. The general solution is $y = c_1e^t + c_2e^{-t}$. If we set $a_0 = c_1 + c_2$ and $a_1 = c_1 - c_2$, so that $c_1 = \tfrac{1}{2}a_0 + \tfrac{1}{2}a_1$ and $c_2 = \tfrac{1}{2}a_0 - \tfrac{1}{2}a_1$, then the general solution becomes

$$y = \left(\frac{1}{2}a_0 + \frac{1}{2}a_1\right)e^t + \left(\frac{1}{2}a_0 - \frac{1}{2}a_1\right)e^{-t} = a_0\frac{e^t + e^{-t}}{2} + a_1\frac{e^t - e^{-t}}{2} = a_0\cosh t + a_1\sinh t$$

If we now replace $\cosh t$ and $\sinh t$ with their Maclaurin-series expansions (see Problems 15.22 and 15.26), we generate the solution obtained in the previous problem.

15.95 Find the general solution near $t = 0$ of $\ddot{y} - 2t\dot{y} - 2y = 0$.

Using the result of Problem 15.68, we have $y = a_0 + a_1t + a_2t^2 + a_3t^3 + \cdots + a_nt^n + \cdots$, with $a_n = \dfrac{2}{n}a_{n-2}$ for $n = 2, 3, 4, \ldots$. Evaluating the recursion formula for successive even integers, we get

$$a_2 = a_0, \quad a_4 = \frac{a_2}{2} = \frac{a_0}{2}, \quad a_6 = \frac{a_4}{3} = \frac{a_0}{3\cdot 2}, \quad a_8 = \frac{a_6}{4} = \frac{a_0}{4\cdot 3\cdot 2}, \text{ and it follows that } a_{2k} = \frac{a_0}{k!}$$

for $k = 0, 1, 2, \ldots$. For the odd integers we have $a_3 = \dfrac{2a_1}{3}$, $a_5 = \dfrac{2a_3}{5} = \dfrac{2^2a_1}{(5)(3)}$, $a_7 = \dfrac{2a_5}{7} = \dfrac{2^3a_1}{(7)(5)(3)}$, and follows that $a_{2k+1} = \dfrac{2^ka_1}{(2k+1)(2k-1)\cdots(5)(3)}$ for $k = 0, 1, 2, \ldots$. Separating the terms defining y into tho[se] indexed by even integers and those indexed by odd integers, we see that the solution can be expressed as

$$y=\sum_{n=0}^{\infty} a_n t^n=\sum_{k=0}^{\infty} a_{2k}t^{2k}+\sum_{k=0}^{\infty} a_{2k+1}t^{2k+1}=a_0\sum_{k=0}^{\infty}\frac{1}{k!}t^{2k}+a_1\sum_{k=0}^{\infty}\frac{2^k}{(2k+1)(2k-1)\cdots(3)(1)}t^{2k+1}$$

As a result of Problem 15.10, the first series in this solution may be simplified to $a_0 e^{t^2}$.

15.96 Find the general solution near $x=0$ of $y''+x^2y'+2xy=0$.

▌ Using the result of Problem 15.70, we have $y=a_0+a_1x+a_2x^2+a_3x^3+\cdots+a_nx^n+\cdots$, with $a_{n+2}=\dfrac{-1}{n+2}a_{n-1}$. Substituting successive values of n into the recursion formula and recalling that $a_{-1}=0$, we find that

$$a_2=a_5=a_8=\cdots=a_{3n-1}=0$$

$$a_3=-\frac{a_0}{3}\qquad a_6=-\frac{a_3}{6}=\frac{a_0}{3^2 2!}\qquad a_9=-\frac{a_6}{9}=-\frac{a_0}{3^3 3!},\ \cdots\qquad a_{3n}=(-1)^n\frac{a_0}{3^n n!}$$

$$a_4=-\frac{a_1}{(1)(4)}\qquad a_7=-\frac{a_4}{7}=\frac{a_1}{(1)(4)(7)}\qquad a_{10}=-\frac{a_7}{10}=-\frac{a_1}{(1)(4)(7)(10)},\cdots$$

$$a_{3n+1}=(-1)^n\frac{a_1}{(1)(4)(7)(10)\cdots(3n+1)}$$

Substituting these values into the power series for y and simplifying, we obtain the solution

$$y=a_0\left(1-\frac{x^3}{3}+\frac{x^6}{3^2 2!}-\frac{x^9}{3^3 3!}+\cdots\right)+a_1\left[x-\frac{x^4}{(1)(4)}+\frac{x^7}{(1)(4)(7)}-\frac{x^{10}}{(1)(4)(7)(10)}+\cdots\right]$$

The power series $y_1=\sum_{k=0}^{\infty}(-1)^k\dfrac{x^{3k}}{3^k k!}$ and $y_2=\sum_{k=0}^{\infty}(-1)^k\dfrac{x^{3k+1}}{(1)(4)(7)(10)\cdots(3k+1)}$ are two linearly independent particular solutions of the given equation, whose complete solution is therefore $y=a_0y_1+a_1y_2$.

15.97 Find the general solution near $x=0$ of $y''-x^2y'-y=0$.

▌ Using the result of Problem 15.72, we have $y=A_0+A_1x+A_2x^2+A_3x^3+\cdots+A_nx^n+\cdots$, with $A_{n+2}=\dfrac{(n-1)A_{n-1}+A_n}{(n+1)(n+2)}$. Evaluating the recursion formula for successive values of n, we get $A_2=-\frac{1}{2}A_{-1}+\frac{1}{2}A_0=\frac{1}{2}A_0$, because A_{-1}, the coefficient of x^{-1}, is zero, along with

$$A_3=\tfrac{0}{6}A_0+\tfrac{1}{6}A_1=\tfrac{1}{6}A_1$$
$$A_4=\tfrac{1}{12}A_1+\tfrac{1}{12}A_2=\tfrac{1}{12}A_1+\tfrac{1}{12}(\tfrac{1}{2}A_0)=\tfrac{1}{12}A_1+\tfrac{1}{24}A_0$$
$$A_5=\tfrac{2}{20}A_2+\tfrac{1}{20}A_3=\tfrac{1}{10}(\tfrac{1}{2}A_0)+\tfrac{1}{20}(\tfrac{1}{6}A_1)=\tfrac{1}{20}A_0+\tfrac{1}{120}A_1$$
$$A_6=\tfrac{3}{30}A_3+\tfrac{1}{30}A_4=\tfrac{1}{10}(\tfrac{1}{6}A_1)+\tfrac{1}{30}(\tfrac{1}{12}A_1+\tfrac{1}{24}A_0)=\tfrac{1}{720}A_0+\tfrac{7}{360}A_1$$
$$A_7=\tfrac{4}{42}A_4+\tfrac{1}{42}A_5=\tfrac{2}{21}(\tfrac{1}{12}A_1+\tfrac{1}{24}A_0)+\tfrac{1}{42}(\tfrac{1}{20}A_0+\tfrac{1}{120}A_1)=\tfrac{13}{2520}A_0+\tfrac{41}{5040}A_1$$

....

Substituting these values and simplifying yield the general solution

$$y=A_0(1+\tfrac{1}{2}x^2+\tfrac{1}{24}x^4+\tfrac{1}{20}x^5+\tfrac{1}{720}x^6+\tfrac{13}{2520}x^7+\cdots)+A_1(x+\tfrac{1}{6}x^3+\tfrac{1}{12}x^4+\tfrac{1}{120}x^5+\tfrac{7}{360}x^6+\tfrac{41}{5040}x^7+\cdots)$$

15.98 Find the general solution near $x=0$ of $(1+x^2)y''+xy'-y=0$.

▌ Using the result of Problem 15.73, we have $y=A_0+A_1x+A_2x^2+A_3x^3+\cdots+A_nx^n+\cdots$, with $A_{n+2}=-\dfrac{n-1}{n+2}A_n$. From the recursion formula it is clear that $A_3=A_5=A_7=\cdots=0$; that is, $A_{n+2}=0$ if n is odd. If n is even $(n=2k)$, then

$$A_{2k}=-\frac{2k-3}{2k}A_{2k-2}=\frac{(2k-3)(2k-5)}{2k(2k-2)}A_{2k-4}=\cdots=(-1)^{k+1}\frac{(1)(3)(5)\cdots(2k-3)}{2^k k!}A_0$$

Thus, the complete solution is

$$y = A_0\left(1 + \frac{1}{2}x^2 - \frac{1}{8}x^4 + \frac{1}{16}x^6 - \frac{5}{128}x^8 + \cdots\right) + A_1 x$$

$$= A_0\left[1 + \frac{1}{2}x^2 + \sum_{k=2}^{\infty}(-1)^{k+1}\frac{(1)(3)(5)\cdots(2k-3)}{2^k k!}x^{2k}\right] + A_1 x$$

$$= A_0\left[1 + \frac{1}{2}x^2 - \sum_{k=2}^{\infty}(-1)^{k}\frac{(1)(3)(5)\cdots(2k-3)}{2^k k!}x^{2k}\right] + A_1 x$$

15.99 Find the general solution near $t = 0$ of $\dfrac{d^2y}{dt^2} + (t-1)\dfrac{dy}{dt} + (2t-3)y = 0.$

▌ Using the result of Problem 15.74, we have $y = a_0 + a_1 t + a_2 t^2 + a_3 t^3 + \cdots + a_n t^n + \cdots,$ with

$$a_{n+2} = \frac{1}{n+2}a_{n+1} - \frac{n-3}{(n+2)(n+1)}a_{n+1} - \frac{2}{(n+2)(n+1)}a_{n-1}$$

Evaluating the recursion formula for successive values of n and noting that $a_{-1} = 0,$ we get

$$a_2 = \tfrac{1}{2}a_1 - \tfrac{-3}{2}a_0 - \tfrac{2}{2}a_{-1} = \tfrac{1}{2}a_1 + \tfrac{3}{2}a_0$$
$$a_3 = \tfrac{1}{3}a_2 + \tfrac{1}{3}a_1 - \tfrac{1}{3}a_0 = \tfrac{1}{3}(\tfrac{1}{2}a_1 + \tfrac{3}{2}a_0) + \tfrac{1}{3}a_1 - \tfrac{1}{3}a_0 = \tfrac{1}{2}a_1 + \tfrac{1}{6}a_0$$
$$a_4 = \tfrac{1}{4}a_3 + \tfrac{1}{12}a_2 - \tfrac{1}{6}a_1 = \tfrac{1}{4}(\tfrac{1}{2}a_1 + \tfrac{1}{6}a_0) + \tfrac{1}{12}(\tfrac{1}{2}a_1 + \tfrac{3}{2}a_0) - \tfrac{1}{6}a_1 = \tfrac{1}{6}a_0$$
$$\cdots\cdots\cdots\cdots\cdots\cdots\cdots\cdots\cdots\cdots\cdots\cdots$$

Thus, the general solution is

$$y = a_0 + a_1 t + (\tfrac{1}{2}a_1 + \tfrac{3}{2}a_0)t^2 + (\tfrac{1}{2}a_1 + \tfrac{1}{6}a_0)t^3 + (\tfrac{1}{6}a_0)t^4 + \cdots$$
$$= a_0(1 + \tfrac{3}{2}t^2 + \tfrac{1}{6}t^3 + \tfrac{1}{6}t^4 + \cdots) + a_1(t + \tfrac{1}{2}t^2 + \tfrac{1}{2}t^3 + 0t^4 + \cdots)$$

15.100 Find the general solution near $v = 0$ of $\dfrac{d^2y}{dv^2} + (v+1)\dfrac{dy}{dv} + y = 0.$

▌ Using the result of Problem 15.76, we have $y = A_0 + A_1 v + A_2 v^2 + A_3 v^3 + \cdots + A_n v^n + \cdots,$ with $A_{n+2} = \dfrac{-1}{n+2}(A_n + A_{n+1}).$ Evaluating the recursion formula for successive values of n, we get

$$A_2 = -\tfrac{1}{2}A_0 - \tfrac{1}{2}A_1$$
$$A_3 = -\tfrac{1}{3}A_1 - \tfrac{1}{3}A_2 = -\tfrac{1}{3}A_1 - \tfrac{1}{3}(-\tfrac{1}{2}A_0 - \tfrac{1}{2}A_1) = \tfrac{1}{6}A_0 - \tfrac{1}{6}A_1$$
$$A_4 = -\tfrac{1}{4}A_2 - \tfrac{1}{4}A_3 = -\tfrac{1}{4}(-\tfrac{1}{2}A_0 - \tfrac{1}{2}A_1) - \tfrac{1}{4}(\tfrac{1}{6}A_0 - \tfrac{1}{6}A_1) = \tfrac{1}{12}A_0 + \tfrac{1}{6}A_1$$
$$A_5 = -\tfrac{1}{5}A_3 - \tfrac{1}{5}A_4 = -\tfrac{1}{5}(\tfrac{1}{6}A_0 - \tfrac{1}{6}A_1) - \tfrac{1}{5}(\tfrac{1}{12}A_0 + \tfrac{1}{6}A_1) = -\tfrac{1}{20}A_0$$
$$A_6 = -\tfrac{1}{6}A_4 - \tfrac{1}{6}A_5 = -\tfrac{1}{6}(\tfrac{1}{12}A_0 + \tfrac{1}{6}A_1) - \tfrac{1}{6}(-\tfrac{1}{20}A_0) = -\tfrac{1}{180}A_0 - \tfrac{1}{36}A_1$$
$$\cdots\cdots\cdots\cdots\cdots\cdots\cdots\cdots\cdots\cdots\cdots\cdots$$

Substituting these values yields the general solution

$$y = A_0 + A_1 v + (-\tfrac{1}{2}A_0 - \tfrac{1}{2}A_1)v^2 + (\tfrac{1}{6}A_0 - \tfrac{1}{6}A_1)v^3 + (\tfrac{1}{12}A_0 + \tfrac{1}{6}A_1)v^4 + (-\tfrac{1}{20}A_0)v^5 + (-\tfrac{1}{180}A_0 - \tfrac{1}{36}A_1)v^6 + \cdots$$
$$= A_0(1 - \tfrac{1}{2}v^2 + \tfrac{1}{6}v^3 + \tfrac{1}{12}v^4 - \tfrac{1}{20}v^5 - \tfrac{1}{180}v^6 + \cdots) + A_1(v - \tfrac{1}{2}v^2 - \tfrac{1}{6}v^3 + \tfrac{1}{6}v^4 - \tfrac{1}{36}v^6 + \cdots)$$

15.101 Find the general solution near $t = 0$ of $\ddot{y} + 2t^2 y = 0.$

▌ Using the result of Problem 15.75, we have $y = a_0 + a_1 t + a_2 t^2 + a_3 t^3 + \cdots + a_n t^n + \cdots,$ with $a_{n+2} = \dfrac{-2}{(n+2)(n+1)}a_{n-2}.$ Evaluating the recursion formula for successive values of n and noting that $a_{-2} = a_{-1} = 0$ because they represent the coefficients of t^{-2} and t^{-1}, we obtain

$$a_2 = -\tfrac{2}{2}a_{-2} = 0 \qquad a_3 = -\tfrac{2}{6}a_{-1} = 0$$
$$a_4 = -\tfrac{2}{12}a_0 = -\tfrac{1}{6}a_0 \qquad a_5 = -\tfrac{2}{20}a_1 = -\tfrac{1}{10}a_1$$
$$a_6 = -\tfrac{2}{30}a_2 = -\tfrac{2}{30}(0) = 0 \qquad a_7 = -\tfrac{2}{42}a_3 = -\tfrac{2}{42}(0) = 0$$
$$a_8 = -\tfrac{2}{56}a_4 = -\tfrac{1}{28}(-\tfrac{1}{6}a_0) = \tfrac{1}{168}a_0 \qquad a_9 = -\tfrac{2}{72}a_5 = -\tfrac{1}{36}(-\tfrac{1}{10}a_1) = \tfrac{1}{360}a_1$$
$$\cdots\cdots\cdots\cdots\cdots\cdots \qquad \cdots\cdots\cdots\cdots\cdots\cdots$$

Substituting these values yields the general solution

$$y = a_0 + a_1 t + (-\tfrac{1}{6}a_0)t^4 + (-\tfrac{1}{10}a_1)t^5 + \tfrac{1}{168}a_0 t^8 + \tfrac{1}{360}a_1 t^9 + \cdots$$
$$= a_0(1 - \tfrac{1}{6}t^4 + \tfrac{1}{168}t^8 + \cdots) + a_1(t - \tfrac{1}{10}t^5 + \tfrac{1}{360}t^9 + \cdots)$$

5.102 Find the general solution near $x = 0$ of *Legendre's equation*, $(1 + x^2)y'' - 2yy' + p(p + 1)y = 0$, where p denotes an arbitrary constant.

▌ The recursion formula for this differential equation was found in Problem 15.82 to be $a_{n+2} = -\dfrac{(p - n)(p + n + 1)}{(n + 1)(n + 2)} a_n$. Thus

$$a_2 = -\frac{p(p+1)}{1 \cdot 2} a_0$$
$$a_3 = -\frac{(p-1)(p+2)}{(2)(3)} a_1$$
$$a_4 = -\frac{(p-2)(p+3)}{(3)(4)} a_2 = \frac{p(p-2)(p+1)(p+3)}{4!} a_0$$
$$a_5 = -\frac{(p-3)(p+4)}{(4)(5)} a_3 = \frac{(p-1)(p-3)(p+2)(p+4)}{5!} a_1$$
$$a_6 = -\frac{(p-4)(p+5)}{(5)(6)} a_4 = -\frac{p(p-2)(p-4)(p+1)(p+3)(p+5)}{6!} a_0$$
$$a_7 = -\frac{(p-5)(p+6)}{(6)(7)} a_5 = -\frac{(p-1)(p-3)(p-5)(p+2)(p+4)(p+6)}{7!} a_1$$
$$\cdots\cdots\cdots\cdots\cdots\cdots\cdots\cdots$$

By inserting these coefficients into the assumed solution $y = \sum a_n x^n$, we obtain

$$y = a_0\left[1 - \frac{p(p+1)}{2!}x^2 + \frac{p(p-2)(p+1)(p+3)}{4!}x^4 - \frac{p(p-2)(p-4)(p+1)(p+3)(p+5)}{6!}x^6 + \cdots\right.$$
$$\left. + a_1\left[x - \frac{(p-1)(p+2)}{3!}x^3 + \frac{(p-1)(p-3)(p+2)(p+4)}{5!}x^5 - \frac{(p-1)(p-3)(p-5)(p+2)(p+4)(p+6)}{7!}x^7 + \cdots\right]\right.$$

.103 Show that whenever p is a positive integer, one solution of Legendre's equation near $x = 0$ is a polynomial of degree n.

▌ The recursion formula of the previous problem contains the factor $p - n$. It follows that when $n = p$, $a_{p+2} = 0$. The recursion formula then implies that $a_{p+4} = a_{p+6} = a_{p+8} = \cdots = 0$. Thus, if p is odd, all odd coefficients a_n $(n > p)$ are zero; if p is even, all even coefficients a_n $(n > p)$ are zero. Thus, one of the bracketed quantities in the solution to the previous problem (depending on whether p is even or odd) will contain only a finite number of terms up to and including x^P; hence it will be a polynomial of degree p.

Since a_0 and a_1 are arbitrary constants, we may choose $a_1 = 0$ when p is even and $a_0 = 0$ when p is odd. If, in addition, we choose the other constant to satisfy the condition $y(1) = 1$, then the resulting solution is a *Legendre polynomial* of degree n. The first few Legendre polynomials are

$$P_0(x) = 1 \qquad P_1(x) = x \qquad P_2(x) = \tfrac{1}{2}(3x^2 - 1) \qquad P_3(x) = \tfrac{1}{2}(5x^3 - 3x) \qquad P_4(x) = \tfrac{1}{8}(35x^4 - 30x^2 + 3)$$

104 Find the general solution near $x = 0$ of $y' = y$.

▌ Using the result of Problem 15.83, we have $y = a_0 + a_1 x + a_2 x^2 + a_3 x^3 + \cdots + a_n x^n + \cdots$, with $a_{n+1} = \dfrac{1}{n+1} a_n$. Evaluating the recursion formula for successive values of n, we get

$$a_1 = a_0 \qquad a_2 = \frac{1}{2}a_1 = \frac{1}{2}a_0$$
$$a_3 = \frac{1}{3}a_2 = \frac{1}{3}\frac{1}{2}a_0 = \frac{1}{3!}a_0 \qquad a_4 = \frac{1}{4}a_3 = \frac{1}{4}\frac{1}{3!}a_0 = \frac{1}{4!}a_0$$
$$a_5 = \frac{1}{5}a_4 = \frac{1}{5}\frac{1}{4!}a_0 = \frac{1}{5!}a_0 \qquad a_6 = \frac{1}{6}a_5 = \frac{1}{6}\frac{1}{5!}a_0 = \frac{1}{6!}a_0$$
$$\cdots\cdots\cdots\cdots \qquad \cdots\cdots\cdots\cdots$$

Substituting these values yields the general solution $y = a_0\left(1 + x + \frac{1}{2!}x^2 + \frac{1}{3!}x^3 + \frac{1}{4!}x^4 + \frac{1}{5!}x^5 + \frac{1}{6!}x^6 + \cdots\right)$ which, as a result of Problem 15.5, may be written as $y = a_0 e^x$. This solution is obtained more easily by the methods described in Chapter 8.

15.105 Discuss alternative procedures for obtaining the power-series expansion of the general solution of $y'' + P(x)y' + Q(x)y = 0$ near a nonzero ordinary point.

▮ If $x_0 \neq 0$ is an ordinary point, then the power-series expansion of the general solution has the form $y = \sum_{n=0}^{\infty} a_n(x - x_0)^n$. The undetermined coefficients $a_2, a_3, a_4, \ldots$ may be obtained in terms of a_0 and a_1 by substituting y and its derivatives into the given differential equation and then equating coefficients of like powers of $(x - x_0)$.

An alternative approach, which often simplifies the algebra, is to first translate the axis so that x_0 becomes the origin. This is effected by setting $t = x - x_0$ and substituting into the original differential equation. Under the substitution, $\frac{dy}{dx} = \frac{dy}{dt}\frac{dt}{dx} = \frac{dy}{dt}(1) = \frac{dy}{dt}$ and

$$\frac{d^2y}{dx^2} = \frac{d}{dx}\left(\frac{dy}{dx}\right) = \frac{d}{dx}\left(\frac{dy}{dt}\right) = \frac{d}{dt}\left(\frac{dy}{dt}\right)\frac{dt}{dx} = \frac{d^2y}{dt^2}\frac{dt}{dx} = \frac{d^2y}{dt^2}(1) = \frac{d^2y}{dt^2}$$

The solution of the differential equation that results can be obtained as a power-series expansion about $t = 0$. Then the solution of the original equation is obtained by substituting for t.

15.106 Find the general solution near $x = -1$ of $y'' + xy' + (2x - 1)y = 0$.

▮ This equation is in standard form. Since $P(x) = x$ and $Q(x) = 2x - 1$ are polynomials, both are analytic everywhere, which implies that every point, including $x = -1$, is an ordinary point. Using the transformation developed in Problem 15.105, we set $t = x - (-1) = x + 1$, $\frac{dy}{dx} = \frac{dy}{dt}$, and $\frac{d^2y}{dx^2} = \frac{d^2y}{dt^2}$. The differential equation becomes $\frac{d^2y}{dt^2} + (t - 1)\frac{dy}{dt} + (2t - 3)y = 0$, and we seek a solution near $t = 0$. Such a solution is found in Problem 15.99 to be

$$y = a_0(1 + \tfrac{3}{2}t^2 + \tfrac{1}{6}t^3 + \tfrac{1}{6}t^4 + \cdots) + a_1(t + \tfrac{1}{2}t^2 + \tfrac{1}{2}t^3 + 0t^4 + \cdots)$$

Since $t = x + 1$, in terms of x this solution is

$$y = a_0[1 + \tfrac{3}{2}(x + 1)^2 + \tfrac{1}{6}(x + 1)^3 + \tfrac{1}{6}(x + 1)^4 + \cdots] + a_1[(x + 1) + \tfrac{1}{2}(x + 1)^2 + \tfrac{1}{2}(x + 1)^3 + 0(x + 1)^4 + \cdots]$$

15.107 Find the general solution near $x = -3$ of $y'' + (2x^2 + 12x + 18)y = 0$.

▮ This equation is in standard form with $P(x) = 0$ and $Q(x) = 2x^2 + 12x + 18$. Since both functions are analytic everywhere, every point (and in particular $x = -3$) is an ordinary point. Using the substitutions developed in Problem 15.105, we set $t = x - (-3) = x + 3$, so $x = t - 3$. The differrential equation becomes $\ddot{y} + [2(t - 3)^2 + 12(t - 3) + 18]y = 0$ or $\ddot{y} + 2t^2y = 0$. Using the result of Problem 15.101, we have as a solution to this last equation near $t = 0$,

$$y = a_0(1 - \tfrac{1}{6}t^4 + \tfrac{1}{168}t^8 + \cdots) + a_1(t - \tfrac{1}{10}t^5 + \tfrac{1}{360}t^9 + \cdots)$$

Substituting $x + 3$ for t, we obtain the solution to the original problem as

$$y = a_0[1 - \tfrac{1}{6}(x + 3)^4 + \tfrac{1}{168}(x + 3)^8 + \cdots] + a_1[(x + 3) - \tfrac{1}{10}(x + 3)^5 + \tfrac{1}{360}(x + 3)^9 + \cdots]$$

15.108 Find the general solution near $x = 1$ of $y'' - xy' + 2y = 0$.

▮ This equation is in standard form with $P(x) = -x$ and $Q(x) = 2$. Since both functions are analytic everywhere, every point (including $x = 1$) is an ordinary point. Using the substitutions developed in Problem 15.105, we set $t = x - 1$, so $x = t + 1$. The differential equation becomes $\frac{d^2y}{dt^2} - (t + 1)\frac{dy}{dt} + 2y = 0$, and we seek a solution near $t = 0$.

Using the result of Problem 15.92, we have $y = a_0(1 - t^2 - \tfrac{1}{3}t^3 - \tfrac{1}{12}t^4 - \tfrac{1}{30}t^5 + \cdots) + a_1(t + \tfrac{1}{2}t^2)$. Since $t = x - 1$, the solution to the original problem is

$$y = a_0[1 - (x - 1)^2 - \tfrac{1}{3}(x - 1)^3 - \tfrac{1}{12}(x - 1)^4 - \tfrac{1}{30}(x - 1)^5 - \cdots] + a_1[(x - 1) + \tfrac{1}{2}(x - 1)^2]$$

15.109 Find the general solution near $x = 5$ of $y'' + (x-5)y = 0$.

▌ Here $x = 5$ is an ordinary point. Using the substitutions developed in Problem 15.105, we set $t = x - 5$, whereupon the differential equation becomes $\ddot{y} + ty = 0$. Using the results of Problem 15.90, we then have

$$y = b_0\left[1 - \frac{1}{(3)(2)}t^3 + \frac{1}{(6)(5)(3)(2)}t^6 + \cdots\right] + b_1\left[t - \frac{1}{(4)(3)}t^4 + \frac{1}{(7)(6)(4)(3)}t^7 + \cdots\right]$$

Substituting $t = x - 5$ into this last equation yields the solution to the original problem as

$$y = b_0\left[1 - \frac{1}{(3)(2)}(x-5)^3 + \frac{1}{(6)(5)(3)(2)}(x-5)^6 + \cdots\right] + b_1\left[(x-5) - \frac{1}{(4)(3)}(x-5)^4 + \frac{1}{(7)(6)(4)(3)}(x-5)^7 + \cdots\right]$$

15.110 Find the general solution near $x = 1$ of $y'' + (2-2x)y' - 2y = 0$.

▌ Here $x = 1$ is an ordinary point. Using the substitutions suggested in Problem 15.105, we set $t = x - 1$, whereupon the differential equation becomes $\ddot{y} - 2t\dot{y} - 2y = 0$. We seek a solution to this new differential equation near $t = 0$.

Using the result of Problem 15.95, we have $y = a_0 e^{t^2} + a_1 \sum_{k=0}^{\infty} \frac{2^k}{(2k+1)(2k-1)\cdots(3)(1)} t^{2k+1}$. Substituting $t = x - 1$ into this last equation, we obtain the solution to the original problem as $y = a_0 e^{(x-1)^2} + a_1 \sum_{k=0}^{\infty} \frac{2^k}{(2k+1)(2k-1)\cdots(3)(1)}(x-1)^{2k+1}$.

15.111 Find the general solution near $x = 2$ of $y'' + (x-1)y' + y = 0$.

▌ Here $x = 2$ is an ordinary point. Using the substitutions developed in Problem 15.105, we set $t = x - 2$, whereupon the differential equation becomes $\frac{d^2y}{dt^2} + (t+1)\frac{dy}{dt} + y = 0$. We seek a solution to this equation near $t = 0$. Using the result of Problem 15.100 (with t replacing v), we have

$$y = A_0(1 - \tfrac{1}{2}t^2 + \tfrac{1}{6}t^3 + \tfrac{1}{12}t^4 - \tfrac{1}{20}t^5 - \tfrac{1}{180}t^6 + \cdots) + A_1(t - \tfrac{1}{2}t^2 - \tfrac{1}{6}t^3 + \tfrac{1}{6}t^4 - \tfrac{1}{36}t^6 + \cdots)$$

Substituting $t = x - 2$ into this last equation, we obtain the solution to the original problem as

$$y = A_0[1 - \tfrac{1}{2}(x-2)^2 + \tfrac{1}{6}(x-2)^3 + \tfrac{1}{12}(x-2)^4 - \tfrac{1}{20}(x-2)^5 - \tfrac{1}{180}(x-2)^6 + \cdots]$$
$$+ A_1[(x-2) - \tfrac{1}{2}(x-2)^2 - \tfrac{1}{6}(x-2)^3 + \tfrac{1}{6}(x-2)^4 - \tfrac{1}{36}(x-2)^6 + \cdots]$$

15.112 Find the general solution near $x = 2$ of $y'' - (x-2)y' + 2y = 0$.

▌ Here $x = 2$ is an ordinary point. Using the substitutions developed in Problem 15.105, we set $t = x - 2$. Then also $\frac{dy}{dx} = \frac{dy}{dt}$ and $\frac{d^2y}{dx^2} = \frac{d^2y}{dt^2}$, and the differential equation becomes $\frac{d^2y}{dt^2} - t\frac{dy}{dt} + 2y = 0$. We seek a solution near $t = 0$.

Using the result of Problem 15.85 (with x replaced by t), we have $y = a_0(1 - t^2) + a_1(t - \frac{1}{6}t^3 - \frac{1}{120}t^5 - \frac{1}{1680}t^7 - \ldots)$. Substituting $t = x - 2$ into this last equation, we obtain the solution to the original problem as

$$y = a_0[1 - (x-2)^2] + a_1[(x-2) - \tfrac{1}{6}(x-2)^3 - \tfrac{1}{120}(x-2)^5 - \tfrac{1}{1680}(x-2)^7 - \cdots]$$

15.113 Find the general solution near $x = -1$ of $(x^2 + 2x)y'' + (x+1)y' - y = 0$.

▌ It follows from Problem 15.49 that $x = -1$ is an ordinary point, so we can find a power-series expansion for y around $x = -1$. We translate -1 to the origin with the transformation $t = x - (-1) = x + 1$, and the differential equation becomes $[(t-1)^2 + 2(t-1)]\ddot{y} + [(t-1) + 1]\dot{y} - y = 0$, or $(t^2 - 1)\ddot{y} + t\dot{y} - y = 0$. We now seek a power-series solution around $t = 0$.

Using the results of Problem 15.88 (with t replacing x), we have $y = a_0(1 - \frac{1}{2}t^2 - \frac{1}{8}t^4 - \frac{1}{16}t^6 - \cdots) + a_1 t$. Substituting $x + 1$ for t, we obtain as the solution to the orginal problem $y = a_0[1 - \frac{1}{2}(x+1)^2 - \frac{1}{8}(x+1)^4 - \frac{1}{16}(x+1)^6 - \cdots] + a_1(x+1)$.

15.114 Find the general solution near $x = 2$ of $(x^2 - 4x + 3)y'' + (x-2)y' - y = 0$.

▌ This differential equation has the standard form $y'' + \frac{x-2}{x^2 - 4x + 3}y' - \frac{1}{x^2 - 4x + 3}y = 0$. Since both

$P(x) = \dfrac{x-2}{x^2-4x+3}$ and $Q(x) = \dfrac{-1}{x^2-4x+3}$ are rational functions with nonzero denominators at $x = 2$, this point is an ordinary point. Accordingly, we can find a power-series expansion for y around $x = 2$.

We translate 2 to the origin with the transformation $t = x - 2$. With this substitution, the differential equation becomes $[(t+2)^2 - 4(t+2) + 3]\dfrac{d^2y}{dt^2} + [(t+2) - 2]\dfrac{dy}{dt} - y = 0$, or $(t^2 - 1)\dfrac{d^2y}{dt^2} + t\dfrac{dy}{dt} - y = 0$. We seek a solution near $t = 0$. The result of Problem 15.88 (with t replacing x) yields $y = a_0(1 - \frac{1}{2}t^2 - \frac{1}{8}t^4 - \frac{1}{16}t^6 - \cdots) + a_1 t$. Substituting $x - 2$ for t, we obtain as the solution to the original problem $y = a_0[1 - \frac{1}{2}(x-2)^2 - \frac{1}{8}(x-2)^4 - \frac{1}{16}(x-2)^6 - \cdots] + a_1(x-2)$.

15.115 Find the general solution near $x = 4$ of $y'' + (4 - x)y = 0$.

▮ This differential equation is in standard form with coefficients $P(x) = 0$ and $Q(x) = 4 - x$; both are analytic everywhere. Thus, every point, and in particular $x = 4$, is an ordinary point.

The substitution $t = x - 4$ translates the point 4 to the origin and transforms the differential equation into $\ddot{y} - ty = 0$. The solution to this equation near $t = 0$ is given in Problem 15.89 (with x replacing t) as $y = a_0(1 + \frac{1}{6}t^3 + \frac{1}{180}t^6 + \cdots) + a_1(t + \frac{1}{12}t^4 + \frac{1}{504}t^7 + \cdots)$. The solution to the original problem is

$$y = a_0[1 + \tfrac{1}{6}(x-4)^3 + \tfrac{1}{180}(x-4)^6 + \cdots] + a_1[(x-4) + \tfrac{1}{12}(x-4)^4 + \tfrac{1}{504}(x-4)^7 + \cdots]$$

15.116 Find the general solution near $t = 2$ for $\ddot{y} + ty = 0$

▮ Using the result of Problem 15.81, we have $y = a_0 + a_1(t-2) + a_2(t-2)^2 + a_3(t-2)^3 + a_n(t-2)^n + \cdots$, with $a_{n+2} = -\dfrac{2a_n + a_{n-1}}{(n+2)(n+1)}$. Evaluating the recursion formula for successive values of n and recalling that $a_{-1} = 0$, we have

$$a_2 = -\frac{2a_0 + a_{-1}}{2} = -a_0 \qquad a_3 = -\frac{2a_1 + a_0}{6} = -\frac{1}{3}a_1 - \frac{1}{6}a_0$$

$$a_4 = -\frac{2a_2 + a_1}{12} = \frac{1}{6}a_0 - \frac{1}{12}a_1 \qquad a_5 = -\frac{2a_3 + a_2}{20} = \frac{1}{30}a_1 + \frac{1}{15}a_0$$

$$\cdots\cdots\cdots\cdots \qquad \cdots\cdots\cdots\cdots$$

Substitution of these values yields the general solution

$$\begin{aligned} y &= a_0 + a_1(t-2) + (-a_0)(t-2)^2 + (-\tfrac{1}{3}a_1 - \tfrac{1}{6}a_0)(t-2)^3 + (\tfrac{1}{6}a_0 - \tfrac{1}{12}a_1)(t-2)^4 + (\tfrac{1}{30}a_1 + \tfrac{1}{15}a_0)(t-2)^5 + \cdots \\ &= a_0[1 - (t-2)^2 - \tfrac{1}{6}(t-2)^3 + \tfrac{1}{6}(t-2)^4 + \tfrac{1}{15}(t-2)^5 \cdots] \\ &\quad + a_1[(t-2) - \tfrac{1}{3}(t-2)^3 - \tfrac{1}{12}(t-2)^4 + \tfrac{1}{30}(t-2)^5 + \cdots] \end{aligned}$$

15.117 Find the general solution near $t = 1$ of $t^2\ddot{y} + 2\dot{y} + (t-1)y = 0$.

▮ Using the results of Problems 15.79 and 15.80, we have $y = a_0 + a_1(t-1) + a_2(t-1)^2 + a_3(t-2)^3 + \cdots$, with $a_{n+2} = -\dfrac{2(n+1)^2 a_{n+1} + n(n-1)a_n + a_{n-1}}{(n+2)(n+1)}$. Evaluating the recursion formula for successive values of n and realizing that $a_{-1} = 0$, we obtain

$$a_2 = -\frac{2a_1 + 0a_0 + a_{-1}}{2} = -a_1 \qquad a_3 = -\frac{8a_2 + 0a_1 + a_0}{6} = \frac{4}{3}a_1 - \frac{1}{6}a_0$$

$$a_4 = -\frac{18a_3 + 2a_2 + a_1}{12} = -\frac{23}{12}a_1 + \frac{1}{4}a_0 \qquad a_5 = -\frac{32a_4 + 6a_3 + a_2}{20} = \frac{163}{60}a_1 - \frac{7}{20}a_0$$

$$\cdots\cdots\cdots\cdots \qquad \cdots\cdots\cdots\cdots$$

Substituting these values yields the solution

$$\begin{aligned} y &= a_0 + a_1(t-1) + (-a_1)(t-1)^2 + (\tfrac{4}{3}a_1 - \tfrac{1}{6}a_0)(t-1)^3 + (-\tfrac{23}{12}a_1 + \tfrac{1}{4}a_0)(t-1)^4 + (\tfrac{163}{60}a_1 - \tfrac{7}{20}a_0)(t-1)^5 + \cdots \\ &= a_0[1 - \tfrac{1}{6}(t-1)^3 + \tfrac{1}{4}(t-1)^4 - \tfrac{7}{20}(t-1)^5 + \cdots] \\ &\quad + a_1[(t-1) - (t-1)^2 + \tfrac{4}{3}(t-1)^3 - \tfrac{23}{12}(t-1)^4 + \tfrac{163}{60}(t-1)^5 + \cdots] \end{aligned}$$

15.118 Solve the previous problem by first transforming $t = 1$ to the origin.

▮ Setting $x = t - 1$, we rewrite the differential equation as $(x+1)^2y'' + 2y' + xy = 0$. We seek a general solution near $x = 0$. A recursion formula for such a solution was developed Problem 15.65; it is identical to the one obtained in the previous problem. Consequently, the coefficients $a_2, a_3, a_4, \ldots$ are identical to those obtained in Problem 15.117, and the solution of the new differential equation is

$$y = a_0(1 - \tfrac{1}{6}x^3 + \tfrac{1}{4}x^4 - \tfrac{7}{20}x^5 + \cdots) + a_1(1 - x^2 + \tfrac{4}{3}x^3 - \tfrac{23}{12}x^4 + \tfrac{163}{60}x^5 + \cdots)$$

If we now replace x with $t - 1$, we obtain the solution of the previous problem.

SOLUTIONS TO NONHOMOGENEOUS DIFFERENTIAL EQUATIONS ABOUT AN ORDINARY POINT

15.119 Find a recursion formula for the coefficients of the general power-series solution near $x = 0$ of $(x^2 + 4)y'' + xy = x + 2$.

▌ Dividing the given equation by $x^2 + 4$, we see that $x = 0$ is an ordinary point and that $\phi(x) = (x + 2)/(x^2 + 4)$ is analytic there. Hence, the power-series method is applicable to the entire equation, which, furthermore, we may leave in the original form. We assume a solution

$$y = a_0 + a_1x + a_2x^2 + a_3x^3 + \cdots + a_nx^n + a_{n+1}x^{n+1} + a_{n+2}x^{n+2} + \cdots \qquad (1)$$

and substitute it and its second derivative into the original differential equation. The result is

$$(x^2 + 4)[2a_2 + 6a_3x + 12a_4x^2 + \cdots + n(n-1)a_nx^{n-2} + (n+1)na_{n+1}x^{n-1} + (n+2)(n+1)a_{n+2}x^n + \cdots]$$
$$+ x[a_0 + a_1x + a_2x^2 + a_3x^3 + \cdots + a_{n-1}x^{n-1} + \cdots] = x + 2$$

or

$$8a_2 + x(24a_3 + a_0) + x^2(2a_2 + 48a_4 + a_1) + x^3(6a_3 + 80a_5 + a_2) + \cdots$$
$$+ x^n[n(n-1)a_n + 4(n+2)(n+1)a_{n+2} + a_{n-1}] + \cdots = 2 + (1)x + (0)x^2 + (0)x^3 + \cdots \qquad (2)$$

Equating coefficients of like powers of x yields

$$8a_2 = 2 \qquad 24a_3 + a_0 = 1 \qquad 2a_2 + 48a_4 + a_1 = 0 \qquad 6a_3 + 80a_5 + a_2 = 0 \qquad \cdots \qquad (3)$$

and, in general, $n(n-1)a_n + 4(n+2)(n+1)a_{n+2} + a_{n-1} = 0$ for $n = 2, 3, \ldots$. This is equivalent to

$$a_{n+2} = -\frac{n(n-1)}{4(n+2)(n+1)}a_n - \frac{1}{4(n+2)(n+1)}a_{n-1} \qquad n = 2, 3, \ldots \qquad (4)$$

The recursion formula is not valid for $n = 0$ or $n = 1$, since the coefficients of x^0 and x^1 on the right side of (1) are not zero.

15.120 Find the general solution near $x = 0$ of $(x^2 + 4)y'' + xy = x + 2$.

▌ Using (2) of the previous problem, we have $a_2 = \frac{1}{4}$ and $a_3 = \frac{1}{24} - \frac{1}{24}a_0$. The recursion formula (4), which is valid for $n = 2, 3, \ldots$, yields

$$a_4 = -\tfrac{1}{24}a_2 - \tfrac{1}{48}a_1 = -\tfrac{1}{24}(\tfrac{1}{4}) - \tfrac{1}{48}a_1 = -\tfrac{1}{96} - \tfrac{1}{48}a_1$$
$$a_5 = -\tfrac{3}{40}a_3 - \tfrac{1}{80}a_2 = -\tfrac{3}{40}(\tfrac{1}{24} - \tfrac{1}{24}a_0) - \tfrac{1}{80}(\tfrac{1}{4}) = \tfrac{-1}{160} + \tfrac{1}{320}a_0$$
$$\cdots\cdots\cdots\cdots\cdots\cdots\cdots\cdots\cdots$$

Thus,
$$y = a_0 + a_1x + \tfrac{1}{4}x^2 + (\tfrac{1}{24} - \tfrac{1}{24}a_0)x^3 + (-\tfrac{1}{96} - \tfrac{1}{48}a_1)x^4 + (\tfrac{-1}{160} + \tfrac{1}{320}a_0)x^5 + \cdots$$
$$= (\tfrac{1}{4}x^2 + \tfrac{1}{24}x^3 - \tfrac{1}{96}x^4 - \tfrac{1}{160}x^5 + \cdots) + a_0(1 - \tfrac{1}{24}x^3 + \tfrac{1}{320}x^5 + \cdots) + a_1(x - \tfrac{1}{48}x^4 + \cdots)$$

The first series is the particular solution. The second and third series together represent the general solution of the associated homogeneous equation $(x^2 + 4)y'' + xy = 0$.

15.121 Find a recursion formula for the coefficients of the general power-series solution near $x = 0$ of $(x^2 + 4)y'' + xy = x^2 - x$.

▌ Since the left side of this differential equation is identical to the left side of the equation in Problem 15.119, much of the work done there remains valid. Equation (2) of that problem becomes

$$8a_2 + x(24a_3 + a_0) + x^2(2a_2 + 48a_4 + a_1) + x^3(6a_3 + 80a_5 + a_2) + \cdots$$
$$+ x^n[n(n-1)a_n + 4(n+2)(n+1)a_{n+2} + a_{n-1}] + \cdots = 0 + (-1)x + (1)x^2 + 0x^3 + 0x^4 + 0x^5 + \cdots$$

reflecting the new right side. Equating coefficients of like powers of x yields

$$8a_2 = 0 \qquad 24a_3 + a_0 = -1 \qquad 2a_2 + 48a_4 + a_1 = 1 \qquad 6a_3 + 80a_5 + a_2 = 0 \qquad \cdots \qquad (1)$$

and, in general, $n(n-1)a_n + 4(n+2)(n+1)a_{n+2} + a_{n-1}$. This may be rewritten as

$$a_{n+2} = -\frac{n(n-1)}{4(n+2)(n+1)}a_n - \frac{1}{4(n+2)(n+1)}a_{n-1} \qquad n = 0, 3, 4, \ldots \qquad (2)$$

This is the same recursion formula as in Problem 15.119, but here it is not valid for $n = 1$ and $n = 2$ because the coefficients of x and x^2 on the right side of the differential equation are not zero.

15.122 Find the general solution near $x = 0$ of $(x^2+4)y'' + xy = x^2 - x$.

▍ Using (*1*) of the previous problem, we have

$$a_2 = 0 \qquad a_3 = -\tfrac{1}{24} - \tfrac{1}{24}a_0 \qquad a_4 = \tfrac{1}{48} - \tfrac{1}{24}a_2 - \tfrac{1}{48}a_1 = \tfrac{1}{48} - \tfrac{1}{48}a_1$$

The recursion formula (*2*) is valid for $n = 3, 4, 5, \ldots$ and yields

$$a_5 = -\tfrac{3}{40}a_3 - \tfrac{1}{80}a_2 = -\tfrac{3}{40}\left(-\tfrac{1}{24} - \tfrac{1}{24}a_0\right) = \tfrac{1}{320} + \tfrac{1}{320}a_0$$
$$a_6 = -\tfrac{1}{10}a_4 - \tfrac{1}{120}a_3 = -\tfrac{1}{10}\left(\tfrac{1}{48} - \tfrac{1}{48}a_1\right) - \tfrac{1}{120}\left(-\tfrac{1}{24} - \tfrac{1}{24}a_0\right) = -\tfrac{1}{576} + \tfrac{1}{480}a_1 + \tfrac{1}{2880}a_0$$
$$\cdots\cdots\cdots\cdots\cdots\cdots\cdots\cdots\cdots\cdots\cdots\cdots$$

The general solution is

$$y = a_0 + a_1x + (0)x^2 + \left(-\tfrac{1}{24} - \tfrac{1}{24}a_0\right)x^3 + \left(\tfrac{1}{48} - \tfrac{1}{48}a_1\right)x^4 + \left(\tfrac{1}{320} + \tfrac{1}{320}a_0\right)x^5 + \left(-\tfrac{1}{576} + \tfrac{1}{480}a_1 + \tfrac{1}{2880}a_0\right)x^6 + \cdots$$
$$= a_0\left(1 - \tfrac{1}{24}x^3 + \tfrac{1}{320}x^5 + \tfrac{1}{2880}x^6 + \cdots\right) + a_1\left(x - \tfrac{1}{48}x^4 + \tfrac{1}{480}x^6 + \cdots\right) + \left(-\tfrac{1}{24}x^3 + \tfrac{1}{48}x^4 + \tfrac{1}{320}x^5 - \tfrac{1}{576}x^6 + \cdots\right)$$

15.123 Find a recursion formula for the coefficients of the general power-series solution near $x = 0$ of $(x^2+4)y'' + xy = e^x$.

▍ Since the left side of this differential equation is identical to the left side of the equation in Problem 15.119, much of the work done there remains valid. Furthermore, it follows from Problem 15.5 that $e^x = \sum_{n=0}^{\infty} \frac{1}{n!}x^n$. Thus, (2) of Problem 15.119 becomes

$$8a_2 + x(24a_3 + a_0) + x^2(2a_2 + 48a_4 + a_1) + x^3(6a_3 + 80a_5 + a_2) + \cdots$$
$$+ x^n[n(n-1)a_n + 4(n+2)(n+1)a_{n+2} + a_{n-1}] + \cdots = 1 + \frac{1}{1!}x + \frac{1}{2!}x^2 + \frac{1}{3!}x^3 + \cdots + \frac{1}{n!}x^n + \cdots$$

reflecting the new right side. Equating coefficients of like powers of x, we get

$$8a_2 = 1 \qquad 24a_3 + a_0 = \frac{1}{1!} \qquad 2a_2 + 48a_4 + a_1 = \frac{1}{2!} \cdots$$

and, in general, $n(n-1)a_n + 4(n+2)(n+1)a_{n+2} + a_{n-1} = \dfrac{1}{n!}$. This may be rewritten as

$$a_{n+2} = \frac{1}{4(n+2)!} - \frac{n(n-1)}{4(n+2)(n+1)}a_n - \frac{1}{4(n+2)(n+1)}a_{n-1} \qquad n = 0, 1, 2, \ldots \qquad (1)$$

15.124 Find the general solution near $x = 0$ of $(x^2+4)y'' + xy = e^x$.

▍ Using the recursion formula (*1*) of the previous problem, we get

$$a_2 = \tfrac{1}{8} \qquad a_3 = \tfrac{1}{24} - \tfrac{1}{24}a_0$$
$$a_4 = \tfrac{1}{96} - \tfrac{1}{24}a_2 - \tfrac{1}{48}a_1 = \tfrac{1}{192} - \tfrac{1}{48}a_1 \qquad a_5 = \tfrac{1}{480} - \tfrac{3}{40}a_3 - \tfrac{1}{80}a_2 = -\tfrac{1}{384} + \tfrac{1}{320}a_0$$
$$\cdots\cdots\cdots\cdots\cdots\cdots \qquad \cdots\cdots\cdots\cdots\cdots\cdots$$

The general solution is

$$y = a_0 + a_1x + \tfrac{1}{8}x^2 + \left(\tfrac{1}{24} - \tfrac{1}{24}a_0\right)x^3 + \left(\tfrac{1}{192} - \tfrac{1}{48}a_1\right)x^4 + \left(-\tfrac{1}{384} + \tfrac{1}{320}a_0\right)x^5 + \cdots$$
$$= a_0\left(1 - \tfrac{1}{24}x^3 + \tfrac{1}{320}x^5 + \cdots\right) + a_1\left(x - \tfrac{1}{48}x^4 + \cdots\right) + x^2\left(\tfrac{1}{8} + \tfrac{1}{24}x + \tfrac{1}{192}x^2 - \tfrac{1}{384}x^3 + \cdots\right)$$

15.125 Find a recursion formula for the coefficients of the general power-series solution near $x = 0$ of $y'' - xy' = e^{-x}$.

▍ Here $x = 0$ is an ordinary point, so we assume a solution of the form

$$y = a_0 + a_1x + a_2x^2 + a_3x^3 + a_4x^4 + \cdots + a_nx^n + a_{n+1}x^{n+1} + a_{n+2}x^{n+2} + \cdots$$

Substituting the first two derivatives of y into the differential equation and noting from Problem 15.5 (with $-x$ replacing x) that the Maclaurin-series expansion for e^{-x} is $\sum_{n=0}^{\infty} \frac{1}{n!}(-x)^n = \sum_{n=0}^{\infty} \frac{(-1)^n}{n!} x^n$, we have

$$[2a_2 + 6a_3x + 12a_4x^2 + \cdots + n(n-1)a_nx^{n-2} + (n+1)(n)a_{n+1}x^{n-1} + (n+2)(n+1)a_{n+2}x^n + \cdots]$$
$$- x[a_1 + 2a_2x + 3a_3x^2 + 4a_4x^3 + \cdots + na_nx^{n-1} + (n+1)a_{n+1}x^n + (n+2)a_{n+2}x^{n+1} + \cdots]$$
$$= 1 - x + \frac{1}{2!}x^2 - \frac{1}{3!}x^3 + \cdots + \frac{(-1)^n}{n!}x^n + \cdots$$

or

$$2a_2 + x(6a_3 - a_1) + x^2(12a_4 - 2a_2) + \cdots + x^n[(n+2)(n+1)a_{n+2} - na_n] + \cdots$$
$$= 1 - x + \frac{1}{2!}x^2 - \frac{1}{3!}x^3 + \cdots + \frac{(-1)^n}{n!}x^n + \cdots \qquad (1)$$

Equating coefficients of like powers of x yields

$$2a_2 = 1 \qquad 6a_3 - a_1 = -1 \qquad 12a_4 - 2a_2 = \frac{1}{2!} \qquad \cdots$$

and, in general, $(n+2)(n+1)a_{n+2} - na_n = (-1)^n/n!$. This may be written as $a_{n+2} = \frac{(-1)^n}{(n+2)!} + \frac{n}{(n+2)(n+1)} a_n$.

15.126 Find the general solution near $x = 0$ of $y'' - xy' = e^{-x}$.

▮ Evaluating the recursion formula of the previous problem for successive values of n, we obtain

$$a_2 = \tfrac{1}{2} + 0a_2 = \tfrac{1}{2} \qquad a_3 = -\tfrac{1}{6} + \tfrac{1}{6}a_1$$
$$a_4 = \tfrac{1}{24} + \tfrac{1}{6}a_2 = \tfrac{1}{8} \qquad a_5 = -\tfrac{1}{120} + \tfrac{3}{20}a_3 = -\tfrac{1}{30} + \tfrac{1}{40}a_1$$
$$\cdots\cdots \qquad \cdots\cdots$$

The general solution is $y = (\frac{1}{2}x^2 - \frac{1}{6}x^3 + \frac{1}{8}x^4 - \frac{1}{30}x^5 + \cdots) + a_0 + a_1(x + \frac{1}{6}x^3 + \frac{1}{40}x^5 + \cdots)$.

15.127 Find a recursion formula for the coefficients of the general power-series solution near $x = 0$ of $y'' - xy' = 12x^3$.

▮ Since the left side of this differential equation is identical to the left side of the equation in Problem 15.125, much of the work done there remains valid. Equation (1) of that problem becomes

$$2a_2 + x(6a_3 - a_1) + x^2(12a_4 - 2a_2) + x^3(20a_5 - 3a_3) + \cdots + x^n[(n+2)(n+1)a_{n+2} - na_n] + \cdots$$
$$= 0 + (0)x + (0)x^2 + 12x^3 + (0)x^4 + \cdots + (0)x^n + \cdots$$

reflecting the new right side. Equating coefficients of like powers of x yields

$$2a_2 = 0 \qquad 6a_3 - a_1 = 0 \qquad 12a_4 - 2a_2 = 0 \qquad 20a_5 - 3a_3 = 12 \qquad \cdots \qquad (1)$$

and, in general, $(n+2)(n+1)a_{n+2} - na_n = 0$. This may be rewritten as $a_{n+2} = \frac{n}{(n+2)(n+1)} a_n$. This recursion formula is valid for all n except $n = 3$; (1) must be used to determine a_5.

15.128 Find the general solution near $x = 0$ of $y'' - xy' = 12x^3$.

▮ It follows from (1) of the preceding problem that

$$a_2 = 0 \qquad a_3 = \tfrac{1}{6}a_1 \qquad a_4 = \tfrac{1}{6}a_2 = \tfrac{1}{6}(0) = 0 \qquad a_5 = \tfrac{3}{5} + \tfrac{3}{20}a_3 = \tfrac{3}{5} + \tfrac{3}{20}(\tfrac{1}{6}a_1) = \tfrac{3}{5} + \tfrac{1}{40}a_1$$

and from the recursion formula that

$$a_6 = a_8 = a_{10} = \cdots = 0$$
$$a_7 = \tfrac{5}{42}a_5 = \tfrac{5}{42}(\tfrac{3}{5} + \tfrac{1}{40}a_1) = \tfrac{3}{42} + \tfrac{1}{336}a_1$$
$$a_9 = \tfrac{7}{72}a_7 = \tfrac{7}{72}(\tfrac{3}{42} + \tfrac{1}{336}a_1) = \tfrac{1}{144} + \tfrac{1}{3456}a_1$$
$$\cdots\cdots$$

The general solution is then

$$y = a_0(1) + a_1(x + \tfrac{1}{6}x^3 + \tfrac{1}{40}x^5 + \tfrac{1}{336}x^7 + \tfrac{1}{3456}x^9 + \cdots) + (\tfrac{3}{5}x^5 + \tfrac{3}{42}x^7 + \tfrac{1}{144}x^9 + \cdots)$$

15.129 Find a recursion formula for the coefficients of the general power-series solution near $x = 0$ of $y'' - 2x^2y' + 4xy = x^2 + 2x + 2$.

▌ The differential equation is in standard form with $P(x) = -2x^2$, $Q(x) = 4x$, and $\phi(x) = x^2 + 2x + 2$. Since all three functions are analytic everywhere, it follows that every point, including $x = 0$, is an ordinary point. The general solution can be expressed in the form $y = A_0 + A_1x + A_2x^2 + A_3x^3 + A_4x^4 + A_5x^5 + \cdots + A_nx^n + \cdots$. Then substitution in the given equation yields

$$y'' - 2x^2y' + 4xy - x^2 - 2x - 2 = (2A_2 - 2) + (6A_3 + 4A_0 - 2)x + (12A_4 + 2A_1 - 1)x^2 + 20A_5x^3 + \cdots$$
$$+ [(n+2)(n+1)A_{n+2} - 2(n-1)A_{n-1} + 4A_{n-1}]x^n + \cdots = 0$$

Equating the coefficients to zero, we obtain

$$A_2 = 1 \qquad A_3 = \tfrac{1}{3} - \tfrac{2}{3}A_0 \qquad A_4 = \tfrac{1}{12} - \tfrac{1}{6}A_1 \qquad A_5 = 0 \qquad \cdots \tag{1}$$

and, in general, $(n+2)(n+1)A_{n+2} - 2(n-3)A_{n-1} = 0$, so that $A_{n+2} = \dfrac{2(n-3)}{(n+1)(n+2)}A_{n-1}$ for $n \geq 3$.

15.130 Find the general solution near $x = 0$ of $y'' - 2x^2y' + 4xy = x^2 + 2x + 2$.

▌ Using (*1*) of the previous problem, we have

$$A_2 = 1 \qquad A_3 = \tfrac{1}{3} - \tfrac{2}{3}A_0 \qquad A_4 = \tfrac{1}{12} - \tfrac{1}{6}A_1 \qquad A_5 = 0$$

Then, use of the recursion formula yields

$$\begin{aligned}
A_6 &= \tfrac{1}{15}A_3 = \tfrac{1}{15}(\tfrac{1}{3} - \tfrac{2}{3}A_0) = \tfrac{1}{45} - \tfrac{2}{45}A_0\\
A_7 &= \tfrac{2}{21}A_4 = \tfrac{2}{21}(\tfrac{1}{12} - \tfrac{1}{6}A_1) = \tfrac{1}{126} - \tfrac{1}{63}A_1\\
A_8 &= A_{11} = A_{14} = \cdots = 0\\
A_9 &= \tfrac{1}{9}A_6 = \tfrac{1}{9}(\tfrac{1}{45} - \tfrac{2}{45}A_0) = \tfrac{1}{405} - \tfrac{2}{405}A_0\\
A_{10} &= \tfrac{1}{9}A_7 = \tfrac{1}{9}(\tfrac{1}{126} - \tfrac{1}{63}A_1) = \tfrac{1}{1134} - \tfrac{1}{567}A_1\\
&\cdots\cdots\cdots
\end{aligned}$$

The complete solution is then

$$y = A_0(1 - \tfrac{2}{3}x^3 - \tfrac{2}{45}x^6 - \tfrac{2}{405}x^9 - \cdots) + A_1(x - \tfrac{1}{6}x^4 - \tfrac{1}{63}x^7 - \tfrac{1}{567}x^{10} - \cdots)$$
$$+ x^2 + \tfrac{1}{3}x^3 + \tfrac{1}{12}x^4 + \tfrac{1}{45}x^6 + \tfrac{1}{126}x^7 + \tfrac{1}{405}x^9 + \tfrac{1}{1134}x^{10} + \cdots$$

15.131 Find a recursion formula for the coefficients of the general power-series solution near $t = 0$ of $d^2y/dt^2 + t^2y = 1 + t + t^2$.

▌ Here $t = 0$ (as well as every other point) is an ordinary point. The general solution has the form

$$y = a_0 + a_1t + a_2t^2 + a_3t^3 + a_4t^4 + a_5t^5 + \cdots + a_{n-2}t^{n-2} + a_{n-1}t^{n-1} + a_nt^n + a_{n+1}t^{n+1} + a_{n+2}t^{n+2} + \cdots$$

Substituting y and its second derivative into the differential equation, we get

$$[2a_2 + 6a_3t + 12a_4t^2 + 20a_5t^3 + \cdots + (n+2)(n+1)a_{n+2}t^n + \cdots]$$
$$+ t^2[a_0 + a_1t + a_2t^2 + \cdots + a_{n-2}t^{n-2} + \cdots] = 1 + t + t^2$$

or
$$2a_2 + t(6a_3) + t^2(12a_4 + a_0) + t^3(20a_5 + a_1) + \cdots + t^n[(n+2)(n+1)a_{n+2} + a_{n-2}] + \cdots$$
$$= 1 + t + t^2 + (0)t^3 + (0)t^4 + \cdots + (0)t^n + \cdots$$

Equating coefficients of like powers of t yields

$$2a_2 = 1 \qquad 6a_3 = 1 \qquad 12a_4 + a_0 = 1 \qquad 20a_5 + a_1 = 0 \qquad \cdots \tag{1}$$

and, in general, $(n+2)(n+1)a_{n+2} + a_{n-2} = 0$. This may be written as $a_{n+2} = \dfrac{-1}{(n+2)(n+1)}a_{n-2}$ for $n \geq 3$.

15.132 Find the general solution near $t = 0$ of $d^2y/dt^2 + t^2y = 1 + t + t^2$.

▌ Using (*1*) of the previous problem, we have $a_2 = \tfrac{1}{2}$, $a_3 = \tfrac{1}{6}$, and $a_4 = \tfrac{1}{12} - \tfrac{1}{12}a_0$. Then the recursion formula produces

$$a_5 = -\tfrac{1}{20}a_1$$
$$a_6 = -\tfrac{1}{30}a_2 = -\tfrac{1}{30}(\tfrac{1}{2}) = -\tfrac{1}{60}$$
$$a_7 = -\tfrac{1}{42}a_3 = -\tfrac{1}{42}(\tfrac{1}{6}) = -\tfrac{1}{252}$$
$$a_8 = -\tfrac{1}{56}a_4 = -\tfrac{1}{56}(\tfrac{1}{12} - \tfrac{1}{12}a_0) = -\tfrac{1}{672} + \tfrac{1}{672}a_0$$
$$a_9 = -\tfrac{1}{72}a_5 = -\tfrac{1}{72}(-\tfrac{1}{20}a_1) = \tfrac{1}{1440}a_1$$
$$\cdots\cdots\cdots\cdots\cdots\cdots\cdots\cdots\cdots\cdots$$

The general solution is

$$y = a_0\left(1 - \frac{t^4}{12} + \frac{t^8}{672} - \cdots\right) + a_1\left(t - \frac{t^5}{20} + \frac{t^9}{1440} - \cdots\right) + \frac{t^2}{2} + \frac{t^3}{6} + \frac{t^4}{12} - \frac{t^6}{60} - \frac{t^7}{252} - \frac{t^8}{672} + \cdots$$

5.133 Find the general solution near $x = -2$ of $y'' + (x^2 + 4x + 4)y = x^2 + 5x + 7$.

▌ We translate the point -2 to the origin with the transformation $t = x - (-2) = x + 2$ (see Problem 15.105). Then $x = t - 2$, $dy/dx = dy/dt$, and $d^2y/dx^2 = d^2y/dt^2$. The differential equation therefore becomes

$$\frac{d^2y}{dt^2} + [(t-2)^2 + 4(t-2) + 4]y = (t-2)^2 + 5(t-2) + 7, \quad \text{or} \quad \frac{d^2y}{dt^2} + t^2y = t^2 + t + 1.$$

The general solution to this equation near $t = 0$ is found in the previous problem. Replacing t with $x + 2$ in that solution, we obtain the solution to the original problem as

$$y = a_0\left[1 - \frac{(x+2)^4}{12} + \frac{(x+2)^8}{672} + \cdots\right] + a_1\left[(x+2) - \frac{(x+2)^5}{20} + \frac{(x+2)^9}{1440} - \cdots\right]$$
$$+ \frac{(x+2)^2}{2} + \frac{(x+2)^3}{6} + \frac{(x+2)^4}{12} - \frac{(x+2)^6}{60} - \frac{(x+2)^7}{252} - \frac{(x+2)^8}{672} + \cdots$$

5.134 Find a recursion formula for coefficients of the general power-series solution near $t = 0$ of $\ddot{y} + ty = e^{t+1}$.

▌ Here $t = 0$ is an ordinary point, so we assume a solution of the form

$$y = a_0 + a_1t + a_2t^2 + a_3t^3 + a_4t^4 + \cdots + a_{n-1}t^{n-1} + a_nt^n + a_{n+1}t^{n+1} + a_{n+2}t^{n+2} + \cdots$$

Substituting y and its second derivative into the differential equation, we get

$$2a_2 + 6a_3t + 12a_4t^2 + \cdots + (n+2)(n+1)a_{n+2}t^n + \cdots + t(a_0 + a_1t + a_2t^2 + \cdots + a_{n-1}t^{n-1} + \cdots) = e^{t+1}$$

Recall that e^{t+1} has the Taylor expansion $e^{t+1} = e\sum_{n=0}^{\infty} t^n/n!$ about $t = 0$. Thus, the last equation can be rewritten as

$$2a_2 + t(6a_3 + a_0) + t^2(12a_4 + a_1) + \cdots + t^n[(n+2)(n+1)a_{n+2} + a_{n-1}] + \cdots$$
$$= \frac{e}{0!} + \frac{e}{1!}t + \frac{e}{2!}t^2 + \cdots + \frac{e}{n!}t^n + \cdots$$

Equating coefficients of like powers of t yields $2a_2 = \dfrac{e}{0!}$, $6a_3 + a_0 = \dfrac{e}{1!}$, $12a_4 + a_1 = \dfrac{e}{2!}, \ldots$. In general, $(n+2)(n+1)a_{n+2} + a_{n-1} = e/n!$ for $n = 1, 2, \ldots,$ or

$$a_{n+2} = -\frac{1}{(n+2)(n+1)}a_{n-1} + \frac{e}{(n+2)(n+1)n!}$$

5.135 Find the general solution near $t = 0$ of $\ddot{y} + ty = e^{t+1}$.

▌ It follows from the previous problem that $a_2 = \dfrac{e}{2}$, and then from the recursion formula that $a_3 = -\dfrac{1}{6}a_0 + \dfrac{e}{6}$, $a_4 = -\dfrac{1}{12}a_1 + \dfrac{e}{24}, \ldots$. Thus,

$$y = a_0 + a_1t + \frac{e}{2}t^2 + \left(-\frac{1}{6}a_0 + \frac{e}{6}\right)t^3 + \left(-\frac{1}{12}a_1 + \frac{e}{24}\right)t^4 + \cdots$$
$$= e\left(\frac{1}{2}t^2 + \frac{1}{6}t^3 + \frac{1}{24}t^4 + \cdots\right) + a_0\left(1 - \frac{1}{6}t^3 + \cdots\right) + a_1\left(t - \frac{1}{12}t^4 + \cdots\right)$$

15.136 Find the general solution near $x = 1$ of $y'' + (x+1)y = e^x$.

▌ We translate the point 1 to the origin with the transformation $t = x - 1$ (see Problem 15.105). Then $x = t + 1$, $\dfrac{dy}{dx} = \dfrac{dy}{dt}$, and $\dfrac{d^2y}{dx^2} = \dfrac{d^2y}{dt^2}$. The differential equation therefore becomes $\ddot{y} + ty = e^{t+1}$. The general solution to this equation near $t = 0$ is given in the previous problem. Replacing t with $x - 1$ in that solution, we obtain the solution to the original problem as

$$y = e[\tfrac{1}{2}(x-1)^2 + \tfrac{1}{6}(x-1)^3 + \tfrac{1}{24}(x-1)^4 + \cdots] + a_0[1 - \tfrac{1}{6}(x-1)^3 + \cdots] + a_1[(x-1) - \tfrac{1}{12}(x-1)^4 + \cdots]$$

15.137 Find a recursion formula for the coefficients of the general power-series solution near $t = 0$ of $y' + 7y = 5t$.

▌ This differential equation is in standard form with $P(t) = 7$ and $\phi(t) = 5t$. Both functions are analytic everywhere, so $t = 0$ is an ordinary point. We assume a solution of the form $y(t) = \sum_{k=0}^{\infty} b_k t^k$, so that $y'(t) = \sum_{k=0}^{\infty} k b_k t^{k-1}$. The power series for $y' + 7y$ is then $y'(t) + 7y(t) = \sum_{k=0}^{\infty} k b_k t^{k-1} + 7\sum_{k=0}^{\infty} b_k t^k$.

The two series are added by adding like powers of t. Since t^3 is the $k = 3$ term in the second series but the $k = 4$ term in the first, we shall change the dummy variable in the series for y', using a double change of variable. We first substitute $k - 1 = m$ or $k = m + 1$ to obtain $y'(t) = \sum_{m+1=0}^{\infty} (m+1)b_{m+1}t^m$, and then let $k = m$, which yields $y'(t) = \sum_{k+1=0}^{\infty} (k+1)b_{k+1}t^k$. Finally, since the $k + 1 = 0$ term contains the factor $k + 1$ whose value is zero, the series may begin with $k = 0$. That is, $y'(t) = \sum_{k=0}^{\infty} (k+1)b_{k+1}t^k$. Then the power series becomes

$$y'(t) + 7y(t) = \sum_{k=0}^{\infty} (k+1)b_{k+1}t^k + 7\sum_{k=0}^{\infty} b_k t^k = \sum_{k=0}^{\infty} [(k+1)b_{k+1} + 7b_k]t^k$$

Since $5t = 0 + 5t + \sum_{k=2}^{\infty} 0t^k$, the differential equation becomes

$$\sum_{k=0}^{\infty} [(k+1)b_{k+1} + 7b_k]t^k = 0 + 5t + \sum_{k=2}^{\infty} 0t^k$$

Equating coefficients of like powers of t now yields $b_1 + 7b_0 = 0$, $2b_2 + 7b_1 = 5$, and, for $k \geq 2$, $(k+1)b_{k+1} + 7b_k = 0$. This last expression may be written as $b_{k+1} = \dfrac{-7}{k+1} b_k$ for $k \geq 2$.

15.138 Find the general solution near $t = 0$ of $y' + 7y = 5t$.

▌ It follows from the previous problem that $b_1 = -7b_0$ and $b_2 = \tfrac{5}{2} - \tfrac{7}{2}b_1 = \tfrac{5}{2} - \tfrac{7}{2}(-7b_0) = \tfrac{1}{2}(5 + 7^2 b_0)$. Then, from the recursion formula,

$$b_3 = (-7)[(-7)^2 b_0 + 5](\tfrac{1}{2})(\tfrac{1}{3})$$
$$b_4 = (-7)^2[(-7)^2 b_0 + 5](\tfrac{1}{2})(\tfrac{1}{3})(\tfrac{1}{4})$$
$$\cdots\cdots\cdots\cdots\cdots\cdots\cdots\cdots\cdots\cdots\cdots\cdots$$

and $$b_k = (-7)^{k-2}\frac{(-7)^2 b_0 + 5}{k!} = (-7)^k \frac{b_0 + 5(-7)^{-2}}{k!} \quad \text{for } k \geq 2$$

Thus, $$y(t) = b_0 + (-7)b_0 t + [b_0 + 5(-7)^{-2}]\sum_{k=2}^{\infty} \frac{(-7t)^k}{k!}$$
$$= b_0 - 7b_0 t + (b_0 + \tfrac{5}{49})\sum_{k=0}^{\infty} \frac{(-7t)^k}{k!} - (b_0 + \tfrac{5}{49})(1 - 7t)$$
$$= b_0 - 7b_0 t + (b_0 + \tfrac{5}{49})e^{-7t} - (b_0 + \tfrac{5}{49})(1 - 7t)$$

where we have used the result of Problem 15.5 with $x = -7t$. If we set $c_0 = b_0 + \tfrac{5}{49}$, we may rewrite the solution as $y = (c_0 - \tfrac{5}{49})(1 - 7t) + c_0 e^{-7t} - c_0(1 - 7t) = c_0 e^{-7t} - \tfrac{5}{49} + \tfrac{5}{7}t$. This solution is obtained more simply with the methods of Chapters 8 and 9 or the method given in Chapter 5, because the differential equation has constant coefficients.

15.139 Find a recursion formula for the coefficients of the general power-series solution near $z = 0$ of $(z + 1)\,dy/dz - y = z + 2$.

▎This differential equation has the standard form $\dfrac{dy}{dz} - \dfrac{1}{z+1}y = \dfrac{z+2}{z+1}$, with $P(z) = -\dfrac{1}{z+1}$ and $\phi(z) = \dfrac{z+2}{z+1}$. Since both functions are analytic at $z = 0$, that point is an ordinary point. We write the general solution to the given differential equation as $y = A_0 + A_1 z + A_2 z^2 + A_3 z^3 + A_4 z^4 + \cdots + A_n z^n + \cdots$. Then $dy/dz = A_1 + 2A_2 z + 3A_3 z^2 + 4A_4 z^3 + \cdots + nA_n z^{n-1} + \cdots$ and

$$\begin{aligned}(z+1)\frac{dy}{dz} - y - z - 2 &= (z+1)(A_1 + 2A_2 z + 3A_3 z^2 + 4A_4 z^3 + \cdots + nA_n z^{n-1} + \cdots) \\ &\quad - z - 2 - (A_0 + A_1 z + A_2 z^2 + A_3 z^3 + \cdots + A_n z^n + \cdots) \\ &= (A_1 - 2 - A_0) + (2A_2 - 1)z + (3A_3 + A_2)z^2 + (4A_4 + 2A_3)z^3 + \cdots + [(n+1)A_{n+1} \\ &\quad + (n-1)A_n]z^n + \cdots = 0\end{aligned}$$

Equating to zero the coefficients of the distinct powers of z yields

$$A_1 = 2 + A_0 \qquad A_3 = -\tfrac{1}{3}A_2 = -\tfrac{1}{6} \qquad A_2 = \tfrac{1}{2} \qquad A_4 = -\tfrac{1}{2}A_3 = \tfrac{1}{12} \qquad \cdots$$

and, in general, $(n+1)A_{n+1} + (n-1)A_n = 0$, so that $A_{n+1} = -\dfrac{n-1}{n+1}A_n$ for $n \geq 2$.

15.140 Find a second form of the recursion formula obtained in the previous problem.

▎We have $A_{n+1} = -\dfrac{n-1}{n+1}A_n$ for $n \geq 2$. Setting $k = n + 1$, we can rewrite this formula as $A_k = -\dfrac{k-2}{k}A_{k-1}$ for $k \geq 3$. Replacing the dummy index k with the dummy index n, we finally obtain $A_n = -\dfrac{n-2}{n}A_{n-1}$ for $n \geq 3$.

15.141 Find the general solution near $z = 0$ of $(z + 1)\,dy/dz - y = z + 2$.

▎We have from Problem 15.139 that $A_1 = 2 + A_0$ and $A_2 = \frac{1}{2}$. For other values of A_n $(n = 3, 4, 5, \ldots)$ we can use either the recursion formula of Problem 15.139 or that of Problem 15.140. We use the latter, which yields

$$\begin{aligned}A_n &= -\frac{n-2}{n}A_{n-1} = \frac{(n-2)(n-3)}{n(n-1)}A_{n-2} = \frac{(n-2)(n-3)(n-4)}{n(n-1)(n-2)}A_{n-3} = \cdots \\ &= (-1)^n\frac{(n-2)(n-3)\cdots(2)(1)}{n(n-1)\cdots(4)(3)}A_2 = (-1)^n\frac{1}{n(n-1)} \qquad \text{for } n \geq 2\end{aligned}$$

Then
$$y = A_0 + (2 + A_0)z + \tfrac{1}{2}z^2 - \tfrac{1}{6}z^3 + \tfrac{1}{12}z^4 - \cdots + (-1)^n\frac{1}{n(n-1)}z^n + \cdots \tag{1}$$

15.142 Find the general solution near $x = 1$ of $xy' - y = x + 1$.

▎We use the transformation $z = x - 1$ to translate the point 1 to the origin (see Problem 15.105). Then $x = z + 1$ and $\dfrac{dy}{dx} = \dfrac{dy}{dz}$. The differential equation therefore becomes $(z + 1)\,dy/dz - y = z + 2$. The general solution to this differential equation near $z = 0$ is given by *(1)* of the previous problem. Substituting $x - 1$ for z, we obtain the solution to the original problem as

$$y = A_0 x + 2(x-1) + \frac{1}{2}(x-1)^2 - \frac{1}{6}(x-1)^3 + \frac{1}{12}(x-1)^4 = A_0 x + 2(x-1) + \sum_{n=2}^{\infty}(-1)^n\frac{1}{n(n-1)}(x-1)^n$$

INITIAL-VALUE PROBLEMS

15.143 Solve $y'' - xy' + 2y = 0$; $y(0) = 2$, $y'(0) = 3$.

▎Since the initial conditions are prescribed at $x = 0$, it is advantageous to obtain the general solution to the differential equation near $x = 0$. This is done in Problem 15.85. Applying the initial conditions to that solution, we find $a_0 = 2$ and $a_1 = 3$. The solution to the initial-value problem is then

$$y = 2(1 - x^2) + 3(x - \tfrac{1}{6}x^3 - \tfrac{1}{120}x^5 - \tfrac{1}{1680}x^7 - \cdots) = 2 + 3x - 2x^2 - \tfrac{1}{2}x^3 - \tfrac{1}{40}x^5 - \tfrac{1}{560}x^7 - \cdots$$

15.144 Solve $y'' - xy' + 2y = 0$; $y(1) = 2$, $y'(1) = 3$.

▮ This is the same differential equation as in the previous problem but with different initial conditions. Since these are specified at $x = 1$, it is advantageous to obtain the general solution to the differential equation near $x = 1$. This is done in Problem 15.108. Applying the initial conditions to that solution, we find $a_0 = 2$ and $a_1 = 3$. The solution to the initial-value problem is then

$$y = 2[1 - (x-1)^2 - \tfrac{1}{3}(x-1)^3 - \tfrac{1}{12}(x-1)^4 - \tfrac{1}{30}(x-1)^5 - \cdots] + 3[(x-1) + \tfrac{1}{2}(x-1)^2]$$
$$= 2 + 3(x-1) - \tfrac{1}{2}(x-1)^2 - \tfrac{2}{3}(x-1)^3 - \tfrac{1}{6}(x-1)^4 - \tfrac{1}{15}(x-1)^5 - \cdots$$

15.145 Discuss the difficulties encountered in attempting to solve the previous problem by first finding the general solution to the differential equation near $x = 0$.

▮ The general solution near $x = 0$ was found in Problem 15.85 to be

$$y = a_0(1 - x^2) + a_1(x - \tfrac{1}{6}x^3 - \tfrac{1}{120}x^5 - \tfrac{1}{1680}x^7 - \cdots) \tag{1}$$

Differentiating this equation term by term yields $y' = a_0(-2x) + a_1(1 - \tfrac{1}{2}x^2 - \tfrac{1}{24}x^4 - \tfrac{1}{240}x^6 - \cdots)$. Then, applying the initial conditions, we get the two equations

$$2 = y(1) = 0a_0 + a_1(1 - \tfrac{1}{6} - \tfrac{1}{120} - \tfrac{1}{1680} - \cdots) \qquad 3 = y'(1) = -2a_0 + a_1(1 - \tfrac{1}{2} - \tfrac{1}{24} - \tfrac{1}{240} - \cdots)$$

Since the limits of the two infinite series are not obvious, the only available representation for a_0 and a_1 is

$$a_1 = \frac{2}{1 - \frac{1}{6} - \frac{1}{120} - \frac{1}{1680} - \cdots} \qquad \text{and} \qquad a_2 = -\frac{3}{2} + \frac{1 - \frac{1}{2} - \frac{1}{24} - \frac{1}{240} - \cdots}{1 - \frac{1}{6} - \frac{1}{120} - \frac{1}{1680} - \cdots}$$

When these quantities are substituted into (1), the result is a much more difficult equation than that obtained in the previous problem.

15.146 Solve $\ddot{y} + ty = 0$; $y(0) = 4$, $\dot{y}(0) = 5$.

▮ Since the initial conditions are prescribed at $t = 0$, it is advantageous to obtain the general solution to the differential equation near $t = 0$. This is done in Problem 15.90, from which

$$y = b_0\left[1 - \frac{1}{(3)(2)}t^3 + \frac{1}{(6)(5)(3)(2)}t^6 - \cdots\right] + b_1\left[t - \frac{1}{(4)(3)}t^4 + \frac{1}{(7)(6)(4)(3)}t^7 - \cdots\right]$$

Differentiating termwise, we obtain $y' = b_0\left[-\frac{1}{2}t^2 + \frac{1}{(5)(3)(2)}t^5 - \cdots\right] + b_1\left[1 - \frac{1}{3}t^3 + \frac{1}{(6)(4)(3)}t^6 - \cdots\right]$.
Applying the initial conditions yields $4 = y(0) = b_0$ and $5 = \dot{y}(0) = b_1$. Thus, the solution to the initial-value problem is

$$y = 4\left[1 - \frac{1}{(3)(2)}t^3 + \frac{1}{(6)(5)(3)(2)}t^6 - \cdots\right] + 5\left[t - \frac{1}{(4)(3)}t^4 + \frac{1}{(7)(6)(4)(3)}t^7 - \cdots\right]$$
$$= 4 + 5t - \frac{4}{(3)(2)}t^3 - \frac{5}{(4)(3)}t^4 + \frac{4}{(6)(5)(3)(2)}t^6 + \frac{5}{(7)(6)(4)(3)}t^7 - \cdots$$

15.147 Solve $\ddot{y} + ty = 0$; $y(2) = 4$; $\dot{y}(2) = 5$.

▮ Since the initial conditions are prescribed at $t = 2$, it is advantageous to obtain the general solution to the differential equation near $t = 2$. Such a solution is found in Problem 15.116 to be

$$y = a_0[1 - (t-2)^2 - \tfrac{1}{6}(t-2)^3 + \tfrac{1}{6}(t-2)^4 + \tfrac{1}{15}(t-2)^5 + \cdots]$$
$$+ a_1[(t-2) - \tfrac{1}{3}(t-2)^3 - \tfrac{1}{12}(t-2)^4 + \tfrac{1}{30}(t-2)^5 + \cdots]$$

Differentiating this equation term by term, we get

$$y' = a_0[-2(t-2) - \tfrac{1}{2}(t-2)^2 + \tfrac{2}{3}(t-2)^3 + \tfrac{1}{3}(t-2)^4 + \cdots] + a_1[1 - (t-2)^2 - \tfrac{1}{3}(t-2)^3 + \tfrac{1}{6}(t-2)^4 + \cdots]$$

Applying the initial conditions, we find that $4 = y(2) = a_0$ and $5 = \dot{y}(2) = a_1$. Thus, the solution to the initial-value problem is

$$y = 4[1 - (t-2)^2 - \tfrac{1}{6}(t-2)^3 + \tfrac{1}{6}(t-2)^4 + \tfrac{1}{15}(t-2)^5 + \cdots]$$
$$+ 5[(t-2) - \tfrac{1}{3}(t-2)^3 - \tfrac{1}{12}(t-2)^4 + \tfrac{1}{30}(t-2)^5 + \cdots]$$
$$= 4 + 5(t-2) - 4(t-2)^2 - \tfrac{7}{3}(t-2)^3 + \tfrac{1}{4}(t-2)^4 + \tfrac{13}{30}(t-2)^5 + \cdots$$

15.148 Show why it is advantageous to solve the previous problem by using the general solution to the differential equation near $t = 2$ rather than near $t = 0$.

If we tried to use the general solution near $t = 0$, we would have, after applying the initial conditions at $t = 2$ to y and y' of Problem 15.146,

$$4 = y(2) = b_0\left[1 - \frac{1}{(3)(2)}2^3 + \frac{1}{(6)(5)(3)(2)}2^6 + \cdots\right] + b_1\left[2 - \frac{1}{(4)(3)}2^4 + \frac{1}{(7)(6)(4)(3)}2^7 + \cdots\right]$$

and
$$5 = \dot{y}(2) = b_0\left[-\frac{1}{2}2^2 + \frac{1}{(5)(3)(2)}2^5 + \cdots\right] + b_1\left[1 - \frac{1}{3}2^3 + \frac{1}{(6)(4)(3)}2^6 + \cdots\right]$$

Since the limits of the four infinite series of numbers are not known, these two equations are algebraically intractable. Thus, in this case the initial conditions do not yield usable information, in contrast to the situation of the previous problem.

15.149 Solve $(x^2 - 1)y'' + xy' - y = 0$; $y(0) = 1$, $y'(0) = 2$.

Since the initial conditions are prescribed at $x = 0$, it is advantageous to obtain the general solution to the differential equation at $x = 0$. This is done in Problem 15.88, which gives us $y = a_0(1 - \frac{1}{2}x^2 - \frac{1}{8}x^4 - \frac{1}{16}x^6 - \cdots) + a_1x$. Differentiating term by term yields $y' = a_0(-x - \frac{1}{2}x^3 - \frac{3}{8}x^5 - \cdots) + a_1$.
Applying the initial conditions to these two equations, we find $1 = y(0) = a_0$ and $2 = y'(0) = a_1$. Then $y = 1 + 2x - \frac{1}{2}x^2 - \frac{1}{8}x^4 - \frac{1}{16}x^6 - \cdots$, which is the solution to the initial-value problem.

15.150 Solve $\ddot{y} + 2t^2y = 0$; $y(0) = 1$, $y'(0) = -1$,

The general solution near $t = 0$ of the differential equation is found in Problem 15.101 as $y = a_0(1 - \frac{1}{6}t^4 + \frac{1}{168}t^8 + \cdots) + a_1(t - \frac{1}{10}t^5 + \frac{1}{360}t^9 + \cdots)$. Differentiating termwise, we get $y' = a_0(-\frac{2}{3}t^3 + \frac{1}{21}t^7 + \cdots) + a_1(1 - \frac{1}{2}t^4 + \frac{1}{40}t^8 + \cdots)$.
Applying the initial conditions to these two equations, we find that $1 = y(0) = a_0$ and $-1 = y'(0) = a_1$. It follows that

$$y = 1(1 - \tfrac{1}{6}t^4 + \tfrac{1}{168}t^8 + \cdots) + (-1)(t - \tfrac{1}{10}t^5 + \tfrac{1}{360}t^9 + \cdots) = 1 - t - \tfrac{1}{6}t^4 + \tfrac{1}{10}t^5 + \tfrac{1}{168}t^8 - \tfrac{1}{360}t^9 + \cdots$$

is a power-series solution to the initial-value problem.

15.151 Solve $y'' - x^2y' - y = 0$; $y(0) = -1$, $y'(0) = 2$.

Since the initial conditions are specified at $x = 0$, it is advantageous to use the general solution to the differential equation near $x = 0$, which is found in Problem 15.97. Applying the initial conditions to that solution, we obtain $A_0 = -1$ and $A_1 = 2$, whereupon the solution becomes

$$\begin{aligned} y &= -1(1 + \tfrac{1}{2}x^2 + \tfrac{1}{24}x^4 + \tfrac{1}{20}x^5 + \tfrac{1}{720}x^6 + \tfrac{13}{2520}x^7 + \cdots) + 2(x + \tfrac{1}{6}x^3 + \tfrac{1}{12}x^4 + \tfrac{1}{120}x^5 + \tfrac{7}{360}x^6 + \tfrac{41}{5040}x^7 + \cdots) \\ &= -1 + 2x - \tfrac{1}{2}x^2 + \tfrac{1}{3}x^3 + \tfrac{1}{8}x^4 - \tfrac{1}{30}x^5 + \tfrac{3}{80}x^6 + \tfrac{1}{90}x^7 + \cdots \end{aligned}$$

15.152 Solve the previous problem if, instead, the initial conditions are $y(0) = 4$ and $y'(0) = 6$.

Since the initial conditions remain specified at $x = 0$, we again apply them to the solution obtained in Problem 15.97. Doing so, we now obtain $A_0 = 4$ and $A_1 = 6$. The solution to the new initial-value problem is

$$\begin{aligned} y &= 4(1 + \tfrac{1}{2}x^2 + \tfrac{1}{24}x^4 + \tfrac{1}{20}x^5 + \tfrac{1}{720}x^6 + \tfrac{12}{2520}x^7 + \cdots) + 6(x + \tfrac{1}{6}x^3 + \tfrac{1}{12}x^4 + \tfrac{1}{120}x^5 + \tfrac{7}{360}x^6 + \tfrac{41}{5040}x^7 + \cdots) \\ &= 4 + 6x + 2x^2 + x^3 + \tfrac{2}{3}x^4 + \tfrac{1}{4}x^5 + \tfrac{11}{90}x^6 + \tfrac{19}{280}x^7 + \cdots \end{aligned}$$

15.153 Solve $y'' + xy' + (2x - 1)y = 0$; $y(-1) = 2$, $y'(-1) = -2$.

Since the initial conditions are prescribed at $x = -1$, it is advantageous to obtain the general solution to the differential equation near $x = -1$. This is done in Problem 15.106. Applying the initial conditions to that solution, we find that $a_0 = 2$ and $a_1 = -2$. The solution to the initial-value problem is then

$$\begin{aligned} y &= 2[1 + \tfrac{3}{2}(x + 1)^2 + \tfrac{1}{6}(x + 1)^3 + \tfrac{1}{6}(x + 1)^4 + \cdots] - 2[(x + 1) + \tfrac{1}{2}(x + 1)^2 + \tfrac{1}{2}(x + 1)^3 + 0(x + 1)^4 + \cdots] \\ &= 2 - 2(x + 1) + 2(x + 1)^2 - \tfrac{2}{3}(x + 1)^3 + \tfrac{1}{3}(x + 1)^4 + \cdots \end{aligned}$$

15.154 Solve the previous problem if, instead, the initial conditions are $y(-1) = 6$ and $y'(-1) = 8$.

▮ Since the initial conditions remain specified at $x = -1$, we again apply them to the solution obtained in Problem 15.106. Doing so, we obtain $a_0 = 6$ and $a_1 = 8$. The solution to the new initial-value problem is

$$y = 6[1 + \tfrac{3}{2}(x+1)^2 + \tfrac{1}{6}(x+1)^3 + \tfrac{1}{6}(x+1)^4 + \cdots] + 8[(x+1) + \tfrac{1}{2}(x+1)^2 + \tfrac{1}{2}(x+1)^3 + 0(x+1)^4 + \cdots]$$
$$= 6 + 8(x+1) + 13(x+1)^2 + 5(x+1)^3 + (x+1)^4 + \cdots$$

15.155 Solve $y'' - xy' = e^{-x}$; $y(0) = 2$, $y'(0) = -3$.

▮ The general solution of the differential equation near $x = 0$ is found in Problem 15.126 to be $y(x) = (\tfrac{1}{2}x^2 - \tfrac{1}{6}x^3 + \tfrac{1}{8}x^4 - \tfrac{1}{30}x^5 + \cdots) + a_0 + a_1(x + \tfrac{1}{6}x^3 + \tfrac{1}{40}x^5 + \cdots)$. Applying the first initial condition to this equation, we get $2 = y(0) = a_0$. Now differentiating term by term, we obtain $y'(x) = (x - \tfrac{1}{2}x^2 + \tfrac{1}{2}x^4 - \tfrac{1}{6}x^5 + \cdots) + a_1(1 + \tfrac{1}{2}x^2 + \tfrac{1}{8}x^4 + \cdots)$. Applying the second initial condition to this equation, we get $-3 = y'(0) = a_1$. With these values of a_0 and a_1, the general solution becomes

$$y = (\tfrac{1}{2}x^2 - \tfrac{1}{6}x^3 + \tfrac{1}{8}x^4 - \tfrac{1}{30}x^5) + 2 + (-3)(x + \tfrac{1}{6}x^3 + \tfrac{1}{40}x^5 + \cdots)$$
$$= 2 - 3x + \tfrac{1}{2}x^2 - \tfrac{2}{3}x^3 + \tfrac{1}{8}x^4 - \tfrac{13}{120}x^5 + \cdots$$

15.156 Solve $y'' - xy' = 12x^3$; $y(0) = 0$, $y'(0) = 0$.

▮ The general solution of the differential equation near $x = 0$ is found in Problem 15.128. Applying the initial conditions to it, we obtain $a_0 = a_1 = 0$. The solution to the initial-value problem is then $y = \tfrac{3}{5}x^5 + \tfrac{3}{42}x^7 + \tfrac{1}{144}x^9 + \cdots$.

15.157 Solve $y'' - xy' = 12x^3$; $y(0) = -1$, $y'(0) = 1$.

▮ Applying these initial conditions to the general solution obtained in Problem 15.128, we get $a_0 = -1$ and $a_1 = 1$. The solution to the initial-value problem is then

$$y = -1(1) + 1(x + \tfrac{1}{6}x^3 + \tfrac{1}{40}x^5 + \tfrac{1}{336}x^7 + \tfrac{1}{3456}x^9 + \cdots) + (\tfrac{3}{5}x^5 + \tfrac{3}{42}x^7 + \tfrac{1}{144}x^9 + \cdots)$$
$$= -1 + x + \tfrac{1}{6}x^3 + \tfrac{25}{40}x^5 + \tfrac{25}{336}x^7 + \tfrac{25}{3456}x^9 + \cdots$$

15.158 Solve $y'' - 2x^2y' + 4xy = x^2 + 2x + 2$; $y(0) = y'(0) = 1$.

▮ The general solution to the differential equation near $x = 0$ is found in Problem 15.130. Applying the initial conditions to it, we obtain $A_0 = A_1 = 1$. The solution to the initial-value problem is then

$$y = 1(1 - \tfrac{2}{3}x^3 - \tfrac{2}{45}x^6 - \tfrac{2}{405}x^9 - \cdots) + 1(x - \tfrac{1}{6}x^4 - \tfrac{1}{63}x^7 - \tfrac{1}{567}x^{10} - \cdots)$$
$$+ (x^2 + \tfrac{1}{3}x^3 + \tfrac{1}{12}x^4 + \tfrac{1}{45}x^6 + \tfrac{1}{126}x^7 + \tfrac{1}{405}x^9 + \tfrac{1}{1134}x^{10} + \cdots)$$
$$= 1 + x + x^2 - \tfrac{1}{3}x^3 - \tfrac{1}{12}x^4 - \tfrac{1}{45}x^6 - \tfrac{1}{126}x^7 - \tfrac{1}{405}x^8 - \tfrac{1}{1134}x^{10} - \cdots$$

15.159 Solve $y'' + (x^2 + 4x + 4)y = x^2 + 5x + 7$; $y(-2) = 3$, $y'(-2) = 1$.

▮ The general solution to the differential equation near $x = -2$ is found in Problem 15.133. Applying the initial conditions to it, we obtain $a_0 = 3$ and $a_1 = 1$. The solution to the initial-value problem is then

$$y = 3[1 - \tfrac{1}{12}(x+2)^4 + \tfrac{1}{672}(x+2)^8 + \cdots] + 1[(x+2) - \tfrac{1}{20}(x+2)^5 + \tfrac{1}{1440}(x+2)^9 + \cdots]$$
$$+ [\tfrac{1}{2}(x+2)^2 + \tfrac{1}{6}(x+2)^3 + \tfrac{1}{12}(x+2)^4 - \tfrac{1}{60}(x+2)^6 - \tfrac{1}{252}(x+2)^7 - \tfrac{1}{672}(x+2)^8 + \cdots]$$
$$= 3 + (x+2) + \tfrac{1}{2}(x+2)^2 + \tfrac{1}{6}(x+2)^3 - \tfrac{1}{6}(x+2)^4 - \tfrac{1}{20}(x+2)^5 - \tfrac{1}{60}(x+2)^6 - \tfrac{1}{252}(x+2)^7 + \tfrac{1}{336}(x+2)^8 + \cdots$$

15.160 Derive an infinite-series expansion for $y = (1+x)^p$, where p denotes an arbitrary constant.

▮ Clearly, $y(0) = (1+0)^p = 1$. Moreover, differentiation yields $y' = p(1+x)^{p-1} = p(1+x)^{-1}(1+x)^p = p(1+x)^{-1}y$. Thus, y is the unique solution to the initial-value problem $(1+x)y' = py$; $y(0) = 1$. If we assume that the solution to the differential equation has a power-series solution to the form

$$y = a_0 + a_1x + a_2x^2 + a_3x^3 + \cdots + a_nx^n + \cdots \tag{1}$$

then it follows from Problem 15.84 that the recursion formula for the coefficients of that solution is $a_{n+1} = \dfrac{p-n}{n+1}a_n$. Applying the initial condition to (1), we obtain $1 = y(0) = a_0$.

Substituting $n = 0, 1, 2, \ldots$ into the recursion formula with $a_0 = 1$ now yields

$$a_1 = p \qquad a_2 = \frac{a_1(p-1)}{2} = \frac{p(p-1)}{2} \qquad a_3 = \frac{a_2(p-2)}{3} = \frac{p(p-1)(p-2)}{2\cdot 3} \qquad \cdots$$

and, in general, $a_n = \dfrac{p(p-1)(p-2)\cdots(p-n+1)}{n!}$. With these coefficients, (*1*) becomes

$$y = 1 + px + \frac{p(p-1)}{2!}x^2 + \frac{p(p-1)(p-2)}{3!}x^3 + \cdots + \frac{p(p-1)(p-2)\cdots(p-n+1)}{n!}x^n + \cdots$$

Because $y = (1+x)^p$, we have

$$(1+x)^p = 1 + px + \frac{p(p-1)}{2!}x^2 + \cdots + \frac{p(p-1)\cdots(p-n+1)}{n!}x^n + \cdots$$

This equality is valid for all x within the interval of convergence of the power series. In this case, that is $|x| < 1$.

15.161 Develop an alternative procedure, using Taylor series, for solving initial-value problems around ordinary points.

▌ An alternative method for solving initial-value problems rests on the assumption that the solution can be expanded in a Taylor series about the initial point x_0; that is, the solution may be written as

$$y = \sum_{n=0}^{\infty} \frac{y^{(n)}(x_0)}{n!}(x-x_0)^n = \frac{y(x_0)}{0!} + \frac{y'(x_0)}{1!}(x-x_0) + \frac{y''(x_0)}{2!}(x-x_0)^2 + \cdots \tag{1}$$

For an nth-order differential equation, $y(x_0)$, $y'(x_0), \ldots, y^{(n-1)}(x_0)$ are given as initial conditions; the terms $y^{(k)}(x_0)$ for $k = n,\ n+1,\ n+2, \ldots$ can be obtained from the differential equation and successive derivatives of that equation.

15.162 Use a Taylor series to solve $y'' + xy' + (2x-1)y = 0$; $y(-1) = 2$, $y'(-1) = -2$.

▌ We seek a solution of the form

$$y(x) = y(-1) + \frac{y'(-1)}{1!}(x+1) + \frac{y''(-1)}{2!}(x+1)^2 + \frac{y'''(-1)}{3!}(x+1)^3 + \cdots \tag{1}$$

We are given $y(-1) = 2$ and $y'(-1) = -2$, as well as the differential equation

$$y''(x) = -xy' - (2x-1)y \tag{2}$$

Substituting $x = -1$ into (*2*), we find

$$y''(-1) = -(-1)y'(-1) - [2(-1) - 1]y(-1) = 1(-2) - (-3)(2) = 4 \tag{3}$$

To obtain $y'''(-1)$, we differentiate (*2*) and then substitute $x = -1$ into the resulting equation. Thus,

$$y'''(x) = -y' - xy'' - 2y - (2x-1)y' = -(-2) + 4 - 2(2) - (-3)(-2) = -4 \tag{4}$$

To obtain $y^{(4)}(-1)$, we differentiate (*4*) and then substitute $x = -1$ into the resulting equation. Thus,

$$y^{(4)}(x) = -xy''' - (2x+1)y'' - 4y' = -4 - (-1)(4) - 4(-2) = 8 \tag{5}$$

This process can be kept up indefinitely. Substituting (*3*), (*4*), (*5*), and the initial conditions into (*1*), we obtain

$$y = 2 + \frac{-2}{1!}(x+1) + \frac{4}{2!}(x+1)^2 + \frac{-4}{3!}(x+1)^3 + \frac{8}{4!}(x+1)^4 + \cdots$$
$$= 2 - 2(x+1) + 2(x+1)^2 - \tfrac{2}{3}(x+1)^3 + \tfrac{1}{3}(x+1)^4 + \cdots$$

(Compare this with the solution to Problem 15.154.)

5.163 Use a Taylor series to solve $y'' - 2xy = 0$; $y(2) = 1$, $y'(2) = 0$.

▌ We assume a solution of the form

$$y(x) = \frac{y(2)}{0!} + \frac{y'(2)}{1!}(x-2) + \frac{y''(2)}{2!}(x-2)^2 + \frac{y'''(2)}{3!}(x-2)^3 + \cdots \tag{1}$$

From the differential equation, $y''(x) = 2xy$; $y'''(x) = 2y + 2xy'$; $y^{(4)}(x) = 4y' + 2xy''$; .... Substituting $x = 2$ into each of these equations and using the initial conditions, we find that

$$y''(2) = 2(2)y(2) = 4(1) = 4$$
$$y'''(2) = 2y(2) + 2(2)y'(2) = 2(1) + 4(0) = 2$$
$$y^{(4)}(2) = 4y'(2) + 2(2)y''(2) = 4(0) + 4(4) = 16$$
$$\cdots\cdots\cdots\cdots\cdots\cdots\cdots\cdots$$

Substituting these results into (*1*) yields $y = 1 + 2(x-2)^2 + \tfrac{1}{3}(x-2)^3 + \tfrac{2}{3}(x-2)^4 + \cdots$.

15.164 Use a Taylor series to solve $y'' - xy' + 2y = 0;\ y(0) = 2,\ y'(0) = 3.$

▌ We seek a solution of the form

$$y(x) = y(0) + \frac{y'(0)}{1!}x + \frac{y''(0)}{2!}x^2 + \frac{y'''(0)}{3!}x^3 + \cdots \qquad (1)$$

From the differential equation and successive differentiation, we have

$$\begin{aligned} y''(x) &= xy' - 2y \\ y'''(x) &= y' + xy'' - 2y' = xy'' - y' \\ y^{(4)}(x) &= y'' + xy''' - y'' = xy''' \\ y^{(5)}(x) &= y''' + xy^{(4)} \\ &\cdots\cdots \end{aligned} \qquad (2)$$

Substituting $x = 0$ into each of these equations and using the initial conditions, we determine that

$$y''(0) = (0)y'(0) - 2y(0) = -4 \qquad y'''(0) = (0)y''(0) - y'(0) = -3$$
$$y^{(4)}(0) = 0y'''(0) = 0 \qquad y^{(5)}(0) = y'''(0) + (0)y^{(4)}(0) = -3$$
$$\cdots\cdots \qquad \cdots\cdots$$

so that (*1*) becomes

$$y = 2 + \frac{3}{1!}x + \frac{-4}{2!}x^2 + \frac{-3}{3!}x^3 + \frac{0}{4!}x^4 + \frac{-3}{5!}x^5 + \cdots = 2 + 3x - 2x^2 - \frac{1}{2}x^3 - \frac{1}{40}x^5 + \cdots$$

(Compare this with the solution to Problem 15.143.)

15.165 Use a Taylor series to solve $y'' - xy' + 2y = 0;\ y(1) = 2,\ y'(1) = 3.$

▌ Since the initial conditions are prescribed at $x = 2$, we assume a solution of the form

$$y(x) = y(2) + \frac{y'(2)}{1!}(x-2) + \frac{y''(2)}{2!}(x-2)^2 + \frac{y'''(2)}{3!}(x-2)^3 + \cdots \qquad (1)$$

Equations (*2*) of the previous problem are valid here. Evaluating them at $x = 1$ and using the new initial conditions, we determine

$$y''(1) = 1y'(1) - 2y(1) = -1 \qquad y'''(1) = 1y''(1) - y'(1) = -4$$
$$y^{(4)}(1) = 1y'''(1) = -4 \qquad y^{(5)}(1) = y'''(1) + 1y^{(4)}(1) = -8$$
$$\cdots\cdots \qquad \cdots\cdots$$

Substituting these values into (*1*), we obtain

$$\begin{aligned} y &= 2 + \frac{3}{1!}(x-2) + \frac{-1}{2!}(x-2)^2 + \frac{-4}{3!}(x-2)^3 + \frac{-4}{4!}(x-2)^4 + \frac{-8}{5!}(x-2)^5 + \cdots \\ &= 2 + 3(x-2) - \tfrac{1}{2}(x-2)^2 - \tfrac{2}{3}(x-2)^3 - \tfrac{1}{6}(x-2)^4 - \tfrac{1}{15}(x-2)^5 + \cdots \end{aligned}$$

(Compare this with the solution to Problem 15.144.)

15.166 Use a Taylor series to solve $y'' + y = 0;\ y(0) = 0,\ y'(0) = 1.$

▌ We assume a solution of the form

$$y(x) = y(0) + \frac{y'(0)}{1!}x + \frac{y''(0)}{2!}x^2 + \frac{y'''(0)}{3!}x^3 + \cdots \qquad (1)$$

The given differential equation and successive differentiation then yield $y'' = -y$, $y''' = -y'$, and, in general, $y^{(n)} = -y^{(n-2)}$, for $n = 2, 3, \ldots$. Evaluating these derivatives at $x = 0$ and using the initial conditions, we have

$$y''(0) = -y(0) = 0 \qquad y'''(0) = -y'(0) = -1$$
$$y^{(4)}(0) = -y''(0) = 0 \qquad y^{(5)}(0) = -y'''(0) = -(-1) = 1$$
$$y^{(6)}(0) = -y^{(4)}(0) = 0 \qquad y^{(7)}(0) = -y^{(5)}(0) = -1$$
$$\cdots\cdots \qquad \cdots\cdots$$

Substituting these values into (1) finally yields

$$y = 0 + \frac{1}{1!}x + \frac{0}{2!}x^2 + \frac{-1}{3!}x^3 + \frac{0}{4!}x^4 + \frac{1}{5!}x^5 + \frac{0}{6!}x^6 + \frac{-1}{7!}x^7 + \cdots = \frac{x}{1!} - \frac{x^3}{3!} + \frac{x^5}{5!} - \frac{x^7}{7!} + \cdots$$

which is shown in Problem 15.20 to be the Maclaurin-series expansion for $\sin x$.

15.167 Use a Taylor series to solve $\ddot{y} + ty = 0$; $y(0) = 4$, $\dot{y}(0) = 5$.

The differential equation and successive differentiation yield

$$\begin{aligned} \ddot{y}(t) &= -ty \\ y^{(3)}(t) &= -y - t\dot{y} \\ y^{(4)}(t) &= -\dot{y} - \dot{y} - t\ddot{y} = -2\dot{y} - t\ddot{y} \\ y^{(5)}(t) &= -2\ddot{y} - \ddot{y} - ty^{(3)} = -3\ddot{y} - ty^{(3)} \\ y^{(6)}(t) &= -3y^{(3)} - y^{(3)} - ty^{(4)} = -4y^{(3)} - ty^{(4)} \\ &\cdots\cdots \end{aligned} \tag{1}$$

Evaluating these equations at $t = 0$ and using the initial conditions, we determine

$$\begin{aligned} y^{(2)}(0) &= -(0)y(0) = 0 & y^{(3)}(0) &= -y(0) - (0)\dot{y}(0) = -4 \\ y^{(4)}(0) &= -2\dot{y}(0) - (0)\ddot{y}(0) = -10 & y^{(5)}(0) &= -3\ddot{y}(0) - 0y^{(3)}(0) = 0 \\ y^{(6)}(0) &= -4y^{(3)}(0) - (0)y^{(4)}(0) = 16 \\ &\cdots\cdots & &\cdots\cdots \end{aligned}$$

The Taylor-series expansion for $y(t)$ around $t = 0$ is then

$$y = 4 + \frac{5}{1!}t + \frac{0}{2!}t^2 + \frac{-4}{3!}t^3 + \frac{-10}{4!}t^4 + \frac{0}{5!}t^5 + \frac{16}{6!}t^6 + \cdots = 4 + 5t - \frac{2}{3}t^3 - \frac{5}{12}t^4 + \frac{1}{45}t^6 + \cdots$$

(Compare this with the solution to Problem 15.146.)

15.168 Use a Taylor series to solve $\ddot{y} + ty = 0$; $y(2) = 4$, $\dot{y}(2) = 5$.

Equations (*1*) of the previous problem are valid here. Evaluating them at $t = 2$ and using the new initial conditions, we obtain

$$\begin{aligned} y^{(2)}(2) &= -2y(2) = -8 & y^{(3)}(2) &= -y(2) - 2\dot{y}(2) = -14 \\ y^{(4)}(2) &= -2\dot{y}(2) - 2\ddot{y}(2) = 6 & y^{(5)}(2) &= -3\ddot{y}(2) - 2y^{(3)}(2) = 52 \\ y^{(6)}(2) &= -4y^{(3)}(2) - 2y^{(4)}(2) = 44 \\ &\cdots\cdots & &\cdots\cdots \end{aligned}$$

The Taylor-series expansion for $y(t)$ around $t = 2$ is then

$$\begin{aligned} y &= 4 + \frac{5}{1!}(t-2) + \frac{-8}{2!}(t-2)^2 + \frac{-14}{3!}(t-2)^3 + \frac{6}{4!}(t-2)^4 + \frac{52}{5!}(t-2)^5 + \frac{44}{6!}(t-2)^6 + \cdots \\ &= 4 + 5(t-2) - 4(t-2)^2 - \tfrac{7}{3}(t-2)^3 + \tfrac{1}{4}(t-2)^4 + \tfrac{13}{30}(t-2)^5 + \tfrac{11}{180}(t-2)^6 + \cdots \end{aligned}$$

(Compare this with the solution to Problem 15.147.)

15.169 Use a Taylor series to solve $\ddot{y} + 2t^2y = 0$; $y(0) = 1$, $\dot{y}(0) = -1$.

The differential equation and successive differentiation yield

$$\begin{aligned} \ddot{y}(t) &= -2t^2y & y^{(3)}(t) &= -4ty - 2t^2\dot{y} \\ y^{(4)}(t) &= -4y - 8t\dot{y} - 2t^2\ddot{y} & y^{(5)}(t) &= -12\dot{y} - 12t\ddot{y} - 2t^2y^{(3)} \\ &\cdots\cdots & &\cdots\cdots \end{aligned}$$

Evaluating these quantities at $t = 0$ and using the given initial conditions, we obtain

$$\begin{aligned} \ddot{y}(0) &= -2(0)^2y(0) = 0 & y^{(3)}(0) &= -4(0)y(0) - 2(0)^2\dot{y}(0) = 0 \\ y^{(4)}(0) &= -4y(0) - 8(0)\dot{y}(0) - 2(0)^2\ddot{y}(0) = -4 & y^{(5)}(0) &= -12\dot{y}(0) - 12(0)\ddot{y}(0) - 2(0)^2y^{(3)}(0) = 12 \\ &\cdots\cdots & &\cdots\cdots \end{aligned}$$

The Taylor-series expansion for $y(t)$ around $t = 0$ is then

$$y = 1 + \frac{-1}{1!}t + \frac{0}{2!}t^2 + \frac{0}{3!}t^3 + \frac{-4}{4!}t^4 + \frac{12}{5!}t^5 + \cdots = 1 - t - \frac{1}{6}t^4 + \frac{1}{10}t^5 + \cdots$$

(Compare this with the solution to Problem 15.150.)

15.170 Use a Taylor series to solve $y'' - x^2y' - y = 0$; $y(0) = -1$, $y'(0) = 2$.

▌ The differential equation and successive differentiation yield

$$\begin{aligned} y'' &= x^2y' + y & y^{(3)} &= (2x+1)y' + x^2y'' \\ y^{(4)} &= 2y' + (4x+1)y'' + x^2y^{(3)} & y^{(5)} &= 6y'' + (6x+1)y^{(3)} + x^2y^{(4)} \\ y^{(6)} &= 12y^{(3)} + (8x+1)y^{(4)} + x^2y^{(5)} \\ &\cdots\cdots & &\cdots\cdots \end{aligned} \tag{1}$$

Evaluating these equations at $x = 0$ and using the given initial conditions, we determine

$$\begin{aligned} y''(0) &= (0)^2y'(0) + y(0) = -1 & y^{(3)}(0) &= [2(0)+1]y'(0) + (0)^2y''(0) = 2 \\ y^{(4)}(0) &= 2y'(0) + [4(0)+1]y''(0) + (0)^2y^{(3)}(0) = 3 & y^{(5)}(0) &= 6y''(0) + [6(0)+1]y^{(3)}(0) + (0)^2y^{(4)}(0) = -4 \\ y^{(6)}(0) &= 12y^{(3)}(0) + [8(0)+1]y^{(4)}(0) + (0)^2y^{(5)}(0) = 27 \\ &\cdots\cdots & &\cdots\cdots \end{aligned}$$

The Taylor-series expansion for $y(x)$ around $x = 0$ is then

$$\begin{aligned} y &= -1 + \frac{2}{1!}x + \frac{-1}{2!}x^2 + \frac{2}{3!}x^3 + \frac{3}{4!}x^4 + \frac{-4}{5!}x^5 + \frac{27}{6!}x^6 + \cdots \\ &= -1 + 2x - \frac{1}{2}x^2 + \frac{1}{3}x^3 + \frac{1}{8}x^4 - \frac{1}{30}x^5 + \frac{3}{80}x^6 + \cdots \end{aligned}$$

(Compare this with the solution to Problem 15.151.)

15.171 Use a Taylor series to solve $y'' - x^2y' - y = 0$; $y(0) = 4$, $y'(0) = 6$.

▌ Equations (*1*) of the previous problem are valid here. Evaluating them at $x = 0$ with the new initial conditions, we get

$$\begin{aligned} y''(0) &= (0)^2y'(0) + y(0) = 4 & y^{(3)}(0) &= [2(0)+1]y'(0) + (0)^2y''(0) = 6 \\ y^{(4)}(0) &= 2y'(0) + [4(0)+1]y''(0) + (0)^2y^{(3)}(0) = 16 & y^{(5)}(0) &= 6y''(0) + [6(0)+1]y^{(3)}(0) + (0)^2y^{(4)}(0) = 30 \\ y^{(6)}(0) &= 12y^{(3)}(0) + [8(0)+1]y^{(4)}(0) + (0)^2y^{(5)}(0) = 88 \\ &\cdots\cdots & &\cdots\cdots \end{aligned}$$

The Taylor-series expansion for $y(x)$ around $x = 0$ is then

$$y = 4 + \frac{6}{1!}x + \frac{4}{2!}x^2 + \frac{6}{3!}x^3 + \frac{16}{4!}x^4 + \frac{30}{5!}x^5 + \frac{88}{6!}x^6 + \cdots = 4 + 6x + 2x^2 + x^3 + \frac{2}{3}x^4 + \frac{1}{4}x^5 + \frac{11}{90}x^6 + \cdots$$

(Compare this with the solution to Problem 15.152.)

15.172 Use a Taylor series to solve $y'' - xy' = e^{-x}$; $y(0) = 2$, $y'(0) = -3$.

▌ The differential equation and successive differentiation yield

$$\begin{aligned} y'' &= xy' + e^{-x} & y^{(3)} &= y' + xy'' - e^{-x} \\ y^{(4)} &= 2y'' + xy^{(3)} + e^{-x} & y^{(5)} &= 3y^{(3)} + xy^{(4)} - e^{-x} \\ &\cdots\cdots & &\cdots\cdots \end{aligned} \tag{1}$$

Evaluating these quantities at $x = 0$ and using the given initial conditions, we find

$$\begin{aligned} y''(0) &= (0)y'(0) + e^{-0} = 1 & y^{(3)}(0) &= y'(0) + 0y''(0) - e^{-0} = -4 \\ y^{(4)}(0) &= 2y''(0) + 0y^{(3)}(0) + e^{-0} = 3 & y^{(5)}(0) &= 3y^{(3)}(0) + 0y^{(4)}(0) - e^{-0} = -13 \\ &\cdots\cdots & &\cdots\cdots \end{aligned}$$

The Taylor-series expansion for $y(x)$ around $x = 0$ is then

$$y = 2 + \frac{-3}{1!}x + \frac{1}{2!}x^2 + \frac{-4}{3!}x^3 + \frac{3}{4!}x^4 + \frac{-13}{5!}x^5 + \cdots = 2 - 3x + \frac{1}{2}x^2 - \frac{2}{3}x^3 + \frac{1}{8}x^4 - \frac{13}{120}x^5 + \cdots$$

(Compare this with the solution to Problem 15.155.)

15.173 Solve the previous problem if the initial conditions are instead $y(1) = 2$ and $y'(1) = -3$.

▌ Equations (*1*) of the previous problem are valid here because the differential equation is the same. Evaluating them at $x = 1$, we get

$$y''(1) = 1y'(1) + e^{-1} = -3 + e^{-1} \qquad y^{(3)}(1) = y'(1) + 1y''(1) - e^{-1} = -6$$
$$y^{(4)}(1) = 2y''(1) + 1y^{(3)}(1) + e^{-1} = -12 + 3e^{-1} \qquad y^{(5)}(1) = 3y^{(3)}(1) + 1y^{(4)}(1) - e^{-1} = -30 + 2e^{-1}$$
$$\cdots\cdots\cdots \qquad \cdots\cdots\cdots$$

The Taylor-series expansion for $y(x)$ around $x = 1$ is then

$$y(x) = 2 + \frac{-3}{1!}(x-1) + \frac{-3+e^{-1}}{2!}(x-1)^2 + \frac{-6}{3!}(x-1)^3 + \frac{-12+3e^{-1}}{4!}(x-1)^4 + \frac{-30+2e^{-1}}{5!}(x-1)^5 + \cdots$$
$$= 2 - 3(x-1) - 1.316(x-1)^2 - (x-1)^3 - 0.454(x-1)^4 - 0.244(x-1)^5 - \cdots$$

5.174 Use a Taylor series to solve $y'' - xy' = 12x^3$; $y(0) = -1$, $y'(0) = 1$.

▌ The differential equation and successive differentiation yield

$$y'' = xy' + 12x^3 \qquad y^{(3)} = y' + xy'' + 36x^2$$
$$y^{(4)} = 2y'' + xy^{(3)} + 72x \qquad y^{(5)} = 3y^{(3)} + xy^{(4)} + 72$$
$$\cdots\cdots\cdots \qquad \cdots\cdots\cdots$$

Evaluating these equations at $x = 0$ and using the given initial conditions, we obtain

$$y''(0) = 0y'(0) + 12(0)^3 = 0 \qquad y^{(3)}(0) = y'(0) + 0y''(0) + 36(0)^2 = 1$$
$$y^{(4)}(0) = 2y''(0) + 0y^{(3)}(0) + 72(0) = 0 \qquad y^{(5)}(0) = 3y^{(3)}(0) + 0y^{(4)}(0) + 72 = 75$$
$$\cdots\cdots\cdots \qquad \cdots\cdots\cdots$$

The Taylor-series expansion for $y(x)$ around $x = 0$ is then

$$y(x) = -1 + \frac{1}{1!}x + \frac{0}{2!}x^2 + \frac{1}{3!}x^3 + \frac{0}{4!}x^4 + \frac{75}{5!}x^5 + \cdots = -1 + x + \frac{1}{6}x^3 + \frac{5}{8}x^5 + \cdots$$

(Compare this with the solution to Problem 15.157.)

5.175 Use a Taylor series to solve $y'' - 2x^2y' + 4xy = x^2 + 2x + 2$; $y(0) = y'(0) = 1$.

▌ The differential equation and successive differentiation yield

$$y'' = 2x^2y' - 4xy + x^2 + 2x + 2 \qquad y^{(3)} = 2x^2y'' - 4y + 2x + 2$$
$$y^{(4)} = 2x^2y^{(3)} + 4xy'' - 4y' + 2 \qquad y^{(5)} = 2x^2y^{(4)} + 8xy^{(3)} \qquad (1)$$
$$y^{(6)} = 2x^2y^{(5)} + 12xy^{(4)} + 8y^{(3)}$$
$$\cdots\cdots\cdots \qquad \cdots\cdots\cdots$$

Evaluating these equations at $x = 0$ and using the given initial conditions, we get

$$y''(0) = 2 \qquad y^{(3)}(0) = -2 \qquad y^{(4)}(0) = -2 \qquad y^{(5)}(0) = 0 \qquad y^{(6)}(0) = -16 \qquad \cdots$$

The Taylor-series expansion for $y(x)$ near $x = 0$ is then

$$y = 1 + \frac{1}{1!}x + \frac{2}{2!}x^2 + \frac{-2}{3!}x^3 + \frac{-2}{4!}x^4 + \frac{0}{5!}x^5 + \frac{-16}{6!}x^6 + \cdots = 1 + x + x^2 - \tfrac{1}{3}x^3 - \frac{1}{12}x^4 - \frac{1}{45}x^6 + \cdots$$

(Compare this with the solution to Problem 15.158.)

.176 Use a Taylor series to solve $y'' - 2x^2y' + 4xy = x^2 + 2x + 2$; $y(1) = y'(1) = 0$.

▌ Equations (*1*) of the previous problem remain valid. Evaluating them at $x = 1$ and using the new initial conditions, we determine

$$y''(1) = 5 \qquad y^{(3)}(1) = 14 \qquad y^{(4)}(1) = 50 \qquad y^{(5)}(1) = 212 \qquad y^{(6)}(1) = 1136 \qquad \cdots$$

The Taylor-series expansion for $y(x)$ around $x = 1$ is then

$$y(x) = 0 + \frac{0}{1!}(x-1) + \frac{5}{2!}(x-1)^2 + \frac{14}{3!}(x-1)^3 + \frac{50}{4!}(x-1)^4 + \frac{212}{5!}(x-1)^5 + \frac{1136}{6!}(x-1)^6 + \cdots$$
$$= \tfrac{5}{2}(x-1)^2 + \tfrac{7}{3}(x-1)^3 + \tfrac{25}{12}(x-1)^4 + \tfrac{53}{30}(x-1)^5 + \tfrac{71}{45}(x-1)^6 + \cdots$$

15.177 Use a Taylor series to solve $y'' + (x^2 + 4x + 4)y = x^2 + 5x + 7;\ y(-2) = 3,\ y'(-2) = 1.$

▌ The differential equation and successive differentiation yield

$$\begin{aligned}
y'' &= -(x^2 + 4x + 4)y + (x^2 + 5x + 7)\\
y^{(3)} &= -(2x + 4)y - (x^2 + 4x + 4)y' + (2x + 5)\\
y^{(4)} &= -2y - (4x + 8)y' - (x^2 + 4x + 4)y'' + 2\\
y^{(5)} &= -6y' - (6x + 12)y'' - (x^2 + 4x + 4)y^{(3)}\\
y^{(6)} &= -12y'' - (8x + 16)y^{(3)} - (x^2 + 4x + 4)y^{(4)}\\
&\cdots\cdots\cdots\cdots\cdots\cdots\cdots\cdots\cdots
\end{aligned}$$

Evaluating these derivatives at $x = -2$ and using the given initial conditions, we obtain

$$y''(-2) = 1 \qquad y^{(3)}(-2) = 1 \qquad y^{(4)}(-2) = -4 \qquad y^{(5)}(-2) = -6 \qquad y^{(6)}(-2) = -12 \qquad \cdots$$

The Taylor-series expansion for $y(x)$ around $x = -2$ is then

$$y(x) = 3 + \frac{1}{1!}(x+2) + \frac{1}{2!}(x+2)^2 + \frac{1}{3!}(x+2)^3 + \frac{-4}{4!}(x+2)^4 + \frac{-6}{5!}(x+2)^5 + \frac{-12}{6!}(x+2)^6 + \cdots$$
$$= 3 + (x+2) + \tfrac{1}{2}(x+2)^2 + \tfrac{1}{6}(x+2)^3 - \tfrac{1}{6}(x+2)^4 - \tfrac{1}{20}(x+2)^5 - \tfrac{1}{60}(x+2)^6 + \cdots$$

(Compare this with the solution to Problem 15.159.)

15.178 Use a Taylor series to solve $xy' - y = x + 1;\ y(1) = 2.$

▌ The differential equation and successive differentiation yield

$$\begin{aligned}
y' &= x^{-1}y + 1 + x^{-1}\\
y'' &= -x^{-2}y + x^{-1}y' - x^{-2}\\
y^{(3)} &= 2x^{-3}y - 2x^{-2}y' + x^{-1}y'' + 2x^{-3}\\
y^{(4)} &= -6x^{-4}y + 6x^{-3}y' - 3x^{-2}y'' + x^{-1}y^{(3)} - 6x^{-4}\\
&\cdots\cdots\cdots\cdots\cdots\cdots\cdots\cdots\cdots
\end{aligned}$$

Evaluating these derivatives at $x = 1$ and using the given initial condition, we obtain

$$y'(1) = 4 \qquad y''(1) = 1 \qquad y^{(3)}(1) = -1 \qquad y^{(4)}(1) = 2 \qquad \cdots$$

The Taylor-series expansion for $y(x)$ around $x = 1$ is then

$$y(x) = 2 + \frac{4}{1!}(x-1) + \frac{1}{2!}(x-1)^2 + \frac{-1}{3!}(x-1)^3 + \frac{2}{4!}(x-1)^4 + \cdots$$
$$= 2 + 4(x-1) + \frac{1}{2}(x-1)^2 - \frac{1}{6}(x-1)^3 + \frac{1}{12}(x-1)^4 + \cdots$$

15.179 Use a Taylor series to solve $y' + (\sin x)y = \cos x;\ y(0) = 2.$

▌ The differential equation and successive differentiation yield

$$\begin{aligned}
y' &= -(\sin x)y + \cos x\\
y'' &= -(\cos x)y - (\sin x)y' - \sin x\\
y^{(3)} &= (\sin x)y - (2\cos x)y' - (\sin x)y'' - \cos x\\
y^{(4)} &= (\cos x)y + (3\sin x)y' - (3\cos x)y'' - (\sin x)y^{(3)} + \sin x\\
&\cdots\cdots\cdots\cdots\cdots\cdots\cdots\cdots\cdots
\end{aligned}$$

Evaluating these derivatives at $x = 0$ and using the given initial condition, we get

$$y'(0) = 1 \qquad y''(0) = -2 \qquad y^{(3)}(0) = -3 \qquad y^{(4)}(0) = 8 \qquad \cdots$$

The Taylor-series expansion for $y(x)$ is then

$$y(x) = 2 + \frac{1}{1!}x + \frac{-2}{2!}x^2 + \frac{-3}{3!}x^3 + \frac{8}{4!}x^4 + \cdots = 2 + x - x^2 - \frac{1}{2}x^3 + \frac{1}{3}x^4 + \cdots$$

15.180 Use a Taylor series to solve $\dfrac{d^3y}{dx^3} - x\dfrac{dy}{dx} + y = 0$; $y(0) = 1$, $y'(0) = 2$, $y''(0) = 3$.

▌ The differential equation and successive differentiation yield

$$y^{(3)} = xy' - y \qquad y^{(4)} = xy''$$
$$y^{(5)} = y'' + xy^{(3)} \qquad y^{(6)} = 2y^{(3)} + xy^{(4)}$$
$$\cdots\cdots\cdots \qquad \cdots\cdots\cdots$$

Evaluating these equations at $x = 0$ and using the three given initial conditions, we find

$$y^{(3)}(0) = 0y'(0) - y(0) = -1 \qquad y^{(4)}(0) = 0y''(0) = 0$$
$$y^{(5)}(0) = y''(0) + 0y^{(3)}(0) = 3 \qquad y^{(6)}(0) = 2y^{(3)}(0) + 0y^{(4)}(0) = -2$$
$$\cdots\cdots\cdots \qquad \cdots\cdots\cdots$$

The Taylor-series expansion for $y(x)$ around $x = 0$ is then

$$y(x) = 1 + \frac{2}{1!}x + \frac{3}{2!}x^2 + \frac{-1}{3!}x^3 + \frac{0}{4!}x^4 + \frac{3}{5!}x^5 + \frac{-2}{6!}x^6 + \cdots = 1 + 2x + \frac{3}{2}x^2 - \frac{1}{6}x^3 + \frac{1}{40}x^5 - \frac{1}{360}x^6 + \cdots$$

15.181 Use a Taylor series to solve $\dfrac{d^3y}{dx^3} - (x^2 + 1)\dfrac{d^2y}{dx^2} + xy = x$; $y(1) = 2$, $y'(1) = 3$, $y''(1) = 4$.

▌ The differential equation and successive differentiation yield

$$y^{(3)} = (x^2 + 1)y'' - xy + x$$
$$y^{(4)} = (x^2 + 1)y^{(3)} + 2xy'' - xy' - y + 1$$
$$y^{(5)} = (x^2 + 1)y^{(4)} + 4xy^{(3)} + (2 - x)y'' - 2y'$$
$$y^{(6)} = (x^2 + 1)y^{(5)} + 6xy^{(4)} + (6 - x)y^{(3)} - 3y''$$
$$\cdots\cdots\cdots$$

Evaluating these equations at $x = 1$ and using the initial conditions, we find

$$y^{(3)}(1) = 7 \qquad y^{(4)}(1) = 18 \qquad y^{(5)}(1) = 62 \qquad y^{(6)}(1) = 255 \qquad \cdots$$

The Taylor-series expansion for $y(x)$ around $x = 1$ is then

$$y(x) = 2 + \frac{3}{1!}(x-1) + \frac{4}{2!}(x-1)^2 + \frac{7}{3!}(x-1)^3 + \frac{18}{4!}(x-1)^4 + \frac{62}{5!}(x-1)^5 + \frac{255}{6!}(x-1)^6 + \cdots$$
$$= 2 + 3(x-1) + 2(x-1)^2 + \tfrac{7}{6}(x-1)^3 + \tfrac{3}{4}(x-1)^4 + \tfrac{31}{60}(x-1)^5 + \tfrac{17}{48}(x-1)^6 + \cdots$$

15.182 Discuss the relative merits of Taylor-series solution.

▌ One advantage of Taylor-series solution, as compared to first finding a power-series solution and then applying the initial conditions, is that the Taylor-series method is easier to apply when only the first few terms of the solution are required. One disadvantage is that a recursion formula cannot be found for the Taylor series, and therefore a general expression for the nth term of the solution cannot be obtained.

THE METHOD OF FROBENIUS

15.183 Find the indicial equation associated with the general solution near $x = 0$ of $8x^2y'' + 10xy' + (x - 1)y = 0$.

▌ Dividing the differential equation by $8x^2$ transforms it to standard form with $P(x) = \dfrac{5}{4x}$ and $Q(x) = \dfrac{1}{8x} - \dfrac{1}{8x^2}$. Neither of these functions is analytic at $x = 0$, but since both $xP(x) = \dfrac{5}{4}$ and

$x^2Q(x) = x/8 - 1/8$ are analytic there, $x = 0$ is a regular singular point. Using the method of Frobenius, we assume that

$$y = x^\lambda \sum_{n=0}^{\infty} a_n x^n = \sum_{n=0}^{\infty} a_n x^{\lambda+n}$$
$$= a_0 x^\lambda + a_1 x^{\lambda+1} + a_2 x^{\lambda+2} + \cdots + a_{n-1}x^{\lambda+n-1} + a_n x^{\lambda+n} + a_{n+1}x^{\lambda+n+1} + \cdots \qquad (1)$$

and, correspondingly, that

$$y' = \lambda a_0 x^{\lambda-1} + (\lambda+1)a_1 x^\lambda + (\lambda+2)a_2 x^{\lambda+1} + \cdots$$
$$+(\lambda+n-1)a_{n-1}x^{\lambda+n-2} + (\lambda+n)a_n x^{\lambda+n-1} + (\lambda+n+1)a_{n+1}x^{\lambda+n} + \cdots \qquad (2)$$

and $y'' = \lambda(\lambda-1)a_0x^{\lambda-2} + (\lambda+1)(\lambda)a_1x^{\lambda-1} + (\lambda+2)(\lambda+1)a_2x^\lambda + \cdots + (\lambda+n-1)(\lambda+n-2)a_{n-1}x^{\lambda+n-3}$
$$+ (\lambda+n)(\lambda+n-1)a_nx^{\lambda+n-2} + (\lambda+n+1)(\lambda+n)a_{n+1}x^{\lambda+n-1} + \cdots \qquad (3)$$

Substituting (1), (2), and (3) into the differential equation and combining, we obtain

$$x^\lambda[8\lambda(\lambda-1)a_0 + 10\lambda a_0 - a_0] + x^{\lambda+1}[8(\lambda+1)\lambda a_1 + 10(\lambda+1)a_1 + a_0 - a_1] + \cdots$$
$$+ x^{\lambda+n}[8(\lambda+n)(\lambda+n-1)a_n + 10(\lambda+n)a_n + a_{n-1} - a_n] + \cdots = 0$$

Dividing by x^λ and simplifying, we have

$$[8\lambda^2 + 2\lambda - 1]a_0 + x[8\lambda^2 + 18\lambda + 9)a_1 + a_0] + \cdots + x^n\{[8(\lambda+n)^2 + 2(\lambda+n) - 1]a_n + a_{n-1}\} + \cdots = 0$$

Factoring the coefficient of a_n and equating the coefficient of each power of x to zero, we find, for $n \ge 1$, that $[4(\lambda+n) - 1][2(\lambda+n) + 1]a_n + a_{n-1} = 0$ or

$$a_n = \frac{-1}{[4(\lambda+n) - 1][2(\lambda+n) + 1]}a_{n-1} \qquad (4)$$

Also, for $n = 0$ we have $(8\lambda^2 + 2\lambda - 1)a_0 = 0$, so either $a_0 = 0$ or

$$8\lambda^2 + 2\lambda - 1 = 0 \qquad (5)$$

It is convenient to keep a_0 arbitrary; therefore, we must choose λ to satisfy (5), which is the indicial equation.

15.184 Find the general solution near $x = 0$ of $8x^2y'' + 10xy' + (x - 1)y = 0$.

▮ We use the results of the previous problem. In that problem, the indicial equation (5) can be factored into $(4\lambda - 1)(2\lambda + 1) = 0$, with roots $\lambda_1 = \frac{1}{4}$ and $\lambda_2 = -\frac{1}{2}$. Since the difference $\lambda_1 - \lambda_2 = \frac{3}{4}$ is not an integer, the two roots will generate different solutions directly.

Substituting $\lambda = \frac{1}{4}$ into the recursion formula (4) and simplifying, we obtain $a_n = \dfrac{-1}{2n(4n+3)}a_{n-1}$ for $n \ge 1$. Thus, $a_1 = -\frac{1}{14}a_0$, $a_2 = -\frac{1}{44}a_1 = \frac{1}{616}a_0, \ldots,$ and

$$y_1(x) = a_0x^{1/4}\left(1 - \tfrac{1}{14}x + \tfrac{1}{616}x^2 + \cdots\right)$$

Substituting $\lambda = -\frac{1}{2}$ into (4) and simplifying, we obtain $a_n = \dfrac{-1}{2n(4n-3)}a_{n-1}$. Thus, $a_1 = -\frac{1}{2}a_0$, $a_2 = -\frac{1}{20}a_1 = \frac{1}{40}a_0, \ldots,$ and

$$y_2(x) = a_0x^{-1/2}\left(1 - \tfrac{1}{2}x + \tfrac{1}{40}x^2 + \cdots\right)$$

The general solution is then

$$y = c_1y_1(x) + c_2y_2(x) = k_1x^{1/4}\left(1 - \tfrac{1}{14}x + \tfrac{1}{616}x^2 + \cdots\right) + k_2x^{-1/2}\left(1 - \tfrac{1}{2}x + \tfrac{1}{40}x^2 + \cdots\right)$$

where $k_1 = c_1a_0$ and $k_2 = c_2a_0$.

15.185 Find the indicial equation associated with the general solution near $x = 0$ of $2x^2y'' + 7x(x+1)y' - 3y = 0$.

▮ Here $P(x) = 7(x+1)/2x$ and $Q(x) = -3/2x^2$; hence, $x = 0$ is a regular singular point and the metho of Frobenius is applicable. Substituting (1), (2), and (3) of Problem 15.183 into the differential equation and combinir we obtain

$$x^\lambda[2\lambda(\lambda-1)a_0 + 7\lambda a_0 - 3a_0] + x^{\lambda+1}[2(\lambda+1)\lambda a_1 + 7\lambda a_0 + 7(\lambda+1)a_1 - 3a_1] + \cdots$$
$$+ x^{\lambda+n}[2(\lambda+n)(\lambda+n-1)a_n + 7(\lambda+n-1)a_{n-1} + 7(\lambda+n)a_n - 3a_n] + \cdots = 0$$

Dividing by x^λ and simplifying yield

$$(2\lambda^2 + 5\lambda - 3)a_0 + x[(2\lambda^2 + 9\lambda + 4)a_1 + 7\lambda a_0] + \cdots + x^n\{[2(\lambda + n)^2 + 5(\lambda + n) - 3]a_n + 7(\lambda + n - 1)a_{n-1}\} + \cdots = 0$$

Factoring the coefficient of a_n and equating each coefficient to zero, we find, for $n \geq 1$, that $[2(\lambda + n) - 1][(\lambda + n) + 3]a_n + 7(\lambda + n - 1)a_{n-1} = 0$, or

$$a_n = \frac{-7(\lambda + n - 1)}{[2(\lambda + n) - 1][(\lambda + n) + 3]} a_{n-1} \qquad (1)$$

For $n = 0$ we have $(2\lambda^2 + 5\lambda - 3)a_0 = 0$, so either $a_0 = 0$ or $2\lambda^2 + 5\lambda - 3 = 0$. It is convenient to keep a_0 arbitrary; therefore, we require λ to satisfy this indicial equation.

15.186 Find the general solution near $x = 0$ of $2x^2y'' + 7x(x + 1)y' - 3y = 0$.

▌ We use the results of the previous problem. The indicial equation found there may be factored into $(2\lambda - 1)(\lambda + 3) = 0$, which has as its roots $\lambda_1 = \frac{1}{2}$ and $\lambda_2 = -3$. Since their difference $\lambda_1 - \lambda_2 = \frac{7}{2}$ is not an integer, each root may be used to generate a linearly independent solution directly.

Substituting $\lambda = \frac{1}{2}$ into (1) of the previous problem and simplifying, we obtain $a_n = \dfrac{-7(2n - 1)}{2n(2n + 7)} a_{n-1}$ for $n \geq 1$. Thus, $a_1 = -\frac{7}{18}a_0$, $a_2 = -\frac{21}{44}a_1 = \frac{147}{792}a_0, \ldots,$ and

$$y_1(x) = a_0x^{1/2}(1 - \tfrac{7}{18}x + \tfrac{147}{792}x^2 + \cdots)$$

Substituting $\lambda = -3$ into (1) and simplifying, we obtain $a_n = \dfrac{-7(n - 4)}{n(2n - 7)} a_{n-1}$ for $n \geq 1$. Thus, $a_1 = -\frac{21}{5}a_0$, $a_2 = -\frac{7}{3}a_1 = \frac{49}{5}a_0$, and $a_3 = -\frac{7}{3}a_2 = -\frac{343}{15}a_0$. The next coefficient, a_4, is zero, so $a_n = 0$ for $n \geq 4$. Thus,

$$y_2(x) = a_0x^{-3}(1 - \tfrac{21}{5}x + \tfrac{49}{5}x^2 - \tfrac{343}{15}x^3)$$

The general solution is then

$$y = c_1y_1(x) + c_2y_2(x) = k_1x^{1/2}(1 - \tfrac{7}{18}x + \tfrac{147}{792}x^2 + \cdots) + k_2x^{-3}(1 - \tfrac{21}{5}x + \tfrac{49}{5}x^2 - \tfrac{343}{15}x^3)$$

where $k_1 = c_1a_0$ and $k_2 = c_2a_0$.

15.187 Find the general solution near $x = 0$ of $3x^2y'' - xy' + y = 0$.

▌ Here $P(x) = -1/3x$ and $Q(x) = 1/3x^2$; hence, $x = 0$ is a regular singular point and the method of Frobenius is applicable. Substituting (1), (2), and (3) of Problem 15.183 into the differential equation and simplifying, we have

$$x^\lambda[3\lambda^2 - 4\lambda + 1]a_0 + x^{\lambda+1}[3\lambda^2 + 2\lambda]a_1 + \cdots + x^{\lambda+n}[3(\lambda + n)^2 - 4(\lambda + n) + 1]a_n + \cdots = 0$$

Dividing by x^λ and equating all coefficients to zero, we find

$$(3\lambda^2 - 4\lambda + 1)a_0 = 0 \qquad (1)$$

and

$$[3(\lambda + n)^2 - 4(\lambda + n) + 1]a_n = 0 \qquad \text{for } n \geq 1 \qquad (2)$$

From (1), we conclude that the indicial equation is $3\lambda^2 - 4\lambda + 1 = 0$, which has roots $\lambda_1 = 1$ and $\lambda_2 = \frac{1}{3}$. We note also that for either value of λ, (2) is satisfied by simply choosing $a_n = 0$ for $n \geq 1$. Thus,

$$y_1(x) = x^1 \sum_{n=0}^{\infty} a_nx^n = a_0x \qquad \text{and} \qquad y_2(x) = x^{1/3} \sum_{n=0}^{\infty} a_nx^n = a_0x^{1/3}$$

and the general solution is $y = c_1y_1(x) + c_2y_2(x) = k_1x + k_2x^{1/3}$, where $k_1 = c_1a_0$ and $k_2 = c_2a_0$.

15.188 Find the indicial equation associated with the general solution near $x = 0$ of $2x^2y'' + x(2x + 1)y' - y = 0$.

▌ If we write this equation in the standard form $y'' + \dfrac{1/2 + x}{x} y' + \dfrac{-1/2}{x^2} y = 0$, then we have $xP(x) = \frac{1}{2} + x$ and $x^2Q(x) = -\frac{1}{2}$, so $x = 0$ is a regular singular point. We assume the Frobenius series solution

$$y = x^m(a_0 + a_1x + a_2x^2 + \cdots) = a_0x^m + a_1x^{m+1} + a_2x^{m+2} + \cdots \qquad (1)$$

with derivatives

$$y' = a_0 m x^{m-1} + a_1(m+1)x^m + a_2(m+2)x^{m+1} + \cdots$$

and
$$y'' = a_0 m(m-1)x^{m-2} + a_1(m+1)mx^{m-1} + a_2(m+2)(m+1)x^m + \cdots$$

To find the coefficients in (*1*), we proceed in essentially the same way as in the case of an ordinary point, with the significant difference that now we must also find the appropriate value (or values) of the exponent m. When the three series above are inserted in the standard-form equation, and the common factor x^{m-2} is canceled, the result is

$$a_0 m(m-1) + a_1(m+1)mx + a_2(m+2)(m+1)x^2 + \cdots$$
$$+ (\tfrac{1}{2} + x)[a_0 m + a_1(m+1)x + a_2(m+2)x^2 + \cdots] - \tfrac{1}{2}(a_0 + a_1 x + a_2 x^2 + \cdots) = 0$$

By inspection, we combine corresponding powers of x and equate the coefficient of each power of x to zero. This yields the system of equations

$$\begin{aligned} a_0[m(m-1) + \tfrac{1}{2}m - \tfrac{1}{2}] &= 0 \\ a_1[(m+1)m + \tfrac{1}{2}(m+1) - \tfrac{1}{2}] + a_0 m &= 0 \\ a_2[(m+2)(m+1) + \tfrac{1}{2}(m+2) - \tfrac{1}{2}] + a_1(m+1) &= 0 \\ \cdots\cdots\cdots\cdots\cdots\cdots\cdots\cdots\cdots\cdots\cdots & \end{aligned} \qquad (2)$$

It is understood that $a_0 \neq 0$. It therefore follows from the first of these equations that $m(m-1) + \frac{1}{2}m - \frac{1}{2} = 0$. This is the required indicial equation.

15.189 Find the general solution near $x = 0$ of $2x^2y'' + x(2x+1)y' - y = 0$.

▌ We use the results of the previous problem. The indicial equation may be written as $2m^2 - m - 1 = 0$, which has the roots $m_1 = 1$ and $m_2 = -\frac{1}{2}$. For each of these values of m, we now use the remaining equations of (*2*) to calculate $a_1, a_2, \ldots$ in terms of a_0. For $m_1 = 1$, we obtain $a_1 = -\frac{2}{5}a_0$, $a_2 = \frac{4}{35}a_0, \ldots$. For $m_2 = -\frac{1}{2}$, we obtain $a_1 = -a_0$, $a_2 = \frac{1}{2}a_0, \ldots$. We therefore have the following two Frobenius series solutions, in each of which we have put $a_0 = 1$: $y_1 = x(1 - \frac{2}{5}x + \frac{4}{35}x^2 + \cdots)$ and $y_2 = x^{-1/2}(1 - x + \frac{1}{2}x^2 + \cdots)$. The general solution is then

$$y = c_1 x(1 - \tfrac{2}{5}x + \tfrac{4}{35}x^2 + \cdots) + c_2 x^{-1/2}(1 - x + \tfrac{1}{2}x^2 + \cdots)$$

15.190 Find a series solution for $2x^2y'' - xy' + (x^2+1)y = 0$.

▌ We assume the solution $y = A_0x^m + A_1x^{m+1} + A_2x^{m+2} + \cdots + A_nx^{m+n} + \cdots$, so that

$$y' = mA_0x^{m-1} + (m+1)A_1x^m + (m+2)A_2x^{m+1} + \cdots + (m+n)A_nx^{m+n-1} + \cdots$$

and $y'' = (m-1)mA_0x^{m-2} + (m+1)mA_1x^{m-1} + (m+1)(m+2)A_2x^m + \cdots + (m+n-1)(m+n)A_nx^{m+n-2} + \cdots$

Substituting these in the given differential equation, we obtain

$$(m-1)(2m-1)A_0x^m + m(2m+1)A_1x^{m+1} + \{[(m+2)(2m+1)+1]A_2 + A_0\}x^{m+2} + \cdots$$
$$+ \{[(m+n)(2m+2n-3)+1]A_n + A_{n-2}\}x^{m+n} + \cdots = 0$$

All terms except the first two will vanish if $A_2, A_3, \ldots$ satisfy the recursion formula

$$A_n = -\frac{1}{(m+n)(2m+2n-3)+1}A_{n-2} \quad \text{for} \quad n \geq 2.$$

The roots of the indicial equation, $(m-1)(2m-1) = 0$, are $m = \frac{1}{2}$ and $m = 1$, and for either value the first term will vanish. Since, however, neither of these values of m will cause the second term to vanish, we take $A_1 = 0$. Then it follows from the recursion formula that $A_1 = A_3 = A_5 = \cdots = 0$. Thus,

$$y = A_0x^m\left\{1 - \frac{1}{(m+2)(2m+1)+1}x^2 + \frac{1}{[(m+2)(2m+1)+1][(m+4)(2m+5)+1]}x^4 - \cdots\right\}$$

When $m = \frac{1}{2}$ and $A_0 = 1$, this yields $y_1 = \sqrt{x}(1 - x^2/6 + x^4/168 - x^6/11{,}088 + \cdots)$, and when $m = 1$ with $A_0 = 1$, we have $y_2 = x(1 - x^2/10 + x^4/360 - x^6/28{,}080 + \cdots)$. The complete solution is then

$$y = Ay_1 + By_2 = A\sqrt{x}\left(1 - \frac{x^2}{6} + \frac{x^4}{168} - \frac{x^6}{11{,}088} + \cdots\right) + Bx\left(1 - \frac{x^2}{10} + \frac{x^4}{360} - \frac{x^6}{28{,}080} + \cdots\right)$$

15.191 Find a series solution for $3xy'' + 2y' + x^2y = 0$.

▌ Substituting y, y', and y'' as in the preceding problem, we obtain

$$m(3m-1)A_0x^{m-1}+(m+1)(3m+2)A_1x^m+(m+2)(3m+5)A_2x^{m+1}$$
$$+[(m+3)(3m+8)A_3+A_0]x^{m+2}+\cdots+[(m+n)(3m+3n-1)A_n+A_{n-3}]x^{m+n-1}+\cdots=0$$

All terms after the third will vanish if $A_3, A_4, \ldots$ satisfy the recursion formula

$$A_n=-\frac{1}{(m+n)(3m+3n-1)}A_{n-3}\quad\text{for}\quad n\geq 3.$$

The roots of the indicial equation $m(3m-1)=0$ are $m=0$ and $m=\frac{1}{3}$. Since neither will cause the second and third terms to vanish, we take $A_1=A_2=0$. Then the recursion formula yields $A_1=A_4=A_7=\cdots=0$ and $A_2=A_5=A_8=\cdots=0$. Thus

$$y=A_0x^m\left[1-\frac{1}{(m+3)(3m+8)}x^3+\frac{1}{(m+3)(m+6)(3m+8)(3m+17)}x^6-\cdots\right]$$

For $m=0$ with $A_0=1$, this yields $y_1=1-x^3/24+x^6/2448-\cdots$, and for $m=1/3$ with $A_0=1$, we obtain $y_2=x^{1/3}(1-x^3/30+x^6/3420-\cdots)$. The complete solution is then

$$y=Ay_1+By_2=A\left(1-\frac{x^3}{24}+\frac{x^6}{2448}-\cdots\right)+Bx^{1/3}\left(1-\frac{x^3}{30}+\frac{x^6}{3420}-\cdots\right)$$

15.192 Find the general solution near $t=0$ of $2ty''+y'-2y=0$.

▌ Here $t=0$ is a regular singular point, so we solve by the method of Frobenius. We assume that $y(t)=t^\sigma\sum_{n=0}^{\infty}a_nt^n=\sum_{n=0}^{\infty}a_nt^{n+\sigma}$. By differentiating this series twice, we obtain series representations for y' and y''. Then, substituting the results into the given differential equation, we obtain

$$2ty''(t)+y'-2y(t)=2\sum_{n=0}^{\infty}(n+\sigma)(n+\sigma-1)a_nt^{n+\sigma-1}+\sum_{n=0}^{\infty}(n+\sigma)a_nt^{n+\sigma-1}-2\sum_{n=0}^{\infty}a_nt^{n+\sigma}$$
$$=\sum_{n=0}^{\infty}[2(n+\sigma)(n+\sigma-1)+(n+\sigma)]a_nt^{n+\sigma-1}-2\sum_{n=1}^{\infty}a_{n-1}t^{n+\sigma-1}$$
$$=[2\sigma(\sigma-1)+\sigma]a_0t^{\sigma-1}+\sum_{n=1}^{\infty}\{[2(n+\sigma)-1][n+\sigma]a_n-2a_{n-1}\}t^{n+\sigma-1}$$

The lowest power of t here is $t^{\sigma-1}$, and hence the indicial equation is $2\sigma(\sigma-1)+\sigma=(2\sigma-1)\sigma=0$. Thus $\sigma=0,\frac{1}{2}$ are the indicial roots. Before substituting the indicial roots, we compute the recursion formula for the a_n's in terms of σ. Setting the terms in the braces equal to zero yields $[2(n+\sigma)-1][n+\sigma]a_n=2a_{n-1}$ for $n=1,2,\ldots$. Hence, $a_n=\dfrac{2a_{n-1}}{(2n+2\sigma-1)(n+\sigma)}$ for $n=1,2,\ldots$.

For $\sigma=0$ we obtain $a_n=\dfrac{2a_{n-1}}{(2n-1)(n)}=\dfrac{4a_{n-1}}{(2n-1)(2n)}$ for $n=1,2,\ldots$, and it follows that a_0 is arbitrary and

$$a_1=\frac{4a_0}{2!}\qquad a_2=\frac{4a_1}{3\cdot 4}=\frac{4^2a_0}{4!}\qquad a_3=\frac{4a_2}{5\cdot 6}=\frac{4^3a_0}{6!}\qquad\cdots$$

It follows now that $a_n=4^na_0/(2n)!$ for all $n\geq 0$. Therefore, a solution to the given equation is

$$y_1(t)=\sum_{n=0}^{\infty}\frac{4^n}{(2n)!}t^n=\sum_{n=0}^{\infty}\frac{1}{(2n)!}(2\sqrt{t})^{2n}=\cosh 2\sqrt{t}$$

For $\sigma=\frac{1}{2}$ we obtain $a_n=\dfrac{2a_{n-1}}{2n(n+\frac{1}{2})}=\dfrac{4a_{n-1}}{2n(2n+1)}$ for $n=1,2,\ldots$, and it follows that a_0 is arbitrary and

$$a_1=\frac{4a_0}{3!}\qquad a_2=\frac{4a_1}{4\cdot 5}=\frac{4^2a_0}{5!}\qquad a_3=\frac{4a_2}{6\cdot 7}=\frac{4^3a_0}{7!}\qquad\cdots$$

It follows now that $a_n=4^na_0/(2n+1)!$ for all $n\geq 0$. Therefore, another solution is

$$y_2(t)=\sum_{n=0}^{\infty}\frac{4^n}{(2n+1)!}t^{n+1/2}=\frac{1}{2}\sum_{n=0}^{\infty}\frac{1}{(2n+1)!}(2\sqrt{t})^{2n+1}=\tfrac{1}{2}\sinh 2\sqrt{t}$$

Hence $y(t)=c_1\cosh 2\sqrt{t}+c_2\sinh 2\sqrt{t}$ is the general solution to the given equation on $(0,\infty)$.

15.193 Find a general expression for the indicial equation associated with the general solution near $x = 0$ of $y'' + P(x)y' + Q(x)y = 0$, if $x = 0$ is a regular singular point.

▮ Since $x = 0$ is a regular singular point, $xP(x)$ and $x^2Q(x)$ are analytic near the origin and can be expanded in Taylor series there. Thus,

$$xP(x) = \sum_{n=0}^{\infty} p_n x^n = p_0 + p_1 x + p_2 x^2 + \cdots \qquad x^2Q(x) = \sum_{n=0}^{\infty} q_n x^n = q_0 + q_1 x + q_2 x^2 + \cdots$$

Dividing by x and x^2, respectively, we obtain

$$P(x) = p_0 x^{-1} + p_1 + p_2 x + \cdots \qquad Q(x) = q_0 x^{-2} + q_1 x^{-1} + q_2 + \cdots$$

Substituting these two results along with (1), (2), and (3) of Problem 15.183 into the differential equation and combining, we obtain $x^{\lambda-2}[\lambda(\lambda-1)a_0 + \lambda a_0 p_0 + a_0 q_0] + \cdots = 0$, which can hold only if $a_0[\lambda^2 + (p_0 - 1)\lambda + q_0] = 0$. Since $a_0 \neq 0$ (a_0 is an arbitrary constant and hence can be chosen to be nonzero), the indicial equation is $\lambda^2 + (p_0 - 1)\lambda + q_0 = 0$.

15.194 Find the indicial equation of $x^2y'' + xe^x y' + (x^3 - 1)y = 0$ for a solution near $x = 0$.

▮ Here $P(x) = e^x/x$ and $Q(x) = x - 1/x^2$, and we have

$$xP(x) = e^x = 1 + x + \frac{x^2}{2!} + \cdots \qquad x^2Q(x) = x^3 - 1 = -1 + 0x + 0x^2 + 1x^3 + 0x^4 + \cdots$$

from which $p_0 = 1$ and $q_0 = -1$. From Problem 15.193, the indicial equation is $\lambda^2 + (1-1)\lambda - 1 = 0$, or $\lambda^2 - 1 = 0$.

15.195 Find the indicial equation for the differential equation of Problem 15.183, using the formula developed in Problem 15.193.

▮ For that differential equation, we found $xP(x) = \frac{5}{4}$ and $x^2Q(x) = -\frac{1}{8} + \frac{1}{8}x$, each of which is its own Maclaurin series. In particular,

$$xP(x) = \tfrac{5}{4} + 0x + 0x^2 + 0x^3 + \cdots \qquad \text{and} \qquad x^2Q(x) = -\tfrac{1}{8} + \tfrac{1}{8}x + 0x^2 + 0x^3 + \cdots$$

Since here $p_0 = \frac{5}{4}$ and $q_0 = -\frac{1}{8}$, the indicial equation is $\lambda^2 + (\frac{5}{4} - 1)\lambda + (-\frac{1}{8}) = 0$, or $8\lambda^2 + 2\lambda - 1 = 0$.

15.196 Find the indicial equation associated with the general solution near $x = 0$ of $x^2y'' + x(-\frac{1}{2} + \frac{1}{2}x)y' + \frac{1}{2}y = 0$.

▮ Dividing this differential equation by x^2, we obtain $P(x) = \dfrac{-1/2 + x/2}{x}$ as the coefficient of y' and $Q(x) = \dfrac{1/2}{x^2}$ as the coefficient of y. Therefore,

$$xP(x) = -\tfrac{1}{2} + \tfrac{1}{2}x = -\tfrac{1}{2} + \tfrac{1}{2}x + 0x^2 + 0x^3 + \cdots \qquad \text{and} \qquad x^2Q(x) = \tfrac{1}{2} = \tfrac{1}{2} + 0x + 0x^2 + 0x^3 + \cdots$$

so each of these two functions is its own Maclaurin series. It follows that $p_0 = -\frac{1}{2}$ and $q_0 = \frac{1}{2}$. Using the result of Problem 15.193, we may write the indicial equation as $\lambda^2 + (-\frac{1}{2} - 1)\lambda + \frac{1}{2} = 0$ or $2\lambda^2 - 3\lambda + 1 = 0$

15.197 Find the general solution near $x = 0$ of $2x^2y'' + (x^2 - x)y' + y = 0$.

▮ Here $x = 0$ is a regular singular point. The indicial equation is given in the previous problem as $2\lambda^2 - 3\lambda + 1 = 0$, which has as its roots $\lambda = 1, \frac{1}{2}$. Since the difference between these roots is not an integer, two linearly independent solutions may be obtained by the method of Frobenius, in the form $y = x^\lambda \sum_{n=0}^{\infty} a_n x^n = \sum_{n=0}^{\infty} a_n x^{n+\lambda}$.

For $\lambda = 1$ (one of the indicial roots), we have $y = \sum_{n=0}^{\infty} a_n x^{n+1}$, $y' = \sum_{n=0}^{\infty} (n+1)a_n x^n$, and $y'' = \sum_{n=1}^{\infty} (n+1)na_n x^{n-1}$. Substituting these into the given equation, we obtain

$$\sum_{n=1}^{\infty} 2(n+1)na_n x^{n+1} + \left[\sum_{n=0}^{\infty} (n+1)a_n x^{n+2} - \sum_{n=0}^{\infty} (n+1)a_n x^{n+1}\right] + \sum_{n=0}^{\infty} a_n x^{n+1} = 0$$

or, by combining the first, third, and fourth sums, $\sum_{n=1}^{\infty} (2n^2 + n)a_n x^{n+1} + \sum_{n=0}^{\infty} (n+1)a_n x^{n+2} = 0$. These two sums can also be combined, if the index of summation in the second is changed from n to $n+1$. Doing this, we obtain

$$\sum_{n=1}^{\infty} (2n^2+n)a_n x^{n+1} + \sum_{n=1}^{\infty} n a_{n-1} x^{n+1} = \sum_{n=1}^{\infty} [(2n^2+n)a_n + n a_{n-1}]x^{n+1} = 0$$

The last equation will be identically satisfied if and only if the coefficient of each power of x is zero. This gives us the recurrence relation $(2n^2+n)a_n + na_{n-1} = 0$, or $a_n = -\dfrac{a_{n-1}}{2n+1}$ for $n = 1, 2, 3, \ldots$. This, in turn, produces $a_1 = -\dfrac{a_0}{3}$, $a_2 = \dfrac{a_0}{(3)(5)}$, $a_3 = -\dfrac{a_0}{(3)(5)(7)}, \ldots$. Hence, taking $a_0 = 1$, we have as one particular solution $y_1 = \sum_{n=0}^{\infty} (-1)^n \dfrac{x^{n+1}}{1 \cdot 3 \cdot 5 \cdots (2n+1)}$.

For $\lambda = \frac{1}{2}$, we have, similarly, $y = \sum_{n=0}^{\infty} a_n x^{n+1/2}$, $y' = \sum_{n=0}^{\infty} (n+\frac{1}{2})a_n x^{n-1/2}$, and $y'' = \sum_{n=0}^{\infty} (n+\frac{1}{2})(n-\frac{1}{2})a_n x^{n-3/2}$ for $x > 0$. Substituting these into the given equation, we obtain

$$\sum_{n=0}^{\infty} 2(n+\tfrac{1}{2})(n-\tfrac{1}{2})a_n x^{n+1/2} + \left[\sum_{n=0}^{\infty} (n+\tfrac{1}{2})a_n x^{n+3/2} - \sum_{n=0}^{\infty} (n+\tfrac{1}{2})a_n x^{n+1/2}\right] + \sum_{n=0}^{\infty} a_n x^{n+1/2} = 0$$

or, by combining the first, third, and fourth sums, $\sum_{n=1}^{\infty} (2n^2-n)a_n x^{n+1/2} + \sum_{n=0}^{\infty} (n+\frac{1}{2})a_n x^{n+3/2} = 0$. If we change the index of summation in the last sum from n to $n+1$ and then combine the two sums, we obtain

$$\sum_{n=1}^{\infty} (2n^2-n)a_n x^{n+1/2} + \sum_{n=1}^{\infty} (n-\tfrac{1}{2})a_{n-1} x^{n+1/2} = \sum_{n=1}^{\infty} [(2n^2-n)a_n + (n-\tfrac{1}{2})a_{n-1}]x^{n+1/2} = 0$$

The last equation will be satisfied if and only if the a's satisfy the recurrence relation $(2n^2-n)a_n + (n-\frac{1}{2})a_{n-1} = 0$, or $a_n = -\dfrac{a_{n-1}}{2n}$. This formula gives us $a_1 = -\dfrac{a_0}{2}$, $a_2 = \dfrac{a_0}{2^2 2!}$, $a_3 = -\dfrac{a_0}{2^3 3!}, \ldots$. Hence, taking $a_0 = 1$, we have a second linearly independent solution, $y_2 = \sum_{n=0}^{\infty} (-1)^n \dfrac{x^{n+1/2}}{2^n n!} = \sqrt{x}e^{-x/2}$ for $x > 0$. The general solution is then

$$y = c_1 y_1(x) + c_2 y_2(x) = c_1 \sum_{n=0}^{\infty} (-1)^n \frac{x^{n+1}}{(1)(3)(5)\cdots(2n+1)} + c_2\sqrt{x}e^{-x/2}$$

where c_1 and c_2 denote arbitrary constants.

15.198 Find the general solution near $x = 0$ of the *hypergeometric equation* $x(1-x)y'' + [C - (A+B+1)x]y' - ABy = 0$, where A and B are any real numbers, and C is any real nonintegral number.

▌ Since $x = 0$ is a regular singular point, the method of Frobenius is applicable. Substituting (*1*), (*2*), and (*3*) of Problem 15.183 into the differential equation, simplifying, and equating the coefficient of each power of x to zero, we obtain the indicial equation and recursion formula

$$\lambda^2 + (C-1)\lambda = 0 \qquad \text{and} \qquad a_{n+1} = \frac{(\lambda+n)(\lambda+n+A+B)+AB}{(\lambda+n+1)(\lambda+n+C)} a_n \tag{1}$$

The roots of the former are $\lambda_1 = 0$ and $\lambda_2 = 1 - C$; hence, $\lambda_1 - \lambda_2 = C - 1$, which is *not* an integer.

Substituting $\lambda = 0$ into the recursion formula, we have $a_{n+1} = \dfrac{n(n+A+B)+AB}{(n+1)(n+C)} a_n$, which is equivalent to $a_{n+1} = \dfrac{(A+n)(B+n)}{(n+1)(n+C)} a_n$. Thus,

$$a_1 = \frac{AB}{C} a_0 = \frac{AB}{1!C} a_0$$

$$a_2 = \frac{(A+1)(B+1)}{2(C+1)} a_1 = \frac{A(A+1)B(B+1)}{2!C(C+1)} a_0$$

$$a_3 = \frac{(A+2)(B+2)}{3(C+2)} a_2 = \frac{A(A+1)(A+2)B(B+1)(B+2)}{3!C(C+1)(C+2)} a_0$$

$$\cdots\cdots\cdots\cdots\cdots\cdots\cdots\cdots\cdots\cdots$$

and $y_1(x) = a_0F(A, B; C; x)$, where

$$F(A, B; C; x) = 1 + \frac{AB}{1!C}x + \frac{A(A+1)B(B+1)}{2!C(C+1)}x^2 + \frac{A(A+1)(A+2)B(B+1)(B+2)}{3!C(C+1)(C+2)}x^3 + \cdots$$

The series $F(A, B; C; x)$ is known as the *hypergeometric series*; it can be shown that this series converges for $-1 < x < 1$. It is customary to assign the arbitrary constant a_0 the value 1. Then $y_1(x) = F(A, B; C; x)$, and the hypergeometric series is a solution of the hypergeometric equation.

To find $y_2(x)$, we substitute $\lambda = 1 - C$ into the recursion formula, obtaining

$$a_{n+1} = \frac{(n+1-C)(n+1+A+B-C)+AB}{(n+2-C)(n+1)}a_n \quad \text{or} \quad a_{n+1} = \frac{(A-C+n+1)(B-C+n+1)}{(n+2-C)(n+1)}a_n$$

Solving for a_n in terms of a_0, and again setting $a_0 = 1$, we find that $y_2(x) = x^{1-C}F(A-C+1, B-C+1; 2-C; x)$. The general solution is $y = c_1y_1(x) + c_2y_2(x)$.

15.199 Find one solution near $x = 0$ of $x^2y'' + xy' + x^2y = 0$.

▌ Here $P(x) = 1/x$ and $Q(x) = 1$, so $x = 0$ is a regular singular point. We assume a solution of the form $y = x^\lambda \sum_{n=0}^{\infty} a_nx^n$. Substituting y and its first two derivatives into the given differential equation and combining, we obtain

$$x^\lambda[\lambda^2a_0] + x^{\lambda+1}[(\lambda+1)^2a_1] + x^{\lambda+2}[(\lambda+2)^2a_2 + a_0] + \cdots + x^{\lambda+n}[(\lambda+n)^2a_n + a_{n-2}] + \cdots = 0$$

Thus, $\lambda^2a_0 = 0$, $(\lambda+1)^2a_1 = 0$, and, for $n \geq 2$, $(\lambda+n)^2a_n + a_{n-2} = 0$, which we write as $a_n = \dfrac{-1}{(\lambda+n)^2}a_{n-2}$ for $n \geq 2$. (The stipulation $n \geq 2$ is required here because a_{n-2} is not defined for $n = 0$ or $n = 1$.) The indicial equation is $\lambda^2 = 0$, which has roots $\lambda_1 = \lambda_2 = 0$.

For $\lambda = 0$ we find that $a_1 = 0$ and $a_n = -(1/n^2)a_{n-2}$. Since $a_1 = 0$, it follows that $0 = a_3 = a_5 = a_7 = \cdots$. Furthermore,

$$a_2 = -\frac{1}{4}a_0 = -\frac{1}{2^2(1!)^2}a_0 \qquad a_6 = -\frac{1}{36}a_4 = -\frac{1}{2^6(3!)^2}a_0$$

$$a_4 = -\frac{1}{16}a_2 = \frac{1}{2^4(2!)^2}a_0 \qquad a_8 = -\frac{1}{64}a_6 = \frac{1}{2^8(4!)^2}a_0$$

and, in general, $a_{2k} = \dfrac{(-1)^k}{2^{2k}(k!)^2}a_0$ for $k = 1, 2, 3, \ldots$. Thus,

$$y_1(x) = a_0x^0\left[1 - \frac{1}{2^2(1!)^2}x^2 + \frac{1}{2^4(2!)^2}x^4 + \cdots + \frac{(-1)^k}{2^{2k}(k!)^2}x^{2k} + \cdots\right] = a_0\sum_{n=0}^{\infty}\frac{(-1)^n}{2^{2n}(n!)^2}x^{2n}$$

15.200 Describe a method for obtaining a second linearly independent solution to a second-order differential equation around a regular singular point when the indicial equation has a single root of multiplicity two.

▌ To find $y_2(x)$ when the indicial roots are equal, we keep the recursion formula in terms of λ and use it to find coefficients a_n $(n \geq 1)$ in terms of both λ and a_0, where the coefficient a_0 remains nonzero. (For convenience a_0 is often set to unity.) Using these coefficients, we can write $y(\lambda, x) = x^\lambda \sum_{n=0}^{\infty} a_n(\lambda)x^n = \sum_{n=0}^{\infty} a_n(\lambda)x^{n+\lambda}$, which depends on the variables λ and x. Then $y_2(x) = \left.\dfrac{\partial y(\lambda, x)}{\partial\lambda}\right|_{\lambda=\lambda_1}$.

15.201 Find a second linearly independent solution near $x = 0$ for the differential equation of Problem 15.199.

▌ We shall use the recursion formula found in Problem 15.199 for $n \geq 2$, and augment it with the equation $(\lambda+1)^2a_1 = 0$ for the special case $n = 1$. It follows from this equation and the fact that $\lambda = 0$ that $a_1 = 0$, which implies that $0 = a_3 = a_5 = a_7 = \cdots$. Then, from the recursion formula, $a_2 = \dfrac{-1}{(\lambda+2)^2}a_0$, $a_4 = \dfrac{-1}{(\lambda+4)^2}a_2 = \dfrac{1}{(\lambda+4)^2(\lambda+2)^2}a_0, \ldots$, and

$$y(\lambda, x) = a_0\left[x^\lambda - \frac{1}{(\lambda+2)^2}x^{\lambda+2} + \frac{1}{(\lambda+4)^2(\lambda+2)^2}x^{\lambda+4} + \cdots\right]$$

Recall that $\dfrac{\partial}{\partial\lambda}(x^{\lambda+k}) = x^{\lambda+k}\ln x$. (In differentiating with respect to λ, x can be considered a constant.) Thus,

$$\frac{\partial y(\lambda, x)}{\partial\lambda} = a_0\left[x^\lambda \ln x + \frac{2}{(\lambda+2)^3}x^{\lambda+2} - \frac{1}{(\lambda+2)^2}x^{\lambda+2}\ln x - \frac{2}{(\lambda+4)^3(\lambda+2)^2}x^{\lambda+4} - \frac{2}{(\lambda+4)^2(\lambda+2)^3}x^{\lambda+4} + \frac{1}{(\lambda+4)^2(\lambda+2)^2}x^{\lambda+4}\ln x + \cdots\right]$$

and

$$y_2(x) = \left.\frac{\partial y(\lambda, x)}{\partial\lambda}\right|_{\lambda=0} = a_0\left(\ln x + \frac{2}{2^3}x^2 - \frac{1}{2^2}x^2\ln x - \frac{2}{4^3 2^2}x^4 - \frac{2}{4^2 2^3}x^4 + \frac{1}{4^2 2^2}x^4 \ln x + \cdots\right)$$

$$= (\ln x)a_0\left[1 - \frac{1}{2^2(1!)}x^2 + \frac{1}{2^4(2!)^2}x^4 + \cdots\right] + a_0\left[\frac{x^2}{2^2(1!)^2}(1) - \frac{x^4}{2^4(2!)^2}\left(\frac{1}{2}+1\right) + \cdots\right]$$

$$= y_1(x)\ln x + a_0\left[\frac{x^2}{2^2(1!)^2}(1) - \frac{x^4}{2^4(2!)^2}\left(1+\frac{1}{2}\right) + \cdots\right]$$

The general solution is $y = c_1y_1(x) + c_2y_2(x)$, where y_1 is found in Problem 15.199.

15.202 Find the general solution near $x = 0$ of $x^2y'' - xy' + y = 0$.

▮ Here $P(x) = -1/x$ and $Q(x) = 1/x^2$, so $x = 0$ is a regular singular point. We assume a solution of the form $y = x^\lambda \sum_{n=0}^{\infty} a_nx^n$. Substituting y and its first two derivatives into the differential equation and combining terms, we obtain

$$x^\lambda(\lambda-1)^2a_0 + x^{\lambda+1}[\lambda^2a_1] + \cdots + x^{\lambda+n}[(\lambda+n)^2 - 2(\lambda+n) + 1]a_n + \cdots = 0$$

Thus, $(\lambda-1)^2a_0 = 0$ and, in general, $[(\lambda+n)^2 - 2(\lambda+n)+1]a_n = 0$. The first of these equations gives the indicial equation as $(\lambda-1)^2 = 0$, which has roots $\lambda_1 = \lambda_2 = 1$. Substituting $\lambda = 1$ into the second, we obtain $n^2a_n = 0$, which implies that $a_n = 0$ for $n \ge 1$. Thus, $y_1(x) = a_0x$.

To find the second solution, we use the method outlined in Problem 15.200. We continue to use the recursion formula $a_n = 0$ for $n \ge 1$, so that $y(\lambda, x) = a_0x^\lambda$. Thus, $\partial y(\lambda, x)/\partial\lambda = a_0x^\lambda \ln x$ and

$$y_2(x) = \left.\frac{\partial y(\lambda, x)}{\partial\lambda}\right|_{\lambda=1} = a_0x\ln x = y_1(x)\ln x$$

The general solution is then $y = c_1y_1(x) + c_2y_2(x) = k_1x + k_2x\ln x$, where $k_1 = c_1a_0$ and $k_2 = c_2a_0$.

15.203 Find the general solution near $x = 0$ of $xy'' + y' + y = 0$.

▮ Here $P(x) = Q(x) = 1/x$, so $x = 0$ is a regular singular point. We assume a solution of the form $y = \sum_{n=0}^{\infty} a_nx^{n+r}$ for $a_0 \ne 0$ (possibly $x > 0$). Substituting this series and its first two derivatives into the differential equation yields

$$\sum_{n=0}^{\infty}(n+r)(n+r-1)a_nx^{n+r-1} + \sum_{n=0}^{\infty}(n+r)a_nx^{n+r-1} + \sum_{n=0}^{\infty}a_nx^{n+r} = 0$$

or, by combining the first two series, $\sum_{n=0}^{\infty}(n+r)^2a_nx^{n+r-1} + \sum_{n=0}^{\infty}a_nx^{n+r} = 0$. If the term corresponding to $n = 0$ is detached from the first series, and if the index of summation is changed from n to $n-1$ in the second series, the two series can be combined to give

$$r^2a_0x^{r-1} + \sum_{n=1}^{\infty}(n+r)^2a_nx^{n+r-1} + \sum_{n=1}^{\infty}a_{n-1}x^{n+r-1} = r^2a_0x^{r-1} + \sum_{n=1}^{\infty}[(n+r)^2a_n + a_{n-1}]x^{n+r-1} = 0$$

The last equation will be satisfied if and only if $r^2a_0 = 0$ while the a's satisfy the recurrence relation $(n+r)^2a_n + a_{n-1} = 0$ or $a_n = -\dfrac{a_{n-1}}{(n+r)^2}$.

The indicial equation is thus $r^2 = 0$. The recurrence relation yields

$a_1 = -\dfrac{a_0}{(r+1)^2}$ and $a_2 = -\dfrac{a_1}{(r+2)^2} = \dfrac{a_0}{(r+1)^2(r+2)^2}$, and, by an obvious induction,

$$a_n = (-1)^n\frac{a_0}{(r+1)^2(r+2)^2\cdots(r+n)^2} \qquad n = 1, 2, \ldots$$

With a_0 arbitrary and all the other a's now determined in terms of a_0 and r, the assumed solution may be written as $y_r = a_0 x^r\left[1 - \frac{x}{(r+1)^2} + \frac{x^2}{(r+1)^2(r+2)^2} - \frac{x^3}{(r+1)^2(r+2)^2(r+3)^2} + \cdots\right]$. Setting $r = 0$ and (for convenience) $a_0 = 1$, we generate as the first solution

$$y_1 = 1 - \frac{x}{(1!)^2} + \frac{x^2}{(2!)^2} - \frac{x^3}{(3!)^2} + \cdots = \sum_{n=0}^{\infty} (-1)^n \frac{x^n}{(n!)^2}$$

To obtain a second linearly independent solution, we first determine

$$\frac{\partial y_r}{\partial r} = x^r \ln|x|\left[1 - \frac{x}{(r+1)^2} + \frac{x^2}{(r+1)^2(r+2)^2} - \frac{x^3}{(r+1)^2(r+2)^2(r+3)^2} + \cdots\right]$$
$$+ x^r\left\{-x\frac{-2}{(r+1)^3} + x^2\left[\frac{-2}{(r+1)^3}\frac{1}{(r+2)^2} + \frac{1}{(r+1)^2}\frac{-2}{(r+2)^3}\right]\right.$$
$$\left. - x^3\left[\frac{1}{(r+2)^2(r+3)^2}\frac{-2}{(r+1)^3} + \frac{1}{(r+1)^2(r+3)^2}\frac{-2}{(r+2)^3} + \frac{1}{(r+1)^2(r+2)^2(r+3)^3}\right] + \cdots\right\}$$

Finally, letting $r = 0$, we have as a second solution of the original equation,

$$y_2 = \left.\frac{\partial y_r}{\partial r}\right|_{r=0} = \ln|x|\left[1 - \frac{x}{(1!)^2} + \frac{x^2}{(2!)^2} - \frac{x^3}{(3!)^2} + \cdots\right] + 2\left[\frac{x}{(1!)^2} - \frac{x^2}{(2!)^2}\left(1 + \frac{1}{2}\right) + \frac{x^3}{(3!)^2}\left(1 + \frac{1}{2} + \frac{1}{3}\right) - \cdots\right]$$
$$= y_1 \ln|x| - 2\sum_{n=1}^{\infty} (-1)^n \frac{x^n}{(n!)^2} H_n \qquad \text{for } x \neq 0$$

where $H_n = \sum_{k=1}^{n} 1/k$ is the nth partial sum of the harmonic series. The general solution is $y = c_1 y_1 + c_2 y_2$, where c_1 and c_2 denote arbitrary constants.

15.204 Find the general solution near $x = 0$ of $xy'' + y' + x^2 y = 0$.

▌ Here $x = 0$ is a regular singular point. We assume a solution of the form $y = x^m \sum_{n=0}^{\infty} A_n x^n = A_0 x^m + A_1 x^{m+1} + A_2 x^{m+2} + \cdots$. Substituting this series and its first two derivatives into the given differential equation and combining terms containing like powers of x, we obtain

$$m^2 A_0 x^{m-1} + (m+1)^2 A_1 x^m + (m+2)^2 A_2 x^{m+1} + [(m+3)^2 A_3 + A_0]x^{m+2} + \cdots$$
$$+ [(m+n)^2 A_n + A_{n-3}]x^{m+n-1} + \cdots = 0$$

Setting the coefficients of powers of x to zero yields $m^2 A_0 = 0$, $(m+1)^2 A_1 = 0, \ldots,$ and, in general, $(m+n)^2 A_n + A_{n-3} = 0$. It follows from the first of these equations that either $A_0 = 0$ (which leads to the trivial solution) or $m^2 = 0$. This last is the indicial equation, and it has $m = 0$ as a root of multiplicity two.

The recursion formula is $A_n = -\frac{1}{(m+n)^2} A_{n-3}$, and it is valid for all positive values of n if we define $A_{-1} = A_{-2} = 0$. Then $A_1 = A_2 = 0$ also, and it follows from the recursion formula that $A_1 = A_4 = A_7 = \cdots = 0$ and $A_2 = A_5 = A_8 = \cdots = 0$. We also have

$$A_3 = -\frac{1}{(m+3)^2} A_0$$
$$A_6 = -\frac{1}{(m+6)^2} A_3 = \frac{1}{(m+3)^2(m+6)^2} A_0$$
$$A_9 = -\frac{1}{(m+9)^2} A_6 = -\frac{1}{(m+9)^2(m+6)^2(m+3)^2} A_0$$
$$\cdots\cdots\cdots\cdots\cdots\cdots\cdots\cdots$$

Substituting these values into the assumed solution for y and setting $A_0 = 1$, we get

$$y = x^m\left[1 - \frac{1}{(m+3)^2}x^3 + \frac{1}{(m+3)^2(m+6)^2}x^6 - \frac{1}{(m+3)^2(m+6)^2(m+9)^2}x^9 + \cdots\right]$$

and $$\frac{\partial y}{\partial m} = y \ln x + 2x^m\left\{\frac{1}{(m+3)^3}x^3 - \left[\frac{1}{(m+3)^3(m+6)^2} + \frac{1}{(m+3)^2(m+6)^3}\right]x^6\right.$$
$$\left. + \left[\frac{1}{(m+3)^3(m+6)^2(m+9)^2} + \frac{1}{(m+3)^2(m+6)^3(m+9)^2} + \frac{1}{(m+3)^2(m+6)^2(m+9)^3}\right]x^9 - \cdots\right.$$

Using the root $m = 0$ of the indicial equation, we obtain

$$y_1 = y|_{m=0} = 1 - \frac{1}{3^2}x^3 + \frac{1}{3^4(2!)^2}x^6 - \frac{1}{3^6(3!)^2}x^9 + \cdots$$

and
$$y_2 = \left.\frac{\partial y}{\partial m}\right|_{m=0} = y_1 \ln x + 2\left[\frac{1}{3^3}x^3 - \frac{1}{3^5(2!)^2}\left(1+\frac{1}{2}\right)x^6 + \frac{1}{3^7(3!)^2}\left(1+\frac{1}{2}+\frac{1}{3}\right)x^9 - \cdots\right]$$

The complete solution is $y = Ay_1 + By_2$, where A and B are arbitrary constants.

15.205 Find one solution near $x = 0$ of $x^2y'' + (x^2 - 2x)y' + 2y = 0$.

▮ Here $P(x) = 1 - 2/x$ and $Q(x) = 2/x^2$, so $x = 0$ is a regular singular point. We solve by the method of Frobenius, assuming a solution of the form $y = x^\lambda \sum_{n=0}^{\infty} a_n x^n = a_0x^\lambda + a_1x^{\lambda+1} + a_2x^{\lambda+2} + \cdots$. Substituting this series and its first two derivatives into the differential equation and simplifying, we have

$$x^\lambda[(\lambda^2 - 3\lambda + 2)a_0] + x^{\lambda+1}[(\lambda^2 - \lambda)a_1 + \lambda a_0] + \cdots$$
$$+ x^{\lambda+n}\{[(\lambda+n)^2 - 3(\lambda+n) + 2]a_n + (\lambda+n-1)a_{n-1}\} + \cdots = 0$$

Dividing by x^λ, factoring the coefficient of a_n, and equating the coefficient of each power of x to zero, we obtain $(\lambda^2 - 3\lambda + 2)a_0 = 0$ and, in general, $[(\lambda+n) - 2][(\lambda+n) - 1]a_n + (\lambda+n-1)a_{n-1} = 0$, or

$$a_n = -\frac{1}{\lambda + n - 2}a_{n-1} \qquad \text{for } n \geq 1 \tag{1}$$

The indicial equation is $\lambda^2 - 3\lambda + 2 = 0$, which has roots $\lambda_1 = 2$ and $\lambda_2 = 1$. Substituting $\lambda = 2$ into (*1*) yields $a_n = -(1/n)a_{n-1}$, from which we obtain $a_1 = -a_0$, $a_2 = -\frac{1}{2}a_1 = \frac{1}{2!}a_0$, and, in general, $a_k = \frac{(-1)^k}{k!}a_0$. Thus, $y_1(x) = a_0x^2 \sum_{n=0}^{\infty} \frac{(-1)^n}{n!}x^n = a_0x^2e^{-x}$.

Observe that we *cannot* find the second solution $y_2(x)$ by repeating the simple method of Frobenius with the *smaller* root λ_2. In fact, if we substitute $\lambda = 1$ into (*1*), we obtain $a_n = -\frac{1}{n-1}a_{n-1}$. Now, however, a_1 is undefined, since the denominator is zero when $n = 1$.

15.206 Describe a method for obtaining a second linearly independent solution to a second-order differential equation around a regular singular point when the indicial equation has two roots that differ by a nonzero integer.

▮ Denote the two roots by λ_1 and λ_2, and order them so that $\lambda_1 > \lambda_2$. The method of Frobenius will always generate a solution corresponding to λ_1, and it *may* generate a solution corresponding to the smaller of the two roots, λ_2. When it does, both solutions will be linearly independent.

If the method of Frobenius does not lead to a solution when $\lambda = \lambda_2$, then we keep the recursion formula in terms of λ and use it to find the coefficients a_n (for $n \geq 1$) in terms of both λ and a_0, where the coefficient a_0 remains nonzero. (For convenience this coefficient is often set equal to 1.) Using these coefficients, we can write $y(\lambda, x) = x^\lambda \sum_{n=0}^{\infty} a_n(\lambda)x^n = \sum_{n=0}^{\infty} a_n(\lambda)x^{n+\lambda}$, which depends on the variables λ and x. Then $y_2(x) = \left.\frac{\partial}{\partial\lambda}[(\lambda - \lambda_2)y(\lambda, x)]\right|_{\lambda=\lambda_2}$.

15.207 Find a second linearly independent solution to the differential equation of Problem 15.205.

▮ We use the recursion formula (*1*) of Problem 15.205 to get

$$a_1 = -\frac{1}{\lambda - 1}a_0 \qquad a_2 = -\frac{1}{\lambda}a_1 = \frac{1}{\lambda(\lambda-1)}a_0 \qquad a_3 = \frac{-1}{(\lambda+1)\lambda(\lambda-1)}a_0 \qquad \cdots$$

Then
$$y(\lambda, x) = a_0\left[x^\lambda - \frac{1}{(\lambda-1)}x^{\lambda+1} + \frac{1}{\lambda(\lambda-1)}x^{\lambda+2} - \frac{1}{(\lambda+1)\lambda(\lambda-1)}x^{\lambda+3} + \cdots\right]$$

and, since $\lambda - \lambda_2 = \lambda - 1$,

$$(\lambda - \lambda_2)y(\lambda, x) = a_0\left[(\lambda-1)x^\lambda - x^{\lambda+1} + \frac{1}{\lambda}x^{\lambda+2} - \frac{1}{\lambda(\lambda+1)}x^{\lambda+3} + \cdots\right]$$

Then $$\frac{\partial}{\partial\lambda}[(\lambda-\lambda_2)y(\lambda,x)] = a_0\left[x^\lambda + (\lambda-1)x^\lambda\ln x - x^{\lambda+1}\ln x - \frac{1}{\lambda^2}x^{\lambda+2} + \frac{1}{\lambda}x^{\lambda+2}\ln x + \frac{1}{\lambda^2(\lambda+1)}x^{\lambda+3} + \frac{1}{\lambda(\lambda+1)^2}x^{\lambda+3} - \frac{1}{\lambda(\lambda+1)}x^{\lambda+3}\ln x + \cdots\right]$$

and $$\begin{aligned} y_2(x) &= \frac{\partial}{\partial\lambda}[(\lambda-\lambda_2)y(\lambda,x)]\Big|_{\lambda=\lambda_2=1} \\ &= a_0(x + 0 - x^2\ln x - x^3 + x^3\ln x + \tfrac{1}{2}x^4 + \tfrac{1}{4}x^4 - \tfrac{1}{2}x^4\ln x + \cdots) \\ &= (-\ln x)a_0(x^2 - x^3 + \tfrac{1}{2}x^4 + \cdots) + a_0(x - x^3 + \tfrac{3}{4}x^4 + \cdots) \\ &= -y_1(x)\ln x + a_0x(1 - x^2 + \tfrac{3}{4}x^3 + \cdots). \end{aligned}$$

The general solution is $y = c_1y_1(x) + c_2y_2(x)$.

15.208 Find the general solution near $x = 0$ of $x^2y'' + xy' + (x^2 - 1)y = 0$.

▮ Here $P(x) = x^{-1}$ and $Q(x) = 1 - x^{-2}$, so $x = 0$ is a regular singular point. Substituting *(1)*, *(2)*, and *(3)* of Problem 15.183 into the differential equation, we obtain

$$x^\lambda[(\lambda^2-1)a_0] + x^{\lambda+1}[(\lambda+1)^2-1]a_1 + x^{\lambda+2}\{[(\lambda+2)^2-1]a_2 + a_0\} + \cdots + x^{\lambda+n}\{[(\lambda+n)^2-1]a_n + a_{n-2}\} + \cdots = 0$$

Thus, $(\lambda^2-1)a_0 = 0$, $[(\lambda+1)^2-1]a_1 = 0$, and, for $n \ge 2$, $[(\lambda+n)^2-1]a_n + a_{n-2} = 0$, or

$$a_n = \frac{-1}{(\lambda+n)^2-1}a_{n-2} \quad \text{for } n \ge 2 \qquad (1)$$

The indicial equation is therefore $\lambda^2 - 1 = 0$, which has roots $\lambda_1 = 1$ and $\lambda_2 = -1$.

For $\lambda = 1$, substitution yields $a_1 = 0$ and $a_n = \dfrac{-1}{n(n+2)}a_{n-2}$ for $n \ge 2$. Since $a_1 = 0$, it follows that $0 = a_3 = a_5 = a_7 = \cdots$. Furthermore,

$$a_2 = \frac{-1}{2(4)}a_0 = \frac{-1}{2^2 1!2!}a_0 \qquad a_4 = \frac{-1}{4(6)}a_2 = \frac{1}{2^4 2!3!}a_0 \qquad a_6 = \frac{-1}{6(8)}a_4 = \frac{-1}{2^6 3!4!}a_0$$

and, in general, $a_{2k} = \dfrac{(-1)^k}{2^{2k}k!(k+1)!}a_0$ for $k = 1, 2, 3, \ldots$. Thus, $y_1(x) = a_0x\sum_{n=0}^{\infty}\dfrac{(-1)^n}{2^{2n}n!(n+1)!}x^{2n}$.

If we substitute $\lambda = \lambda_2 = -1$ in *(1)*, we obtain $a_n = \dfrac{-1}{n(n-2)}a_{n-2}$, which fails to define a_2. Thus, the simple method of Frobenius does not provide the second solution $y_2(x)$, and we must use the modification described in Problem 15.206. Again $0 = a_1 = a_3 = a_5 = \cdots$, and now *(1)* yields

$$a_2 = \frac{-1}{(\lambda+3)(\lambda+1)}a_0 \qquad a_4 = \frac{1}{(\lambda+5)(\lambda+3)^2(\lambda+1)}a_0 \qquad \cdots$$

Thus, $$y(\lambda,x) = a_0\left[x^\lambda - \frac{1}{(\lambda+3)(\lambda+1)}x^{\lambda+2} + \frac{1}{(\lambda+5)(\lambda+3)^2(\lambda+1)}x^{\lambda+4} + \cdots\right]$$

and, since $\lambda - \lambda_2 = \lambda + 1$,

$$(\lambda-\lambda_2)y(\lambda,x) = a_0\left[(\lambda+1)x^\lambda - \frac{1}{(\lambda+3)}x^{\lambda+2} + \frac{1}{(\lambda+5)(\lambda+3)^2}x^{\lambda+4}\cdots\right]$$

Then $$\frac{\partial}{\partial\lambda}[(\lambda-\lambda_2)y(\lambda,x)] = a_0\left[x^\lambda + (\lambda+1)x^\lambda\ln x + \frac{1}{(\lambda+3)^2}x^{\lambda+2} - \frac{1}{(\lambda+3)}x^{\lambda+2}\ln x - \frac{1}{(\lambda+5)^2(\lambda+3)^2}x^{\lambda+4} - \frac{2}{(\lambda+5)(\lambda+3)^3}x^{\lambda+4} + \frac{1}{(\lambda+5)(\lambda+3)^2}x^{\lambda+4}\ln x + \cdots\right]$$

and $$\begin{aligned} y_2(x) &= \frac{\partial}{\partial\lambda}[(\lambda-\lambda_2)y(\lambda,x)]\Big|_{\lambda=\lambda_2=-1} \\ &= a_0(x^{-1} + 0 + \tfrac{1}{4}x - \tfrac{1}{2}x\ln x - \tfrac{1}{64}x^3 - \tfrac{2}{32}x^3 + \tfrac{1}{16}x^3\ln x + \cdots) \\ &= -\tfrac{1}{2}(\ln x)a_0x(1 - \tfrac{1}{8}x^2 + \cdots) + a_0(x^{-1} + \tfrac{1}{4}x - \tfrac{5}{64}x^3 + \cdots) \\ &= -\tfrac{1}{2}(\ln x)y_1(x) + a_0x^{-1}(1 + \tfrac{1}{4}x^2 - \tfrac{5}{64}x^4 + \cdots) \end{aligned}$$

The general solution is $y = c_1y_1(x) + c_2y_2(x)$.

15.209 Find the general solution near $x = 0$ of $xy'' - 3y' + xy = 0$.

▌ Here $P(x) = -3/x$ and $Q(x) = 1$, so $x = 0$ is a regular singular point. We assume a solution of the form $y = x^m \sum_{n=0}^{\infty} A_n x^n = A_0 x^m + A_1 x^{m+1} + A_2 x^{m+2} + \cdots$. Substituting this series and its first two derivatives into the given differential equation and simplifying, we get

$$(m-4)mA_0 x^{m-1} + (m-3)(m+1)A_1 x^m + [(m-2)(m+2)A_2 + A_0]x^{m+1} + \cdots$$
$$+ [(m+n-4)(m+n)A_n + A_{n-2}]x^{m+n-1} + \cdots = 0$$

It follows that the indicial equation is $(m-4)m = 0$, which has roots 0 and 4. The recursion formula, left in terms of m, is $(m+n-4)(m+n)A_n + A_{n-2} = 0$, or $A_n = -\dfrac{1}{(m+n-4)(m+n)} A_{n-2}$ for $n \geq 2$. This formula is also valid for $n = 2$ if we define $A_{-1} = 0$.

Evaluating the recursion formula for successive values of n, we determine $A_1 = -\dfrac{1}{(m+1-4)(m+1)} A_{-1} = 0$, from which we conclude that $A_3 = A_5 = A_7 = \cdots = 0$. Also,

$$A_2 = -\frac{1}{(m-2)(m+2)} A_2 \qquad A_4 = -\frac{1}{m(m+4)} A_4 = \frac{1}{m(m-2)(m+2)(m+4)} A_0 \qquad \cdots$$

Substituting these quantities into the assumed solution, we obtain

$$y = A_0 x^m \left[1 - \frac{1}{(m-2)(m+2)} x^2 + \frac{1}{m(m-2)(m+2)(m+4)} x^4 - \frac{1}{m(m-2)(m+2)^2(m+4)(m+6)} x^6 \right.$$
$$\left. + \frac{1}{m(m-2)(m+2)^2(m+4)^2(m+6)(m+8)} x^8 - \cdots \right]$$

One solution is obtained by evaluating this expression for $m = 4$, the larger of the two indicial roots. This yields $y_1 = y|_{m=4} = A_0 \left[x^4 - \dfrac{1}{(2)(6)} x^6 + \dfrac{1}{(2)(4)(6)(8)} x^8 - \dfrac{1}{(2)(4)(6^2)(8)(10)} x^{10} + \cdots \right]$.

If we try to use $m = 0$, the smaller of the indicial roots, we find that $A_4, A_6, A_8, \ldots$ are undefined, because their denominators all contain m as a factor. Following the procedure described in Problem 15.206, we form

$$(m-0)y = my = A_0 x^m \left[m - \frac{m}{(m-2)(m+2)} x^2 + \frac{1}{(m-2)(m+2)(m+4)} x^4 - \frac{1}{(m-2)(m+2)^2(m+4)(m+6)} x^6 \right.$$
$$\left. + \frac{1}{(m-2)(m+2)^2(m+4)^2(m+6)(m+8)} x^8 - \cdots \right]$$

and then differentiate with respect to m to obtain

$$\frac{\partial}{\partial m}[(m-0)y] = A_0 x^m (\ln x) \left[m - \frac{m}{(m-2)(m+2)} x^2 + \frac{1}{(m-2)(m+2)(m+4)} x^4 \right.$$
$$\left. - \frac{1}{(m-2)(m+2)^2(m+4)(m+6)} x^6 + \cdots \right]$$
$$+ A_0 x^m \left\{ 1 + \frac{m^2+4}{[(m-2)(m+2)]^2} x^2 - \frac{1}{(m-2)(m+2)(m+4)} \left[\frac{1}{m-2} + \frac{1}{m+2} + \frac{1}{m+4} \right] x^4 \right.$$
$$+ \frac{1}{(m-2)(m+2)^2(m+4)(m+6)} \left[\frac{1}{m-2} + \frac{2}{m+2} + \frac{1}{m+4} + \frac{1}{m+6} \right] x^6$$
$$\left. - \frac{1}{(m-2)(m+2)^2(m+4)^2(m+6)(m+8)} \left[\frac{1}{m-2} + \frac{2}{m+2} + \frac{2}{m+4} + \frac{1}{m+6} + \frac{1}{m+8} \right] x^8 + \cdots \right\}$$

Then
$$y_2 = \frac{\partial}{\partial m}(my)\bigg|_{m=0} = -16A_0 y_1 \ln x + A_0 \left\{ 1 + \frac{1}{2^2} x^2 + \frac{1}{(2^5)(2!)} x^4 - \frac{1}{(2^6)(3!)(1!)} \left(1 + \frac{1}{2} + \frac{1}{3} \right) x^6 \right.$$
$$+ \frac{1}{(2^8)(4!)(2!)} \left[\left(1 + \frac{1}{2} + \frac{1}{3} + \frac{1}{4} \right) + \frac{1}{2} \right] x^8$$
$$\left. - \frac{1}{(2^{10})(5!)(3!)} \left[\left(1 + \frac{1}{2} + \frac{1}{3} + \frac{1}{4} + \frac{1}{5} \right) + \left(\frac{1}{2} + \frac{1}{3} \right) \right] x^{10} + \cdots \right\}$$

The general solution is $y = c_1 y_1 + c_2 y_2$.

15.210 Find the general solution near $x = 0$ of $x^2y'' + (x^2 + 2x)y' - 2y = 0$.

▌ Here $P(x) = 1 + 2/x$ and $Q(x) = -2/x^2$, so $x = 0$ is a regular singular point. We assume a solution of the form $y = a_0x^\lambda + a_1x^{\lambda+1} + a_2x^{\lambda+2} + a_3x^{\lambda+3} + \cdots$. Substituting y and its first two derivatives (see Problem 15.183) into the given differential equation and simplifying, we obtain

$$x^\lambda[(\lambda^2 + \lambda - 2)a_0] + x^{\lambda+1}[(\lambda^2 + 3\lambda)a_1 + \lambda a_0] + \cdots + x^{\lambda+n}\{[(\lambda + n)^2 + (\lambda + n) - 2]a_n + (\lambda + n - 1)a_{n-1}\} + \cdots = 0$$

Dividing by x^λ, factoring the coefficient of a_n, and equating to zero the coefficient of each power of x, we then obtain $(\lambda^2 + \lambda - 2)a_0 = 0$ and, for $n \geq 1$, $[(\lambda + n) + 2][(\lambda + n) - 1]a_n + (\lambda + n - 1)a_{n-1} = 0$. This is equivalent to $a_n = -\dfrac{1}{\lambda + n + 2}a_{n-1}$ for $n \geq 1$.

The indicial equation is $\lambda^2 + \lambda - 2 = 0$, which has roots $\lambda_1 = 1$ and $\lambda_2 = -2$. Substituting $\lambda = 1$ into the recursion formula, we obtain $a_n = [-1/(n + 3)]a_{n-1}$, which in turn yields

$$a_1 = -\frac{1}{4}a_0 = -\frac{3!}{4!}a_0 \qquad a_2 = -\frac{1}{5}a_1 = \frac{3!}{5!}a_0 \qquad a_3 = -\frac{1}{6}a_2 = -\frac{3!}{6!}a_0$$

and, in general, $a_k = \dfrac{(-1)^k 3!}{(k + 3)!}a_0$. Hence

$$y_1(x) = a_0x\left[1 + 3!\sum_{n=1}^{\infty}\frac{(-1)^n x^n}{(n + 3)!}\right] = a_0x\sum_{n=0}^{\infty}\frac{(-1)^n 3!x^n}{(n + 3)!}$$

which can be simplified to $y_1(x) = \dfrac{3a_0}{x^2}(2 - 2x + x^2 - 2e^{-x})$.

Although $\lambda_1 - \lambda_2 = 3$, a positive integer, let us try to find $y_2(x)$ by repeating the method of Frobenius with $\lambda = \lambda_2$. Substituting $\lambda = -2$ into the recursion formula, we obtain $a_n = (-1/n)a_{n-1}$ for $n \geq 1$. This does not cause any a_n to be undefined for $n \geq 1$, so the unmodified method of Frobenius can be used to find $y_2(x)$. We obtain $a_1 = -a_0 = -\dfrac{1}{1!}a_0$, $a_2 = -\dfrac{1}{2}a_2 = \dfrac{1}{2!}a_0$, and, in general, $a_k = (-1)^k a_0/k!$. Therefore,

$$y_2(x) = a_0x^{-2}\left[1 - \frac{1}{1!}x + \frac{1}{2!}x^2 + \cdots + \frac{(-1)^k}{k!}x^k + \cdots\right] = a_0x^{-2}\sum_{n=0}^{\infty}\frac{(-1)^n x^n}{n!} = a_0x^{-2}e^{-x}$$

The general solution is $y = c_1y_1(x) + c_2y_2(x)$.

15.211 Find the general solution near $x = 0$ of $(x - x^2)y'' - 3y' + 2y = 0$.

▌ Here $xP(x) = 1/(1 - x^2)$ and $x^2Q(x) = 2x/(1 - x)$, both of which are analytic at $x = 0$, so $x = 0$ is regular singular point. We assume $y = x^m\sum_{n=0}^{\infty}A_nx^n = \sum_{n=0}^{\infty}A_nx^{m+n}$. Substituting for y, y', and y'' in the differential equation, we obtain

$$(m - 4)mA_0x^{m-1} + [(m - 3)(m + 1)A_1 - (m - 2)(m + 1)A_0]x^m + [(m - 2)(m + 2)A_2 - (m - 1)(m + 2)A_1]x^{m+1}$$
$$+ \cdots + [(m + n - 4)(m + n)A_n - (m + n - 3)(m + n)A_{n-1}]x^{m+n-1} + \cdots = 0$$

The recursion formula is then $A_n = \dfrac{m + n - 3}{m + n - 4}A_{n-1}$, so that

$$y = A_0x^m\left(1 + \frac{m - 2}{m - 3}x + \frac{m - 1}{m - 3}x^2 + \frac{m}{m - 3}x^3 + \frac{m + 1}{m - 3}x^4 + \frac{m + 2}{m - 3}x^5 + \frac{m + 3}{m - 3}x^6 + \cdots\right)$$

The roots $m = 0, 4$ of the indicial equation differ by an integer. However, when $m = 0$ the expected vanishing of the denominator in the coefficient of x^4 does not occur, since the factor m appears in both numerator and denominator. Note that the coefficient of x^3 is zero when $m = 0$. Thus, with $A_0 = 1$.

$$y_1 = y|_{m=0} = 1 + \frac{2x}{3} + \frac{x^2}{3} + 0 - \frac{x^4}{3} - \frac{2x^5}{3} - \frac{3x^6}{3} - \frac{4x^7}{3} - \cdots$$

and

$$y_2 = y|_{m=4} = x^4(1 + 2x + 3x^2 + 4x^3 + \cdots)$$

so that $y_1 = (1 + 2x/3 + x^2/3) - y_2/3$. The complete solution is $y = C_1y_1 + C_2y_2$, which may be simplified to $A(x^2 + 2x + 3) + B\dfrac{x^4}{(1 - x)^2}$.

BESSEL FUNCTIONS

15.212 Use the method of Frobenius to find one solution of Bessel's equation of order p: $x^2y'' + xy' + (x^2 - p^2)y = 0$.

▌ We assume a solution of the form $y = x^\lambda \sum_{n=0}^{\infty} a_n x^n = \sum_{n=0}^{\infty} a_n x^{n+\lambda}$. Substituting this series and its first two derivatives into the differential equation and combining coefficients of like powers of x, we obtain

$$x^\lambda(\lambda^2 - p^2)a_0 + x^{\lambda+1}[(\lambda+1)^2 - p^2]a_1 + x^{\lambda+2}\{[(\lambda+2)^2 - p^2]a_2 + a_0\} + \cdots$$
$$+ x^{\lambda+n}\{[(\lambda+n)^2 - p^2]a_n + a_{n-2}\} + \cdots = 0$$

Thus, $(\lambda^2 - p^2)a_0 = 0$, $[(\lambda+1)^2 - p^2]a_1 = 0$, and, in general, $[(\lambda+n)^2 - p^2]a_n + a_{n-2} = 0$, or $a_n = -\dfrac{1}{(\lambda+n)^2 - p^2} a_{n-2}$.

The indicial equation is $\lambda^2 - p^2 = 0$, which has the roots $\lambda_1 = p$ and $\lambda_2 = -p$ (for p nonnegative). For $\lambda = p$, we find that $a_1 = 0$ and $a_n = -\dfrac{1}{n(2p+n)} a_{n-2}$ $(n \ge 2)$. Hence, $0 = a_1 = a_3 = a_5 = a_7 = \cdots$ and

$$a_2 = \frac{-1}{2^2 1!(p+1)} a_0$$

$$a_4 = -\frac{1}{2^2 2(p+2)} a_2 = \frac{1}{2^4 2!(p+2)(p+1)} a_0$$

$$a_6 = -\frac{1}{2^2 3(p+3)} a_4 = \frac{-1}{2^6 3!(p+3)(p+2)(p+1)} a_0$$

and, in general, $a_{2k} = \dfrac{(-1)^k}{2^{2k} k!(p+k)(p+k-1)\cdots(p+2)(p+1)} a_0$ for $k \ge 1$. Thus,

$$y_1(x) = x^\lambda \sum_{n=0}^{\infty} a_n x^n = x^p\left[a_0 + \sum_{k=1}^{\infty} a_{2k} x^{2k}\right]$$
$$= a_0 x^p \left[1 + \sum_{k=1}^{\infty} \frac{(-1)^k x^{2k}}{2^{2k} k!(p+k)(p+k-1)\cdots(p+2)(p+1)}\right]$$

It is customary to choose the arbitrary constant $a_0 = \dfrac{1}{2^p \Gamma(p+1)}$. Then

$$y_1(x) = \frac{1}{2^p\Gamma(p+1)} x^p + \sum_{k=1}^{\infty} \frac{(-1)^k x^{2k+p}}{2^{2k+p} k!\Gamma(p+k+1)} = \sum_{k=0}^{\infty} \frac{(-1)^k x^{2k+p}}{2^{2k+p} k!\Gamma(p+k+1)} \equiv J_p(x)$$

15.213 Find the general solution to Bessel's equation of order zero.

▌ For $p = 0$, the equation is $x^2y'' + xy' + x^2y = 0$, which was solved in Problem 15.201. By the result of Problem 15.199, one solution is $y_1(x) = a_0 \sum_{n=0}^{\infty} \dfrac{(-1)^n x^{2n}}{2^{2n}(n!)^2}$. If we change n to k and let $a_0 = \dfrac{1}{2^0\Gamma(0+1)} = 1$ as indicated in Problem 15.212, it follows that $y_1(x) = J_0(x)$.

A second solution is (see Problem 15.201, with a_0 again chosen to be 1)

$$y_2(x) = J_0(x)\ln x + \left[\frac{x^2}{2^2(1!)^2}(1) - \frac{x^4}{2^4(2!)^2}\left(1 + \frac{1}{2}\right) + \frac{x^6}{2^6(3!)^2}\left(1 + \frac{1}{2} + \frac{1}{3}\right) - \cdots\right]$$

which is usually designated $N_0(x)$. Thus, the general solution to Bessel's equation of order zero is $y = c_1 J_0(x) + c_2 N_0(x)$.

Another common form of the general solution is obtained when the second linearly independent solution is not taken to be $N_0(x)$, but a combination of $N_0(x)$ and $J_0(x)$. In particular, if we define $Y_0(x) = \dfrac{2}{\pi}[N_0(x) + (\gamma - \ln 2)J_0(x)]$, where γ is the *Euler constant* defined by

$$\gamma = \lim_{k \to \infty}\left(1 + \frac{1}{2} + \frac{1}{3} + \cdots + \frac{1}{k} - \ln k\right) \approx 0.57721566$$

then the general solution to Bessel's equation of order zero can be given as $y = c_1 J_0(x) + c_2 Y_0(x)$.

15.214 Find the Laplace transform of $J_0(t)$.

▎ We have
$$J_0(t) = 1 - \frac{t^2}{2^2} + \frac{t^4}{(2^2)(4^2)} - \frac{t^6}{(2^2)(4^2)(6^2)} + \cdots$$

Then
$$\begin{aligned}\mathscr{L}\{J_0(t)\} &= \frac{1}{s} - \frac{1}{2^2}\frac{2!}{s^3} + \frac{1}{(2^2)(4^2)}\frac{4!}{s^5} - \frac{1}{(2^2)(4^2)(6^2)}\frac{6!}{s^7} + \cdots \\ &= \frac{1}{s}\left(1 - \frac{1}{2}\frac{1}{s^2} + \frac{(1)(3)}{(2)(4)}\frac{1}{s^4} - \frac{(1)(3)(5)}{(2)(4)(6)}\frac{1}{s^6} + \cdots\right) \\ &= \frac{1}{s}\left(1 + \frac{1}{s^2}\right)^{-1/2} = \frac{1}{\sqrt{s^2+1}}\end{aligned}$$

where we have used the binomial expansion $(1+x)^{-1/2} = 1 - \frac{1}{2}x + \frac{(1)(3)}{(2)(4)}x^2 - \frac{(1)(3)(5)}{(2)(4)(6)}x + \cdots$.

15.215 Prove that $\displaystyle\sum_{k=0}^{\infty} \frac{(-1)^k(2k)x^{2k-1}}{2^{2k+p}k!\Gamma(p+k+1)} = -\sum_{k=0}^{\infty} \frac{(-1)^k x^{2k+1}}{2^{2k+p+1}k!\Gamma(p+k+2)}$.

▎ Writing the $k=0$ term separately, we have
$$\sum_{k=0}^{\infty} \frac{(-1)^k(2k)x^{2k-1}}{2^{2k+p}k!\Gamma(p+k+1)} = 0 + \sum_{k=1}^{\infty} \frac{(-1)^k(2k)x^{2k-1}}{2^{2k+p}k!\Gamma(p+k+1)}$$

which, under the change of variables $j = k-1$, becomes
$$\begin{aligned}\sum_{j=0}^{\infty} \frac{(-1)^{j+1}2(j+1)x^{2(j+1)-1}}{2^{2(j+1)+p}(j+1)!\Gamma(p+j+1+1)} &= \sum_{j=0}^{\infty} \frac{(-1)(-1)^j 2(j+1)x^{2j+1}}{2^{2j+p+2}(j+1)!\Gamma(p+j+2)} \\ &= -\sum_{j=0}^{\infty} \frac{(-1)^j 2(j+1)x^{2j+1}}{2^{2j+p+1}(2)(j+1)(j!)\Gamma(p+j+2)} \\ &= -\sum_{j=0}^{\infty} \frac{(-1)^j x^{2j+1}}{2^{2j+p+1}j!\Gamma(p+j+2)}\end{aligned}$$

The desired result follows if we change the dummy variable in the last summation from j to k.

15.216 Prove that $\displaystyle -\sum_{k=0}^{\infty} \frac{(-1)^k x^{2k+p+2}}{2^{2k+p+1}k!\Gamma(p+k+2)} = \sum_{k=0}^{\infty} \frac{(-1)^k(2k)x^{2k+p}}{2^{2k+p}k!\Gamma(p+k+1)}$.

▎ We make the change of variables $j = k+1$:
$$\begin{aligned}-\sum_{k=0}^{\infty} \frac{(-1)^k x^{2k+p+2}}{2^{2k+p+1}k!\Gamma(p+k+2)} &= -\sum_{j=1}^{\infty} \frac{(-1)^{j-1}x^{2(j-1)+p+2}}{2^{2(j-1)+p+1}(j-1)!\Gamma(p+j-1+2)} \\ &= \sum_{j=1}^{\infty} \frac{(-1)^j x^{2j+p}}{2^{2j+p-1}(j-1)!\Gamma(p+j+1)}\end{aligned}$$

Now we multiply numerator and denominator in the last summation by $2j$, noting that $j(j-1)! = j!$ and $2^{2j+p-1}(2) = 2^{2j+p}$. The result is $\displaystyle\sum_{j=1}^{\infty} \frac{(-1)^j(2j)x^{2j+p}}{2^{2j+p}j!\Gamma(p+j+1)}$. Because the factor j appears in the numerator, th infinite series is not altered if the lower limit in the sum is changed from $j=1$ to $j=0$. Once this is done, the desired result is achieved by simply changing the dummy index from j to k.

15.217 Prove that $\dfrac{d}{dx}[x^{p+1}J_{p+1}(x)] = x^{p+1}J_p(x)$.

▎ We may differentiate the series for the Bessel function term by term. Thus,
$$\begin{aligned}\frac{d}{dx}[x^{p+1}J_{p+1}(x)] &= \frac{d}{dx}\left[x^{p+1}\sum_{k=0}^{\infty} \frac{(-1)^k x^{2k+p+1}}{2^{2k+p+1}k!\Gamma(k+p+1+1)}\right] \\ &= \frac{d}{dx}\left[\sum_{k=0}^{\infty} \frac{(-1)^k x^{2k+2p+2}}{2^{2k+p}(2)k!\Gamma(k+p+2)}\right] = \sum_{k=0}^{\infty} \frac{(-1)^k(2k+2p+2)x^{2k+2p+1}}{2^{2k+p}k!2\Gamma(k+p+2)}\end{aligned}$$

Noting that $2\Gamma(k+p+2) = 2(k+p+1)\Gamma(k+p+1)$ and that the factor $2(k+p+1)$ cancels, we have

$$\frac{d}{dx}[x^{p+1}J_{p+1}(x)] = \sum_{k=0}^{\infty} \frac{(-1)^k x^{2k+2p+1}}{2^{2k+p}k!\Gamma(k+p+1)} = x^{p+1}J_p(x)$$

For the particular case $p = 0$, it follows that $\dfrac{d}{dx}[xJ_1(x)] = xJ_0(x)$.

15.218 Prove that $xJ'_p(x) = pJ_p(x) - xJ_{p+1}(x)$.

▌ We have

$$pJ_p(x) - xJ_{p+1}(x) = p\sum_{k=0}^{\infty} \frac{(-1)^k x^{2k+p}}{2^{2k+p}k!\Gamma(p+k+1)} - x\sum_{k=0}^{\infty} \frac{(-1)^k x^{2k+p+1}}{2^{2k+p+1}k!\Gamma(p+k+2)}$$

$$= \sum_{k=0}^{\infty} \frac{(-1)^k p x^{2k+p}}{2^{2k+p}k!\Gamma(p+k+1)} - \sum_{k=0}^{\infty} \frac{(-1)^k x^{2k+p+2}}{2^{2k+p+1}k!\Gamma(p+k+2)}$$

Using the result of Problem 15.216 in the last summation, we find

$$pJ_p(x) - xJ_{p+1}(x) = \sum_{k=0}^{\infty} \frac{(-1)^k p x^{2k+p}}{2^{2k+p}k!\Gamma(p+k+1)} + \sum_{k=0}^{\infty} \frac{(-1)^k (2k) x^{2k+p}}{2^{2k+p}k!\Gamma(p+k+1)}$$

$$= \sum_{k=0}^{\infty} \frac{(-1)^k (p+2k) x^{2k+p}}{2^{2k+p}k!\Gamma(p+k+1)} = xJ'_p(x)$$

For the particular case $p = 0$, it follows that $xJ'_0(x) = -xJ_1(x)$, or $J'_0(x) = -J_1(x)$.

5.219 Prove that $xJ'_p(x) = -pJ_p(x) + xJ_{p-1}(x)$.

▌ We have

$$-pJ_p(x) + xJ_{p-1}(x) = -p\sum_{k=0}^{\infty} \frac{(-1)^k x^{2k+p}}{2^{2k+p}k!\Gamma(p+k+1)} + x\sum_{k=0}^{\infty} \frac{(-1)^k x^{2k+p-1}}{2^{2k+p-1}k!\Gamma(p+k)}$$

Multiplying the numerator and denominator in the second summation by $2(p+k)$ and noting that $(p+k)\Gamma(p+k) = \Gamma(p+k+1)$, we find

$$-pJ_p(x) + xJ_{p-1}(x) = \sum_{k=0}^{\infty} \frac{(-1)^k (-p) x^{2k+p}}{2^{2k+p}k!\Gamma(p+k+1)} + \sum_{k=0}^{\infty} \frac{(-1)^k 2(p+k) x^{2k+p}}{2^{2k+p}k!\Gamma(p+k+1)}$$

$$= \sum_{k=0}^{\infty} \frac{(-1)^k [-p+2(p+k)] x^{2k+p}}{2^{2k+p}k!\Gamma(p+k+1)}$$

$$= \sum_{k=0}^{\infty} \frac{(-1)^k (2k+p) x^{2k+p}}{2^{2k+p}k!\Gamma(p+k+1)} = xJ'_p(x)$$

5.220 Use Problems 15.218 and 15.219 to derive the recursion formula $J_{p+1}(x) = \dfrac{2p}{x}J_p(x) - J_{p-1}(x)$.

▌ Subtracting the results of Problem 15.219 from the results of Problem 15.218, we find that $0 = 2pJ_p(x) - xJ_{p-1}(x) - xJ_{p+1}(x)$. Solving for $J_{p+1}(x)$ produces the desired result. As an example of the use of this formula, for $p = 1$ we have $J_2(x) = (2/x)J_1(x) - J_0(x)$.

.221 Show that $y = xJ_1(x)$ is a solution of $xy'' - y' - x^2J'_0(x) = 0$.

▌ We first note that $J_1(x)$ is a solution of Bessel's equation of order one:

$$x^2J''_1(x) + xJ'_1(x) + (x^2 - 1)J_1(x) = 0 \tag{1}$$

Now substituting $y = xJ_1(x)$ into the left side of the given differential equation yields

$$x[xJ_1(x)]'' - [xJ_1(x)]' - x^2J'_0(x) = x[2J'_1(x) + xJ''_1(x)] - [J_1(x) + xJ'_1(x)] - x^2J'_0(x)$$

But $J'_0(x) = -J_1(x)$ (see Problem 15.218), so that the right-hand side becomes

$$x^2J''_1(x) + 2xJ'_1(x) - J_1(x) - xJ'_1(x) + x^2J_1(x) = x^2J''_1(x) + xJ'_1(x) + (x^2 - 1)J_1(x) = 0$$

where the last equality follows from (*1*).

15.222 Show that $y = \sqrt{x}J_{3/2}(x)$ is a solution of $x^2y'' + (x^2 - 2)y = 0$.

▌ Observe that $J_{3/2}(x)$ is a solution of Bessel's equation of order $\frac{3}{2}$:

$$x^2 J''_{3/2}(x) + xJ'_{3/2}(x) + (x^2 - \tfrac{9}{4})J_{3/2}(x) = 0 \tag{1}$$

Now we substitute $y = \sqrt{x}J_{3/2}(x)$ into the left side of the given differential equation, obtaining

$$\begin{aligned} &x^2[\sqrt{x}J_{3/2}(x)]'' + (x^2 - 2)\sqrt{x}J_{3/2}(x) \\ &\quad = x^2[-\tfrac{1}{4}x^{-3/2}J_{3/2}(x) + x^{-1/2}J'_{3/2}(x) + x^{1/2}J''_{3/2}(x)] + (x^2 - 2)x^{1/2}J_{3/2}(x) \\ &\quad = \sqrt{x}[x^2 J''_{3/2}(x) + xJ'_{3/2}(x) + (x^2 - \tfrac{9}{4})J_{3/2}(x)] = 0 \end{aligned}$$

where the last equality follows from (*1*). Thus $\sqrt{x}J_{3/2}(x)$ satisfies the given differential equation.

15.223 Find the Laplace transform of $J_1(t)$.

▌ It follows from Problem 15.218 (with x replaced by t) that $J'_0(t) = -J_1(t)$. Then, using the result of Problem 15.214, we have

$$\mathscr{L}\{J_1(t)\} = -\mathscr{L}\{J'_0(t)\} = -(s\mathscr{L}\{J_0(t)\} - 1) = 1 - \frac{s}{\sqrt{s^2+1}} = \frac{\sqrt{s^2+1} - s}{\sqrt{s^2+1}}$$

CHAPTER 16

Eigenfunction Expansions

STURM-LIOUVILLE PROBLEMS

16.1 Define the second-order Sturm-Liouville problem.

▌ A *second-order Sturm-Liouville problem* is a homogeneous boundary-value problem of the form

$$[p(x)y']' + q(x)y + \lambda w(x)y = 0; \qquad \alpha_1 y(a) + \beta_1 y'(a) = 0, \qquad \alpha_2 y(b) + \beta_2 y'(b) = 0$$

where $p(x)$, $p'(x)$, $q(x)$, and $w(x)$ are continuous on $[a, b]$, and both $p(x)$ and $w(x)$ are positive on $[a, b]$. The constant λ is arbitrary.

16.2 Determine whether the boundary-value problem $e^x y'' + e^x y' + \lambda y = 0$; $y(0) = 0$, $y'(1) = 0$ is a Sturm-Liouville problem.

▌ The equation can be rewritten as $(e^x y')' + \lambda y = 0$; hence $p(x) = e^x$, $q(x) \equiv 0$, and $w(x) \equiv 1$. This is a Sturm-Liouville problem.

16.3 Determine whether the boundary-value problem $xy'' + y' + (x^2 + 1 + \lambda)y = 0$; $y(0) = 0$, $y'(1) = 0$ is a Sturm-Liouville problem.

▌ The equation is equivalent to $(xy')' + (x^2 + 1)y + \lambda y = 0$; hence $p(x) = x$, $q(x) = x^2 + 1$, and $w(x) \equiv 1$. Since $p(x)$ is zero at a point in the interval $[0, 1]$, this is not a Sturm-Liouville problem.

16.4 Determine whether the boundary-value problem $(xy')' + (x^2 + 1 + \lambda e^x)y = 0$; $y(1) + 2y'(1) = 0$, $y(2) - 3y'(2) = 0$ is a Sturm-Liouville problem.

▌ This is a Sturm-Liouville problem with $p(x) = x$, $q(x) = x^2 + 1$, and $w(x) = e^x$. Note that on $[1, 2]$, which is the interval of interest, both $p(x)$ and $w(x)$ are positive.

16.5 Determine whether the boundary-value problem $\left(\frac{1}{x} y'\right)' + (x + \lambda)y = 0$; $y(0) + 3y'(0) = 0$, $y(1) = 0$ is a Sturm-Liouville problem.

▌ Here $p(x) = 1/x$, $q(x) = x$, and $w(x) \equiv 1$. Since $p(x)$ is not continuous in $[0, 1]$, in particular at $x = 0$, this is not a Sturm-Liouville problem.

16.6 Determine whether the boundary-value problem $y'' + \lambda(1 + x)y = 0$; $y'(0) = 0$, $y(2) + y'(2) = 0$ is a Sturm-Liouville problem.

▌ The equation can be rewritten as $(y')' + \lambda(1 + x)y = 0$; hence $p(x) \equiv 1$, $q(x) \equiv 0$, and $w(x) = 1 + x$. This is a Sturm-Liouville problem.

16.7 Determine whether the boundary-value problem $y'' + \lambda y = 0$; $y(0) = 0$, $y(1) = 0$ is a Sturm-Liouville problem.

▌ This is a Sturm-Liouville problem with $p(x) = w(x) = 1$ and $q(x) = 0$. Both $p(x)$ and $w(x)$ are positive everywhere, so they are positive on the interval of interest, $[0, 1]$.

16.8 Determine whether the boundary-value problem $y'' + \lambda y = 0$; $y(0) = 0$, $Y'(\pi) = 0$ is a Sturm-Liouville problem.

▌ It is, for the reasons given in the previous problem.

16.9 Show how to convert the second-order differential equation $a_2(x)y'' + a_1(x)y' + a_0(x)y + \lambda r(x)y = 0$ into a differential equation having the form required for a Sturm-Liouville problem when $a_2(x)$ and $r(x)$ are positive in the interval of interest.

▌ Multiplying the differential equation by $I(x) = e^{\int [a_1(x)/a_2(x)]\,dx}$, we obtain $I(x)a_2(x)y'' + I(x)a_1(x)y' + I(x)a_0(x)y + \lambda I(x)r(x)y = 0$, which can be rewritten as $a_2(x)[I(x)y']' + I(x)a_0(x)y + \lambda I(x)r(x)y = 0$. To obtain the desired form, we divide this equation by $a_2(x)$ and then set $p(x) = I(x)$, $q(x) = I(x)a_0(x)/a_2(x)$, and $w(x) = I(x)r(x)/a_2(x)$. Note that since $I(x)$ is an exponential and $a_2(x)$ does not vanish, $I(x)$ is positive.

16.10 Transform $y'' + 2xy' + (x + \lambda)y = 0$ into the form required for a Sturm-Liouville problem.

▌ Here $a_2(x) \equiv 1$ and $a_1(x) = 2x$; hence $a_1(x)/a_2(x) = 2x$ and $I(x) = e^{\int 2x\,dx} = e^{x^2}$. Multiplying the given differential equation by $I(x)$, we obtain $e^{x^2}y'' + 2xe^{x^2}y' + xe^{x^2}y + \lambda e^{x^2}y = 0$, which can be rewritten as $(e^{x^2}y')' + xe^{x^2}y + \lambda e^{x^2}y = 0$ with $p(x) = e^{x^2}$, $q(x) = xe^{x^2}$, and $w(x) = e^{x^2}$.

16.11 Transform $(x + 2)y'' + 4y' + xy + \lambda e^x y = 0$ into the form required for a Sturm-Liouville problem.

▌ Here $a_2(x) = x + 2$ and $a_1(x) \equiv 4$; hence $a_1(x)/a_2(x) = 4/(x + 2)$ and $I(x) = e^{\int [4/(x+2)]\,dx} = e^{4\ln|x+2|} = e^{\ln(x+2)^4} = (x + 2)^4$. Multiplying the given differential equation by $I(x)$ yields

$$(x + 2)^5 y'' + 4(x + 2)^4 y' + (x + 2)^4 xy + \lambda(x + 2)^4 e^x y = 0$$

which can be rewritten as $(x + 2)[(x + 2)^4 y']' + (x + 2)^4 xy + \lambda(x + 2)^4 e^x y = 0$ or $[(x + 2)^4 y']' + (x + 2)^3 y + \lambda(x + 2)^3 e^x y = 0$. This last equation has $p(x) = (x + 2)^4$, $q(x) = (x + 2)^3$, and $w(x) = (x + 2)^3 e^x$. Note that since we divided by $a_2(x)$, it is necessary to restrict $x \neq -2$. Furthermore, in order that both $p(x)$ and $w(x)$ be positive, we must require $x > -2$.

16.12 Find the eigenvalues and eigenfunctions of the Sturm-Liouville problem $y'' + \lambda y = 0$; $y(0) = 0$, $y(1) = 0$.

▌ The characteristic equation is $m^2 + \lambda = 0$. We consider the cases $\lambda = 0$, $\lambda < 0$, and $\lambda > 0$ separately, since they lead to different solutions.

For $\lambda = 0$, the solution is $y = c_1 + c_2 x$. Applying the boundary conditions, we obtain $c_1 = c_2 = 0$, which results in the trivial solution $y \equiv 0$.

For $\lambda < 0$, the solution is $y = c_1 e^{\sqrt{-\lambda}x} + c_2 e^{-\sqrt{-\lambda}x}$, where $-\lambda$ and $\sqrt{-\lambda}$ are positive. Applying the boundary conditions, we obtain $c_1 + c_2 = 0$ and $c_1 e^{\sqrt{-\lambda}} + c_2 e^{-\sqrt{-\lambda}} = 0$. The only solution to this set of equations is $c_1 = c_2 = 0$, which leads to the trivial solution $y \equiv 0$.

For $\lambda > 0$, the solution is $y = A \sin\sqrt{\lambda}x + B\cos\sqrt{\lambda}x$. Applying the boundary conditions, we obtain $B =$ and $A \sin\sqrt{\lambda} = 0$. Note that $\sin\theta = 0$ if and only if $\theta = n\pi$, where $n = 0, \pm 1, \pm 2, \ldots$. Furthermore, if $\theta > 0$, then n must be positive. To satisfy the boundary conditions, $B = 0$ and either $A = 0$ or $\sin\sqrt{\lambda} = 0$. This last equation is equivalent to $\sqrt{\lambda} = n\pi$, where $n = 1, 2, 3, \ldots$. The choice $A = 0$ results in the trivial solution; the choice $\sqrt{\lambda} = n\pi$ results in the nontrivial solution $y_n = A_n \sin n\pi x$. Here the notation A_n signifies that the arbitrary constant A_n can be different for different values of n.

Collecting the results of all three cases, we conclude that the eigenvalues are $\lambda_n = n^2\pi^2$ and the corresponding eigenfunctions are $y_n = A_n \sin n\pi x$, for $n = 1, 2, 3, \ldots$.

16.13 Find the eigenvalues and eigenfunctions of the Sturm-Liouville problem $y'' + \lambda y = 0$; $y(0) = 0$, $y'(\pi) = 0$.

▌ The cases $\lambda = 0$, $\lambda < 0$, and $\lambda > 0$ must be considered separately.

For $\lambda = 0$, the solution is $y = c_1 + c_2 x$. Applying the boundary conditions, we obtain $c_1 = c_2 = 0$ an hence $y \equiv 0$.

For $\lambda < 0$, the solution is $y = c_1 e^{\sqrt{-\lambda}x} + c_2 e^{-\sqrt{-\lambda}x}$, where $-\lambda$ and $\sqrt{-\lambda}$ are positive. Applying the boundary conditions, we obtain $c_1 + c_2 = 0$ and $c_1\sqrt{-\lambda}e^{\sqrt{-\lambda}\pi} - c_2\sqrt{-\lambda}e^{-\sqrt{-\lambda}\pi} = 0$. The only solution to these equations is $c_1 = c_2 = 0$; hence $y \equiv 0$.

For $\lambda > 0$, the solution is $y = A \sin\sqrt{\lambda}x + B\cos\sqrt{\lambda}x$. Applying the boundary conditions, we obtain $B = 0$ and $A\sqrt{\lambda}\cos\sqrt{\lambda}\pi = 0$. For $\theta > 0$, $\cos\theta = 0$ if and only if θ is a positive odd multiple of $\pi/2$; that is, for $\theta = (2n - 1)\pi/2 = (n - \frac{1}{2})\pi$, where $n = 1, 2, 3, \ldots$. Therefore, to satisfy the boundary conditions we must have $B = 0$ and either $A = 0$ or $\cos\sqrt{\lambda}\pi = 0$. This last equation is equivalent to $\sqrt{\lambda} = n - \frac{1}{2}$. The choice $A = 0$ results in the trivial solution; the choice $\sqrt{\lambda} = n - \frac{1}{2}$ results in the nontrivial solution $y_n = A_n \sin(n - \frac{1}{2})x$.

Collecting all three cases, we conclude that the eigenvalues are $\lambda_n = (n - \frac{1}{2})^2$ and the corresponding eigenfunctions are $y_n = A_n \sin(n - \frac{1}{2})x$, where $n = 1, 2, 3, \ldots$.

FOURIER SERIES

16.14 Prove $\int_{-L}^{L} \cos\frac{m\pi x}{L}\cos\frac{n\pi x}{L}\,dx = \begin{cases} 0 & m \neq n \\ L & m = n \end{cases}$, where m and n can assume the values 1, 2, 3, ...

From trigonometry, $\cos A \cos B = \frac{1}{2}[\cos(A-B) + \cos(A+B)]$. Hence, when $m \neq n$, we have

$$\int_{-L}^{L} \cos\frac{m\pi x}{L}\cos\frac{n\pi x}{L}\,dx = \frac{1}{2}\int_{-L}^{L}\left[\cos\frac{(m-n)\pi x}{L} + \cos\frac{(m+n)\pi x}{L}\right]dx$$
$$= \frac{L}{2(m-n)\pi}\sin\frac{(m-n)\pi x}{L}\Bigg|_{-L}^{L} + \frac{L}{2(m+n)\pi}\sin\frac{(m+n)\pi x}{L}\Bigg|_{-L}^{L} = 0$$

Furthermore, $\cos^2 A = \frac{1}{2}(1 + \cos 2A)$, so when $m = n$, we have

$$\int_{-L}^{L} \cos\frac{m\pi x}{L}\cos\frac{n\pi x}{L}\,dx = \frac{1}{2}\int_{-L}^{L}\left(1 + \cos\frac{2n\pi x}{L}\right)dx = \frac{1}{2}\left[x + \frac{L}{2n\pi}\sin\frac{2n\pi x}{L}\right]_{-L}^{L} = L$$

16.15 Prove $\int_{-L}^{L} \sin\frac{m\pi x}{L}\sin\frac{n\pi x}{L}\,dx = \begin{cases} 0 & m \neq n \\ L & m = n \end{cases}$, where m and n can assume the values 1, 2, 3, ...

From trigonometry, $\sin A \sin B = \frac{1}{2}[\cos(A-B) - \cos(A+B)]$ and $\sin^2 A = \frac{1}{2}(1 - \cos 2A)$. When $m \neq n$,

$$\int_{-L}^{L} \sin\frac{m\pi x}{L}\sin\frac{n\pi x}{L}\,dx = \frac{1}{2}\int_{-L}^{L}\left[\cos\frac{(m-n)\pi x}{L} - \cos\frac{(m+n)\pi x}{L}\right]dx = 0$$

When $m = n$, $\int_{-L}^{L} \sin\frac{m\pi x}{L}\sin\frac{n\pi x}{L}\,dx = \frac{1}{2}\int_{-L}^{L}\left(1 - \cos\frac{2n\pi x}{L}\right)dx = L.$

16.16 Prove $\int_{-L}^{L} \sin\frac{m\pi x}{L}\cos\frac{n\pi x}{L}\,dx = 0$, where m and n can assume the values 1, 2, 3, ...

We have $\sin A \cos B = \frac{1}{2}[\sin(A-B) + \sin(A+B)]$. Then, if $m \neq n$,

$$\int_{-L}^{L} \sin\frac{m\pi x}{L}\cos\frac{n\pi x}{L}\,dx = \frac{1}{2}\int_{-L}^{L}\left[\sin\frac{(m-n)\pi x}{L} + \sin\frac{(m+n)\pi x}{L}\right]dx = 0$$

If $m = n$, $\int_{-L}^{L} \sin\frac{m\pi x}{L}\cos\frac{n\pi x}{L}\,dx = \frac{1}{2}\int_{-L}^{L}\sin\frac{2n\pi x}{L}\,dx = 0.$

16.17 If the series $f(x) = A + \sum_{n=1}^{\infty}\left(a_n\cos\frac{n\pi x}{L} + b_n\sin\frac{n\pi x}{L}\right)$ converges uniformly in $(-L, L)$, show that $a_n = \frac{1}{L}\int_{-L}^{L} f(x)\cos\frac{n\pi x}{L}\,dx$, for $n = 1, 2, 3, \ldots$.

Multiplying $f(x) = A + \sum_{n=1}^{\infty}\left(a_n\cos\frac{n\pi x}{L} + b_n\sin\frac{n\pi x}{L}\right)$ by $\cos\frac{m\pi x}{L}$ and integrating from $-L$ to L (using Problems 16.14 and 16.16), we have

$$\int_{-L}^{L} f(x)\cos\frac{m\pi x}{L}\,dx = A\int_{-L}^{L}\cos\frac{m\pi x}{L}\,dx + \sum_{n=1}^{\infty}\left(a_n\int_{-L}^{L}\cos\frac{m\pi x}{L}\cos\frac{n\pi x}{L}\,dx + b_n\int_{-L}^{L}\cos\frac{m\pi x}{L}\sin\frac{n\pi x}{L}\,dx\right)$$
$$= a_m L \qquad \text{for } m \neq 0$$

Thus, $a_m = \frac{1}{L}\int_{-L}^{L} f(x)\cos\frac{m\pi x}{L}\,dx$ for $m = 1, 2, 3, \ldots$.

16.18 For the series in Problem 16.17, show that $b_n = \frac{1}{L}\int_{-L}^{L} f(x)\sin\frac{n\pi x}{L}\,dx.$

Multiplying $f(x) = A + \sum_{n=1}^{\infty}\left(a_n\cos\frac{n\pi x}{L} + b_n\sin\frac{n\pi x}{L}\right)$ by $\sin\frac{m\pi x}{L}$ and integrating from $-L$ to L (using

Problems 16.15 and 16.16), we have

$$\int_{-L}^{L} f(x)\sin\frac{m\pi x}{L}\,dx = A\int_{-L}^{L}\sin\frac{m\pi x}{L}\,dx + \sum_{n=1}^{\infty}\left(a_n\int_{-L}^{L}\sin\frac{m\pi x}{L}\cos\frac{n\pi x}{L}\,dx + b_n\int_{-L}^{L}\sin\frac{m\pi x}{L}\sin\frac{n\pi x}{L}\,dx\right)$$
$$= b_m L$$

Thus, $b_m = \frac{1}{L}\int_{-L}^{L} f(x)\sin\frac{m\pi x}{L}\,dx$ for $m = 1, 2, 3, \ldots$.

16.19 For the series in Problem 16.17, show that $A = a_0/2$, where a_0 is the extension of a_m to $m = 0$.

▌ Integrating $f(x) = A + \sum_{n=1}^{\infty}\left(a_n\cos\frac{n\pi x}{L} + b_n\sin\frac{n\pi x}{L}\right)$ from $-L$ to L gives $\int_{-L}^{L} f(x)\,dx = 2AL$, so that $A = \frac{1}{2L}\int_{-L}^{L} f(x)\,dx$. By putting $m = 0$ in the result of Problem 16.17, we find $a_0 = \frac{1}{L}\int_{-L}^{L} f(x)\,dx$. Thus $A = a_0/2$.

16.20 Determine formulas for the Fourier coefficents for $f(x)$ on the interval $(c, c + 2L)$ for any real number c.

▌ The results of Problems 16.17 through 16.19 remain valid if the limits of integration $-L$ and L are replaced, respectively, with c and $c + 2L$. Thus, if the infinite series of Problem 16.17 converges uniformly to $f(x)$ in $(c, c + 2L)$, then $f(x) = \frac{a_0}{2} + \sum_{n=1}^{\infty} a_n\cos\frac{n\pi x}{L} + b_n\sin\frac{n\pi x}{L}$, with

$$a_0 = \frac{1}{2L}\int_c^{c+2L} f(x)\,dx \qquad a_n = \frac{1}{L}\int_c^{c+2L} f(x)\cos\frac{n\pi x}{L}\,dx \qquad b_n = \frac{1}{L}\int_c^{c+2L} f(x)\sin\frac{n\pi x}{L}\,dx$$

16.21 State sufficient conditions for the convergence of the series $\frac{a_0}{2} + \sum_{n=1}^{\infty}\left(a_n\cos\frac{n\pi x}{L} + b_n\sin\frac{n\pi x}{L}\right)$ to $f(x)$ on $(c, c + 2L)$ when a_0, a_n, and b_n are determined by the formulas found in Problem 16.20.

▌ The Dirichlet conditions are sufficient for convergence:

1. $f(x)$ is defined and single-valued except possibly at a finite number of points in $(c, c + 2L)$.
2. $f(x)$ is periodic with period $2L$.
3. $f(x)$ and $f'(x)$ are piecewise continuous in $(c, c + 2L)$.

Then the series converges to $f(x)$ if x is a point of continuity, or to $\frac{f(x+0) + f(x-0)}{2}$ if x is a point of discontinuity.

By this result we can write $f(x) = \frac{a_0}{2} + \sum_{n=1}^{\infty}\left(a_n\cos\frac{n\pi x}{L} + b_n\sin\frac{n\pi x}{L}\right)$ at any point of continuity x. However, if x is a point of discontinuity, then the left side must be replaced with $\frac{1}{2}[f(x+0) + f(x-0)]$, so that the series converges to the mean of $f(x+0)$ and $f(x-0)$.

16.22 Graph the function $f(x) = \begin{cases} 3 & 0 < x < 5 \\ -3 & -5 < x < 0 \end{cases}$, period = 10.

▌ Since the period is 10, that portion of the graph in $-5 < x < 5$ (heavy in Fig. 16.1) must be extended periodically outside this range (indicated dashed). Note that $f(x)$ is not defined at $x = 0, 5, -5, 10, -10, 15, -15$, etc. These values are the *discontinuities* of $f(x)$.

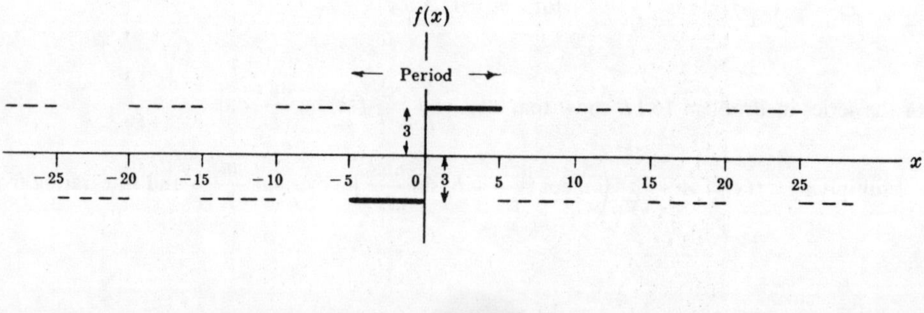

Fig. 16.

16.23 Determine whether the function $f(x)$ defined in Problem 16.22 satisfies the Dirichlet conditions.

▌ Here $c = -5$ and $L = 5$. The function is single-valued everywhere on the interval $(-5, 5)$ except at $x = 0$. At the end points of the subintervals on which $f(x)$ is continuous, we have

$$f(-5+0) = \lim_{\substack{x \to -5 \\ x > -5}} f(x) = -3 \qquad f(0-0) = \lim_{\substack{x \to 0 \\ x < 0}} f(x) = -3$$

$$f(0+0) = \lim_{\substack{x \to 0 \\ x > 0}} f(x) = 3 \qquad f(5-0) = \lim_{\substack{x \to 5 \\ x < 5}} f(x) = 3$$

so $f(x)$ is piecewise continuous on $(-5, 5)$.

In addition, $f'(x) = 0$ everywhere on $(-5, 5)$ except at $x = 0$, where it is undefined. For this derivative, the right-hand limit at $x = -5$, the right-hand and left-hand limits at $x = 0$, and the left-hand limit at $x = 5$ are zero. Since these limits exist, $f'(x)$ is piecewise continuous.

Finally, since $f(x)$ has period 10, it satisfies the Dirichlet conditions.

16.24 Graph the function $f(x) = \begin{cases} \sin x & 0 \le x \le \pi \\ 0 & \pi < x < 2\pi \end{cases}$, period $= 2\pi$.

▌ The graph is shown in Fig. 16.2.

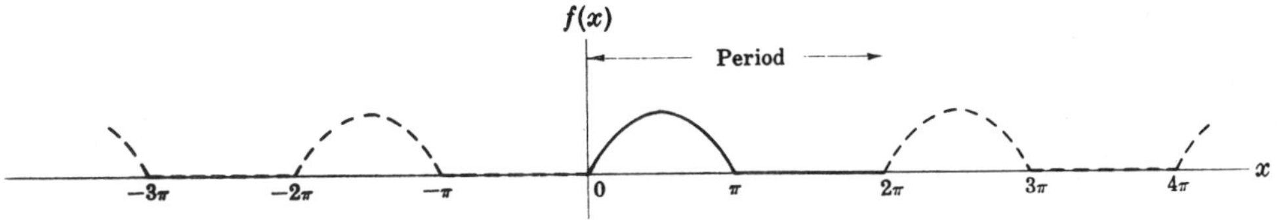

Fig. 16.2

16.25 Determine whether the function $f(x)$ defined in Problem 16.24 satisfies the Dirichlet conditions.

▌ Here $c = 0$ and $L = \pi$. The function is single-valued and continuous everywhere. On the interval $(0, 2\pi)$, its derivative is $f'(x) = \begin{cases} \cos x & 0 < x < \pi \\ 0 & \pi < x < 2\pi \end{cases}$, which is single-valued and continuous on the two subintervals $(0, \pi)$ and $(\pi, 2\pi)$. At the end points of these intervals,

$$f'(0+0) = \lim_{\substack{x \to 0 \\ x > 0}} f'(x) = \lim_{\substack{x \to 0 \\ x > 0}} \cos x = 1 \qquad f'(\pi-0) = \lim_{\substack{x \to \pi \\ x < \pi}} f'(x) = \lim_{\substack{x \to \pi \\ x < \pi}} \cos x = -1$$

$$f'(\pi+0) = \lim_{\substack{x \to \pi \\ x > \pi}} f'(x) = \lim_{\substack{x \to \pi \\ x > \pi}} 0 = 0 \qquad f'(2\pi-0) = \lim_{\substack{x \to 2\pi \\ x < 2\pi}} f'(x) = \lim_{\substack{x \to 2\pi \\ x < 2\pi}} 0 = 0$$

Thus, all required limits exist, and $f'(x)$ is piecewise continuous. Since $f(x)$ also has period 2π, it satisfies the Dirichlet conditions.

16.26 Graph the function $f(x) = \begin{cases} 0 & 0 \le x < 2 \\ 1 & 2 \le x < 4, \\ 0 & 4 \le x < 6 \end{cases}$ period $= 6$.

▌ The function is graphed in Fig. 16.3. Note that $f(x)$ is defined for all x and is discontinuous at $x = \pm 2, \pm 4, \pm 8, \pm 10, \pm 14, \ldots$.

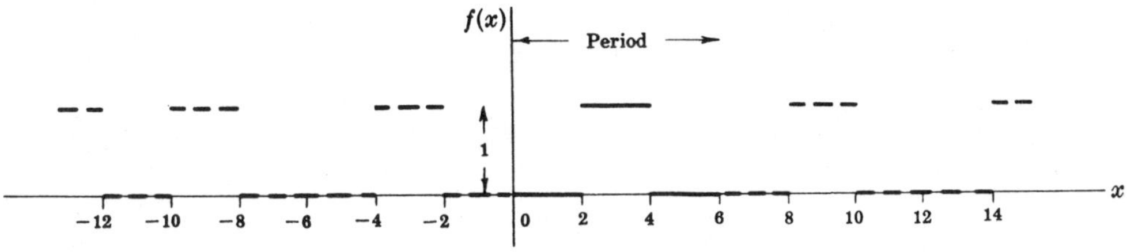

Fig. 16.3

16.27 Determine whether the function defined in Problem 16.26 satisfies the Dirichlet conditions.

▌ Here $c = 0$ and $L = 3$. The function is single-valued and continuous everywhere on the subintervals (0, 2), (2, 4), (4, 6). The right-hand limit at 0, the left-hand limit at 2, the right-hand limit at 4, and the left-hand limit at 6 are zero. In addition, both the right-hand limit at 2 and the left-hand limit at 4 are 1, so all the required limits exists and $f(x)$ is piecewise continuous.

In addition, $f'(x) = 0$ everywhere on (0, 6) except at $x = 2$ and $x = 4$. The right- and left-hand limits of $f'(x)$ at these two points are zero, as are the right-hand limit at $x = 0$ and the left-hand limit at $x = 6$. Thus $f'(x)$ is also piecewise continuous on (0, 6).

Since $f(x)$ is periodic with period 6, $f(x)$ satisfies the Dirichlet conditions.

16.28 Determine whether $f(x) = 1/x$, $0 < x < 4$ with period 4, satisfies the Dirichlet conditions.

▌ Here $c = 0$ and $L = 2$. The function is single-valued and continuous everywhere on (0, 4). At the left-hand end point, however, $\lim_{\substack{x\to 0\\ x>0}} f(x) = \lim_{\substack{x\to 0\\ x>0}} \frac{1}{x} = \infty$. Since one of the required limits does not exist, $f(x)$ is not piecewise continuous on (0, 4) and, therefore, does not satisfy the Dirichlet conditions.

16.29 Determine whether $f(x) = \sqrt{x-2}$, $2 < x < 5$ with period 3, satisfies the Dirichlet conditions.

▌ Here $c = 2$ and $L = 1.5$. The function $f(x)$ is single-valued and continuous everywhere on (2, 5), and it has a right-hand limit at $x = 2$ and a left-hand limit at $x = 5$ (namely 0 and $\sqrt{3}$, respectively), so $f(x)$ is piecewise continuous.

The derivative, $f'(x) = \frac{1}{2}(x-2)^{-1/2}$, is also continuous everywhere on (2, 5); however, at the left-hand end point, $\lim_{\substack{x\to 2\\ x>2}} f(x) = \lim_{\substack{x\to 2\\ x>2}} \frac{1}{2\sqrt{x-2}} = \infty$. Since one of the required limits does not exist, $f'(x)$ is not piecewise continuous and, therefore, $f(x)$ does not satisfy the Dirichlet condtions.

16.30 Determine whether the function $f(x) = (x^2 - 4)/(x - 2)$, defined everywhere on the interval (1, 5) except at $x = 2$ and having period 4, satisfies the Dirichlet conditions.

▌ Here $c = 1$ and $L = 2$. The function is single-valued everywhere on (1, 5) except at $x = 2$, where it is undefined. Thus, $f(x)$ is single-valued on $(c, c + 2L)$ except at a single point. The function is continuous everywhere on (1, 2) and (2, 5), with $f(2 + 0) = \lim_{\substack{x\to 2\\ x>2}} \frac{x^2-4}{x-2} = \lim_{\substack{x\to 2\\ x>2}} (x + 2) = 4$ and similarly with $f(2 - 0) = 4$, $f(1 + 0) = 3$, and $f(5 - 0) = 7$. Thus, $f(x)$ is piecewise continuous.

Its derivative $f'(x) = 1$ everywhere on (1, 5) except at $x = 2$, where it is undefined. Since $f'(1 + 0) = f'(2 - 0) = f'(2 + 0) = f(5 - 0) = 1$, $f'(x)$ is also piecewise continuous. Since $f(x)$ has period 4, it satisfies the Dirichlet conditions.

16.31 Find the Fourier series corresponding to the function $f(x) = \begin{cases} 3 & 0 < x < 5 \\ -3 & -5 < x < 0 \end{cases}$, period = 10.

▌ The graph of this function is shown in Fig. 16.1. With $c = -5$ and $L = 5$, it follows from Problem 16.20 that

$$a_n = \frac{1}{L}\int_{-L}^{L} f(x)\cos\frac{n\pi x}{L}\,dx = \frac{1}{5}\int_{-5}^{5} f(x)\cos\frac{n\pi x}{5}\,dx = \frac{1}{5}\left[\int_{-5}^{0}(-3)\cos\frac{n\pi x}{5}\,dx + \int_{0}^{5} 3\cos\frac{n\pi x}{5}\,dx\right]$$

$$= \frac{1}{5}\left(-\frac{15}{n\pi}\sin\frac{n\pi x}{5}\bigg|_{-5}^{0} + \frac{15}{n\pi}\sin\frac{n\pi x}{5}\bigg|_{0}^{5}\right) = 0 \quad \text{for } n \neq 0$$

If $n = 0$, then $a_0 = \frac{2}{L}\int_{-L}^{L} f(x)\,dx = \frac{2}{5}\int_{-5}^{5} f(x)\,dx = \frac{2}{5}\int_{-5}^{0}(-3)\,dx + \int_{0}^{5} 3\,dx = 0$. Also,

$$b_n = \frac{1}{L}\int_{-L}^{L} f(x)\sin\frac{n\pi x}{L}\,dx = \frac{1}{5}\int_{-5}^{5} f(x)\sin\frac{n\pi x}{5}\,dx = \frac{1}{5}\left[\int_{-5}^{0}(-3)\sin\frac{n\pi x}{5}\,dx + \int_{0}^{5} 3\sin\frac{n\pi x}{L}\,dx\right]$$

$$= \frac{1}{5}\left(\frac{15}{n\pi}\cos\frac{n\pi x}{5}\bigg|_{-5}^{0} - \frac{15}{n\pi}\cos\frac{n\pi x}{5}\bigg|_{0}^{5}\right) = \frac{6}{n\pi}(1 - \cos n\pi)$$

With $a_n = a_0 = 0$, the corresponding Fourier series is

$$\frac{a_0}{2} + \sum_{n=1}^{\infty} a_n \cos\frac{n\pi x}{L} + b_n \sin\frac{n\pi x}{L} = \sum_{n=1}^{\infty} \frac{6}{n\pi}(1 - \cos n\pi)\sin\frac{n\pi x}{5} = \frac{12}{\pi}\left(\sin\frac{\pi x}{5} + \frac{1}{3}\sin\frac{3\pi x}{5} + \frac{1}{5}\sin\frac{5\pi x}{5} + \cdots\right)$$

2 How should the function in Problem 16.31 be defined at $x = -5$, $x = 0$, and $x = 5$ so that its Fourier series will converge to $f(x)$ everywhere on $[-5, 5]$?

▌ It is shown in Problem 16.23 that $f(x)$ satisfies the Dirichlet conditions. Thus, the Fourier series obtained in Problem 16.31 converges to $f(x)$ at all points of continuity, and at points of discontinuity the series converges to $\frac{1}{2}[f(x+0) + f(x-0)]$. In $[-5, 5]$, the points of discontinuity are -5, 0, and 5. We can see from the graph (Fig. 16.1) or from the results of Problem 16.23 that at each point of discontinuity the series converges to $(-3+3)/2 = 0$. Thus the series will converge to $f(x)$ everywhere on $[-5, 5]$ if we redefine $f(x)$ as follows:

$$f(x) = \begin{cases} 0 & x = -5 \\ -3 & -5 < x < 0 \\ 0 & x = 0 \\ 3 & 0 < x < 5 \\ 0 & x = 5 \end{cases} \qquad \text{period} = 10$$

33 Find the Fourier series corresponding to the function $f(x) = \begin{cases} 0 & -5 < x < 0 \\ 3 & 0 < x < 5 \end{cases}$, period = 10.

▌ Here
$$a_n = \frac{1}{L}\int_{-L}^{L} f(x)\cos\frac{n\pi x}{L}\,dx = \frac{1}{5}\int_{-5}^{5} f(x)\cos\frac{n\pi x}{5}\,dx$$
$$= \frac{1}{5}\left\{\int_{-5}^{0}(0)\cos\frac{n\pi x}{5}\,dx + \int_{0}^{5}(3)\cos\frac{n\pi x}{5}\,dx\right\} = \frac{3}{5}\int_{0}^{5}\cos\frac{n\pi x}{5}\,dx$$
$$= \frac{3}{5}\left(\frac{5}{n\pi}\sin\frac{n\pi x}{5}\right)\Bigg|_{0}^{5} = 0 \qquad \text{for } n \neq 0$$

If $n = 0$ $a_0 = \frac{3}{5}\int_0^5 \cos\frac{0\pi x}{5}\,dx = \frac{3}{5}\int_0^5 dx = 3$. Also,

$$b_n = \frac{1}{L}\int_{-L}^{L} f(x)\sin\frac{n\pi x}{L}\,dx = \frac{1}{5}\int_{-5}^{5} f(x)\sin\frac{n\pi x}{5}\,dx$$
$$= \frac{1}{5}\left\{\int_{-5}^{0}(0)\sin\frac{n\pi x}{5}\,dx + \int_{0}^{5}(3)\sin\frac{n\pi x}{5}\,dx\right\} = \frac{3}{5}\int_{0}^{5}\sin\frac{n\pi x}{5}\,dx$$
$$= \frac{3}{5}\left(-\frac{5}{n\pi}\cos\frac{n\pi x}{5}\right)\Bigg|_{0}^{5} = \frac{3(1 - \cos n\pi)}{n\pi}$$

The corresponding Fourier series is

$$\frac{a_0}{2} + \sum_{n=1}^{\infty}\left(a_n\cos\frac{n\pi x}{L} + b_n\sin\frac{n\pi x}{L}\right) = \frac{3}{2} + \sum_{n=1}^{\infty}\frac{3(1-\cos n\pi)}{n\pi}\sin\frac{n\pi x}{5}$$
$$= \frac{3}{2} + \frac{6}{\pi}\left(\sin\frac{\pi x}{5} + \frac{1}{3}\sin\frac{3\pi x}{5} + \frac{1}{5}\sin\frac{5\pi x}{5} + \cdots\right)$$

.34 How should the function $f(x)$ in Problem 16.33 be defined at $x = -5$, $x = 0$, and $x = 5$ so that its Fourier series will converge to $f(x)$ everywhere on $[-5, 5]$?

▌ Since $f(x)$ satisfies the Dirichlet conditions, we can say that the series converges to $f(x)$ at all points of continuity and to $\frac{1}{2}[f(x+0) + f(x-0)]$ at points of discontinuity. At $x = -5, 0,$ and 5, which are points of discontinuity, the series converges to $(3+0)/2 = 3/2$. The series will converge to $f(x)$ for $-5 \le x \le 5$ if we redefine $f(x)$ as follows:

$$f(x) = \begin{cases} 3/2 & x = -5 \\ 0 & -5 < x < 0 \\ 3/2 & x = 0 \\ 3 & 0 < x < 5 \\ 3/2 & x = 5 \end{cases} \qquad \text{period} = 10$$

16.35 Expand $f(x) = x^2$, $0 < x < 2\pi$, in a Fourier series having period 2π.

▌ Here, the period is $2L = 2\pi$, so $L = \pi$. Choosing $c = 0$, we have

$$a_n = \frac{1}{L}\int_c^{c+2L} f(x)\cos\frac{n\pi x}{L}\,dx = \frac{1}{\pi}\int_0^{2\pi} x^2\cos nx\,dx$$

$$= \frac{1}{\pi}\left(x^2\frac{\sin nx}{n} - 2x\frac{-\cos nx}{n^2} + 2\frac{-\sin nx}{n^3}\right)\Bigg|_0^{2\pi} = \frac{4}{n^2} \qquad \text{for } n \neq 0$$

If $n = 0$, $a_0 = \dfrac{1}{\pi}\displaystyle\int_0^{2\pi} x^2\,dx = \frac{8\pi^2}{3}$. Also,

$$b_n = \frac{1}{L}\int_c^{c+2L} f(x)\sin\frac{n\pi x}{L}\,dx = \frac{1}{\pi}\int_0^{2\pi} x^2\sin nx\,dx$$

$$= \frac{1}{\pi}\left[(x^2)\left(-\frac{\cos nx}{n}\right) - (2x)\left(-\frac{\sin nx}{n^2}\right) + (2)\left(\frac{\cos nx}{n^3}\right)\right]_0^{2\pi} = \frac{-4\pi}{n}$$

Then $f(x) = x^2 = \dfrac{4\pi^2}{3} + \displaystyle\sum_{n=1}^{\infty}\left(\frac{4}{n^2}\cos nx - \frac{4\pi}{n}\sin nx\right)$ for $0 < x < 2\pi$.

16.36 How should the function $f(x)$ in Problem 16.35 be defined at $x = 0$ and $x = 2\pi$ so that its Fourier series will converge everywhere on $[0, 2\pi]$?

▌ At all points of continuity, the graph of the Fourier series obtained in Problem 16.35 is given in Fig. 16.4. If we redefine $f(x)$ at both 0 and 2π as $\frac{1}{2}(0 + 4\pi^2) = 2\pi^2$, then the Fourier series will converge to $f(x)$ everywhere on $[0, 2\pi]$.

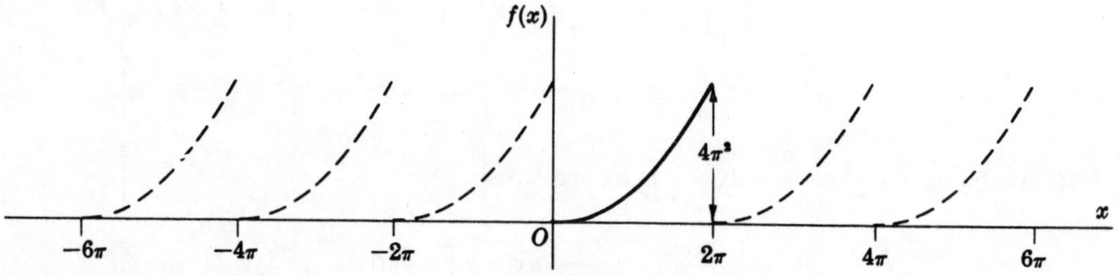

Fig. 16.4

16.37 Expand $f(x) = x^2$, $-\pi < x < \pi$, in a Fourier series having period 2π.

▌ As in Problem 16.35, $L = \pi$, but now $c = -\pi$, so

$$a_n = \frac{1}{\pi}\int_{-\pi}^{\pi} x^2\cos nx\,dx$$

$$= \frac{1}{\pi}\left(x^2\frac{\sin nx}{n} - 2x\frac{-\cos nx}{n^2} + 2\frac{-\sin nx}{n^3}\right)\Bigg|_{-\pi}^{\pi} = \frac{4}{n^2}\cos n\pi \qquad \text{for } n \neq 0$$

If $n = 0$, then $a_0 = \dfrac{1}{\pi}\displaystyle\int_{-\pi}^{\pi} x^2\,dx = \frac{2\pi^2}{3}$, Also,

$$b_n = \frac{1}{\pi}\int_{-\pi}^{\pi} x^2\sin nx\,dx = \frac{1}{\pi}\left\{x^2\frac{-\cos nx}{n} - 2x\frac{-\sin nx}{n^2} + 2\frac{\cos nx}{n^3}\right)\Bigg|_{-\pi}^{\pi} = 0$$

Thus, $x^2 = \dfrac{\pi^2}{3} + \displaystyle\sum_{n=1}^{\infty}\left(\frac{4}{n^2}\cos n\pi\cos nx\right)$ for $-\pi < x < \pi$.

16.38 Graph the Fourier series obtained in Problem 16.37.

▌ This series will converge to x^2 at every point in the interval $(-\pi, \pi)$ for which x is continuous, which is throughout the interval. The Fourier series is also periodic with period 2π. At the end points $\pm n\pi$, for $n = 1, 2, 3, \ldots,$ the series must converge to the average value of its left- and right-hand limits, which in each case is $(\pi^2 + \pi^2)/2 = \pi^2$. The graph is shown in Fig. 16.5. Because the Fourier series is continuous at both end points, we can extend the result of Problem 16.37 to include these end points; hence

$$x^2 = \frac{\pi^2}{3} + \sum_{n=1}^{\infty} \left(\frac{4}{n^2} \cos n\pi \cos nx \right) \quad \text{for} \quad -\pi \le x \le \pi.$$

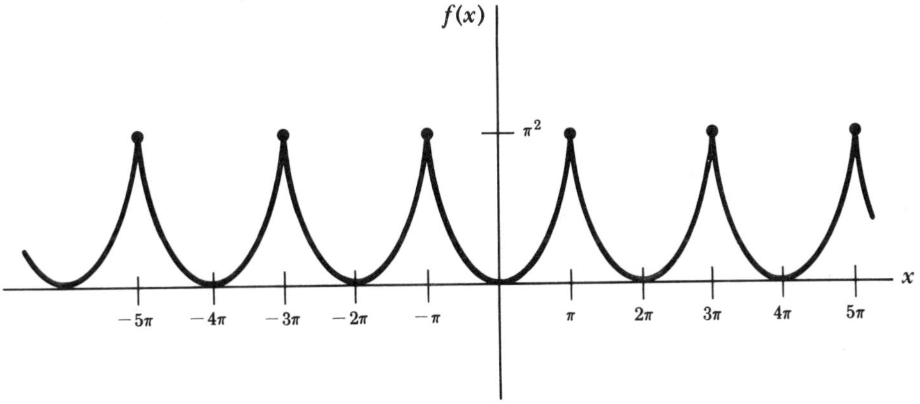

Fig. 16.5

6.39 Find the Fourier expansion of the periodic function whose definition in one period is

$$f(t) = \begin{cases} 0 & -\pi \le t \le 0 \\ \sin t & 0 < t \le \pi \end{cases}.$$

▮ The half-period of the given function is $L = \pi$. Hence, taking $c = -\pi$, we have

$$a_n = \frac{1}{\pi} \int_{-\pi}^{\pi} f(t) \cos nt\, dt = \frac{1}{\pi} \int_{-\pi}^{0} 0 \cos nt\, dt + \frac{1}{\pi} \int_0^{\pi} \sin t \cos nt\, dt = \frac{1}{\pi} \left\{ -\frac{1}{2} \left[\frac{\cos(1-n)t}{1-n} + \frac{\cos(1+n)t}{1+n} \right] \right\}_0^{\pi}$$

$$= -\frac{1}{2\pi} \left[\frac{\cos(\pi - n\pi)}{1-n} + \frac{\cos(\pi + n\pi)}{1+n} - \left(\frac{1}{1-n} + \frac{1}{1+n} \right) \right] = -\frac{1}{2\pi} \left(\frac{-\cos n\pi}{1-n} + \frac{-\cos n\pi}{1+n} - \frac{2}{1-n^2} \right)$$

$$= \frac{1 + \cos n\pi}{\pi(1-n^2)} \quad \text{for} \quad n \ne 1$$

$$a_1 = \frac{1}{\pi} \int_0^{\pi} \sin t \cos t\, dt = \frac{\sin^2 t}{2\pi} \bigg|_0^{\pi} = 0$$

$$b_n = \frac{1}{\pi} \int_{-\pi}^{\pi} f(t) \sin nt\, dt = \frac{1}{\pi} \int_{-\pi}^{0} 0 \sin nt\, dt + \frac{1}{\pi} \int_0^{\pi} \sin t \sin nt\, dt$$

$$= \frac{1}{\pi} \left\{ \frac{1}{2} \left[\frac{\sin(1-n)t}{1-n} - \frac{\sin(1+n)t}{1+n} \right] \right\}_0^{\pi} = 0 \quad \text{for} \quad n \ne 1$$

$$b_1 = \frac{1}{\pi} \int_0^{\pi} \sin^2 t\, dt = \frac{1}{\pi} \left(\frac{t}{2} - \frac{\sin 2t}{4} \right)_0^{\pi} = \frac{1}{2}$$

Since $f(t)$ is continuous everywhere as a periodic function of period 2π and satisfies the Dirichlet conditions on $(0, 2\pi)$ (see Problems 16.24 and 16.25), it follows that $f(t)$ equals its Fourier series everywhere; hence

$$f(t) = \frac{1}{\pi} + \frac{\sin t}{2} - \frac{2}{\pi} \left(\frac{\cos 2t}{3} + \frac{\cos 4t}{15} + \frac{\cos 6t}{35} + \frac{\cos 8t}{63} + \cdots \right).$$

.40 Sketch the graphs of the first few partial sums for the Fourier series computed in Problem 16.39.

▮ Plots showing the accuracy with which the first n terms of this series represent the given function are shown in Fig. 16.6 for $n = 1, 2, 3$. For $n = 4, 5, 6, \ldots$ the graphs of the partial sums are almost indistinguishable from the graph of $f(t)$.

.41 Find the Fourier expansion of the periodic function whose definition in one period is

$$f(t) = \begin{cases} -t & -3 \le t \le 0 \\ t & 0 < t \le 3 \end{cases}$$

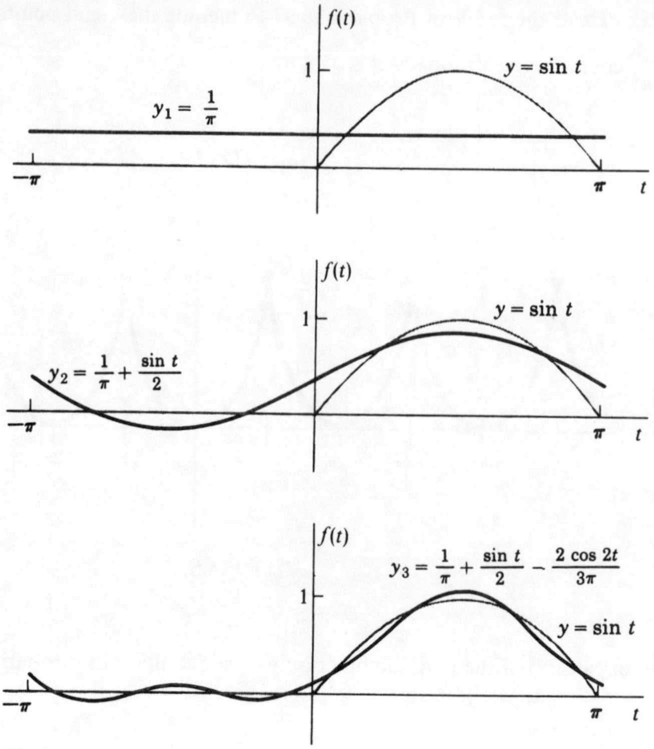

Fig. 16.6

▌ The period of this function is 6. Hence $L = 3$ and $c = -3$, so

$$a_n = \frac{1}{3}\int_{-3}^{0} -t\cos\frac{n\pi t}{3}\,dt + \frac{1}{3}\int_0^3 t\cos\frac{n\pi t}{3}\,dt$$

$$= -\frac{1}{3}\left(\frac{9}{n^2\pi^2}\cos\frac{n\pi t}{3} + \frac{3t}{n\pi}\sin\frac{n\pi t}{3}\right)\Bigg|_{-3}^{0} + \frac{1}{3}\left(\frac{9}{n^2\pi^2}\cos\frac{n\pi t}{3} + \frac{3t}{n\pi}\sin\frac{n\pi t}{3}\right)\Bigg|_0^3$$

$$= -\frac{3}{n^2\pi^2}(1-\cos n\pi) + \frac{3}{n^2\pi^2}(\cos n\pi - 1) = \frac{6}{n^2\pi^2}(\cos n\pi - 1) \qquad \text{for } n \neq 0$$

$$a_0 = \frac{1}{3}\int_{-3}^{0} -t\,dt + \frac{1}{3}\int_0^3 t\,dt = -\frac{t^2}{6}\Bigg|_{-3}^{0} + \frac{t^2}{6}\Bigg|_0^3 = \frac{3}{2} + \frac{3}{2} = 3$$

$$b_n = \frac{1}{3}\int_{-3}^{0} -t\sin\frac{n\pi t}{3}\,dt + \frac{1}{3}\int_0^3 t\sin\frac{n\pi t}{3}\,dt$$

$$= -\frac{1}{3}\left(\frac{9}{n^2\pi^2}\sin\frac{n\pi t}{3} - \frac{3t}{n\pi}\cos\frac{n\pi t}{3}\right)\Bigg|_{-3}^{0} + \frac{1}{3}\left(\frac{9}{n^2\pi^2}\sin\frac{n\pi t}{3} - \frac{3t}{n\pi}\cos\frac{n\pi t}{3}\right)\Bigg|_0^3$$

$$= \frac{3}{n\pi}\cos(-n\pi) - \frac{3}{n\pi}\cos n\pi = 0$$

Thus, $f(t) = \frac{3}{2} - \frac{12}{\pi^2}\left(\frac{1}{1}\cos\frac{\pi t}{3} + \frac{1}{9}\cos\frac{3\pi t}{3} + \frac{1}{25}\cos\frac{5\pi t}{3} + \cdots\right).$

16.42 Find the Fourier series for $f(x) = \begin{cases} -x & -4 \le x \le 0 \\ x & 0 \le x \le 4 \end{cases}$, with period 8.

▌
$$a_n = \frac{1}{4}\int_{-4}^{0} -x\cos\frac{n\pi x}{4}\,dx + \frac{1}{4}\int_0^4 x\cos\frac{n\pi x}{4}\,dx$$

$$= -\frac{1}{4}\int_4^0 y\cos\frac{n\pi y}{4}\,dy + \frac{1}{4}\int_0^4 x\cos\frac{n\pi x}{4}\,dx \qquad \text{where } y = -x$$

$$= \frac{2}{4}\int_0^4 x\cos\frac{n\pi x}{4}\,dx = \left(\frac{8}{n^2\pi^2}\cos\frac{n\pi x}{4} + \frac{2x}{n\pi}\sin\frac{n\pi x}{4}\right)\Bigg|_0^4 = \frac{-8(1-\cos n\pi)}{n^2\pi^2} \qquad \text{for } n \neq 0$$

$$a_0 = \tfrac{1}{4}\int_{-4}^{0} -x\,dx + \tfrac{1}{4}\int_0^4 x\,dx = 4$$

$$b_n = \frac{1}{4}\int_{-4}^{0} -x\sin\frac{n\pi x}{4} + \frac{1}{4}\int_0^4 x\sin\frac{n\pi x}{4}\,dx$$

$$= \frac{1}{4}\int_4^0 y\sin\frac{n\pi y}{4}\,dy + \frac{1}{4}\int_0^4 x\sin\frac{n\pi x}{4}\,dx \qquad \text{where} \quad y=-x$$

$$= 0$$

Thus, the Fourier series is $2 - \dfrac{8}{\pi^2}\displaystyle\sum_{n=1}^{\infty}\frac{1-\cos n\pi}{n^2}\cos\frac{n\pi x}{4}$.

16.43 Find the Fourier series for $f(t)=t$ on the interval (0, 6) with period 6.

▌ The period is $2L=6$, so $L=3$. Also, $c=0$, so

$$a_n = \frac{1}{3}\int_0^6 t\cos\frac{n\pi t}{3}\,dt = \frac{1}{3}\left(\frac{9}{n^2\pi^2}\cos\frac{n\pi t}{3} + \frac{3t}{n\pi}\sin\frac{n\pi t}{3}\right)_0^6 = 0 \qquad \text{for} \quad n \neq 0$$

$$a_0 = \frac{1}{3}\int_0^6 t\,dt = 6$$

$$b_n = \frac{1}{3}\int_0^6 t\sin\frac{n\pi t}{3}\,dt = \frac{1}{3}\left(\frac{9}{n^2\pi^2}\sin\frac{n\pi t}{3} - \frac{3t}{n\pi}\cos\frac{n\pi t}{3}\right)_0^6 = \frac{-6}{n\pi}$$

Thus, $t = \dfrac{6}{2} + \displaystyle\sum_{n=1}^{\infty}\frac{-6}{n\pi}\sin\frac{n\pi t}{3}$, for $0<t<6$.

16.44 Graph the Fourier series obtained in Problem 16.43.

▌ This series converges to t at every point in the interval (0, 6), because the function is continuous there. The Fourier series is also periodic with period 6. At the points $\pm 6n$, $n = 0, 1, 2, \ldots,$ the series must converge to the average of its left- and right-hand limits, which in each case is $(6+0)/2=3$. The graph of the Fourier series is shown in Fig. 16.7.

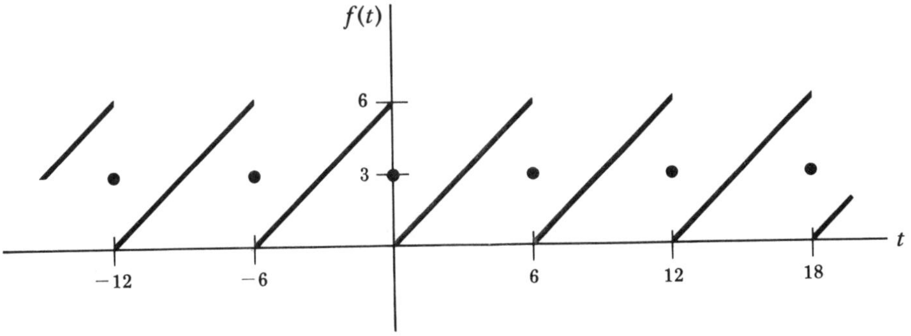

Fig. 16.7

16.45 Find the Fourier expansion of $f(t)=t$ on the interval $(-3, 3)$ with period 6.

▌ Here $L = 6/2 = 3$ and $c=-3$, so

$$a_n = \frac{1}{3}\int_{-3}^{3} t\cos\frac{n\pi t}{3}\,dt = \frac{1}{3}\left(\frac{9}{n^2\pi^2}\cos\frac{n\pi t}{3} + \frac{3t}{n\pi}\sin\frac{n\pi t}{3}\right)_{-3}^{3} = 0 \qquad \text{for} \quad n \neq 0$$

$$a_0 = \frac{1}{3}\int_{-3}^{3} t\,dt = 0$$

$$b_n = \frac{1}{3}\int_{-3}^{3} t\sin\frac{n\pi t}{3}\,dt = \frac{1}{3}\left(\frac{9}{n^2\pi^2}\sin\frac{n\pi t}{3} - \frac{3t}{n\pi}\cos\frac{n\pi t}{3}\right)_{-3}^{3} = \frac{-6}{n\pi}\cos n\pi = \frac{-6}{n\pi}(-1)^n$$

Therefore, $t = \displaystyle\sum_{n=1}^{\infty}\frac{-6}{n\pi}(-1)^n\sin\frac{n\pi t}{3}$ for $-3<t<3$.

16.46 Graph the Fourier series obtained in Problem 16.45.

▌ The series converges to t at every point in the interval $(-3, 3)$ because the function is continuous there. At the points $3 \pm 6n$, $n = 0, 1, 2, \ldots$, $f(t) = [3 + (-3)]/2 = 0$. The graph is shown in Fig. 16.8.

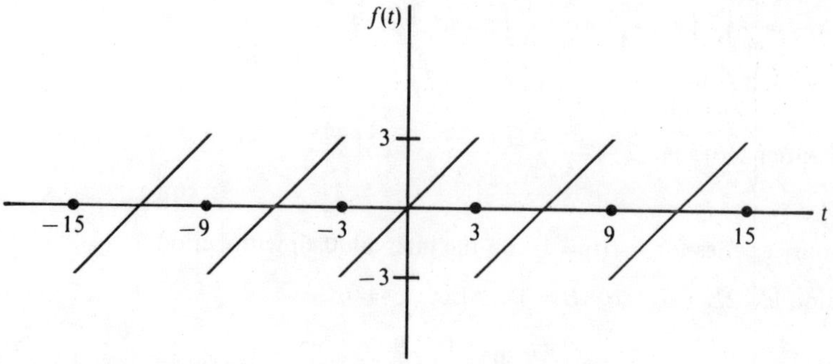

Fig. 16.8

16.47 Find the Fourier series for $f(t) = t$ on the interval $(-1, 5)$ with period 6.

▌ Here $L = 3$ and $c = -1$, so

$$a_n = \frac{1}{3}\int_{-1}^{5} t\cos\frac{n\pi t}{3}\,dt = \frac{1}{3}\left(\frac{9}{n^2\pi^2}\cos\frac{n\pi t}{3} + \frac{3t}{n\pi}\sin\frac{n\pi t}{3}\right)\Bigg|_{-1}^{5}$$

$$= \frac{1}{3}\left[\frac{9}{n^2\pi^2}\left(\cos\frac{5n\pi}{3} - \cos\frac{n\pi}{3}\right) + \frac{3}{n\pi}\left(5\sin\frac{5n\pi}{3} - \sin\frac{n\pi}{3}\right)\right] \qquad \text{for } n \neq 0$$

$$a_0 = \frac{1}{3}\int_{-1}^{5} t\,dt = 4$$

$$b_n = \frac{1}{3}\int_{-1}^{5} t\sin\frac{n\pi t}{3}\,dt = \frac{1}{3}\left(\frac{9}{n^2\pi^2}\sin\frac{n\pi t}{3} - \frac{3t}{n\pi}\cos\frac{n\pi t}{3}\right)\Bigg|_{-1}^{5}$$

$$= \frac{1}{3}\left[\frac{9}{n^2\pi^2}\left(\sin\frac{5n\pi}{3} + \sin\frac{n\pi}{3}\right) - \frac{3}{n\pi}\left(5\cos\frac{5n\pi}{3} + \cos\frac{n\pi}{3}\right)\right]$$

But
$$\cos\frac{5n\pi}{3} = \cos\left(2n\pi - \frac{n\pi}{3}\right) = \cos 2n\pi\cos\frac{n\pi}{3} + \sin 2n\pi\sin\frac{n\pi}{3} = \cos\frac{n\pi}{3}$$

and
$$\sin\frac{5n\pi}{3} = \sin\left(2n\pi - \frac{n\pi}{3}\right) = \sin 2n\pi\cos\frac{n\pi}{3} - \sin\frac{n\pi}{3}\cos 2n\pi = -\sin\frac{n\pi}{3}$$

so that $a_n = \dfrac{-6}{n\pi}\sin\dfrac{n\pi}{3}$ and $b_n = \dfrac{-6}{n\pi}\cos\dfrac{n\pi}{3}$ for $n \neq 0$. Thus,

$$t = 2 - \frac{6}{\pi}\sum_{n=1}^{\infty}\frac{1}{n}\left(\sin\frac{n\pi}{3}\cos\frac{n\pi t}{3} + \cos\frac{n\pi}{3}\sin\frac{n\pi t}{3}\right) \qquad -1 < t < 5$$

16.48 Graph the Fourier series obtained in Problem 16.47.

▌ The graph is shown in Fig. 16.9.

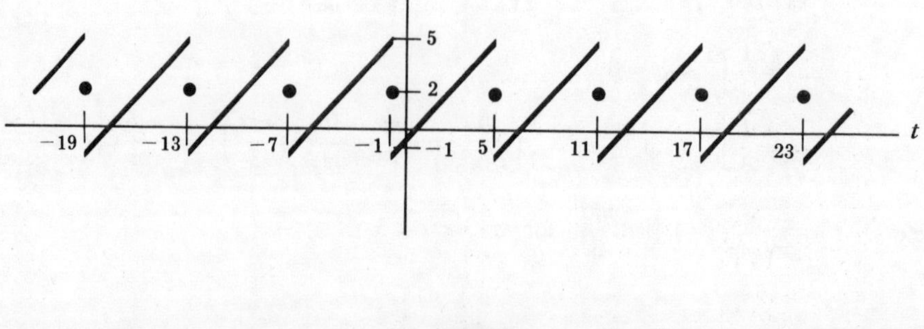

Fig. 16.9

16.49 Find the Fourier series for $f(y) = e^{-y}$ on the interval (0, 2) with period 2.

Here $c = 0$ and $2L = 2$, so $L = 1$. Then

$$a_n = \frac{1}{1}\int_0^2 e^{-y}\cos n\pi y\,dy = \frac{e^{-y}}{1+n^2\pi^2}(-\cos n\pi y + n\pi \sin n\pi y)\Big|_0^2 = \frac{1-e^{-2}}{1+n^2\pi^2}$$

$$b_n = \frac{1}{1}\int_0^2 e^{-y}\sin n\pi y\,dy = \frac{e^{-y}}{1+n^2\pi^2}(-\sin n\pi y - n\pi \cos n\pi y)\Big|_0^2 = \frac{n\pi(1-e^{-2})}{1+n^2\pi^2}$$

Therefore, on (0, 2), $e^{-y} = (1-e^{-2})\left(\frac{1}{2} + \sum_{n=1}^{\infty}\frac{\cos n\pi y}{1+n^2\pi^2} + \sum_{n=1}^{\infty}\frac{n\pi \sin n\pi y}{1+n^2\pi^2}\right)$.

16.50 Determine the values to which the Fourier series obtained in Problem 16.49 converges.

Since $f(y) = e^{-y}$ is continuous on (0, 2) and satisfies the Dirichlet conditions there, the Fourier series converges to e^{-y} everywhere on (0, 2). The Fourier series is also periodic with period 2. At each of the discontinuities $y = 2n$, for $n = 0, \pm 1, \pm 2, \ldots,$ the Fourier series converges to the average of its left- and right-hand limits, which is $(1 + e^{-2})/2 = 0.568$.

16.51 Find the Fourier series for $f(x) = \begin{cases} 2x & 0 \le x \le 3 \\ 0 & -3 < x < 0 \end{cases}$ with period 6.

$$a_n = \frac{1}{3}\int_{-3}^{0}(0)\cos\frac{n\pi x}{3}\,dx + \frac{1}{3}\int_0^3 2x\cos\frac{n\pi x}{3}\,dx$$

$$= \left(\frac{6}{n^2\pi^2}\cos\frac{n\pi x}{3} + \frac{2x}{n\pi}\sin\frac{n\pi x}{3}\right)\Big|_0^3 = \frac{6(\cos n\pi - 1)}{n^2\pi^2} \quad \text{for} \quad n \ne 0$$

$$a_0 = \frac{1}{3}\int_{-3}^{0} 0\,dx + \frac{1}{3}\int_0^3 2x\,dx = 3$$

$$b_n = \frac{1}{3}\int_{-3}^{0}(0)\sin\frac{n\pi x}{3} + \frac{1}{3}\int_0^3 2x\sin\frac{n\pi x}{3}\,dx = \left(\frac{6}{n^2\pi^2}\sin\frac{n\pi x}{3} - \frac{2x}{n\pi}\cos\frac{n\pi x}{3}\right)\Big|_0^3 = \frac{-6\cos n\pi}{n\pi}$$

The Fourier series is thus $\frac{3}{2} + \sum_{n=1}^{\infty}\left[\frac{6(\cos n\pi - 1)}{n^2\pi^2}\cos\frac{n\pi x}{3} - \frac{6\cos n\pi}{n\pi}\sin\frac{n\pi x}{3}\right]$.

6.52 Find the Fourier series of period 2 that converges to $f(t) = \sin 3t$ on (0, 2).

$$a_n = \frac{1}{1}\int_0^2 \sin 3t\cos n\pi t\,dt = -\frac{1}{2}\left[\frac{\cos(3-n\pi)t}{3-n\pi} + \frac{\cos(3+n\pi)t}{3+n\pi}\right]_0^2$$

$$= -\frac{1}{2}\left[\frac{\cos(6-2n\pi)}{3-n\pi} + \frac{\cos(6+2n\pi)}{3+n\pi}\right] + \frac{1}{2}\left(\frac{1}{3-n\pi} + \frac{1}{3+n\pi}\right)$$

But $\cos(6-2n\pi) = \cos 6\cos 2n\pi + \sin 6\sin 2n\pi = \cos 6$

and $\cos(6+2n\pi) = \cos 6\cos 2n\pi - \sin 6\sin 2n\pi = \cos 6$

so $a_n = \frac{3(1-\cos 6)}{9-n^2\pi^2}$. Also,

$$b_n = \frac{1}{1}\int_0^2 \sin 3t\sin n\pi t\,dt = -\frac{1}{2}\left[\frac{\sin(3+n\pi)t}{3+n\pi} - \frac{\sin(3-n\pi)t}{3-n\pi}\right]_0^2 = -\frac{1}{2}\left[\frac{\sin(6+2n\pi)}{3+n\pi} - \frac{\sin(6-2n\pi)}{3-n\pi}\right]$$

But here $\sin(6+2n\pi) = \sin(6-2n\pi) = \sin 6$, so $b_n = \frac{n\pi\sin 6}{9-n^2\pi^2}$. Since the expression for a_n may be used for the special case $n = 0$, we have $a_0 = \frac{1-\cos 6}{3}$. Then the Fourier series is

$$\frac{1-\cos 6}{6} + \sum_{n=1}^{\infty}\frac{3(1-\cos 6)}{9-n^2\pi^2}\cos n\pi t + \sum_{n=1}^{\infty}\frac{n\pi\sin 6}{9-n^2\pi^2}\sin n\pi t$$

.53 Find a Fourier series of period 2π that converges to $f(t) = \sin 3t$ on $(0, 2\pi)$.

▌ The function $\sin 3t$ is its own Fourier series with period 2π. That is,

$$\sin 3t = \frac{a_0}{2} + \sum_{n=1}^{\infty} a_n \cos\frac{n\pi t}{\pi} + \sum_{n=1}^{\infty} b_n \sin\frac{n\pi t}{\pi}$$

where $a_n = 0$ for $n = 0, 1, 2, \ldots$; $b_n = 0$ for $n \neq 3$; and $b_3 = 1$.

16.54 Find a Fourier series of period 4 that converges to $f(t) = 5$ on $(-2, 2)$.

▌ The function $f(t) = 5$ is its own Fourier series over any interval with any period. In particular, $5 = \frac{a_0}{2} + \sum_{n=1}^{\infty} a_n \cos\frac{n\pi t}{L} + \sum_{n=1}^{\infty} b_n \sin\frac{n\pi t}{L}$, where $a_0 = 10$ and $a_n = b_n = 0$ for $n = 1, 2, 3, \ldots$.

16.55 Find the Fourier series of period 2 that converges to $f(t) = t - t^2$ on $(-1, 1)$.

▌
$$a_n = \frac{1}{1}\int_{-1}^{1} (t - t^2)\cos\frac{n\pi t}{1}\,dt = \left[\frac{\cos n\pi t}{n^2\pi^2} + \frac{t}{n\pi}\sin n\pi t - \left(\frac{2t}{n^2\pi^2}\cos n\pi t - \frac{2}{n^3\pi^3}\sin n\pi t + \frac{t^2}{n\pi}\sin n\pi t\right)\right]_{-1}^{1}$$
$$= -\frac{4\cos n\pi}{n^2\pi^2} \quad \text{for } n \neq 0$$

$$a_0 = \frac{1}{1}\int_{-1}^{1} (t - t^2)\,dt = \left(\frac{t^2}{2} - \frac{t^3}{3}\right)_{-1}^{1} = -\frac{2}{3}$$

$$b_n = \frac{1}{1}\int_{-1}^{1} (t - t^2)\sin\frac{n\pi t}{1}\,dt$$
$$= \left[\frac{\sin n\pi t}{n^2\pi^2} - \frac{t}{n\pi}\cos n\pi t - \left(\frac{2t}{n^2\pi^2}\sin n\pi t + \frac{2}{n^3\pi^3}\cos n\pi t - \frac{t^2}{n\pi}\cos n\pi t\right)\right]_{-1}^{1} = -\frac{2\cos n\pi}{n\pi}$$

Thus, the Fourier series is

$$-\frac{1}{3} + \frac{4}{\pi^2}\left(\frac{\cos\pi t}{1} - \frac{\cos 2\pi t}{4} + \frac{\cos 3\pi t}{9} - \frac{\cos 4\pi t}{16} + \cdots\right) + \frac{2}{\pi}\left(\frac{\sin\pi t}{1} - \frac{\sin 2\pi t}{2} + \frac{\sin 3\pi t}{3} - \frac{\sin 4\pi t}{4} + \cdots\right)$$

16.56 Show that $\frac{1}{(1)(3)} - \frac{1}{(3)(5)} + \frac{1}{(5)(7)} - \frac{1}{(7)(9)} + \cdots = \frac{\pi - 2}{4}$.

▌ Evaluating the result of Problem 16.39 at $t = \pi/2$, we find

$$f\left(\frac{\pi}{2}\right) = 1 = \frac{1}{\pi} + \frac{1}{2} - \frac{2}{\pi}\left(-\frac{1}{3} + \frac{1}{15} - \frac{1}{35} + \frac{1}{63} - \cdots\right) \quad \text{or} \quad \frac{1}{(1)(3)} - \frac{1}{(3)(5)} + \frac{1}{(5)(7)} - \frac{1}{(7)(9)} + \cdots = \frac{\pi - 2}{4}$$

16.57 Show that $\sum_{n=1}^{\infty} \frac{1}{n^2} = \frac{\pi^2}{6}$.

▌ Evaluating the result of Problem 16.37 at $x = \pi$, we obtain $\pi^2 = \frac{\pi^2}{3} + \sum_{n=1}^{\infty} \frac{4}{n^2}\cos^2 n\pi = \frac{\pi^2}{3} + 4\sum_{n=1}^{\infty} \frac{1}{n^2}$, from which the desired result follows.

16.58 Show that $\sum_{n=1}^{\infty} \frac{(-1)^{n+1}}{n^2} = \frac{\pi^2}{12}$.

▌ Evaluating the result of Problem 16.37 at $x = 0$, we find $0 = \frac{\pi^2}{3} + \sum_{n=1}^{\infty} \frac{4}{n^2}\cos n\pi = \frac{\pi^2}{3} - 4\left(\frac{1}{1^2} - \frac{1}{2^2} + \frac{1}{3^2} - \frac{1}{4^2} + \cdots\right)$, from which the desired result follows immediately

PARSEVAL'S IDENTITY

16.59 Assuming that the Fourier series corresponding to $f(x)$ converges uniformly to $f(x)$ in $(-L, L)$, prove Parseval's identity, $\frac{1}{L}\int_{-L}^{L} [f(x)]^2\,dx = \frac{a_0^2}{2} + \sum_{n=1}^{\infty} (a_n^2 + b_n^2)$, where the integral is assumed to exist.

▌ If $f(x) = \frac{a_0}{2} + \sum_{n=1}^{\infty}\left(a_n \cos\frac{n\pi x}{L} + b_n \sin\frac{n\pi x}{L}\right)$, then multiplying by $f(x)$ and integrating term by term from −

to L (which is justified since the series is uniformly convergent) yield

$$\int_{-L}^{L} [f(x)]^2\,dx = \frac{a_0}{2}\int_{-L}^{L} f(x)\,dx + \sum_{n=1}^{\infty}\left[a_n\int_{-L}^{L} f(x)\cos\frac{n\pi x}{L}\,dx + b_n\int_{-L}^{L} f(x)\sin\frac{n\pi x}{L}\,dx\right]$$
$$= \frac{a_0^2}{2}L + L\sum_{n=1}^{\infty}(a_n^2+b_n^2) \qquad (1)$$

where we have used the results

$$\int_{-L}^{L} f(x)\cos\frac{n\pi x}{L}\,dx = La_n \qquad \int_{-L}^{L} f(x)\sin\frac{n\pi x}{L}\,dx = Lb_n \qquad \int_{-L}^{L} f(x)\,dx = La_0$$

obtained from the Fourier coefficients. The required result follows on dividing both sides of (*1*) by L.

16.60 Write Parseval's identity corresponding to the Fourier series of Problem 16.42.

▮ Here $L=4$, $a_0=4$, $a_n = \dfrac{-8(1-\cos n\pi)}{n^2\pi^2}$ for $n\neq 0$, and $b_n=0$. Then $\frac{1}{4}\int_{-4}^{4}[f(x)]^2\,dx = \frac{1}{4}\int_{-4}^{4}|x|^2\,dx = \frac{1}{4}\int_{-4}^{4}x^2\,dx = \frac{32}{3}$, and

$$\frac{a_0^2}{2} + \sum_{n=1}^{\infty}(a_n^2+b_n^2) = \frac{16}{2} + \sum_{n=1}^{\infty}\left[\frac{64(1-\cos n\pi)^2}{n^4\pi^4} + 0^2\right] = 8 + \frac{64}{\pi^4}\left(\frac{4}{1^4}+\frac{4}{3^4}+\frac{4}{5^4}+\frac{4}{7^4}+\cdots\right)$$

Parseval's identity thus becomes $\frac{32}{3} = 8 + \frac{256}{\pi^4}\left(\frac{1}{1^4}+\frac{1}{3^4}+\frac{1}{5^4}+\frac{1}{7^4}+\cdots\right)$, which may be rewritten as $\frac{1}{1^4}+\frac{1}{3^4}+\frac{1}{5^4}+\frac{1}{7^4}+\cdots = \frac{\pi^4}{96}$.

16.61 Determine the sum $\frac{1}{1^4}+\frac{1}{2^4}+\frac{1}{3^4}+\frac{1}{4^4}+\cdots$.

▮ Using the result of Problem 16.60, we have

$$S = \frac{1}{1^4}+\frac{1}{2^4}+\frac{1}{3^4}+\cdots = \left(\frac{1}{1^4}+\frac{1}{3^4}+\frac{1}{5^4}+\cdots\right) + \left(\frac{1}{2^4}+\frac{1}{4^4}+\frac{1}{6^4}+\cdots\right)$$
$$= \left(\frac{1}{1^4}+\frac{1}{3^4}+\frac{1}{5^4}+\cdots\right) + \frac{1}{2^4}\left(\frac{1}{1^4}+\frac{1}{2^4}+\frac{1}{3^4}+\cdots\right) = \frac{\pi^4}{96}+\frac{S}{16}$$

from which $S=\pi^4/90$.

16.62 Write Parseval's identity corresponding to the Fourier series of Problem 16.45.

▮ With $L=3$, $a_0=a_n=0$, and $b_n = \dfrac{-6}{n\pi}(-1)^n$, we have $\frac{1}{3}\int_{-3}^{3}t^2\,dt = 6$ and $\frac{a_0^2}{2}+\sum_{n=1}^{\infty}(a_n^2+b_n^2) = \frac{0^2}{2}+\sum_{n=1}^{\infty}\left(0^2+\frac{36}{n^2\pi^2}\right)$. Parseval's identity then becomes $6 = \frac{36}{\pi^2}\sum_{n=1}^{\infty}\frac{1}{n^2}$, which can be simplified to $\sum_{n=1}^{\infty}\frac{1}{n^2} = \frac{\pi^2}{6}$. (Compare this with Problem 15.57.)

16.63 Write Parseval's identity corresponding to the Fourier series of Problem 16.31.

▮ Here $L=5$, $a_0=a_n=0$, and $b_n = \dfrac{6}{n\pi}(1-\cos n\pi)$. Thus we have $\frac{1}{5}\int_{-5}^{5}[f(x)]^2\,dx = \frac{1}{5}\int_{-5}^{5}|3|^2\,dx = \frac{9}{5}\int_{-5}^{5}dx\ 18$, and Parseval's identity becomes

$$18 = \frac{0^2}{2}+\sum_{n=1}^{\infty}\left[0^2+\frac{36}{n^2\pi^2}(1-\cos n\pi)^2\right] \qquad \text{or} \qquad 18 = \frac{36}{\pi^2}\left(\frac{4}{1^2}+\frac{4}{3^2}+\frac{4}{5^2}+\frac{4}{7^2}+\cdots\right)$$

which may be rewritten as $\frac{1}{1^2}+\frac{1}{3^2}+\frac{1}{5^2}+\frac{1}{7^2}+\cdots = \frac{\pi^2}{8}$.

16.64 Write Parseval's identity corresponding to the Fourier series of Problem 16.33.

▌ Here $L = 5$, $a_0 = 3$, $a_n = 0$ for $n = 1, 2, 3, \ldots$, and $b_n = \dfrac{3(1 - \cos n\pi)}{n\pi}$. Thus we have $\frac{1}{5}\int_{-5}^{5} [f(x)]^2\,dx = \frac{1}{5}\int_{-5}^{0} (0)^2\,dx + \frac{1}{5}\int_0^5 (3)^2\,dx = 9$, and Parseval's identity becomes

$$9 = \frac{(3)^2}{2} + \sum_{n=1}^{\infty}\left[0^2 + \frac{9}{n^2\pi^2}(1 - \cos n\pi)^2\right] \quad \text{or} \quad 9 = \frac{9}{2} + \frac{9}{\pi^2}\left(\frac{4}{1^2} + \frac{4}{3^2} + \frac{4}{5^2} + \frac{4}{7^2} + \cdots\right)$$

which may be rewritten as $\displaystyle\sum_{n=1}^{\infty} \frac{1}{(2n-1)^2} = \frac{\pi^2}{8}$. Compare this result with that of the previous problem.

16.65 Determine the sum $\dfrac{1}{2^2} + \dfrac{1}{4^2} + \dfrac{1}{6^2} + \cdots$.

▌ Using the result of Problem 16.62, we have

$$\frac{1}{2^2} + \frac{1}{4^2} + \frac{1}{6^2} + \cdots = \sum_{n=1}^{\infty} \frac{1}{(2n)^2} = \sum_{n=1}^{\infty} \frac{1}{4n^2} = \frac{1}{4}\sum_{n=1}^{\infty} \frac{1}{n^2} = \frac{1}{4}\frac{\pi^2}{6} = \frac{\pi^2}{24}$$

Alternatively, using the results of Problems 16.62 and 16.64 and denoting the sum of interest by S, we find

$$\frac{\pi^2}{6} = \frac{1}{1^2} + \frac{1}{2^2} + \frac{1}{3^2} + \frac{1}{4^2} + \cdots = \left(\frac{1}{1^2} + \frac{1}{3^2} + \frac{1}{5^2} + \cdots\right) + \left(\frac{1}{2^2} + \frac{1}{4^2} + \frac{1}{6^2} + \cdots\right) = \frac{\pi^2}{8} + S$$

so that $S = \dfrac{\pi^2}{6} - \dfrac{\pi^2}{8} = \dfrac{\pi^2}{24}$.

16.66 Write Parseval's identity corresponding to the Fourier series of Problem 16.39.

▌ In this case, $\dfrac{1}{L}\int_{-L}^{L} [f(t)]^2\,dt = \dfrac{1}{\pi}\int_{-\pi}^{0} 0^2\,dt + \dfrac{1}{\pi}\int_0^{\pi} \sin^2 t\,dt = \dfrac{1}{\pi}\left(\dfrac{t}{2} - \dfrac{\sin 2t}{4}\right)\Big|_0^{\pi} = \dfrac{1}{2}$. In addition, $a_1 = 0$; $a_n = \dfrac{1 + \cos n\pi}{\pi(1 - n^2)}$ for $n \neq 1$, $b_1 = \frac{1}{2}$, and $b_n = 0$ for $n \neq 1$; thus,

$$\frac{a_0^2}{2} + \sum_{n=1}^{\infty}(a_n^2 + b_n^2) = \frac{a_0^2}{2} + (a_1^2 + b_1^2) + \sum_{n=2}^{\infty}(a_n^2 + b_n^2) = \frac{(2/\pi)^2}{2} + \sum_{n=2}^{\infty} \frac{(1 + \cos n\pi)^2}{\pi^2(1 - n^2)^2}$$

$$= \frac{2}{\pi^2} + \frac{1}{4} + \frac{1}{\pi^2}\left(\frac{4}{3^2} + \frac{4}{15^2} + \frac{4}{35^2} + \frac{4}{63^2} + \cdots\right)$$

and Parseval's identity becomes $\dfrac{1}{2} = \dfrac{2}{\pi^2} + \dfrac{1}{4} + \dfrac{4}{\pi^2}\left[\dfrac{1}{(1\cdot 3)^2} + \dfrac{1}{(3\cdot 5)^2} + \dfrac{1}{(5\cdot 7)^2} + \dfrac{1}{(7\cdot 9)^2} + \cdots\right]$,

which may be rewritten as $\dfrac{1}{(1\cdot 3)^2} + \dfrac{1}{(3\cdot 5)^2} + \dfrac{1}{(5\cdot 7)^2} + \dfrac{1}{(7\cdot 9)^2} = \dfrac{\pi^2 - 8}{16}$.

EVEN AND ODD FUNCTIONS

16.67 Determine whether any of the following functions are even: (a) $\sin 3x$, (b) $\cos 5x$, and (c) e^x.

▌ A function $f(x)$ is even if and only if $f(-x) = f(x)$.

(a) For $f(x) = \sin 3x$, we have $f(-x) = \sin(-3x) = -\sin 3x$. Since $f(-x) \neq f(x)$, the function is not even.

(b) For $f(x) = \cos 5x$, we have $f(-x) = \cos(-5x) = \cos 5x$. Since $f(-x) = f(x)$, the function is even.

(c) For $f(x) = e^x$, we have $f(-x) = e^{-x} = 1/e^x$. Since $f(x) \neq f(-x)$, the function is not even.

16.68 Determine whether any of the functions given in the previous problem are odd.

▌ A function $f(x)$ is odd if and only if $f(-x) = -f(x)$. Using the results of the previous problem, we deduce

(a) $f(-x) = -\sin 3x = -f(x)$, so $\sin 3x$ is odd.

(b) $f(-x) = \cos 5x \neq -f(x)$, so $\cos 5x$ is not odd.

(c) $f(-x) = 1/e^x \neq -e^x$, so e^x is not odd.

16.69 Describe the graph of $y = f(x)$ when the function is even.

▌ The graph of an even function is symmetric with respect to the y axis. If the point (x, y) is on the graph, then so too is the point $(-x, y)$.

16.70 Determine whether $f(x) = x(10 - x)$, $0 < x < 10$, with period 10 is even.

▌ The graph of this function, shown in Fig. 16.10, is symmetric with respect to the y axis, so the function is even.

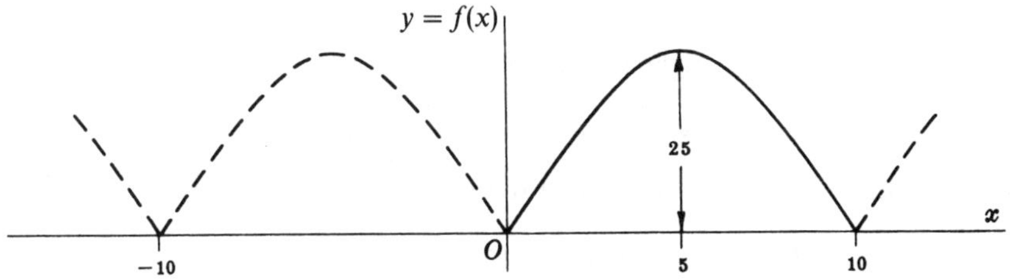

Fig. 16.10

16.71 Determine whether $f(x) = \begin{cases} \cos x & 0 < x < \pi \\ 0 & \pi < x < 2\pi \end{cases}$, period $= 2\pi$, is even.

▌ The graph of this function, shown in Fig. 16.11, is not symmetric with respect to the y axis; hence the function is not even.

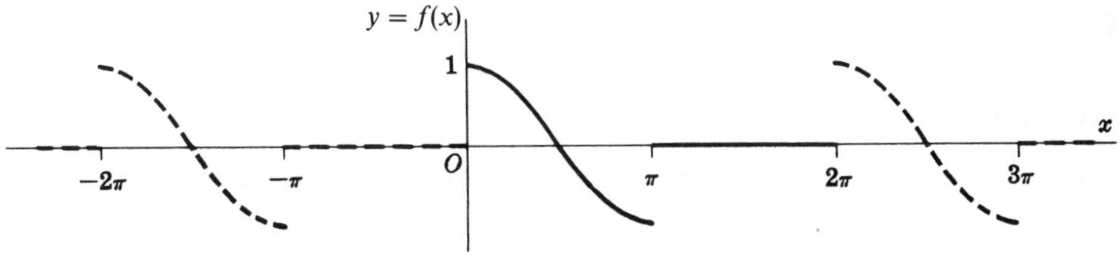

Fig. 16.11

6.72 Describe the graph of $y = f(x)$ when the function is odd.

▌ The graph of an odd function is symmetric with respect to the origin. If the point (x, y) is on the graph, then so too is the point $(-x, -y)$.

6.73 Determine whether $f(x) = \begin{cases} 2 & 0 < x < 3 \\ -2 & -3 < x < 0 \end{cases}$, period $= 6$, is odd.

▌ The graph of this function, shown in Fig. 16.12, is symmetric with respect to the origin, so the function is odd.

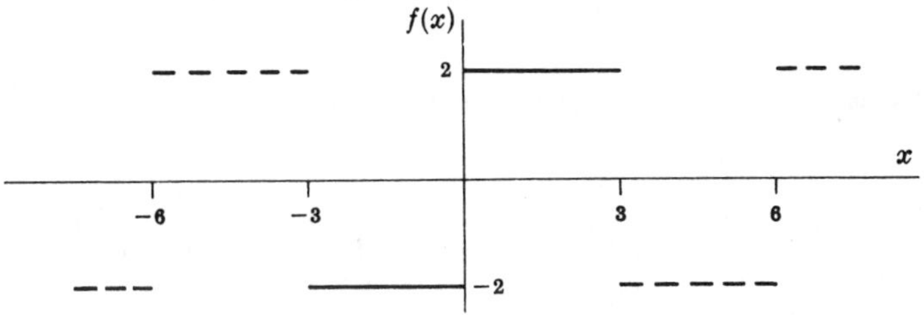

Fig. 16.12

.74 Determine whether the function defined in Problem 16.71 is odd.

▌ The graph of this function is shown in Fig. 16.11. Since the graph is not symmetric with respect to the origin, the function is not odd.

16.75 Determine whether the function shown in Fig. 16.2 is even or odd.

▌ Since the graph is not symmetric with respect to either the vertical axis or the origin, the function is neither even nor odd.

16.76 Determine whether the function shown in Fig. 16.3 is even or odd.

▌ Since the graph is symmetric with respect to the vertical axis, the function is even.

16.77 Determine whether the function shown in Fig. 16.4 is even or odd.

▌ The graph is not symmetric with respect to the vertical axis, so the function is not even; the graph is not symmetric with respect to the origin, so the function is not odd.

16.78 Determine whether the function shown in Fig. 16.5 is even or odd.

▌ Since the graph is symmetric with respect to the vertical axis, the function is even.

16.79 Determine whether the function shown in Fig. 16.7 is even or odd.

▌ It is neither even nor odd because its graph is not symmetric with respect to the vertical axis or the origin.

16.80 Determine whether the function shown in Fig. 16.8 is even or odd.

▌ It is odd because its graph is symmetric with respect to the origin.

16.81 Determine whether the function shown in Fig. 16.9 is even or odd.

▌ It is neither.

SINE AND COSINE SERIES

16.82 Show that an even function can have no sine terms in its Fourier expansion of period $2L$ over the interval $(-L, L$

▌ No sine terms will appear if $b_n = 0$ for $n = 1, 2, 3, \ldots$. To show that this is the case, we write

$$b_n = \frac{1}{L}\int_{-L}^{L} f(x)\sin\frac{n\pi x}{L}\,dx = \frac{1}{L}\int_{-L}^{0} f(x)\sin\frac{n\pi x}{L}\,dx + \frac{1}{L}\int_{0}^{L} f(x)\sin\frac{n\pi x}{L}\,dx \qquad (1$$

If we make the transformation $x = -u$ in the first integral on the right of (1), we obtain

$$\frac{1}{L}\int_{-L}^{0} f(x)\sin\frac{n\pi x}{L}\,dx = \frac{1}{L}\int_{0}^{L} f(-u)\sin\left(-\frac{n\pi u}{L}\right)du = -\frac{1}{L}\int_{0}^{L} f(-u)\sin\frac{n\pi u}{L}\,du$$

$$= -\frac{1}{L}\int_{0}^{L} f(u)\sin\frac{n\pi u}{L}\,du = -\frac{1}{L}\int_{0}^{L} f(x)\sin\frac{n\pi x}{L}\,dx \qquad ($$

where we have used the fact that for an even function $f(-u) = f(u)$ and, in the last step, that the dummy variable of integration u can be replaced with any other symbol, in particular x. Thus, from (*1*) and using (*2*), we ha

$$b_n = -\frac{1}{L}\int_{0}^{L} f(x)\sin\frac{n\pi x}{L}\,dx + \frac{1}{L}\int_{0}^{L} f(x)\sin\frac{n\pi x}{L}\,dx = 0$$

Alternative Method: Assuming convergence, we have $f(x) = \frac{a_0}{2} + \sum_{n=1}^{\infty}\left(a_n\cos\frac{n\pi x}{L} + b_n\sin\frac{n\pi x}{L}\right)$. If $f(x)$ is eve then $f(-x) = f(x)$. Hence

$$\frac{a_0}{2} + \sum_{n=1}^{\infty}\left(a_n\cos\frac{n\pi x}{L} + b_n\sin\frac{n\pi x}{L}\right) = \frac{a_0}{2} + \sum_{n=1}^{\infty}\left(a_n\cos\frac{n\pi x}{L} - b_n\sin\frac{n\pi x}{L}\right)$$

which implies that $\sum_{n=1}^{\infty} b_n\sin\frac{n\pi x}{L} = 0$. That is, $f(x) = \frac{a_0}{2} + \sum_{n=1}^{\infty} a_n\cos\frac{n\pi x}{L}$, in which no sine terms appear. This method is weaker than the first, since convergence is assumed.

16.83 If $f(x)$ is even, show that its Fourier series has $a_n = \frac{2}{L}\int_{0}^{L} f(x)\cos\frac{n\pi x}{L}\,dx$.

$$a_n = \frac{1}{L}\int_{-L}^{L} f(x)\cos\frac{n\pi x}{L}\,dx = \frac{1}{L}\int_{-L}^{0} f(x)\cos\frac{n\pi x}{L}\,dx + \frac{1}{L}\int_{0}^{L} f(x)\cos\frac{n\pi x}{L}\,dx$$

Letting $x = -u$, we obtain

$$\frac{1}{L}\int_{-L}^{0} f(x)\cos\frac{n\pi x}{L}\,dx = \frac{1}{L}\int_{0}^{L} f(-u)\cos\left(\frac{-n\pi u}{L}\right)du = \frac{1}{L}\int_{0}^{L} f(u)\cos\frac{n\pi u}{L}\,du$$

since by the definition of an even function, $f(-u) = f(u)$. Then

$$a_n = \frac{1}{L}\int_{0}^{L} f(u)\cos\frac{n\pi u}{L}\,du + \frac{1}{L}\int_{0}^{L} f(x)\cos\frac{n\pi x}{L}\,dx = \frac{2}{L}\int_{0}^{L} f(x)\cos\frac{n\pi x}{L}\,dx$$

Observe that when $n = 0$, $\cos\frac{n\pi x}{L} = \cos 0 = 1$, so the formula for a_n is reduced to $a_0 = \frac{2}{L}\int_0^L f(x)\,dx$.

16.84 Assume that $f(x)$ and $f'(x)$ are piecewise continuous on (0(L). Discuss how to expand $f(x)$ in a Fourier series of period $2L$ that contains only a_0 and cosine terms and converges to $f(x)$ at all points in $(0, L)$ where $f(x)$ is continuous.

Define a new function $F(x)$ on $(-L, L)$ as follows:

$$F(x) = \begin{cases} f(x) & 0 < x < L \\ f(x+0) & x = 0 \\ f(-x) & -L < x < 0 \end{cases}$$

Then $F(x)$ is even on $(-L, L)$. If we now extend $F(x)$ so it equals $f(L-0)$ at $x = L$ and is periodic with period $2L$, then $F(x)$ is defined everywhere on $(-\infty, \infty)$, it is even, and it satisfies the Dirichlet conditions.

Since $F(x)$ is even, its Fourier series must contain only cosine terms and a_0 (see Problem 16.83). Moreover, since $F(x)$ satisfies the Dirichlet conditions, its Fourier series must converge on $(0, L)$ to $F(x) = f(x)$ at all points in $(0, L)$ where the function $f(x)$ is continuous. $F(x)$ is called the *even periodic extension* of $f(x)$.

16.85 Graph the even periodic extension of $f(x) = \sin x$, $0 < x < \pi$.

The graph is shown in Fig. 16.13. On $(0, \pi)$, the graph is the graph of $f(x)$ (drawn as a solid curve). On $(-\pi, 0)$ the graph is chosen to be the reflection around the y axis of $f(x)$ as drawn on $(0, \pi)$. At all other points, the graph is drawn to be periodic with period 2π.

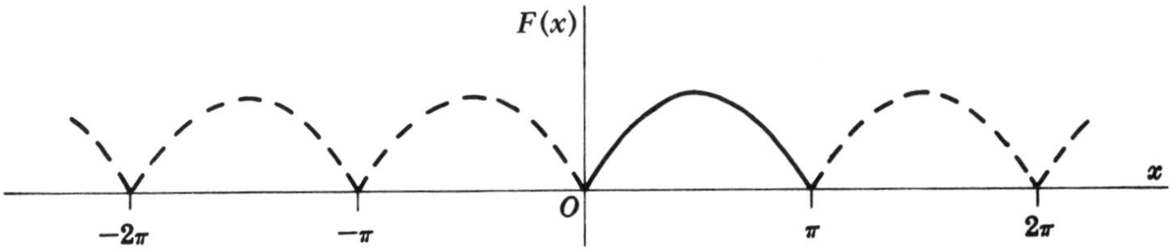

Fig. 16.13

16.86 Expand $f(x) = \sin x$, $0 < x < \pi$, in a Fourier cosine series.

A Fourier series consisting of cosine series alone is obtained only for an even function. We therefore find the Fourier series for $F(x)$, the even periodic extension of $f(x)$ as determined in Problem 16.85. The Fourier series for $F(x)$ will converge to $F(x)$ wherever $F(x)$ is continuous. Since $F(x) = f(x)$ on $(0, \pi)$, the Fourier series for $F(x)$ will converge to $f(x) = \sin x$ everywhere on $(0, \pi)$. Here $L = \pi$, $b_n = 0$ [because $F(x)$ is even], and

$$a_n = \frac{2}{L}\int_0^L F(x)\cos\frac{n\pi x}{L}\,dx = \frac{2}{\pi}\int_0^\pi \sin x\cos nx\,dx$$

$$= \frac{1}{\pi}\int_0^\pi [\sin(x+nx) + \sin(x-nx)]\,dx = \frac{1}{\pi}\left[-\frac{\cos(n+1)x}{n+1} + \frac{\cos(n-1)x}{n-1}\right]_0^\pi$$

$$= \frac{1}{\pi}\left[\frac{1-\cos(n+1)\pi}{n+1} + \frac{\cos(n-1)\pi - 1}{n-1}\right] = \frac{1}{\pi}\left(-\frac{1+\cos n\pi}{n+1} - \frac{1+\cos n\pi}{n-1}\right)$$

$$= \frac{-2(1+\cos n\pi)}{\pi(n^2-1)} \quad \text{for } n \neq 1$$

For $n = 1$, $a_1 = \frac{2}{\pi}\int_0^\pi \sin x \cos x\,dx = \frac{2}{\pi}\frac{\sin^2 x}{2}\Big|_0^\pi = 0$. Then

$$f(x) = \frac{2}{\pi} - \frac{2}{\pi}\sum_{n=2}^{\infty}\frac{1+\cos n\pi}{n^2-1}\cos nx = \frac{2}{\pi} - \frac{4}{\pi}\left(\frac{\cos 2x}{2^2-1} + \frac{\cos 4x}{4^2-1} + \frac{\cos 6x}{6^2-1} + \cdots\right)$$

16.87 Graph the even periodic extension of $f(x) = x$, $0 < x < 2$.

▌ The graph is shown in Fig. 16.14. On (0, 2) the graph of $F(x)$ (drawn as a solid line) is identical to the graph of $f(x)$. On $(-2, 0)$ the graph of $F(x)$ is the reflection around the y axis of $f(x)$ as drawn on (0, 2). At all other points, the graph is drawn to be periodic with period 4.

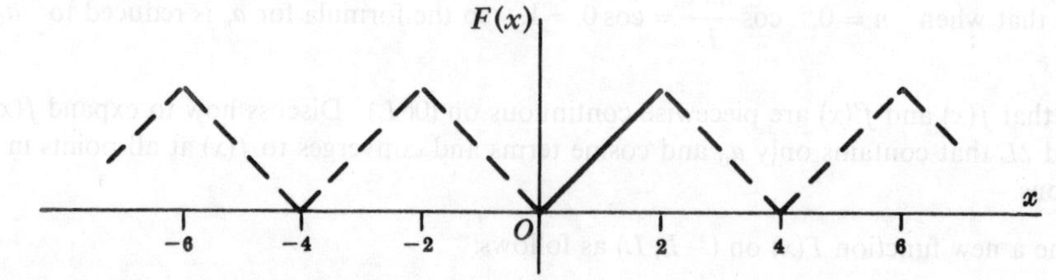

Fig. 16.14

16.88 Expand $f(x) = x$, $0 < x < 2$, in a Fourier cosine series.

▌ We shall determine the Fourier series for the periodic extension of $f(x)$, denoted $F(x)$, as found in Problem 16.87. The Fourier series for $F(x)$ will converge to $F(x)$ everywhere, and since $F(x) = f(x)$ on (0, 2), the Fourier series for $F(x)$ will converge to $f(x)$ on (0, 2). Here $L = 2$, $b_n = 0$ [because $F(x)$ is even], and

$$a_n = \frac{2}{L}\int_0^L F(x)\cos\frac{n\pi x}{L}\,dx = \frac{2}{2}\int_0^2 x\cos\frac{n\pi x}{2}\,dx = \left(x\frac{2}{n\pi}\sin\frac{n\pi x}{2} - \frac{-4}{n^2\pi^2}\cos\frac{n\pi x}{2}\right)_0^2$$

$$= \frac{-4}{n^2\pi^2}(\cos n\pi - 1) \qquad \text{for } n \neq 0$$

If $n = 0$, then $a_0 = \int_0^2 x\,dx = 2$. Thus,

$$f(x) = 1 + \sum_{n=1}^{\infty}\frac{4}{n^2\pi^2}(\cos n\pi - 1)\cos\frac{n\pi x}{2} = 1 - \frac{8}{\pi^2}\left(\cos\frac{\pi x}{2} + \frac{1}{3^2}\cos\frac{3\pi x}{2} + \frac{1}{5^2}\cos\frac{5\pi x}{2} + \cdots\right)$$

16.89 Show that an odd function can have no cosine terms in its Fourier expansion of period $2L$ over the interval $(-L, L)$.

▌ No cosine terms will appear if $a_n = 0$ for $n = 0, 1, 2, \ldots$. To show that this is the case, we write

$$a_n = \frac{1}{L}\int_{-L}^{L} f(x)\cos\frac{n\pi x}{L}\,dx = \frac{1}{L}\int_{-L}^{0} f(x)\cos\frac{n\pi x}{L}\,dx + \frac{1}{L}\int_0^L f(x)\cos\frac{n\pi x}{L}\,dx$$

Making the substitution $u = -x$ in the first integral on the right of the last equality and noting that for an odd function $f(-u) = -f(u)$, we obtain

$$a_n = \frac{1}{L}\int_L^0 f(-u)\cos\frac{-n\pi u}{L}(-du) + \frac{1}{L}\int_0^L f(x)\cos\frac{n\pi x}{L}\,dx$$

$$= \frac{1}{L}\int_L^0 f(u)\cos\frac{n\pi u}{L}\,du + \frac{1}{L}\int_0^L f(x)\cos\frac{n\pi x}{L}\,dx$$

$$= -\frac{1}{L}\int_0^L f(u)\cos\frac{n\pi u}{L}\,du + \frac{1}{L}\int_0^L f(x)\cos\frac{n\pi x}{L}\,dx = 0$$

This last result follows because the two integrals are identical, albeit with different dummy variables of integration. Note that all equalities remain valid when $n = 0$.

16.90 If $f(x)$ is odd, show that the coefficients of the sine terms in its Fourier expansion of period $2L$ on $(-L, L)$ can be written $b_n = \frac{2}{L}\int_0^L f(x)\sin\frac{n\pi x}{L}\,dx$.

▮
$$b_n = \frac{1}{L}\int_{-L}^{L} f(x)\sin\frac{n\pi x}{L}\,dx = \frac{1}{L}\int_{-L}^{0} f(x)\sin\frac{n\pi x}{L}\,dx + \frac{1}{L}\int_{0}^{L} f(x)\sin\frac{n\pi x}{L}\,dx$$

Making the substitution $u = -x$ in the first integral on the right of the last equality and noting that $f(-u) = -f(u)$, we find

$$b_n = \frac{1}{L}\int_{L}^{0} f(-u)\sin\frac{-n\pi u}{L}(-du) + \frac{1}{L}\int_{0}^{L} f(x)\sin\frac{n\pi x}{L}\,dx = \frac{1}{L}\int_{0}^{L} f(u)\sin\frac{n\pi u}{L}\,du + \frac{1}{L}\int_{0}^{L} f(x)\sin\frac{n\pi x}{L}\,dx$$

Since the last two integrals are identical, albeit with different dummy variables of integration, the desired equality follows.

16.91 Assume that $f(x)$ and $f'(x)$ are piecewise continuous on $(0, L)$. Discuss how to expand $f(x)$ in a Fourier series of period $2L$ that contains only sine terms and converges to $f(x)$ at all points in $(0, L)$ where $f(x)$ is continuous.

▮ Define a new function $F(x)$ on $(-L, L)$ as follows:

$$F(x) = \begin{cases} f(x) & 0 < x < L \\ 0 & x = 0 \\ -f(-x) & -L < x < 0 \end{cases}$$

Then $F(x)$ is odd on $(-L, L)$. If we now extend $F(x)$ so that it is zero at $x = L$ and periodic with period $2L$, then $F(x)$ is defined everywhere on $(-\infty, \infty)$, it is odd, and it satisfies the Dirichlet conditions.

Since $F(x)$ is odd, its Fourier series must contain only sine terms (see Problem 16.89). Moreover, since $F(x)$ satisfies the Dirichlet conditions, its Fourier series must converge on $(0, L)$ to $F(x) = f(x)$ at all points where $f(x)$ is continuous. $F(x)$ is called the *odd periodic extension* of $f(x)$.

6.92 Graph the odd periodic extension of $f(x) = \cos x$, $0 < x < \pi/2$.

▮ The graph is shown in Fig. 16.15. On $(0, \pi/2)$, the graph is the graph of $f(x) = \cos x$ (drawn as a solid curve). On $(-\pi/2, 0)$ the graph is drawn to be the reflection around the origin of $f(x)$ as drawn on $(0, \pi/2)$; that is, on $(-\pi/2, 0)$, $F(x) = -\cos(-x) = -\cos x$. Also, $F(0) = 0$ and $F(\pi) = 0$. At all other points, the graph is drawn to be periodic with period π.

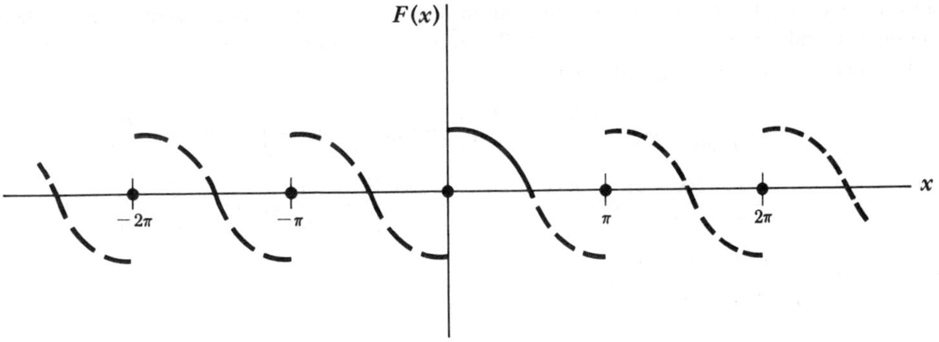

Fig. 16.15

.93 Expand $f(x) = \cos x$, $0 < x < \pi/2$, in a Fourier sine series.

▮ A Fourier series consisting of sine terms alone is obtained only for an odd function. We therefore find the Fourier series for $F(x)$, the odd periodic extension of $f(x)$ as determined in Problem 16.92. The Fourier series for $F(x)$ will converge to $F(x)$ wherever $F(x)$ is continuous; since $F(x) = f(x) = \cos x$ on $(0, \pi/2)$, the Fourier series for $F(x)$ will converge to $f(x)$ on this interval.

We have $L = \pi/2$, $a_n = 0$ [because $F(x)$ is odd], and

$$b_n = \frac{2}{L}\int_{0}^{L} F(x)\sin\frac{n\pi x}{L}\,dx = \frac{2}{L}\int_{0}^{L} f(x)\sin\frac{n\pi x}{L}\,dx = \frac{4}{\pi}\int_{0}^{\pi/2}\cos x\sin 2nx\,dx$$

$$= -\frac{2}{\pi}\left[\frac{\cos(2n+1)x}{2n+1} + \frac{\cos(2n-1)x}{2n-1}\right]_{0}^{\pi/2} = -\frac{2}{\pi}\left(-\frac{1}{2n+1} - \frac{1}{2n-1}\right) = \frac{8n}{\pi(4n^2-1)}$$

Thus, $f(x) = \dfrac{8}{\pi}\displaystyle\sum_{n=1}^{\infty}\frac{n}{4n^2-1}\sin 2nx.$

16.94 Graph the odd periodic extension of $f(x) = x + 1, \quad 0 < x < 2$.

▮ The graph is shown in Fig. 16.16. On (0, 2) the graph of $F(x)$ (drawn as a solid line) is identical to the graph of $f(x) = x + 1$. On $(-2, 0)$ the graph of $F(x)$ is the reflection around the origin of $f(x)$ as drawn on (0, 2). Furthermore, $F(0) = F(2) = 0$. At all other points, the graph is drawn to be periodic with period 4.

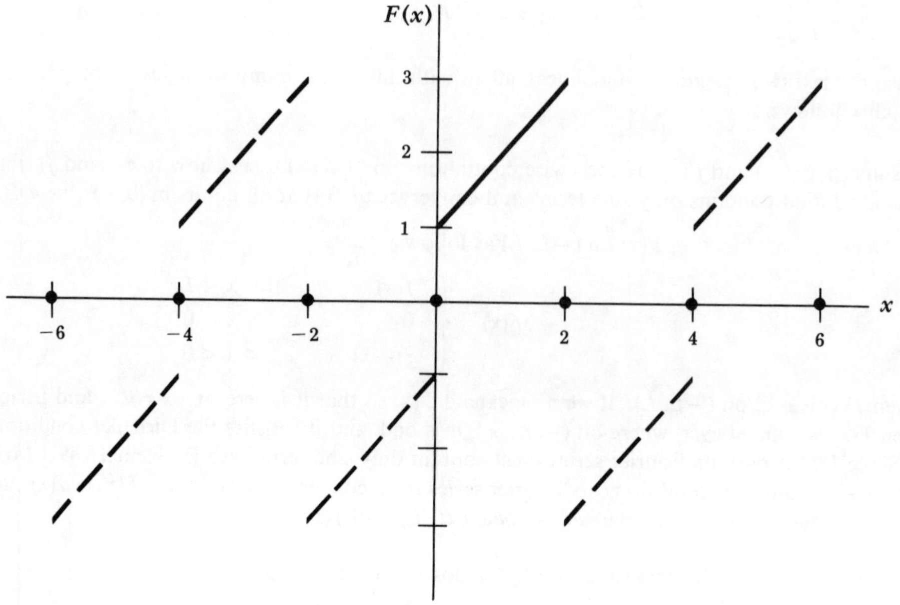

Fig. 16.1

16.95 Expand $f(x) = x + 1, \quad 0 < x < 2$, in a Fourier sine series.

▮ We calculate the Fourier series for $F(x)$, the odd periodic extension of $f(x)$, as determined in Problem 16.94. Since $F(x)$ is even, the Fourier series will contain only sine terms; it will converge to $F(x)$ wherever $F(x)$ is continuous; and, since $F(x) = f(x)$ on (0, 2), the Fourier series will converge to $f(x)$ there.

We have $L = 2, \quad a_0 = a_n = 0$, and

$$b_n = \frac{2}{L}\int_0^L F(x)\sin\frac{n\pi x}{L}\,dx = \frac{2}{L}\int_0^L f(x)\sin\frac{n\pi x}{L}\,dx = \int_0^2 (x+1)\sin\frac{n\pi x}{2}\,dx$$

$$= \left[-\frac{2(x+1)}{n\pi}\cos\frac{n\pi x}{2} + \frac{4}{n^2\pi^2}\sin\frac{n\pi x}{2}\right]_0^2 = -\frac{6}{n\pi}\cos n\pi + \frac{2}{n\pi} = \frac{2}{n\pi}(1 - 3\cos n\pi)$$

Thus, $f(x) = \dfrac{2}{\pi}\displaystyle\sum_{n=1}^{\infty}\frac{1 - 3\cos n\pi}{n}\sin\frac{n\pi x}{2}$.

16.96 Graph the even periodic extension of $f(t) = t - t^2, \quad 0 < t < 1$.

▮ The graph is shown in Fig. 16.17.

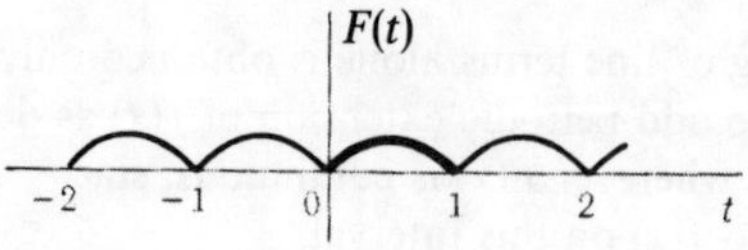

Fig. 16.17

16.97 Graph the odd periodic extension of $f(t) = t - t^2, \quad 0 < t < 1$.

▮ The graph is shown in Fig. 16.18.

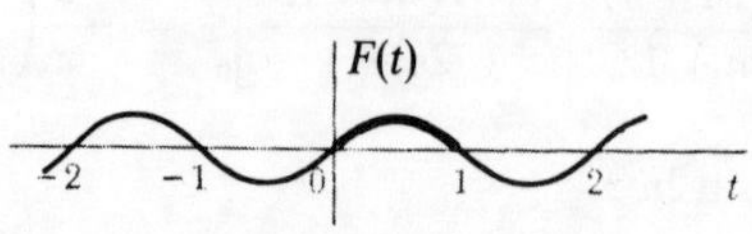

Fig. 16.18

16.98 Find the Fourier sine series of period 2 that converges to $f(t) = t - t^2$ on (0, 1).

▌ We have $a_n \equiv 0$ and

$$b_n = \frac{2}{1}\int_0^1 (t - t^2)\sin\frac{n\pi t}{1}\,dt = 2\left[\frac{\sin n\pi t}{n^2\pi^2} - \frac{t}{n\pi}\cos n\pi t - \left(\frac{2t}{n^2\pi^2}\sin n\pi t + \frac{2}{n^3\pi^3}\cos n\pi t - \frac{t^2}{n\pi}\cos n\pi t\right)\right]_0^1$$

$$= 2\left[\left(-\frac{\cos n\pi}{n\pi}\right) - \left(\frac{2(\cos n\pi - 1)}{n^3\pi^3} - \frac{\cos n\pi}{n\pi}\right)\right] = \frac{4(1 - \cos n\pi)}{n^3\pi^3}$$

Hence, for $0 < t < 1$, $f(t) = \dfrac{8}{\pi^3}\left(\dfrac{\sin \pi t}{1} + \dfrac{\sin 3\pi t}{27} + \dfrac{\sin 5\pi t}{125} + \dfrac{\sin 7\pi t}{343} + \cdots\right)$.

16.99 Find the Fourier cosine series of period 2 that converges to $f(t) = t - t^2$ on (0, 1).

▌ We have $b_n \equiv 0$ and

$$a_n = \frac{2}{1}\int_0^1 (t - t^2)\cos\frac{n\pi t}{1}\,dt = 2\left[\frac{\cos n\pi t}{n^2\pi^2} + \frac{t}{n\pi}\sin n\pi t - \left(\frac{2t}{n^2\pi^2}\cos n\pi t - \frac{2}{n^3\pi^3}\sin n\pi t + \frac{t^2}{n\pi}\sin n\pi t\right)\right]_0^1$$

$$= 2\left(\frac{\cos n\pi - 1}{n^2\pi^2} - \frac{2\cos n\pi}{n^2\pi^2}\right) = -\frac{2(1 + \cos n\pi)}{n^2\pi^2} \qquad \text{for } n \neq 0$$

$$a_0 = \frac{2}{1}\int_0^1 (t - t^2)\,dt = 2\left[\frac{t^2}{2} - \frac{t^3}{3}\right]_0^1 = \frac{1}{3}$$

Hence, for $0 < t < 1$, $f(t) = \dfrac{1}{6} - \dfrac{4}{\pi^2}\left(\dfrac{\cos 2\pi t}{4} + \dfrac{\cos 4\pi t}{16} + \dfrac{\cos 6\pi t}{36} + \dfrac{\cos 8\pi t}{64} + \cdots\right)$.

16.100 Find a Fourier sine series for $f(x) = \begin{cases} x & x \le 1 \\ 2 & x > 1 \end{cases}$ on (0, 2).

▌ Here $L = 2$, $a_n = 0$, and

$$b_n = \frac{2}{L}\int_0^L f(x)\sin\frac{n\pi x}{L}\,dx = \frac{2}{2}\int_0^1 x\sin\frac{n\pi x}{2}\,dx + \frac{2}{2}\int_1^2 2\sin\frac{n\pi x}{2}\,dx$$

$$= \left(\frac{4}{n^2\pi^2}\sin\frac{n\pi x}{2} - \frac{2x}{n\pi}\cos\frac{n\pi x}{2}\right)_0^1 + \left(\frac{-4}{n\pi}\cos\frac{n\pi x}{2}\right)_1^2 = \frac{4}{n^2\pi^2}\sin\frac{n\pi}{2} + \frac{2}{n\pi}\cos\frac{n\pi}{2} - \frac{4}{n\pi}\cos n\pi$$

Thus, the sine series is $\displaystyle\sum_{n=1}^{\infty}\left(\frac{4}{n^2\pi^2}\sin\frac{n\pi}{2} + \frac{2}{n\pi}\cos\frac{n\pi}{2} - \frac{4}{n\pi}\cos n\pi\right)\sin\frac{n\pi x}{2}$.

16.101 Graph the sine series found in Problem 16.100.

▌ The sine series for $f(x)$ on (0, 2) is the Fourier series of period 4 for the odd periodic extension of $f(x)$. Thus, the series converges to $f(x)$ at all points on (0, 2) where $f(x)$ is continuous, it is an odd function, and it has period 4. At all points of discontinuity, the series converges to the average of its left- and right-hand limits. The graph is shown in Fig. 16.19.

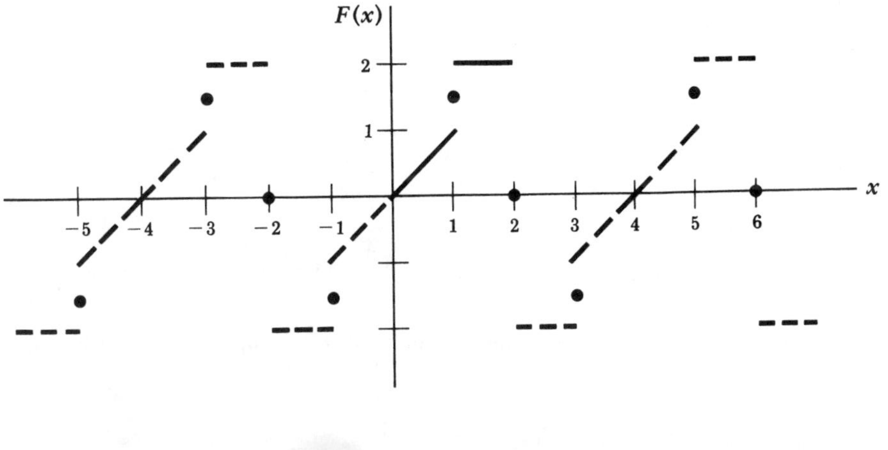

Fig. 16.19

16.102 Find a Fourier cosine series for $f(x) = \begin{cases} x & x < 1 \\ 2 & x \geq 1 \end{cases}$ on (0, 2).

▌ Here $L = 2$, $b_n = 0$, and

$$a_n = \frac{2}{L}\int_0^2 f(x)\cos\frac{n\pi x}{L}\,dx = \frac{2}{2}\int_0^1 x\cos\frac{n\pi x}{2}\,dx + \frac{2}{2}\int_1^2 2\cos\frac{n\pi x}{2}\,dx$$

$$= \left(\frac{4}{n^2\pi^2}\cos\frac{n\pi x}{2} + \frac{2x}{n\pi}\sin\frac{n\pi x}{2}\right)_0^1 + \left(\frac{4}{n\pi}\sin\frac{n\pi x}{2}\right)_1^2$$

$$= \frac{4}{n^2\pi^2}\cos\frac{n\pi}{2} - \frac{4}{n^2\pi^2} - \frac{2}{n\pi}\sin\frac{n\pi}{2} \qquad \text{for } n \neq 0$$

Since $a_0 = \frac{2}{2}\int_0^1 x\,dx + \frac{2}{2}\int_1^2 2\,dx = \frac{1}{2} + 2 = \frac{5}{2}$, the cosine series is

$$\frac{5}{4} + \sum_{n=1}^{\infty}\left(\frac{4}{n^2\pi^2}\cos\frac{n\pi}{2} - \frac{4}{n^2\pi^2} - \frac{2}{n\pi}\sin\frac{n\pi}{2}\right)\cos\frac{n\pi x}{2}$$

16.103 Graph the cosine series found in Problem 16.102.

▌ The cosine series for $f(x)$ on (0, 2) is the Fourier series of period 4 for the even periodic extension of $f(x)$. Thus, the series converges to $f(x)$ at all points on (0, 2) where $f(x)$ is continuous, it is an even function, and it has period 4. At all points of discontinuity, the series converges to the average of its left- and right-hand limits. The graph is shown in Fig. 16.20.

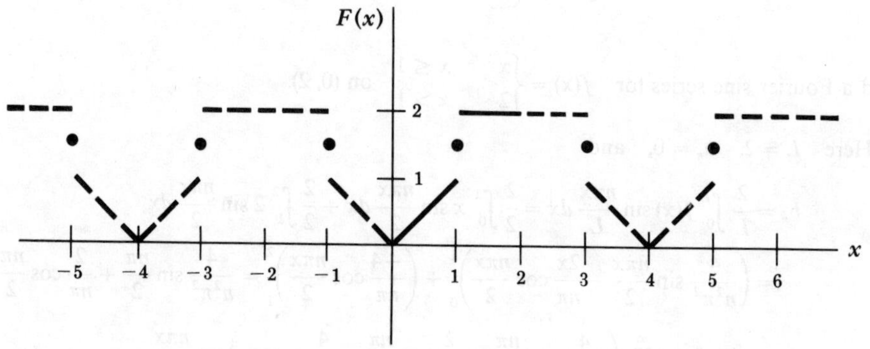

Fig. 16.20

16.104 Find a Fourier cosine series for $f(x) = 1$ on (0, 5).

▌ The function $f(x) = 1$ is its own even periodic extension and its own Fourier cosine series over any interval with any period. In particular, $1 = \frac{a_0}{2} + \sum_{n=1}^{\infty} a_n \cos\frac{n\pi x}{L}$ with $a_0 = 2$ and $a_n = 0$ for $n = 1, 2, 3, \ldots$.

16.105 Find a Fourier sine series for $f(x) = 1$ on (0, 5).

▌ With $L = 5$, we have

$$b_n = \frac{2}{L}\int_0^L f(x)\sin\frac{n\pi x}{L}\,dx = \frac{2}{5}\int_0^5 (1)\sin\frac{n\pi x}{5}\,dx$$

$$= \frac{2}{5}\left(-\frac{5}{n\pi}\cos\frac{n\pi x}{5}\right)_{x=0}^{x=5} = \frac{2}{n\pi}(1 - \cos n\pi) = \frac{2}{n\pi}[1 - (-1)^n]$$

Thus, $$1 = \sum_{n=1}^{\infty}\frac{2}{n\pi}[1 - (-1)^n]\sin\frac{n\pi x}{5} = \frac{4}{\pi}\left(\sin\frac{\pi x}{5} + \frac{1}{3}\sin\frac{3\pi x}{5} + \frac{1}{5}\sin\frac{5\pi x}{5} + \cdots\right)$$

Since both $f(x)$ and $f'(x) = 0$ are continuous on (0, 5), this result is valid for all x in (0, 5).

16.106 Find a Fourier cosine series for $f(x) = x$ on (0, 3).

▌ With $L = 3$, we have

$$a_0 = \frac{1}{L}\int_0^L f(x)\,dx = \frac{1}{3}\int_0^3 x\,dx = \frac{3}{2}$$

$$a_n = \frac{2}{L}\int_0^L f(x)\cos\frac{n\pi x}{L}\,dx = \frac{2}{3}\int_0^3 x\cos\frac{n\pi x}{3}\,dx = \frac{2}{3}\left(\frac{3x}{n\pi}\sin\frac{n\pi x}{3} + \frac{9}{n^2\pi^2}\cos\frac{n\pi x}{3}\right)\Bigg|_{x=0}^{x=3}$$

$$= \frac{2}{3}\left(\frac{9}{n^2\pi^2}\cos n\pi - \frac{9}{n^2\pi^2}\right) = \frac{6}{n^2\pi^2}[(-1)^n - 1]$$

Thus $$x = \frac{3}{2} + \sum_{n=1}^{n} \frac{6}{n^2\pi^2}[(-1)^n - 1]\cos\frac{n\pi x}{3} = \frac{3}{2} - \frac{12}{\pi^2}\left(\cos\frac{\pi x}{3} + \frac{1}{9}\cos\frac{3\pi x}{3} + \frac{1}{25}\cos\frac{5\pi x}{3} + \cdots\right)$$

Since $f(x)$ and $f'(x) = 1$ are both continuous on (0, 3), this result is valid for all x in (0, 3).

16.107 Find a Fourier sine series for $f(x) = e^x$ on $(0, \pi)$.

▌ With $L = \pi$, we obtain

$$b_n = \frac{2}{\pi}\int_0^\pi e^x \sin\frac{n\pi x}{\pi}\,dx = \frac{2}{\pi}\left[\frac{e^x}{1+n^2}(\sin nx - n\cos nx)\right]_{x=0}^{x=\pi}$$

$$= \frac{2}{\pi}\left(\frac{n}{1+n^2}\right)(1 - e^\pi \cos n\pi)$$

Thus $e^x = \frac{2}{\pi}\sum_{n=1}^{\infty}\frac{n}{1+n^2}[1 - e^\pi(-1)^n]\sin nx$ for all x in $(0, \pi)$.

16.108 Find a Fourier cosine series for $f(x) = e^x$ on $(0, \pi)$.

▌ With $L = \pi$, we have

$$a_0 = \frac{1}{\pi}\int_0^\pi e^x\,dx = \frac{1}{\pi}(e^\pi - 1)$$

$$a_n = \frac{2}{\pi}\int_0^\pi e^x\cos\frac{n\pi x}{\pi}\,dx = \frac{2}{\pi}\left[\frac{e^x}{1+n^2}(\cos nx + n\sin nx)\right]_{x=0}^{x=\pi}$$

$$= \frac{2}{\pi}\left(\frac{1}{1+n^2}\right)(e^\pi\cos n\pi - 1)$$

Thus $e^x = \frac{1}{\pi}(e^\pi - 1) + \frac{2}{\pi}\sum_{n=1}^{\infty}\frac{1}{1+n^2}[(-1)^n e^\pi - 1]\cos nx$ for all x in $(0, \pi)$.

16.109 Find a Fourier sine series for $f(x) = \begin{cases} 0 & x \le 2 \\ 2 & x > 2 \end{cases}$ on (0, 3).

▌ With $L = 3$, we obtain

$$b_n = \frac{2}{3}\int_0^3 f(x)\sin\frac{n\pi x}{3}\,dx = \frac{2}{3}\int_0^2 (0)\sin\frac{n\pi x}{3}\,dx + \frac{2}{3}\int_2^3 2\sin\frac{n\pi x}{3}\,dx$$

$$= 0 + \frac{4}{3}\left(-\frac{3}{n\pi}\cos\frac{n\pi x}{3}\right)\Bigg|_{x=2}^{x=3} = \frac{4}{n\pi}\left(\cos\frac{2n\pi}{3} - \cos n\pi\right)$$

Thus $f(x) = \sum_{n=1}^{\infty}\frac{4}{n\pi}\left[\cos\frac{2n\pi}{3} - (-1)^n\right]\sin\frac{n\pi x}{3}$. Furthermore, $\cos\frac{2\pi}{3} = -\frac{1}{2}$, $\cos\frac{4\pi}{3} = -\frac{1}{2}$, $\cos\frac{6\pi}{3} = 1, \ldots,$

so that $f(x) = \frac{4}{\pi}\left(\frac{1}{2}\sin\frac{\pi x}{3} - \frac{3}{4}\sin\frac{2\pi x}{3} + \frac{2}{3}\sin\frac{3\pi x}{3} - \cdots\right)$.

Since $f(x)$ and $f'(x)$ are continuous everywhere on (0, 3) except at $x = 2$, it follows that this result is valid for all x in (0, 2) and (2, 3). At $x = 2$, the sine series converges to the average of its left- and right-hand limits, or $(0 + 2)/2 = 1$.

16.110 Find a Fourier cosine series for $f(x) = \begin{cases} 0 & x \le 2 \\ 2 & x > 2 \end{cases}$ on (0, 3).

▌ With $L = 3$, we obtain

$$a_n = \frac{2}{3}\int_0^3 f(x)\cos\frac{n\pi x}{3}\,dx = \frac{2}{3}\int_0^2 (0)\cos\frac{n\pi x}{3}\,dx + \frac{2}{3}\int_2^3 2\cos\frac{n\pi x}{3}\,dx$$

$$= 0 + \frac{2}{3}\left(\frac{6}{n\pi}\sin\frac{n\pi x}{3}\right)\Bigg|_2^3 = \frac{-4}{n\pi}\sin\frac{2n\pi}{3} \qquad \text{for } n \ne 0$$

$$a_0 = \frac{2}{3}\int_0^3 f(x)\,dx = \frac{2}{3}\int_0^2 (0)\,dx + \frac{2}{3}\int_2^3 2\,dx = \frac{4}{3}$$

Thus, the cosine series is $\dfrac{2}{3} - \dfrac{4}{\pi}\displaystyle\sum_{n=1}^{\infty}\frac{1}{n}\sin\frac{2n\pi}{3}\cos\frac{n\pi x}{3}$.

16.111 Find a Fourier cosine series for $f(x) = x^2$ on $(0, \pi)$.

▌ Here $L = \pi$, $b_n = 0$, and

$$a_n = \frac{2}{\pi}\int_0^\pi x^2\cos nx\,dx = \frac{2}{\pi}\left(-\frac{2}{n^3}\sin nx + \frac{2x}{n^2}\cos nx + \frac{x^2}{n}\sin nx\right)\Bigg|_0^\pi$$

$$= \frac{2}{\pi}\frac{2\pi}{n^2}\cos n\pi = \frac{4}{n^2}\cos n\pi = \frac{4}{n^2}(-1)^n \qquad \text{for } n \ne 0$$

Since $a_0 = \dfrac{2}{\pi}\displaystyle\int_0^\pi x^2\,dx = \frac{2}{\pi}\frac{\pi^3}{3} = \frac{2\pi^2}{3}$, the cosine series is $\dfrac{1}{3}\pi^2 + 4\displaystyle\sum_{n=1}^{\infty}\frac{(-1)^n}{n^2}\cos nx$.

16.112 Explain why the result of Problem 16.111 is identical to the result of Problem 16.37.

▌ The cosine series for x^2 is the Fourier series for the even periodic extension of x^2 on $(0, \pi)$. Since x^2 is itself an even function, its even periodic extension is identical to x^2 on $(-\pi, \pi)$. Thus, both series are Fourier series for x^2 on $(-\pi, \pi)$.

16.113 Find a Fourier sine series for $f(x) = x^2$ on $(0, \pi)$.

▌ Here $L = \pi$, $a_n = 0$, and

$$b_n = \frac{2}{\pi}\int_0^\pi x^2\sin nx\,dx = \frac{2}{\pi}\left(\frac{2}{n^3}\cos nx + \frac{2x}{n^2}\sin nx - \frac{x^2}{n}\cos nx\right)\Bigg|_0^\pi$$

$$= \frac{2}{\pi}\left[\frac{2}{n^3}(\cos n\pi - 1) - \frac{\pi^2}{n}\cos n\pi\right] = \frac{2}{\pi}\left\{\frac{2}{n^3}[(-1)^n - 1] - \frac{\pi^2}{n}(-1)^n\right\}$$

The sine series is $\dfrac{2}{\pi}\displaystyle\sum_{n=1}^{\infty}\left\{\frac{2}{n^3}[(-1)^n - 1] - \frac{\pi^2}{n}(-1)^n\right\}\sin nx$.

16.114 Find a Fourier sine series for $f(t) = t$ on (0, 3).

▌ Here $L = 3$, $a_n = 0$, and

$$b_n = \frac{2}{3}\int_0^3 t\sin\frac{n\pi t}{3}\,dt = \frac{2}{3}\left(\frac{9}{n^2\pi^2}\sin\frac{n\pi t}{3} - \frac{3t}{n\pi}\cos\frac{n\pi t}{3}\right)\Bigg|_0^3 = \frac{2}{3}\left[-\frac{3(3)}{n\pi}\cos n\pi\right] = \frac{-6}{n\pi}(-1)^n$$

The sine series is $\displaystyle\sum_{n=1}^{\infty}\frac{-6}{n\pi}(-1)^n\sin\frac{n\pi t}{3}$.

16.115 Explain why the result of Problem 16.114 is identical to the result of Problem 16.45.

▌ The sine series for $f(t) = t$ is the Fourier series for the odd periodic extension of $f(t)$ on $(-3, 3)$. Since $f(t) =$ is itself odd, it equals its odd periodic extension on $(-3, 3)$. Thus, both series are Fourier series for $f(t) = t$ on $(-3, 3)$.